"Most studies on energy transitions emphasize the need for non-fossil fuels to solve the current problems of the world's energy problems. This book gives a refreshingly broader view: an energy transition with innovation and changes of behavior as key factors".

    – **Professor José Goldemberg**, *University of Sao Paulo, Brazil; Lead Author for the* World Energy Assessment *and Co-Chair of the* Global Energy Assessment

"This is an important book for those interested in the crucial aspect of the energy transition to a decarbonized economy. As climate change has become a threat multiplier for global insecurities, we need to move as rapidly as possible, informed by sound analysis such as what these authors provide, to transition our energy systems".

    – **Sherri Goodman**, *Secretary General, International Military Council on Climate and Security; Vice Chair, Secretary of State's International Security Advisory Board; Formerly, First Deputy Undersecretary of Defense, Environmental Security, United States*

"This timely reference should come as a welcome relief to any policymaker, investor, or practitioner trying to navigate the complex energy transition now underway – a transition we have no choice but to accelerate. Many useful blueprints can be drawn from its comprehensive coverage of topics, and the crosscutting way they are presented".

    – **Mark Radka**, *Chief, Energy and Climate Branch, Economy Division, United Nations Environment Program*

"This volume provides a comprehensive overview of all aspects of the ongoing energy transition. It provides an excellent introduction for students and practitioners alike".

    – **Jim Skea**, *Professor of Sustainable Energy, Centre for Environmental Policy, Imperial College, United Kingdom; IPCC Working Group III Chair*

"The great energy transition of the 21st century is well underway – and it is accelerating. Those of us working in the field hope to see by mid-century a fundamentally new global energy system built on a backbone of renewable energy. Understanding this transition is key to realizing this vision and this work provides a critical view from a multidisciplinary perspective addressing the economic, policy, innovation, and management aspects of the energy transition".

    – **Katherine Dykes**, *Head of Section, Systems Engineering and Optimization, DTU Wind Energy, Technical University of Denmark*

"Energy transitions are frequently mentioned but poorly understood, with too much focus on utopian technical perspectives without considering the wider socio-economical and philosophical impacts that will affect all communities. This *Handbook* helps to address this imbalance through careful curation to provide a comprehensive overview of the essential aspects needing to be considered when devising energy policies and assessing their effectiveness. The *Handbook* should be essential reading for everyone who wants to enable just and effective energy transitions for all".

    – **Martin Young**, *Former Senior Director, World Energy Council*

"Humanity needs to do something unprecedented: foster a complete transformation of a vital technological infrastructure in less than a decade. And not just any infrastructure but the world's energy system. *The Routledge Handbook of Energy Transitions* sets a foundation for everyone to better understand the choices in all aspects of the challenge, making it a must-read".

> – **Greg Unruh**, *Arison Chair in Values Leadership; Associate Professor,*
> *Sustainable Business Strategy and Social Innovation, George Mason University;*
> *Strategic Partner of the United Nations Global Compact*

"With considerable accumulated expertise, Dr. Araújo has produced important international work on energy transitions. The *Routledge Handbook of Energy Transition*s continues this tradition. This book highlights regional energy insights from Latin America, Europe, Asia, Africa, Australia, and North America, with valuable takeaways for global energy policies. More broadly, this *Handbook* brings attention to diverse aspects of energy transitions, primarily triggered by the necessity of the increasing use of low-carbon energy sources. As such, this book serves as a valuable reference for energy researchers and policymakers".

> – **Jose R. Moreira**, *Senior Professor at the Institute of Energy*
> *and Environment, University of Sao Paulo, Sao Paulo, Brazil; Lead Author,*
> *IPCC Special Report, Methodological and Technological Issues in Technology Transfer*

"This *Routledge Handbook of Energy Transitions* is a guide to the ongoing transformations taking place and the possible pathways to an – as yet – unrealized low-carbon energy future. Each of the proposed technological options is covered comprehensively and authoritatively. Of equal importance is the attention given to the requirements for an accompanying economic and societal transition that will eventually stabilize atmospheric CO_2 and other greenhouse gases at concentrations that avoid dangerous anthropogenic interference with the climate system. As this volume makes clear, the transition to wondrous low-carbon technologies can and must be affordable and acceptable to individuals and societies. This is a recommended reading for anyone wanting to understand energy transitions".

> – **Bill Moomaw**, *Member, National Advisory Board for the*
> *Union of Concerned Scientists; Chair of the Board of Climate Group North America;*
> *Professor Emeritus of International Environmental Policy, the Fletcher School, Tufts University*

"A timely resource that goes beyond the obvious topics to illuminate the links between energy system transformation and other key policy domains, including food systems, equity concerns, trade, and macroeconomic stability".

> – **Stephen Hammer**, *Adviser, Global Partnerships and Strategy, World Bank*

ROUTLEDGE HANDBOOK OF ENERGY TRANSITIONS

The *Routledge Handbook of Energy Transitions* draws upon a unique and multidisciplinary network of experts from around the world to explore the expanding field of energy transitions.

This *Handbook* recognizes that considerable changes are underway or are being developed for the modes in which energy is sourced, delivered, and utilized. Employing a sociotechnical approach that accounts for economics and engineering, as well as more cross-cutting factors, including innovation, policy and planning, and management, the volume considers contemporary ideas and practices that characterize the field. The book explores pressing issues, including choices about infrastructure, the role of food systems and materials, sustainability, and energy democracy. Disruption is a core theme throughout, with the authors examining topics such as digitalization, extreme weather, and COVID-19, along with regional similarities and differences. Overall, the *Routledge Handbook of Energy Transitions* advances the field of energy transitions by connecting ideas, taking stock of empirical insights, and challenging how we think about the theory and practice of energy systems change.

This innovative volume functions as an authoritative roadmap with both regional and global relevance. It will be an essential resource for students, policymakers, researchers, and practitioners researching and working in the fields of energy transitions, planning, environmental management and policy, sustainable business, engineering, science and technology studies, political science, geography, design anthropology, and environmental justice.

Kathleen M. Araújo is Director of the CAES Energy Policy Institute and Associate Professor at Boise State. She earned her PhD at MIT and specializes in energy transitions, regional diversification, innovation, and resilience. She is the author of *Low Carbon Energy Transitions: Turning Points in National Policy and Innovation*.

ROUTLEDGE HANDBOOK OF ENERGY TRANSITIONS

Edited by Kathleen M. Araújo

Routledge
Taylor & Francis Group

LONDON AND NEW YORK

earthscan
from Routledge

Designed cover image: jonathanfilskov-photography / Getty Images

First published 2023
by Routledge
4 Park Square, Milton Park, Abingdon, Oxon OX14 4RN

and by Routledge
605 Third Avenue, New York 10158

Routledge is an imprint of the Taylor & Francis Group, an informa business

British Library Cataloguing-in-Publication Data
A catalogue record for this book is available from the British Library

ISBN: 978-1-032-02350-2 (hbk)
ISBN: 978-1-032-02402-8 (pbk)
ISBN: 978-1-003-18302-0 (ebk)

DOI: 10.4324/9781003183020

The Open Access version of chapter 10 was funded by the University of Birmingham and chapter 26 was funded through the ENCLUDE project, supported by the EU's Horizon 2020 Research and Innovation Program, under grant agreement no. 101022791.

This book is for those who want to understand the complex dynamics of energy transitions.

CONTENTS

Contents

FIGURES

BOXES

TABLES

CONTRIBUTORS

Madhav Acharya is VP, Commercialization at Syzygy Plasmonics. He previously served as Technology to Market Adviser at ARPA-E, where he helped innovative energy technologies transition from lab to market. He has a PhD in chemical engineering from the University of Delaware and worked in technical and managerial roles at ExxonMobil for 17 years.

Kathleen M. Araújo directs the CAES Energy Policy Institute and is Associate Professor at Boise State. She specializes in the social and technical aspects of energy transitions, regional diversification, innovation, and resilience. She earned her PhD at MIT and completed post-doctoral research at the Harvard Kennedy School of Government. She is the author of *Low Carbon Energy Transitions: Turning Points in National Policy and Innovation* (Oxford University Press).

Fabian Barrera works at Agora Energiewende as Project Manager PtX for Latin America. Previously, he worked at the International Renewable Energy Agency, consulted for Ecofys and worked with the International Energy Agency and the Colombian energy sector. He completed his master's in energy and process engineering.

Tobias Bischof-Niemz is Head of Division for the New Energy Solutions at Enertrag AG and Director at Enertrag South Africa. He is responsible for leading market entry into new geographies and commercializing new business models around green hydrogen.

Saurabh Biswas is a scientist in the Energy Policy and Economics Group at the Pacific Northwest National Laboratory. He studies the energy-poverty nexus, marginalization, and designing multi-stakeholder transitions for sustainable development. Prior to this role, he was a postdoctoral researcher at Arizona State University.

Wouter Botzen is Professor of Economics of Climate Change and Natural Disasters and Head of the Department of Environmental Economics, Institute for Environmental Studies, VU Amsterdam. In addition, he is Professor of Economics of Global Environmental Change at the Utrecht University School of Economics, Utrecht University.

Joshua Brinkman currently teaches and writes in the areas of the history of technology and engineering, US history, STS, and legal studies. His work often examines how technology and identity are co-constructed, as well as the relationship between law, policy, science, and technology.

Karla Graciela Cedano-Villavicencio is Head of Technology Management and Liaison at the Renewable Energy Institute of the Universidad Nacional Autonoma de Mexico, as well as President of the Solar Energy National Association in Mexico. She holds an MSc in computer sciences and a PhD in applied sciences.

Terence Creamer is Editor of *Engineering News* and Deputy Editor of *Mining Weekly*. He has editorial responsibility at Polity.org.za, a legal and political website, and is a journalist with writing that focuses primarily on energy, as well as techno-economic and policy developments, in South Africa.

Bruno S.L. Cunha is a postdoctoral researcher in the Energy Planning Program at Coppe/UFRJ. His main scientific interests include energy-economic modeling, energy economics, and the role of food systems in global climate stabilization.

Yanelys Delgado-Triana is Professor of the Law Department of the Marta Abreu de Las Villas Central University (UCLV). She has a PhD in legal sciences. She is the president of the chapter of the Cuban Society of Constitutional, Administrative, and Environmental Law and the director of the national project Legal Bases for Sustainable Energy Development in Cuba.

Lorenzo De Marinis is an electronic engineer and a PhD candidate in photonic technologies at Sant'Anna School of Advanced Studies in Pisa, Italy. His main research concerns the development of photonic circuits for artificial intelligence, with a focus on energy-efficient optical computing.

David Diaz-Florian is a full-time professor in the Economy Department at Universidad del Norte in Barranquilla, Colombia. His research interests are currently related to how climatic events like El Niño can influence energy system vulnerability in Colombia, which is heavily dependent on hydropower. In addition, David's research interests are also linked to an ecological economics approach to the study of climate change.

Gary Dirks is Senior Director of the Global Futures Laboratory and Director of the LightWorks Initiative at Arizona State University. He also serves as Chairman of Carbon Collect, a company commercializing direct air capture technology. Dr. Dirks has been at ASU since 2009.

Niall P. Dunphy is CPPU Director and a senior research fellow with the School of Engineering and Architecture and the Environmental Research Institute, University College Cork, Ireland. His research intersects the social sciences with science and engineering, focused on the theme of society, sustainability, and energy.

Angel L. Echevarria Barreto is a research associate in Arizona State University's Center for Energy and Society. His research brings together issues of sustainability, energy systems, and how peoples relate to both in society. He focuses on Puerto Rico's energy system transformation.

Megan Farrelly is Associate Professor of Geography within the School of Social Sciences, Monash University. She is as an interdisciplinary researcher whose research examines how sociotechnical and governance experimentation informs sustainable urban transitions and the governance and policy levers to underpin transitions in practice.

Robert Ferry is a LEED accredited licensed architect whose practice centers on regenerative design and urban energy landscapes. As founding co-director of the nonprofit Land Art Generator Initiative, he supports the role of design as part of a comprehensive solution to climate change.

Timothy J. Foxon is Professor of Sustainability Transitions at the Science Policy Research Unit, University of Sussex. His research explores the technological and social factors relating to the innovation of new energy technologies, the co-evolution of technologies and institutions for a sustainable low-carbon economy, and relations between energy use and economic growth.

Raghu Garud is Professor of Management and Organization and the Farrell Chair in Innovation and Entrepreneurship, Pennsylvania State University. Raghu earned a PhD degree in strategic management and organization from the University of Minnesota.

Deborah Gordon is Senior Principal in RMI's Climate Intelligence Program. Her work focuses on oil, gas, and climate change, with positions at Chevron, Yale, Brown, Union of Concerned Scientists, and Carnegie Endowment. Gordon published her third book, *No Standard Oil* (Oxford University Press), in 2021.

Nihit Goyal is Assistant Professor at the Faculty of Technology, Policy, and Management, Delft University of Technology. His research interests lie in analyzing and informing the governance of the energy transition, with a focus on comparative public policy, energy and sustainable development, and the use of computational social science.

Arnulf Grubler is Emeritus Research Scholar at the International Institute for Applied Systems Analysis (IIASA) in Austria having until recently served as Acting Director of the Transitions to New Technologies Program. He also holds an honorary professorship at Montan University, Leoben, Austria.

Richard Hirsh, Professor of the History of Technology and Science and Technology Studies at Virginia Tech, has written extensively on the history of the electric power system. His newest book, *Powering American Farms*, available in 2022, reinterprets the history of rural electrification since the 1920s.

Michael Howlett, FRSC, is Burnaby Mountain Professor and Canada Research Chair (Tier 1) in the Department of Political Science at Simon Fraser University in Vancouver, British Columbia, Canada. He specializes in resource and environmental policymaking.

Faheem Hussain is Clinical Associate Professor at Arizona State University and Chair of the MS in Global Technology and Development (GTD) program. He advises the United Nations and USAID as a technology policy specialist for community development.

Peter Karnøe is Professor in the Department of Planning at Aalborg University in Denmark.

Jennifer Keahey is Assistant Professor at Arizona State University. As a critical development scholar, Keahey is interested in questions pertaining to sustainability, social justice, and social change. Her current scholarship is focused on the social and cultural influences shaping transitions to democracy and sustainability.

Donald V. Kingsbury is Assistant Professor in Political Science and Latin American Studies at the University of Toronto.

Jack Kiruja is an Associate Program Officer for Geothermal Energy at the International Renewable Energy Agency (IRENA). He holds a master's degree in sustainable energy engineering from Reykjavik University, Iceland and is a fellow of the Geothermal Training Program in Iceland (GRO GTP).

Breffní Lennon is a research fellow at the Environmental Research Institute, University College Cork, Ireland. He is a human geographer researching the social and economic dimensions of sustainability issues, with a focus on the human aspects of climate change and the energy transition.

Nairo Ruperto Leon-Rodriguez is a PhD candidate at the Renewable Energy Institute of the Universidad Nacional Autonoma de Mexico. His research explores innovation in solar energy technologies, specifically in bifacial photovoltaics and thermal photovoltaics systems.

Beatriz Lorenzo-Yera is Professor at the Law Department of the Marta Abreu de Las Villas Central University (UCLV). She is a member of the chapter of the Cuban Society of Constitutional, Administrative, and Environmental Law and a member of the national project Legal Bases for Sustainable Energy Development in Cuba.

José Grabiel Luis-Cordova, a PhD candidate in legal sciences, is Professor of Energy Law at the Marta Abreu de Las Villas Central University (UCLV). He is an associate researcher at the Center for Private and Economic Law of the Free University of Brussels and member of the national project Legal Bases for Sustainable Energy Development in Cuba.

Arian Mahzouni is Senior Lecturer in Urban Energy Transitions at ISBA University in Freiburg, Germany. In his work, he addresses energy transition pathways, policies, and practices in housing and mobility sectors in several European cities: Stockholm (Sweden), Freiburg (Germany), and Basel and Sion (Switzerland).

Jochen Markard is Privatdozent (Habilitation) at ETH Zurich and a senior researcher at the Zurich University of Applied Sciences. Jochen works at the intersection of innovation and transition studies, policy analysis, and management studies. He is Chair of the Sustainability Transitions Research Network.

Masahiro Matsuura (Masa) is Professor at the Graduate School of Governance Studies, Meiji University (Tokyo, Japan) focusing on the practice of transition management, consensus building, and negotiation in the sustainable urban and regional planning contexts. He has a PhD and master's in city planning (MIT).

Adolfo Mejia-Montero is a research fellow in energy vulnerability at the University of Birmingham, in the UK. He earned an MSc in sustainable energy systems and a PhD in energy justice and wind power development from the University of Edinburgh.

Clark Miller is a theorist of techno-human design and transformation. His work seeks to accelerate decarbonization as a project of human progress and uplift. He is Professor and Director of the Center for Energy and Society at Arizona State University.

Elizabeth Monoian (MFA Carnegie Mellon University) is the founding co-director of the Land Art Generator, a nonprofit working to provide models of community-centered sustainable infrastructures that inspire people about the beauty of a world without fossil fuel and educate the next generation of designers.

Majia Nadesan is Professor at Arizona State University. Her research examines the biopolitics (or politics of life) of energy and other key assemblages, seeking greater transparency, and democratic reform of extractive processes and effects.

Mary Jane C. Parmentier is Clinical Professor at Arizona State University and Co-Associate Director for Academic Programs at the School for the Future of Innovation in Society (SFIS). She focuses on global development and technology theory and policy, with an emphasis on Latin America and North Africa.

Martin J. Pasqualetti is Professor in the School of Geographical Sciences and Urban Planning and Senior Sustainability Scientist in the Julie Ann Wrigley Global Futures Laboratory, both at Arizona State University. He has served as an adviser to the US Department of Energy, the National Renewable Energy Laboratory, the US Office of Technical Assessment, and the National Academy of Sciences.

Sarah Pink is Professor and Director of the Emerging Technologies Research Lab at Monash University and Associate Director of Monash Energy Institute. She is a design anthropologist who specializes in developing new approaches to understanding and working toward futures.

Joana Portugal-Pereira is Assistant Professor of the Energy Planning Program of COPPE/UFRJ and a visiting fellow at the Centre for Environmental Policy of Imperial College London. Her research focuses on sustainability of low-carbon innovation technologies. She earned a PhD and MSc in environmental engineering (from UTokyo and ULisbon).

Guilherme Pratti is a PhD candidate in the philosophy of law at Sant'Anna School of Advanced Studies, Italy, and a member of its Center for Inter-legality Research. Currently a visiting PhD at the Copenhagen Business School. He investigates how the green industrial revolution impacts legal rationality.

Alex Putzer is a UN Harmony with Nature expert and a PhD student in the Rights of Nature at Sant'Anna School of Advanced Studies in Pisa, Italy. He researches the urban implications of non-anthropocentric ethics and how a changing environment affects the human condition.

Rob Raven is an interdisciplinary scholar and Professor of Sustainability Transitions and Deputy Director (Research) at the Monash Sustainable Development Institute. His research broadly addresses questions on how sociotechnical experimentation, institutional change, and incumbent urban regimes shape transitions toward sustainable cities.

Clare Richardson-Barlow is a research fellow at the University of Leeds, where she examines industrial decarbonization from a political economy perspective and teaches Asia-Pacific political economy and energy transition topics. She is also a fellow for the National Bureau of Asian Research, where she provides expertise related to energy policy, energy transitions, and the political economy of decarbonization.

Yiamar Rivera-Matos is a doctoral student in the College of Global Futures at Arizona State University, studying the Puerto Rican energy transition. Her research focuses on sustainable futures, energy transitions, and grassroots community solar energy projects.

Tiare Robles-Bonilla is a PhD candidate at the Renewable Energy Institute of the Universidad Nacional Autonoma de Mexico. Her research explores ways to understand and measure energy systems resilience and energy poverty.

Pedro Rochedo is Assistant Professor in the Energy Planning Program at Coppe/UFRJ and a certified trainer for the IAEA's analytical tool for energy modeling. Pedro has contributed to the development of global and national integrated assessment models for energy and land systems analysis.

Juan Manuel Romero-Bravo is an MSc student at the Renewable Energy Institute of the Universidad Nacional Autonoma de Mexico. His research is on energy vulnerability at the systems level for the case of Mexico.

Daniel Rosenbloom is a SSHRC postdoctoral fellow at the University of Toronto. Daniel is also a board member of the Sustainability Transitions Research Network. Drawing on transition and political perspectives, his research explores the intersection of climate change, energy, and societal transformation.

Roberto Schaeffer is Full Professor at the Energy Planning Program of the Federal University of Rio de Janeiro. He has been a coordinating lead author, lead author, and review editor of various IPCC reports since 1998. He is Associate Editor of Energy at International Journal.

Darren Sharp (PhD) is a research fellow at Monash Sustainable Development Institute where he is Research Coordinator of the ARC Linkage project Net Zero Precincts: an interdisciplinary approach to decarbonizing cities. Darren is a sustainability transitions researcher interested in urban experimentation, living labs, multi-stakeholder governance, grassroots innovations, and the sharing economy.

Swati Singh is an energy analyst with ADI Analytics, a boutique oil and gas consulting firm based in Houston, Texas. Swati holds an MS in energy management from University of Houston, ME in renewable energy Planning from NTNU, Norway, and BE in mechanical engineering from Tribhuwan University, Nepal.

Fereidoon Sioshansi is President of Menlo Energy Economics, a consulting firm based in San Francisco, California, advising clients on the rapid transformation of the electricity sector and emerging business models. He has over four decades of experience covering all aspects of the electricity power sector.

Alexandre Szklo is Associate Professor of the Energy Planning Program of Coppe/UFRJ. He is the author of numerous books and around 150 peer-reviewed articles and has supervised over 150 theses. Alexandre leads the development of integrated assessment models and optimization models for oil refineries.

Araz Taeihagh (DPhil, Oxon) is Head of the Policy Systems Group at the Lee Kuan Yew School of Public Policy and Principal Investigator at the Centre for Trusted Internet and Community at the National University of Singapore. His research interest has been on the interface of technology and society.

Harriet Thomson is Associate Professor in Global Social Policy and Sociology and Director of Postgraduate Taught Programs at the University of Birmingham, in the UK. Her research interests broadly concern the role of public policy and policymaking processes, access to affordable and clean forms of energy, and indicators for measuring the complex realities of energy poverty.

Uday Turaga is the founder and CEO of ADI Analytics. Through 20 years of industry experience gained at ExxonMobil, ConocoPhillips, Booz, and ADI, he brings deep commercial and technical expertise in energy, chemical, and industrial markets. He holds a PhD in fuel science from Penn State and an MBA from the University of Texas at Austin.

Jeroen van den Bergh is the ICREA Professor in the Institute of Environmental Science and Technology, Universitat Autònoma de Barcelona (since 2007), and Professor of Environmental and Resource Economics at Vrije Universiteit Amsterdam (since 1997). He was founding Editor in Chief of *Environmental Innovation and Societal Transitions* (2011–2021).

Charlie Wilson is Professor of Energy and Climate Change in the Environmental Change Institute (ECI) at the University of Oxford in the UK, and a Visiting Research Scholar at the International Institute for Applied Systems Analysis (IIASA) in Austria.

Rebecca Windemer is a lecturer in environmental planning at the University of the West of England. Her existing research has focused on planning for the future of existing renewable energy infrastructure, including the opportunities and challenges of repowering, life extension, and decommissioning.

Caroline Zimm is a research scholar in the Transformative Institutional and Social Solutions (TISS) research group at IIASA. Currently, she works on the international research initiatives of the Earth Commission of the Global Commons Alliance, and Just Transitions to Net Zero-Carbon Emissions for All.

Marianne Zotin is a doctoral student and research assistant in the Energy Planning Program at Coppe/UFRJ. Her main research interests include renewable energy policymaking, industry sector decarbonization, critical materials, and integrated assessment modeling.

PREFACE

This *Handbook* was an idea for many years. It became a reality during the pandemic.

Over the course of the past two years, I had the pleasure of partnering with nearly 70 collaborators from around the world on this project. We worked primarily remotely. As we emailed draft chapters and discussed the nuances of the writing in web-based meetings, we challenged each other's thinking about the state of knowledge.

In the process, we witnessed unprecedented disruptions in energy and the world at large. We also sought out junior members of the field, particularly from underrepresented groups, to partner on a number of chapters.

Recognizing that improved understanding of energy is crucial for society's choices, this *Handbook* is designed to be accessible to a broad group of policymakers, students, interested citizens, and experts.

We welcome continued refinements in the field and look forward to contributing not only to the ideas, but to good choices.

– Kathleen M. Araújo, 2022

ACKNOWLEDGMENTS

Credits are acknowledged for the following:

- The International Energy Agency, World Bank, IRENA, the Web of Science, and Japan's Ministry of Economy, Trade, and Industry for use of their data.
- Marianne Zotin for permission to reproduce "Main linkages between materials and energy sources/technologies".
- Jochen Markard and Daniel Rosenbloom for permission to reproduce: "Global CO_2 emissions across sectors and projected path to net-zero", "Non-linear development and different stages of a transition", "Multiple innovations and two transitions affecting each other", "Low-carbon innovations and individual transitions build toward net-zero", and "Diffusion of new renewable energy technologies".
- The Carnegie Endowment for International Peace for permission to reprint Deborah Gordon and Madhav Acharya's "Oil Shake-Up: Refining Transitions in a Low-Carbon Economy".
- Bruno Cunha for permission to reproduce "Global primary energy consumption and food production by human society" and "GHG emissions of the global food system by activities in 2015".
- Taylor and Francis for permission to reproduce and update Peter Karnøe and Raghu Garud's "Path Creation: Co-creation of Heterogeneous Resources in the Emergence of the Danish Wind Turbine Cluster", *European Planning Studies*.
- Claire Richardson for permission to reproduce "Geographical context", "ASEAN interconnections summary forecast", "Current ASEAN Interconnections", and "BIMP-EAGA forecast and current interconnections".
- Tobias Bischof-Niemz and Terence Creamer for permission to reprint "Structure of the future energy system based on solar and wind electricity as the primary energy", "The four broad categories of tradeable products: green ammonia, green methanol, green kerosene, and green steel", "The buildup of solar and wind capacity under the hypothetical scenario whereby South Africa would produce 25 to 30 million tons of green hydrogen per year", and "A comparison of the present value of potential electricity production from South Africa's coal resources versus its combined solar and wind resources".

- Jack Kiruja and Fabian Barrera for permission to reproduce "Global installed capacity (MWt) of geothermal direct use in the last 25 years" and "Direct applications of geothermal energy".
- Rebecca Windemer for permission to reproduce the photograph (authors own) showing the scale of the transformer substation.
- Masa Matsuura for permission to reproduce "Japan's electricity generation portfolio by fuel type (2010–2019)", "Cumulative installed solar power generation capacity and the feed-in tariff rate", and "Cumulative installed wind power generation capacity in Japan".
- Elsevier for permission to reprint the figure "The transition management cycle" from Loorbach, Derk, and Jan Rotmans' "The practice of transition management: Examples and lessons from four distinct cases".
- Nihit Goyal for permission to reprint "The number of publications on energy transitions over time", "Inter-institutional collaboration in the research on energy transitions", and "Correlation network of terms associated with policy".
- Arian Mahzouni for permission to reprint "An integrated socio-technical approach to low-carbon urban mobility".
- Charlie Wilson for permission to reproduce "Transforming how services are provided to meet human needs with fewer resource inputs" and "Global resource efficiency cascades throughout the service provisioning system for energy".
- Fereidoon Sioshansi for permission to reproduce "Annual energy cost outcomes for three scenarios for residential solar, battery, and water heater system in Queensland, Australia" and excerpts from his company newsletter.
- Jennifer Keahey for permission to reproduce "Map of renewable energy approaches".
- Saurabh Biswas for permission to reproduce "Social value of energy mapping (process schematic)".
- Robert Ferry and Elizabeth Monoian for permission to reproduce "Solar canopy provides shade over the Orange Mall of the Arizona State University campus in Tempe, Arizona".
- Penelope Boyer for permission to reproduce a photo of *La Monarca*.
- Robert Flottemesch for permission to reprint *Nest*, by Robert Flottemesch.
- Martin Heide for permission to reprint *Light Up*, by Martin Heide, Dean Boothroyd, Emily Van Monger, David Allouf, Takasumi Inoue, Liam Oxlade, Michael Strack, and Richard Le (NH Architecture); Mike Rainbow and Jan Talacko (Ark Resources); John Bahoric (John Bahoric Design); and Bryan Chung, Chea Yuen Yeow Chong, Anna Lee, and Amelie Noren (RMIT students).

1

THE EVOLVING FIELD OF ENERGY TRANSITIONS

A World of Change

Kathleen M. Araújo

1. Introduction

To paraphrase an old expression, there is no constant like change.[1] Digitalization is dramatically altering the way we work, communicate and live (WEF, n.d.). The global pandemic has led to more than a half billion cases of infection and over six million deaths (WHO, n.d.), considerable economic disruption (IMF, n.d.), and scientific innovations (Fuller et al., 2021; Szabo, 2022). The frequency of extreme weather events is also increasing. (WMO, 2021). Weather that was once deemed as a '100-year occurrence' is now occurring much more regularly.

Alongside the above disruption, the Russian war on Ukraine that began in early 2022 has had a profound effect on society and energy. In addition to the toll of human tragedy, the risks associated with energy import dependence have prompted countless decision-makers to adopt energy strategies that shift away from reliance on Russian energy supplies (Tollefson, 2022).[2] Outside of the war, shocks, including negative oil prices, have occurred recently in energy markets (Hansen, 2020; Gencer and Akcura, 2022). Technologies that have largely been on drafting boards for years, like hydrogen and small modular reactors, are now appearing in energy deployment policies and projects (WEC, 2022; Trinomics, 2020; Deign, 2021; IEA, 2019; Shabenah, 2020). Sources of energy, like wind and solar, that were deemed niche or alternative forms of energy in recent decades are now the fastest-growing areas for new power generation globally, often in cost-competitive terms without subsidies, and at the best levelized cost in specific areas (EIA, 2022a; Lazard, 2021; Marcacci, 2020; Engel, 2021, see also Section 4.6).

Shifts that occur in energy systems can have a profound influence on the functioning of our human, built, and natural systems.[3] One only needs to consider today's debates and discussions about energy to find disruptive thinking about ways that energy should managed and in what timeframe (OECD, n.d.; WEC, n.d.a, n.d.b; IRENA, 2022; Smil, 2016; Sovacool, 2016; Sovacool and Geels, 2016; Grubler, Wilson and Nemet, 2016; Araújo, 2013, 2017).

The goal of this *Handbook* is to provide an overview of energy transitions, by drawing upon insights from a unique and multidisciplinary network of experts from around the world. In doing so, the *Handbook* goes beyond a single approach, sector, or lens. It considers ideas, influences, and approaches that shape the current state of knowledge. The *Handbook* also explores pressing issues, including choices about infrastructure, the role of materials and food systems, sustainability, and energy democracy. Key questions include the following: How is

DOI:10.4324/9781003183020-1

planning anticipating new urban forms of mobility, retirements in electricity infrastructure, and regional security changes? What types of policies appear to be pivotal in energy transitions of different types? In what ways are countries and regions managing their strategies for net-zero change? To what extent should geopolitics, resilience and net-zero carbon priorities be considered together? Finally, where do key interdependencies exist across systems that are relevant to energy? These are simply a point of departure for a long list of topics that this *Handbook* will cover.

The current chapter builds on core, earlier work on energy transitions (Araújo, 2014, 2017) while taking stock of the playing field, based on discussions in the field, new research, and writing, such as that found in the *Routledge Studies in Energy Transitions* book series.[4] The rest of this chapter is organized as follows. Section 2 reviews terminology and publication developments. Section 3 discusses key inflections in energy systems from the last decade. Section 4 looks at longer-term trends and indicators, discussing implications for critical materials and different scopes for costs and benefits. Section 5 considers developments in data and reporting from the last decade, then offers ideas on areas to watch in terms of carbon accounting. Section 6 closes with a review of the rest of the *Handbook*.

2. Definition Refinement and Priorities

The use of the concept of energy transition has varied since the early 1900s (Araújo, 2014). Two prominent writers in the field in more recent times have employed distinctly different definitions. Arnulf Grubler defined an energy transition as "a change in the state of an energy system as opposed to a change in an individual energy technology or fuel source" (Grubler, 1991). Somewhat differently, Vaclav Smil referred to an energy transition and the study of the subject matter as centering on the changing composition (structure) of the primary energy supply (Smil, 2017). This includes the time elapsed between the introduction of a new primary energy source and its rise, with attention given to absolute quantities and qualitative changes that result in wider availability. According to Smil, devices which convert primary energy are also to be considered (ibid.).

Recognizing key elements of definitions such as the above, as well as the advances in the field, this *Handbook* puts forward a more comprehensive definition:

> A considerable shift in the nature or pattern of how energy is used within a system, including the type, quantity, or quality of how energy is sourced, delivered, or utilized. This can be a planned or unplanned change that encompasses the emergence and decline of an energy industry, together with geopolitical, economic, social, and ecological factors that connect to all stages of energy utilization.

Today, energy transitions are now common priorities within government and industry, as well as broader communities (G20, 2022; Office of International Affairs, 2022; WEC, n.d.a; Yazdani, 2020; Shaver, 2021; Smart Cities World Team, 2021). At any level from the local to the international, one can find reporting or related writing on the need or planning for energy system change (ibid.). A close inspection of recent writing on the subject also reveals a considerable increase in publications, as the topic has become more mainstream (Figure 1.1).

In tandem with widespread adoption of energy transition priorities and increased writing on the subject, the range of relevant topics has expanded to include more coverage of regional dynamics, governance and institutional aims, the environment and life cycle, markets and their

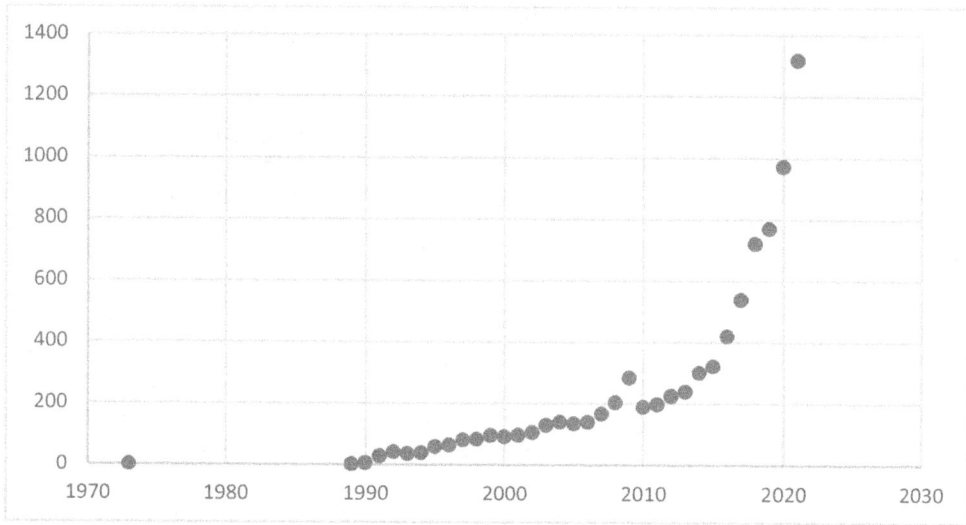

Figure 1.1 Publications on energy transitions.

Source: Web of Science, accessed April 4, 2022, based on the keyword "energy transition".

conversions, infrastructure and technology in all phases of maturity and adoption, and system interactions, among the points of coverage.[5] Participation and fairness in energy decision-making and facility siting are now also among the core themes (Keahey et al., Dunphy and Lennon, Sharp et al., Wilson et al., and Sioshansi, all within this volume).

Many research and policy initiatives tied to energy transitions currently center on the aim to decarbonize the planet. When people to refer *the* energy transition, they are typically speaking about a goal-driven shift in the world's energy balance or mix away from the status quo. Specifically, it is a move toward an energy mix that is without carbon and related greenhouse gas emissions by or before 2050. This is done to keep the global average surface temperature from exceeding a 1.5°C increase over the pre-industrial period (IPCC, 2021, 2022). This objective is a critical priority for many working on energy in the context of the international environment or investment (World Bank, 2021; UN, 2021; IRENA, 2022; S&P Global, 2020). From local efforts through to global ones, this particular kind of transition is also often referred to as net-zero (carbon or greenhouse gas emissions), carbon-neutral, or decarbonized. While net-zero change is an important priority for many who are working on energy transitions, it is important to recognize it is not the only focus. Energy transitions can include intentional or unintended change. Planned changes could entail moves toward greater security, modernization or urbanization, for example. Unplanned energy transitions can include changes for instance, that are triggered by extreme events, such as with the Deep Water Horizon oil spill or the Fukushima Daiichi nuclear meltdown. Following accidents such as these, energy transitions may be reframed by aims centered on recovery, safety or resilience. Today's conditions highlight critical needs to account for decarbonization of energy with resilience and security, where energy security incorporates access to affordable and reliable energy.

On balance, it is a valuable time to be evaluating the field and the state of knowledge for energy transitions. The following section details key inflections in energy with historical and contemporary examples.

3. Then and Now

The strategic importance of energy and aspects of its systemic change can be seen in decision-making and reporting over the past roughly 200 years. In World War I, for example, First Admiral of the Navy Winston Churchill is famously known for changing his position on energy dependence to strategically enhance the speed of the British navy (Araújo, 2017). In doing so, the British navy realigned its fuel along with social and technical systems from domestic coal use to Persian oil (Churchill, 1928, 1968; Churchill and Heath, 1965; Yergin, 1991, 2011), with spillover effects in energy security and geopolitics, economics and jobs, industries, infrastructure, and other aspects.

Early energy reporting also enabled analysts to learn about historical dimensions of commercial and non-commercial energy use, including fuel displacement and associated technologies. Energy company, BP (formerly known as British Petroleum), has extensive reporting on global primary energy since 1965 (BP, 2022), with industrial reporting on coal production that extends back to the 1800s (Communications with Ellen Williams, 2011). Palmer Cosslett Putnam provided some of the earliest national and global energy accounts in the 1900s (Grubler, 2012, citing Putnam, 1953), while Darmstadter et al. wrote about energy in the world economy since 1925 (1971). Mid-20th-century, futures-based research provided tools and methods for envisioning and forecasting alternative outlooks that enabled planning and policy to advance in ways that are now common in energy transitions research (Seefried, 2014; see also journal series, *Technological Forecasting and Social Change,* that started in 1969). At least since the 1970s, energy companies, like Royal Dutch Shell, have employed scenario planning in their decision-making (Jefferson, 1982, 2012),[6] a fundamental tool for energy transitions analysis. For more discussion of energy data, reporting, and accounting, see Section 5.

Turning to more recent times next, one can see critical shifts or inflections since 2010 in the energy system.

3a. 2010–Recent Past

At the start of the last decade, influences were seen with inflation, energy opportunities, and climate. Responses would in some cases evolve throughout the decade.

- Starting with the time period around 2010, a global economic recession was finishing (began in 2007), in which many governments put forward substantial green stimulus packages to facilitate investment in renewable and low-carbon energy development, along with other sectors (see Fox, this volume).
- A boom in unconventional oil and gas associated with hydraulic fracturing and horizontal drilling was underway, with concerns about peak oil subsiding. US shale gas and oil production, for example, rose from near-negligible levels in 2008 to 65 billion cubic feet per day (Bcf/d) of natural gas (70% of total US dry gas production) and about 7 million barrels per day (b/d) of crude oil (60% of total US oil production) in 2018 alone (EIA, 2019; Blasi, 2017).
- A nuclear renaissance was also widely being discussed. The Fukushima Daiichi accident that occurred the following year altered this focus. The ensuing period included near nuclear phase-outs in countries, such as Germany (Rankine, 2022).
- Dissatisfaction with the negotiation deadlock of the 2009 Conference of Parties Summit (Dimitrov, 2010) permeated international climate discussions. This sentiment paved the way for the Paris Accords, which were agreed upon in 2015 (UNF-CCC, n.d.).

3b. Today

By the 2020s, shifts with energy producers, war and pandemic-related risks, as well as attention to an emerging energy economy reflect the more current playing field. Among these conditions, inflation and climate influences are still evident.

- With the ramp-up of unconventional oil and gas in the past decade or so, the leverage held by Russia and countries in the Middle East, as lead producers, has shifted to some degree. Other countries, like the US, have become pivotal players, transforming from net importers of oil and gas to net exporters and leading producers (BP, 2021; EIA, n.d.a and n.d.b, 2022b)
- New risks of energy supply constraints alongside high energy prices tied to the war in Ukraine and COVID-19 economic recovery have prompted a reprioritization of energy security (World Economic Forum, 2022), with deep moves away from dependence on Russian energy exports (Tollefson, 2022; see also Araújo, Chapter 2, this volume).
- The global economy continues to recover from the pandemic, which has persisted for more than two years, and inflation. Like the green stimulus packages from a decade ago, many national governments developed new stimulus packages to build on the progress and lessons from the earlier, global recession, focusing on new levels of support for low-carbon energy jobs, and infrastructure (see Fox, this volume).
- There is greater attention to the "new" energy economy that emphasizes electrifying everything, leveraging digitalization, decarbonization, and the role that users have (IEA, 2022c, 2021d; see also Sioshansi and Wilson et al., this volume). Specific to users, actors such as universities are prioritizing sustainability and in some cases carbon reductions (AASHE, n.d.), as students are influencing the global discourse and agenda (Tollefson, 2019; Long, 2019).

The shifts from the last decade highlight critical junctures in energy transitions that directly frame today's choices. These shifts also inform us more fully about dynamic relationships within and between energy systems. The next section details key trends and notable developments of relevance to energy systems change.

4. Key Trends and Indicators for Energy

Highlights: Broadly, trends are evident in decarbonization, growth in electricity utilization, and sharp declines in the costs for certain power generation and storage technologies, as energy market flux for oil and gas reflects considerable volatility.

Such change is happening as the rate of global economic growth diverges from the pace of growth for population, CO_2 emissions, and energy consumption.

4.1 The Energy Sector and Greenhouse Gases

Today, roughly three-fourths of greenhouse gas (GHG) emissions are generated by the energy sector (IEA, 2021a), with oil, gas, and coal – the main carbon-intensive fuels – representing 81% of the world energy supply in 2019 (IEA, 2021b). If only carbon is considered from fossil fuel combustion, the global emissions have increased by 64% since 1990, with the non-OECD country emissions surpassing those from the OECD country emissions in 2004 (Figure 1.2).[7]

In line with an increased interest in decarbonizing the energy mix, it is now common to see commitments to move away from carbon-intensive and other GHG-intensive energy (UN, n.d.a, Carlin, 2022) toward net-zero energy that emits no carbon or other GHGs. Following the

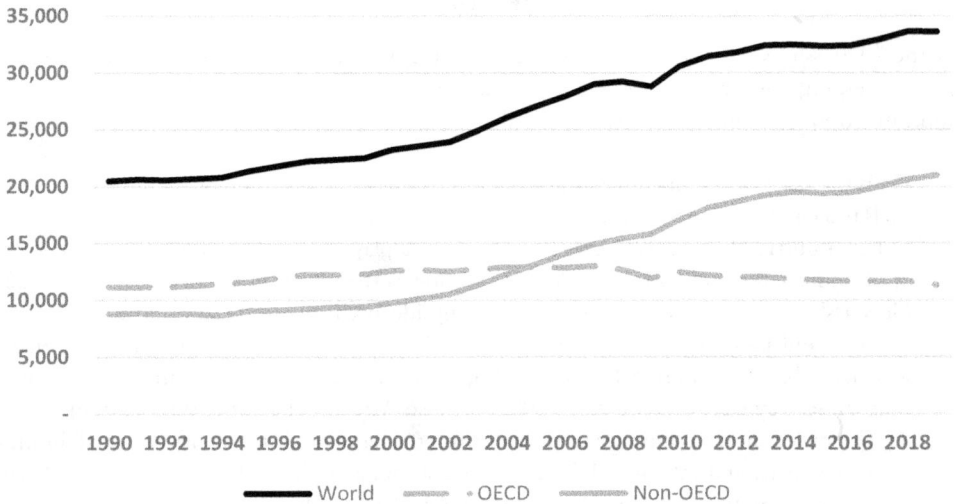

Figure 1.2   Carbon dioxide from fuel combustion, 1990 to 2019 (Mt CO_2).

Source: IEA, 2022b.

Note: Mt is million tonnes.

Conference of Parties Summit in Glasgow, Scotland, in 2021, more than 100 governments have set decarbonization targets, (Carlin, 2022; see also Araújo, Chapter 2). The UN also identifies a growing group of actors across sectors that are pledging to reach net-zero aims. Among these actors are 1,200+ companies, 1,000+ cities, 1,000+ educational institutions, and 400+ financial institutions (UN, n.d.a).

4.2 Global Power Generation Growth and Related Cost Declines

Global power generation and related technology trends reflect additional patterns of relevance for energy transitions. In the period between 1990 and 2020, for example, power generation or electricity totals doubled worldwide, and projections indicate the growth may continue robustly (Raimi et al., 2022). By 2050, such projections suggest that the global power generation will increase by more than 50%, as the share of fossil fuels in power generation drops from 60% to less than 40% (ibid.).

Alongside the above growth is a considerable decline in the costs of electricity and storage for certain technologies. Between 2010 and 2020 alone, for example, costs fell for utility-scale solar photovoltaics (PV) by 85%, concentrating solar power (CSP) by 68%, onshore wind by 56%, and offshore wind by 48% (IRENA, 2021). Real prices for lithium batteries similarly declined by 89% for the same period (BNEF, 2020), which has significance for storage and electrifying transport with electric vehicles. All of these mentioned technologies may be key enablers for decarbonization in electricity and transport.

4.3 Market, Fuel Pricing, and Supply Flux

In line with a theme of disruptive change, energy markets, fuel pricing, and supplies are experiencing considerable flux in recent years (Gould and Atkinson, 2020; IEA, 2022a; World

Economic Forum, 2022). In the first four months of 2020 (with the early spread of the COVID-19 pandemic), Brent crude spot prices, an international benchmark, dropped from $70 per barrel to under $10 per barrel (Raimi et al., 2022). West Texas Intermediate crude oil, another international benchmark, was also trading at negative futures prices for a period (Hansen, 2020), which was largely unheard of in traditional economics training. Brent crude spot prices rose from $42 on average per barrel in 2020 to $71 per barrel on average in 2021 (Raimi et al., 2022). By March 2022, with the Russian war on Ukraine underway, oil prices reached more than $120 per barrel (ibid.).

Similar to oil, natural gas prices, particularly in European markets, have been under major stress. Dutch prices for natural gas, for example, grew by more than a factor of 10 by early March 2022 compared to the 2020 average (ibid.).[8]

Global uranium markets are also an area of flux in connection with the Russian war on Ukraine. Price increases of roughly 20% were evident between February 24, 2022, the day of the Russian invasion, and March 8, 2022 (Erickson, 2022). With Russia and Ukraine among the world's top ten uranium producers in 2021, the geopolitical unrest tied to this nuclear fuel type may spur uranium buyers to diversify their supply chains (ibid.).

4.4 Longer Time Horizons and Energy-Related Indicators

Trends in broad energy-related indicators provide added context for understanding energy systems. Specific to the period between 1971 and 2019, the total energy supply, population, and carbon dioxide (CO_2) emissions more than doubled worldwide, as the global economy increased by more than a factor of 5 (Figure 1.3). Closer scrutiny of the chart shows a decoupling occurring between (1) economic growth (GDP) and (2) energy as well as emissions, starting in the 1980s and then amplifying. The drivers affecting this divergence represent a good subject for additional study. Along with the above shifts, the shares of the global energy supply from fossil fuels shifted from 87% to 81% (IEA, 2022b) (Figure 1.4) – a relatively limited degree of decarbonization as the overall amount of fossil fuels and total energy increased by roughly a factor of 2.5.

Together with the above trends, the rural-urban breakdown in the global population has skewed increasingly with urbanization rising from 37% to 56% during the same period (World Bank, n.d.). This shift has many implications for energy efficiency, the environment, governance,

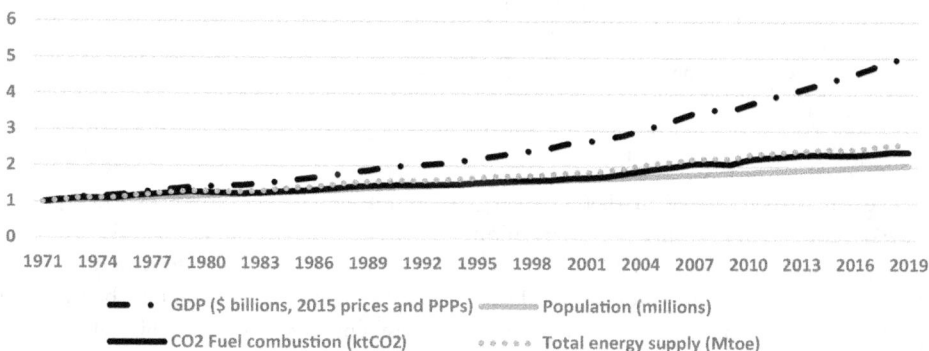

Figure 1.3 Global indicator change (base year: 1971).
Source: IEA, 2022b.

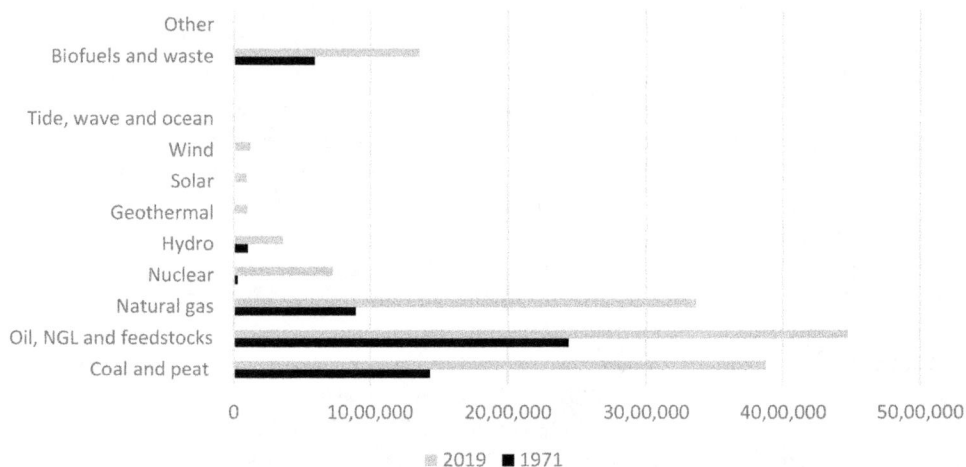

Figure 1.4 Global energy supply (kilotons of oil equivalent).
Source: IEA, 2022b.

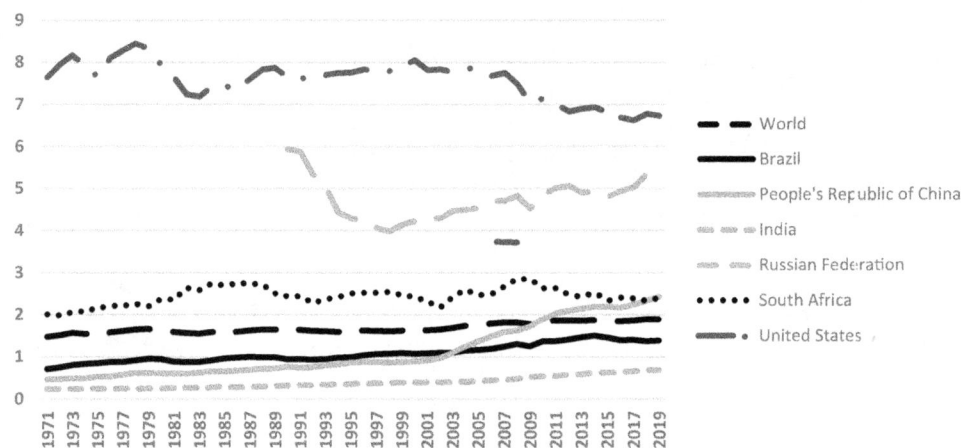

Figure 1.5 Total energy supply per person (TOE per capita).
Source: IEA, 2022b.

resilience, infrastructure, and markets. If energy consumption occurs in concentrated demand hubs, like cities, planning may focus on minimizing energy losses (UNDP et al., 2000) and reducing the land use footprint for the infrastructure. Governance and infrastructure choices may also work out differently in urban and rural energy settings. Rural users may be more self-sufficient and choose options such as microgrids for resilient coverage, especially when market choices may be limited. By contrast, urban energy use may be more regulated. The distinctions between rural and energy decision-making reflect a critical area for additional study.

Another way to consider the long-term trends is to look at the total energy supply per capita for various regions. This characterizes the energy intensity of a country or region's population.

Table 1.1 Global Energy Imports

Fuel Type	1971 (TJ)	2019 (TJ)	Growth Factor 1971–2019
Electricity	.3	2.6	9.8
Coal and peat products	5.4	35.6	6.6
Natural gas	2.0	43.0	21.1
Crude oil, NGL, feedstock, and oil products	67.1	159.5	2.4

Source: IEA (2022b). Accessed June 17, 2022.

Note: NGL refers to natural gas liquids.

Figure 1.5 shows how the US outpaces other major regions and the global average, by using a considerably larger amount of energy per person at 6.7 tons of oil equivalent (TOE) per capita, followed by Russia at 5.4 TOE per capita in 2019. The direction of change for this trend is also instructive. China, Russia, and India (to a lesser extent) show notable increases in the last two decades, as they have industrialized. Meanwhile, the world trend has remained largely flat. By contrast, the US is trending downward, reflecting changes from its heavy intensity that may be tied to efficiency, conservation, and population growth.

Changes in the scale and relative shares of energy imports provide another means to evaluate the global energy system. Table 1.1 shows fossil fuels and electricity imports over the course of roughly five decades. Crude oil and related oil fuel products were and remain the largest share of energy that is imported worldwide, rising from 67.1 to 159.5 terajoules (TJ) between 1971 and 2019. Growth in crude oil and related products may be a reflection of the ease of transport, energy density, and diversity of applications, as well as the market and infrastructure lock-in (see Chapter 2 for a discussion of lock-in) and other barriers to decarbonize. In the same period, natural gas surpassed coal as the second most imported fuel type, reflecting a factor increase of 21 – the largest overall for the period. This reflects a build-out of natural gas infrastructure, including pipelines and liquified natural gas facilities, as well as market expansion and other dynamics.

4.5 Net-Zero Ambitions and Critical Materials

If one were to look ahead in the energy playing field, and consider net-zero ambitions that include increased adoption of wind power, solar photovoltaic energy, and electric vehicles (EVs) using low carbon energy, then important access considerations will need to be addressed in terms of critical minerals (IEA, 2021c; Cunha et al., this volume). Current wind turbine and EV engine designs, for example, use rare earths in their permanent magnets (ibid.). In addition, EVs and other applications for storage rely on battery performance, longevity, and energy density that require minerals, such as lithium, cobalt, nickel, manganese, and graphite. Electricity systems also need copper and aluminum (ibid.).

Cunha et al. (this volume) elaborate on crucial connections between materials and energy transitions. Geopolitical implications of such connections represent security challenges that require strategies today, as well as continued oversight going forward. Such resource challenges can be expected to differ from current fossil fuel issues in terms of the resource geography, infrastructure, actors, and technologies (Araújo, Chapter 2, this volume). Yet similar dynamics may emerge, unless recycling and other use pathways are developed.

4.6 *Costs and Benefits in Energy Systems*

The scope for costs and benefits, as well as assumptions, can vary widely for energy systems. The following outlines a number of basic approaches for measuring energy shifts at micro- and more macro-levels.

The levelized cost of electricity (LCOE) is a common way to compare different types of energy plants in a mostly apples-to-apples manner.[9] The International Energy Agency and OECD Nuclear Energy Agency jointly report on power plant-level costs for baseload electricity generated from fossil fuel, nuclear power, and a range of renewable generation plants (IEA and NEA, 2020).[10] In 2020, the levelized costs of electricity generation for low-carbon generation technologies reflected declines and increasingly were below the costs of conventional fossil fuel generation (ibid.).

Viewed at a broader level, Bloomberg New Energy Finance (BNEF) reported that $755 billion was invested in energy in 2021 (BNEF, 2022). Investments grew in renewable energy, energy storage, electrified transport, electrified heat, nuclear, hydrogen, and sustainable materials, and declined in carbon capture and storage, although new projects were announced (ibid.). Renewables, including wind, solar, plus other types of renewable energy, reflected the largest energy sector of commitment at $366 billion, followed by electrified transport $273 billion (ibid.).

Since energy is the largest contributor to GHG emissions, one can also factor the cost of the changing climate with energy system change. The World Bank produced a study in 2019 for the necessary global infrastructure investment by 2030 to meet stated climate aims, finding that $90 trillion is needed (U.N. n.d.b). The World Bank study went further, by also analyzing the potential to recoup such investments in a transition to a green economy with benefits in jobs and other economic opportunities. It found that there can be a yield of $4 in benefits for every $1 of investment (ibid.).

While these studies had distinct scopes and assumptions, the larger takeaway is the enormous cost of inaction, and opportunity with action relative to the changing climate. Another critical insight is that benefits may be not only in employment and industrial advance, but also in improvements to the quality of life, among possibilities. A challenge for planning or evaluating energy system change, and for making the case for change is that some attributes are difficult to quantify. Nonetheless, they need to be part of the decision-making.

Adopting ways to gauge more qualitative benefits can be challenging for gains such as a cleaner and healthier environment, safety, and resilience outcomes. One way to evaluate a cleaner environment with better health conditions is to assess air quality. An estimated 97% of low- and middle-income countries with more than 100,000 inhabitants fail to meet the World Health Organization's air quality standards (IASS, n.d.). Changing to renewables from fossil fuels could minimize pollutant emissions, which are measurable. Such a shift can then also translate to less disease and burden on health systems (ibid.).

Turning to water, scarcity considerations are a major challenge in different regions worldwide. Shifting to energy strategies that have a reduced water footprint is a co-benefit that can be critical for system resilience and evaluating energy system change (ibid.).[11] This metric may be combined with costs to form net co-benefits, an important approach that has emerged and should be more fully developed in the field.

5. Data, Reporting, and Accounting

Planning, analysis, and tracking of energy systems naturally depend on good data. Critical energy data and reporting sources at the international level include those produced by the

International Energy Agency, BP, International Renewable Energy Agency, OECD, International Atomic Energy Agency, United Nations, World Bank, International Monetary Fund, Renewable Energy Policy Network for the 21st Century, and Nuclear Energy Agency, as well as more industry-based associations reflecting the various energy technologies.[12]

Over the past decade, notable developments in international reporting are evident. These include breaking out renewables more by energy type rather than consolidating them all into an aggregate total, systematic reporting on critical minerals or materials that are relevant to net-zero systems, and more systematic reporting on the subsidies for fossil fuels and renewables at the international level as well as on access to energy (BP, 2021; IEA, n.d.a and IEA, n.d.b).

An important challenge for today's analysis is in standardizing accounting and reporting. Fossil fuel reserves and resources, for example, have been tracked in diverse ways by different countries and organizations (WEC, 2010; EC, n.d.; UNDP et al., 2000/2004). Combustible fuels and assumptions for converting renewables and nuclear can also vary in reporting, based on a number of factors, such as high or low assumed heat values (Araújo, 2014).

An area where additional and critical disconnect exists is in carbon accounting. There is not one standardized approach to represent carbon emissions in an era when net-zero targets and carbon policies are increasingly adopted and viewed as vital worldwide (see Chapter 2, Araujo, and Chapter 16, van den Bergh and Botzen, this volume). Reporting differences may stem, for example, from scoping that centers on the geographic source of the emissions versus the location where a carbon-intensive product may be consumed (EPI, 2022). The UN reporting system also allows one set of standards for developed countries and another for developing countries, with flexibility to decide what, how, and when to report (Mooney et al., 2021). This reflects the fact that traditionally industrialized countries have historically contributed the majority of cumulative GHGs in the atmosphere since the Industrial Revolution. This reporting circumstance also recognizes that the same industrialized countries often have more technical capacity to evaluate their emissions relative to poorer countries (ibid.).

A recent review of country-level reports – the nationally determined contributions agreed to in the Paris Accords – identified major inconsistencies and errors in GHG reporting. The study found significant under-reporting on GHG that ranges from 8.5 billion tons to 13.3 billion tons, with the majority of the gap stemming from the way in which countries account for carbon dioxide emissions from land (ibid.). One of the challenges in correcting this is that the UN does not require atmospheric or satellite measurements; moreover, there are high stakes in the reporting.

One area where carbon accounting is particularly complicated is with bioenergy, where the science shows competing evidence about carbon storage that can depend on the specific feedstock, land use, and management practices (EPI, 2022). How carbon is factored in terms of the selected method, assumptions, sector, geographic scope, boundary, and timing of actual carbon benefits will influence the final accounting (ibid., Mackey et al., 2022; Sterman et al., 2022; IEA Bioenergy, 2018).

As carbon pricing and border carbon adjustments become more prevalent (see Chapter 2), such accounting questions will require resolution.

6. Coverage in the Rest of this *Handbook*

Having highlighted different aspects in the evolving field of energy transitions, the rest of this *Handbook* is organized as follows.

> **Part I** begins with an examination of *concepts and theory* relating to energy transitions. In Chapter 2, Kathleen M. Araújo reviews *influential ideas* that shape energy transitions

knowledge, as well as specific *disruptions* that are of relevance for energy systems. In Chapter 3, Timothy J. Foxon considers the economic framing of *green and post growth*, asking whether these initiatives can effectively integrate carbon reduction with jobs creation, while benefitting disadvantaged communities. Marianne Zotin and her co-authors in Chapter 4 explore critical connections between the *transition in materials and energy*. In Chapter 5, Josh Brinkman and Richard Hirsh reflect on underlying logic in energy choices that may not be evident in broader voting patterns, with *unarticulated identities*. Taking a more unique direction in Chapter 6, Guilherme Pratti and co-authors explore legal conceptualization and related ideas on the *non-human right to energy*. Rounding out this section, Jochen Markard and Daniel Rosenbloom provide an extended perspective in Chapter 7 on *phases of a net-zero energy transition* and how to attain it.

Part II focuses on *systems and geographic dynamics* of energy transitions. The section begins in Chapter 8 with review of challenges and adaptations for the *oil and gas refining* system in a decarbonizing world by Madhav Acharya and co-authors. In Chapter 9, Bruno Cunha and co-authors explore the intersection of the *food system* and decarbonizing energy transition from the standpoint of behavior change. In Chapter 10, Adolfo Mejia-Montero and co-authors detail *vulnerability and resilience in energy system shifts* with cases from Latin America. Chapter 11 changes direction with Peter Karnøe and Raghu Garud examining actors and networks in *path creation of the Danish wind turbine industry*. In Chapter 12, Clare Richardson-Barlow delves into *international political economy aspects of cross-border electricity trade in East Asia*. That is followed by an exploration of *reactive decarbonization in Cuba and Venezuela* in Chapter 13 by Don Kingsbury. The potential for *South Africa to become a hydrogen products superpower* is considered in Chapter 14 by Tobias Bischof-Niemz and Terence Creamer. The section ends with a review in Chapter 15 of micro-cases in *Africa, China, and other regions that are leveraging geothermal energy* to decarbonize heating systems and related applications by Jack Kiruja and Fabian Barrera.

Part III explores *policy, politics, and behavior* in relation to energy system change. Jeroen van den Bergh and Wouter Botzen begin in Chapter 16 with an in-depth review of the debates on *carbon pricing*. Rebecca Windemer focuses next in Chapter 17 on the need to more fully account for *energy infrastructure retirements in planning* by examining onshore wind farms in the United Kingdom. In Chapter 18, Masahiro Matsuura highlights the *politics and policies of a missed opportunity* for Japan to shift its energy strategy, following the Fukushima Daiichi accident. People, politics, and place are considered, then, by Daniel Sharp and co-authors in Chapter 19 in relation to an interdisciplinary agenda for *governance of urban energy* transitions. *Public policy theories are mapped* next by Nihit Goyal and co-authors with bibliometric and computational text analysis in Chapter 20.

Part IV outlines elements of *Strategic and Deliberate Transitions*. Chapter 21 begins with an argument by Clark Miller and co-authors for regional leaders to collaborate on *planned transitions with responsible design*. *Urban mobility planning for future cities* is examined next in Chapter 22 by Arian Mahzouni. Charlie Wilson and co-authors turn, then, in Chapter 23 to deep strategic change to attain global sustainability aims through *energy-services-led transformation*. This is followed in Chapter 24 by Fereidoon Sioshansi's examination of the emerging *demand-side paradigm in the power sector*. Jennifer Keahey and co-authors next explore *conventional notions about energy, energy system change, and its governance* with new ways to frame in Chapter 25. Niall Dunphy and Breffní Lennon offer related insight on citizen participation in energy choices in Chapter 26. The *human development paradigm and social value of energy* are then considered in Chapter 27 by Saurabh Biswas and co-authors. Robert Ferry and Elizabeth Monoian review ways to *plan and design solar energy infrastructure* in Chapter 28 with an aim to be more inclusive.

The *Handbook* concludes with perspectives by transition researchers Kathleen M. Araújo and a number of co-authors, providing ideas on how the *next frontier of energy transitions may be reconsidered in terms of priorities and approaches.*

Notes

1 This is attributed to Greek Philosopher Heraclitus. See Kahn (1979).
2 See also Araújo, Chapter 2 (this volume).
3 Energy systems are defined as "interconnected networks of people and institutions engaged in processes of energy exploration, production, transformation, delivery and use within an enabling environment or ecosystem. These systems include inputs (i.e., fuel resources) and outputs or energy services that are linked by infrastructure and management systems, typically within a market" (Araújo, 2017). *Energy system change* and *energy transition* will be used interchangeably in this chapter.
4 See the series at www.routledge.com/Routledge-Studies-in-Energy-Transitions/book-series/RSENT.
5 This is based on a review of topics that are within the data for Figure 1.1. See also Goyal et al., this volume).
   A follow-up query in the Web of Science on June 17, 2022, for the same author keyword phrase (energy transition) revealed that the top country affiliations for 8,187 publications are China (18.1%), US (15.9%), Germany 10.9%, England (7.4%), and India (6.5%).
6 The author appreciates numerous discussions and communications with Michael Jefferson on this topic (Communications with Michael Jefferson, Spring 2021). He was the first Group Chief Economist for Royal Dutch Shell Group and a member of Shell's scenario team from 1974 to 1979, later holding senior roles within the company associated planning, oil supply, and pricing through to 1990.
7 OECD and non-OECD terms, based on the intergovernmental Organization for Economic Cooperation and Development, refer loosely to industrialized and industrializing countries. Newer terms include the Global North and Global South.
8 In the September to October 2021 period, gas, coal, and electricity prices all reached the highest levels they had been in decades due to a combination of factors (Alvarez and Molnar, 2021).
9 Unlike numbers for simply equipment costs, the levelized cost of electricity reflects the cost to build and operate a power-generating plant, accounting for its assumed financial life and performance cycle. LCOE represents the costs associated with equipment for the power plant, plus operations and maintenance, as well as fuel and financing, based on an assumed use rate over an expected project lifetime (Araújo, 2016). This indicator is often shown in terms of the local currency – for instance, dollars per megawatt-hour or cents per kilowatt-hour. LCOE accounts for some uncertainty with respect to price and discount rate scenarios (ibid.). An advantage to this approach is that it highlights differences in the relative cost structures of energy technologies, reflecting sensitivities to different assumptions about price and discount rate (ibid.).
   An alternative indicator is the levelized avoided costs (LACE) which incorporates expected grid costs to generate power that would otherwise be displaced by a new generation asset or project (Araújo, 2016, citing EIA, 2015; Pentland, 2014).
10 Baseload power is seen as a dispatchable or stable power source predicated on technology characteristics and usually cost.
11 Coal plants and nuclear plants often consume vast amounts of potable water (IASS, n.d.; Araújo, 2017).
12 IEA: www.iea.org/data-and-statistics; BP: www.bp.com/en/global/corporate/energy-economics/statistical-review-of-world-energy.html; IRENA: www.irena.org/statistics; https://stats.oecd.org/; IAEA: https://www-nds.iaea.org/; UN: https://data.un.org/; World Bank: https://data.worldbank.org/; IMF: www.imf.org/en/Data; REN21: www.ren21.net/reports/global-status-report/; NEA: www.oecd-nea.org/jcms/rni_6525/data-bank; etc.

References

Alvarez, C. and Molnar, G. (2021). What Is Behind Soaring Energy Prices and What Is Next? *IEA*, October 12, www.iea.org/commentaries/what-is-behind-soaring-energy-prices-and-what-happens-next.

Araújo, K. (2013). Energy at the Frontier: Low Carbon Energy System Transitions and Innovation in Prime Mover Countries. PhD Dissertation, Cambridge, MA: Massachusetts Institute of Technology.

Araújo, K. (2014). The Emerging Field of Energy Transitions: Progress, Challenges, and Opportunities, *Energy Research and Social Science*, 1: 112–121.

Araújo, K. (2016). Truer Costs in Energy Systems Change, *Papers in Energy*, 141–173 (in Spanish, Los Costes Reales del Cambio del Sistema Energético, *Papeles de Energía*, 43–79). https://www.funcas.es/wp-content/uploads/Migracion/Articulos/FUNCAS_PE/001art07.pdf

Araújo, K. (2017). *Low Carbon Energy Transitions: Turning Points in National Policy and Innovation*. New York: Oxford University Press.

Association for the Advancement of Sustainability in Higher Education (AASHE) (n.d.). www.aashe.org/, Accessed April 5, 2021.

Blasi, A. (2017). Witnessing the Ongoing Transformation of the Oil and Gas Industry, *IEA Commentary*, July 28, www.iea.org/commentaries/witnessing-the-ongoing-transformation-of-the-oil-and-gas-industry.

BNEF (2020). Battery Pack Prices Cited Below $100/kWh for the First Time in 2020, While Market Average Sits at $137/kW, December 16, https://about.bnef.com/blog/battery-pack-prices-cited-below-100-kwh-for-the-first-time-in-2020-while-market-average-sits-at-137-kwh.

BNEF (2022). Global Investment in Low-Carbon Energy Transition Hit $755 Billion in 2021, January 27, 2022, https://about.bnef.com/blog/global-investment-in-low-carbon-energy-transition-hit-755-billion-in-2021.

BP (2022). Statistical Review of World Energy, 1965–2021, www.bp.com/en/global/corporate/energy-economics/statistical-review-of-world-energy.html.

Carlin, D. (2022). How Net Zero Became Our Global Climate Goal and Why We Need It, *Forbes*, April 18, www.forbes.com/sites/davidcarlin/2022/04/18/how-net-zero-became-our-global-climate-goal/?sh=4a69334b2ac9.

Churchill, R. (1968). *Winston Churchill, Volume 2, Young Statesman, 1901–1914*. London: Heiman.

Churchill, W. (1928). *The World Crisis, 1911–1918*. London: Penguin.

Churchill, W. and Heath, F. (1965). *Great Destiny: Sixty Years of the Memorable Events in the Life of the Man of the Century Recounted in His Own Incomparable Words*. New York: Putnam.

Communications with Ellen Williams, Former Chief Scientist for BP (2011).

Communications with Michael Jefferson (Spring 2021).

Darmstadter, J., Teitelbaum, P. and Polack, J. (1971). *Energy in the World Economy*. Baltimore, MD: Johns Hopkins Press.

Deign, J. (2021). Nuclear: These Countries Are Investing in Small Modular Reactors, *World Economic Forum*, January 13, www.weforum.org/agenda/2021/01/buoyant-global-outlook-for-small-modular-reactors-2021.

DiChristopher, T. (2019). Climate Disasters Cost the World $650 Billion Over 3 Years – Americans Are Bearing the Brunt: Morgan Stanley, February 14, www.cnbc.com/2019/02/14/climate-disasters-cost-650-billion-over-3-years-morgan-stanley.html.

Dimitrov, R. (2010). Inside UN Climate Change Negotiations: The Copenhagen Conference, https://onlinelibrary.wiley.com/doi/10.1111/j.1541-1338.2010.00472.x.

Energy Information Administration (EIA) (2019). EIA Adds New Play Production Data to Shale Gas and Tight Oil Reports, February 15, 2019, https://www.eia.gov/todayinenergy/detail.php?id=38372#.

Energy Information Administration (EIA) (2022a). Levelized Costs of New Generation Resources in the Annual Energy Outlook 2022, March 2022, www.eia.gov/outlooks/aeo/pdf/electricity_generation.pdf.

Energy Information Administration (EIA) (2022b). The United States Has Been an Annual Net Total Energy Exporter Since 2019, June 10, 2022, www.eia.gov/outlooks/aeo/pdf/electricity_generation.pdf, Accessed June 11, 2022.

Energy Information Administration (EIA) (n.d.a). Natural Gas Exports Reached a Record High in 2021, www.eia.gov/energyexplained/natural-gas/imports-and-exports.php, Accessed May 13, 2022.

Energy Information Administration (EIA) (n.d.b). Oil and Petroleum Products Explained, www.eia.gov/energyexplained/oil-and-petroleum-products/imports-and-exports.php#:~:text=The%20United%20States%20was%20a,row%20since%20at%20least%201949, Accessed April 23, 2022.

Energy Policy Institute (2022). Carbon Accounting for Forest-based Bioenergy, *Power Talk, Panel*, June 2.

Engel, J. (2021). With or Without Subsidies, Wind and Solar Are Still Cost-Competitive, Report Says, *Renewable Energy World*, October 29, www.renewableenergyworld.com/solar/with-or-without-subsidies-wind-and-solar-are-still-cost-competitive-report-says/#gref.

Erickson, C. (2022). Uranium Buyers, Ukraine Conflict [sic] Drive Spot Price to 10-Year High, *S&P Global Market Intelligence*, March 11, www.spglobal.com/marketintelligence/en/news-insights/latest-news-headlines/uranium-buyers-ukraine-conflict-drive-spot-price-to-10-year-high-69251651.

European Council (EC) (n.d). National Reporting, https://ec.europa.eu/assets/jrc/minventory/national-reportingc1c0.html?field_cs_country_lexique_tid=&order=name&sort=desc, Accessed February 10, 2021.

Fuller, D., Titus, A. and Krogan, N. (2021). Three Medical Innovations Fueled By Covid-19 that Will Outlast the Pandemic, March 9, https://theconversation.com/3-medical-innovations-fueled-by-covid-19-that-will-outlast-the-pandemic-156464.

G20 (2022). G20 Needs to Provide Inclusive Energy System to Accelerate Global Energy Transition, February 21, www.g20-insights.org/2022/02/28/g20-needs-to-provide-inclusive-energy-system-to-accelerate-global-energy-transition.

Gencer, D. and Akcura, E. (2022). Amid Energy Price Shocks, Five Lessons to Remember on Energy Subsidies, May 13, https://blogs.worldbank.org/energy/amid-energy-price-shocks-five-lessons-remember-energy-subsidies#:~:text=The%20World%20Bank's%20latest%20Commodity,and%20economies%20from%20price%20shocks.

Gould, T. and Atkinson, N. (2020). The Global Oil Industry Is Experiencing a Shock Like No Other in its History, *IEA*, April 1, www.iea.org/articles/the-global-oil-industry-is-experiencing-shock-like-no-other-in-its-history.

Grubler, A. (1991). Diffusion: Long-term Patterns and Discontinuities, *Technological Forecasting and Social Change*, 39(1–2): 159–180.

Grubler, A. (2012). Energy Transitions Research: Insights and Cautionary Tales, *Energy Policy*, 50: 8–16.

Grubler, A., Wilson, C. and Nemet, G. (2016). Apples, Oranges, and Consistent Comparisons of the Temporal Dynamics of Energy Transitions, *Energy Research & Social Science*, 22: 18–25.

Hansen, S. (2020). Here's What Negative Oil Prices Really Mean, *Forbes*, April 21, www.forbes.com/sites/sarahhansen/2020/04/21/heres-what-negative-oil-prices-really-mean/?sh=782e67505a85.

IEA Bioenergy (2018). Is Energy from Woody Biomass Positive for the Climate? January, www.ieabioenergy.com/wp-content/uploads/2018/01/FAQ_WoodyBiomass-Climate_final-1.pdf

Institute for Advanced Sustainability Studies (IASS) (n.d.). Focusing on the Co-benefits of the Energy Transition, www.iass-potsdam.de/en/news/focusing-co-benefits-energy-transition, Accessed December 18, 2021.

Intergovernmental Panel on Climate Change (IPCC) (2021). Global Warming of 1.5 °C, Special Report, www.ipcc.ch/sr15.

Intergovernmental Panel on Climate Change (IPCC) (2022). IPCC Sixth Assessment Report: Impacts, Adaptation and Vulnerability, www.ipcc.ch/report/ar6/wg2.

International Energy Agency (IEA) (2019). The Future of Hydrogen, www.iea.org/reports/the-future-of-hydrogen.

International Energy Agency (IEA) (2021a). Net Zero by 2050: A Roadmap for the Global Energy Sector, May 2021, www.iea.org/reports/net-zero-by-2050.

International Energy Agency (IEA) (2021b). Key World Energy Statistics 2021, www.iea.org/reports/key-world-energy-statistics-2021/energy-balances.

International Energy Agency (IEA) (2021c). The Role of Critical Materials in Clean Energy Transitions, https://iea.blob.core.windows.net/assets/ffd2a83b-8c30-4e9d-980a-52b6d9a86fdc/TheRoleofCriticalMineralsinCleanEnergyTransitions.pdf

International Energy Agency (IEA) (2021d). Smart Grids, Report, November 2021, www.iea.org/reports/smart-grids.

International Energy Agency (IEA) (2022a). Emergency Measures Can Quickly Cut Global Oil Demand by 2.7 Million Barrels a Day, Reducing the Risk of a Damaging Supply Crunch, www.iea.org/news/emergency-measures-can-quickly-cut-global-oil-demand-by-2-7-million-barrels-a-day-reducing-the-risk-of-a-damaging-supply-crunch.

International Energy Agency (IEA) (2022b). Data, Subscription, www.iea.org/.

International Energy Agency (IEA) (2022c). The Potential of Digital Business Models in the New Energy Economy, January 7, 2022, www.iea.org/articles/the-potential-of-digital-business-models-in-the-new-energy-economy.

International Energy Agency (IEA) (n.d.a). Energy Subsidies, www.iea.org/topics/energy-subsidies, Accessed August 1, 2021.

International Energy Agency (IEA) (n.d.b). Access to Electricity, www.iea.org/reports/sdg7-data-and-projections/access-to-electricity, Accessed August 1, 2021.

International Energy Agency (IEA) and Nuclear Energy Agency (NEA) (2020). Projected Costs of Electricity 2020, www.iea.org/reports/projected-costs-of-generating-electricity-2020.

International Monetary Fund (IMF) (n.d.). Policy Responses to COVID 19, www.imf.org/en/Topics/imf-and-covid19/Policy-Responses-to-COVID-19, Accessed April 30, 2022.

IRENA (2021). Power Generation Costs in 2020, www.irena.org/-/media/Files/IRENA/Agency/Publication/2021/Jun/IRENA_Power_Generation_Costs_2020.pdf.

IRENA (2022). Energy Transition Holds Key to Tackle Global Energy and Climate Crisis, March 29, www.irena.org/newsroom/pressreleases/2022/Mar/Energy-Transition-Holds-Key-to-Tackle-Global-Energy-and-Climate-Crisis.

Jefferson, M. (1982). Historical Perspectives of Societal Change and the use of Scenarios at Shell, Chapter 8 in: Twiss, B. (Ed.), *Social Forecasting for Company Planning*. London: Macmillan.

Jefferson, Michael (2012). Shell Scenarios: What Really Happened in the 1970s and What May be Learned for Current World Prospects. *Technological Forecasting & Social Change*, 79: 192.

Kahn, C. (Ed.) (1979). *The Art and Thought of Heraclitus*. Cambridge: Cambridge University Press.

Lazard (2021). Levelized Cost of Energy, Levelized Cost of Storage, and Levelized Cost of Hydrogen, October 28, www.lazard.com/perspective/levelized-cost-of-energy-levelized-cost-of-storage-and-levelized-cost-of-hydrogen.

Long, C. (2019). The 'Greta Effect' on Student Activism and Climate Change, *National Education Association*, September 19, www.nea.org/advocating-for-change/new-from-nea/greta-effect-student-activism-and-climate-change.

Mackey, B., Moomaw, W., Lindemayer, D. and Keith, H. (2022). Net Carbon Accounting and Reporting are a Barrier to Understanding the Mitigation Value of Forest Protection in Developed Countries, *Environmental Research Letters*, April 28, https://iopscience.iop.org/article/10.1088/1748-9326/ac661b/meta.

Marcacci, S. (2020). Renewable Energy Prices Hit Record Lows: How Can Utilities Benefit from Unstoppable Solar and Wind? *Forbes*, January 21, www.forbes.com/sites/energyinnovation/2020/01/21/renewable-energy-prices-hit-record-lows-how-can-utilities-benefit-from-unstoppable-solar-and-wind/?sh=27bc7a7c2c84.

Mooney, C., Eilperin, J., Butler, D., Muyskens, J., Narayanswamy, A. and Ahmed, N. (2021). Countries' Climate Pledges Built on Flawed Data, Post Investigation Finds, *The Washington Post*, November 7, www.washingtonpost.com/climate-environment/interactive/2021/greenhouse-gas-emissions-pledges-data.

OECD (n.d.). Managing Environmental and Energy Transitions for Regions and Cities, www.oecd.org/greengrowth/energy-environment-transition.htm, Accessed March 1, 2022.

Office of International Affairs (2022). U.S. Department of Energy, Joint Statement between the Ministry of the Economy, Trade and Industry of Japan, and the U.S. Department of Energy on Cooperation toward Energy Security and Clean Energy Transition, May 4, www.energy.gov/ia/articles/joint-statement-between-ministry-economy-trade-and-industry-japan-and-united-states.

Raimi, D., Campbell, E., Newell, R., Preset, B., Villanueva, S. and Wingenroth, J. (2022). Global Energy Outlook 2022: Turning Points and Tension in the Energy Transition, RFF, Report 22–04, April, https://media.rff.org/documents/Report_22-04_v1.pdf.

Rankine, A. (2022). Why Europe Could be Heading for a "Nuclear Renaissance" With Energy Prices at Record Highs, *Money Week*, April 15, https://moneyweek.com/investments/commodities/energy/604701/why-europe-could-be-heading-for-a-nuclear-renaissance-with.S&P Global (2020). What is Energy Transition? February 24, www.spglobal.com/en/research-insights/articles/what-is-energy-transition.

Seefried, E. (2014). Steering the Future, *European Journal of Futures Research*, 15: 29.

Shabenah, R. (2020). Adaptive Policy to Leverage Hydrogen in the Energy Transition, *G20 Insights*, November 20, www.g20-insights.org/policy_briefs/adaptive-policy-to-leverage-hydrogen-in-the-energy-transition.

Shaver, L. (2021). To Advance a Clean Energy Transition, US Cities and Corporations Should Collaborate, *The City Fix*, September 2, https://thecityfix.com/blog/to-advance-a-clean-energy-transition-us-cities-and-corporations-should-collaborate.

Smart Cities World Team (2021). COP26: C40 Cities Launches Clean Construction Coalition to Halve Emissions by 2030, November 11, www.smartcitiesworld.net/commercial-buildings/commercial-buildings/cop26-c40-cities-launches-clean-construction-coalition-to-halve-emissions-by-2030-7124.

Smil, V. (2016). Examining Energy Transitions: A Dozen Insights Based on Performance, *Energy Research and Social Science*, 22: 194–197.

Smil, V. (2017). *Energy Transitions: Global and National Perspectives* (2nd ed.). Westport, CT: Praegar.

Sovacool, B. (2016). How Long Will It Take? *Energy Research and Social Science*, 13: 202–215.

Sovacool, B. and Geels, F. (2016). Further Reflections on the Temporality of Energy Transitions, *Energy Research and Social Science*, 22: 232–237.

Sterman, J., Moomaw, W., Rooney-Varga, J. and Siegel, L. (2022). Does Wood Bioenergy Help or Harm the Climate? *The Bulletin of Atomic Scientists*, 78(3): 128–138, www.tandfonline.com/doi/full/10.1080 /00963402.2022.2062933.

Szabo, L. (2002). Covid's 'Silver Lining': Research Breathroughs for Chronic Disease, Cancer and the Common Flu, March 17, www.medscape.com/viewarticle/970503.

Tollefson, J. (2019). The Hard Truths of Climate Change–by the Numbers, *Nature*, 573: 325–326, September 19, www.nature.com/immersive/d41586-019-02711-4/index.html.

Tollefson, J. (2022). What the War in Ukraine Means for Energy, *Climate and Food, Nature*, 604: 232–233, April 5, www.nature.com/articles/d41586-022-00969-9.

Trinomics (2020). Opportunities for Hydrogen Energy Technologies Considering the National Energy & Climate Plans, August 31, www.fch.europa.eu/publications/opportunities-hydrogen-energy-technologies-considering-national-energy-climate-plans.

United Nations (2021). Theme Report on Energy Transition, www.un.org/sites/un2.un.org/files/2021-twg_2-062321.pdf.

United Nations (n.d.a). Climate Action, www.un.org/en/climatechange/net-zero-coalition, Accessed January 10, 2022.

United Nations (n.d.b). Financing Climate Action, www.un.org/en/climatechange/raising-ambition/climate-finance, Accessed March 2, 2022.

UN Development Programme (UNDP), UN Dept of Econ and Social Affairs (UNDESA), World Energy Council (WEC) (2000/2004 update). World Energy Assessment, www.nzdl.org/cgi-bin/library.cgi?e=d-00000-00 – off-0edudev – 00-0-0-10-0-0-0direct-10-4-0-11 – 11-en-50–20-about – 00-0-1-00-0-0-11–1–0utfZz-8–00&a=d&c=edudev&cl=CL3.159&d=HASH01a07b9e87963817e770 4dde.3; https://sustainabledevelopment.un.org/content/documents/2420World_Energy_Assessment_Overview_2004_Update.pdf

UNFCCC (n.d.). What is the Paris Agreement? https://unfccc.int/process-and-meetings/the-paris-agreement/the-paris-agreement, Accessed January 5, 2022.

Web of Science, 'Energy Transition', Search of 'Author Keyword', Accessed April 4, 2022.

World Bank (2021). Transitions at the Heart of Climate Change, May 24, 2021, www.worldbank.org/en/news/feature/2021/05/24/transitions-at-the-heart-of-the-climate-challenge.

World Bank (n.d.). World Development Indicators, https://databank.worldbank.org/source/world-development-indicators, Accessed March 1, 2022.

World Energy Council (WEC) (2010). World Energy Resources, Report, www.worldenergy.org/publications/entry/world-energy-resources-2010-survey.

World Energy Council (WEC) (2022). Regional Insights into Low-Carbon Hydrogen Scale Up, www.worldenergy.org/assets/downloads/World_Energy_Insights_Working_Paper_Regional_insights_into_low-carbon_hydrogen_scale_up.pdf?v=1653898629.

World Energy Council (n.d.a). Creating Insight for [a] Successful Energy Transition, www.worldenergy.org/transition-toolkit#fullpage1, Accessed April 5, 2022.

World Energy Council (n.d.b). World Energy Transition Radar, www.worldenergy.org/transition-toolkit/world-energy-scenarios/covid19-crisis-scenarios/world-energy-transition-radar, Accessed February 3, 2022.

World Economic Forum (WEF) (2022). Fostering (an) Effective Energy Transition 2022, May 11, www.weforum.org/reports/fostering-effective-energy-transition-2022/in-full/1-1-economic-development-and-growth.

World Economic Forum (WEF) (n.d.). Understanding the Impact of Digitalization on Society, https://reports.weforum.org/digital-transformation/understanding-the-impact-of-digitalization-on-society.

World Health Org (WHO). Coronavirus Dashboard, https://covid19.who.int/, Accessed October 9, 2022.

World Meteorological Organization (WMO) (2021). Weather-Related Disasters Increase Over Past 50 Years, Causing More Damage but Fewer Deaths, August 31, 2021, https://public.wmo.int/en/media/press-release/weather-related-disasters-increase-over-past-50-years-causing-more-damage-fewer.

Yazdani, H. (2020). The Great Reset: How Cities Are Leading the Energy Transition, *World Economic Forum*, August 21, 2020, www.weforum.org/agenda/2020/08/role-of-cities-in-the-energy-transition.

Yergin, D. (1991). *The Prize: The Epic Quest for Oil, Money & Power*. New York: Free Press.

Yergin, D. (2011). *The Quest: Energy, Security, and the Remaking of the Modern World*. New York: Penguin.

PART I

Concepts and Theory

2

A ROADMAP FOR CONCEPTS AND THEORY OF ENERGY TRANSITIONS

Kathleen M. Araújo

1. Introduction

There continues to be a strong sense today that we need to manage energy differently (Araújo, 2017). Forces at work, including decarbonization, geopolitics, digitalization, and COVID-19 conditions, among other disruptions, are transforming our energy systems.

To advance thinking about energy system change and deepen related discussions, this *Handbook* provides insight into lessons from around the world. As a foundation for the remaining chapters within the *Handbook*, this chapter examines key ideas that shape or explain elements of energy transitions. These concepts and theory may be applicable at different scales, as well as in varied timeframes and geographic scopes.

For the purposes here, an energy transition is defined as a considerable shift in the nature or pattern of how energy is used within a system, including the type, quantity, or quality of how energy is sourced, delivered, or utilized. This can be a planned or unplanned change that encompasses the emergence and decline of an energy industry, together with geopolitical, economic, social, and ecological factors that connect to all stages of energy utilization (Araújo, Chapter 1, this volume).

The rest of this chapter is structured as follows: Section 2 builds on the coverage in Chapter 1 to review a wider set of influential ideas. Section 3 reviews similarities and differences between the focus of energy transitions and sustainability transitions fields of study. Section 4 details a range of disruptive conditions, with an emphasis on their relevance for current energy systems change. Section 5 looks more deeply at advanced concepts and theory that relate to energy systems and associated change. The chapter closes with takeaways for continued work in the field and the choices that are associated with it.

2. A Deeper Review of Early Influences in the Writing

One might say there is always some form of energy transition underway. A review of influences in the writing and developments relating to energy provides a basis for a fuller perspective on this topic.

In 1778, Thomas Malthus wrote "An Essay on the Principle of Population", emphasizing how constraints in our natural system (including energy) limit the growth and quality of life

DOI:10.4324/9781003183020-3

for society (Malthus, 1999). That same idea was examined nearly 200 years later in the *Limits to Growth* study by the Club of Rome (Meadows et al., 1972), which used system dynamics modeling to show projections of *human-natural system interactions* over time for five key variables: population, consumption of nonrenewable natural resources (including petroleum), industrialization, pollution, and food production (ibid.). Consistent with Malthus' earlier conceptual work, the *Limits* study reinforced the point that *unsustainable practices* in natural resource use, along with other exponential growth, can lead to society ultimately exceeding the carrying capacity of the planet.[1] The notion of society overshooting sustainable limits of natural resources was developed further in the articulation of *sustainable development* by the World Commission on Environment and Development, in *Our Common Future* (1987) – what became known as the Brundtland Report. The concept of sustainable development integrates three interdependent and mutually reinforcing priorities that value the environment, economics, and society in an equitable manner across generations (ibid.).

Box 2.1: Differences with Sustainable, Low-Carbon, Renewable, and Clean Energy

Different interpretations of the term *sustainability* complicate its use (Daly, 1996). In lay terms, sustainability refers to a quality of durability. In the context of energy, *sustainable* is often used interchangeably with *low-carbon*, *renewable*, and *clean*. However, important distinctions exist. *Sustainable* implies durability or (in more advanced terms) enduring intergenerationally in a manner that does not unduly compromise society, the economy, or the environment. By contrast, *low-carbon* generally means less carbon is emitted into the atmosphere relative to conventional, fossil-fuel-based options. *Renewable* typically refers to sources of energy that naturally regenerate, and *clean* typically refers to energy that does not pollute.[2]

During the 1970s and early 1980s, *energy security* and *resilience* measures were both evident. Energy security, as discussed in Chapter 1, is defined as access to affordable and reliable energy, whereas resilience refers to the ability to withstand or recover from low-frequency, high-impact events (Araújo and Shropshire, 2021). Two oil crises of that era challenged the pivotal role that energy plays in the global and national economies, with new capacity-building occurring to minimize disruption (IEA, 1994a, 1994b; Araújo, 2017). In the first crisis of 1973–1974, an international embargo was imposed on oil exports to select oil-importing countries, together with a reduction in oil production (Office of the Historian, n.d.). These actions contributed to a quadrupling of oil prices between 1973 and 1978 (BP, 2021). A second crisis occurred in 1979–1980, following the Iranian Revolution, as concern for another global oil disruption took hold (Graefe, 2013). The latter resulted in more than a doubling of the oil prices between 1978 and 1980, on top of the earlier price increase from the previous crisis (BP, 2021). Responses to the shocks included the formation of the International Energy Agency and the strategic petroleum reserve to provide energy-importing nations with energy and institutional capacity to lessen the risk of energy market flux (IEA, 1994a, 1994b). During this time, it was not unusual to also see the establishment of a national cabinet-level or ministry-level agency to oversee national energy security (Office of Legacy Management, n.d.).

In the same period, Amory Lovins wrote *Soft Energy Paths* and *The Road Not Taken* (1977, 1976), combining ideas about natural resource limits and energy security with analysis that compared a conventional energy approach, including fossil fuels and nuclear energy, relative to an alternative approach that incorporated renewables, energy efficiency, and conservation. Consistent with the heightened energy security concerns of the times, Lovins' alternative path mapped a course for society to adopt an informed shift away from the status quo to a novel strategy that would be more sustainable. The practice of producing alternative energy plans became a more substantive step in decision-making during these times (Araújo, 2017). A critical example of such an alternative strategy was in the *Energiewende (Energy Transition)* study, produced by Germany's Institute for Applied Ecology in the early 1980s (Krause et al., 1982). Unlike the *Limits to Growth* study, which did not propose specific alternative options, *Energiewende* reflects some of the earliest writing to propose a holistic approach to strategically alter an energy system. It argued that economic growth was possible with lower energy consumption that could be accomplished through renewable energy and efficiency (ibid.).

In addition to the oil crises and capacity-building that occurred during the above time period, key analysis of trends in global, national, and related energy substitution was completed by researchers, such as those with the International Institute of Applied Systems Analysis.[3]

Employing Fisher and Pry's simple substitution model for technology change (1971), Cesare Marchetti and Nabojsa Nakicenovic produced market penetration analysis for energy sources (1979). Evaluating 300 cases of energy systems, Marchetti and Nakicenovic concluded, among their findings, that the *timescale* to shift a specific energy source in the global energy mix was 50–100 years (ibid.).

More recently, others have examined timescales for energy transitions, such as that with national change. Smil (2016), Sovacool (2016), Sovacool and Geels (2016), Grubler et al. (2016), and Araújo (2013, 2017) have produced a range of analytical insights, including, for example, the potential for considerable national energy system change in less than 15 years. Distinctions across the findings exist, due in part to differences in methodology, assumptions, and scope. The *pace of change (or temporality)* to shift an energy system remains an important area for continued study. General timelines to attainment may factor critically in societal buy-in and evaluating net costs and benefits, as more priorities converge around altering existing energy systems in connection with the carbon intensity, security and resilience, affordability, and access, among aims.

Timescales or temporality for change connects importantly to the concepts of *diffusion* and *the S-curve,* which explain adoption modes for a specific change (Griliches, 1957; Mansfield, 1961). Everett Rogers' breakthrough writing on diffusion of innovations sheds light on how adoption of change occurs with early versus late adopters, and what factors can affect this process (1995). Represented by the S-curve, diffusion begins gradually, according to Rogers, as early adopters overcome barriers. The pace of diffusion accelerates, then, as more people learn, seeing the value of the benefits and feasibility of use. The process eventually tapers off, as the market saturates. Rogers then elaborated on key factors that can affect the pace of change including: the relative advantage of a change compared to the status quo that it supersedes; the compatibility and/or coherence of change with norms, needs, and understanding; the complexity of a change; opportunities for testing a change; and the observability or clarity of positive results (ibid.). Such ideas are useful for all sectors as people aim to catalyze or leverage change.

In addition to evolving ideas on sustainability, security, timescales, and diffusion, forms of change and lock-in are also key for conceptualizing energy transitions. Technology writing characterizes basic forms of change in terms of *incremental versus disruptive* types that generally imply a slight modification compared to substantial adaptation (Araújo, 2017). Closer inspection reveals nuances even for this simple classification. For example, changes can be at varied scales, including at the end user level or at the macro-system level. If considering disruptive change at the end user level, such a shift could be from riding a horse to a car. At the macro-system level, an example of disruptive change could be in shifting regional electricity from centralized coal generation to distributed power from renewables and storage, which is not necessarily apparent to all users.

A refinement to the change categories could include the introduction of an intermediate stage of *continuous* change between incremental (end-of-pipe) and disruptive (discontinuous) forms (Unruh, 2000, 2002). This could be accomplished by repositioning of an incumbent system with an upgrade, without fundamentally changing the technology or institutions (ibid.; Berkhout, 2002).

Frank Geels and Johan Schot contributed a related classification of technology shifts, based on the processes that are involved. These include *substitution, reconfiguration, transformation, and realignment/dealignment* (2007). Among the four shifts, two common examples with energy systems are found in substitution and reconfiguration. Substitution involves one fuel type replacing another, for example, in energy production with natural gas replacing coal. By contrast, reconfiguration could entail a shift from distributed, onshore wind power with small turbine groupings to installations of large offshore wind farms that are centralized. With reconfiguration, the form of electricity generation could be the same, as was the case in the wind power example, but the system to produce and supply it (together with its geographic footprint) can be restructured quite differently.

In 2012, Geert Verbong and Derk Loorbach put forward one of the earliest edited volumes on energy transitions, with a focus on governance. Their book led a period of exponential growth in writing on energy transitions, which continues to today (see Chapter 1, Figure 1.1).

Going beyond the above concepts and developments, one may also want to consider how to adapt an energy system or technology with ideas on *leapfrogging* and *hybridization*. Leapfrogging in terms of energy system change refers to the potential to sidestep early lessons and issues, by directly adopting a more superior option or system (Goldemberg, 1998). By contrast, hybridization conceives of energy system change somewhat differently by envisioning incumbent and novel technologies being used concurrently rather than sequentially. This latter approach permits ongoing learning and incremental refinements, reflecting Unruh's continuous category of transitioning (2000, 2002).

Closely connected to these ideas are those on *path dependence and carbon lock-in*, as well as *path creation*. Path dependence is the notion that earlier choices limit the range of later options, based on barriers such as sunk costs, increased returns from continuing on the existing path, or the interrelatedness of technologies (Arthur, 1989; David, 1985). Essentially, early steps crowd out future options. Carbon lock-in takes this idea further, as a special case of path dependence, by describing how industrial economies become entrenched in fossil-fuel-based systems with related developments in technology and institutions (Unruh, 2000; Seto et al., 2016). Ideas, like path dependence, may be criticized for sounding *technologically deterministic*, as if the evolution of technology occurs autonomously without society having opportunity to intervene (Kline, 2001). *Path creation* represents an alternative to more technologically deterministic scoping, by indicating that there are opportunities throughout evolution to alter the status quo with "mindful deviation" (Garud and Karnøe, 2001; Karnøe and Garud, this volume). Path creation thinking is more *socially constructive*, seeing human development and the advance of knowledge, as being based on social interaction (Detel, 2001).

3. Energy Transitions Versus Sustainability Transitions

When approaching ideas about energy transitions, it is important to understand how the field of study relates to and yet also differs from that for *sustainability transitions*. Commonality exists in the scopes of focus that prioritize sustainability and in the researchers who work on both.

Where these two fields of study differ is in their expanded areas of coverage. The field of sustainability transitions can focus, for instance, on subjects that go beyond energy, such as with water, food, or materials. The field of energy transitions may approach these subjects as key, related systems (see Cunha et al. and Zotin et al., this volume). The energy transitions field of study also evaluates aims that can differ from sustainability, such as ones like regional diversification, affordability, resilience, technology leadership, jobs, or security.[4] Scenario development that is carried out by fossil fuel companies, for example, may focus on anticipating market change or shifts in demand (Jefferson, 1982, 2012), irrespective of environmental impacts. Energy transition analysis may also be reflected in strategies of cities, states, countries, supranational regions, companies, and universities that are adopting plans to alter their approach to carbon intensity, industry, workforce diversity, affordability, or fairness, etc. These aims may include sustainability, but could also translate to distinct strategies that may be at odds with each other, especially during periods of tight budgets. The key, naturally, is to identify areas for synergy.

4. Disruptions

Having considered some of the conceptual and development playing field, this section examines disruptions to current energy systems in more depth. Forms of disruption include decarbonization, geopolitics, digitalization, and the COVID-19 pandemic.

4a. Decarbonization and Carbon Policies

Priorities to shift the global energy mix away from its carbon-intensive form that relies primarily on fossil fuels, to one that no longer emits carbon and related greenhouse gasses (GHGs) into the atmosphere, are now evident in many public and private sector agendas (Araújo, Chapter 1, this volume; UN, n.d.). Whether one calls this strategy *decarbonizing, net-zero, or carbon-neutral*, the disruption translates in such a way that carbon and other GHGs are not being generated when the energy is produced, or the emissions are captured and stored, in order not to escape into the atmosphere. If GHGs are emitted into the atmosphere, *radiative forcing* creates a heating effect on the planet.

Decarbonization has been quantified with a range of metrics, including the following:

- Absolute emissions of carbon dioxide (CO_2 or CO_2 equivalents[5] in tonnes
- CO_2 per total primary energy produced (tonnes per TJ, MTOE, or BTU)
- CO_2 per gross domestic product (tonnes per \$GDP)
- CO_2 per person (tonnes per capita)
- CO_2 per square mile (tonnes per square mile)

The last four metrics represent the carbon intensity of certain subtopics, such as in the carbon intensity of the economy. The variation in these metrics can mean that actors, like countries that are parties to a treaty, may measure their decarbonization performance in different ways (see Chapter 1, Section 5). This diversity complicates shared international efforts and can in the end lead to an increase in absolute carbon emissions, when all may indicate they are decarbonizing.

In recent years, the terminology and form of measurement appear to be converging around a more simplified term of *net-zero*. The net-zero phrase has become a mainstream way to characterize carbon reduction efforts, largely since the Paris Agreement of 2015. In the legally binding international treaty known as the Paris Accords, 196 countries adopted the aim to reduce GHG emissions to as close to zero as possible (UNFCCC, 2015). Taking effect in 2016, the

Table 2.1 Net-Zero Targets – Worldwide, as of 2021

Net-Zero Target Status	Share (%) of Total Global Energy Supply	Share (%) of Global CO_2 Emissions from Fuel Combustion	Share (%) of Total Global Gross Domestic Product (Nominal)
Laws	14	12	0
Policy documents	44	50	50
Declaration/pledges	21	21	12
Under discussion	5	4	4
Other	16	13	34

Sources: World Economic Forum, 2022, citing Energy and Climate Intelligence Unit; IEA; World Bank.

treaty focuses on reducing global warming and building resilience to climate change, limiting warming to no more than 1.5°C (UN, n.d.). The energy sector is crucial for avoiding the worst effects of climate change, as noted in Chapter 1, with science emphasizing a diminishing window of actionable time to reduce carbon emissions and reach net-zero by 2050 (ibid.; IPCC, 2021). To keep warming to no more than 1.5°C above pre-industrial levels, countries must cut emissions by at least 45% by 2030 compared to 2010 levels, with the transition to net-zero emissions fully complete by 2050 (UN, n.d.).

To support net-zero efforts, net-zero targets, carbon pricing, and border carbon adjustments are three key policy-based measures that governments utilize.[6] Table 2.1 outlines *net-zero targets* worldwide, as of 2021. As opposed to being codified in law, the majority of targets are found in policy documents. These represent 44% of the total global energy supply, 50% of global CO_2 emissions from fuel combustion, and 50% of the total global nominal gross domestic product. Whether targets are voluntary or mandated, they can signal a direction but not carry much weight without the fortitude of implementation by the public and private sectors.

A second form of carbon policy is *carbon pricing*. This accounts for the external costs of carbon – basically, what the public pays for in other ways, such as in health care costs (World Bank, n.d.a). The cost of carbon raises the price of a product in relation to the amount of carbon involved in the creation of the given product. Carbon pricing introduces accountability for carbon-related damage to those who are responsible for it and who could minimize it (ibid.).[7] Such pricing can take the form of a *cap-and-trade system* in which a predetermined quantity of allowable carbon is set by a rule (i.e., capped), allowing the market to determine the price. Alternatively, a *premium/tax* has a price set by a rule, with the quantity that is traded left to the market forces. Countries have been continuing to expand their consideration of carbon pricing more intensely, as IPCC report conclusions increasingly emphasize the importance of action (ibid.; de Coninck et al., 2018; IPCC, 2022).

Challenges exist in carbon pricing with respect to *leakage*, where pricing may not be equitably applied across domestic or international markets. Some industries, such as steel, concrete, and aluminum production, are more carbon-intensive than other industries. With carbon pricing implemented in one region but not another, carbon-intensive industries could relocate their operations to regions/jurisdictions that do not tax carbon, creating an unfair advantage for non-regulated industry competitors. This process of carbon leakage can undermine domestic production and production capability while globally increasing carbon emissions, in effect weakening the attainment of aims behind carbon pricing.

Border carbon adjustments seek to address the issue of carbon leakage in two ways: (1) a tariff or tax on imported goods and (2) a rebate credit for exported goods. The Carbon Border

Adjustment Mechanism (CBAM) for the European Union is arguably the boldest version of a border carbon adjustment put forward to date. It is designed to be gradually phased in with an increasing commitment to more restrictive international carbon policy, and a commitment to equitable policies in light of unequal international competition (European Council, 2022). CBAM targets imports of carbon-intensive products: cement, aluminum, fertilizers, electric energy production, iron, and steel (ibid.). It aims to accelerate the decarbonization of European industries while also protecting these industries from companies in regions with less ambitious climate goals. Additionally, CBAM is designed to incentivize other countries to also be more sustainable and emit less carbon (ibid.). It will function in parallel with the EU's Emissions Trading System, an emissions cap-and-trade system, and will create a new registry of CBAM importers that is to be centralized at EU level (ibid.).

4b. Geopolitics, Energy Security, and Associated Considerations

Geopolitics represent another critical influence in energy transitions. In simplest terms, geopolitics reflect power relationships tied to geography and/or geostrategy. The control of natural resources, such as energy, has figured prominently in geopolitical conflicts in history. One need only look to Hitler's strategic aims to obtain the oil fields in the Caucasus in World War II, the Iraqi-Kuwait dispute over oil in the Gulf War, or Arctic territorial claims for untapped oil and gas reserves to see energy's role in geopolitical power plays (Yergin, 1991; Hayes, 1990; Walker, 2010).

The events in 2022 underscore critical connections between geopolitics and energy. The Russian invasion of Ukraine in February 2022 triggered considerable disruption in energy markets, driving oil and gas prices to their highest in nearly a decade, with cascading security effects for the global economy (Tollefson, 2022; IEA, n.d.a). As peak winter conditions played out in the Northern Hemisphere with a tight energy supply market and high prices (Tollefson, 2022), the war exacerbated uncertainties, catalyzing reassessments of energy dependence. Critical responses included sanctions being imposed on Russian energy imports, the suspension of the Nord Stream 2 gas pipeline project between Russia and Germany (and subsequent bankruptcy filing by the project's company), and the strategic refocus to options, such as liquified natural gas, renewables, efficiency or conservation, and nuclear energy (ibid., BBC, 2022; European Commission, 2022). IEA member countries agreed to two emergency releases of oil stocks equaling 240 million barrels in a period of six months equal to more than one million barrels per day in the global market (IEA, 2022).

Among regions that were affected in energy terms beyond the two battling countries, the EU was, arguably, the most dependent region on Russian energy. The EU imported roughly 40% of its natural gas, about half of its coal, and a quarter of its oil from Russia in 2019 (Tollefson, 2022). To address the energy system disruption, the European Commission announced plans to reduce Russian gas imports by approximately two-thirds by the end of the year (ibid.). The plan included realignment with suppliers from other regions for roughly 60% of the energy and the scaling of renewables and conservation for the remainder (ibid.).

As noted earlier in the discussion of the 1970s and early 1980s oil shocks, *energy security* is closely tied to geopolitics and can take numerous forms.[8] A very basic way to gauge energy security quantitatively is in terms of the share of imports that are based on specific fuels. Table 2.2 shows the top ten countries for fuel imports as a share of merchandise imports in 2020, with higher dependence reflecting roughly 20–40% of country-level merchandise imports. Brunei, a small nation on the island of Borneo that is surrounded by Malaysia and the South China Sea, had the highest fuel import dependence at over 36% of its merchandise imports in 2020, up roughly 3 percentage points over 2019. Importantly, the time period reflected in the table

Table 2.2 Top Countries for Fuel Imports in 2020 as a Percentage of Merchandise Imports

Country	Fuel Imports in 2020 as a % of Merchandise Imports	Percentage Point Change vs. 2019
Brunei	36.5	2.8
Cape Verde	32.6	20.7
Mauritania	28.9	−0.7
Lebanon	28.4	−5.5
India	28.4	−3.5
Burkina Faso	26.6	−1.6
Senegal	23.2	−2.7
Pakistan	22.5	−6.3
Belarus	20.4	−4.4
Greece	19.9	−7.0

Source: World Bank, n.d.b, Accessed May 5, 2022.

Note: Simple variances are listed, based on rounding.

Table 2.3 Countries with Top Fossil Fuel Rents in 2019, as a Share of Gross Domestic Product

Rank	Natural Gas	Rent as a % of GDP	Petroleum	Rent as a % of GDP	Coal	Rent as a % of GDP
1	Timor-Leste	29.0	Libya	43.9	Mongolia	6.9
2	Brunei	10.5	Congo, Rep.	43.4	Mozambique	3.4
3	Uzbekistan	6.9	Kuwait	42.1	South Africa	1.9
4	Equatorial Guinea	6.9	Iraq	39.6	India	0.8
5	Papua New Guinea	5.2	Angola	25.1	Kazakhstan	0.8
6	Trinidad and Tobago	4.6	Oman	24.9	Indonesia	0.7
7	Qatar	3.8	Saudi Arabia	24.2	Australia	0.7
8	Azerbaijan★	3.6	Equatorial Guinea	22.3	Colombia	0.5
9	Russian Federation★	2.8	Azerbaijan★	21.9	China	0.4
10	Myanmar	2.5	Gabon	18.8	Russian Federation★	0.4

Source: World Bank, n.d.b, Accessed May 5, 2022.

★Countries top ranked for more than one fossil fuel.

represents an early stage of the COVID-19 pandemic, which produced considerable flux in economic activity and energy use (IEA, 2021a).

Another way to consider energy security is in the form of *energy rents*, which generally reflect the difference between energy revenues and the costs of extraction (van de Graaf and Sovacool, 2020). Rents represent a means to charaterize energy dependence within the economy. Table 2.3 shows countries ranked by top fuel rents as a share of gross domestic product in 2019, with nations such as Russia and Azerbaijan appearing in the table for more than one fuel type. Looking across Table 2.3, the highest rents are for petroleum at 40–45% in countries, including

Table 2.4 Top Countries with Limited Access to Electricity in 2020

Country	Access to Electricity in 2020 (% of Population)	Percentage Point Change vs. 2019
South Sudan	7.2	0.5
Chad	11.1	2.7
Burundi	11.7	0.4
Malawi	14.9	3.7
Central African Republic	15.5	1.2
Burkina Faso	19.0	0.6
Congo, Dem. Rep.	19.1	0
Niger	19.3	0.3
Sierra Leone	26.2	3.5
Liberia	27.5	4.4

Source: World Bank, n.d.b, Accessed May 5, 2022.

Note: Simple variances are listed, based on rounding.

Libya, the Republic of Congo, and Kuwait. For natural gas, the highest rent is roughly 30% for Timor-Leste, and for coal, the top rent share is considerably lower at approximately 7% for Mongolia. Looking more fully across Tables 2.2 and 2.3, one sees that Brunei had the highest share of merchandise imports from fossil fuels in 2020 and ranked the second highest rent for natural gas at +10% in 2019. This highlights forms of energy dependence and security concerns.

The share of a country's population that has access to electricity characterizes another aspect of energy security. Table 2.4 shows the countries with the lowest share of access by the population in 2020, based on World Bank reporting.

South Sudan represents the country with the lowest share of its population having access to power at roughly 7%. Importantly, limited access to electricity can be a function of institution-building or vestiges of war, both of which are part of South Sudan's recent history (BBC, 2016). Depending on how one operationalizes "having access to reliable and affordable energy", the countries in Table 2.4 could be considered energy insecure. Others may argue, however, that these countries do not have high levels of access to electricity and historically have met energy needs through other forms of energy. When used, this indicator requires explanation.

While geopolitics and concerns over energy security can be major drivers of energy transitions, other concepts that relate to geopolitics and energy security can, by extension, also relate to energy transitions. *Resource nationalism* refers to a process or action in which a government or a select group of people asserts an ownership claim over natural resources within their territory. This was evident, for example, in 2006 when Bolivia nationalized its oil and gas reserves, giving foreign investors a six-month deadline to comply with demands or leave (Zissis, 2006). In doing so, Bolivia appropriated ownership from foreign investors, including natural gas fields under contract to produce for Brazil. This process brought energy security concerns to the foreground in Brazil, as Brazil has been more reliant on natural gas, when droughts minimize its hydropower generation (ibid.; Trevisani and Lewis, 2021). Distinctly different is the concept of the *resource curse*, coined by Richar Auty, which refers to a condition in which a country produces natural resource wealth, such as from oil or gas, but does not attain economic growth (Mittleman, 2017). This curse can be represented as a negative association between raw-material exports and economic growth, which may tie to conflict, corruption, or poverty. Closely related to the resource curse is the concept of the *Dutch disease*, which was named for a situation that emerged in the Netherlands in the 1960s after natural gas was discovered in the North Sea (ibid.). The commodity boom for the Netherlands from natural gas, increased the value of the country's currency, in turn making other Dutch goods less competitive in international markets (ibid.). A resource curse or Dutch disease conditions are examples of what countries aim to avoid in energy shifts that include new production of energy.

4c. Digitalization in Energy Systems: A Focus on Cyber Threats

Like geopolitics, digitalization is another disruptive force that is affecting energy systems. Rapid advances in data analytics, remote sensing, artificial intelligence, and machine learning are enabling energy systems to be managed in new ways (IEA, n.d.b, 2017, 2021b). If done well, digitalization allows energy systems to be more productive and less costly while improving safety, reliability, resilience, and potentially, stability, among factors (ibid.).

In addition to possible gains, the digital transformation is also widening the surface area for cybercriminals to attack (MIT Technology Review, 2021). This points to the importance of strengthening energy systems resilience (i.e., hardening) against cyber threats, as energy companies modernize their operations with digital and cloud-based technology. Cyber threats may emerge from adversaries' deliberate actions, as well as unintended conditions that are transferred by digital systems. Cyber risks could include the disabling of physical protection measures, compromising of facility instrumentation and control systems, or a breach with the loss of security-sensitive information, enabling future attacks, among possibilities. For energy systems, this may entail vulnerability to industrial or consumer control of the smart grids, homes, and cars as well as delivery systems, such as that for oil.

A recent example of a cyberattack on an energy system was the May 2021 ransomware attack on the Colonial Pipeline in the United States, the largest publicly disclosed cyberattack against critical infrastructure in the US (Kerner, 2022). As a delivery system for oil, the Pipeline is one of the largest in the US, spanning over 5,500 miles from the states of Texas to New Jersey to supply almost half the fuel for the East Coast (ibid., Vasquez, 2022). This includes refined fuel for gasoline and diesel, home heating, and jet fuel. In the attack, hackers entered the Pipeline's information technology (IT) network with an exposed virtual private network (VPN) password, stealing 100 gigabytes of data in a two-hour timeframe (Kerner, 2022). Subsequent to this step, ransomware infected the Colonial Pipeline IT network. The Pipeline operational technology systems, which transport the oil, were taken offline by the Pipeline operators to reduce the risk of exposure for the operational network (Shea, 2021). Highlighting the quick impact that such an incident can have on a region, lines at the gas pumps formed over concerns about energy shortages (Vasquez, 2022). The Colonial Pipeline management paid a ransom of $4.4 million to obtain a decryption key, enabling the company's IT staff to regain control of its systems (ibid.). Ultimately, the Pipeline was offline from May 7–12, 2021. The US Congress subsequently mandated that companies operating infrastructure must report to federal authorities when they are hacked. The Biden administration also imposed mandates to secure oil and gas pipelines (ibid.). This example highlights how energy system change often entails trade-offs and learning that must be factored with the risks that exist alongside the benefits.

4d. COVID-19

The onset of the COVID-19 pandemic and associated recovery periods represent additional forms of disruption and learning that have significance for energy transitions.[9]

During the pandemic period, supply chains reflected bottlenecks and considerable disorder (Sultan, 2022). Crude oil prices collapsed in March 2020, complicated by a price war that was waged by Russia and Saudi Arabia (Meredith, 2020). While the early pandemic proved to be unfavorable for oil and gas (IEA, n.d.a; 2021c), renewable generation earned more stable revenue and continued to grow (ibid., IEA, 2021d). Estimated renewable energy capacity additions that became operational in 2021 (270 GW) and will become operational in 2022 (280 GW) reflect a more than 50% increase in average annual additions over the record period of 2017–2019 (IEA, 2021d). Behavioral shifts during the pandemic also represent considerable influences of a remote workforce and consumer actions (Cruickshank, 2020) that have implications for energy.

Government attempts to protect communities from the spread of the disease with economic shutdowns had cascading effects on the energy sector. An unprecedented decline in mobility was evident in reduced transport, trade, and economic activity (IEA, n.d.b). This, in turn, was reflected in a 4% decline in energy use in 2020 (ibid.). Recovery packages (see Fox, this volume) and vaccines have helped to facilitate rebounds, with energy demand rising 4.6% in 2021, above the pre-pandemic level (IEA, n.d.c). Government spending focused on sectors for sustainable recovery, including clean energy (IEA, 2021c). As of October 2021, roughly $400 billion on average per year is estimated as potentially being mobilized in clean energy and sustainable recovery spending (ibid.). Lockdowns also cut daily carbon emissions by over 15% in 2020 versus the 2019 average (le Quere et al., 2020), revealing that global carbon emissions can be drastically reduced in a short period of time, albeit with economic paralysis.

5. Advancing the Concepts and Theory

Looking more fully across concepts and theory provides policymakers, students, and scholars with ways to further explain energy transitions. Meta-systems scoping and ideas, like the circular economy, offer frames of reference with which to elaborate on phases and elements of energy transitions.

5a. Systems

When considering the complex interdependencies of energy, systems thinking is often key for more comprehensive analysis. System approaches may use boundaries, stocks and flow, interactions, and feedback loops as well as consider overshoot (Meadows, 2008) as a way to conceptually distill the areas of necessary focus. System-of-systems theory-building then goes even further with examples, such as socio-ecological systems and sociotechnical systems that are discussed next.

Socio-ecological systems (SES) is a meta-level conceptual framework that envisions nested system hierarchies, with variables that may interact and affect outcomes (Ostrom, 2007, 2009; Ostrom and Cox, 2010). SES captures interdependent linkages between *social and environmental change* and how those connections influence the attainment of sustainability aims across different systems, levels, and scale. This framework has evolved from advancing collective action ideas to be a widely cited tool to diagnose sustainability in a social-ecological system (Partelow, 2018).

A related meta-level framework is that associated with *sociotechnical systems* (STS). This conceptual framework also allows for nested system hierarchies, highlighting coevolution and multidimensional interactions, such as those between industry, technology, markets, policy culture, and civil society (Geels, 2012). In contrast to SES, STS centers primarily on the interdependent linkages and interactions between *society and technology* across different systems, levels, and scale.

For energy-specific STS – namely, with electricity, heating, and mobility – analysis may cover internal processes, innovations, and activities that can change the elements of the system entities and the relationship between them. Instead of viewing system transitions as a unitary process of disruption, STS may include dispersed processes that cumulatively modify the elements and the architecture of the systems (Geels and Turnheim, 2022). Drawing on evolutionary economics, sociology, and science, technology and society studies, the *multi-level perspective* (MLP) is a regularly employed theoretical lens within the STS domain, conceiving of sociotechnical transitions as a multidimensional interplay between layers of emergent niche-based innovations with established systems, and developments within the broader context (i.e., landscape) (ibid.; Geels, 2006). Considerable change is achieved when trajectory realignment occurs within and between the levels. This *Handbook* covers sample landscape elements of energy systems in terms of infrastructure integration in communities (Ferry and Monoian, and Sharp et al., this volume), retrofitting or decommissioning (Windemer, this volume), and urban mobility planning (Mazouni, this volume).

Associated with MLP are *strategic niche management* and *transitions management*, which more deeply explore how to nurture socially desired aims and technological innovation (Schot and Geels, 2008; Raven et al., 2010). Experimentation is critical and highlighted with the blending of transitions management and design anthropology in the chapter by Sharp et al. (this volume) related to urban infrastructural systems at the urban precinct level.

Another theoretical lens that informs current energy transitions work is that on *large technical systems* (LTS). Derived primarily from history of science writing by Thomas Hughes, such as *The Seamless Web* (1986) and *Networks of Power* (1983), this meta-level framing integrates physical infrastructure with economic, legal, and social aspects within a system. Key to this framing is the coherence of technical systems within the social environment. Resistance to change can occur as a system matures with elements that are out of sync. When not corrected, the circumstances can become radical, with a novel, alternative system emerging (Hughes, 2012). This meta-level framing is highly instructive for today's practitioners and analysts of energy transitions who aim to more fully account for the scope of relevant factors and their interactions. Questions have been raised about the seamless potential of LTS (Cherp et al., 2018). LTS contributions, however, underscore that systems and the environment mutually influence the other with momentum, reinforcing factors, and resistance potentially playing a role (Hughes, 1983, 1986, 2012).

Innovation systems (IS) represents another theoretical framing to shed light on energy transitions, based on three kinds of learning – innovation, research and development, and competence building (e.g., training and education) (Edquist, 2006). At the national level, Chris Freeman used IS to assess Japan's economy (Freeman, 1987). The scope captures the natural idiosyncrasies of language, culture, and institutions within a given country. Freeman's approach evolved into the *national innovation systems* (NIS/NSI) framing that views the interactions among institutions and other elements of national significance as a basis for comparison. Critical writing in this area has emphasized national research and development capacity (Edquist, 2006, citing Nelson, 1993) and the role of user-producer interactions and the domestic market (Lundvall, 1992). *Technological innovation systems* (TIS), in contrast to NIS, adopts a functional lens for the way in which innovations are generated, disseminated, and utilized (Jacobsson and Bergek, 2004), with seven subfunctions (Hekkert et al., 2007). In this framing, a transition may occur, contingent on the level of subfunction completion and interaction with other subfunctions. Markard and Truffer (2008) highlighted early connections between MLP and TIS. *Regional innovation systems* (RIS) takes a slightly different lens, recognizing that knowledge production and the innovation process exhibit a distinctive locational value in territorially based IS (Asheim and Gertler, 2006). Combined with the grounded nature of local knowledge, the systemic nature of the RIS is reinforced by the prevalence of a set of attitudes, norms, values, routines, and expectations in a region (ibid.).

Techno-economic paradigms (TEP) represent another critical means to explain system change that draws from long-wave theories of business cycles. This theoretical framing views interlocking technologies, beliefs, processes, and economic structures as sustaining until a technological revolution occurs (Freeman and Perez, 1988, Perez, 2009, Freeman and Louca, 2002). Fluctuations are seen as tempered by connections between business, political, and cultural paths. New logic can shape changes over the course of five to six decades with incumbents and new entrants (ibid.).

Across these framings, one can consider the *readiness* of a country for an energy transition (Araújo, 2017). This concept can be applied to any society, as it aims, for example, to rapidly shift based on energy security and net zero aims. Readiness may generally be evaluated in terms of the availability of a given natural resource (if fuel switching); the presence of appropriate infrastructure, markets, and industry; and the existence of indigenous experts/expertise. Additional considerations that may have a strong bearing include historical familiarity for the adopting community in terms of the planned energy option. Recent examples of national energy transitions show that countries can adapt fairly quickly by scaling up more from an existing strategy or repurposing existing markets and infrastructure (ibid.).

5b. Related Concepts

Related ideas that support thinking about energy system change include those on bounded rationality, the role and types of knowledge, and the circular economy. Each provides a basis for policy and practice today.

Bounded rationality reflects the idea that one's logic is shaped by the individual's experiences, cognitive capacity, and circumstances in which to make a decision (Simon, 1957, 1990, 1991). Based on this, individuals may respond differently to the same set of conditions and are not fully predictable. Bounded rationality connects to Brinkman and Hirsh's chapter (this volume) on unarticulated identities, which highlights how a person's underlying logic for energy adoption may not necessarily align with their voting pattern.

The pivotal role of *knowledge* represents still another critical lens for understanding energy transitions, especially in terms of regional and national readiness, mentioned earlier. What we know, how, and when can shape acceptance of change. Sources of knowledge can be roughly broken down into two general groups, which capture the process of how knowledge emerges. Established (or expert) knowledge is typically tested through mainstream disciplinary investigation, then is reviewed and accepted by scientific peers (Araújo, 2017). By contrast, local knowledge does not owe its origin, testing, degree of verification, truth, status, or currency to distinctive professional techniques but rather to common sense, causal empiricism, or thoughtful speculation and analysis (ibid., citing Lindblom and Cohen, 1979). It is gained in ways such as experience and not controlled experimentation that is vetted by peer review. Fields such as environmental planning and ecology; science, technology, and society; and broader technology writing recognize the value of local knowledge, as an area where early indicators and important, untapped insight lie. Harnessing both forms of knowledge presents opportunities for more comprehensive and potentially timely learning amid complex and at times disruptive dynamics of an energy transition.

Recognizing that systemic relationships exist between society, the economy, and the natural environment, *industrial ecology* focuses on the stages of the production of goods and services from a point of view of nature, seeing ways to mimic a natural system through conservation and reuse of resources (Chertow, 2008). The *circular economy* is a related model of production and consumption, which involves sharing, leasing, reusing, repairing, refurbishing, and recycling existing materials and products as long as possible (EP, 2015). It is a departure from a linear economy, which is based on a produce-consume-throw away pattern, that generally includes planned obsolescence.[10] Taken together, industrial ecology and circular economy ideas point to an industrial ecosystem in which the waste of one industry may be the feedstock for another, and so on. As decision-makers evaluate energy systems change, efficiency and conservation can more fully support objectives, like decarbonization, energy security, and resilience, exemplifying the principles of industrial ecology and the circular economy.

6. Looking ahead

Energy transitions by nature are dynamic and may be evolutionary or revolutionary. (Araújo, 2017). In taking stock of where writing and developments are on energy transitions, a number of takeaways are offered.

A critical area to watch includes the simple ways that transitions are framed. Thinking of energy transitions as an opportunity for modernizing/repurposing infrastructure and advancing the quality of life or of the workforce to sidestep old issues translates not only for trade-offs and buy-in but for the fuller set of solutions that are considered. As the next set of energy choices are navigated, decision-makers and research funders are encouraged to think differently about how they frame requests for analysis and public discussion. The aim is to minimize the recreation of lock-in to old incumbencies. Recognizing that energy systems are linked not only by regional interdependencies and technologies but also to other systems such as food, materials, and water presents a more realistic and savvy way to evaluate the choices at hand.

For there to be a more engaged society that contributes to the transition, decision-makers can go far by acknowledging that different types of knowledge from various sectors can deepen the learning and buy-in. Moreover, discussing energy transitions in ways that look beyond simply one-for-one technology substitution can enlarge the window of opportunity not only for planners but for adopters who will grapple with real implications and the trade-offs. Windows of opportunity may emerge from disruption, such as with the Russian war on Ukraine, in which countless decision-makers have been prioritizing shifts in their energy strategies.

When considering the costs of system change, it is important for analysts, decision-makers, and impacted communities to also account for the range of co-benefits as well. This includes co-benefits that may not be readily quantified but resonate with the local values.

Key questions to ask about divergent views or approaches in the field include the following:

- How do policymakers and researchers account for differences in the science on timescales for energy change and the carbon storage capacity, such as that of bioenergy?
- What are the appropriate forms of carbon pricing to use with net-zero systems, and how is carbon leakage effectively addressed?
- In what way(s) should carbon accounting and reporting be standardized, bearing in mind the need for accuracy and buy-in?
- Can the distinct areas of energy transitions and sustainability transitions research continue to mutually inform, or will they ultimately converge?

While some may debate prioritization between energy security, resilience, and net-zero priorities, it may serve everyone to instead consider how such aims can amplify each other with a multiplier effect, enabling fuller buy-in during challenging times.

There is no constant like change, but there is also no time like the present to become more informed and proactive in our choices.

Notes

1 The study was criticized for, among reasons, the sensitivity of the simulations to assumptions (Cole et al., 1973). Updated analysis of the study in the ensuing decades with refinements to assumptions and data shows that the broad trends generally held true (Hall and Day, 2009; Turner, 2008; Meadows et al., 2004).
2 For more discussion and examples, see Araújo, 2017.
3 Later energy-related studies by IIASA considered a broad array of research on energy, including diffusion of decarbonization, technology change and innovation, sustainability and access, and energy-services-led transformation, among others (Grubler and Nakicenovic, 1996; GEA, 2012; Grubler, 2012; Nakicenovic, 1996; Wilson et al., this volume).
4 These objectives often relate to sustainability but are not identical to it.
5 CO_2 *equivalent* measures the emissions from various GHGs in terms of their *global warming potential* (GWP). GWP characterizes the potency of a GHG, accounting for the duration it is in the atmosphere.
6 See also Chapter 1 (Araújo, this volume) for a discussion of issues with carbon accounting and international reporting, and Chapter 16 (van den Bergh and Botzen, this volume) for more in-depth discussion of the debates and implications of carbon pricing.
7 Carbon pricing and policies in some respects complicate Harold Hotelling's more traditional thinking about exhaustible natural resources, which indicates that owners of nonrenewable energy resources will leave their resource untapped if the financial yield of production is less than the cost of production (Hotelling, 1931). This is an area for fuller examination in theory and practice.
8 It can relate to a diversity of suppliers in the case of an energy importer. For an energy exporter, this definition may also be rethought in terms of having a diversity of buyers.
9 See also Foxon's chapter (this volume).
10 *Planned obsolescence* is an approach that assumes at the design phase that a product will have a limited lifespan, encouraging consumers to more than once.

References

Araújo, K. (2013). Energy at the Frontier: Low Carbon Energy System Transitions and Innovation in Prime Mover Countries. PhD Dissertation, Massachusetts Institute of Technology: Cambridge, MA.
Araújo, K. (2017). *Low Carbon Energy Transitions: Turning Points in National Policy and Innovation*. Oxford University Press: New York.
Araújo, K. and Shropshire, D. (2021). A Meta-Level Framework for Evaluating Resilience in Net-Zero Carbon Power Systems with Extreme Weather Events in the United States, *Energies*, 14(14): 4243.
Arthur, W. B. (1989). Competing Technologies, Increasing Returns, and Lock-In by Historical Events, *The Economic Journal*, 99(394): 116–131.
Asheim, B., and Gertler, M. (2006). The Geography of Innovation: Regional Innovation Systems. in: J. Fagerburg, D. Mowery and R. Nelson (Eds.). *Oxford Handbook of Innovation*. Oxford University Press: Oxford.
BBC (2016). South Sudan Profile—Overview, April 27, www.bbc.com/news/world-africa-14019208, Accessed February 2, 2021.
BBC (2022). What Sanctions Are Being Imposed on Russia over Ukraine Invasion? May 1, www.bbc.com/news/world-europe-60125659, Accessed June 1, 2022.
Berkhout, F. (2002). Technological Regimes, Path Dependency and the Environment, *Global Environmental Change*, 12(1): 1–4.
BP (2021). Statistical Review of World Energy, 1965–2020, www.bp.com/en/global/corporate/energy-economics/statistical-review-of-world-energy.html
Cherp, A., Vinichenk, V., Jewell, J., Brutschin, E., Sovacool, B. (2018). Integrating techno-Economic, Socio-Technical and Political Perspectives on National Energy Transitions: A Meta-Theoretical Framework, *Energy Research & Social Science*, 37: 175–190.
Chertow, M. (2008). Industrial Ecology in a Developing Context, in: C. Clini, I. Musu and M. L. Gullino (Eds.). *Sustainable Development and Environmental Management*, Springer Netherlands: Dordrecht, 335–349.

Cole, H., Freeman, C., Jahoda, M., Pavitt, K. (Eds.) (1973). *Models of Doom: A Crtique of the Limits to Growth* (1st Hardcover ed.). Universe Publishing: New York.

Cruickshank, A. (2020). COVID Pandemic-19 Shows Telecommuting Can Help Fight Climate Change, *Scientific American*, July 22, www.scientificamerican.com/article/covid-19-pandemic-shows-telecommuting-can-help-fight-climate-change.

Daly, H. (1996). Beyond Growth: The Economics of Sustainable Development. Beacon Press: Boston, MA.

David, P. (1985). Clio and the Economics of QWERTY, *American Economic Review (Papers and Proceedings)*, 75: 332–337.

de Coninck, H., Revi, A., Babiker, M., Bertoldi, P., Buckeridge, M., Cartwright, A., Dong, W., Ford, J., Fuss, S., Hourcade, J.-C., Ley, D., Mechler, R., Newman, P., Revokatova, A., Schultz, S., Steg, L., Sugiyama, T. (2018). Strengthening and Implementing the Global Response. in: V. Masson-Delmotte, P. Zhai, H.-O. Pörtner, D. Roberts, J. Skea, P. R. Shukla, A. Pirani, W. Moufouma-Okia, C. Péan, R. Pidcock, S. Connors, J. B. R. Matthews, Y. Chen, X. Zhou, M. I. Gomis, E. Lonnoy, T. Maycock, M. Tignor and T. Waterfield (Eds.) *Global Warming of 1.5°C. An IPCC Special Report on the Impacts of Global Warming of 1.5°C Above Pre-Industrial Levels and Related Global Greenhouse Gas Emission Pathways, in the Context of Strengthening the Global Response to the Threat of Climate Change, Sustainable Development, and Efforts to Eradicate Poverty*. Cambridge University Press: Cambridge and New York. doi: 10.1017/9781009157926.

Detel, W. (2001). Social Constructivism, in: N. Smelser and P. Baltes (Eds.) *International Encyclopedia of the Social & Behavioral Sciences*, www.sciencedirect.com/topics/psychology/social-constructivism

Edquist, C. (2006). Systems of Innovation: Perspectives and Challenges, in: J. Fagerburg, D. Mowery and R. Nelson (Eds.) *Oxford Handbook of Innovation*. Oxford University Press: Oxford.

European Commission (2022). REPowerEU, May 18, https://ec.europa.eu/info/strategy/priorities-2019-2024/european-green-deal/repowereu-affordable-secure-and-sustainable-energy-europe_en, Accessed May 19, 2022.

European Council (2022). Council Agrees on the Carbon Border Adjustment Mechanism (CBAM), March 15, www.consilium.europa.eu/en/press/press-releases/2022/03/15/carbon-border-adjustment-mechanism-cbam-council-agrees-its-negotiating-mandate/#:~:text=The%20Commission%20presented%20its%20proposal,than%20those%20of%20the%20EU).

European Parliament (EP) (2015). Circular Economy: Definition, Importance and Benefits, February 12, www.europarl.europa.eu/news/en/headlines/economy/20151201STO05603/circular-economy-definition-importance-and-benefits, Accessed October 2, 2021.

Fisher, J. and Pry, R. (1971). A Simple Substitution Model of Technological Change, *Technological Forecasting and Social Change*, 3(1): 75–78.

Freeman, C. (1987). *Technology Policy and Economic Performance: Lessons from Japan*. Pinter: London.

Freeman, C. and Louca, F. (2002). *As Time Goes By: From the Industrial Revolutions to the Information Revolution*. Oxford University Press: New York.

Freeman, C. and Perez, C. (1988). Structural Crises of Adjustment, Business Cycles and Investment Behaviour, in: G. Dosi (Ed.) *Technical Change and Economic Theory*. Pinter Publisher: London.

Garud, R. and Karnøe, P. (2001). Path Creation as a Process of Mindful Deviation, in: R. Garud and P. Karnøe (Eds.) *Path Dependence and Creation in the Danish Wind Turbine Field*. Chapters 1–2, LEA Associates Publishers: London.

Geels, F. (2006). Multi-level Perspective on System Innovation: Relevance for Industrial Transformation, Chapter 9 in: X. Olshoorn and A. Wieczorek (Eds.) *Understanding Industrial Transformation: Views from Different Disciplines*, 163–186.

Geels, F. (2012). A Socio-Technical Analysis of Low-carbon Transitions, *Journal of Transport Geography*, 24: 471–482.

Geels, F. and Schot, J. (2007). Typology of Socio-Technical Transition Pathways, *Research Policy*, 36(3): 399–417.

Geels, F. and Turnheim, B. (2022). *The Great Reconfiguration*. Cambridge University Press: Cambridge.

Global Energy Assessment (GEA) (2012). *Global Energy Assessment – Toward a Sustainable Future*, Cambridge University Press: Cambridge and New York, USA and the International Institute for Applied Systems Analysis, Laxenburg, Austria.

Goldemberg, J. (1998). Leapfrog Energy Technologies, *Energy Policy*, 26(10): 729–742.

Graefe, L. (2013). Oil Shock of 1978–79, Federal Reserve of Atlanta, November 22, www.federalreserve-history.org/essays/oil-shock-of-1978-79#:~:text=Oil%20prices%20began%20to%20rise,of%20oil%20during%20the%20crisis, Accessed February 1, 2021.

Griliches, Z. (1957). Hybrid Corn: An Exploration in the Economics of Technical Change, *Econometrica*, 25: 501–522.

Grubler, A. (2012). Energy Transitions Research: Insights and Cautionary Tales, *Energy Policy*, 50: 8–16.

Grubler, A. and Nakicenovic, N. (1996). Decarbonizing the Global Energy System, *Technological Forecasting and Social Change*, 53(1): 97–110.

Grubler, A., Wilson, C. and Nemet, G. (2016). Apples, Oranges, and Consistent Comparisons of the Temporal Dynamics of Energy Transitions, *Energy Research & Social Science*, 22: 18–25.

Hall, C. and Day, J. (2009). Revisiting the Limits to Growth After Peak Oil, *American Scientist*, 97(3): 230–237.

Hayes, T. (1990). Confrontation in the Gulf: The Oilfield Lying below the Iraq-Kuwait Dispute, *New York Times*, September 3, www.nytimes.com/1990/09/03/world/confrontation-in-the-gulf-the-oilfield-lying-below-the-iraq-kuwait-dispute.html

Hekkert, M., Suurs, R., Negro, S., Kuhlmann, S. and Smits, R. (2007). Functions of Innovation Systems: A New Approach for Analysing Technological Change, *Technological Forecasting and Social Change*, 74(4): 413–432.

Hotelling, H. (1931). The Economics of Exhaustible Resources, *Journal of Political Economy* 39: 137–175.

Hughes, T. (1983). *Networks of Power: Electrification in Western Society*. Johns Hopkins University Press: Baltimore, MD.

Hughes, T. (1986). The Seamless Web: Technology, Science, Etcetera, Etcetera, *Social Studies of Science*, 16(2): 281–292.

Hughes, T. (2012). The Evolution of Large Technological Systems, in: W. Bijker, T. Hughes, and T. Pinch (Eds.) *The Social Construction of Technological Systems*. MIT Press: Cambridge, MA.

IEA (1994a). The History of the International Energy Agency—The First 20 Years, Report, Volume I, https://iea.blob.core.windows.net/assets/9f9fe7d6-223a-40db-8725-d8d2539b0e3f/TheHistory-oftheInternationalEnergyAgency-TheFirst20Years.pdf.

IEA (1994b). The History of the International Energy Agency—Major Policies and Actions, Report, Volume II, https://iea.blob.core.windows.net/assets/3d2118d4-db47-40ad-ada5-3facda90c53e/2ieahistory.pdf.

IEA (2017). Digitalization and Energy, Report, www.iea.org/reports/digitalisation-and-energy.

IEA (2021a). COVID 19 Impact on Electricity, www.iea.org/reports/covid-19-impact-on-electricity, Accessed February 2, 2021.

IEA (2021b). Smart Grids, Report, November 2021, www.iea.org/reports/smart-grids.

IEA (2021c). Global Energy Review 2021, April 2021, www.iea.org/reports/global-energy-review-2021.

IEA (2021d). *Renewable Energy Market Update 2021*, IEA, www.iea.org/reports/renewable-energy-market-update-2021.

IEA (2022). IEA Confirms Member Country Contributions to Second Collective Action to Release Oil Stocks in Response to Russia's Invasion of Ukraine, April 7, 2022, www.iea.org/news/iea-confirms-member-country-contributions-to-second-collective-action-to-release-oil-stocks-in-response-to-russia-s-invasion-of-ukraine.

IEA (n.d.a). Russia's War on Ukraine, www.iea.org/topics/russia-s-war-on-ukraine, Accessed June 1, 2022.

IEA (n.d.b). Digitalisation, Making Energy Systems Smarter, More Connected, Efficient and Resilient, www.iea.org/topics/digitalisation, Accessed February 1, 2022.

IEA (n.d.c). COVID-19, www.iea.org/topics/covid-19, Accessed February 1, 2022.

Intergovernmental Panel on Climate Change (IPCC) (2021). Global Warming of 1.5 °C, Special Report, www.ipcc.ch/sr15.

Intergovernmental Panel on Climate Change (IPCC) (2022). *IPCC Sixth Assessment Report: Impacts, Adaptation and Vulnerability*, www.ipcc.ch/report/ar6/wg2.

Jacobsson, S. and Bergek, A. (2004). Transforming the Energy Sector: The Evolution of Technological Systems in Renewable Energy Technology, *Industrial and Corporate Change*, 13(5): 815–849.

Jefferson, M. (1982). Historical Perspectives of Societal Change and the use of Scenarios at Shell, Chapter 8 in: B. Twiss (Ed.) *Social Forecasting for Company Planning*. Macmillan: London.

Jefferson, Michael (2012). Shell Scenarios: What Really Happened in the 1970s and What May Be Learned for Current World Prospects, *Technological Forecasting & Social Change*, 79: 192.

Kerner, S. (2022). Colonial Pipeline Hack Explained: Everything you Need to Know, April 26, www.techtarget.com/whatis/feature/Colonial-Pipeline-hack-explained-Everything-you-need-to-know

Kline, R. (2001). Technological Determinism, chapter in: N. Smelser, and P. Baltes (Eds.) *International Encyclopedia of the Social & Behavioral Sciences*, www.sciencedirect.com/topics/computer-science/technological-determinism.

Krause, F., Bossel, H., Muller-Reißmann, K-F. (1982). *Energiewende* (*Energy Transition: Growth and Prosperity without Oil and Uranium*). Fischer Verlag: Frankfurt, Germany.

Le Quéré, C., Jackson, R. B., Jones, M. W. et al. (2020). Temporary Reduction in Daily Global CO_2 Emissions during the COVID-19 Forced Confinement, *Nature Climate Change*, 10: 647–653. https://doi.org/10.1038/s41558-020-0797-x.

Lovins, A. (1976). Energy Strategy: The Road Not Taken? *Foreign Affairs*, www.foreignaffairs.com/articles/united-states/1976-10-01/energy-strategy-road-not-taken.

Lovins, A. (1977). *Soft Energy Paths*. Ballinger Pub. Co: Pensacola, FL.

Lundvall, B. (Ed.) (1992). *National Systems of Innovation. Towards a Theory of Innovation and Interactive Learning*. Pinter Publishers: London.

Malthus, T. (1999). An Essay on the Principle of Population, in: G. Gilbert (Ed.), *Oxford World's Classics*. Oxford University Press: Oxford.

Mansfield, E. (1961). Technical Change and the Rate of Imitation. *Econometrica*, 29(4).

Marchetti, N. and Nakicenovic, C. (1979). The Dynamics of Energy Systems and the Logistic Substitution Model. RR-79-13. Laxenburg, *International Institute for Applied Systems Analysis*, https://pure.iiasa.ac.at/id/eprint/1024/1/RR-79-013.pdf.

Markard, J. and Truffer, B. (2008). Technological Innovation Systems and the Multi-level Perspective: Towards an Integrated Framework, *Research Policy*, 37: 596–615.

Meadows, D., Meadows, D., Randers, J. and Behrens III, W. (1972). *The Limits to Growth*. Universe Books: New York.

Meadows, Donella (2008). *Thinking in Systems*, Ed. Diana Wright. Chelsea Green Publishing: White River Junction, VT.

Meadows, Donella, Randers, J. and Meadows, Dennis (2004). *Limits to Growth: The 30-Year Update*. Chelsea Green Publishing Co: White River Junction, VT.

Meredith, S. (2020). The Losers—and Even Bigger Losers – of An Oil Price War Between Saudi Arabia and Russia, *CNBC*, March 12, www.cnbc.com/2020/03/12/oil-the-losers-of-the-price-war-between-saudi-arabia-and-russia.html

MIT Technology Review (2021). Transforming the Energy Industry with AI, Report, https://wp.technologyreview.com/wp-content/uploads/2021/01/Transforming-the-energy-industry-with-AI.pdf

Mittleman, M. (2017). The Resource Curse, *Bloomberg*, May 19, www.bloomberg.com/quicktake/resource-curse.

Nakicenovic, N. (1996). Decarbonization: Doing More with Less, *Technological Forecasting and Social Change*, 51: 1–17.

Office of the Historian, Department of State, United States Government (n.d.). Oil Embargo, 1973–1974, https://history.state.gov/milestones/1969-1976/oil-embargo, Accessed August 1, 2021.

Office of Legacy Management, United States Government (n.d.). A Brief History of the Department of Energy, www.energy.gov/lm/doe-history/brief-history-department-energy, Accessed November 2, 2021.

Ostrom, E. (2007). A Diagnostic Approach for Going Beyond Panaceas, *Proceedings of the National Academy of Sciences of the United States of America,* 104(39): 15181–15187.

Ostrom, E. (2009). A General Framework for Analyzing Sustainability of Social-Ecological Systems, *Science,* 325(5939): 419–422.

Ostrom, E. and Cox, M. (2010). Moving Beyond Panaceas: A Multi-Tiered Diagnostic Approach for Social-Ecological Analysis, *Environmental Conservation*, 37(4): 451–463.

Partelow, S. (2018). A Review of the Social-Ecological Systems Framework: Applications, Methods, Modifications, and Challenges, *Ecology and Society*, 23(4): 36.

Perez, C. (2009). Technological Revolutions and Techno-economic Paradigms, *Cambridge Journal of Economics*, 34(1): 185–202.

Raven, R., van den Bosch, S. and Weterings. R. (2010). Transitions and Strategic Niche Management, *International Journal of Technology Management*, 51(1): 57–74.

Rogers, E. (1995). *Diffusion of Innovations* (4th ed.). Free Press: New York.

Schot, J. and Geels, F. (2008). Strategic Niche Management and Sustainable Innovation Journeys: Theory, Findings, Research Agenda, and Policy, *Technology Analysis & Strategic Management,* 20(5): 537–554.

Seto, K., Davis, S., Mitchell, R., Stokes, E., Unruh, G., and Urge-Versatz, D. (2016). Carbon Lock-in: Types, Causes, and Policy Implications, *Annual Review of Environment Resources*, 41: 425–452.

Shea, D. (2021). Lessons From the Colonial Pipeline Attack: Heading Off Cyberthreats, *Natl Conf of State Legislatures*, October 26, www.ncsl.org/research/energy/lessons-from-the-colonial-pipeline-attack-heading-off-cyberthreats-magazine2021.aspx?msclkid=4fae3484cf0311ec8df90201bbe929d2.

Simon, H. (1957). A Behavioral Model of Rational Choice, in: H. Simon (Ed.) *Models of Man, Social and Rational: Mathematical Essays on Rational Human Behavior in a Social Setting*. Wiley: New York.

Simon, H. (1990). A Mechanism for Social Selection and Successful Altruism, *Science*, 250: 4988, 1665–1668.

Simon, H. (1991). Bounded Rationality and Organizational Learning, *Organization Science*, 2(1): 125–134.

Smil, V. (2016). Examining Energy Transitions: A Dozen Insights Based on Performance, *Energy Research and Social Science*, 22: 194–197.

Sovacool, B. (2016). How Long Will it Take? *Energy Research and Social Science*, 13: 202–215.

Sovacool, B. and Geels, F. (2016). Further Reflections on the Temporality of Energy Transitions, *Energy Research and Social Science*, 22: 232–237.

Sultan, T. (2022). 5 Ways the COVID-19 Pandemic Has Changed the Supply Chain, January 14, www.weforum.org/agenda/2022/01/5-ways-the-covid-19-pandemic-has-changed-the-supply-chain.

Tollefson, J. (2022). What the War in Ukraine Means for Energy, Climate and Food, Nature, 604: 232–233, www.nature.com/articles/d41586-022-00969-9.

Trevisani, P. and Lewis, J. (2021). Brazil's Drought Pressures Power Grid, Boosting Case for Renewables—and Fossil Fuels, *Wall Street Journal*, October 11, www.wsj.com/articles/brazils-drought-pressures-power-grid-boosting-case-for-renewablesand-fossil-fuels-11633946401?mod=trending_now_video_4.

Turner, G. (2008). A Comparison of 'The Limits to Growth' with Thirty Years of Reality, *Socio-Economics and the Environment in Discussion (SEED). CSIRO Working Paper Series*. Commonwealth Scientific and Industrial Research Organisation (CSIRO).

UNFCCC (2015). The Paris Agreement, https://unfccc.int/process-and-meetings/the-paris-agreement/the-paris-agreement, Accessed January 2, 2022.

United Nations (UN) (n.d.). Climate Action, www.un.org/en/climatechange/net-zero-coalition, Accessed May 1, 2022.

Unruh, G. (2000). Understanding Carbon Lock-in, *Energy Policy*, 28(12): 817–830.

Unruh, G. (2002). Escaping Carbon Lock-in, *Energy Policy*, 30(4): 317–325.

Van de Graaf, T. and Sovacool, B. (2020). *Global Energy Politics*. Polity Press: Camnbridge, UK.

Vasquez, C. (2022). How the Colonial Pipeline Hack Galvanized a Nation at Risk, *Energy Wire*, May 9, www.eenews.net/articles/how-the-colonial-pipeline-hack-galvanized-a-nation-at-risk.

Verbong, G. and Loorbach, D. (2012). *Governing the Energy Transition?* Routledge: Oxfordshire.

Walker, S. (2010). Countries Lay Claim to Arctic in Battle for Oil and Gas Reserves, *The Independent*, September 23, www.independent.co.uk/climate-change/news/countries-lay-claim-to-arctic-in-battle-for-oil-and-gas-reserves-2087040.html.

World Bank (n.d.a). Carbon Pricing Dashboard, https://carbonpricingdashboard.worldbank.org/what-carbon-pricing, Accessed April 10, 2022.

World Bank (n.d.b). World Development Indicators, https://databank.worldbank.org/source/world-development-indicators, Accessed May 5, 2022.

World Commission on Environment and Development (1987). Our Common Future, https://sustainabledevelopment.un.org/content/documents/5987our-common-future.pdf

Yergin, D. (1991). *The Prize*. Free Press: New York.

Zissis, C. (2006). Bolivia's Nationalization of Oil and Gas, Council of Foreign Relations, May 12, www.cfr.org/backgrounder/bolivias-nationalization-of-oil-and-gas

Zotin, M., Rochedo, P., Portugal-Pereira, J., Szklo, A., and Shaeffer, R. (this volume). Critical Connections in Material Transitions and Energy Transitions. in: K. Araújo (Ed.) *Routledge Handbook of Energy Transitions*. Routledge: Abingdon.

3

GREEN GROWTH AND POST GROWTH

Economic Framing in Low-Carbon Energy Transitions

Timothy J. Foxon

A. Introduction

At the COP26 Climate Summit in Glasgow, all major emitting countries, including the US, China, India, the European Union, and the UK agreed to decarbonize their economies by the middle of the 21st century, though commitments for action by 2030 still fall short of putting the world on a path to limiting global warming to 1.5°C above pre-industrial levels (WRI, 2021). This means that all major economies will now need to act rapidly to enable a transition in national and global energy systems toward net-zero carbon emissions. This implies a green industrial revolution in technologies, institutions, and practices on a scale equivalent to or faster than previous energy-industrial revolutions (Foxon, 2018). The rate and directions of linked energy transitions in different countries will therefore depend on the economic and social priorities in those countries, and in gaining public legitimacy for rapid changes, which will inevitably produce winners and losers.

This chapter considers the macro-economic framing of low-carbon energy transitions in this context. It will review the current and emerging theoretical debates about this framing and consider how these could inform and challenge current political consensus. The current dominant political framing is that of green growth. As President Biden remarked on launching his climate action plan, "a key plank of our Build Back Better Recovery Plan is building a modern, resilient climate infrastructure and clean energy future that will create millions of good-paying union jobs" (Biden, 2021). Linked to this is the idea of a just transition, in providing opportunities for retraining and reskilling for communities dependent on high-carbon industries, which is part of the European Union's Green Deal. This approach has a clear political saliency for building a wide political coalition for action, particularly in countries with democratic political systems.

However, some scientists, theorists, and political activists argue that this framing will not generate a low-carbon energy transition that is fast enough to achieve the rapid reductions in carbon emissions needed to reach net-zero by mid-century while also ensuring that the costs and benefits of the transition are shared equitably across societies (Jackson, 2017; Speth, 2008). Instead, they argue that climate change represents a symptom of a more fundamental contradiction between current capitalist economic systems and the ability of the planet Earth to provide the resources and services needed by a growing human population in an equitable way (Hickel, 2020). So they argue for a transition to a post-growth economy, which would prioritize social

DOI:10.4324/9781003183020-4

and environmental goals, based on more local and small-scale economic activity and increased community values (Jackson, 2017; Speth, 2008). This builds on a body of thinking on ecological economics, pioneered by Nicholas Georgescu-Roegen and former World Bank economist Hermann Daly, exploring the implications of the dependence of all economic activity on flows of energy and resources from and to natural ecosystems. Daly (2015) argued that the scale of economic activity has moved from an "empty world" to a "full world", which is supported by recent scientific analysis that human-induced activity is now exceeding crucial planetary boundaries, including climate change, biodiversity loss, and nutrient cycling (Steffen et al., 2015).

This chapter will review these two political and economic framings and discuss the implications of contestation between these framings for the rate and direction of a transition in national and global energy systems to net-zero carbon.

B. Economic Framing of a Low-Carbon Energy Transition

Green Growth Framing

The scale of national economic activity is usually measured by GDP (gross domestic product), which represents the value of all traded goods and services. Due to the COVID-19 pandemic and the associated slowdown in economic activity, global GDP fell by 3.2% in 2020 (IMF, 2021). Linked to this, governments in industrialized countries significantly increased their public borrowing to support health care and other services and to provide additional support to workers who would otherwise have lost their jobs. As a result of the public health measures and rollout of vaccines in these countries enabling economic activity to return, the global economy is projected to grow by 6% in 2021 and by 4.9% in 2022 (IMF, 2021). However, the form of this economic recovery will be crucial if the rapid transition in energy systems toward net-zero carbon emissions by around 2050 is to be achieved. This will require deep decarbonization across all sectors, including power, heating and cooling, transport, and industry, complemented by negative emissions options, including biomass with carbon capture and storage and direct air capture. The green growth position that it is feasible and desirable to achieve high levels of economic growth while rapidly reducing carbon emissions. This is often referred to as decoupling of economic growth from carbon emissions.

The green growth framing then focuses on achieving this decoupling through the innovation and deployment of low-carbon technologies, including renewable energy, electric vehicles, nuclear power, and carbon capture and storage, plus negative emissions options, and development of infrastructure and skills to support these. This will require high levels of both public and private investment, which, it is argued, rely on high levels of economic growth (Stern, 2021). The pursuit of economic growth has been the central plank of all industrialized countries since the calculation of GDP was formulated as a way of enhancing economic output during the World War II (Coyle, 2017). This measures private consumption, investment, government spending and net exports of a country. Sustained economic growth is usually understood by economists as being driven by innovation, which enables the production of goods and services at lower cost and hence enhances the range and quality of goods and services available to consumers (Beinhocker, 2006). So a green growth perspective argues that this innovation has to be directed toward reducing the costs and improving the quality of green technologies (Acemoglu et al., 2012). A focus on smart, inclusive green growth follows from the recognition that such growth needs to be enabled by investment in digital technologies and that the benefits of growth need to include poorer and disadvantaged communities (Perez, 2019). This perspective argues that continuing green growth and decoupling is possible, as more of economic activity shifts to an unbounded "intellectual economy" rather than the bounded "material economy" (Hepburn and Bowen, 2013).

Proponents of this view point to the success in achieving rapid cost reductions for renewable electricity generation technologies, including solar photovoltaics and wind power (IRENA, 2016). Though the cost of these technologies was initially much higher than the cost of coal or gas-powered electricity generation, a combination of public support for deployment of these technologies in countries and regions, such as Germany and California, and learning effects and economies of scale in manufacturing of these technologies, particularly in China, led to these rapid cost reductions (Geels et al., 2016). However, some low-carbon technologies are still seen as high-risk investments by many private investors, leading to calls for mission-oriented innovation to deliver desired outcomes, such as smart cities, which have been taken forward by the European Union and other countries (Mazzucato, 2018).

Redirecting private investment toward low-carbon technologies is thus seen as requiring a combination of regulatory requirements and fiscal incentives to change the balance of risks and rewards associated with investing in these technologies. In economic terms, carbon emissions are an "externality" – a cost that is borne by society at large rather than by buyers and sellers of fossil-fuel-based products. The simplest way to "internalize" this externality is by putting a price on carbon emissions, either through a carbon tax or by a cap-and-trade scheme that establishes tradable permits in carbon. However, efforts to impose a carbon tax have met with resistance from current industries who argue that this would constrain overall economic activity. The European Union instituted a carbon Emissions Trading Scheme (ETS) in 2005, but following the economic crash in 2008/2009, the price of tradable carbon permits under the ETS shrank to below €15 per tonne of carbon emissions (Borghesi and Montini, 2016). This is only high enough to incentivize incremental changes in emissions, such as switching from coal to gas-fired electricity generation, and not to stimulate more transformative change. China has also tested regional carbon trading schemes but plans for a national scheme have been put on hold by the pandemic.

Thus, governments have largely fallen back on regulations designed to promote innovation. Examples include the UK government's decision to prohibit the sale of new internal combustion engine cars from 2030 onward in order to promote innovation and deployment of electric cars (HM Government, 2020) and targeted support for particular technology sectors, such as portfolio standards or feed-in tariffs for renewable electricity generation. Such measures still also require substantial public investment in areas where it is difficult for private investors to make an adequate return. Infrastructure, such as national electricity grids, research, development, and demonstration of new technologies, including hydrogen production and carbon capture and sequestration, and skills training and development are among such areas.

From a macroeconomic perspective, the issue of the sources and scale of public funding needed for such measures, and the impact that this effort could have on private investment raises questions. In relation to the second point, critics of government spending often argue that such an initiative will "crowd out" private investment, creating a drag on economic growth. However, proponents such as Mariana Mazzucato, point to the key role of public investment in earlier successful technological innovations, such as the internet, global positioning systems (GPS), and voice recognition systems, and public health initiatives, including vaccine development, arguing that public investment can mobilize private investment in technologies, such as renewable electricity generation (Deleidi et al., 2020).

In relation to the first point, long-standing and unresolved debates about the role of public debt in modern economies resonate. Each year, national governments spend money on health, education, defense, social welfare, and direct or indirect support for industry. They raise money through taxes on income, capital gains, sales, and private profits. If government spending is higher than tax receipts, that leads to a deficit for that year. This adds to the stock of national debt, which is financed by governments issuing interest-paying bonds that are purchased by private investors, who see this as a safe investment as governments are unlikely to go bankrupt. To alleviate the effects of the COVID-19 pandemic, governments around the world have increased borrowing to provide support for families, businesses, and communities affected, such as the $1.9 trillion American Rescue Plan, signed into law in March 2021 by the new Biden administration (Biden, 2021). The US Senate also passed an Innovation and Competition Act in June 2021 to support industries such as semiconductor manufacturing and the Infrastructure Investment and Jobs Act, with an additional $550 billion of newly authorized spending on infrastructure. The Inflation Reduction Act, including £391 billion of investment in clean energy and climate change mitigation measures, was passed in August 2022. Until recently, governments in industrialized economies have been able to finance additional borrowing, as interest rates have been historically low. However, some critics point to the dangers of inflation (CRFB, 2021), which occurs when too much money in the economy is chasing too little goods and services, leading to the price of those goods and services rising, and central banks to increase interest rates to slow down economic activity and prevent further price rises. Proponents of green stimulus investment, such as Nicholas Stern, argue that the danger is exaggerated, provided that an economy grows faster than the rate of interest charged on its debt, then its debt to GDP ratio will fall and borrowing can be securely financed (Stern and Zenghelis, 2021). Thus, under the current economic system, green growth is essential to enable governments to safely simulate a transition to a net-zero carbon energy system by around 2050.

Post-Growth Framing

More radical critics of the current economic system doubt whether the green growth framing will be enough to achieve the necessary rapid transition to net-zero energy systems and whether it will promote achieving this in a just and equitable way (Hickel and Kallis, 2019). They argue that it depends on an unrealistic confidence in the ability of new technologies to be deployed and deliver expected carbon reductions rather than addressing excessive levels of consumption in richer countries (Anderson and Peters, 2016; Wiedmann et al., 2020) and downplays other carbon reduction options, including energy efficiency and energy demand reduction measures, such as insulating all homes and buildings and promoting active travel, such as walking and cycles in towns and cities to replace short car journeys (CREDS, 2019, 2021), which may not deliver private economic returns. These options would instead be justified by the wider wellbeing benefits that they could deliver, including positive health benefits, social cohesion, and local environmental benefits, including reducing air pollution (Creutzig et al., 2022; Barrett et al., 2022).

The dominant political economy model of economic growth over the last 40 years has been underpinned by neoclassical economic theory. This has held that growth is mainly driven by private investment leading to productivity improvements, that private firms should mainly or exclusively focus on providing returns to their shareholders, and that the benefits of this growth will largely "trickle down" to workers, with a minimal public safety net for those unable to work. However, income and wealth inequality has risen significantly in many industrialized countries, as shown by Piketty (2014). This means that the benefits of productivity improvements and economic growth have largely been captured by the wealthiest in society, with average wages for workers static or falling. Arguably, this rising inequality has been one of the triggers for rising populism, seen in countries from USA and Brazil to Russia and Hungary (Engler and Weisstanner, 2021).

Proponents of a post-growth position[1] argue that a continued green growth strategy is dangerous, as this would lead to the benefits of public and private investment in low-carbon technologies being largely captured by wealthy individuals and firms, as has already happened with internet-based industries, such as electronic search and online sales. They argue that this would slow the diffusion of socially beneficial low-carbon technologies that do not give rise to high private returns, such as forms of distributed electricity generation, including small-scale solar PV. If green growth policies lead to an inequitable distribution of benefits, this would be seen as unjust and could lead to populist backlashes, such as the Gilets Jaunes protests in France against the imposition of higher fuel taxes (Tainturier, 2020).

The post-growth position argues for the need to transform current economic systems to address these challenges and enable a rapid and just low-carbon energy transition (Jackson, 2018). This would involve (1) changing the political economy goals of economic activity to address the climate and ecological emergency, (2) having a greater role for both public- and community-based economic activity compared to private economic activity, (3) prioritizing interests of the most vulnerable people, and (4) instituting democratic reforms to give more people a greater say in decision-making toward these goals. This is based on the view that this transformation is both necessary because of the scale and urgency of the climate and ecological emergencies, including biodiversity loss, and that action to transfer power to more people to achieve these goals would be popular.

Within this view, a network of scholars and activists are advocating for "degrowth" as a radical social and political transformation to reduce energy and resource use while improving people's quality of life or well-being (Kallis et al., 2018). They argue for the need to directly challenge the dominant economic growth framing, so as to highlight the need for social and political change well as technological change, and to avoid co-option by the mainstream, which they argue happened with previous concepts, such as sustainable development.

Other scholars argue instead for an "a-growth" position, under which social and environmental goals are prioritized, but with a neutral perspective on whether GDP would fall or rise as a result (van den Bergh, 2011, 2017). This view aims to propose a more pragmatic way forward by avoiding the polarization often associated with debates between green growth and degrowth.

From a post-growth perspective, "renegade economist" Kate Raworth has proposed a doughnut economy model (Raworth, 2017). This argues that the goals of economic activity should be changed to fitting within the "safe and just space" inside the doughnut. This would be achieved by reducing the scale of economic activity to inside the planetary boundaries, such as climate change, biodiversity loss, and nutrient cycling (Steffen et al., 2015), while exceeding minimum thresholds for social goals, such as health and poverty alleviation, defined by the UN Sustainable Development Goals for 2030 (UN, n.d.). Unfortunately, the scale of global inequality means that no country is currently living within the doughnut. Richer countries largely deliver on social goals for their citizens but at the expense of exceeding their fair share of planetary boundaries, whereas poorer countries tend to be below their fair share of these planetary boundaries but at the expense of not delivering on social goals for their citizens (O'Neill et al., 2018). Post-growth economists generally recognize the need for economic growth in developing countries to help achieve these social goals. They also argue that if these developing countries follow the economic model of industrialized countries, even in a green growth version, they will rapidly exceed their fair share of planetary boundaries, as emerging economies such as China are already doing (O'Neill et al., 2018). Moreover, they argue that there are limits to the potential for decoupling of economic growth from resource use, as all economic activity relies on flows of energy and materials into the economy and wastes back into ecosystems, which have historically been closely coupled with levels of economic growth (Haberl et al., 2020). This does not prove that decoupling is impossible if efforts are focused on achieving this goal but leads to a justifiable skepticism about the rates of decoupling assumed (Ward et al., 2016).

Critics of the post-growth position argue that we have to work within the current economic system, as systems change tends to be slow and highly contested, and that is not clear that a post-growth economy could really deliver on its noble aims, even if these could be agreed upon. They argue that economic growth is what has delivered the rise in living standards in industrialized countries and so should not be discarded as a goal without a clear understanding of what should replace it. In particular, the creation of well-paying jobs enables the spread of economic prosperity, and under the current system, these are largely created by private sector investment to deliver productivity improvements and growth in new industries to replace declining industries. With such a view, a post-growth economy would only gain widespread popular support if it leads to job creation and the spreading of wealth.

Insights from the Pandemic

The COVID-19 pandemic and associated disruption of economic activity has brought these considerations into a new focus but also perhaps offered some new insights. It is clear that, under the current economic system, if demand for goods and services dramatically declines, then this trend would lead to high levels of unemployment, without rapid government action. Critics of a post-growth perspective tend to critique this position as being similar to an economic recession under the current system, which would have severe detrimental effects on the jobs and livelihoods of the poorest in society. However, post-growth proponents respond that we already have enough to meet everyone's basic needs for food, shelter, mobility, and social interactions – the problem is that these resources are inequitably distributed (Steinberger et al., 2020). If resources were used in a more equitable way and strategies aimed at sufficiency in high-income countries, then it would be possible to provide final energy required to meet decent living standards for the entire global population (Oswald et al., 2021; Millward-Hopkins et al., 2020). This would require radical changes to current economic and social provisioning systems to meet end-use demands, including demand-side changes to reduce consumption of energy and embodied energy in products, and measures to significantly reduce inequalities, both within and between countries (Creutzig et al., 2018; Vogel et al., 2021; Fanning et al., 2020). So how do we move to an economic system that is more "regenerative and distributive by design" (Raworth, 2017)?

The first insight from the pandemic is that governments can intervene in economies to achieve social goals. The governments that have been most successful at dealing with the pandemic prioritized the health and well-being of their citizens by intervening to prevent activities likely to spread the disease but also by supporting the wages of those who would otherwise have lost their jobs through furlough schemes. This was possible through an increase in government borrowing, enabled by low interest rates. Furthermore, as many have pointed out, central banks in the US, the UK, and Eurozone have been injecting money into the economy through so-called quantitative easing since the 2008/2009 economic crisis. This involves central banks creating new money from nothing and using it to buy financial assets from commercial banks and other financial institutions, including government bonds and corporate bonds. This means that much of the new debt created through government borrowing ends up being owned by the relevant central bank and could, in principle, be written off without creating inflationary pressures (Kelton, 2020). Efforts have also begun to "green" this quantitative easing by putting conditions on what the money is used to purchase so that it systemically favors investment in low-carbon and other green technologies. A new mandate on the Bank of England in this regard was imposed by the UK Chancellor of the Exchequer in his April 2021 budget (Bank of England, 2021).

A second broader insight from the pandemic relates to what people value in terms of extreme stress. Surveys have shown that people value their own and their family's health, and the sense of community to which they belong, higher than material or economic goals. Ecological economists Tim Jackson and Peter Victor have argued that this could provide a basis for moving away from a society focused on economic growth to one based on goals of individual, social, and planetary health, which emphasize healthy balance rather than increasing growth (Jackson and Victor, 2021).

Reconciling Green Growth and Post-Growth Views

Both green growth and post-growth framings thus agree on the need for rapid public and private action to achieve a transition to a net-zero carbon energy system by at least the mid-2050s and that this needs to be done in a way which is socially inclusive and equitable. However, these different framings can lead to different priorities for action to achieve such goals, which could affect the political feasibility and likelihood of success.

The green growth perspective argues that the key to achieving a sustainable energy transition is the scale-up of redirection of investment from the further exploitation of fossil fuel sources to the innovation and diffusion of low-carbon alternatives, including renewables, nuclear power, and carbon capture and sequestration. Under the current economic system, such an approach means changing the relative risks and rewards for relevant investments. Imposing a carbon price through carbon taxes or trading schemes, with reasonable expectations that this price would keep increasing into the future, would make fossil fuel investments increasingly risky. Government support for low-carbon innovation and diffusion through incentives such as feed-in tariffs and regulatory measures, including bans on the sale of internal combustion engine vehicles, would increase rewards for alternatives. This leads to the idea of a policy mix and the need for policymakers to strive to improve both the consistency of elements of the policy mix and the coherence of policy processes, in order to minimize any adverse economic impacts and improve effectiveness in redirecting innovation (Rogge and Reichardt, 2016). In this view, generalized growth still remains necessary to create the savings needed for this investment, and to generate tax returns for governments to fund these support measures and to address the social challenges relating to inequity. However, such an approach leads to less emphasis on the need for rapid scale-down of fossil fuel industries in order to achieve stringent carbon reduction targets.

The post-growth perspective emphasizes the need for taxes and regulatory measures to rapidly reduce investment and use of high-carbon energy technologies and argues that this can be done while maintaining health and well-being of citizens. This perspective notes that the world is not short of the savings needed for investment, and governments can support additional investment through borrowing. Trillions of dollars are already invested annually, mainly for short-term financial gains, and savings by private individuals and companies have increased due to the pandemic. Governments can easily finance additional borrowing when interest rates are low (Kelton, 2020). Up to the end of 2021, central banks were continuing to create additional money through quantitative easing to stimulate economic recovery from the pandemic despite increasing concerns about price inflation.

More generally, the post-growth perspective argues that, since the adoption of neoliberal[2] economic policies designed to promote businesses, free trade, and economic growth, most of the benefits of this growth have gone to the top 1% most wealthy, with relatively little or no "trickle down" of the benefits to the majority of populations (Jackson, 2018). In some emerging economies, such as China, the pursuit and achievement of high levels of economic growth has been successful in reducing the numbers of people in absolute poverty but at the expense of massive environmental impacts, including rapidly rising greenhouse gas emissions, and rising levels of inequality (Piketty et al., 2019).

In the short term, there is some overlap between the policies arising from the green growth and post-growth perspectives. Both would agree on the need and feasibility of government incentives to promote low-carbon energy choices, particularly renewable energy options. However, a green growth perspective would tend to also favor incentives and direct support for new nuclear power and carbon capture and sequestration (CCS) technologies (including negative emissions options, such as biomass energy with CCS), whereas a post-growth perspective would tend to prefer further support for energy-efficiency and energy-demand-reduction measures. This is because the latter measures are less likely to deliver private economic returns, whereas investment in new technologies would deliver high private economic returns if these technologies are successfully deployed. However, energy-efficiency and energy-demand-reduction measures may be justified in terms of delivering other social, health, and local environmental benefits (Creutzig et al., 2022).

These types of approaches are already being put into practice. In the US, the Inflation Reduction Act contains support for renewables and household energy efficiency measures, and in the UK, smaller amounts are being provided for renewables, small modular nuclear reactors, and hydrogen production and carbon sequestration demonstration projects under the prime minister's 10 Point Plan (HM Government, 2020). There is also the recognition that, in capitalist, democratic societies, such measures need to be justified on the basis of the net jobs that are expected to be created and the need to support communities dependent on high-carbon jobs that are expected to lose out. For example, the European Union's Green Deal includes provisions for a "just transition", focused on providing support and reskilling for communities in coal-mining regions that are being affected (European Commission, 2021). Thus, elements of a Green New Deal model are starting to be put into place, emphasizing public investment in low-carbon energy options and support for communities losing out. Academic analysis shows that, as renewables and energy efficiency are more labor-intensive than fossil fuel alternatives, they are likely to lead to net job creation in the short term, provided that economies are operating at less than full capacity. However, the long-term impacts on job creation depend on whether this investment puts economies on to a more economically efficient pathway, taking into account externalities including environmental impacts and energy security (Blyth et al., 2014).

Drawing on elements of the post-growth perspective, many proponents of a Green New Deal in the US, the UK, and Europe would argue for going much further. The core of this thinking would be the need to redesign Western economies so that environment and social goals are built in from the start rather than being seen as "externality" problems that need to be addressed as unfortunate adverse consequences of a growth-oriented model. Indeed, the Green New Deal, non-binding resolution introduced in the US Congress, did not mention economic growth but prioritized "achieving net-zero greenhouse gas emissions through a fair and just transition . . . creating millions of good high-wage jobs and ensuring prosperity and economic security . . . and promoting justice and equity" (Ocasio-Cortez and Markey, 2018). These objectives would be achieved through mobilization to achieve systems change for decarbonization in electricity, buildings, transport, and manufacturing systems, while supporting vulnerable and deindustrialized communities.

Clearly, redesigning economic systems is a huge challenge, especially in democratic countries. Strong resistance to perceived threats to their interests has already come from fossil fuel companies and other interests in business and the media. Though outright denial of the existence of human-induced climate change is on the wane, tactics designed to promote doubt and delay to the low-carbon transition have been identified (Lamb et al., 2020).

The challenge posed to a post-growth perspective is to go beyond a critique of the shortcomings of the current growth-based economic system and propose an alternative that is still consistent with values of individual freedom and choice (while recognizing that there are limits on these freedoms). One more specific post-growth approach that is now starting to inform policy in some countries is that of a well-being economy. In particular, the government in New Zealand has already developed a Well-Being Budget, which is being used to assess the contribution of individual policies to overall goals. This Well-Being Budget argues for the need for an integrated approach to social, economic, and environmental objectives, with high level goals of *taking mental health seriously, improving child well-being, supporting indigenous and disadvantaged people, building a productive nation,* and *transforming the economy* (New Zealand Treasury Department, 2019). In this way, investment in innovation, skills, and incentives for a transition to a low-carbon energy system is seen as part of a wider transformation of the economy. The crucial change in perspective is that all government spending, investment, taxation, and regulation should be assessed in this holistic way, with social and environmental impacts being assessed on their own terms rather than, at best, being given notional monetary values. Thus, rather than social and environmental policies being required to clean up the negative impacts of economic growth on nature and excluded groups, the aim is that by integrating these goals from the start, a more sustainable and inclusive form of economic prosperity can be achieved. Needless to say, it would be a huge political challenge in many countries to move to this more integrated and holistic approach to government budgeting, but useful lessons can be learned from countries and regions adopting well-being budgeting, including New Zealand, Iceland, Ireland, Scotland, and Wales (Wellbeing Economy Alliance, 2020).

C. Conclusion

What can we conclude about the relevance of these competing economic framings for how energy is sourced, delivered, and utilized? First, conventional macroeconomic theory has neglected the important role of access to useful energy in stimulating economic growth (Warr and Ayres, 2012; Foxon, 2018). This means that there is a danger that assessments based on such theory may underestimate the scale of the economic challenge in a transition to a net-zero energy system. The UK Committee on Climate Change estimates that the additional economic costs of the transition only amount to around 1% of UK GDP per year by 2050 (CCC, 2020). This means that if the UK economy continues to grow at 2.5% per annum, it would reach the same level of economic output only a few months later than if it had not invested in the transition (CCC, 2019). Clearly, this is an attractive message for politicians, in arguing that the costs of the transition are relatively low. However, there remain questions as to whether investments in energy efficiency improvements in buildings, industry, and transport will deliver the expected energy savings or whether some of the associated economic savings will be reinvested in carbon-intensive activities – the so-called rebound effect, which has been argued to be a significant driver of past economic growth (Brockway et al., 2021). The point is important, as it would imply that higher levels of investment in low-carbon energy supply would be needed to reach the same level of carbon reduction. Furthermore, questions also remain as to whether renewable energy sources, with lower net energy returns than fossil fuels, could sustain equivalent levels of economic activity (Jackson and Jackson, 2021). Some of this could be counterbalanced by the fact that renewables enable more widespread electrification of transport and heating end-uses with high efficiency of electric motors and heat pumps, compared to internal combustion engines and gas boilers.

Second, the response to the COVID-19 pandemic has shown that governments can act rapidly and decisively in the face of a perceived emergency to mobilize desired actions, such as rolling out vaccinations, and to restrict individual freedoms for a common benefit, including preventing meeting others indoors in order to reduce collective rates of infection. This has been made possible through huge injections of public financing, both for funding health services and for alleviating pressure on those who would otherwise lose out (e.g., by furloughing workers in affected industries). This suggests that the scale of public financing for a more rapid low-carbon transition is dependent on the extent to which dangerous climate change is seen as a similar emergency requiring rapid government action.

The framing of the economic recovery from the impacts of the enforced economic contraction due to the pandemic will also have important implications for the scale of public funding available for the low-carbon transition. In the US, under the Biden administration, efforts are being made to mobilize trillions of dollars for infrastructure spending, including that on expanded renewable generation and upgrading electricity networks. These efforts recognize the argument discussed earlier that countries issuing their own sovereign currency can increase government spending as much as they like, as long as this is not likely to lead to higher price inflation – that is, the level of national public debt is not a problem per se. However, in other countries such as the UK, there are signs that a renewed framing of austerity – that government spending has to be reduced in order to control the level of public debt – is threatening the level of government funding available to support the net-zero transition. As happened after the 2008 financial crisis, further imposition of austerity measures could substantially reduce support available to kick-start a rapid energy transition, as well as reduce support for other public services and welfare measures that would mitigate the transitional effects on those working in industries that would lose out from the transition.

Finally, though the need for a rapid energy transition to meet net-zero goals in order to address the threat of climate change is now widely accepted in principle by governments around the world, the actions to stimulate this are still lacking in practice. This chapter has argued that this lack of radical action is at least partly related to the economic framing under which governments act. The dominant green growth framing argues that generalized economic growth is necessary to stimulate private investment and to provide the funds through taxation for public investment on innovation and deployment of low-carbon energy technologies. The point implies that the key role for governments is to promote the conditions for higher economic growth, while steering such growth toward green technologies. Critics of this view argue that, while this provides a basis for government support for innovation in new technologies, such as solar and wind generation, and hydrogen as an energy carrier, it does not give rise to policies that seek to reduce the scale of high-carbon energy and energy-intensive industries. These more radical policies could include eliminating subsidies to fossil fuel exploration or imposing taxes on high levels of flying. In essence, a green growth framing leads to the perspective that innovation in clean energy technologies will occur rapidly enough, leading to cost reductions so that they outcompete fossil-fuel-based energy technologies fast enough to avert dangerous climate change. This situation is exemplified by the UK government's promotion of Jet Zero, technologies for net-zero aviation can be rolled out by the 2040s so that there is no need to impose restrictions on the scale of aviation in the meantime (HM Government, 2020).

The post-growth perspective argues that this view is a dangerous delusion. The recent IPCC Sixth Assessment Report and the evidence of the impacts of climate change already being felt, from heat waves and forest fires in Canada, Australia, and Mediterranean countries to floods in India, China, and Central Europe, point to the need for a more rapid energy transition, including action to promptly scale down the use of fossil-fuel-energy technologies through fiscal and regulatory measures. To avoid the risk of these impacts having a severely adverse impact on people's welfare, particularly the poorest in society, post-growth ideas advocate for the need for wider change to current economic systems in order to ensure that more of the present benefits of economic activity flow to the less well-off in society. Measures proposed to enable this include expansion of free or reduced cost public services, such as health, social care, and basic amenities; a universal basic or citizens income; and community wealth-building approaches to retain more wealth generated within local communities. Such measures would be justified on a well-being economy basis, in which the wider social and environmental benefits, such as improved health outcomes and cleaner air, would offset any reduction in the overall level of economic activity (Wellbeing Economy Alliance, 2020). Though such measures could generate widespread public support, any such major economic transformation would need to overcome resistance from powerful industries and individuals that may lose out from this transformation (Lamb et al., 2020). As a start, post-growth advocates argue that these types of measures should be included in climate mitigation scenarios to avoid overreliance on assumptions of unfeasible rates of innovation in new technological and negative emissions options (Hickel et al., 2021). Critics of a post-growth perspective tend to view such proposals as utopian with little guarantee of success, compared with direct economic incentives for technological innovation which have been seen to work in other contexts.

So where does that leave us? Both the green growth and post-growth economic framings support more rapid action to promote innovation and deployment of renewable energy technologies, such as solar and wind power for electricity generation, and the rapid electrification of other end-uses, such as heating and mobility through heat pumps and electric vehicles. These approaches are justified economically on the basis of stimulus effect of such investment in promoting new jobs and increasing revenues from taxation, as well as longer-term benefits to countries that are able to generate cost savings by reducing the imports of fossil-fuel-energy sources. However, a green growth framing also puts more faith in the ability of new energy technologies, such as hydrogen, small-scale nuclear power, biofuels, and carbon capture and sequestration, to come down in cost so that they can outcompete or offset the impacts of existing fossil-fuel-energy technologies, alongside renewables. Green growth framing also implies that negative emissions technology options can be developed and implemented fast enough to offset remaining emissions from hard to decarbonize sectors, such as aviation and cement production.

A post-growth framing leads to more emphasis on rapidly reducing public and private investment in fossil-fuel-energy technologies, and supporting energy efficiency and energy demand reduction measures, such as insulating all homes and buildings and promoting active travel, such as walking and cycles in towns and cities to replace short car journeys (CREDS, 2019, 2021). Under a green growth framing, such measures are likely to be blocked or decreased as they would be seen to reduce the level of value-added and economic growth compared with alternatives. The post-growth framing would counter that such measures would be assessed as beneficial to well-being if all the positive social and environmental benefits of these measures were properly valued (Creutzig et al., 2022).

The most desirable and effective economic framing for a low-carbon energy transition is the subject of ongoing public and intellectual debate, and any positive changes will depend on

the ability of progressive social movements to convince public opinion and democratic governments to overcome resistance from vested interests and act rapidly and decisively. The extent to
which a transition to a net-zero carbon energy system is achieved fast enough to avoid the most
catastrophic environmental and social impacts of climate change, and in a way that is equitable
and inclusive to the least well-off in society, will depend on the outcomes of these political and
economic debates, at least as much as on the pace of innovation in desired low-carbon energy
technologies.

Notes

1 Here, we are including a "degrowth" position (Kallis et al., 2018) within a post-growth perspective, as
  both involve a radical critique of the emphasis on economic growth as the main goal of public policy.
2 Neoliberalism is used to refer to market-oriented reform policies, such as "eliminating price controls,
  deregulating capital markets, lowering trade barriers" and reducing, especially through privatization
  and austerity, state influence in the economy (Wikipedia, accessed February 6, 2022).

References

Acemoglu, D., Aghion, P., Bursztyn, L. and Hamous, D. (2012). 'The environment and directed technical
  change', *American Economic Review* 102(1): 131–166.
Anderson, K. and Peters, G. (2016). 'The trouble with negative emissions', *Science* 354: 182–183.
Bank of England (2021). 'Options for greening the Bank of England's Corporate Bond Purchasing Scheme',
  Discussion Paper, May 2021 www.bankofengland.co.uk/paper/2021/options-for-greening-the-bank-
  of-englands-corporate-bond-purchase-scheme
Barrett, J., Pye, S., Betts-Davies, S. *et al.* (2022), 'Energy demand reduction options for meeting national
  zero-emission targets in the United Kingdom', *Nature Energy* 7: 726–735.
Beinhocker, E. (2006). *The Origin of Wealth: Evolution, Complexity and the Radical Remaking of Economics.*
  London: Random House.
Biden, J. R. (2021). 'Remarks by President Biden', January 2021 www.whitehouse.gov/briefing-room/
  speeches-remarks/2021/01/27/remarks-by-president-biden-before-signing-executive-actions-on-
  tackling-climate-change-creating-jobs-and-restoring-scientific-integrity.
Blyth, W., Gross, R., Speirs, J., Sorrell, S., Nichols, J., Dorgan, A. and Hughes, N. (2014). *Low Carbon Jobs:
  The Evidence for Net Job Creation from Policy Support for Energy Efficiency and Renewable Energy.* London:
  UK Energy Research Centre. https://ukerc.ac.uk/publications/low-carbon-jobs-the-evidence-for-
  net-job-creation-from-policy-support-for-energy-efficiency-and-renewable-energy.
Borghesi, S. and Montini, M. (2016). 'The best (and worse) of GHG emissions trading systems: Comparing
  the EU ETS with its follower', *Frontiers in Energy Research* 4: 27.
Brockway, P. E., Sorrell, S., Semieniuk, G., Heun, M. K. and Court, V. (2021). 'Energy efficiency and
  economy-wide rebound effects: A review of the evidence and its implications', *Renewable and Sustainable Energy Reviews* 141: 110781.
Centre for Research on Energy Demand Solutions (CREDS) (2019). *Shifting the Focus: Energy Demand
  in a Net Zero Carbon UK*, Oxford, July 2019, www.creds.ac.uk/publications/shifting-the-focus-
  energy-demand-in-a-net-zero-carbon-uk.
Centre for Research on Energy Demand Solutions (CREDS) (2021). *The Role of Energy Demand Reduction
  in Achieving Net Zero in the UK*, Oxford, October 2021, https://low-energy.creds.ac.uk/the-report.
Committee for a Responsible Federal Budget (CRFB) (2021). 'What will build back better mean for inflation?', *CRFB Blog*, December 6, www.crfb.org/blogs/what-will-build-back-better-mean-inflation
Committee on Climate Change (CCC) (2019). *Net Zero: The UK's Contribution to Stopping Global
  Warming*, London, May 2019, www.theccc.org.uk/publication/net-zero-the-uks-contribution-to-
  stopping-global-warming.
Committee on Climate Change (CCC) (2020). *The Sixth Carbon Budget: The UK's Path to Net Zero*, Lon-
  don, December 2020, www.theccc.org.uk/publication/sixth-carbon-budget.
Coyle, D. (2017). 'Rethinking GDP', *IMF Finance and Development*, March 2017, 16–19.
Creutzig, F., Niamir, L., Bai, X. et al. (2022). 'Demand-side solutions to climate change mitigation consist-
  ent with high levels of well-being', *Nature Climate Change* 12: 36–46.

Creutzig, F., Roy, J., Lamb, W. F., Azevedo, I. M. L., Bruine De Bruin, W., Dalkmann, H., Edelenbosch, O. Y., Geels, F. W., Grubler, A., Hepburn, C., Hertwich, E. G., Khosla, R., Mattauch, L., Minx, J. C., Ramakrishnan, A., Rao, N. D., Steinberger, J. K., Tavoni, M., Urge-Vorsatz, D., Weber, E. U. (2018). 'Towards demand-side solutions for mitigating climate change', *Nature Climate Change* 8: 260–263.

Daly, H. E. (2015). 'Economics for a full world', *Great Transition Initiative* (June 2015). www.greattransition.org/publication/economics-for-a-full-world

Deleidi, M., Mazzucato, M. and Semeniuk, G. (2020). 'Neither crowding in nor out: Direct public investment mobilizing private investment in renewable electricity projects', *Energy Policy* 140: 111195.

Engler, S. and Weisstanner, D. (2021). 'The threat of social decline: Income inequality and the radical right support', *Journal of European Public Policy* 28(2): 153–173.

European Commission (2021). Delivering the European Green Deal, Brussels https://ec.europa.eu/info/strategy/priorities-2019-2024/european-green-deal/delivering-european-green-deal_en

Fanning, A.L., O'Neill, D. W. and Buchs, M. (2020). 'Provisioning systems for a good life within planetary boundaries', *Global Environmental Change* 64: 102135.

Foxon, T. J. (2018). *Energy and Economic Growth: Why we Need a New Pathway to Prosperity*. Abingdon and New York: Routledge.

Geels, F. W., Kern, F., Fuchs, G., Hinderer, N., Kungl, G., Mylan, J., Neukirch, M. and Wassermann, S. (2016). 'The enactment of socio-technical transition pathways: A reformulated typology and a comparative multi-level analysis of the German and UK low-carbon electricity transitions (1990–2014)', *Research Policy* 45: 896–913.

Haberl, H, Wiedenhofer, D, Virag, D, Kalt, G, Plank, B, Brockway, P, Fishman, T, Hausknost, D, Krausmann, F, Leon-Gruchalski, B, Mayer, A, Pichler, M, Schaffartzik, A, Sousa, T, Streeck, J and Creutzig, F (2020). 'A systematic review of the evidence on decoupling of GDP, resource use and GHG emissions, part II: Synthesizing the insights', *Environmental Research Letters* 15: 065003.

Hepburn, C. and Bowen, A. (2013). 'Prosperity with growth: Economic growth, climate change and environmental limits', in Fouquet, R. (Ed.) *Handbook of Energy and Climate Change*, London: Edward Elgar.

Hickel, J. (2020). 'Quantifying national responsibility for climate breakdown: An equality-based attribution approach for carbon dioxide emissions in excess of the planetary boundary', *Lancet Planetary Health* 4: e399–404.

Hickel, J., Brockway, P., Kallis, G., Keyßer, L., Lenzen, M., Slameršak, A., Steinberger, J. and Ürge-Vorsatz, D. (2021). 'Urgent need for post-growth climate mitigation scenarios', *Nature Energy* 6: 766–768.

Hickel, J. and Kallis, G. (2019). 'Is green growth possible?', *New Political Economy* 25(4): 469–486.

HM Government (2020). *The Ten Point Plan for a Green Industrial Revolution*, London, November 2020 www.gov.uk/government/publications/the-ten-point-plan-for-a-green-industrial-revolution

International Monetary Fund (IMF) (2021). 'World Economic Output Update July 2021', www.imf.org/en/Publications/WEO/Issues/2021/07/27/world-economic-outlook-update-july-2021.

International Renewable Energy Agency (IRENA) (2016). *The Power to Change: Solar and Wind Cost Reduction Potential to 2025*. Bonn, Germany: IRENA.

Jackson, A. and Jackson, T. (2021). 'Modelling energy transition risk: The impact of declining energy return on investment (EROI)', *Ecological Economics* 185: 107023.

Jackson, T. (2017). *Prosperity Without Growth: Foundations for the Economy of Tomorrow*. Abingdon and New York: Routledge.

Jackson, T. (2018). 'The post-growth challenge: Secular stagnation, inequality and the limits to growth', CUSP Working Paper 12, University of Surrey www.cusp.ac.uk/wp-content/uploads/WP-12-The-Post-Growth-Challenge-1.2MB.pdf

Jackson, T. and Victor, P. (2021). 'Confronting inequality in the "new normal": Hyper-capitalism, proto-socialism, and post-pandemic recovery', *Sustainable Development* 2021: 1–13.

Kallis, G., Kostakis, V., Lange, S., Muraca, B., Paulson, S. and Schmelzer, D. (2018). 'Research on degrowth', *Annual Review of Environment and Resources* 43: 291–316.

Kelton, S (2020). *The Deficit Myth: Modern Monetary Theory and the Birth of the People's Economy*. New York: PublicAffairs.

Lamb, W. F. et al. (2020). 'Discourses of climate delay', *Global Sustainability* 3: e17, 1–5.

Mazzucato, M (2018). *Mission-oriented Research and Innovation in the European Union*, European Commission. https://ec.europa.eu/info/sites/default/files/mazzucato_report_2018.pdf

Millward-Hopkins, J., Steinberger, J. K., Rao, N. D. and Oswald, Y. (2020). 'Providing decent living with minimum energy: A global scenario', *Global Environmental Change* 65: 102168.

New Zealand Treasury Department (2019). The Wellbeing Budget, Auckland. www.treasury.govt.nz/sites/default/files/2019-05/b19-wellbeing-budget.pdf

Ocasio-Cortez, A. and Markey, W. (2018). *Resolution on a Green New Deal*. US House of Representatives.

O'Neill, D., Fanning, A., Lamb, W. and Steinberger, J. (2018). 'A good life for all within planetary boundaries', *Nature Sustainability* 1: 88–95.

Oswald, Y., Steinberger, J. K., Ivanova, D. and Millward-Hopkins, J. (2021). 'Global redistribution of income and household energy footprints: A computational thought experiment', *Global Sustainability* 4: e4, 1–13.

Perez, C. (2019). 'Transitioning to smart green growth: Lessons from history' in: Fouquet, R. (Ed.) *Handbook on Green Growth*. Elgar: Cheltenham, pp. 447–463.

Piketty, T. (2014). *Capital in the 21st Century*. Cambridge MA: Harvard University Press.

Piketty, T., Yang, L. and Zuchman, G. (2019). 'Income inequality is growing fast in China and making it look more like the US', *LSE Blog*, April 1, https://blogs.lse.ac.uk/businessreview/2019/04/01/income-inequality-is-growing-fast-in-china-and-making-it-look-more-like-the-us.

Raworth, K. (2017). *Doughnut Economics: Seven Ways to Think Like a 21st Century Economist*. London: Random House Business.

Rogge, K. and Reichardt, K. (2016). 'Policy mixes for sustainability transitions: An extended concept and framework for analysis', *Research Policy* 45: 1620–1635.

Speth, J. G. (2008). *The Bridge at the Edge of the World: Capitalism, the Environment, and Crossing from Crisis to Sustainability*. New Haven: Yale University Press.

Steffen, W. et al. (2015). 'Planetary boundaries: Guiding human development on a changing planet', *Science* 347: 6223.

Steinberger, J. K., Lamb, W. F. and Sakai, M. (2020). 'Your money or your life? The carbon-development paradox', *Environmental Research Letters* 15: 044016.

Stern, N. (2021). 'G7 leadership for sustainable, resilient and inclusive economic recovery and growth', *LSE Grantham Institute*, www.lse.ac.uk/granthaminstitute/publication/g7-leadership-for-sustainable-resilient-and-inclusive-economic-recovery-and-growth.

Stern, N. and Zenghelis, D. (2021). 'Fiscal responsibility in advanced economies through investment for economic recovery from the COVID-19 pandemic', *LSE Grantham Institute*, www.lse.ac.uk/granthaminstitute/publication/fiscal-responsibility-in-advanced-economies-through-investment-for-economic-recovery-from-the-COVID-19-pandemic.

Tainturier, B. (2020). 'To explain, to understand, or to tell the Gilet Jaunes', *Europe Now*, Issue 31, Council for European Studies.

UN (n.d.). Sustainable Development Goals for 2030, https://sdgs.un.org/goals

van den Bergh, J. C. J. M. (2011). 'Environment versus growth – A criticism of "degrowth" and a plea for "a-growth"', *Ecological Economics* 70(5): 881–890.

van den Bergh, J. C. J. M. (2017). 'A third option for climate policy within potential limits to growth', *Nature Climate Change* 7(Feb.): 107–112.

Vogel, J., Steinberger, J. K., O'Neill, D. W., Lamb, W. F. and Krishnakumar, J. (2021). 'Socio-economic conditions for satisfying human needs at low energy use: An international analysis of social provisioning', *Global Environmental Change* 69: 102287.

Ward, J. D., Sutton, P. C., Werner, A. D., Costanza, R., Mohr, S. H. and Simmons, C. T. (2016). 'Is decoupling GDP growth from environmental impact possible?', *PLoS One* 11(10): e0164733.

Warr, B. and Ayres, R. U. (2012). 'Useful work and information as drivers of growth', *Ecological Economics* 73: 93–102.

Wellbeing Economy Alliance (2020). *Wellbeing Economy Policy Guide*, https://weall.org/policyguide

Wiedmann, T., Lenzen, M., Keyser, L. and Steinberger, J. (2020). 'Scientists' warning on affluence', *Nature Communications* 11: 3107.

World Resources Institute (WRI) (2021). COP26: Key outcomes from the UN Climate Talks in Glasgow, November 17, 2021, www.wri.org/insights/cop26-key-outcomes-un-climate-talks-glasgow

4

CRITICAL CONNECTIONS IN MATERIAL TRANSITIONS AND ENERGY TRANSITIONS

Marianne Zotin, Pedro Rochedo, Joana Portugal-Pereira, Alexandre Szklo, and Roberto Schaeffer

A. Introduction

As non-linear, complex, far-reaching, and inherently uncertain processes, global energy transitions comprise a number of transformations inextricably linked to the diffusion of new prime movers – defined here as machines or devices that convert primary and/or secondary energy sources into energy services – and the structural change of global primary energy supply. Energy transitions in different scales may be driven by techno-economic aspects, natural resources limitations, and power relations in energy systems, among other influences. The implications of these processes echo in all other human systems (see also Wilson et al., 2023, this volume).

Throughout history, energy transitions have been closely related to materials demand in quantitative and qualitative terms, determining the material foundations of societies. Likewise, advances in materials science and technology, as well as inexpensive availability of materials with desirable properties, have been necessary though not sufficient conditions for new energy systems to emerge, stabilize, and expand. As a result, societal use has evolved from 13 elements known in the periodic table prior to 1750, to the material basis of the current global economy consisting of 80 stable chemical elements in a wide spectrum of natural and synthetic materials. Metal alloys, synthetic polymers, high-performance composites, fine chemicals, and ceramics have been invented, and their properties have been improved to meet material services demand.

While energy transition studies have substantially contributed to the understanding of how such processes unfold and shape (or are shaped by) economic development, technological innovation, and social change (Araujo, 2017; Geels et al., 2017; Grubler and Wilson, 2013), socio-metabolic research has similarly provided a conceptual basis to understand society–nature interactions, offering valuable insights into how socioeconomic activity relates to biophysical flows and stocks use (Haberl et al., 2019). Furthermore, a fragmented literature on the role of biomaterials, materials efficiency strategies, circular economy, industry sector decarbonization, and critical materials in energy transitions has gained relevance in recent years (Allwood et al., 2011; Hertwich et al., 2020; IEA, 2020a; Kermeli et al., 2021; Oliveira et al., 2021; van Sluisveld et al., 2021). Nevertheless, how energy transitions shape materials transitions and vice versa has not been well analyzed in a systematic way.

Thus, the aim of this chapter is twofold: first, to present energy-materials interlinkages in past energy transitions (Section C) and, second, to discuss potential materials-related issues that

DOI:10.4324/9781003183020-5

may play a role in shaping the current energy transition and the material foundations of the 21st century (Section D). Section E draws concluding remarks.

B. Materials-Energy Linkages over Transitions

This section aims to highlight how the provision of material and energy services[1] have been intertwined through technical, socioeconomic, and environmental linkages throughout history. It is not our goal to fully cover materials use and technology history or the chronology of events that have led to profound shifts in energy production and consumption globally. Instead, the objective is to shed light on when and how energy and materials developments have been interdependent in such a way that when one shifted the other was also altered. While the historical trajectories of energy and materials consumption have been geographically and culturally diverse, we focus on how materials transitions shaped global energy transitions and vice versa in the past. We aim to contribute to the debate on sustainability transitions by reviewing and analyzing how materials can play the roles of constraints or opportunities in the ongoing energy transition to a low-carbon economy.

B.1 The Transition to Agrarian Societies

In the search for water, food, and firewood, hunter-gatherers were limited by the distances traveled, the productivity of the region, the inventiveness of the people and the efficiency of human metabolism to convert food into mechanical power (Cunha et al., 2023). Energy use (i.e., food intake and wood burning) in prehistoric times was therefore limited to physiological needs (energy and materials), movement, and human reproduction (Simmons, 2004). In turn, prehistoric innovations were driven by survival needs. Readily available natural materials, such as stones, bones, branches, and vines, were used to tailor better tools and weapons that ultimately delivered a necessary means for defense, for foraging and/or for hunting larger animals (Diamond, 2005).

By the late Paleolithic, wooden vessels also facilitated transportation, expanding distance limits that constrained food access, and the deployment of animal hides was extensively used for resisting harsh winters and building tents. Small kilns or pits were also used in some regions to fire ceramics and shape them into tools for cooking and storage.

The increased availability of energy surpluses through the domestication of grains and animals around 10,000 BC in the Fertile Crescent region[2] provided the comparative advantages of sedentary societies over hunter-gatherers, raising population density on the order of a 100 times (Burger et al., 2017). Establishing, maintaining, and expanding towns required a significant quantity of energy to find and transport construction materials. From simple huts to sophisticated hydraulic systems, the supply of common natural materials, such as stones, wood, clay, limestone, and bamboo, required planning and basic empirical knowledge on materials' properties. Thus, energy surpluses could be shifted to activities beyond the simple maintenance and reproduction of existing systems, increasing the complexity of human societies (Hall and Klitgaard, 2012).

The use of metals depended on early energy developments. Leveraging mining and metallurgy fire-based techniques, such as fire-setting and smelting, was the necessary precondition to create access to metallic compounds and to shape them into useful objects (Perlin, 2005). In premodern agrarian societies, metals in limited availability were used when, as an alternative for or in combination with wood or stone, they presented a better cost-benefit ratio in materials services provision, such as shelter (nails), transport (horseshoes), production (tools in general), and security (weaponry) (Smil, 2014).

Even though these events unfolded differently across the world, in Europe these advances had such a remarkable impact on materials culture that historical periods were named after materials. For some applications, stone could be replaced by copper (3200–2300 BC), whose malleability enabled humans to shape tools according to their needs. Bronze, an alloy of copper and tin, replaced copper for the same reason (2300–700 BC): it had greater hardness, a lower melting point and greater resistance to corrosion. Iron, despite its higher melting point, was more abundant than copper and tin. Triggered mainly by tin production shortages, iron replaced bronze (1200–300 BC), although it would only be a better performing material when further processed to steel alloys, centuries later (Debeir et al., 1992). Naturally, previous knowledge of ceramic kilns and high-temperature heat was the pillar of early metallurgy, driving the former two materials transitions with performance opportunities. The latter materials transition, on the other hand, was driven by lack of materials availability due to trade disruptions.

Among the inorganic and more commonly used natural materials, stone use was generally oriented to durable infrastructure and ceramic bricks eventually came into use for construction where clay was easily found and fired. The Romans developed the *opus caementicium*, a concrete-like material resultant of mixing rock or ceramic tiles as aggregates with gypsum, quicklime, or pozzolana as binders, rendering one of the most time-resistant construction materials.

B.2 Industrial Revolution and the Transition to Coal: Steel and Construction Materials

From ancient times to the 18th century in Europe (and still today in several regions of the Global South), biomass was the main source for both materials and energy services provisioning. Though metals have better mechanical and chemical properties (e.g., malleability, specific stiffness, corrosion resistance) than wood. Their utilization was limited by mining productivity, smelting capacity, transportation distances, and costs. Wood, on the other hand, was vastly used for buildings, vessels, furniture, and artisanal manufacturing, not to mention its energy uses for heating, lighting, and cooking. In other words, economic activities were constrained by energy resources (firewood and food), materials (timber, stone, and clay for housing, transportation, and tools), energy converters (human and animal power), and innovation in traditional biomass energy systems (Smil, 2010).

In the slow transition to coal within Britain, a number of factors preceded Watt's engine in defining the turning point. The expansion of international trade for Great Britain with its colonies through the hegemony of the British merchant fleet, as well as the development of rural textiles manufacturing industries, gave rise to population growth and urbanization. London's population more than doubled from 1520 to 1550 (Rutter and Keirstead, 2012). The expanding trade also introduced news forms of knowledge. In addition, the price revolution (1560–1620) increased real wages, particularly high in Britain, leading to higher material well-being (Allen, 2012). Taken together, the demand for energy and materials grew, pressing the renewability of the agrarian system.

If, on the one hand, deforestation meant more timber and wood production and more land available to feed the growing population, on the other, it implied the gradual reduction of forest stocks, leading to wood transportation prohibiting costs (Perlin, 2005). Coal was considered of inferior quality due to the bad odor when combusted, but it offered greater availability and became cost-competitive relative to depleting forest wood (Araujo, 2017), given the lower costs of open-pit mining and transportation via cabotage (sea coal). In addition, it had higher energy density and reduced the pressure on the land, as it was found underground (Sieferle, 2001). The price advantage led to the early use of coal for thermal energy provision in glassworks, breweries, pottery, and ultimately metallurgy. As early as 1700, coal use was already over 48% of the total energy consumed in England and Wales (Warde, 2007).

The utilization of coke instead of charcoal in blast furnaces reduced the specific consumption of the reducing agent from around 20 m³ of softwood/tonne of hot metal (t HM) in the 18th century to around 1 m³ of coking coal/t HM (Babich and Senk, 2018). However, waterwheels remained operating blowers, hammers and rolling cylinders until James Watt's steam engine invention (1776) – which would be impossible without the use of iron cylinders.

Watt's steam engine was based on Newcomen's engine (1708) for water removal of the increasingly deeper coal mines. Therefore, coal mining provided not only a new source of energy but also set the stage for the development of the first prime mover to convert the chemical energy of coal into mechanical power – that is, a device universally applicable in industry. The innovation initially driven by the repeated shortages of energy, materials, and land not only relaxed the competition on wood for material and energy services but, mostly, led to unprecedented gains of productivity and cost reduction in transportation, metals, and construction materials production (Smil, 2014).

Indirectly, coal use expansion also drove the reduction of wood demand in the construction sector by driving cost reductions of brick production since it provided cheaper high-temperature heat than charcoal. The use of coking coal enabled the construction of larger blast furnaces due to its greater mechanical strength compared to charcoal, reducing costs of iron products, machinery, and railroads. This reinforced the positive feedback on materials production and the further expansion of land and maritime transport prime movers and networks, as well as urbanization. Later on, Portland cement (1824) and Bessemer steel (1856) processes were patented, enabling mass production of construction materials production to supply the rapidly growing demand in cities.

By removing the energy constraints set by the traditional organic system on one hand and by creating demand to expand transportation and urban infrastructure on the other, the Industrial Revolution supported large-scale and diversified delivery of materials services. The expansion of energy surpluses was the essence of the transition from organic to inorganic materials (Wrigley, 1962), including ferrous and non-ferrous metals, glass, and bricks. These were, in turn, materials requirements for both producing and transporting raw materials and final products for a vast array of material and energy services.

B.3 The Transition to Oil: Automobiles, Polymers, and Fertilizers

From its modern birth in the 19th century to the development of Henry Ford's Model T by 1907, the emergent petroleum industry was mostly based on providing kerosene use as a cheap illuminant substitute to the whale oil and coal town gas, although a minor use also was evident as a lubricant to reduce friction in machinery parts, as asphalt and as other product (Yergin, 1991). As the coal-based energy system expanded to the rest of Europe and to North America, urbanization strengthened and cities widened, changing the urban transport regime (Geels, 2005). The increased pressure on the horse-based transport system created favorable conditions to the diffusion of electric trams during the 1888–1914 period and, later, to the establishment of gasoline automobiles as a cheap and practical means of transportation.

The advent of petroleum-based liquid fuels overcame the shortcomings of steam engines to transform the land mobility landscape with internal combustion engines (ICE), smaller and with more power per volume or mass than the steam engine. The automobile's diffusion demanded increasing amounts of steel (used in the body structure and other parts) and concrete to expand roads and bridges networks. The emergent automobile industry both benefitted from and reinforced the existing oil-based infrastructure: not only gasoline but also asphalt for road pavements and eventually synthetic rubber feedstock for tires production (in substitution for natural rubber) were inexpensively co-produced from oil refining and processing to fulfill the energy and materials requirements of the automobile industry.

The influence of warfare in energy technology innovation and in shaping energy and materials transitions cannot be overlooked. Though scientific and technological breakthroughs were achieved in the pre–World War I period in the organic chemical industry, there was considerable uncertainty over the potential market for synthetic materials (Smith, 1988). Major demand for such materials was created during the two world wars (mostly World War II), which spurred further innovation and massive investments in the oil-based materials industry. Large-scale production of polyvinyl chloride (electrical insulation in wires), neoprene, and other synthetic rubbers (tires, tubes, wear, and oil resistant applications), nylon (parachutes and hammocks), Teflon (proximity fuses and uranium isotopes separation), and polyethylene (insulator for radar cables) was driven not only by the opportunities to substitute metals whenever possible but also by their unique properties, which yielded tremendous advantages in warfare (ibid.). After the war, organic chemical technology co-evolved with refining technologies, grasping opportunities for achieving economies of scope through diversification and verticalization.

Oil and gas also became central to fertilizer production in industrial scale, essential to feed the rapidly growing population in the 20th century. The first ammonia plant in Oppau (Germany) in 1914, based on coal feedstock in the Haber-Bosch process, initially aimed at improving agricultural productivity until it was diverted to HNO_3 explosives and munitions production when the World War I broke out (Smil, 2001a). In general, the organic chemical industry back then – located mainly in Germany (IG Farben), France (Rhône-Poulenc), and the UK (ICI) – was based on coal tars. After World War II, cheap availability of petroleum rapidly displaced coal as a chemical feedstock (Bennett and Pearson, 2009), and natural gas became the main source of hydrogen for ammonia synthesis. Hence, the improvement of average diets in growing populations after 1950 was enabled by large-scale ammonia production for fertilizer use, which in turn depended on the cheap feedstock co-production in refineries (Cunha et al., 2023; Smil, 2001b).

Relative to the transition to coal, the transition to oil also represented an even more abundant energy surplus to support societal materials needs. The complexity achieved by modern industrial civilization is revealed in the material abundance and diversity that flooded global markets in the postwar period. New synthetic materials with a peerless combination of properties, such as chemical and thermal resistance, durability, strength, and low cost, underwent competition in several different markets providing a better combination of cost and performance than incumbent materials, such as ivory, glass, tin, lead, steel, and many others. In the 1930s, global plastics production was around 50,000 t, reached 6 Mt in 1960 and 380 Mt in 2015, as a result of the expansion of oil production worldwide (Geyer et al., 2017; Smil, 2014). The fast growth of plastics, fertilizers, lubricants, solvents, rubbers, fibers, dyes, waxes, surfactants, detergents, asphalt, and other organic fossil-fuel-based materials production mainly after 1950 not only reveals the diversity of oil-based products and their pervasiveness in the global economy but also the extent to which the last energy transition radically transformed the materials culture worldwide.

The technology landscape of the postwar period also represented a shift from military to civil purposes. The reconstruction of the infrastructure destroyed by the war and supporting the growing middle class required intensive mining activity of widely known metals such as iron, aluminum, and copper. After the war, the demand for by-product (minor) metals – such as cobalt, titanium, vanadium, and many others – also increased as aviation, communications, rocketry, and nuclear power technologies were gradually introduced in modern civil society (Kleijn, 2012).

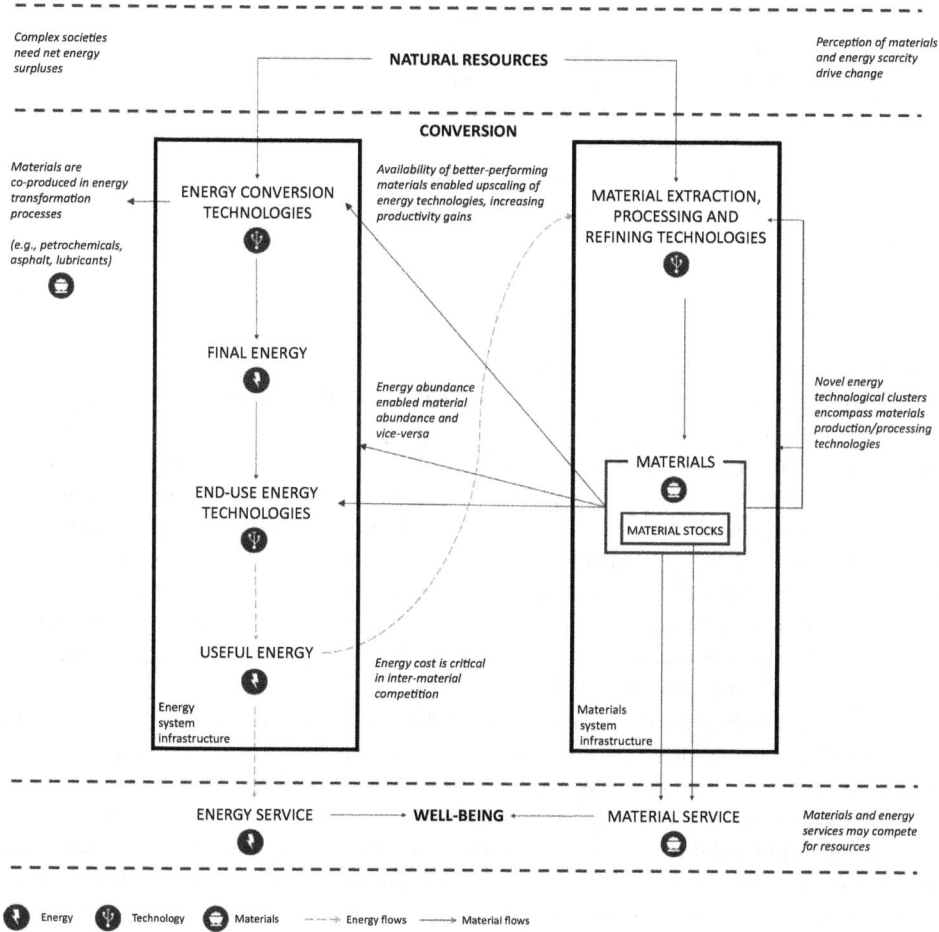

Figure 4.1 Main linkages between materials and energy sources/technologies.

Source: Authors.

Thus, oil became a key feedstock in the 20th century (so far, also in the 21st). The combining effects of expanding urban and agriculture frontiers and increasing the production of inorganic and organic materials led to human-made materials surpassing global living biomass in 2020 (Elhacham et al., 2020).

C. Materials Opportunities and Constraints in the 21st-Century Energy Transition

The conclusions drawn in the previous section add decisive questions for the unfolding energy transition. As greenhouse gas (GHG) emissions reduction targets become stricter every year due to insufficient climate action, an unprecedented and far-reaching global energy transition is urgently required so that all parties can meet their commitments under the Paris Agreement (IEA, 2021a; IPCC, 2018). In spite of the potential role of degrowth (Keyßer and Lenzen,

2021) or low-energy-demand scenarios (Grubler et al., 2018), the required energy transition will entail expanding the diffusion of solar PV panels, wind turbines, and electric vehicles worldwide, as well as extensive transmission lines and industrial-scale carbon dioxide removal (CDR) strategies. This engineered energy transition (Sovacool and Geels, 2016), at unprecedented rates to deal with the climate emergency, has implications to materials use and production. Materials can act as constraints or opportunities to energy transition pathways. The following subsections summarize key issues underlying how different groups of materials may influence energy transition pathways.

C.1 Organic Bulk Materials

Organic bulk materials are those natural or synthetic materials (mostly made from petroleum, natural gas, and coal) containing covalently bonded carbon and hydrogen – also, oxygen, nitrogen, sulfur, chlorine, and other elements can be present – and produced in large quantities, usually associated with the co-production of fuels. Fossil fuels can be used for combusted and non-combusted uses. In a typical refinery, some 75% of an oil barrel's total volume is converted to gasoline, diesel, and jet fuel as its main products, and another 10% to other fuels. The remaining 15% is mostly petrochemical feedstock, non-combusted and cheaply co-produced with fuels, which feeds 90% of the modern chemical industry today (IEA, 2018). They are subsequently transformed into primary chemicals, which are further processed into intermediates and, finally, converted to a vast array of final products among polymers, agrochemicals, and specialty chemicals in a complex supply chain.

Primary chemicals account for two-thirds of the energy demand in the chemical industry (IEA, 2018). These are ethylene, propylene, butadiene, and aromatics (benzene, toluene, and xylenes, or BTX) – also known as high-value chemicals (HVCs) – as well as methanol and ammonia. HVCs are the building blocks of plastics production. Ethylene and propylene are produced in larger volumes worldwide (around 165 Mt/year and 120 Mt/year, respectively) and are the key raw materials for thermoplastics production, such as polyethylene and polypropylene, vastly used for a number of different applications from food packaging to electrical insulation (Oil & Gas Journal, 2015). Butadiene (around 10 Mt/year) is critical to elastomers and synthetic rubbers production, essential in automotive tires. In turn, aromatics (~100 Mt/year) are ring-structured molecules largely used for increasing fuel performance as a high-octane blend component but also as intermediates for polystyrene and polyethylene terephthalate (PET) production (Bender, 2014). HVCs are conventionally produced in the steam cracking of (oil-based) naphtha or (natural-gas-based) ethane, although there is a significant co-production of propylene and aromatics in the fluidized catalytic cracking and catalytic reforming units in refineries, respectively. Also, the last decade saw an increase in the dedicated development of propylene production capacity based on propane dehydrogenation in the Middle East and coal-to-olefins in China, as well as a green ethylene plant based on bioethanol dehydration in Brazil, although in a much smaller scale. This was distinct from traditional steam-cracking, multi-product facilities.

Ammonia (~170 Mt/year) and methanol (~110 Mt/year) production processes are based on the production of syngas, a mixture of hydrogen and carbon monoxide, through steam methane reforming, coal gasification (especially in China and South Africa), or partial oil oxidation (IEA, 2018). Although ammonia is strictly an inorganic molecule, it requires large amounts of hydrocarbons to generate hydrogen, which further reacts with nitrogen. Around 80% of ammonia is consumed as nitrogen fertilizer (mostly in the form of urea and ammonium nitrate) (IEA, 2019a), whereas methanol is used in a variety of energy (mostly as gasoline blending or substitute and in biodiesel production but also in other forms) and non-energy (intermediate for formaldehyde, acetic acid, and silicone production, to name a few) applications. Although in ammonia-urea integrated facilities, CO_2 is captured and used within the process, in standalone ammonia facilities, process CO_2 emissions can increase from 2.1 to 4.6 tCO_2/tNH_3, depending on the feedstock (IPCC, 2006). On the other hand, limited process emissions are generated in methanol production.

Hence, from high-performance engineering composites to grocery single-use plastic bags, these hydrocarbon-based raw materials constitute the core of synthetic materials production today. Therefore, the fates of the fossil fuel and fossil-fuel-derived materials industries in the energy transition are thoroughly interdependent. An energy transition away from fossil fuels does not necessarily imply a materials transition that overthrows the non-energy use of petroleum, natural gas, and coal. Indeed, in the near- to medium-term, we shall expect an increasing refinery-petrochemical integration. As gasoline and diesel demands reduce due to passenger fleet electrification and carbon pricing, both refining capacity and cheap conventional feedstock free up for materials production. Intermediate streams from petroleum refineries are increasingly diverted toward petrochemicals production. For instance, IEA (2021a) indicates that at a deep decarbonization scenario petroleum refinery throughput would reduce 85% in 2050 while its yield in petrochemicals would increase to 70%, almost leading to the survival of only refineries integrated to petrochemicals. Also, the increasing demand for low-carbon and energy-efficient technologies is already driving up the demand for lightweight, stiff, durable, fatigue-resistant, and strong materials, such as polymer-based composites and carbon fibers (Mishnaevsky et al., 2017). The latter is at least four times lighter and its tensile strength three times higher than in steels (Liu et al., 2015). This, coupled with the capital-intensive nature of chemical industries, make it very unlikely that a paradigm shift happens in the organic chemical technology and infrastructure anytime soon.

In the long term, however, trends and uncertainties in the demand and supply sides may redefine the sector. From a demand-side perspective, plastic marine pollution and microplastics contamination concerns are driving single-use plastics bans and changes in consumer profile toward circular economy, which might lead the industry to shift end-use markets (Jambeck et al., 2015). At least 127 countries announced different laws and measures to limit single-use plastics use and manufacturing, though the COVID-19 pandemic delayed implementation (Excell et al., 2018). Furthermore, the substitution of single-use non-degradable plastics for biodegradable alternatives has been considered but performance efficacy and potential for global scale substitution are yet to be proven.

In fact, large-scale plastic waste management methods must consider climate change mitigation, material efficiency, and circular economy principles. Mechanical recycling plays a role in reducing GHG emissions and the use of virgin raw materials by reconverting plastic waste back into valuable – though inevitably lower-grade – products. Chemical recycling, on the other hand, can be a game changer in plastics demand if ever economically feasible since it enables remanufacturing waste plastic to the original polymer with the same quality, other synthetic materials, or fuels, avoiding investment needs in new capacity. This material advantage comes with an energy penalty, however, due to heat requirements of pyrolysis and gasification processes, as well as purification processes.

On the supply side, however, stringent climate targets can open opportunities to the chemical sector to develop and implement deep decarbonization measures Three general approaches arise to decarbonize the organic chemical industry. The first one consists of keeping up with oil, gas, and coal as raw materials but combining it with carbon capture and storage (CCS) to mitigate residual process and energy emissions. The second one counts on using CO_2 captured from other industrial processes or directly from air as the carbon (and oxygen) feedstock for chemical industrial processes (CCU). In the CCU case, cheap renewable electricity would play a central role in green hydrogen production and in promoting highly non-spontaneous reactions[3]. As renewable electricity costs decline, the conversion of electrolytic hydrogen (i.e., power-to-X) to both feedstock and energy carrier would become increasingly viable, enabling scaling-up production. To date, green hydrogen technologies have not reached maturity. Policy coordination to expand hydrogen demand and develop the necessary infrastructure for large-scale transport and storage will be necessary. However, though a hydrogen-based strategy would deliver better climate and air pollution outcomes, concerns on safety and water requirements are still a challenge.

The third approach is to feed renewable biomass in direct chemical, biochemical, and thermochemical processes. Though large-scale biomass harvesting may imply adverse side effects, this approach could benefit from infrastructure, technological, and logistics synergies with similar bio-based strategies in other hard-to-abate sectors, such as aviation and shipping, which consider low-carbon fuels to comply with the International Maritime Organization (IMO) 50% reduction in carbon emissions goal by 2050 (Carvalho et al., 2021; Oliveira et al., 2021). Also, existing fossil fuel infrastructure can play a role in smoothing the transition in these sectors and reducing large-scale capital risks by co-processing bio-based feedstock in conventional refinery units aiming at producing biofuels and bio-derived chemicals.

Away from the chemical sector, organic bulk materials also play an essential role as reducing agents in the iron, steel, and ferroalloys industries, mainly coking coal in blast furnaces and natural gas in direct reduced iron (DRI) facilities to a lesser extent. Hydrogen and charcoal are potential substitute candidates; also, scrap recycling in electric arc furnaces certainly plays a role by reducing mineral ore demand.

C.2 Inorganic Bulk Materials

Inorganic bulk materials are metallic (ferrous and non-ferrous) and non-metallic minerals produced in large volumes and whose production is energy- and/or emission-intensive. The emphasis of this section lies on basic inorganic raw materials that are inextricably linked to the expansion of new electric and digital infrastructure for low-carbon technology deployment and to the intensification of potassium and phosphorous fertilizer use for bioenergy development. Thus, they influence energy transition pathways in (1) the increment of energy inputs and direct and indirect GHG emissions, (2) the material intensity of renewable energy technologies, and (3) the potential reduction of the energy return on investment (EROI) due to declining metal ore grade.

Today, some 4,000 Mt of cement are produced every year, accounting for around 7% of both global direct GHG emissions and final energy consumption (GCCA, 2019). For iron and steelmaking processes, these figures are 6% and 7%, respectively, and yearly global production reached around 1,800 Mt of crude steel (IEA, 2020b). With very few cost-competitive substitutes, especially for structural applications, cement and steel demand have strong relationship with economies' GDP, industrialization, and in the future, probably the new capital stock associated with renewable energy expansion as well – recognized as more material-intensive than traditional fossil fuel technologies. Actually, estimates show that global demand for steel in the power sector could grow by a factor of 2.6 in 2050, driven by the deployment of renewable electricity technologies to limit global average temperature increase to 2°C (Deetman et al., 2021).

Cement and steel are enablers of transitions, as traditional low-cost materials with the desired properties to be used in foundations for wind turbines, concentrated solar power towers, transmission lines, and CCS infrastructure. However, they are also constraints, as hard-to-abate and emissions-intensive sectors. Though energy efficiency played a central role in the past decades within these industrial sectors, replacing traditional with innovative technologies is unavoidable to limit emissions. Furthermore, shelter, transportation, and storage services demands will probably increase as populations grow, urbanize, and gain affluence worldwide. However, innovative low-carbon technologies are still on early stages of development, and around 90% of the traditional existing capacity will reach the end of investment cycle only by 2040 (IEA, 2021a).

Similar to the chemical industry, these industries are heavy, hard-to-abate, capital-intensive, and with long-lasting equipment. They demand high-temperature heat in metallurgical (i.e., reduction of hematite and magnetite ores in temperatures as high as 1400–1,500°C in blast furnaces and up to 1,000°C in direct reduction) and calcination (i.e., decomposition of limestone, releasing carbon at 850–900°C) processes, which are energy services not easily electrified. It has been advocated that nuclear energy can be particularly useful in achieving these high temperatures or producing hydrogen as an energy carrier, thus playing a key role in hard-to-abate sectors as a mature, reliable, and low-carbon technology. But this comes with the acknowledged safety challenges faced by the nuclear energy itself and its toxic waste disposal, failing to meet circular economy principles (Costanza et al., 2011; Thomas, 2010). Nuclear plants are also high material consuming (Peterson et al., 2005), with materials performance issues as well (Zinkle and Was, 2013).

Furthermore, these processes release not only energy- but also process-related emissions, thus requiring mitigation strategies, such as clinker substitution, the substitution of coking coal as a reducing agent in iron production, and carbon capture utilization and storage (CCUS) in cement and steel production. Increasing the share of alternative materials into cement and using hydrogen as a reductant in the direct reduction of iron ore have been identified as key measures to achieve global net-zero emissions by 2050 (IEA, 2021a). However, permanently capturing CO_2 emissions and storing/using them is critical to address residual process emissions mitigation in these sectors. In this sense, while carbon capture and geological storage technologies are commercially available (though costly), using CO_2 as a feedstock could deliver additional economic revenues that would make the whole process economically viable (IEA, 2019b).

The material-intensive aspect of energy transition also affects the supply and demand of other inorganic materials. Phosphate, potassium, and nitrogen fertilizers will be required to meet expanding needs for food and bioenergy. Aluminum is a light, malleable, corrosion-resistant, and ductile metal used for transportation, packaging, and transmission lines. Copper, as an essential metal in the electric and electronic industry, will be extensively used in electricity networks. Most of the nickel produced today is used to make stainless steel and other steel alloys, though increasingly demanded by the batteries industry. Zinc is widely used in steel galvanization and for anticorrosion protection in turbine types. Beyond their traditional uses and new energy applications, most of these metals are also part of the broader digital and electric technology cluster, which include mobile phones, ICTs, EVs, and new lighting technologies, among others.

These metals' demands are expected to increase in unprecedented levels and most assessments assume an increase of around 215% for aluminum, 140% for copper, 140% for nickel, and 46% for zinc by 2050 (Watari et al., 2021). Though abundant and produced in distributed regions, these materials usually present a cheaply available range of properties, making them hard to substitute. Hence, the more we delay large-scale deployment of low-carbon technology, much less densely packed than fossil-fuel-based ones, the faster will be the ramp-up of material extraction and production rates needed. Also, the stock of materials in use, thus prevented from being recycled, will also be higher, which can hamper the efficacy of material efficiency measures (Vidal et al., 2013).

As a consequence, speculations over potential exhaustion of metal reserves emerge, in a way revisiting historical concerns over the scarcity of natural resources. While many researches point out evidence that mined ore grades have been declining over the years (Calvo et al., 2016), others suggest that misconceptions over mining concepts and economics, as well as inappropriate analogies with petroleum reserves, have created a false idea of causality between observed declining mined ores and the depletion of higher-grade deposits (Northey et al., 2018; West, 2011). The question, though, seems to be less about the physical limits of metal endowments and more about the implications of increasing energy intensity of metal extraction, how fast can metal extraction increase to respond to demand escalation, and the socio-environmental impacts that may arise from mining intensification.

Whether higher-quality deposits are indeed facing depletion or technological improvements enabled lower-grade mining in high-demand periods, it remains true that energy consumption and GHG emissions increased with decreasing ore grades (Azadi et al., 2020). This effect may be amplified with increasing depth of deposits and impurities concentration due to mining activity intensification. This underlies a key concern over the reduction of the net energy available to society if the energy return on investment (EROI) decreases, from which a vicious cycle would result: the energy transition to metal-intensive technologies would increase metal demand, gradually increasing the energy intensity of extraction, ultimately decreasing the EROI of the system and increasing the need for renewable technologies deployment (Fizaine and Court, 2015). Though only copper, zinc, lead, and nickel have shown declining grade trends as of today, managing environmental impacts of mining and processing of all metals would add to the energy cost accountability.

Moreover, mining and metallurgical activities entail serious human health and environmental impacts. For instance, mining extraction provokes land-use change – affecting biodiversity and displacement of communities – as well as large water use and waste generation, which are increased with lower ore grades. For example, water use in brine lithium and copper production can be as high as around 1 m^3/kg and 0.05 m^3/kg, respectively (IEA, 2021b). Depending on the ore, toxic pollutants may be emitted, such as SO_2 in copper chalcopyrite and HF in aluminum processing. Naturally, appropriate technology was developed for decades to minimize these impacts, not without affecting prices and leading to offshoring of mining activities to countries with less stringent environmental regulations. Thus, if the social and environmental burden of mining, metallurgical, and refining activities are not rigorously handled, energy transition pathways may result in extensive local impacts.

C.3 Critical Materials

While inorganic and organic bulk materials directly relate to CO_2 emissions and energy transition pathways, critical materials neither have a rigorous physicochemical nature nor have its impacts on the basis of energy/CO_2 emissions accountability. The concept of critical materials has been evolving since its first use in the 1930s–1940s to represent national security concerns over resource access: when the national security concept changed – from mainly military defense in war times to a vast array of economic securities – so did the critical materials concept to encompass material access concerns over energy transition strategies (Haglund, 1986). Since the late 2000s, this concept has been used in reference to materials that present a high risk of supply and are considered important in a given economy, corporation, or supply chain (DoE, 2011). Criticality is, therefore, not an inherent feature of a material but a dynamic concept that changes depending on the perspective and historical time.

Naturally, the diversity of perspectives in criticality assessments often results in contrasting outcomes regarding which materials are and which are not critical (Schrijvers et al., 2020). For example, even bulk materials such as copper and phosphates are understood as critical in some assessments (e.g., Sokhna et al., 2020; Cordell et al., 2009). However, our focus here lies on the materials more frequently considered critical, which usually (but not always), (1) are used in small mass per product unit, for which its property is essential; (2) are minor metals, mined in much smaller quantities compared to bulk or industrial metals (as a consequence, they usually cannot be considered mineral commodities); (3) have increasing demand rate prospects; (4) are co- or by-products of a major metal production; (5) have very low price elasticity of supply; (6) require complicated (intensive in energy, chemicals, and water use) extractive metallurgy; (7) have low recycling rates; and/or (8) have no substitutes with similar performance in the short/medium term. The fundamental issue is that some of these materials could delay or slow down energy technologies deployment on both supply (e.g., renewables) and demand (e.g., digitalization) sides as a consequence of geopolitical, economic, technological, or socio-environmental aspects. Hence, if organic and inorganic bulk materials are the macronutrients of the socio-economic metabolism, critical materials are the micronutrients (Pauliuk and Hertwich, 2015).

Examples can be drawn. Lithium's lowest standard reduction potential and cobalt's low electric and thermal conductivity make them suitable for long-lasting and high-energy-density batteries. Rare earth elements (REE), such as dysprosium, neodymium, terbium, and praseodymium, make up NdFeB magnets, which are capable of creating a powerful magnetic field without undergoing demagnetization in high temperatures, thus suitable for wind turbines and electric motors. Platinum group metals (PGMs) are extensively used in heterogeneous and electrochemical catalysis – also, potentially in biorefineries and hydrogen electrolyzers – given their high chemical resistance, high catalytic activity, and stability properties.

The claim that the ongoing energy transition is one from an oil-based to a mineral-based economy finds evidence not only in the increasing dependence on material-intensive, digital, and low-carbon technologies but also in the changing energy geopolitical landscape that emerges from it. The high geographical concentration of reserves, production, and/or processing capacity of some materials was what triggered off materials supply concerns and multiple criticality assessments in the last decade, after the 2010 rare earth crisis. China raised export taxes and decreased export quotas of REE in that year, which increased prices from 5 to 13 times (Kiggins, 2015). By that time, 97% of REE production was in China, and despite the diversification and recycling measures taken by industry and governments, it reduced to 65% (90% of the processing capacity) in 2019 (IEA, 2021b). China also holds around 60% and 50% of the processing capacity of cobalt (whose global production is 70% in the Democratic Republic of the Congo) and lithium, respectively, and 40% of copper refining capacity (IEA, 2021b). China has been able to expand access to key resources of the energy transition so as to convert it to industrial advantage, fulfilling its own materials requirements for strategic emerging industries, such as wind turbines, electronic equipment, and electric vehicles manufacturing (Kenderdine, 2017).

Of particular importance is the limited capability of the mining industry to respond to such crises and general market fluctuations. New project development, from the discovery of a new deposit to the production, takes on average 17 years (IEA, 2021b). To further process the metal extracted, capital-intensive metallurgical and refining processes are required, and such investments can be hampered due to the characteristic price volatility of these materials. Price volatility is commonly observed for metals that are recovered as by- or co-products of a primary desired metal. They are not necessarily scarce in the earth's crust but sparsely distributed in specific ore bodies and sometimes in such low grades that remain as waste products unless proven economical. Hence, co- and by-products pricing is highly dependent on the market dynamics of the main metal, which can make supply insensitive to high prices.

Besides geopolitical and economic factors, mining, recovering, and refining minor metals inflict major socio-environmental impacts. As ore grades decrease, more waste is generated and more water, energy, and chemicals are required to extract and purify the desired metals, frequently also exposing workers to toxic chemicals. It is estimated that over 50% of lithium and copper production today is carried out in high-water-stress regions (IEA, 2021b). While steel production consumes around 23 MJ/kg, cobalt production ranges from 140 to 2,100 MJ/kg and platinum group metals can reach up to 254,860 MJ/kg (EU Commission, 2018). Child labor, unsafe work conditions in artisanal mining sites, and political instability in the Democratic Republic of the Congo has been also documented as critical aspects of cobalt production in the country. Hence, measures to ensure material access to energy transition pathways must not overlook potential trade-offs between local and global impacts.

D. Concluding Remarks

All social and economic activities are mediated by the metabolic exchange of both matter and energy within nature, inextricably conditioned by the laws of thermodynamics (Georgescu-Roegen, 1986). Energy transition conceptualizations are diverse; they converge, though, in that materials aspects are often underrated as part of the transformations that go by with structural changes in energy production, transformation and consumption globally. As a result, energy transitions modeling that supports energy policy do not fully consider the materials-energy nexus as constraints or opportunities for deep decarbonization pathways.

From this perspective, an energy transition from fossil to renewable energy, not unusually envisioned as one from finite to infinite energy resources, may underestimate energy sources' and/or prime movers' substitutions triggered by unavailability of materials, much similar to that of the depletion of forest stocks in Europe. However, while back then material shortage revealed itself as a driver for harnessing the increasing availability of coal energy, as of today material shortage can hamper GHG emissions mitigation strategies to avoid irreversible climate change impacts. As the IEA (2021b) recently ascertained, investment plans for expanding mineral supplies to the extent required by an accelerated transition are still insufficient, not to mention geopolitical and environmental risks that could disrupt the operating capacity.

Likewise, materials-related opportunities may be sidelined as well. Efforts to establish a fully circular economy can both strengthen energy security and reduce the pressure on local communities and the environment. Moreover, material efficiency strategies – such as material substitution, recycling, remanufacturing, design, repair, and lifetime expansion – can also become relevant to loosen the pressure over supplies and further reduce impacts and improve energy efficiency. Nevertheless, primary material supply is unavoidable to fulfill the increasing demand for materials in emerging countries where the buildup of new material stocks for infrastructure and buildings is required (Krausmann et al., 2017). In addition, materials are part of the transition system and thus can affect energy technology choices (for example, facilities that co-produce materials and fuels from renewable sources may become systemically preferable compared to energy-focused facilities). Therefore, energy transition strategies need a holistic and systemic approach so as to consider the life cycles of materials and the industrial ecology in the cascading use of resources. Materials not only add up to energy and emissions accounting but can abruptly disrupt energy pathways and represent an underexplored potential to minimize socio-environmental impacts of an inevitably fast transition on account of decades of global climate inaction. To make ends meet, a just global energy transition must be based on the sustainability materials' life cycles, taking stock of the materials-energy linkages revealed in past transitions.

Notes

1 We build on the conceptualization adapted from Whiting et al. (2020) that defines material services as "those functions that materials contribute to personal or societal activity with the purpose of obtaining or facilitating desired end goals or states, regardless of whether or not a material flow or stock is supplied by the market". The authors consider all energy flows (excluding non-material energy flows, such as solar, wind, and hydro) as materials; likewise, all energy services are also material services (e.g., illumination, transport). However, in our text, "materials" and "material services" will be used in reference to flows/stocks and their respective services only when they provide exclusively material functions (e.g., shelter, packaging) to emphasize how natural resources used for energy and non-energy purposes interacted in past transitions.
2 This spans modern-day Iraq, Syria, Lebanon, Jordan, Palestine, Israel, and parts of Turkey and Egypt.
3 In 2020, electrolysis-based hydrogen production reached 5% of the 8 Mt of global capacity (IEA 2021a).

References

Allen, R. C., 2012. Backward into the future: The shift to coal and implications for the next energy transition. *Energy Policy* 50, 17–23. https://doi.org/10.1016/j.enpol.2012.03.020.

Allwood, J. M., Ashby, M. F., Gutowski, T. G., Worrell, E., 2011. Material efficiency: A white paper. *Resour. Conserv. Recycl.* 55, 362–381. https://doi.org/10.1016/j.resconrec.2010.11.002.

Araujo, K., 2017. *Low Carbon Energy Transitions: Turning Points in National Policy and Innovation.* New York: Oxford University Press.

Azadi, M., Northey, S. A., Ali, S. H., Edraki, M., 2020. Transparency on greenhouse gas emissions from mining to enable climate change mitigation. *Nat. Geosci.* 13, 100–104. https://doi.org/10.1038/s41561-020-0531-3.

Babich, A., Senk, D., 2018. *Coke in the Iron and Steel Industry, New Trends in Coal Conversion: Combustion, Gasification, Emissions, and Coking.* Elsevier. https://doi.org/10.1016/B978-0-08-102201-6.00013-3.

Bender, M., 2014. An Overview of industrial processes for the production of olefins – C4 hydrocarbons. *Chem. Bio. Eng. Rev.* 1, 136–147. https://doi.org/10.1002/cben.201400016.

Bennett, S. J., Pearson, P. J. G., 2009. From petrochemical complexes to biorefineries? The past and prospective co-evolution of liquid fuels and chemicals production in the UK. *Chem. Eng. Res. Des.* 87, 1120–1139. https://doi.org/10.1016/j.cherd.2009.02.008.

Burger, J. R., Weinberger, V. P., Marquet, P. A., 2017. Extra-metabolic energy use and the rise in human hyper-density. *Sci. Rep.* 7, 1–5. https://doi.org/10.1038/srep43869.

Calvo, G., Mudd, G., Valero, Alicia, Valero, Antonio, 2016. Decreasing ore grades in global metallic mining: A theoretical issue or a global reality? *Resources* 5, 1–14. https://doi.org/10.3390/resources5040036.

Carvalho, F., Müller-Casseres, E., Poggio, M., Nogueira, T., Fonte, C., Wei, H. K., Portugal-Pereira, J., Rochedo, P. R. R., Szklo, A., Schaeffer, R., 2021. Prospects for carbon-neutral maritime fuels production in Brazil. *J. Clean. Prod.* 326, 129385. https://doi.org/10.1016/j.jclepro.2021.129385.

Cordell, D., Drangert, J. O., White, S., 2009. The story of phosphorus: Global food security and food for thought. *Glob. Environ. Chang.* 19, 292–305. https://doi.org/10.1016/j.gloenvcha.2008.10.009.

Costanza, R., Cleveland, C., Cooperstein, B., Kubiszewski, I., 2011. Can nuclear power be part of the Solution? *Solut. J.* 2, 29–31.

Cunha, B. S. L., Portugal-Pereira, J., Szklo, A., Schaeffer, R., 2023. The Role of Food Systems in Energy Transitions, in: Araújo, K. (Ed.), *Routledge Handbook of Energy Transitions* (1st ed.). Routledge: Abingdon.

Debeir, J.-C., Deléage, J.-P., Hémery, D., 1992. Les servitudes de la puissance. Une histoire de l'energie. *Flammarion.* https://doi.org/10.2307/3768761.

Deetman, S., de Boer, H. S., Van Engelenburg, M., van der Voet, E., van Vuuren, D. P., 2021. Projected material requirements for the global electricity infrastructure – generation, transmission and storage. *Resour. Conserv. Recycl.* 164. https://doi.org/10.1016/j.resconrec.2020.105200.

Diamond, J., 2005. *Guns, Germs, and Steel: The Fates of Human Societies.* Norton: New York.

DoE, 2011. *Critical Materials Strategy.* U.S. Department of Energy: Washington, DC.

Elhacham, E., Ben-Uri, L., Grozovski, J., Bar-On, Y. M., Milo, R., 2020. Global human-made mass exceeds all living biomass. *Nature* 588, 442–444. https://doi.org/10.1038/s41586-020-3010-5.

EU Commission (Directorate-General for Internal Market, Industry, Entrepreneurship and SMEs, Bobba, S., Claudiu, P., Huygens, D., et al.), 2018. Report on critical raw materials and the circular economy, Publications Office, https://data.europa.eu/doi/10.2873/167813

Excell, C., Salcedo-La Viña, C., Worker, J., Moses, E., 2018. Legal limits on single-use plastics and microplastics: A global review of national laws and regulation. United Nations Environ. Program. Nairobi, Kenya, 1–118.

Fizaine, F., Court, V., 2015. Renewable electricity producing technologies and metal depletion: A sensitivity analysis using the EROI. *Ecol. Econ.* 110, 106–118. https://doi.org/10.1016/j.ecolecon.2014.12.001.

GCCA, 2019. GNR database – Getting the Numbers Right.

Geels, F., 2005. The dynamics of transitions in socio-technical systems: A multi-level analysis of the transition pathway from horse-drawn carriages to automobiles. *Technol. Anal. Strateg. Manag.* 17, 445–476.

Geels, F. W., Sovacool, B. K., Schwanen, T., Sorrell, S., 2017. The socio-technical dynamics of low-carbon transitions. *Joule* 1, 463–479. https://doi.org/10.1016/j.joule.2017.09.018.

Georgescu-Roegen, N., 1986. The entropy law and the economic process in retrospect. *East. Econ. J.* XII, 3–25.

Geyer, R., Jambeck, J. R., Law, K. L., 2017. Production, use, and fate of all plastics ever made. *Sci. Adv.* 3, 19–24. https://doi.org/10.1126/sciadv.1700782.

Grubler, A., Wilson, C. (Eds.), 2013. *Energy Technology Innovation: Learning from Historical Successes and Failures.* Cambridge University Press: Cambridge. https://doi.org/10.1017/CBO9781139150880.

Grubler, A., Wilson, C., Bento, N., Boza-Kiss, B., Krey, V., McCollum, D. L., Rao, N. D., Riahi, K., Rogelj, J., De Stercke, S., Cullen, J., Frank, S., Fricko, O., Guo, F., Gidden, M., Havlík, P., Huppmann, D., Kiesewetter, G., Rafaj, P., Schoepp, W., Valin, H., 2018. A low energy demand scenario for meeting the 1.5 °c target and sustainable development goals without negative emission technologies. *Nat. Energy* 3, 515–527. https://doi.org/10.1038/s41560-018-0172-6.

Haberl, H., Wiedenhofer, D., Pauliuk, S., Krausmann, F., Müller, D. B., Fischer-Kowalski, M., 2019. Contributions of sociometabolic research to sustainability science. *Nat. Sustain.* 2, 173–184. https://doi.org/10.1038/s41893-019-0225-2.

Haglund, D. G., 1986. The new geopolitics of minerals. An inquiry into the changing international significance of strategic minerals. *Polit. Geogr. Q.* 5, 221–240. https://doi.org/10.1016/0260-9827(86)90035-2.

Hall, C. A. S., Klitgaard, K. A., 2012. *Energy and the Wealth of Nations: Understanding the Biophysical Economy.* Springer: New York (407 pp.).

Hertwich, E., Lifset, R., Pauliuk, S., Heeren, N., Ali, S., Tu, Q., Ardente, F., Berrill, P., Fishman, T., Kanaoka, K., Makov, T., Masanet, E., Wolfram, P., Acheampong, E., Beardsley, E., Calva, T., Ciacci, L., Clifford, M., Eckelman, M., Hashimoto, S., Hsiung, S., Huang, B., Kallmyer, F., Kul, J., Khursid, N., Klose, S., Mainhart, D., Michalowska, K., Reed, T., Myers, R., Asghari, F. N., Olivetti, E., Pamenter, S., Pearson, J., Stocker, A., Verma, S., Vollmer, P., Williams, E., Zabel, J., Zheng, S., Zhu, B., 2020. Resource efficiency and climate change. https://doi.org/10.5281/zenodo.3542680.

IEA, 2018. The future of petrochemicals, towards more sustainable plastics and fertilisers. https://doi.org/10.1016/B978-0-12-394381-1.00002-7

IEA, 2019a. The future of hydrogen. https://doi.org/10.1787/1e0514c4-en

IEA, 2019b. Transforming industry through CCUS. https://doi.org/10.1787/09689323-en

IEA, 2020a. The challenge of reaching zero emissions in heavy industry: Heavy industries both facilitate and complicate the transition to a net-zero emissions energy system [https://www.iea.org/articles/the-challenge-of-reaching-zero-emissions-in-heavy-industry].

IEA, 2020b. Iron and steel technology roadmap. Iron Steel Technol. Roadmap. https://doi.org/10.1787/3dcc2a1b-en

IEA, 2021a. Net zero by 2050: A roadmap for the global energy sector. Paris, 222 pp.

IEA, 2021b. The role of critical minerals in clean energy transitions. Paris. IPCC, 2006. Chemical industry emissions. *IPCC Guidelines for National Greenhouse Gas Inventories. Volume 3: Industrial Processes and Product Use* 110.

IPCC, 2018. Summary for policymakers. Global warming of 1.5°C. An IPCC Special Report on the Impacts of Global Warming of 1.5°C above Pre-Industrial Levels and Related Global Greenhouse Gas Emission Pathways, in the Context of Strengthening the Global Response to the Threat of Climate Change. https://doi.org/10.1017/CBO9781107415324.

Jambeck, J. R., Ji, Q., Zhang, Y.-G., Liu, D., Grossnickle, D. M., Luo, Z.-X., 2015. Plastic waste inputs from land into the ocean. *Science* 347, 764–768. https://doi.org/10.1126/science.1260879.

Kenderdine, T., 2017. China's industrial policy, strategic emerging industries and space law. *Asia Pacific Policy Stud.* 4, 325–342. https://doi.org/10.1002/app5.177.

Kermeli, K., Edelenbosch, O. Y., Crijns-Graus, W., van Ruijven, B. J., van Vuuren, D. P., Worrell, E., 2021. Improving material projections in integrated assessment models: The use of a stock-based versus a flow-based approach for the iron and steel industry. *Energy* 122434. https://doi.org/10.1016/j.energy.2021.122434.

Keyßer, L. T., Lenzen, M., 2021.1.5 °C degrowth scenarios suggest the need for new mitigation pathways. *Nat. Commun.* 12, 1–16. https://doi.org/10.1038/s41467-021-22884-9.

Kiggins, R. D., 2015. The political economy of rare earth elements: Rising powers and technological change. https://doi.org/10.1057/9781137364241.

Kleijn, R., 2012. *Materials and Energy: A Story of Linkages.* Ph.D. Thesis. Leiden University: The Netherlands.

Krausmann, F., Wiedenhofer, D., Lauk, C., Haas, W., Tanikawa, H., Fishman, T., Miatto, A., Schandl, H., Haberl, H., 2017. Global socioeconomic material stocks rise 23-fold over the 20th century and require half of annual resource use. *Proc. Natl. Acad. Sci. U. S. A.* 114, 1880–1885. https://doi.org/10.1073/pnas.1613773114.

Liu, Y., Zwingmann, B., Schlaich, M., 2015. Carbon fiber reinforced polymer for cable structures-a review. *Polymers (Basel)* 7, 2078–2099. https://doi.org/10.3390/polym7101501.

Mishnaevsky, L., Branner, K., Petersen, H. N., Beauson, J., McGugan, M., Sørensen, B. F., 2017. Materials for wind turbine blades: An overview. *Materials (Basel)* 10, 1–24. https://doi.org/10.3390/ma10111285.

Northey, S. A., Mudd, G. M., Werner, T. T., 2018. Unresolved complexity in assessments of mineral resource depletion and availability. *Nat. Resour. Res.* 27, 241–255. https://doi.org/10.1007/s11053-017-9352-5.

OGJ, 2015. International survey of ethylene from steam crackers – 2015. *Oil & Gas Journal.*

Oliveira, C. C. N., Angelkorte, G., Rochedo, P. R. R., Szklo, A., 2021. The role of biomaterials for the energy transition from the lens of a national integrated assessment model. *Clim. Change* 167. https://doi.org/10.1007/s10584-021-03201-1.

Pauliuk, S., Hertwich, E. G., 2015. Socioeconomic metabolism as paradigm for studying the biophysical basis of human societies. *Ecol. Econ.* 119: 83–93. https://doi.org/10.1016/j.ecolecon.2015.08.012.

Perlin, J., 2005. *A Forest Journey: The Story of Wood and Civilization.* The Countryman Press: Woodstock, Vermont.

Peterson, P. F., Zhao, H., Petroski, R., 2005. *Metal and Concrete Inputs for Several Nuclear Power Plants.* University of California: Berkeley.

Rutter, P., Keirstead, J., 2012. A brief history and the possible future of urban energy systems. *Energy Policy* 50, 72–80. https://doi.org/10.1016/j.enpol.2012.03.072.

Schrijvers, D., Hool, A., Blengini, G. A., Chen, W. Q., Dewulf, J., Eggert, R., van Ellen, L., Gauss, R., Goddin, J., Habib, K., Hagelüken, C., Hirohata, A., Hofmann-Amtenbrink, M., Kosmol, J., Le Gleuher, M., Grohol, M., Ku, A., Lee, M. H., Liu, G., Nansai, K., Nuss, P., Peck, D., Reller, A., Sonnemann, G., Tercero, L., Thorenz, A., Wäger, P. A., 2020. A review of methods and data to determine raw material criticality. *Resour. Conserv. Recycl.* 155, 104617. https://doi.org/10.1016/j.resconrec.2019.104617.

Sieferle, R. P., 2001. *The Subterranean Forest. Energy Systems and the Industrial Revolution.* The White Horse Press: Cambridge.

Simmons, I. G., 2004. Environmental Change and Energy, in: Cleveland, C. (Ed.), *Encyclopedia of Energy.* Elsevier Science: San Diego, CA.

Smil, V., 2010. *Energy Transitions: History, Requirements, Prospects.* Praeger Publishers: Santa Barbara, CA.

Smil, V., 2014. *Making the Modern World – Materials & Dematerialization.* John Wiley and Sons: New York, NY.

Smith, J. K., 1988. World War II and the transformation of the American chemical industry. *Sci. Technol. Mil.* XII, 307–322. https://doi.org/10.1007/978-94-017-2958-1_2.

Sokhna, G., Hache, E., Bonnet, C., Simoën, M., Carcanague, S., 2020. Copper at the crossroads : Assessment of the interactions between low-carbon energy transition and supply limitations. *Resour. Conserv. Recycl.* 163, 105072. https://doi.org/10.1016/j.resconrec.2020.105072.

Sovacool, B. K., Geels, F. W., 2016. Further reflections on the temporality of energy transitions: A response to critics. *Energy Res. Soc. Sci.* 22, 232–237. https://doi.org/10.1016/j.erss.2016.08.013.

Thomas, S., 2010. Competitive energy markets and nuclear power: Can we have both, do we want either? *Energy Policy* 38, 4903–4908. https://doi.org/10.1016/j.enpol.2010.04.051.

van Sluisveld, M. A. E., de Boer, H. S., Daioglou, V., Hof, A. F., van Vuuren, D. P., 2021. A race to zero—Assessing the position of heavy industry in a global net-zero CO2 emissions context. *Energy Clim. Chang.* 2, 100051. https://doi.org/10.1016/j.egycc.2021.100051.

Vidal, O., Goffé, B., Arndt, N., 2013. Metals for a low-carbon society. *Nat. Geosci.* 6, 894–896. https://doi.org/10.1038/ngeo1993.

Warde, P., 2007. Energy consumption in England & Wales 1560–2000.

Watari, T., Nansai, K., Nakajima, K., 2021. Major metals demand, supply, and environmental impacts to 2100: A critical review. *Resour. Conserv. Recycl.* 164, 105107. https://doi.org/10.1016/j.resconrec.2020.105107.

West, J., 2011. Decreasing metal ore grades: Are they really being driven by the depletion of high-grade deposits? *J. Ind. Ecol.* 15, 165–168. https://doi.org/10.1111/j.1530-9290.2011.00334.x

Whiting, K., Carmona, L. G., Brand-Correa, L., Simpson, E., 2020. Illumination as a material service: A comparison between Ancient Rome and early 19th century London. *Ecological Economics.* 169, 106502. https://doi.org/10.1016/J.ECOLECON.2019.106502.

Wilson, C., Grubler, A., Zimm, C., 2023. Energy Services-Led Transformation, in Araújo, K. (Ed.), *Routledge Handbook of Energy Transitions* (1st ed.). Routledge: Oxfordshire.

Wrigley, E. A., 1962. The supply of raw materials in the industrial revolution. *Econ. Hist. Rev. New Ser.* 15, 1–16.

Yergin, D., 1991. *The Prize: The Epic Quest for Oil, Money and Power.* Blackwell: Oxford.

Zinkle, S. J., Was, G. S., 2013. Materials challenges in nuclear energy. *Acta Mater.* 61, 735–758. https://doi.org/10.1016/j.actamat.2012.11.004.

5

THE EFFECT OF UNARTICULATED IDENTITIES AND VALUES ON ENERGY POLICY

Joshua T. Brinkman and Richard F. Hirsh

Introduction

The lead author spent many consecutive summers of his childhood in rural Iowa, helping his uncle and grandfather, who had built a small family farm into a 1,500-acre (approximately 600-hectare) business. College and work interrupted the annual family tradition, but in 2004 he visited again with family. The first activity there, as always, was to admire the new tractor, combine, grain elevator, machine shed, or semitruck that had been recently purchased. Next, his uncle drove him in a pickup truck through a grid pattern of roads traversing identical flat cornfields. Then, in the distance, a huge white wind turbine jutted above the flat blue horizon. It had all the magnificence of a futuristic spacecraft. Stopping next to the base of the "monument", his uncle declared, "This is my new turbine!" The massive device seemed to physically and conceptually indicate, "This is the future and it's unstoppable!" As with his uncle's farm, everything appeared clean and new. The turbine itself had sleek lines, and its blades rotated almost effortlessly, the perfectly square concrete base separated from the rich black Iowa dirt by a border of raked gravel. The field from which the turbine tower sprouted featured, like all of the nearby fields, rows of uniform corn plants in straight lines. Everything within sight appeared neat, ordered, and prosperous. Then in his late 70s, his grandfather looked up at the huge blades passing in front of the sun and smiled, with an expression as if to say, "I can't believe we've come this far".

The farm family's enthusiasm for the wind turbine reveals valuable insights about how people interact with technology that go beyond purely economic explanations (see, for example, Pasqualetti et al. 2004; Sowers 2006). It suggests, as we argue, that unarticulated values help create identities of technological users that underlie rational arguments offered to justify the acceptance or rejection of new energy technologies. Moreover, this formation of identity through technological use constitutes a historical and cultural practice forming an important aspect of rural modernity that has existed in the American Midwest at least since the 1920s (Brinkman 2017a; Brinkman and Hirsh 2017). Finally, we suggest that this unspoken construction and reflection of identity through the daily use of material objects influences energy policy and discourses,[1] especially on the local- and state-government levels, and may affect the success or failure of energy transitions.

In the rural Corn Belt region of the US, the national debate of economic-versus-environmental interests becomes reframed to fit into a production ethos that often results in unexpected policy

DOI:10.4324/9781003183020-6

outcomes. This bottom-up approach to policy, which emphasizes unexpressed identity, helps explain why Republican-leaning Midwestern states surprisingly lead the nation in policies promoting the transition to renewable energy. These policies often appear popular among "progressive", Democratic-leaning states such as those in the Far West, New York, and New England, and unpopular in other Republican-dominated states.[2]

This chapter offers a novel approach for understanding the cultural dimensions of energy transitions for several reasons. First, our argument shifts the focus from simple economic incentives or environmental concerns of stakeholders to deeply embedded and unexpressed ideas about identity (see, for example, Graff et al. 2018). Rather than relying on articulated reasons for embracing or rejecting new technologies, our approach views such statements as proxy arguments in favor of underlying cultural views of the self and technology. As such, this chapter offers a different methodology to anyone seeking to understand energy transitions. Instead of simply looking at financial gains from energy sources, we concentrate on the discourse employed by users within a broader cultural context.[3] Our use of discourse to identify underlying values shaping energy adoption draws from social constructivist methodologies employed by historians of technology, such as Thomas Hughes (1994), who demonstrates how social and cultural factors lead to different technological styles (see also Bijker et al. 1987). In a similar fashion, we draw from Gabrielle Hecht's (1998) work highlighting how technologies and politics often co-construct one another to shape national identities.

Second, this chapter contributes to energy transition studies by proposing a nonobvious conception in which stances on policy are determined from the bottom-up rather than imposed top-down by government officials (see Brinkman 2017b; Hossain 2016). Hecht's study of nuclear power in France, for instance, reveals how a desire to regain national pride and independence after the country's humiliation during World War II drove technological boosterism (Hecht 1998). But Hecht's work focused on elite officials at national agencies in forming discourse around technology rather than on users (ibid.). Our bottom-up approach grants agency to farmers using technology in forming technological discourse, combining Hecht's insightful theoretical approach with perspectives from other scholars such as Kathleen M. Araújo (2017), whose study of Danish wind power examined citizens' and scientists' encouragement of widespread change in systems.

Further, this chapter distinguishes itself from other bottom-up studies of energy policy by moving away from investigating grassroots citizens' movements. Rather, it highlights how the daily use of technology helps form identities and discourses that find political expression by policymakers (Hess 2018; Oteman et al. 2017). Policy frameworks, such as Paul Sabatier and Hank C. Jenkins-Smith's advocacy coalition approach, for example, assumes that if citizens influence policy, they do so through formal groups struggling to achieve favorable outcomes by using well-articulated ideas and belief systems. While this advocacy coalition approach makes room for bottom-up implementation of ideas, it does not offer a means for understanding how and why beliefs form among groups in the first place (Sabatier and Jenkins-Smith 1993). Our approach proposes that citizens construct their sense of self by using technology and by forming a discourse around material objects. In other words, our methodology views energy policy as the end result of cultural values and identities formed through technological use.[4] We suggest that energy transition policy – especially that dealing with the employment of renewable technologies – does not always begin at the governor's mansion with the signing of a bill but in the shed when a farmer starts a state-of-the-art machine and forms an identity as a modern producer.

Approaches for Understanding Technology and Energy Choices

Traditional academic and mainstream rhetoric often views technology as neutral and devoid of social, cultural, and political characteristics (Balabanian 2006). Such neutrality assumes that technology results from rational design on the part of inventors to construct the most efficient object to fulfill a need. This tendency toward technological neutrality arises out of a progress narrative in the US and other Western nations representing an unquestioned cultural practice that considers technology as always good (or at least as the most important factor in determining history) (Smith and Marx 1994; Hecht 1998). Such determinism typically inspires innovation-focused narratives of technology that emphasize the new, the big, and the rich; it places the most importance on great inventors rather than on the experience of users with everyday objects. Instead of framing technology as simply material artifacts arising from human interactions, the American progress narrative sees technology as a panacea coming out of nowhere to "change the world".

Such "rehashed futurism", a phrase coined by the historian David Edgerton (2007), has proven remarkably persistent throughout US history (at least since the late 19th century), and it becomes particularly evident during times of perceived energy transitions. In 1917, for example, when less than 35% of American households enjoyed electricity, Fern Van Bramer wrote a prize-winning essay in the *Electric City Magazine* predicting that new electric lighting in the home would improve the American family by keeping husbands and sons out of saloons. By taking advantage of the services of the Commonwealth Edison Company, wives could dissuade their husbands and sons from frequenting these well-lit yet "unwholesome" places. Electric toys would also teach children about electricity, allowing them to "enter something better than the 'blind alley' jobs that lead nowhere" (Van Bramer, 1917). In Van Bramer's view, the new electricity radiates from almost a spiritual place to render the children smarter, the mother less stressed, and the father more sober, in a way that demonstrates the power of technology to improve people's lives (ibid.).

Under this unquestioned faith in technology, considerations of the causes of energy transitions remain simple and straightforward, often based on how the best technologies naturally win in the marketplace. Adoption results from rational choices among developers and users to embrace or resist new energy sources, and groups obtain coveted technologies by leveraging economic power. Supporting this mindset, a 2019 report in *Wallaces Farmer*, a farm journal promoting farmers' interests, attributed the lack of local opposition to a proposed wind farm in Cherokee County, Iowa, simply to the rational economic expectations among residents who looked forward to increased tax revenue (Flaugh 2019).

Scholarship in the history of technology since the late 1980s, however, refutes technological neutrality, contending that social and cultural factors embed themselves within technologies and systems. (See Bijker 2007, Bijker et al. 1987, and Winner 1986 for examples of a social constructivist approach.) Electricity's widespread use did not eradicate alcoholism nor end unhappy marriages, as Van Bramer hoped, because the design of electric systems and the use of electricity arose out of the same social, economic, cultural, and political factors that produced these social problems. Additionally, scholars have highlighted the impact technological users have on shaping the design, impact, and social meaning of material objects (Cowan 1983; Fouché 2006). Historian David Nye (1992, 1990), for example, has shown how users and producers of electricity in the late 19th and early 20th centuries facilitated an energy transition in the US by creating an "electrical sublime" or a sense of awe (and at times dread) around electricity. Further, Nye points out that electricity became a "visual discourse" promoting a whole set of corporate and middle-class values. In short, technologies are social constructs, not objects devoid of culture (Noble 1999; Schatzberg 1994).

Social constructivist perspectives on technology have important policy implications. Most significantly, if one accepts that social and cultural factors shape the design and use of energy technologies, then the number of factors influencing energy transition policy expands and becomes more complex. Policy choices do not result only from economic considerations of policymakers and stakeholders but also from less obvious social meanings of technology. The focus of policy analysis should therefore extend, at least sometimes, beyond actions of politicians, regulators, or interest groups to regular folk who use technology every day. As people adopt and employ technology, we argue, they create social and cultural meanings that can ultimately become reflected as energy policy.

Renewable Energy as Part of an Ultramodern Rural Identity

Taking farmers in the Corn Belt as a case study of this bottom-up process of influencing policymaking, we contend that rural Midwesterners have formed an ultramodern rural identity through use of state-of-the-art technologies. Identity for these agrarians includes notions of "inborn" innovation inherited from a family legacy of farming that benefits from the use of the newest forms of technologies (Giddens 1990). It frames farmers as the ultimate experts, combining practical experience with a high level of scientific and technological know-how that exceeds that of their urban cousins. Consequently, ultramodern farmers often celebrate older farming equipment not as a form of nostalgia but to highlight how far their families have progressed since the "bad old days" (Adams 1994; Mowitz 1992; Thompson 1983).

Rural ultramodernity starts with the image of farmers heroically overcoming the challenges presented by the frontier and reframes the concept to meet the threat of world hunger in a globalized marketplace. The set of ideas often attributed to Thomas Jefferson in the 18th century views small yeoman farmers as chosen by God to civilize the American continent and to preserve democratic ideals. The prosperity of the cities, according to Jefferson, rested on this group of independent farmers overcoming obstacles through inventiveness and creativity, providing the food necessary for larger urban populations.[5] Other American figures, such as politician William Jennings Bryan (1896) and historian Frederick Jackson Turner (1894) in the late 19th and early 20th centuries, promoted this Jeffersonian perspective that lauds rural denizens.[6] Even today, farmers view themselves as the backbone of American democratic society who work as moral, independent producers and who distrust the outside and less righteous, urban "others" (Brinkman and Hirsh 2017). But instead of acting as 18th-century yeomen, they see themselves as modern, savvy businesspeople.

Farmers rarely express this ultramodern rural identity overtly, preferring to offer rational economic explanations for their business decisions. Of course, rural denizens do not employ words such as "modernity" in their rhetoric, nor do farmers self-examine their identities. But an examination of discourse dealing with farmers' daily use of sophisticated technology reveals a rural ultramodern sense of self. In other words, through language and actions, agrarians perform an identity, which we call rural ultramodernity, featuring a particular set of ideas shaped by the historical experiences of the American Midwest.[7] From such discourse, we learn that rural ultramodern farmers view land and technologies used for family-based production as the only proper symbols of wealth, and they display their identities by using the latest equipment. For example, a farm journal editor, Lon Tonneson, signaled his celebration of the ultramodern agrarian by first describing a group of Illinois soybean growers who visited Brazil in 2011 as "definitely high tech". The remainder of Tonneson's article debunks urban stereotypes of rural people as backward bumpkins, stating that the soybean farmers "use some of biggest, most sophisticated farming equipment in the world"; they "are at ease with Global Positioning Systems, variable rating seed equipment and auto steer technology" (Tonneson 2011, Inside Dakota Ag). Back home in the Midwest, many farmers perform their ultramodern identities by forming limited liability companies (LLCs) to erect their own turbines, the editor observed, keeping their production of wind energy independent from developers even though such methods expose farmers to more financial risk. This ultramodern identity often remains unspoken behind expressed economic motives (ibid.).

Farmers' business decisions also come with discourse emphasizing the newest technologies. Such objects represent the modern progress of rural communities or regions. "Midwest farmers and landowners are getting involved in the fastest-growing electricity source in the world: wind energy", one farm journal article bragged in 2006 (Sauck 2006, 2). "Wind power has the potential to supply more than one-and-a-half times the current electricity consumption of the United States", the piece continued, "and the Midwest has the potential to lead the way" (ibid.). After discussing financial pros and cons of installing wind power, the farm journal editor appealed to ultramodernity and progress, concluding, "The winds of change are blowing through the Midwest" (ibid. 9). Other farm journals delighted in "the wind-power revolution", a form of discourse typical of rural ultramodernity ("Wind Turbines Power Cooper Farms" 2012). Not surprisingly perhaps, Iowa in 2020 obtained almost 60% of its electrical energy from wind turbines, the highest percentage in the country.[8]

Ideas of rural ultramodernity helps explain why many Midwest farmers support Republican political candidates who resist environmental regulation while embracing renewable energy technologies (Swoboda 2018). Several farm journal surveys prior to the 2020 presidential election showed that more than 80% of farmers in upper Midwestern states (such as Iowa and Minnesota) supported Donald Trump's position on eliminating environmental regulations (Bunge 2020). Yet farmers in these same states often championed renewable energy technologies and disapproved of the national Republican Party's efforts to downplay them (see, for example, "Putting Wind in The Rotation" 2012).[9] Among expressions of rural support, the traditionally red, Republican (and intensely rural[10]) state of Nebraska celebrates wind technology through outreach such as the Nebraska Public Power District's online hosting of "watch the wind" sessions, which began in 2014 ("NPPD Introduces Online 'Wind Watch'" 2014).

In other words, something other than the rational economic reasons given by farmers for their support of Trump must explain this apparent disconnect. We contend that a dislike of environmental regulation coupled with an enthusiasm for renewable energy makes sense only in the context of ultramodern rural identity and values. Objects such as solar panels or wind turbines, when viewed through this rural ultramodern lens, do not represent environmentalism or other "liberal" or progressive causes. Rather, farmers appreciate energy transition technologies as just one part of a suite of state-of-the-art artifacts controlled by the farmer to maintain independent family-based production. Farmers see little difference between using the latest combine or a technology operated by wind energy since both help them form an ultramodern identity. For example, in the article about Illinois soybean farmers who visited Brazil, Tonneson noted that when farmers encountered a replica of an old Dutch windmill, they "poured over the structure" and examined its marvelous mechanical components. When asked why the group showed such fascination in the old windmill, one rural denizen did not express an interest in the economic efficiency of the technology but, instead, stated proudly, "We're farmers . . . We just like knowing how things work" (Tonneson 2011, Inside Dakota Ag). By figuring out how the machine worked, the farmers sought to project an ultramodern image to observers as experts with savvy mechanical knowledge, thereby debunking yokel stereotypes about rural people as being backward.

Interestingly, farmers in Midwestern states who overwhelmingly supported anti-environmental Republican candidates often cited concerns for the environment and sustainability when discussing renewable energy technologies, but they did so when such devices allowed for the reinforcement of ultramodern identity. One 2013 article in *Wallaces Farmer* recalled how Dean and Betty Van Kooten, who raised corn and soybeans near Kellogg, Iowa, installed the state's largest privately owned solar structure, consisting of 60 arrays, after seeing similar systems in Europe. While the article hailed "sustainability", the Van Kootens built the project to allow for the preservation of family-based independent production under the farmer's control, the central rationale of rural ultramodernity, by "ensuring the farm produces its own energy when Joey and Michelle [the son and his wife] assume control of the farm" (Harris 2013, 3). The article's author used discourse typical of an agrarian producer ethic to describe the family's motives by declaring, "the Van Kootens began harvesting another crop this fall – the sun". By doing so, the farmers viewed the renewable energy technology as morally positive because it produced a desirable product and enabled the family to maintain control of farming, not because of a desire to "save" the environment. Only after fitting into this family-based production process does sustainability itself become an articulated rationale for promoting energy transition (ibid.).

A rural ultramodern identity similarly led Nebraska farmer Mike Zakrzewski to highlight environmental concerns when discussing the reasons for embracing wind turbines on his property. "You see it [climate change effect] firsthand on agriculture", he stated, "because we're working with nature every day" (Bergen 2020, 1). Zakrzewski even encouraged his neighbors to reject "the trend in national politics" that opposes "anything green". Explicitly, he opposed President Trump's anti-environmental rhetoric, and he urged other farmers to view wind turbines as a means of preserving independent family farms by mitigating the effects of climate change on crop production. To place his plea further within a production ethos, Zakrzewski pointed out his turbines allowed him to make improvements on his land that enhanced cattle grazing while also improving his farm's infrastructure. More importantly, he saw the tall devices as the means to continue a family legacy of farming land by mitigating the negative impact of climate change and countering the heavy carbon footprint of modern agriculture. Zakrzewski described growing environmental threats to his farm output: "Our growing seasons have extended and shifted a little bit. They're a little later into the fall . . . it's the new norm to have wider extremes in weather – drier drys, wetter wets, windier winds" (ibid.). When farmers see turbines through an underlying ultramodern discourse – as the latest equipment maintaining the farmer's control of production, improving the land, and preserving a moral family legacy – they express environmental and economic benefits as rational reasons for embracing renewable energy technologies. In the process, they discount political rhetoric (even from the party they support) that could be viewed as antithetical to their ultramodern identity.

Other farmers employed a similar ultramodern discourse of production and independence when installing renewable energy systems. For example, in 2013, dairy farmers Francis and Susan Thicke in Fairfield, Iowa, hosted a Practical Farmers Field Day event so their neighbors could learn about renewable energy systems. Rather than describing the system in the language of environmentalism or organic farming, the Thickes emphasized energy independence and using the newest technology as ultramodern dairy producers. The couple harvested energy with a 40-kilowatt wind turbine matched with a solar-thermal water heater, a solar-powered pumping system, and a geothermal heating and cooling system. The field demonstration included consultants from energy companies as participants, but farmers led the event as the ultimate practical experts who offered a tour of the farm's several state-of-the-art facilities ("Harvest Energy from Wind and Solar to Power Your Farm" 2013). In a similar fashion, some self-described innovators separately collected big data from their fields of solar panels not because they sought to demonstrate economic or environmental benefits but because they appreciated being on the next wave of technological progress (Vogel and Bechman 2014).[11]

The Impact of Ultramodern Rural Identity on Policy Choices in the Midwest

An understanding of unarticulated values among Midwestern farmers helps explain several energy policy initiatives. In particular, the ultramodern rural identity that was formed and reinforced using technology leads to surprising policies promoting renewable energy in the very communities supporting Trump and other Republican candidates promising to eradicate environmental regulations. This ultramodern discourse around renewable energy flows upward to public officials from farmers themselves, as well as from policy and business organizations such as the American Wind Energy Association (AWEA), the Energy Foundation, and Harvesting Clean Energy ("25X'25 Helping Wind Energy Spread" 2009). Reflecting this form of influence, for example, Iowa senator Charles Grassley, a conservative Republican, proposed the first production tax credit in 1992, giving a tax incentive to producers using wind, solar, biomass, and geothermal sources to generate electric power. Although the conservative Heritage Foundation opposed the tax credit in favor of letting the market decide energy use, he and the entire state's congressional delegation strongly supported the measure. Grassley also attempted to extend the wind energy tax credit for ten years in 2009 as part of an economic stimulus bill, at a time when normally pro-environmental Democrats in the Senate rejected it ("Grassley's Wind Energy Proposals Rejected" 2009).[12]

This outcome appears nonsensical unless one considers rural ultramodernity as driving policy. Iowa farmers themselves embraced these pro-wind polices, often taking the lead in installing turbines on their farms ("Wind Powered Ethanol" 2007). While Grassley often cited economic reasons for supporting renewable energy, such as creating manufacturing jobs, he also appealed to ultramodernity when discussing energy legislation; he expressed concern, for example, for "rural families", and he framed farming as a moral and noble profession (Offshore Production and Safety Act 2011). His claims gain credence because, as a farmer himself (running an operation in New Hartford, where he recycles everything, collects rainwater, and conserves energy at every chance), the senator exemplifies rural ultramodernity while also protecting ethanol and biofuel tax credits. In 2012, when the tax credit for wind turbines threatened to expire, Grassley even allied with President Barack Obama to push its renewal through Congress. When asked to describe Grassley's overall political philosophy, Ferd Hoefner of the National Sustainable Agriculture Coalition stated, "I think he . . . believes we should have a country of family farms" (Looker 2019, 6). Thus, the long-time senator clearly echoes ultramodern discourse that associates wind turbines with family-based production.[13]

Iowa representative Steve King also championed the renewable energy tax break extension. A Republican, he and fellow policymakers bolstered this and other incentives through a Treasury grant program under the 2009 Recovery Act, which gave farmers cash grants for 30% of a wind project's total capital costs ("Existing Government Policies Spur Wind Power Development" 2010). According to one wind industry executive, "the wind energy tax credit isn't stimulus money or a grant. It's an offset to tax liability only when turbines are spinning and electricity is being generated" (Swoboda 2012, 7). In other words, the fact that the wind turbines *produce* a desirable product alters the notion of the credit from an immoral "handout" to a moral "offset" ("The Wind Energy Developer Question" 2010). This subtle use of production to moralize technology reflects an ultramodern rural discourse. By 2014, many Midwestern politicians celebrated the region's expected status as "world leader in renewable energy".

Republican Kansas governor Jim Brownback appealed to rural ultramodernity at the time, stating, "The moment is approaching when our nation must decide how it's going to power the future" (Paulos 2014, 5).

In 2022, the tax credit for renewable energy remains part of the US tax code (United States Environmental Protection Agency, n.d.). Since 2010, Iowa lawmakers have advocated a national energy goal that utilities obtain 25% of their electricity from renewable energy ("Governors Take Stand on Wind Energy" 2010). In 2018, the Center for Rural Affairs conducted a survey of elected officials and landowners in Iowa to gauge support for wind and solar power, concluding, "We were happy to see strong support for wind and solar among local elected leadership" and "We were pleased to see the results shared by the landowners" (Swoboda 2018, 4). *Wallaces Farmer* observed that the growth of wind power "presents a major opportunity for Iowa landowners" and encouraged them to enter into wind turbine contracts. While the article detailed rational disadvantages of wind turbines – from noise to shadow flicker – an interviewed land manger recounted the most common objection to wind energy among farmers: "A very real sentiment, but more difficult to pinpoint, is that these projects are perceived to be brokered by power companies and politicians not living in the local area". In short, "There is an arm's length transaction aspect that feeds into a rural vs. urban divide" ("Farmland Ownership and wind energy" 2019, 5). Thus, the same rural-urban conflict that contributed to the basis of rural modern identities since the 1920s controls farmers' perception of renewable energy to the present time, notwithstanding rational economic or environmental proxy arguments for or against energy policies (Brinkman and Hirsh 2017). Studies of energy transitions in local communities in France, Denmark, and Pennsylvania have noted similar objections among residents who perceived the new systems as primarily benefiting outside interests (Araújo 2017). Despite these rare objections, Iowa led the nation in its wind energy mix, even though about 82% of the state's farmers expressed their intention to vote for Trump in 2020. And after the election (in March 2021), the Iowa Environmental Council (IEC) proudly noted that its state produced the highest percentage of electricity by wind in the country – 60% – with a total wind turbine capacity of 11,660 MW (Iowa Environmental Council 2021).

Other Midwestern states featuring prominent agricultural sectors have begun to embrace the idea of an energy transition within an ultramodern rural discourse that highlights independent family-based production and high-tech leadership. In Illinois, for example, the Illinois Institute for Rural Affairs established a Wind for Schools program from 2011 to 2014, in which participating rural educational institutions received a classroom set of experimental model wind turbines, weather balloons, and model wind tunnels. The wind energy curriculum included discussions of energy and electricity, turbines and engineering, and "environmental considerations" ("Illinois Wind for Schools Picks Six for Wind Energy Curriculum" 2013; "Illinois Wind for Schools", n.d.). Such a program reflects neither an environmental ethos that rejects development of large-scale projects nor a right-wing discourse seeking to preserve older nonrenewable energy sources. Rather, it expresses an ultramodern rural discourse that frames wind energy as the latest in a series of productive technologies. Illinois further joined several other Midwestern states in 2017 in an alliance of governors calling on Congress to pass a renewable energy stimulus package (Calma 2020). Most recently in 2021, Indiana's Republican governor, Eric J. Holcomb, joined his counterparts from Illinois, Minnesota, Michigan, and Wisconsin to sign the Regional Electric Vehicle for the Midwest Memorandum of Understanding (REV Midwest MOU) to accelerate vehicle electrification. Holcomb's reasoning for signing the MOU reflected a desire, typical of rural ultramodern discourse, to associate his state with the latest technologies ("Governor Walz Joins Midwest Governors to Announce Partnership to Expand Electric Vehicle Charging Infrastructure, Grow Jobs, Futureproof Regional Commerce" 2021).

Midwestern politicians and policy advocates have similarly responded to rural ultramodern identities not only in the form of promoting an energy transition but also to echo agrarian discourse. As early as 2004, for example, Iowa senator Tom Harken advocated for tax incentives for farm installations of turbines by saying, "Wind power is a new 'cash crop' for farmers and rural communities, and we know it holds great potential for Iowa" (Wenzel 2004). Likewise, Michigan senator Debbie Stabenow, while chairing the Agriculture Committee in 2004, appealed to rural ultramodern notions of "inborn innovation" and farmers' control over production to justify solar grants through the US Department of Agriculture's Rural Energy for America Program (REAP). Stabenow argued that under the program, farmers "will be empowered to harness the energy potential of American grown agriculture – by developing innovative systems" ("Federal Funding Goes to Rural Solar Renewable Energy Projects" 2004, 5). Far from distrusting REAP as an instance of "big government", farmers throughout the Midwest applied for grants of up to $18,000 or more each to install solar- and wind-energy systems ("REAP Program Supports Energy Efficiency" 2015; "Reduced Energy Costs for Rural Businesses" 2011). Another policy advocate for the Central Iowa Power Cooperative and Center for Rural Affairs, Johnathan Hladik, similarly appealed in 2016 to rural ultramodernity to advocate that utility cooperatives install more solar generation, stating, "Rural Americans have always been innovative, self-reliant and pragmatic. Solar energy is a tangible example of those values" ("Iowa rural electric cooperative leads the way on utility solar" 2016). Thus, Hladik appealed primarily to the identities of rural people as independent innovators and practical experts in the latest technologies, not to rational economic benefits.

These rural ultramodern discourses have proven remarkably resilient in red Midwestern states despite President Trump's promise to discourage a transition to sustainable energy systems. Ultramodern discourses also reject the anti-environmental rhetoric of Trump because rural identity is rooted in Jeffersonian agrarianism that sees land as representing family cohesion and sustenance. "Iowa farmers are known for their strong legacy of stewardship – taking care of the natural resources to reap the rewards they offer in return", one wind energy executive explained. "Wind energy is no different". Even after Trump's anti-renewable energy messages, Iowans and Midwestern energy companies still used rural ultramodern discourse to promote wind energy as "a cash crop for Iowa" and "the future of energy" (Fehr 2020, 7). Iowa agriculturalists and farm journals even lamented Trump's climate change denial during the 2016 election campaign because of its implications for energy policy, and many in the Midwest openly asked if clean energy was "Trump-proof". In a strange reversal of political priorities, Trump, not an environmentally minded Democrat, had framed wind turbines at the time as "bird killers", a characterization opposed by Republicans in the Midwest (Cama 2016; Foley and Young 2017; Usher 2016).

Many Republican politicians appealed to modern discourse in breaking with the president, calling his views of wind turbines "outdated". In a subtle reference to the rural-urban conflict, Des Moines City Council candidate Josh Mandelbaum countered that tall buildings in the city, like the ones owned by Trump in New York City, kill more birds than wind turbines in Iowa; he quipped, "If the president's concerned, maybe he should take a look at his own portfolio" (Foley and Young 2017, 2). By contrast, a commercial supporting Grassley's 2016 re-election bid featured the senator atop a wind turbine donning a climbing harness and hard hat, flashing a victory sign, and smiling at the camera. Behind him, wind turbines stretched as far as the eye could see over hundreds of acres of productive farmland in a way that connected the technology to the high morality of the ultramodern farm family. The advertisement even included Mike Garland, a former chairman of the AWEA, who exclaimed, "Senator Grassley is literally the father of the modern wind industry". The ad rejects a retreat to old energy sources, such as coal, and projects the image of Iowa farmers and their senator as innovators boldly leading the nation toward the technology of the future. Grassley's slogan, "Grassley works", equally appealed to the production ethos of rural ultramodernity (Vander Hart 2016).

Even before President Trump took office in 2017, farm journal editors noted he "bashed wind energy" on the campaign trail and feared that he would "spell trouble for Iowa" by rejecting the Renewable Fuel Standard and tax incentives supporting wind, solar, ethanol, and biodiesel production (Swoboda 2016). When asked to comment on Trump's plans to reverse federal backing for wind energy, Senator Grassley responded, "Over my dead body" (Longman 2019, 2). In 2017, renewable energy industry representatives – many Republican – urged the Trump administration to "get out of the coal mines and look to the sky". Iowa's Republican governor, Terry Branstad, as well as Grassley, remained fervent promotors of renewable energy including wind and solar, for example. After Trump gave a speech alleging that noise from "windmills" causes cancer, Grassley publicly called the comment "idiotic" and issued a bipartisan press release (with Democratic Senators Bernie Sanders, Kamala Harris, Elizabeth Warren, and Amy Klobuchar, among others) calling for more federal funding for wind energy (ibid.). Toward the end of 2020, farmers and farm journal editors had hung onto their ultramodern view of wind energy in spite of the Trump administration becoming even more aggressive in its denial of climate change, with one industry spokesman declaring, "Simply put, wind energy is good for Iowa", an attitude shared by many Republican lawmakers.[14]

In the end, the production tax credit that promotes a cleaner energy transition in rural Iowa survived the Trump presidency, and farmers (along with utility companies such as MidAmerican Energy) constructed wind turbines across the state for a total investment of $19 billion by 2020 (Webber 2020). Similarly, conservative advocacy groups on the federal level have failed to persuade most Republican Midwestern states to overturn legislation containing ambitious renewable energy targets (Williams 2019). In the few cases in which states have rescinded such targets, such as in Kansas in 2015 (Walton 2015), the prevalence of rural ultramodernity arguably remained a powerful force that drove continued investment in renewable energy systems by landowners and power companies (Habeeb 2021). While many commentators have attributed the popularity of renewable energy in "Trump country" to expressed economic benefits, the use of the newest energy technologies also reflects an adherence to an ultramodern rural identity. Iowans, for instance, delight not only in the economic boon of wind power but in the fact that the state leads the world in the state-of-the-art energy production technologies and in its status as a destination for the world's major energy companies. Midwestern politicians appeal to rural ultramodernity by placing energy within a production ethos by framing it as simply another "crop" supplied by farmers in a globalized marketplace (Brinkman and Hirsh 2017). The Republican governor of Kansas, Sam Brownback, for example, bragged in 2011, "We export lots of things, and in our future, I want us to export a lot of wind power" (Gillis and Popovich 2017, 2).

Conclusion

Regarding technological use as a means of forming identity suggests that an element of energy transition policy starts at the user level, not necessarily in the halls of state legislatures, in Congress, or in the Oval Office. Thus, energy transition policy may manifest as simply bottom-up adoption. For rural Midwesterners, the values underlying policymaking form in farmers' positive attitudes about the newest technologies. These rural denizens created an identity of being technologically adept, continuing a legacy of employing state-of-the-art equipment to yield increasing amounts of food and other products (such as electricity) for an increasingly global community. A focus on these farmers' ultramodern identities also comports with the social constructivist view of technology, which implies that policy choices do not occur in a vacuum dependent on narrow economic interests. Rather, such choices reflect identities and discourses drawn from still-widely-held Jeffersonian ideals and a sense of the moral center of the country. Farmers do not simply profit economically from "harvesting the wind"; embracing wind energy technologies reinforces their ultramodern sense of self, an identity that policymakers promote even in an age of partisan division.

In this chapter, we have highlighted a way in which discourse reveals unarticulated identities shaping attitudes about technology. Consequently, these socially constructed relationships between people and artifacts sometimes translate into political action. Our contribution emphasizes how an appreciation of such unexpressed notions adds to the toolbox of those studying energy policy. Interest group analyses and bottom-up and top-down approaches have great value in certain circumstances, and scholars have profitably used them in their analyses. But we argue that those approaches do not capture the entire breadth of policy considerations, and we have sought to offer something novel in explaining people's attitudes about energy technologies.[15] More broadly, an understanding of unspoken identities underlying policy preferences could aid in breaking through the political impasses so prevalent in contemporary politics. In the case of the agrarian Midwest, framing technology within a rural ultramodern discourse may yield positive outcomes in terms of promoting renewable energy systems as part of contemporary energy transitions.

Notes

1 By "discourse", we mean more than simply rhetoric or messaging. As we state in Brinkman and Hirsh (2017, 344–345), "An examination of rhetoric traces the use of specific words, while discourse analysis employs language in a contextual framework to more fully understand the motivations of historical actors. Importantly, actions, such as decisions about the use of various technologies, can also constitute discourse while rhetoric limits its focus on language". One would not expect farmers or politicians to use the term "ultramodernity" to describe why they adopt a technology such as wind turbines. Instead, Midwesterners use language that reveal an underlying ultramodern identity "such as 'efficiency' and 'progress' to signify an up-to-date and future-oriented sensibility, or they simply acted in ways consistent with such notions" (ibid.). This conception of discourse draws on the work of Ernesto Laclau and Chantal Mouffe (Jorgensen 2002).

2 For example, in 2017, 69% of the wind power produced in the United States came from states that Donald Trump won in the 2016 presidential election (Gillis and Popovich 2017; Hess, Mai, and Brown 2016).

3 This chapter will build on studies of unarticulated reasons for accepting or rejecting energy technologies in Brinkman and Hirsh (2017) and Hirsh and Sovacool (2013).

4 For more traditional top-down policy studies, see Stokes and Breetz (2018); Nordensvard, and Urban (2015); Solomon and Krishna (2011). With our novel form of analysis, we do not aim to discount political science or other approaches to understand energy policy. For example, Stokes (2020) uses interest group theory to persuasively explain why an early-21st-century movement stalled in advocating for renewable energy policies.

5 For a more detailed description of Jeffersonian agrarianism, see Peterson (1999).

6 Importantly, rural ultramodern discourse jealously protects the moral image of the farmer as engaged in family-based production and resents characterizations of large Midwest farms as part of corporate agribusiness, or "big ag". Rural women, who have historically engaged in production, often find urban critiques of modern agriculture particularly offensive because they imply a "male" domination of nature that discounts contributions made by other family members (Brinkman 2017a).

7 For a more detailed discussion of the historical formation of this rural ultramodern identity, see Brinkman and Hirsh (2017).

8 Percent calculated from US Department of Energy, Energy Information Administration, data for 2020, which showed that of the 57,463,000 megawatt-hours of electricity generated in Iowa, 34,178,000

megawatt-hours were produced from wind sources ("Electricity Data Browser", n.d.). Also see Kiernan (2020).

9 On the other hand, some farming communities in India have resisted solar power farms. See Bechman (2020) and Brinkman and Hirsh (2017).

10 In 2021, farms and ranches utilized 92% of Nebraska's total land area (United States Department of Agriculture 2021).

11 Of course, not all Iowa farmers hold the same belief system. Notably, the ultramodern discourse around renewable energy differs significantly from organic discourses among a minority of Midwest farmers when discussing the same technology ("Learn How to Harness Solar Energy on the Farm" 2013). For examples of this organic discourse, see Pollan (2008), Kingsolver (2007), Berry (1995), and Seymour and Sutherland (2009).

12 The release prominently notes, "Grassley is the father of the wind energy tax credit, having sponsored the first-ever provision enacted in 1992" ("Grassley Works to Bolster Wind Energy Production and Generate Job Creation" 2009, 1).

13 This view of wind turbines contrasts with environmental discourses that emphasize climate change (Watson 2019).

14 Lawmakers were joined in their opposition to President Trump's anti-wind rhetoric by Abigail Hopper, chief executive officer of the Solar Energy Industries Association. She challenged Trump's assertion that solar power threatened the coal industry by contending, "I reject the idea that there has to be a winner and a loser". Tom Kiran of the Wind Energy Association utilized ultramodern rural discourses that positively highlighted family-based independent production more directly in order to push back on Trump's pro-fossil-fuel-energy policy. He stated, "We're helping families keep farms they've held for generations. The lifeblood of our industry is in rural America". Indeed, by 2017, the top ten congressional districts for wind energy all remained in Republican-dominated rural areas in the Corn Belt (Martin 2017).

15 Explicit acknowledgment of ultramodern sensibilities can aid business actors as well, particularly those seeking to market renewable energy more effectively to certain groups. Instead of just focusing on the economic benefits that farmers can accrue by placing wind turbines on their properties, company representatives can appeal to farmers' positive thinking about using the most up-to-date technologies to boost overall productivity and maintain the ability of future generations to use their families' land.

Bibliography

"25X'25 Helping Wind Energy Spread." *Wallaces Farmer*, Oct. 5, 2009.

Adams, Jane. *The Transformation of Rural Life*. Chapel Hill, NC: University of North Carolina Press, 1994.

Araújo, Kathleen M. *Low Carbon Energy Transitions: Turning Points in National Policy and Innovation*. New York: Oxford University Press, 2017.

Balabanian, Norman. "On the Presumed Neutrality of Technology." *IEEE Technology and Society Magazine*, Winter, 2006: 15–25.

Bechman, Tom J. "Will You Grow Corn Under Solar Panels Someday?" *Wallaces Farmer*, Feb. 4, 2020, Crops.

Bergen, Molly. "How Wind Turbines Are Providing a Safety Net for Rural Farmers." *World Resources Institute*, Oct. 13, 2020.

Berry, Wendell. *Collected Poems*. New York: North Point Press, 1995.

Bijker, Wiebe E. "American and Dutch Coastal Engineering: Differences in Risk Conception and Differences in Technological Culture." *Social Studies of Science* 37, no. 1 (2007), 143–151.

Bijker, E., Thomas P. Hughes, and Trevor J. Pinch. *The Social Construction of Technological Systems: New Directions in the Sociology and History of Technology*. Cambridge, MA: MIT Press, 1987.

Brinkman, Joshua T. "From 'Hicks' to High Tech: Performative Use in the American Corn Belt" PhD diss., Virginia Polytechnic Institute and State University, 2017a, https://vtechworks.lib.vt.edu/handle/10919/81991 (accessed 11/9/21).

———. ""Thinking Like a lawyer" in an Uncertain World: The Politics of Climate, Law, and Risk Governance in the United States." *Energy Research & Social Science* 34 (2017b), 104–121.

Brinkman, Joshua T. and Richard F. Hirsh. "Welcoming Wind Turbines and the PIMBY ("Please in My Backyard") Phenomenon: The Culture of the Machine in the Rural American Midwest." *Technology and Culture* 58, no. 2 (April 2017), 335–367.

Bryan, William, Jennings. "A Cross of Gold-Democratic National Convention." July 8, 1896. Buffalo, NY. www.americanrhetoric.com/speeches/williamjenningsbryan1896dnc.htm (accessed 10/27/21).

Bunge, Jacob. "Farmers Favor Trump Despite Trade Woes." *Wall Street Journal*, Oct. 19, 2020.

Calma, Justine. "How Governors Are Fighting for Clean Energy Jobs." *The Verge*, Dec. 10, 2020.

Cama, Timothy. "Trump: Wind Power 'Kills all your Birds," *The Hill*, Aug. 2, 2016.

Cowan, Ruth Schwartz. *More Work for Mother: The Ironies of Household Technology from the Open Hearth to the Microwave*. New York: Basic Books, 1983.

Edgerton, David. *Shock of the Old*. Oxford: Oxford University Press, 2007.

"Electricity Data Browser." *U.S. Energy Information Administration*, www.eia.gov/electricity/data/browser/#/topic/0?agg=2,0,1&fuel=vtvv&geo=00000g&sec=008&freq=A&start=2001&end=2020&ctype=linechart&ltype=pin&rtype=s&pin=&rse=0&maptype=0 (accessed 11/4/21).

"Existing Government Policies Spur Wind Power Development." *Wallaces Farmer*, March 19, 2010.

"Farmland Ownership and Wind Energy." *Wallaces Farmer*, Feb. 8, 2019, Farm Business.

"Federal Funding Goes to Rural Solar Renewable Energy Projects." *Wallaces Farmer*, Sept. 18, 2004, Farm Business.

Fehr, Mike. "Wind Energy Good for Iowa." *Wallaces Farmer*, Sept. 22, 2020, Farm Business.

Flaugh, Loren G. "Landowners Welcome Wind Farm." *Wallaces Farmer*, Jul. 17, 2019, Farm Business.

Foley, Ryan and Aaron Young. "Trump's Putdown of Wind Energy Whips up a Backlash in Iowa." *Des Moines Register*, Jun. 22, 2017.

Fouché, Rayvon. "Say it Loud, i'm Black and i'm Proud: African Americans, American Artifactual Culture, and Black Vernacular Technological Creativity." *American Quarterly* 58 (2006): 639–661.

Garud, Raghu and Peter Karnøe. "Path Creation Processes Underlying Energy Transitions: The Case of Danish Wind Energy," in: Araújo, K. (Ed.) *The Routledge Handbook of Energy Transitions*. Routledge: Abingdon.

Giddens, Anthony. *The Consequences of Modernity*. Stanford, CA: Stanford University Press, 1990.

Gillis, Justin and Nadja Popovich. "In Trump Country, Renewable Energy Is Thriving," *New York Times*, June 6, 2017.

"Governors Take Stand on Wind Energy." *Wallaces Farmer*, Mar. 21, 2010.

"Governor Walz Joins Midwest Governors to Announce Partnership to Expand Electric Vehicle Charging Infrastructure, Grow Jobs, Futureproof Regional Commerce," *Office of Governor Tim Walz & Lt. Governor Peggy Flanagan*, https://mn.gov/governor/news/?id=1055-501265 (accessed 10/27/21).

Graff, Michelle et al. "Stakeholder Perceptions of the United States Energy Transition: Local-level Dynamics and Community Responses to National Politics and Policy." *Energy Research & Social Science* 43 (Sept. 2018): 144–157.

"Grassley's Wind Energy Proposals Rejected." *Wallace's Farmer*, Jan. 28, 2009.

"Grassley Works to Bolster Wind Energy Production and Generate Job Creation." *News Release on Senator Grassley's Senate Web Site*, Jan. 26, 2009, www.grassley.senate.gov/news/news-releases/grassley-worksbolster-wind-energy-production-and-generate-job-creation (accessed 11/28/21).

Habeeb, Gabrielle. "The Kansas Energy Story – and Opportunity," *NRDC*, Apr. 22, 2021.

Harris, Tyler. "European Influence in the Midwest." *Wallaces Farmer*, Dec. 19, 2013.

"Harvest Energy from Wind and Solar to Power Your Farm." *Wallaces Farmer*, September 11, 2013.

Hecht, Gabrielle. *The Radiance of France: Nuclear Power and National Identity after World War II*. Cambridge, MA: MIT Press, 1998.

Hess, David J. "Energy Democracy and Social Movements: A Multi-coalition Perspective on the Politics of Sustainability Transitions." *Energy Research & Social Science* 40 (2018): 177–189.

Hess, David J. Quan D. Mai, and Kate Pride Brown. "Red States, Green Laws: Ideology and Renewable Energy Legislation in the United States." *Energy Research & Social Science* 11 (2016): 19–28.

Hirsh Richard F. and Benjamin K. Sovacool. "Wind Turbines and Invisible Technology: Unarticulated Reasons for Local Opposition to Wind Energy." *Technology and Culture* 54, no. 4 (2013): 705–34.

Hossain, Mokter. "Grassroots Innovation: A Systematic Review of Two Decades of Research." *Journal of Cleaner Production* 137 (2016): 973–981.

Hughes, Thomas P. "Technological Momentum." In Merritt Roe Smith and Leo Marx (eds). *Does Technology Drive History? The Dilemma of Technological Determinism*. Cambridge, MA: MIT Press, 1994, 101–113.

"Illinois Wind for Schools." *Illinois Institute for Rural Affairs*, www.illinoiswind.org/active-projects/illinois-wind-for-schools/ (accessed 1/13/21).

"Illinois Wind for Schools Picks Six for Wind Energy Curriculum." *Prairie Farmer*, April 30, 2013.

Iowa Environmental Council. "Wind Energy: Wind Energy Is Clean, Affordable, and Provides Proven Economic and Environmental Benefits Throughout Iowa," www.iaenvironment.org/our-work/clean-energy/wind-energy (accessed 10/27/21).

"Iowa Rural Electric Cooperative Leads the Way on Utility Solar." *Wallaces Farmer*, Apr. 5, 2016.

Jorgensen, Marianne, and Louis Phillips. *Discourse Analysis as Theory and Method*. London: Sage, 2002.

Kiernan, Tom. "Six Facts to Know About American Wind Power During National Clean Energy Week, Climate Week NYC." *Into the Wind: The AWEA Blog*, Sept. 21, 2020, www.aweablog.org/six-facts-about-american-wind-power-during-ncew-climatenyc/ (accessed 2/4/21).

Kingsolver, Barbara. *Animal, Vegetable, Miracle*. New York: HarperCollins Publishers, 2007.

"Learn How to Harness Solar Energy on the Farm." *Wallaces Farmer*, Jun. 17, 2013, Farm Business.

Looker, Dan. "SF Special: Chuck Grassley, a Senate Force for Farmers." *Successful Farming*, May 29, 2019.

Longman, Martin. "Chuck Grassley Has Lost His Patience with Trump's Wind Lies." *Washington Monthly*, April 4, 2019.

Martin, Christopher. "Renewable Energy Industry Advises Trump to Look to the Sky." *Wallaces Farmer*, Feb. 2, 2017, Farm Business.

Mowitz, Dave. "Ageless iron: Reunions that Revive the Past." *Successful Farming* 90, no. 3 (mid-February 1992).

Noble, David F. "Social Choice in Machine Design: The Case of Automatically Controlled Machine Tools." In Donald MacKenzie and Judy Wajcman, eds., *The Social Shaping of Technology*, 2nd ed. Buckingham: Open University Press, 1999, 161–176.

Nordensvard, Johan and Frauke Urban. "The Stuttering Energy Transition in Germany: Wind Energy Policy and Feed-In Tariff Lock-In." *Energy Policy* 82 (2015): 156–165.

"NPPD Introduces Online 'Wind Watch.'" *Nebraska Farmer*, Jan. 7, 2014.

Nye, David E. *Electrifying America: Social Meaning of a New Technology*. Cambridge, MA: MIT Press, 1992.

———. "Republicanism and the Electrical Sublime," *ATQ: 19th Century American Literature and Culture*, 4, no. 3 (1990).

Offshore Production and Safety Act of 2011 – Motion to Proceed, Senator Charles Grassley, May 17, 2011: S3034, www.govinfo.gov/content/pkg/CREC-2011-05-17/pdf/CREC-2011-05-17-pt1-PgS3013-4.pdf (accessed 2/5/21).

Oteman, Marieke et al. "Pioneering Renewable Energy in an Economic Energy Policy System: The History and Development of Dutch Grassroots Initiatives." *Sustainability* 9 (2017): 550.

Pasqualetti, Martin J., Robert Righter, and Paul Gipe. "Rejuvenated North America." In: Cutler J. Cleveland (Ed.). *History of Wind Energy*, 419–433. Vol. 6 of *Encyclopedia of Energy*, Amsterdam: Elsevier, 2004.

Paulos, Ben. "The United States Midwest: A Leader in Clean Energy." *Renewable Energy World*, June 10, 2014.

Peterson, Tarla Rai. "Jefferson's Yeoman Farmer as Frontier Hero: A Self-Defeating Mythic Structure." *Agriculture and Human Values* (Winter 1999): 9–19.

Pollan, Michael. *In Defense of Food*. New York, Penguin Press, 2008.

"Putting Wind in The Rotation." *Ohio Farmer*, Apr. 11, 2012, Farm Business.

"REAP Program Supports Energy Efficiency." *Wallaces Farmer*, Sept. 28, 2015, Farm Business.

"Reduced Energy Costs for Rural Businesses. *Wallaces Farmer*, Nov. 9, 2011, Farm Business.

Sabatier, Paul A. and Hank C. Jenkins-Smith. *Policy Change and Learning: An Advocacy Coalition Approach*. Boulder, CO: Westview Press, 1993.

Sauck, Katie. "Wind Power Your Future." *Wallaces Farmer*, Dec. 1, 2006.

Schatzberg, Eric. "Ideology and Technical Choice: The Decline of the Wooden Airplane in the United States, 1920–1945." *Technology and Culture* 35, no. 1 (1994): 34–69.

Seymour, John and Will Sutherland. *The Self-Sufficient Life and How to Live It*. New York: DK Publishing, Inc., 2009.

Smith, Merritt Roe and Leo Marx, Introduction: "Technological Determinism in American Culture." In *Does Technology Drive History? The Dilemma of Technological Determinism*. Cambridge, MA: MIT Press, 1994, 1–37.

Solomon, Barry D. and Karthik Krishna. "The Coming Sustainable Energy Transition: History, Strategies, and Outlook." *Energy Policy* 39 (2011): 7422–7431.

Sowers, Jacob. "Fields of Opportunity: Wind Machines Return to the Plains." *Great Plains Quarterly* 26, no. 2 (2006): 99–112.

Stokes, Leah Cardamore. *Short Circuiting Policy: Interest Groups and the Battle Over Clean Energy and Climate Policy in the American States*. New York: Oxford University Press, 2020.

Stokes, Leah C. and Hannah L. Breetz. "Politics in the US Energy Transition: Case Studies of Solar, Wind, Biofuels and Electric Vehicles Policy." *Energy Policy* (February 1, 2018): 113–156.

Swoboda, Rod. "Iowans Cite Mostly Positive Experiences with Renewable Energy." *Wallaces Farmer*, Feb. 6, 2018, Farm Business.

_____. "What Will Happen to RFS Under President Trump?" *Wallaces Farmer*, Dec. 5, 2016, Iowa Farm Scene.

_____. "Wind Energy Progress is in Jeopardy," *Wallaces Farmer*, Jun. 4, 2012.

"The Wind Energy Developer Question." *EHS Energy*, Feb. 17, 2010.

Thompson, Carl. "The Model T." In Carol Burke (eds). *Plain Talk*. West Lafayette, IN: Purdue University Press, 1983.

Tonneson, Lon. "Blowing in the Wind." *Dakota Farmer*, Feb. 5, 2011, Inside Dakota Ag.

Turner, Frederick Jackson. *"Significance of the Frontier in American History."* Madison, WI: State Historical Society of Wisconsin, 1894.

United States Department of Agriculture. "Nebraska Agricultural Fact Card," Feb., 2021. https://nda.nebraska.gov/facts.pdf (accessed 11/15/21).

United States Environmental Protection Agency. "Renewable Electricity Production Tax Credit Information: Renewable Electricity Production Tax Credit." www.epa.gov/lmop/renewable-electricity-production-tax-credit-information (accessed 12/28/20).

Usher, Anne. "How Obama is 'Trump-Proofing' His Climate Pact," *Politico Magazine*. June 6, 2016.

Van Bramer, Fern. "How Electricity Effects Economy in the Home and Adds to the Happiness of the Family." *Electric City Magazine,* April 1917: 9.

Vander Hart, Shane. "Grassley Touts Wind Energy in New Campaign Ad." *Caffeinated Thoughts*, Sept. 19, 2016, https://caffeinatedthoughts.com/2016/09/grassley-touts-wind-energy-new-campaign-ad/ (accessed 2/6/21).

Vogel, John and Tom J. Bechman. "Solar Panels, Field Data." *Wallaces Farmer*, Mar. 31, 2014, Crops.

Walton, Robert. "Kansas ends 20% Renewables Mandate, Replaces it with Voluntary Goal." *Utility Dive*, Jun. 1, 2015.

Watson, Sarah. "What Iowa's Political Leaders have Said About Climate Change." *The Daily Iowan*, Oct. 7, 2019.

Webber, Tim. "Wind Blows By Coal to Become Iowa's Largest Source of Electricity." *Des Moines Register*, Apr. 16, 2020.

Wenzel, Wayne. "Harvest the Wind." Wallaces Farmer, Mar. 1, 2004.

Williams, Samantha. "Midwest Renewables Surge Forward: But Where Is Ohio?" *NRDC*, Mar. 29, 2019.

"Wind Powered Ethanol." *Wallaces Farmer*, Feb. 1, 2007.

"Wind Turbines Power Cooper Farms." *Ohio Farmer*, Feb. 3, 2012.

Winner, Langdon. "Do Artifacts have Politics?" In *The Whale and the Reactor: A Search for Limits in an Age of High Technology*. Chicago, IL: University of Chicago Press, 1986.

6

ADVANCING LEGAL AND PRACTICAL RECOGNITION OF THE NON-HUMAN RIGHT TO *ENERGY*

Guilherme Pratti, Alex Putzer, and Lorenzo De Marinis[1*]

1. The Dimensions of Energy Transition

There is a need for a rapid, just, and effective energy transition (DESA 2021). This is easier said than done, as such a transition includes several dimensions. Some dimensions relate to regulatory frameworks, and others tie to the deployment of new technologies. Transitions are inherently dynamic, a feature that Talus and Pami (2020, 560) stress by stating, "transitions are a *status quo* for the energy world".

We believe that, in addition to various mainstream approaches, a philosophical reassessment of the energy in the energy transition is necessary. This stands in contrast with most of the past 200 years, where "the study of energy has been the domain of the natural sciences and engineering" (Frigo 2018a, 178). Such a traditional energy paradigm, however, "hinders more nuanced, complex understandings of what energy might be" (Frigo 2017). As a result of the energy crises of the 1970s, this one-sided perspective began to change. Scholars and practitioners started to consider energy within "economic, socio-political and ecological dimensions" (Frigo 2020, 150). The philosophy of energy, including so-called energy ethics, represents one of these emerging fields.

Mitcham and Frigo distinguish between type I and type II energy ethics.[2] Type I identifies a linear relation between energy use and human well-being, where energy use is seen as fundamentally good. Type II proposes a skepticism to this linear relation, questioning such an assumption (Mitcham 2013). In this chapter, we will address this second type. We are not the first to do so: Meinhold (2018, 296) explores the idea of a "holistic understanding of energy", Paterson (2020) its legal conceptualization, and Frigo (2020, 151) reimagines what he calls the "human-energy-nature nexus according to an eco-centric perspective".

We aim to provide a twofold reassessment. The first one relates to the meanings and use of the term energy; the second builds upon this and proposes the non-human right to *Energy*. To illustrate the meanings regarding energy, we rely on a metaphor drawn from Shakespeare's 16th-century play, *The Merchant of Venice* (Shakespeare 1600). Specifically, at the moment of the plot in which Shylock, the play's antagonist, was about to extract a pound of flesh from Antonio's body, as forfeiture for an unpaid bond. He was unable to do so, because said bond only mentioned flesh but not blood. The scene's captivating force comes from this juxtaposition of

DOI:10.4324/9781003183020-7

tangible and intangibleness: for one cannot *be* without the other. This excerpt is capable of illustrating the so-called Anthropocene's narrative[3] and its respective role in the ongoing energy and environmental crises, which are at the center of the global energy transition and, in allegorical terms, are caused by the carving of the Earth in a manner which is careless of its consequences. We take this powerful image to be an innovative reading-key of the intertwinement between the worldwide ever-increasing energy production and the corresponding ecological crises it causes.

The main claim of the first part of this chapter is that the legal realm has hitherto recognized the tangible but not quite properly the intangible aspects of Nature when it comes to the production and use of energy and environmental protection. An energy transition is a well-timed opportunity to reassess the said lack of proper recognition. With the Shakespearean juxtaposition in mind, we analyze the main international legal instruments on environmental and energy production matters and pin down the overall lack of proper recognition regarding the intangible aspects of Nature within the realm of the law. To clarify our claim, we consequently distinguish between an uppercase *Energy* and a lowercase *energy* (always evidenced in italics). Whereas the former represents the energy within Nature, the latter is the one so incessantly used by humans.[4]

Building on this first section, we introduce the Rights of Nature, a legal and ethical theory that seeks to recognize the value of Nature independent of human use. In particular, we analyze the relation between the Rights of Nature and the *Energy* within natural processes, proposing the so-called non-human right to *Energy*.

Natural processes (or cycles) can be defined in two ways. The *Cambridge Dictionary* sees them as "a series of actions that [one takes] in order to achieve a result". Examples include the annual harvest or the generation of electric energy. Nevertheless, according to Rescher (1996), such a purpose-oriented vision is only one possible interpretation. In another, the process itself matters. Instead of leading up toward a particular goal, these processes represent continuous and dynamic interactions. Throughout this chapter, when we refer to processes or cycles, we will thus talk about this continuing interpretation, hence processes detached from instrumentalized purposes.[5]

Combining both parts into a coherent framework (i.e., reassessing the concept of energy and bridging the identified language gap through the non-human right to *Energy*) allows us to recognize Nature in a more sophisticated way. We believe the result creates the philosophical foundation for a more effective energy transition.

2. Reassessing the Concept of Energy

Both "reassessing" and "energy" in "reassessing the energy" have a dual dimension. We first reassess two dimensions of energy, one of which we claim as not being sufficiently distinguished. The unrecognized type of energy we are trying to bring to light is the dimension commonly ignored by the law, as for example, the energy that perpetuates hydrological cycles throughout Earth's natural history and that, ultimately, allows us to "generate" electric energy by placing dams and turbines along riverbeds or by diverting a river's flow to power plants.

We claim that this relinquishing is a major reason for the growing global environmental and climate repercussions, which, as shown by the 2012 IPCC Special Report, result in extreme weather and in an alteration of the frequency, intensity, timing, and duration of climate events.

As we will show, the law usually covers the regulation, production, protection, use, and continuity of human-use energy, while neglecting the regulation, protection, use, and continuity of the energies within Nature. The latter we frequently consider as a mere resource. These are the energies that allow us to count on their cycles as the Heideggerian "standing-reserve" (Heidegger 1977, 17).[6] They are crucial for energy grids and for the commodities-dependent global market (cotton, soy, corn, wheat, sugarcane, meat, etc.).

The aforementioned dual dimension is thus as follows: we reassess the concept of energy in the energy transition, all the while reassessing the dimensions of energies that ought to be considered by the law. If the energy transition is the passage from the hegemonic use of fossil fuels to a worldwide sustainable power grid, the law should be as clear as possible with regard to these definitions and to the relation between human-use energy and within-Nature energies.

2a. Energy *Contra* energy

Scientifically, energy represents the ability of a body to perform work on another body (Müller 2007). Energy is an *indirectly observable* quantity, meaning that it is measured through its effects, such as the speed of a car or the temperature of a lit match. Energy comes in many different forms, but today it is commonly associated with its electrical form, an association deriving from its versatility. As opposed to other forms, electrical energy can be relatively easily transformed into other forms, such as heat, and then be stored and transferred. In short, electrical energy has become so widespread because it can be easily instrumentalized for human use.

Describing all forms of energy is not necessary for the scope of this chapter, as a few basic underlying concepts suffice for understanding our energy systems. To begin, energy can be stored. Wood, for instance, stores the radiant energy transformed through photosynthesis by the sun in the form of chemical energy. Different forms store energy differently: for example, kinetic energy represents the energy associated with motion. It is the form of energy that makes a river flow. The waterway gains its kinetic energy from transforming potential energy, the "energy of heights", which derives from Earth's gravitational field, thus flowing downstream. Another example is found in water droplets that rise by the conversion of solar energy into thermal energy, called the process of vaporization. This continuous transformation of energy creates the Earth's water cycles. Humans then, by use of a hydroelectric power plant, extract the kinetic energy of a river from the water cycle and transform it via turbines into electric energy.

Importantly for us, the law of conservation states that energy can neither be created nor destroyed but only transformed. As a result, talking about the "production" or "generation" of energy is misleading. Throughout this chapter, we will apply this terminology exclusively for the conventional usage of the electrical grid, while using the more appropriate "conversion" and "transformation" when describing our proposed conceptualization.[7] Through this, we aim to create awareness for the energies that we rely upon as a standing-reserve for human use. In other words, these are the within-Nature energies that we take for granted and that make us perceive trees as wood logs and hence as energy sources or rivers as mere "water power suppliers" (Heidegger 1977, 16).

In order to distinguish between the human-converted form of energy and the one that maintains natural processes, we emphasize vocabulary caution by introducing lowercase *energy* and uppercase *Energy*.[8] The former refers to instrumentalized-for-human-use energy, the latter to within-Nature energy. This distinction is done for linguistic accuracy, describing *energy*, on one hand, in its human-converted or human-transformed electrical form, and *Energy*, on the other hand, as "nature-energy" (Frigo 2020, 157) or for what Heidegger uses the plural "energies of nature" (1977, 18).

Crucially, when talking about *Energy*, we have to keep in mind both the tangibility of the natural realm and the intangibleness of its ethereal equilibrium. Such a causal relationship is clearly illustrated by the IPCC's AR6, when it states that the rapid "chemical and biological changes in the Earth system such as rapid ocean acidification" are due to "massive destruction of tropical forests, a worldwide loss of biodiversity and the sixth mass extinction of species" (Chen et al. 2021, 1–17).[9] All mentioned elements, including the oceans, forests, and biodiversity but also the Earth system are but prime examples of the intrinsic relation between the tangible and intangible aspects of Nature.[10]

In this context, the Shakespearean juxtaposition, as conveyed at the beginning of this chapter, envisions a systemic equilibrium between the tangible and intangible aspects of Nature. *Energies* relate to Antonio's intangible traits as nature's tangibility relates to his body. The Earth system's equilibrium, the natural processes' continuous transformation of *Energy*, and the roles that trees, oceans, insects, animals, bacteria, and so on play in it are all altered by the disturbances of one another. And humankind's huge demand for one particular tangible aspect, that is, fossil fuels' converted *energy* disturbs the balance of the whole system.

With these physical and vernacular distinctions of energies in mind, we turn our analysis to the legal language spoken by these mentioned disturbances to highlight its lack of sufficient consideration for the intangible aspects of Nature. We will show the apposing of these two concepts of energy as visible in the language of the law.

2b. *The Legal Traces of* Energy *and* energy

In this analysis, we consider both nature-oriented and energy-oriented international laws. Due to the variety of legal definitions and in order to avoid any misunderstandings, we will generally refer to these as "legal instruments". We will begin by looking into a set of environmental legal instruments, followed by ones concerned about energy production. We chose the most important instruments in the field since we believe they are the most representative of the dominant dimension of energy.

The first legal instruments we consider emerged from the 1972 United Nations Conference on the Human Environment. This "foundational moment of modern international environmental law" (Dupuy and Viñuales 2018, 9) had several outcomes: the Stockholm Declaration (named after the Conference's location), the establishment of the United Nations Environment Program (UNEP), and an Action Plan for the Human Environment. We focus on the Declaration and the Action Plan, which represent the beginning of global conversations regarding the intertwining of environmental and development through industrialization concerns.

The Declaration's 26 principles frequently mention "natural resources", "natural ecosystems", and Earth's capacity "to produce vital renewable resources" (principles 2, 3, 6, 12, and 13, in particular).[11] The Action Plan recommends the "environmental aspects of natural resources management" (rec. 19 to 69), such as "soil potentialities, capabilities, use and restoration" (rec. 19 to 21) and the need to "compile, register and catalogue genetic resources" and "develop comprehensive inventories of" microorganisms culture collections (rec. 40 and 41). These can all be interpreted as a way of preserving the Heideggerian standing-reserve (Heidegger 1977, 16). While naming *energy* only in its instrumental form,[12] it becomes evident that these first-of-their-kind legal instruments show traces of *Energy* through the reference to natural processes.

In ten-year intervals, the United Nations subsequently adopted the 1982 World Charter for Nature (WCN) and the 1992 Rio Declaration on Environment and Development. These two legal instruments, although of soft law,[13] deepened the reference to *Energy* since they put forward an overarching conceptualization regarding the tangible and intangible aspects of Nature (Handl 2012). In the WCN's preamble and the annex to it, this distinction is shown through the pairings of "natural and ecological processes and diversity of life forms" and "natural and life support systems, balance, and quality of nature and natural resources". In a strongly worded declaration of awareness, the General Assembly wrote that "[m]ankind is a part of nature and life depends on the uninterrupted functioning of natural systems which ensure the supply of energy and nutrients", to then define as a general principle that "Nature shall be respected and its essential processes shall not be impaired". The supply of energy is to be here understood as a supply of *Energy*, for it is the result of the "uninterrupted functioning of natural systems". Principle seven of the Rio Declaration summed up this vocabulary in a call for States to "cooperate in spirit of global partnership to conserve, protect and restore the health and integrity of the Earth's ecosystem".

The richness of the *Energy*-oriented vocabulary of these legal instruments is not followed by subsequent developments. The 1992 United Nations Framework Convention on Climate Change, the 1992 Convention on Biological Biodiversity, the 1994 Convention on Desertification, and the 1997 Kyoto Protocol considerably weaken these provisions,[14] with the most recent 2015 Paris Agreement referring to "environment" and "ecosystems" but not offering any additional elaborations.

We thus identify only an indirect reference to *Energy*. There are clear references to the tangible aspects of Nature, although with a strong emphasis on protecting these exclusively for human benefit, as well as with a clear dispersion of vernacular precision throughout time. Even though a substantial vocabulary was built up between 1972 and 1992, subsequent legal instruments diluted the provisions into widely generic terms.

A different trend can be observed in the major energy-oriented legal instruments. To begin with, the 1991 European Energy Charter, in its preamble, states the "essential importance of efficient energy systems in the production, conversion, transport, distribution and use of energy for security of supply and for the protection of the environment". Interesting is the juxtaposition of "efficient *energy* systems" that are used "for the protection of the environment", where environmental protection is always mentioned in the context of "energy efficiency" (Titles I and II).

The 1994 Energy Charter Treaty, aiming to implement the guidelines of the European Energy Charter, mentions "environmental aspects" in article 19, where it states that all parties shall strive to minimize harmful environmental impacts, from all operations within the "[e]nergy [c]ycle", in a cost-effective manner. This entails precautionary measures "to prevent or minimise environmental degradation" by taking account of environmental dangers "throughout the formulation and implementation of their energy policies". The Treaty defines energy cycle as "the entire energy chain", including "exploration, production, conversion, storage, transport, distribution and consumption of the various forms of energy". This legal instrument outlines a complete separation of *Energy* and *energy*. Put differently, human-use energy is set apart from Nature and its processes.

More than this, the Energy Charter Treaty set forth an additional split. Indeed, it divides the "natural" from "resources" in "natural resources" by translating it into "energy resources". It does so by transposing the "principle of sovereignty over natural resources",[15] originally linked to the right of peoples to self-determination, to the realm of *energy* law, now worded as "sovereign rights over energy resources".[16] This can be seen as understanding Nature "as the chief storehouse of the standing energy reserve" (Heidegger 1977, 21) – that is, in its availability to be instrumentalized for human use as mentioned earlier.

The 2015 International Energy Charter is in line with the language division and reasoning put forward by the Energy Charter Treaty and the European Energy Charter. A singular shining beacon in the gloom is presented by the 2009 International Renewable Energies Agency's Statute. It mentions the role renewable energy sources can play in the stabilization of the "climate system" and on "environmental preservation", for it can limit the pressure on "natural resources" and reduce deforestation, desertification, and biodiversity loss and contribute to "climate protection".

The analysis showed that, on the one hand, international environmental legal instruments reflect a strong anthropocentric view when it comes to the protection of Nature and the legal vernacular distinction regarding its tangible and intangible aspects. This was intensified by the diachronic dilution of the legal traces of *Energy*. The initial vocabular richness of the environmental legal instruments was further reduced to the term "environment". On the other hand, the reading of energy-oriented legal instruments showed not only the separation of *Energy* and *energy* but also the illusion of a split between energy and natural resources.

In this regard, we can paraphrase Viñuales (2013, 6) to say,

> *Energy* is seen as an "immigrant" in the land of *energy* law and as such, it would only have the scope of action that is consistent with *energy*. If *Energy* protection were to collide with *energy* cycle considerations, the latter would likely prevail.[17]

Comparing *Energy* and *energy* shows a contrast within the two realms of the law. This leads us to the following observations: (1) these realms do not share the vocabulary when referring to the same "object", although they share the same conceptualization of nature as a standing-reserve at their basis; (2) the nature-oriented legal instruments mention the need to preserve the environment to protect *energy* production, while the energy-oriented legal instruments urge for the improvement of *energy* systems as a way to enhance the protection of the "environment" (as long as it is on an acceptable, cost-effective, economic basis). This can be said to be an oddity in the form of reciprocity that, at worst, can be further deepened if not straightforwardly addressed in the energy transition. The first step to counter this trajectory has been minding the language gap between these two legal realms. The second step is to bridge it.

3. The Non-Human Right to *Energy*

We have described the lack of shared vocabulary between energy-oriented and nature-oriented international legal documents. By differentiating *Energy* and *energy*, we have problematized a limited understanding of energy. In this section, we want to build on these findings and suggest a way forward. Inspired by Rights of Nature scholarship, we propose the non-human right to *Energy* – that is, the attempt of bridging a language gap between legal fields. The Rights of Nature are particularly apt for such an endeavor as they, too, represent a bridge between differing normative systems. First, we briefly contextualize the idea and identify its main characteristics. In particular, we reflect on the aim and composition of rights-holders and see a multifaceted relation between natural entities and processes. With this analysis in mind, we highlight the non-human right to *Energy*, in particular the right to flow, in our case study – the Flying Rivers of the Amazon. We conclude this section by envisioning some implications that go beyond the field of energy.

3a. The Rights of Nature

Nature-oriented and energy-oriented international legal instruments are one way of addressing the manifold ecological and energy crises we face today. They form part of a wider, growing field of unconventional legal tools that question practices that ignore planetary and other boundaries. Among these constructs, both national and international, are also the human right to a healthy environment, ecocide,[18] and climate change litigation. The Rights of Nature go one step further and question the supremacy of human interests. They are "calling for acknowledgement of the fact that [non-human Nature should have] rights that humans [would be] morally [and legally] obligated to respect and protect" (Boyd 2017, 219). Nature should not be seen through the lens of human usefulness but be accepted and respected for its intrinsic value. According to Berry's principles of Earth Jurisprudence, these rights are supposed to be overlapping and role-specific. "Rivers have river rights. Birds have bird rights. Insects have insect rights. Humans have human rights. [The] [d]ifference in rights is qualitative, not quantitative" (Berry 1999, 161).

Christopher Stone's 1972 article "Should Trees Have Standing?" founded the theoretical interest of the Rights of Nature.[19] Several decades later, the idea has led to over 400 initiatives in at least 39 countries across all continents.[20] Arguably, the most basic argument for the attribution of rights to non-human Nature is a call for coherence between the natural and social, but also within social sciences. At least since Darwin's and Wallace's breakthroughs in evolutionary theory, biology is increasingly defeating the so-called Cartesian dualism and accepts that humans have derived from the same lineage as other animals and are, in fact, as much part of the natural world, as any other entity (or process, for that matter).[21] Opposing these findings are religious and legal perceptions of human superiority. While it is certainly true that human talents are distinctive, other parts of Nature are special in their own meaning. To use the words of Keeley (2015, 9), "while humans are unique, they are not *uniquely unique*". Beyond the realization that the attribution of rights is based on a threshold of arbitrarily chosen characteristics, another direction seeks coherence within the legal realm itself. Following it, if such abstract entities such as organizations are able to be rights-holders, Nature shall too. While we do not elaborate here on the ontological and epistemological justification of these claims (see Kurki 2021), we want to recognize that the Rights of Nature are criticized with regard to their purpose, use, and content of legal terminology (rights, personhood, legal standing) and, at times, their blurry division between morality and legality (Corrigan and Oksanen 2021).[22]

For this chapter, in particular, we are concerned about the relation between the Rights of Nature and natural processes. We are trying to answer the question of *who* or what is supposed to have *which* rights. In order to answer this question, we need to distinguish between the composition of rights-holders and the rights themselves. As for the former, a vision of natural entities or objects is complemented by one of natural processes. As for the latter, we distinguish between an explicit right to processes and one without. Similar to the previous section, we will carry out an analysis of legal instruments. In particular, informed by the respective literature, we will analyze the language of the most complete dataset of Rights of Nature initiatives to date (Putzer et al. 2022).

The field of environmental ethics generally distinguishes between holistic and individualistic approaches of protection. While proponents of the former aim at a collective, the latter identify the value of individuals. Such a division can already cause dilemmas: a focus on individualism, on one hand, does not allow the culling of one animal in order to save an entire species. A holistic approach, on the other hand, is frequently applied to entire ecosystems, with legal Rights of Nature predominantly following such a pattern. In fact, more than two-thirds of all legally implemented initiatives refer to Nature indistinctively. They thus align, consciously or not, with a holistic approach. Building on the previous section, these initiatives would consequently take a process approach. Corrigan and Oksanen (2021, 9), however, argue that an indistinctive reference to Nature or ecosystems keeps referring to objects since natural entities remain the dominant element. For them, a process approach looks different, "for example, whether natural selection and evolution have a right to continue in plants and animals or whether the great wildebeest migration on the West-African savannah has a right to take place without human interference" (ibid., 9). They thus make a distinction between an initiative that regards an entire ecosystem that includes both entities and processes and one that focuses exclusively on the process itself. When considering the composition of rights-holders, such an explicit reference to processes has indeed gone unnoticed. There are currently no initiatives that refer explicitly to the specific rights of ocean or wind currents, none that consider water cycles, weather phenomena, oxygen production, or soil formation. Put differently, while both are crucial for efficient protection efforts, we need to distinguish a reference to ecosystems (the tangible) from a reference to natural processes (the intangible).

In addition to that, we need to examine which rights these rights-holders have. As opposed to the rights *of* processes, what do exist are the rights *to* processes. Close to 8% of all initiatives include such a right (Putzer et al. 2022).[23] Specifically, the initiatives include references to the right to naturally evolve, to maintain vital cycles and natural processes. Natural ecosystems have a right to remain intact, while rivers have a right to flow. A prime example of such rights to processes is the 2008 Constitution of Ecuador. Paragraph 1 of Article 71 states, "Nature, or Pacha Mama, where life is reproduced and occurs, has the right to integral respect for its existence and for the maintenance and regeneration of its life cycles, structure, functions and evolutionary processes". Here, we can identify the right *of* an ecosystem *to* natural processes. This constellation is also rather similar to the aforementioned WCN's preamble: "Nature shall be respected and its essential processes shall not be impaired".

Up until now, the most elaborated reference to the right to processes is an initiative from Canada. On February 16 and 18, 2021, respectively, the Minganie Regional County Municipality and the Innu Council of Ekuanitshit, both located in the province of Quebec, adopted twin resolutions for the rights of the Mutehekau Shipu, or Magpie River. Through them, the 200 km long waterway has become a "legal person" and possesses now

> the right to life, to exist and to flow; the right to respect of its natural cycles, the right to naturally evolve, to be preserved and protected; the right to maintain its natural biodiversity; the right to maintain its integrity; the right to fulfil its essential ecosystem functions; the right to be free from pollution; the right to regeneration and restoration; [as well as] legal standing.

The language of this initiative implies that a river without its flow ceases to be a river. Put differently, a river without its *Energy* ceases to be a river. Tangible Nature is incomplete without its intangible characteristics. The right to natural processes is thus seen as key to its efficient protection.

Even though we consider both entities and processes when talking about an entire system, it is crucial to specifically identify the lacking protection of the latter. Just like the tangible and intangible aspects of Nature, the rights of entities and processes are connected. Similar to our analogy, the latter is frequently taken for granted. We therefore propose not only to acknowledge and name *Energy* but also to recognize and protect the non-human right to it.

A clear distinction becomes increasingly significant in times when transnational environmental harm is on the rise. Nature does not consider the boundaries of countries. This increasingly becomes a problem since the protective jurisdiction in one country does not apply to another one, consequently representing a major hurdle in the protection of Nature. Border rivers, mountain ranges, oceans, or the seasonal migration of animals are just some examples of such transnationality, continental, and multidimensional processes just as water or weather cycles are another. Since current legal Rights of Nature face the same problems as conventional environmental protection efforts, with international environmental initiatives mostly representing forms of soft law,[24] we believe that a focus on natural processes offers an additional argument to the advancement of not only a just energy transition but also transnational environmental law.

3b. *The Flying Rivers Phenomenon*

The ever-increasing demand for energy has transformed considerable parts of Nature: landscapes have been altered for both renewable and nonrenewable *energy* sources. The Earth system is under the constant and imminent perils of, among others, climate-change-induced extreme weather events, such as extended droughts, which occur due to (and are perpetuated by) the disturbances in hydrological cycles around the globe. A prime example of the interrelation of these events with human-related landscape-altering activities can be illustrated by the aerial rivers phenomenon, which "connects moisture-donating regions with moisture-receiving locations" (Nobre 2014, 19).

This phenomenon is also known by the name of Flying Rivers and can be defined to be of paramount importance to the hydrological cycle of many of the world's regions, including South America, Siberia, Scandinavia, and China (Pearce 2019, 42).[25] We will focus on the Amazonian phenomenon, which derives from a combination of natural events that connect "the moisture-laden trade winds from the equatorial Atlantic with the winds that blow over the Amazon as far as the Andes, and from there, depending on the season, over to the southern part of South America" (Nobre 2014, 19).[26] This results in the Amazon exporting water vapor to the southern regions of the continent and thus nourishing river basins along the way (Salati et al. 1979, 1250; Nobre 2014, 18). In this process, the trees in the Amazon function as water vapor pumps due to a process called evapotranspiration (Marengo et al. 2004, 2262), which basically describes the recycling of the humidity caused by rainfalls through the absorption of soil water by the roots, which is then transpired by the leaves back into the air and subsequently directed toward the south of the continent (Nobre 2014, 12). Antonio Donato Nobre, one of the leading scientists on this phenomenon,[27] illustrates this process by comparing evapotranspiration to plants' "sweat" and defining the Amazon aerial rivers' cycle as fresh water through airborne "arteries" (Nobre 2014, 12–14).

The Flying Rivers of the Amazon, however, are threatened. They are being greatly impacted by landscape alterations, mainly due to deforestation occurring in the Amazon rainforest, hence causing the alteration of rain patterns and a great decrease in precipitation (Nobre 2014, 21–22), directly affecting the River Plate Basin countries such as Brazil, Bolivia, Paraguay, Uruguay, and Argentina (Pearce 2019, 42). As for Brazil, in particular, deforestation's impacts on this phenomenon are being clearly felt. Despite having 12% of the world's fresh water and 53% of the continent's water supply, throughout 2021, the country experienced its worst drought in 90 years, the second since 2014, and the third since the beginning of this century (Brasil 2021). These facts raise *Energy* security concerns in all River Plate Basin countries, especially so in Brazil, for more than 60% of its energy matrix comes from hydropower plants located in its southern region. Put differently, much of Brazil's *energy* is sourced from the region to where the Flying Rivers flow. And to where they are to flow ever less.

The situation, as illustrated, is "local" in the sense that it immediately affects "only" one of the world's regions but should nonetheless urge for *Energy* security concerns everywhere. It cannot be left out of transition-related conversations.[28]

In the current efforts to curb deforestation, the protection of the Flying Rivers is not directly addressed. Or to put it differently, it is so done only in an incidental manner. The traditional priority lies on the tangible parts of Nature, thus leaving only a secondary emphasis on intangible aspects. This perception of forests becomes clear within the concept of sustainable forestry. This practice defines the strategic clearing and restoration of forests and is championed by, among others, the Rainforest Alliance.[29] For this "international non-profit organization working at the intersection of business, agriculture, and forests to make responsible business the new normal", forests serve as "fuel for cooking and warmth, medicinal plants, food, wildlife habitat, clean water, spiritual and cultural touchstones" (Rainforest Alliance 2021).[30] All elements, thus, that can be summarized within the idea of ecosystem services – that is, "[t]he benefits people obtain from ecosystems" (Millennium Ecosystem Assessment 2005, 9).

In 1993, the Rainforest Alliance co-founded the Forest Stewardship Council, or FSC, which describes itself as promoting "environmentally appropriate, socially beneficial, and economically viable management of the world's forests". The FSC developed ten principles that serve as a guide to action. Principle 6 is concerned about environmental values and impact. Subprinciple 6.7 states that businesses should implement "[m]easures to maintain natural hydrological patterns and stream flows" (FSC 2021, 38).[31] With a focus on rivers, this principle could suggest an emphasis on the importance of natural processes and, as such, a reference to (the non-human right to) *Energy*. However, throughout the documents, the repeated emphasis on ecosystem services and the strong focus on "natural resources" and "management" contradicts and weakens such a conclusion.

Beyond the field of sustainable forestry, natural processes play a more prominent role. In fact, the global importance of the Amazon rainforest is reflected, among others, through a focus on both its tangible and intangible characteristics. The analogy of the Amazon as one of the "green lungs" of the planet depicts a widespread understanding of a natural process called the oxygen cycle. A related and increasingly popular idea is the vision of the Amazon as a carbon sink (i.e., the carbon cycle).[32]

While these developments represent progress toward more complete environmental protection efforts, we see that other processes, among them the Flying Rivers, fly largely under the radar. By highlighting the non-human right to *Energy*, we can build upon existing narratives, and bring lesser-known processes to the forefront. The non-human right to *Energy* has the ability to offer an addition for the explicit protection of natural processes, as well as for a more complete energy transition in general.

Natural processes are manifold and interconnected. The mentioned oxygen cycle is interrelated with both the carbon and hydrological cycles. Such connectivity becomes even more complex when analyzing processes. Evapotranspiration is a process on its own, as are the wind currents pushing the water droplets toward the west and south of the South American continent. Rain is a process and the accumulation and flowing of the river downstream is as well. While it can be argued that each one of these processes would have its own rights, when we talk about the Flying Rivers' right to flow (i.e., the non-human right to *Energy*), we include all of these overlapping and multidimensional forms of natural processes. The goal is to offer a more complete form of environmental protection that can enter the public debate about the energy transition and help to bridge the gap that has existed largely to this point.

3c. Implications Beyond Energy

The focus of this *Handbook* is on energy transitions, but the framework we propose can also be applied beyond this field. In this chapter, we outline some implications for both a conceptual and a governance level. By searching for traces of *Energy* in a variety of international and national legal documents, we evaluated the explicit use of natural processes. We proposed the *non-human* right to *Energy* to legally recognize our reassessment. This approach contrasted the conventional and dominant *human* right to *energy*, which considers an energy that is focused on the human-exclusive conversion and use of the natural standing-reserve.

Reassembling these elements, we can envision two additional applications. One of them is the *non-human* right to *energy*. This inverts the traditional relationship since natural processes would be able to access human-use *energy* for their benefit. With the 1991 European Energy Charter using such an approach, an example of this dynamic would be the use of (heavy) machinery for the renaturation of a river. Such an approach is similar to Nature's "right to restoration", for which some Rights of Nature scholars advocate (see Luuppala 2021). The second relationship is the *human* right to *Energy*. Rather than converting *Energy* into *energy* to complete a task, this application would act as a guide to access and interact with *Energy* directly. An anthropocentric vision would interpret this as "letting Nature do the work". However, a non-anthropocentric vision would see a re-rapprochement between humankind and Nature, thus a re-emerging of the human in natural processes (i.e., *Energy*). An established example of this dynamic is drying clothes in the sun, as opposed to using a laundry dryer. A considerably more elaborated illustration is the use of the so-called gravitational assist, which uses orbital mechanics to propel a spacecraft more efficiently.

In addition to these conceptual reflections, the outcome of this chapter can be used to enhance the worldwide governance policies that go beyond the energy transition. Here, we specifically refer to Kotzé's two suggestions (2014, 19–32) on how to improve global governance on biodiversity matters. Those include, on one hand, re-envisioning the current institutional design and procedures and, on the other hand, the legal instruments capable of dealing with the ever more complex and connected global environmental challenges.

We agree with Kotzé on the possibility to create a "hybrid mix of legal instruments that apply transnationally and that are sensitive to the integrated Earth-system . . . in a holistic setting" (Kotzé 2014, 33). Although there might be a variety of ways to do so, we put forward a third essential suggestion alongside the ones mentioned, Kotzé's urging to reimagine normative frameworks would arguably benefit from correction of the language gap we identified. At the same time, this chapter's proposal would gain considerable leverage if taken into deliberation in the "transboundary governance" backdrop of Kotzé's stance. This could lead energy transitions into building up an *Energized* grid of "connectivity norms" (Kjaer 2018, 126), including a balanced account of the (in)tangible aspects of nature being shared by different realms of the law, thus increasing its possibilities of global success.[33]

To summarize, both the distinction between *Energy* and *energy* and the resulting non-human right to *Energy* create manifold opportunities to advance not only a just energy transition but also a tool to enhance and widen a variety of other conceptual and policy provisions, overcoming currently outdated normative frameworks.

4. A More Complete Energy Transition

Global energy crises create the need for and emergence of various dimensions of a global energy transition. By reassessing the concept of energy itself, we set out to add a currently missing dimension. We distinguish between uppercase *Energy*, which represents the independent, within-Nature form that maintains natural processes, and lowercase *energy*, which is the instrumentalized, human-converted form, most commonly associated with the electric power grid. With this distinction in mind, we identified traces of the two energies in both nature- and energy-oriented international legal instruments and established that *Energy* is not properly addressed. As such, we set out to further clarify but also bridge this language gap.

Inspired by the scholarship and legal implementation of the Rights of Nature, we developed the traces into two conceptualizations of processes, one as a rights-holder and one as a right. We thus proposed the non-human right to *Energy*. Whereas the non-human is seen as a process, *Energy* enables a clear legal recognition of this neglected, intangible aspect of Nature. South America's biggest water cycle, and one of the most important natural processes in the world, served as a case study to illustrate what we coined the Flying Rivers' right to flow. We finished our reflections by outlining some of the concept's possible implications beyond the energy realm.

This chapter serves as an exercise to underline the importance of language in regulating not only energy transitions. It is crucial to identify, name, and recognize the elements that need to be regulated. Whereas *energy* focuses on the important human-used concept, *Energy* recognizes the frequently neglected (and insufficiently addressed) intangible aspects of Nature. As a first step, such a reassessment helps to create a more complete legal representation of Nature. Just like Shakespeare's point in *The Merchant of Venice*, natural entities without their processes are an insufficient depiction of Nature. At the same time as keeping them linguistically apart, we want to emphasize that both are inseparable. In a second step, we used these findings to support a new legal vision of Nature, which explicitly refers to natural processes. Eventually, such a linguistic turn leads to an energy transition in every sense of the word.

Notes

1* We would like to thank Katja Schechtner for her thorough support throughout all phases of our writing process and Kathy Araújo for the backing and encouragement during the editing.
2 In an interview, he speaks of "type zero" as not thinking about the ethical implication of energy at all (Mitcham and Frigo 2018, 303).
3 The Anthropocene narrative claims that human beings have become a geologic force, responsible for substantial modifications of the Earth's system. Regarding the Anthropocene narrative, please see Zalasiewicz et al. (2010), Kotzé (2014), and Viñuales (2018). For a counternarrative, please see Gibbard et al. (2021).
4 We follow the policy of various scholars and organizations, including the United Nations Harmony with Nature Program, which write Nature in uppercase to establish it also linguistically as a subject (see, for instance, footnote 1 of the UN GA Resolution A/75/266). This policy is further emphasized by a contrast to a lowercase anthropocentric object vision of nature.
5 "Processes detached from a purpose" refers to a factual state of natural processes which occur regardless of human interpretation and instrumentalization.
6 We are here relying upon the astute insight Heidegger put forward in *Die Frage nach der Technik* regarding humankind's domination of Nature through technology. In this sense, "standing-reserve" (*Bestand*) should be comprehended as the fact that technology has allowed us to render Nature something orderable and always available to be used.
7 Power is a quantity often misinterpreted as a synonym for energy (electrical power and electrical energy are used in the same way). Power quantifies the amount of energy transferred from a body to another per time; its unit is the watt (i.e., joule transferred per second). For instance, charging a smartphone with a 30 W power charger means that there are 30 J of energy flowing in the battery every second.
8 Throughout the chapter, we italicize *energy* and *Energy*, when we talk about the dual dimensions, and keep energy in Roman format, when we talk about its everyday usage.
9 More recently, the IPCC's Working Group II, in their contribution to the AR6, established a direct link between the Earth system's changes and energy systems compromise due to lack of stability and rise of infrastructure maintenance costs related to climate change disruptions (Pörtner et al. 2022 11–15).
10 We are aware that our take on this particular matter can be said to be ethnocentric – and indeed it is – for the ontological distinction of "energies" we take stock from physics of energy, continental philosophy, and international law. The understanding of the inseparable quality of *Energy* and *energy* (or so the tangible and intangible aspects of Nature, especially on how these relate in the Amazon forest) can also be achieved through what Viveiros de Castro (1996) calls "amerindian perspectivism", which offers access to the "amerindian cosmologies"; through Descola's take on humans and non-humans joint contribution to the "general equilibrium of the cosmos" (1996, 89); the Yanomami shaman Kopenawa's critique of the carelessness of "Merchant People" for the non-human (Kopenawa and Albert 2015); and Krenak's (2019) critique of how human beings behave as if they were separated from Nature.

11 Declaration of the United Nations Conference on the Human Environment. Stockholm. June 16, 1972.

12 While the Declaration does not name "energy", the Action Plans does, although always in reference to the collection of data regarding its environmental effects, consequences, and management mechanisms (recommendations 57, 58, and 59). The opposite occurs in the Statute of the International Atomic Energy Agency, which, albeit in force much prior to the arising of strong environmental concerns in the 1970s, does not mention natural processes or the environment whatsoever (its article XX contains the expression "occurring in nature", as in "naturally occurring", when referring to uranium depletion). This has not changed with subsequent revisions and additional documents. In their current wording, the treaties regarding nuclear safety, signed under the IAEA's Statute, present environmental concerns under the embodiment of the sign "environment" – for instance, the 1994 Convention on Nuclear Safety's preamble and articles 1(iii), 6, and 17(i) and the Convention on the Physical Protection of Nuclear Material, as amended in 2005, in its preamble and articles 1(e) and 7(a, e, g).

13 On the intrinsic relation between soft law and international environmental law, see Dupuy (1990) and Dupuy and Viñuales (2018). On its role in lawmaking procedures, see Brunnée (2002, 1–52).

14 For instance, these refer to "climate system" as "the totality of the atmosphere, hydrosphere, biosphere and geosphere and their interactions" (UNFCCC, articles 1 and 3; KP, art. 10d), "systems of the biosphere" (CBB, preamble), and "bio-productive system", composed of "soil, vegetation, other biota, and the ecological and hydrological processes that operate within" it (CD, articles 1(e) and 2). In the Paris Agreement, the earlier vocabulary precision is substituted by the loose terms "ecosystem" (preamble, article 7), "ecological systems and natural resources" (article 7), and "environmental integrity" (articles 4, 6).

15 United Nations document A/RES/1803. Permanent Sovereignty over Natural Resources. December 14, 1962.

16 This phrase can be found in article 18 of the Energy Charter Treaty. This wording was already present in the 1991 European Charter (e.g., Preamble, Titles I and II), but the binding force of the 1994 Energy Charter Treaty is its milestone. The same concept is present in the 2015 International Energy Charter (Preamble, Titles I and II).

17 Viñuales was referring to the relationship between "environmental protection" and "development/ growth". We substituted these with *Energy* and *energy*, respectively.

18 While some argue that the concept of *ecocide* aims at questioning the supremacy of human interests, its legal implementation is currently still anthropocentric. See, for instance, the definition put forward by the Independent Expert Drafting Panel.

19 Indigenous norms are oftentimes cited as a crucial source and/or point of departure for the Rights of Nature. Whereas close to 10% of all initiatives do refer to indigenous norms, such a generalized *ex post* framing is debatable and criticized (at times, by indigenous groups themselves). The first explicit conceptualization was carried out by Stone, who argued about the legal standing of Nature to weigh in the decision of *Sierra Club v. Morton*. While ultimately unsuccessful, his ideas were cited in the dissenting opinion.

20 The Rights of Nature in a legal setting have been elaborated by Berry's "Earth Jurisprudence" (1999) and Cullinan's "Wild Law" (2011). Every statistic on the legal implementation of the Rights of Nature is taken from Putzer et al. (2022).

21 In the late 1850s, both Charles Darwin and Alfred Russel Wallace independently developed a theory of evolution, which showed that humans underwent the same processes of natural selection as every other living being. This discovery presupposes and explains the comparable features and characteristics of all of life. Such a biological leveling clashed with the culturally dominant idea of human exceptionalism, as promoted by, among others, René Descartes. His vision, dubbed Cartesian dualism, presents humans as the only rational beings, standing in contrast to animals, which were seen as mere machines, guided by reflexes without any consciousness.

22 For introductions, see Boyd (2017), Kauffman and Martin (2021), and Tănăsescu (2022).

23 We were taking into account only passed original legislation, thus excluding policies, ongoing initiatives, and those that merely refer to other initiatives (Putzer et al., 2022).

24 Nine out of ten initiatives are taking place within the borders of a single nation-state. Corrigan and Oksanen (2021) call these domestic Rights of Nature.

25 China, in particular, is aware of the importance of such a hydrological cycle that they are experimenting to emulate what they call "sky rivers" (Dockrill 2018).

26 For an in-depth analysis of how this phenomenon unfolds, please see Marengo et al. (2004, 2262).

27 The most comprehensive research done regarding this phenomenon is entitled "The Flying Rivers Project". In it, Nobre calculated the daily basis amount of water released by the Amazon trees to be around 20 billion tonnes (circa 20 trillion liters). See http://riosvoadores.com.br/english.

28 In this volume, for instance, the contributions by J. Keahey et al. share, to some extent, our *Energy* concerns.

29 We look at NGOs in the assumption that they are the most advanced in rainforest protection efforts.

30 "The economic value of these ecosystem services has been estimated at $33 trillion per year, twice the GDP of the United States. . . . The complete answer is complex, but if we had to reduce it to one word, it would be 'balance'" (Rainforest Alliance 2021).

31 For them, all of the remaining Amazon rainforest has "high conservation value". As such, their principles apply to the case study.

32 The effects of the recent growth in deforestation have already impacted the carbon cycle, for a portion of the Amazon Forest now releases more carbon than it absorbs, as shown in the recent study by Qin et al. (2021, 442–8).

33 Kjær sustains that "connectivity norms" facilitate the transfer of legal components between different legal structures and spheres of legality. In this sense, an *Energized* grid of connectivity norms is a major step toward bridging the tangible-intangible gap between the realms of international environmental and energy laws, as this chapter identified. This is precisely what was recommended by the WGII in the Summary for Policymakers of the IPCC's AR6, by calling for an integration of the energy-oriented normative infrastructures "with ecological and social approaches" (Pörtner et al. 2022, 33). See endnote 9.

Bibliography

Berry, Thomas. 1999. *The Great Work: Our Way Into the Future*. New York: Harmony/Bell Tower.

Boyd, David R. 2017. *The Rights of Nature: A Legal Revolution That Could Save the World*. Toronto: ECW Press.

Brasil. 2021. "Conjuntura Recursos Hídricos Brasil, 2021". *Relatório Pleno – Agência Nacional de Águas e Saneamento Básico*. https://relatorio-conjuntura-ana-2021.webflow.io.

Brunnée, Jutta. 2002. "COPing with Consent: Law-Making under Multilateral Environmental Agreements." *Leiden Journal of International Law* 15, no. 1: 1–52.

Chen, D., M. Rojas, B. H. Samset, K. Cobb, A. Diongue Niang, P. Edwards, S. Emori, S. H. Faria et al. 2021. "2021: Framing, Context, and Methods". In *Climate Change 2021: The Physical Science Basis. Contribution of Working Group I to the Sixth Assessment Report of the Intergovernmental Panel on Climate Change*, edited by V., Masson-Delmotte, P. Zhai, A. Pirani, S. L. Connors, C. Péan, S. Berger, N. Caud et al. Cambridge and New York: Cambridge University Press. www.ipcc.ch/assessment-report/ar6.

Corrigan, Daniel P., and Markku Oksanen. 2021. "Rights of Nature: Exploring the Territory." In *Rights of Nature: A Re-examination*. London: Routledge.

Cullinan, Cormac. 2011. *Wild Law: A Manifesto for Earth justice*. White River Junction: Chelsea Green Publishing.

DESA. 2021. *Theme Report on Energy Transition*. United Nations Department of Economic and Social Affairs. Accessed December 22, 2021. www.un.org/sites/un2.un.org/files/2021-twg_2-062321.pdf.

Descola, Philippe. 1996. "Constructing Natures: Symbolic Ecology and Social Practice". In *Nature and Society – Anthropological Perspectives*, edited by Philippe Descola and Gísli Pálsson. 82–102. London: Routledge.

Dockrill, Peter. 2018. *China's 'Sky River' Will Be the Biggest Artificial Rain Experiment Ever*. Accessed December 22, 2021. www.sciencealert.com/how-china-s-sky-river-will-be-the-biggest-artificial-rain-experiment-ever-cloud-seeding.

Dupuy, Pierre-Marie. 1990. "Soft Law and the International Law of the Environment." *Michigan Journal of International Law* 12, no. 2: 420–435.

Dupuy, Pierre-Marie, and Jorge E. Viñuales. 2018. *International Environmental Law*. Second edition. Cambridge: Cambridge University Press.

Frigo, Giovanni. 2017. "Energy Ethics, Homogenization, and Hegemony: A Reflection on the Traditional Energy Paradigm." *Energy Research & Social Science* 30: 7–17. https://doi.org/10.1016/j.erss.2017.06.030.

Frigo, Giovanni. 2018a. "Energy Ethics: a Literature Review." In *Relations Beyond Anthropocentrism—Energy Ethics: Emerging Perspectives in a Time of Transition*. Special issue edited by Giovanni Frigo. Part II, 177–214.

Frigo, Giovanni. 2020. "Beyond the Capitalocene: An Ecocentric Perspective for the Energy Transition". In *Ethics and Politics of Space for the Anthropocene*, edited by Anu Valtonen, Outi Rantala, and Paolo D. Farah, 150–174. Cheltenham: Edward Elgar Publishing.

FSC. 2021. *International Generic Indicators.* Accessed December 22, 2021. https://fsc.org/en/document-centre/documents/resource/262.

Gibbard, Philip L., Andrew M. Bauer, Matthew Edgeworth, William F. Ruddiman, Jacquelyn L. Gill, Dorothy J. Merritts, Stanley C. Finney et al. 2021. "A Practical Solution: The Anthropocene Is a Geological Event, Not a Formal Epoch." *Episodes* 1: 1–9. https://doi.org/10.18814/epiiugs/2021/021029.

Handl, Günther. 2012. "Declaration of the United Nations Conference on the Human Environment (Stockholm Declaration), 1972 and the Rio Declaration on Environment and Development, 1992." *United Nations Audiovisual Library of International Law.* Accessed December 22, 2021. https://legal.un.org/avl/ha/dunche/dunche.html.

Heidegger, Martin. 1977. *The Question Concerning Technology and Other Essays.* Translated by William Lovitt. New York and London: Garlan Publishing Inc.

Kauffman, Craig M., and Pamela L. Martin. 2021. *The Politics of Rights of Nature: Strategies for Building a More Sustainable Future.* Cambridge: MIT Press.

Keeley, Brian L. 2015. "Nonhuman Animal Senses." In *The Oxford Handbook of Philosophy of Perception*, edited by Mohan Matten, 853–870. Oxford: Oxford University Press.

Kjaer, Poul F. 2018. "Constitutionalizing Connectivity: The Constitutional Grid of World Society." *Journal of Law and Society* 45, no. 1: 114–134.

Kopenawa, Davi, and Bruce Albert. 2015. *A queda do céu – Palavras de um xamã Yanomami.* Translated by Beatriz Perrone-Moisés. São Paulo: Companhia das Letras.

Kotzé, Louis J. 2014. "Transboundary Environmental Governance of Biodiversity in the Anthropocene." In *Transboundary Governance of Biodiversity*, edited by Louis J. Kotzé and Thilo Marauhn, 12–33. Leiden-Boston: Brill Nihoff.

Krenak, Aílton. 2019. *Ideias Para Adiar O Fim Do Mundo.* São Paulo: Companhia das Letras.

Kurki, Visa. 2021. "Can Nature Hold Rights? It's Not as Easy as You Think." *Helsinki Legal Studies Research Paper,* no 66. http://dx.doi.org/10.2139/ssrn.3853708.

Luuppala, Linnea. 2021. "Rights-Based Restoration." In *Right of Nature*, edited by Daniel P. Corrigan and Markku Oksanen. London: Routledge. Chapter 8.

Marengo, J., W. Soares, C. Saulo, and M. Cima. 2004. "Climatology of the Low-Level Jet East of the Andes as Derived from the NCEP-NCAR Reanalyses: Characteristics and Temporal Variability." *Journal of Climate* 17, no. 12: 2261–2280.

Meinhold, Roman. 2018. "Human Energy: Philosophical-Anthropological Pressupositions of Anthropogenicc Energy, Movement, and Activity and Their Implications for Well-being." In *Relations Beyond Anthropocentrism – Energy Ethics: Emerging Perspectives in a Time of Transition.* Special issue edited by Giovanni. Frigo: Part II, 287–298.

Millennium Ecosystem Assessment. 2005. *Ecosystems and Human Well-being: Synthesis.* Washington, DC: Island Press.

Mitcham, Carl. 2013. "Energy Constraints". *Science and Engineering Ethics* 19, no. 2: 313–319.

Mitcham, Carl, and Giovanni Frigo. 2018. "Energy Ethics Outside the Box: Carl Mitcham in Conversation with Giovanni Frigo." In *Relations Beyond Anthropocentrism – Energy Ethics: Emerging Perspectives in a Time of Transition.* Special issue edited by Giovanni Frigo: Part II, 301–312. http://dx.doi.org/10.7358/rela-2018-002-mitc.

Müller, Ingo. 2007. *A History of Thermodynamics: The Doctrine of Energy and Entropy.* Berlin: Springer Science & Business Media.

Nobre, Antonio Damato. 2014. *The Future Climate of Amazonia. Scientific Assessment Report.* São Paulo. ARA, CCST-INPE and INPA.

Paterson, John. 2020. "Reconceptualising Energy Security from a Legal Perspective in the Context of Climate Change." In *Routledge Handbook of Energy Law*, edited by Tina Soliman Hunter et al., 58–73. Milton Park: Routledge.

Pearce, Fred. 2019. "Rivers in the Sky." *New Scientist Magazine*: 40–43.

Pörtner, H.-O., D. C. Roberts, E. S. Poloczanska, K. Mintenbeck, M. Tignor, A. Alegría, M. Craig et al. 2022. "Climate Change 2022: Impacts, Adaptation, and Vulnerability". In *Summary for Policymakers: Contribution of Working Group II to the Sixth Assessment Report of the Intergovernmental Panel on Climate Change*, edited by H.-O. Pörtner, and D. C. Roberts, M. Tignor, E. S. Poloczanska, K. Mintenbeck, A. Alegría et al. Cambridge University Press. In Press. https://report.ipcc.ch/ar6wg2.

Putzer, Alex, Tineke Lambooy, Ronald Jeurissen, and Eunsu Kim. 2022. "Putting the Rights of Nature on the Map. A Quantitative Analysis of Rights of Nature Initiatives Across the World." *Journal of Maps* 18: 1–8. https://doi.org/10.1080/17445647.2022.2079432.

Qin, Yuanwei, Xiangming Xiao, Jean-Pierre Wigneron, Philippe Ciais, Martin Brant, Lei Fan, Xiaojun Li et al. 2021. "Carbon Loss from Forest Degradation Exceeds that from Deforestation in the Brazilian Amazon." *Nature Climate Change* 11: 442–448. https://doi.org/10.1038/s41558-021-01026-5.

Rainforest Alliance. 2021. *What Is Sustainable Forestry?* Accessed December 22, 2021. www.rainforest-alliance.org/insights/what-is-sustainable-forestry.

Rescher, Nicholas. 1996. *Process Metaphysics: An Introduction to Process Philosophy*. New York: SUNY Press.

Salati, Eneas, Attilio Dall'Olio, Eiichi Matsui, and Joel R. Gat. 1979. "Recycling of Water in the Amazon Basin: An Isotopic Study." *Water Resources Research* 15, no. 5: 1250–1258. https://doi.org/10.1029/WR015i005p01250.

Shakespeare, William. 1600. *The Merchant of Venice*. Accessed December 22, 2021. http://shakespeare.mit.edu/merchant/full.html.

Stone, Christopher D. 1972. "Should Trees Have Standing? Toward Legal Rights for Natural Objects." *Southern California Law Review* 45: 450–501.

Talus, Kim, and Aalto Pami. 2020. "Energy Transitions and the Law." In *Routledge Handbook of Energy Law*, edited by Tina Soliman Hunter et al., chapter 32. London: Routledge.

Tănăsescu, Mihnea. 2022. *Understanding the Rights of Nature: A Critical Introduction*. Bielefeld: Transcript.

Viñuales, Jorge E. 2013. "The Rise and Fall of Sustainable Development." *RECIEL* 22, no. 1: 3–13.

Viñuales, Jorge E. 2018. *The Organisation of the Anthropocene: In Our Hands?* Leiden: Brill.

Viveiros de Castro, Eduardo. 1996. "Os Pronomes Cosmológicos e o Perspectivismo Ameríndio". *Mana* 2, no. 2: 115–144.10.1590/S0104-93131996000200005.

Zalasiewicz, Jan, Mark Williams, Will Steffen, and Paul Crutzen. 2010. "The New World of the Anthropocene". *Environmental Science & Technology* 44, no. 7: 2228–2231.10.1021/es903118j.

7

PHASES OF THE NET-ZERO ENERGY TRANSITION AND STRATEGIES TO ACHIEVE IT

Jochen Markard and Daniel Rosenbloom

1. Introduction

The low-carbon energy transition has entered a new phase of development as more and more governments, and private firms, are making pledges to reduce their greenhouse gas emissions to net-zero. As of 2021, over 120 countries, which together represent 61% of the global greenhouse gas (GHG) emissions, had announced commitments to reach net-zero by mid-century or soon after (ECIU, 2021). Major emitters such as the United States, China, the European Union, the United Kingdom, Canada, and Japan are also on board.

With the rise of net-zero targets, societal and policy discourse surrounding climate change has shifted fundamentally. Framing the climate challenge in terms of *net-zero* foregrounds the deep changes that will be required. It will not suffice to make improvements in some sectors (e.g., phasing out coal-fired power generation in favor of renewable energy). Instead, *all GHG emissions in all sectors and places* will need to be cut or compensated for. To achieve this, far-reaching and economy-wide changes in production and consumption systems will be necessary, including the transformation of difficult-to-decarbonize industries, such as aviation, shipping, or cement production (Davis et al., 2018; Miller et al., 2021).

This analysis focuses on the energy dimensions of this challenge as energy-related CO_2 emissions account for about three quarters of all emissions, or 34 Gt in 2020 (IEA, 2021). Herein, the associated technological, organizational, political, and institutional change processes toward eliminating or compensating for these emissions will be referred to as the *net-zero energy transition*. It includes multiple interconnected sociotechnical transitions across a broad range of sectors, from electricity to transport, buildings, and industry.

Much of what we know about change processes of this sort has come from the field of transition studies (Markard et al., 2012). This research has uncovered the multiple interacting factors that shape transitions in multiple societal domains, such as transport (Geels, 2005; Geels et al., 2012) and electricity (Foxon, 2013; Verbong and Geels, 2007). However, this body of work has predominantly focused on single sectors or the emergence of one major innovation. Take, for example, the shift from sailing to steam-powered ships or from propeller to jet engines (Geels, 2002, 2006). Key frameworks in transition studies such as the multi-level perspective (Geels, 2019) or technological innovation systems (Markard, 2020) have been developed to explain transformation processes around a focal innovation or focal sector, but they are not

DOI:10.4324/9781003183020-8

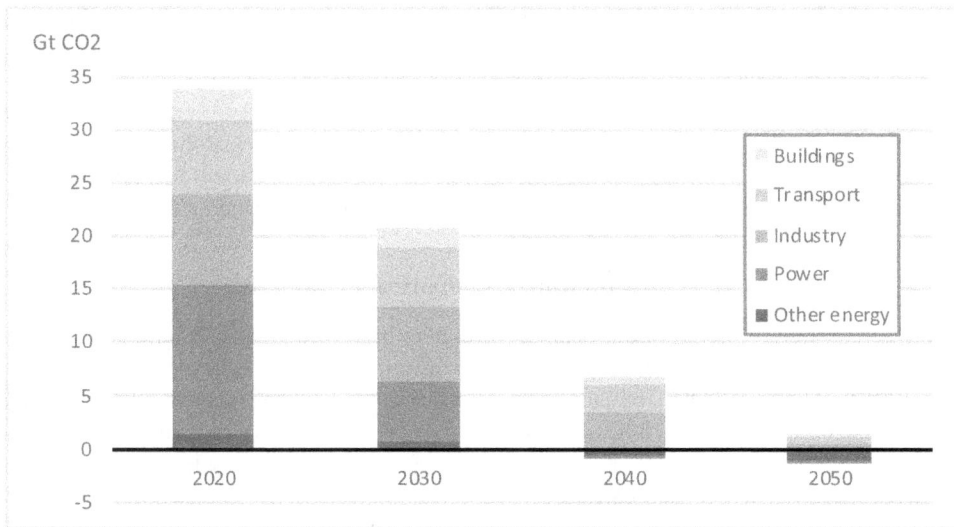

Figure 7.1 Global CO_2 emissions across sectors and projected path to net-zero.
Source: IEA, 2021.

geared toward the complexity of multiple innovations and multiple transforming sectors. While scholars have identified specific deficits in existing approaches (Papachristos et al., 2013; Rosenbloom, 2020) and made first suggestions of how to overcome them (Andersen and Markard, 2020; Geels, 2018; Schot and Kanger, 2018), we still lack an overarching framework to capture transitions as complex as the net-zero energy transition. In particular, we still have a limited understanding of the new policy challenges that arise as the transition enters new phases of development, especially as change processes accelerate, accumulate, and broaden in scope.

In the following, we address some of these shortcomings. We develop a simple schematic model that conceptualizes the net-zero energy transition as a set of interdependent sectoral transitions and distinguishes four qualitatively different phases of development. The distinction of phases is important because each comes with new, additional challenges for policymaking. Next to the phases and policy challenges, we discuss different decarbonization strategies, such as low-carbon electrification, new types of fuels, and lifestyle changes. These are an important element in a framework for the net-zero energy transition because the challenges that lie ahead will, most likely, require a mix of approaches to tackle decarbonization.

Our arguments build on insights from the field of sustainability transitions research, which highlights that large-scale transformations are systemic and non-linear and involve sociopolitical and techno-economic processes (Geels et al., 2017; Köhler et al., 2019; Markard et al., 2012). The phases and strategies we discuss will be of a broad, partly stylized nature to provide general guidance. Our propositions are inspired by future necessities and past developments in some places. Countries such as Germany, the UK, or Denmark, for example, which have traditionally very much relied on electricity from fossil fuels, exhibit certain commonalities in their low-carbon transition of the electricity sector and beyond (Geels et al., 2016; Markard, 2018). We acknowledge that there is no general blueprint of the net-zero energy transition but that it very much depends on specific contextual conditions (e.g., emerging vs. industrialized economies, availability of natural resources, political priorities, industry structure, societal preferences). Nonetheless, we believe that the reflections we share are valuable in situating individual transition processes in terms of an overarching pattern of development we refer to as the net-zero transition.

In Section 2, we briefly introduce the perspective of sustainability transitions. Section 3 represents the core of the paper, in which we discuss the particularities of the net-zero energy transition, including four main phases (3.1), key policy challenges (3.2), and five generic decarbonization strategies (3.3). Section 4 generates lessons and offers concluding remarks.

2. A Sustainability Transition Perspective

Grand sustainability challenges, from climate change to biodiversity loss and societal inequality, are pervasive and seemingly intractable crises that resist conventional policy approaches (Levin et al., 2012). They are highly complex, span-across sectors and places (requiring a high degree of coordination) facing conflicting views and interests of stakeholders (regarding problem definition and potential solutions) and change over time (moving target). In addition, they undermine the very basis of our (co)existence.

Due to these particularities, grand sustainability challenges call for novel approaches in research and policy. Grounded in a growing evidentiary base, the field of sustainability transitions offers such an approach (Köhler et al., 2019). In particular, this body of research (1) demonstrates that fundamental changes in existing sociotechnical systems (around electricity, transportation, and buildings, for instance) are needed to address sustainability challenges and (2) provides lessons for accelerating transition processes (Markard et al., 2020).

Transition studies take *sociotechnical systems* as the primary unit undergoing change. These systems consist of different kinds of elements, including actors (e.g., firms, NGOs), institutions (e.g., policies, societal norms), technologies, and infrastructures (Rip and Kemp, 1998). We look at large-scale sociotechnical systems[1] that provide key societal functions such as energy supply, transportation, or housing (Konrad et al., 2008). In established sociotechnical systems, the various elements have co-evolved over long time spans. Dominant designs, sunk costs and vested interests reinforce a particular "way of doing things" – a phenomenon often referred to as *lock-in* (Berkhout, 2002; Unruh, 2000). As a consequence, mature sociotechnical systems are highly resistant to radical changes. In the energy sectors, we currently see how difficult it is to break up the lock-in around fossil fuels (Trencher et al., 2020).

Nonetheless, systems can and do change. Fundamental changes of sociotechnical systems are referred to as *sociotechnical transitions* (Kemp et al., 2001). Examples of past transitions include the transition from an oceanic shipping system based around sail to one relying on steam-powered vessels (Geels, 2002) or a transport system based on horse-drawn carriages to one anchored around internal combustion automobiles and the use of oil (Geels, 2005). And each of these involved co-evolving changes in infrastructure (ports and roads), business models (more accurate shipping times and automotive maintenance services), practices (recreational travel), rules (speed limits), and so on.

Innovations are central to sociotechnical transitions and as a response to grand challenges. Innovations include new technologies but also non-technical novelties (e.g., changes in policies, practices, or lifestyles). Often, technical and non-technical changes are intertwined. This is why the term *sociotechnical configurations* is often used to embrace innovations (e.g., solar PV) but also their interconnected business models, regulations, or practices (e.g., self-consumption of electricity generated at home). Emerging sociotechnical configurations such as those taking hold around renewable electricity production are expected to be the seeds for the formation of alternative and potentially more sustainable sociotechnical systems (e.g., around low-carbon electricity).

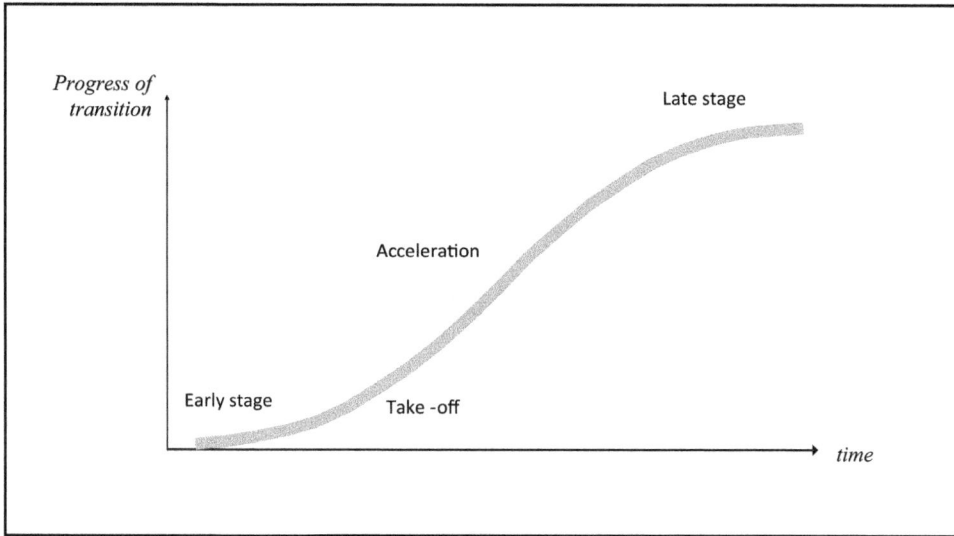

Figure 7.2 Non-linear development and different stages of a transition.

Research highlights that transitions unfold in a non-linear way in the form of an S-curve (Rotmans et al., 2001; Figure 7.2). At an early stage, progress is slow, and changes are minor and confined to niches, in which multiple innovations develop (Kemp et al., 2001; Smith and Raven, 2012). In the take-off phase, one or more innovations start to diffuse. The diffusion stimulates further improvements, which then speed the uptake of the novel innovation. During the acceleration phase, changes accumulate and eventually transform many elements of the sociotechnical system (Markard et al., 2020; McMeekin et al., 2019). Finally, dynamics slow down again as a new, reconfigured system emerges and stabilizes (Geels, 2002). While this general pattern of development has mostly been associated with transitions of single sociotechnical systems, we also see a similar dynamic in transitions that span across several sectors (Schot and Kanger, 2018).

In the context of climate change and other pressing sustainability challenges, transitions involve multiple innovations, and *multiple sociotechnical systems* (Andersen and Markard, 2020; Papachristos et al., 2013; Rosenbloom, 2020; Schot and Kanger, 2018). For example, the transition toward low-carbon e-mobility is not just about electric vehicles but also depends on the availability of low-carbon electricity, innovations in battery technology, or the rollout of a charging infrastructure (Köhler et al., 2020). It involves major changes in the automobile sector, the energy sector, the battery industry, and public infrastructure. Importantly, progress in one technology (e.g., batteries) can stimulate further progress in another (e.g., electric vehicles), and vice versa. Similarly, changes in one system (e.g., transformation of electricity supply based on renewable energy sources) can lead to changes in another (e.g., utility companies investing into vehicle charging stations), and vice versa. Over time, transitions in different sectors may begin to co-evolve, potentially reinforcing each other, leading to the buildup of an even larger transition that involves multiple sociotechnical systems. One part of the larger transition might be the formation of an overarching strategy, or *paradigm*, for decarbonization, similar to what Schot and Kanger (2018) have called meta-rules. One such paradigmatic feature we observe at the moment is the approach to electrify as many energy applications as possible, a strategy that

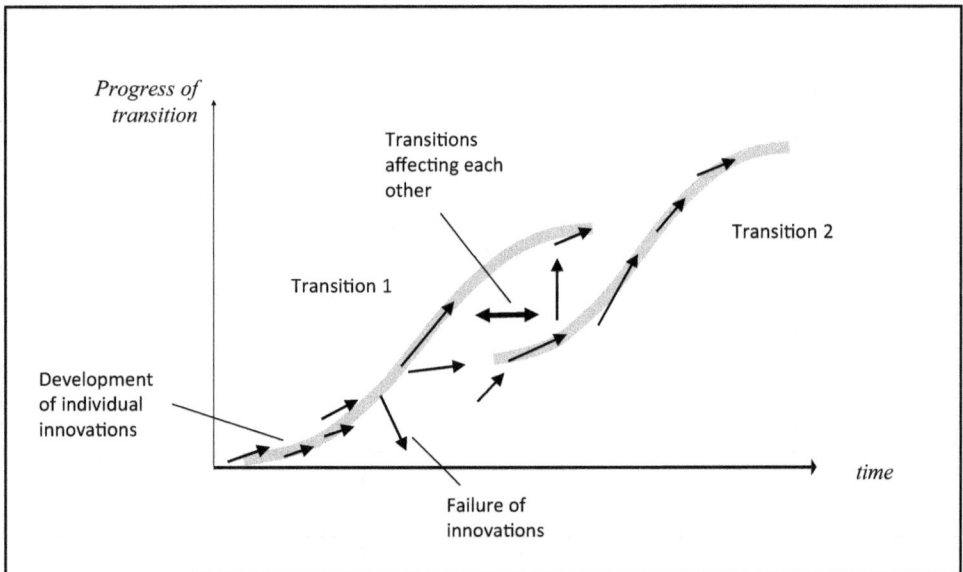

Figure 7.3 Multiple innovations and two transitions affecting each other.

is very much driven by rapidly falling costs of renewable electricity generation. We will come back to this in the following.

Figure 7.3 shows two transitions that interact, each involving a variety of technologies. At a general level, we can distinguish complementary interactions and competition (Markard and Hoffmann, 2016; Rosenbloom, 2019; Sandén and Hillman, 2011). Returning to the electric vehicle example, a transition to low-carbon electricity supply in the electricity sector enables the electrification and simultaneous decarbonization of transport (arguably a complementary interaction). At the same time, electric mobility drives up demand for additional low-carbon electricity, which could rationalize greater investment in the build-out of renewables (complementary) but may also result in tensions among energy end-uses (e.g., mobility and indoor temperature regulation) for limited electricity supply (competitive).

3. The Net-Zero Energy Transition

The net-zero energy transition is a particularly complex and demanding transition. Complex because it entails multiple, partly simultaneous transitions of different sociotechnical systems. The transformation will affect almost every sector and part of society, there will be many interactions, unexpected developments, and setbacks. Demanding because it must be swift, radical, and actively pushed forward by a number of key societal actors with governments taking a central role. While past transitions often unfolded over the course of decades (Araújo, 2017; Grubler et al., 2016; Sovacool, 2016) and there was not much need for coordination between them, we only have about 30 years left to complete the net-zero transition. Different sectoral transitions, therefore, have to unfold in parallel rather than one at a time. Also, it will not suffice to make incremental improvements (e.g., to just increase the fuel-efficiency of cars or to replace coal-fired power generation by natural gas). Instead, radical changes capable of bringing about new net-zero sociotechnical configurations are needed. While some innovations are readily available or in early stages of

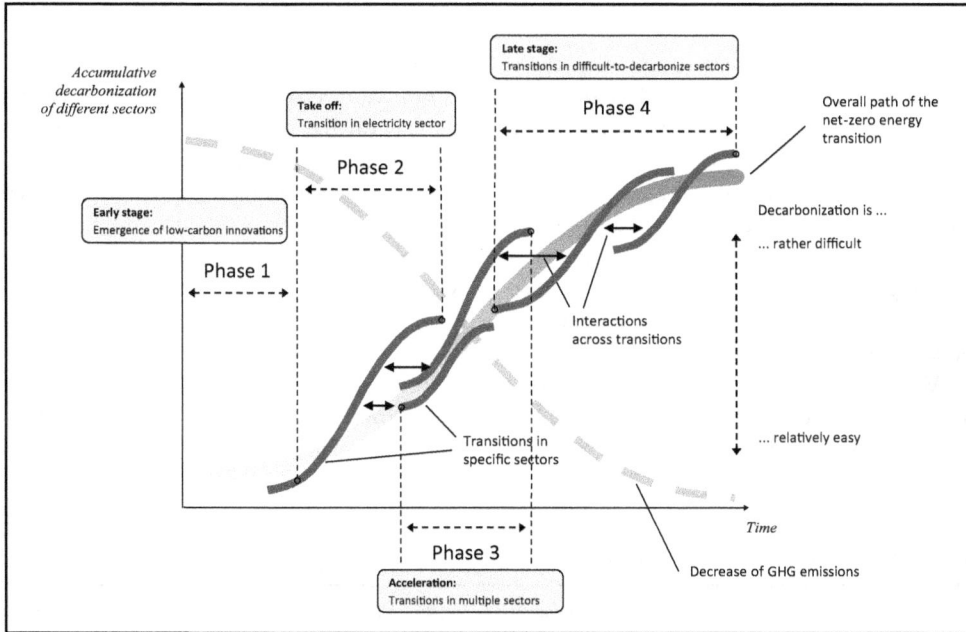

Figure 7.4 Low-carbon innovations and individual transitions build toward net-zero: schematic model of interdependent innovation dynamics and sectoral transition. Dynamics of decline have been omitted for the sake of simplicity.

development[2], others may not even be known yet. As a consequence, the net-zero energy transition inherently involves the challenge of realizing rapid action in the context of deep uncertainty.

We understand the net-zero energy transition as an assemblage of interdependent and potentially complementary transitions that unfold in different sectors (and places) at different times and with different dynamics (Figure 7.4). It is a cumulative process involving multiple sociotechnical transitions (gray S-curves) that together move society toward complete decarbonization. Each individual sociotechnical transition entails the development and diffusion of multiple innovations (small arrows in Figure 7.3, not depicted in Figure 7.4). These innovations may stimulate other innovations or otherwise enable transitions within adjacent sociotechnical system (e.g., a disruptive business model that emerged in one sector is taken up and adapted to another sector). In other words, one transition may benefit from another (e.g., electric vehicles using low-carbon electricity), and vice versa (black arrows in Figure 7.4). We contend that the cumulative nature and directionality of multiple transitions is key for the ultimate realization of net-zero in a timescale needed to avert serious climate disruption (dashed gray S-curve in the background of Figure 7.4).

This is a simple model of a larger transition that spans multiple systems and involves multiple innovations. It highlights that the overall outcome depends on how individual transitions complement each other – that is, how their effects accumulate (e.g., the second and the third transition can use low-carbon electricity so they already start at a higher level) and how innovations, policy approaches, business models, or practices cross over from one transition to another (horizontal black arrows).[3] It also conveys the message that the transition moves from easier targets to more difficult ones, which means that policymaking will need to continually adapt and develop new solution strategies as the transition unfolds.

3.1 Phases to Net-Zero

To capture the increasing complexity of the net-zero energy transition, we distinguish four qualitatively different phases of development. This distinction is based on sectoral scope (which was wide in the first phase, then narrow with a focus on the electricity sector, then widened again) and the relative weight placed on specific decarbonization approaches. Renewable energy electrification, for instance, has emerged as such a dominant approach:[4] it is now used to decarbonize applications in the building and transport sectors, and it might also be increasingly applied elsewhere. The phases are understood as mutually reinforcing and partly overlapping. To be sure, this exercise is necessarily stylized, and we acknowledge that specific places or sectors may show different patterns (e.g., due to geographic, economic, or political particularities).[5]

With each phase, new processes come on top of those that are already underway (Table 7.1). This increasing complexity has repercussions for the political dynamics that unfold and for the resulting policy challenges. Next, we discuss each phase in detail.

Table 7.1 Phases of the Net-Zero Energy Transition

	Phase 1 Early Stage: Emergence of Low-Carbon Innovations	Phase 2 Take-off: Transition in Electricity Sector	Phase 3 Acceleration: Transitions in Multiple Sectors	Phase 4 Late Stage: Transitions in Difficult-to-Decarbonize Sectors
Estimated time interval	~1980–2010	Since ~2010	Since ~2015	Since ~2020
Phase description	Emergence of climate change issue and various low-carbon innovations	New renewable energy sources drive the transition in the electricity sector	Low-carbon electricity diffusion and complementary innovations (e.g., EVs) drive transitions in buildings and transport	Development and diffusion of new innovations (e.g., around hydrogen) to tackle applications low-carbon electricity cannot reach
Sectoral scope	Multiple sectors	Focus on the electricity sector	Multiple sectors	All sectors (also, redefinition of sectoral boundaries)
Dominant decarbonization strategy	None	Low-carbon electrification emerges as a dominant approach for decarbonization	"Electrify everything": cross-sector diffusion of dominant strategy	Alternative strategies needed as limits of electrification are reached

Table 7.1 (Continued)

	Phase 1 Early Stage: Emergence of Low-Carbon Innovations	Phase 2 Take-off: Transition in Electricity Sector	Phase 3 Acceleration: Transitions in Multiple Sectors	Phase 4 Late Stage: Transitions in Difficult-to-Decarbonize Sectors
Key transition processes	Emergence of innovations	Change of entire sociotechnical system, including decline of established configurations	Accelerated diffusion and multi-system interaction	Multiple overlapping transitions and strategies for decarbonization
Political dynamics	New entrants (e.g., start-ups) and incumbent actors developed new technologies and business models. First climate policies introduced. Disagreement over the long-term viability of alternatives and the adequate reaction to climate change	Incumbent actors face increasing pressure; decline of (some) established business models (e.g., coal phase-outs). Policies for deployment and decline in electricity proliferate. Actors struggle over renewable energy deployment, policy ambition, and the pace of the transition	Organizations increasingly active across sectoral boundaries. Policies to drive electrification of buildings and transport proliferate. Opposition to renewables and climate policy continues. Increasing cross-sectoral struggles among incumbents to capture market share	Organizations formulate individual net-zero targets and strategies. Net-zero policy targets rapidly expand. Potential contests surround role of different energy carriers and import dependencies
Policy challenges	Advancing innovations through research, development, and demonstration	Diffusing innovations and promoting the decline of carbon-intensive arrangements	Policy coordination and coherence to ensure alignment across sectors, files, and mandates	Preventative measures, policies to monitor new energy demands

Installed capacity (GW)

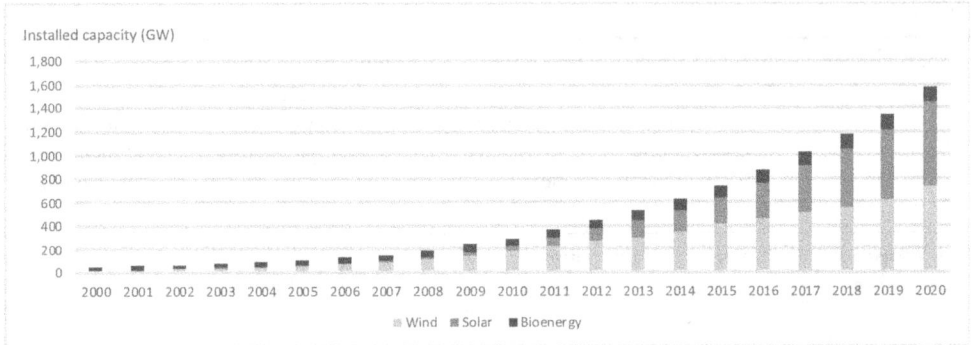

Figure 7.5 Diffusion of new renewable energy technologies (2000–2020).

Source: irena.org.[6]

The *first phase* started around the 1980s and stretched into the 2000s. It was characterized by the emergence of (1) climate change as a topic of societal concern (e.g., foundation of the Intergovernmental Panel on Climate Change (IPCC); first IPCC report in 1990) and (2) a broad range of low-carbon technologies in sectors such as electricity, buildings, and transport. Examples include power generation technologies such as wind or solar PV, new heating technologies such as solar heating or heat pumps, building efficiency technologies, biofuels, and first-generation electric and fuel cell vehicles in transport. The first phase also saw changes in regulations, such as grid access and feed-in regulations for independent power producers, stricter building codes, and fuel efficiency standards for vehicles. Various policies were implemented to support innovation, technology development and diffusion. New actors emerged, including project developers, energy service providers, and technology producers (e.g., for wind and solar). There was contestation around whether the emerging innovations would become viable alternatives and about the gravity of climate change as a policy issue. Altogether, phase 1 did not upset the equilibrium of established systems. Lock-ins remained strong and incumbents successfully weakened attempts at stringent climate policies (Meckling, 2011). Many innovations were not successful (e.g., electric vehicles) or remained confined to niches (e.g., biofuels). However, some innovations matured and started to diffuse more widely. In the electricity sector, especially wind and solar began to present promise.

The *second phase* began around 2010 when wind and solar PV and other renewables started to diffuse rapidly and in many places around the world (Figure 7.5). Solar PV grew from 40 GW of installed capacity in 2010 to more than 700 GW in 2020 (a growth rate of more than 30% p.a.), while onshore wind diffused from around 178 GW to 698 GW in the same period (irena.org). Momentum around these innovations helped promote a shift in focus toward the low-carbon transition of the electricity sector (Nemet, 2019). While there were also improvements in other sectors (e.g., efficiency improvements and biomass use in heating),[7] they were far outpaced by the changes in electricity. In fact, the transition in electricity toward renewables has become an initial case, perhaps a prototype even, of a sectoral transition toward net-zero (Markard, 2018). It includes major changes in power supply (including distributed generation), system balancing, sector-specific policies, and new entrants and new business models (Geels et al., 2016). Incumbent actors have increasingly recognized that this transition threatens their established assets and business practices and have adopted political strategies to slow down the pace of transformation (Lauber and Jacobsson, 2016; Hess, 2014). Intense contests over the legitimacy and future of coal, for example, have ensued in this phase (Isoaho and Markard, 2020; Rosenbloom, 2018;

Stutzer et al., 2021). A new phenomenon of phase 2 is that renewable electrification has started to emerge as a dominant strategy for decarbonization: wind, solar PV, and battery technology have seen dramatic cost and performance improvements (IRENA, 2021), which is why they might become dominant configurations for the net-zero energy transition. The second phase is still ongoing. It will end when electricity is fully decarbonized.

Phase 2 has come with several new challenges, which may also provide lessons for sustainability transitions in other sectors (Markard et al., 2020). These include whole systems change and decline. Once transitions take off and build momentum, important interdependencies among sociotechnical system components are revealed (Andersen et al., 2022; McMeekin et al., 2019). This creates a need for complementary innovations (e.g., grid infrastructures, balancing technologies, demand-side innovations) to ensure the functioning of the whole system (Markard and Hoffmann, 2016). Another key challenge is decline. To effectively respond to climate change, there is not only a need to expand low-carbon innovation but also to phase out existing carbon-intensive technologies, such as coal-fired power generation (Rosenbloom and Rinscheid, 2020; Turnheim and Geels, 2012). Policy needs to cope with increasing resistance (Geels, 2014; Wells and Xenias, 2015) and to address, such as regions or workers on the losing side (Johnstone and Hielscher, 2017).

The *third phase* began around 2015 and is characterized by the acceleration of electrification to decarbonize multiple sectors. The directionality of this phase is underpinned both by progress in low-carbon electricity supply (phase 2) and by complementary innovations that *use electricity* for purposes such as heating (e.g., heat pumps) or transport (e.g., batteries and electric vehicles). Renewable electrification has become a dominant paradigm and the associated decarbonization strategy is to "electrify everything" (Roberts, 2017). Also, complementary innovations are a key feature of the third phase. Phase 3 is in an early stage of development. It started to take off around 2015 with the Paris climate agreement, and it will end when the potential for electrification in a broad variety of sectors has been exhausted.

In the third phase, transition dynamics are becoming more complex as they increasingly span multiple sectors or systems (Rosenbloom, 2020). These multi-sector interactions create new opportunities (e.g., cross-sectoral balancing of electricity supply and demand), but they also come with new challenges (Mäkitie et al., 2020). Norwegian utilities, for example, face a trade-off of whether to build more transmission lines to export hydroelectricity or whether to use their hydroelectric base to drive the domestic decarbonization of transport and industry (Moe et al., 2021). The implications for incumbents are yet to become clear, though there are signs that previous incumbent strategies backing the status quo may no longer be appropriate (e.g., in light of changing consumer preferences, policy shifts, or threats from new entrants). In the building sector, for instance, there is still quite some resistance (e.g., suppliers seeking to frame natural gas as a green fuel), whereas in transport, more and more automakers are announcing ambitious targets or the end of gasoline cars (Bullard, 2021). Rising electrification will also mean the erosion of lock-ins as old assets (e.g., conventional fueling and automotive servicing stations or natural gas distribution networks) are displaced or re-envisioned. Deepening disruptions of this nature are expected to intensify political struggles where new entrants increasingly compete with incumbent service providers. Underlying these tensions are the long-term prospects for specific actor networks, business models, and social practices. And many of these debates have become increasingly polarized against a backdrop of rising right-wing populism.

The *fourth phase* started very recently when more and more jurisdictions such as the European Union introduced net-zero emission targets. The overriding emphasis during this phase is on tackling those sectors and applications that are difficult to decarbonize (Davis et al., 2018) because of the challenges facing electrification (e.g., due to insufficient power storage capacity, remoteness of grid interconnection, or cost-efficiency considerations). Examples include aviation, shipping, or long-distance trucking (Gray et al., 2021). Even though we know relatively little about phase 4 as of today, we expect that it will be characterized by a new set of low-carbon innovations. One such innovation, which is currently receiving considerable attention, is hydrogen, a general-purpose energy carrier, which can be produced in different ways and used for a broad variety of applications (Gray et al., 2021; Staffell et al., 2019). Hydrogen has the potential to play an indispensable role in driving the net-zero transition as it shows promise in replacing fossil fuels as a low-carbon feedstock in industrial processes (e.g., in chemical production and steelmaking) and as a general-purpose energy carrier (e.g., as a fuel for heavy equipment and transport). It can also interconnect with electricity to store surplus power (over long durations even) and balance intermittent renewables, potentially solidifying hydrogen and electricity as dominant energy carriers of the future (Dowling et al., 2020; van Renssen, 2020). However, hydrogen may also prolong the use of fossil fuels if natural gas is used as an input for its production. Tensions surrounding hydrogen or synthetic fuels center around production (e.g., from renewables or natural gas), use (e.g., only for difficult-to-decarbonize applications) or imports, and geopolitical implications. We might also see struggles around competing alternative fuels, such as biofuels and hydrogen, and the reach of electrification, such as full-battery trucks, catenary trucks, and fuel-cell trucks (Mäkitie et al., 2020). And while reframing the challenge in terms of net-zero along with pandemic recovery efforts appear poised to accelerate climate action, it remains too early to tell how political dynamics will unfold.

In addition to predominant supply side innovations, phase 4 might also bring about major changes on the demand side, including more serious engagement with demand reduction and changes in lifestyles (Spaargaren and Cohen, 2020). These could become increasingly necessary as the limits of supply side innovations are encountered (e.g., a reduction in overseas travel becomes a favorable solution to decrease emissions from aviation) or continued concentration on "one side of the equation" necessitates increasingly extreme options (e.g., widescale direct air capture or bioenergy with carbon capture utilization and sequestration) with their own sustainability challenges (e.g., land use conflicts, regional environmental pollution and resource depletion). Phase 4 will end when net-zero technologies and energy carriers become dominant and targets are in sight.

Similar to phase 3, the fourth phase will be characterized by a multi-transition setting. But here the interactions among previously siloed and novel energy systems are becoming even more imperative. On the one hand, electrification is continuing (and will continue) to spread rapidly across sectors and industries where fossil fuels previously dominated. And on the other hand, alternative fuels (partly enmeshed with electricity) are now reaching nearly all parts of the economy, including difficult-to-decarbonize industries. Integrated energy planning is, therefore, essential. And even the notion of distinct energy systems (electricity, transport, hydrogen, and so on) may become obsolete. Decline will also become increasingly critical during this phase to ensure the complete displacement of fossil fuels and their associated end-use technologies. For example, communities relying on natural gas for heating will need to retire distribution networks and end-use technologies so that alternative fuel options can take their place. In contrast, blended heating systems that continue to rely on natural gas with supplemental alternative fuel injection represents a potential dead-end pathway.

3.2 Policy Challenges

As the net-zero energy transition unfolds, policymaking becomes more and more complex. With each phase, new policy challenges come on top of the already existing ones. This evolution is comparable and, to some extent, related to how problem framings, conceptual frameworks, and approaches for innovation policy changed since the 1950 from R&D policy to systemic innovation policy and to transformative innovation policy (Schot and Steinmueller, 2018; Smith et al., 2010).

In the first phase, the focus of policymaking was on innovation. The policy challenge was to stimulate technology development and the formation of early niche markets (Hoogma et al., 2002; Kemp et al., 1998; Schot and Geels, 2008). The emphasis was on research, development, and demonstration of alternative sociotechnical configurations, and scholars and policymakers highlighted the importance of systems approaches such as technological innovation systems (Bergek et al., 2008; Markard et al., 2015). Low-carbon innovations were supported in specific sectors and places (e.g., wind in Denmark or solar in Germany), and policy programs were geared toward knowledge generation, building an early industry base and creating local markets (Bergek and Jacobsson, 2003; Dewald and Truffer, 2011; Garud and Karnøe, 2003). Even though in these early stages, there was comparatively little orientation toward transforming sectors or the entire economy, systemic innovation policy approaches will remain important in the future. The IEA states, "Reaching net zero by 2050 requires . . . widespread use of technologies that are not on the market yet" (IEA, 2021; p. 15), and estimates a need for public funding for R&D and demonstration of around $90 billion until 2030 (ibid.).

In the second phase, there was an increasing shift from innovation to deployment policies to further improve existing technologies (e.g., realizing learning effects and reaping economies of scale) and help accelerate their diffusion (e.g., through technology-specific deployment policies) (Hoppmann et al., 2013; Jacobsson and Bergek, 2011; Sandén and Azar, 2005). As a complementary element in the policy mix, decline policies, particularly phase-outs (e.g., targeting coal-fired power, incandescent bulbs or internal combustion engines), are garnering increasing attention as a way to promote decarbonization (Markard and Rosenbloom, 2020; Rosenbloom and Rinscheid, 2020). These raise new issues for policy development and implementation given their implications for targeted industries and associated communities. As part of this, affected regions (around coal mining, for instance) might need to be supported or compensated (Johnstone and Hielscher, 2017; Rinscheid et al., 2021). Overall, the policy focus is shifting toward embracing an entire sector or whole system (McMeekin et al., 2019), thereby including a broad range of system elements (e.g., transmission grids or storage technologies in electricity) and tackling potential bottlenecks that may arise during the transition (Andersen et al., 2022; Haley, 2018).

A key policy challenge for phase 3 is cross-sectoral policy coordination (Markard et al., 2020) and multi-system interaction (Rosenbloom, 2020). This requires close exchange across governmental departments that were designed to operate individually. In earlier phases, it was possible to design decarbonization policies for one sector largely independent of those for other sectors. In the third phase, however, more and more interactions emerge across sectors, and there might also be competition between different sectors (e.g., for low-carbon electricity). Consider, for instance, how integrated energy system planning is increasingly being called for and attempted, bringing together previously siloed files on electricity, transport, buildings, and novel energy carriers (hydrogen). Phase 3 will also see an increasing need for investments into new infrastructures (e.g., public charging for electric vehicles), and public funding will play a key role in this (IEA, 2021). In fact, we might see a much stronger role of the state, and the contestation that comes along with it (Roberts and Geels, 2019), emerging in the course of phase 3.

To facilitate phase 4, continued support for innovations, sectoral transition policies and cross-sectoral policies will be needed. However, a new challenge emerges around novel technologies around hydrogen (van Renssen, 2020) or carbon capture and storage (Martin-Roberts et al.,

2021) and the necessity to build up new infrastructures, e.g., to transport hydrogen or captured CO_2 at large scales and at a high pace (IEA, 2021). This will require massive investments, both from public and private parties (ibid.), and it will also come with a risk of potential (new) lock-ins once these infrastructures are in place.

A very different policy challenge but also very crucial for the fourth phase will be to address major changes in lifestyles and consumption patterns. This is a domain where policymaking has seen challenges in the past and, therefore, has mostly shied away from (Saujot et al., 2020). An additional policy challenge is to develop (and diffuse) a new set of low-carbon innovations (and the associated infrastructures) to tackle difficult-to-decarbonize industries and applications. Due to a rapidly growing demand for low-carbon energy carriers, we can also expect increasing political conflicts, which require careful deliberations. Conflicts may occur in energy generation (e.g., land use conflicts in the case of wind energy) and energy use (e.g., whether some applications or sectors have priority access to clean hydrogen).

3.3 Strategies for Net-Zero

There are many different approaches, or strategies, of how to decarbonize existing sectors and practices. One such strategy, which we already mentioned due to its current dominance in several sectors, centers around low-carbon electrification. In the following, we discuss net-zero strategies in some more detail. We group the large number of options into strategies and sub-strategies. Each strategy centers around a guiding principle (e.g., reducing demand, substituting fuels, changing lifestyles) and encourages the development of various sociotechnical configurations that correspond to this principle (e.g., renewable power sources and electric vehicles go together with low-carbon electrification). In the following paragraphs, we identify five main strategies. Three strategies have already been deployed to varying degrees: energy efficiency, low-carbon electrification, and low-carbon fuels. Two additional strategies are in an early stage of development: negative emissions and a broader category with more radical demand-side approaches that have remained largely unexplored so far (untapped approaches).

The *first strategy* focuses on reducing energy demand through energy efficiency measures. This strategy has a long history and encompasses many mature innovations (e.g., LED lighting). Conservation (e.g., adjusting temperature settings for heating and cooling) and technological efficiency (e.g., building materials with higher insulation values) represent the dominant ways in which to realize this strategy. Energy efficiency measures can target all applications and sectors, though specific technologies (e.g., LED light bulbs) are often needed. The success of this strategy strongly depends on user involvement. That is, actors need to adopt more efficient technologies and change certain practices to realize gains. And of equal importance, the efficiency strategy cannot drive net-zero in isolation as there will always be some energy demand left (residual demand). While the efficiency strategy is relevant across all phases, it has not been given significant weight so far and largely remains underutilized.

The *second strategy* concentrates on low-carbon electrification and has emerged as a dominant approach to reach net-zero. It is composed of two sub-strategies: (2.1) substitution of carbon-intensive power generation and (2.2) extension of electrification to use cases that were traditionally served by fossil fuels (e.g., in buildings and transport). Both include a variety of sociotechnical configurations, the choice of which depends on techno-economic performance improvements and sociopolitical conditions (e.g., a preference for or against nuclear for low-carbon electricity). These choices will have important implications for the operation of electricity systems. A greater reliance on distributed and intermittent technologies, for instance, may necessitate the emergence of a more flexible power system, extensive interconnections, and/

Table 7.2 Five Main Strategies for Decarbonization Toward Net-Zero[8]

	Principle(s)	*Examples*	*Maturity*	*Particularities*	*Limitations*
Energy efficiency (1)	Conservation (1.1)	Switch off unused loads	Many mature technologies and services	Hinges on energy users	Decarbonization gap: residual energy demand cannot be addressed by efficiency
	Technological efficiency (1.2)	LED light bulbs; higher-fuel-economy gas engines		Variety of approaches and technologies tailored to different use cases	
Low-carbon electrification (2)	Substitute carbon-intensive with low-carbon electricity generation (2.1)	Renewable power generation; nuclear; storage; flexibility technologies	Mostly mature (but still much potential for further diffusion)	Variable renewables require more flexible grid	Land use; minerals and resource needs; many sectors (e.g., heavy transport, shipping, air travel) currently defy electrification
	Electrify additional use cases (2.2)	Electric vehicles; air-source heat pumps	Diffusing rapidly in transport, more slowly in buildings	Significant expansion of low-carbon electricity needed	
Low-carbon fuels (3)	Alternative fuels based on biomass (3.1)	Ethanol and biodiesel, methane and biogas, biomass power	Mature	Potential dead-end pathway	Secondary GHG emissions, land use, monocultures, soil degradation, limited feedstock, etc.
	Direct hydrogen use or synthetic fuels based on hydrogen (3.2)	Hydrogen, ammonia	Very early stage	Produced through a variety of approaches, including fossil fuels Geopolitical implications	Conversion losses, high costs, potential fossil fuel lock-ins
Negative emissions (4)	CCS-based technologies (4.1)	Bioenergy and CCS, direct air capture and CCS	Early stage/ some experience	Societal acceptance of CCS might be a barrier	Energy-intensive, requires renewable electricity

(Continued)

Table 7.2 (Continued)

	Principle(s)	Examples	Maturity	Particularities	Limitations
	Other negative emissions technologies (4.2)	Enhanced weathering, carbon sequestration in soil, reforestation and afforestation, change in agricultural practices	Early stage	Many nature-based strategies	Processes not (yet) fully understood; not commercially viable; politically challenging
Untapped demand-side approaches (5)	Major changes in lifestyles and work practices (5.1)	Car-free lifestyle; restrictions to regional air travel; telework and conferencing	Very early stage to emergence (e.g., some car-free communities)	Strong user and industry resistance, low political feasibility	Politically and administratively challenging; many other institutional changes required (e.g., standards, building codes)
	Radical substitution of carbon-intensive products (5.2)	Replace cement or steel with plant-based structural materials	Very early stage	Requires radical changes in business models	Hinges on viable alternatives that generate sufficient interest and resources
	Restrict emergence of new carbon-intensive practices (5.3)	SUVs; space tourism; outdoor heating	Very early stage	Requires societal debate about needs and values	Institutional capacity building needed to shift from firefighting to fireproofing

or energy storage capabilities. This may also involve different ways to value energy and even flexible loads. While the second strategy has successfully been implemented, there are challenges in other sustainability dimensions such as land use or depletion of critical minerals (van den Bergh et al., 2015; Sovacool et al., 2020) and for its use in difficult-to-decarbonize sectors (Davis et al., 2018).

A *third strategy* focuses on low-carbon fuels. It reflects a growing recognition of the limitations of low-carbon electrification and seeks to substitute carbon-intensive fuels with low-carbon fuels. There are two sub-strategies. The first (3.1) is based on biofuels, which are produced from biomass such as wood, energy crops, or organic waste. Biofuels gained some traction in the 2000, showing promise in realizing GHG reductions and generating new income streams in agriculture. However, unwanted effects have also become visible – such as land use conflicts for food production, feedstock limitations, energy crop monocultures, or previously overlooked GHG emissions from soils or nitrous oxide (Scharlemann and Laurance, 2008). Today, biofuels remain in niche applications and have yet to play a major role in the net-zero challenge. And given their limitations, biofuels may turn out as a dead-end pathway (Hillman and Sandén, 2008). Indeed, there is a risk that biofuels will help extend the life of internal combustion vehicles, for instance, by incrementally reducing emissions but failing to yield the radical change needed to reach net-zero.

The second sub-strategy (3.2) involves the use of hydrogen and hydrogen-based fuels. Like biofuels, hydrogen received considerable attention in the early 2000s. There were high expectations around fuel cell technology in transport (Budde et al., 2012). These expectations, however, have yet to be borne out and interest largely faded until recently. Net-zero targets have created renewed interest in hydrogen as an energy carrier that might be used to address many difficult-to-decarbonize sectors (e.g., shipping, aviation, and others). Currently, many hydrogen-based technologies are in an early stage of development (except hydrogen vehicles), and it heavily depends on political support. Major limitations include energy losses in production and conversion and high costs. There are also risks that hydrogen – when produced from natural gas – locks in continued reliance on fossil fuels. The contribution of hydrogen to net-zero will, therefore, hinge on the uncertain prospects of carbon capture utilization and sequestration or a widespread expansion of low-carbon electricity to produce hydrogen through electrolysis.

A *fourth strategy* centers around carbon dioxide removal and "negative emission technologies" (Haszeldine et al., 2018). It is based on the idea that CO_2 needs to actively be removed from the atmosphere because other strategies are not sufficient to reach net-zero goals in time. There are two sub-strategies: approaches based on carbon capture and storage (CCS) (4.1) and other negative emissions strategies (4.2).

CCS technology is already used to capture and store CO_2 from fossil fuels before it is released into the atmosphere through various methods (pre-combustion, post-combustion, oxyfuel) and can be paired with other technologies to achieve negative emissions (Martin-Roberts et al., 2021). For example, when pairing bioenergy with CCS (BECCS), biomass can act as a carbon sink while it grows, and when converted to bioenergy, CCS comes in handy. A second approach can be through direct air capture (DAC) and CCS. DAC technology is one of few that can remove CO_2 directly from the atmosphere. CCS has been under development for more than 25 years, and so far, it has fallen short compared both to earlier expectations (e.g., in terms of costs and performance) and future necessities (e.g., in terms of required capacities) (Martin-Roberts et al., 2021).

Other negative emissions strategies include reforestation (planting trees where forests used to be) and afforestation (planting trees where there were previously none), methods like enhanced weathering or ocean alkalinity enhancement (spreading fine basalt rock over large land/sea areas to accelerate chemical weathering reactions leading to CO_2 removal and storage in, e.g., solid carbonate minerals, Bach et al., 2019), and carbon sequestration by changing agricultural practices (away from conventional farming toward, e.g., agroforestry). These strategies are in an early stage of development and many of the underlying processes and implications are not yet fully understood.

While strategies 1–4 may contribute significantly to the pathway to net-zero, it is not yet possible to discern all approaches needed for full decarbonization by the middle or end of century. Acknowledging this, we offer three illustrative examples of relatively untapped approaches that can complement more established solutions (e.g., should limitations be encountered around electrification or low-carbon fuels). These are major lifestyle changes (5.1), radical substitution of products (5.2), and restricting new carbon-intensive practices (5.3).

While strategy 1 already targets demand-side changes, much more profound changes in lifestyles may be required to reduce energy demand to levels compatible with net-zero emissions. Consider, for instance, car-free housing projects (supported by new approaches in city planning), telework and virtual conferencing to reduce travel, low-carbon diets, local vacations, or extensive changes in consumption patterns (e.g., a shift away from fast fashion). A second approach (5.2) centers around the substitution of carbon-intensive, difficult-to-decarbonize products, such as cement, steel, or aluminum, with alternative materials (e.g., wood or other plant-based building materials). A third approach (5.3) focuses on preventing or downscaling unwanted developments. The ongoing diffusion of pickup trucks and SUVs, for example, significantly increases energy demand in transportation and sets even higher hurdles for electrifying transport. There are also entirely new energy uses emerging. Think of space tourism: if current pilot projects scale up and – eventually – diffuse more widely (as air travel did some decades ago), they will make achieving net-zero emission targets all the more difficult.

So far, these strategies have remained largely unexplored or restricted to very small niches. However, it may become increasingly vital to mobilize a broader range of approaches and to gain experience with policies to support radical change. Efforts to advance these strategies can expect to encounter serious political challenges such as when the German Green Party suggested a "veggie day" in staff canteens or a restriction of suburban single-family homes (The Guardian, 2021). In Switzerland, a policy initiative to ban SUVs and off-road vehicles was rejected by the parliament and government some years ago, but stricter emission regulations were implemented instead (Swiss Confederation, 2008). While political feasibility may still be a major hurdle for these more far-reaching approaches to decarbonization, pressure to widen the repertoire of decarbonization approaches is likely to increase.

4. Conclusions

Reducing GHG emissions to net-zero is one of the biggest challenges of our times. Building on research in the field of sustainability transitions, we have proposed four major phases of change. After an early period with the emergence of a first generation of low-carbon technologies (phase 1), we are currently witnessing the acceleration of the low-carbon transition in the electricity sector (phase 2). This will provide the basis for electrifying further energy uses and sectors such as buildings and automobility, which have thus far relied on fossil fuels (phase 3). While low-carbon electrification has become the dominant strategy for decarbonization, it is also clear that this will not suffice. There are difficult-to-decarbonize sectors (aviation, shipping, cement), for which it is not a feasible option.

Engagement with climate change over the past few decades suggests that complementary strategies are beginning to move us toward net-zero, including energy efficiency, low-carbon electrification, and low-carbon fuels. However, there is also room to consider more radical strategies, from major changes in lifestyles to the substitution of difficult-to-decarbonize materials (e.g., steel, cement) and a restriction of new, carbon-intensive practices (e.g., space tourism).

To be sure, the phases and strategies we suggested are not meant to capture all possible pathways. Indeed, different developments are possible, such as rapid advancements in negative emission technologies (engineered or natural), to either draw down historical emissions or offset ongoing emissions. Net-zero pathways can also be expected to vary considerably across contexts (e.g., a jurisdiction with an already near-decarbonized electric power system versus another with mixed or carbon-intensive electricity). And there may be unwanted developments that could lead society along dead-end pathways (with unsuccessful strategies) or dystopian futures in which society fails to reach net-zero emissions.

Taken together, our analysis points to several principles for sustainability transition policy aimed at the pursuit of net-zero. First, it is important to recognize climate change as a grand challenge and systems problem, which requires fundamental transformations instead of piecemeal adaptation (Rosenbloom et al., 2020). For this, public policies have to foster radical innovations, which include new sociotechnical configurations (e.g., around hydrogen) and non-technical innovations (e.g., low-carbon lifestyles). Second, given the inherent uncertainty in the transition toward net-zero, it will be vital to carefully monitor ongoing developments, in particular to avoid dead-end pathways (e.g., biofuels, natural gas use), which create further delays and sunk investments (Hillman and Sandén, 2008; Meadowcroft et al., 2019). Third, policymaking has to manage conflicting interests (e.g., by compensating losers), forge strong coalitions in favor of change (Hess, 2019; Meckling et al., 2015), and carefully attend to those political strategies that seek to undermine stringent decarbonization (e.g., the new politics of delay, Lamb et al., 2020). Fourth, it will be necessary to develop and apply a mix of policies that reflect the particularities of different phases of development and sector- and country-specific conditions (Meckling et al., 2017; Rosenbloom et al., 2020). It will not be possible to devise one-size-fits-all approaches. Finally, policymaking has to prepare to tackle the inconvenient policy areas, including established but unsustainable consumption practices and lifestyles (Spaargaren and Cohen, 2020). These will evoke much opposition and resistance, but they may prove to play a key role in our endeavor to successfully navigate the net-zero energy transition.

Acknowledgments

We thank Allan Dahl Anderson, Leticia Müller, Dianne Hondeborg, and our colleagues in the Pathways project and at the ETH Group for Sustainability and Technology for their comments and inputs to earlier versions of this manuscript. Jochen Markard acknowledges funding from the Norwegian Research Council (Conflicting Transition Pathways for Deep Decarbonization, Grant number 295062/E20) and from the Swiss Federal Office of Energy (SWEET program, PATHFNDR consortium). Daniel Rosenbloom would like to acknowledge the financial support of the Social Sciences and Humanities Research Council of Canada.

Notes

1 For specific systems, we also use the term *sector* (e.g., transport sector) because it is very common.
2 The IEA estimates that more than 40% of the low-carbon technologies needed by 2050 are only in a developmental stage at the moment (IEA, 2021. Net Zero by 2050: A Roadmap for the Global Energy Sector. International Energy Agency, Paris, p. 224.).
3 Of course, there might also be unwanted developments, such as new, energy-intensive products or sectors emerging, such as SUVs or space tourism (Markard et al., 2021). Neglected developments undermining sustainability transitions. Environmental Innovation and Societal Transitions.).
4 The IEA estimates that by 2050, two-thirds of total energy supply will be from renewable energy sources and that low-carbon electricity will be a key pillar in the pathway to net-zero: "Ever-cheaper renewable energy technologies give electricity the edge in the race to zero" (IEA 2021, p. 14).
5 Also, countries such as Norway, Switzerland, and Iceland, whose electricity systems were traditionally based on low-carbon electricity, will have no need for a low-carbon transition in electricity (phase 2).
6 www.irena.org/Statistics/View-Data-by-Topic/Capacity-and-Generation/Statistics-Time-Series, accessed Jan-16, 2022.
7 There were hardly any in transportation.
8 Note that we do not include carbon dioxide removal (CDM) approaches here as they are not strictly related to the energy transition. Bioenergy plus carbon capture and storage (BECCS) is another technological option (perhaps to be subsumed under 3.1) we do not address here in any further detail.

References

Andersen, A.D., Markard, J., 2020. Multi-technology interaction in socio-technical transitions: How recent dynamics in HVDC technology can inform transition theories. *Technological Forecasting and Social Change* 151, 119802.

Andersen, A.D., Markard, J., Bauknecht, D., Korpås, M., 2022. *Architectural Change in Accelerating Transitions—Insights from the German energy transition.* Trondheim, Norway.

Araújo, K.M., 2017. *Low Carbon Energy Transitions: Turning Points in National Policy and Innovation.* Oxford University Press: New York.

Bach, L.T., Gill, S.J., Rickaby, R.E.M., Gore, S., Renforth, P., 2019. CO2 removal with enhanced weathering and ocean alkalinity enhancement: Potential risks and co-benefits for marine pelagic ecosystems. *Frontiers in Climate* 1.

Bergek, A., Jacobsson, S., 2003. The Emergence of a Growth Industry: A Comparative Analysis of the German, Dutch and Swedish Wind Turbine Industries, in: Metcalfe, J.S., Cantner, U. (Eds.), *Change, Transformation and Development.* Physica-Verlag (Springer), Heidelberg, pp. 197–228.

Bergek, A., Jacobsson, S., Carlsson, B., Lindmark, S., Rickne, A., 2008. Analyzing the Functional dynamics of technological innovation systems: A scheme of analysis. *Research Policy* 37, 407–429.

Berkhout, F., 2002. Technological regimes, path dependency and the environment. *Global Environmental Change* 12, 1–4.

Budde, B., Alkemade, F., Weber, K.M., 2012. Expectations as a key to understanding actor strategies in the field of fuel cell and hydrogen vehicles. *Technological Forecasting and Social Change* 79, 1072–1083.

Bullard, N., 2021. *Automakers are Investing in EVs Like They Mean It.* Bloomberg Green.

Davis, S.J., Lewis, N.S., Shaner, M., Aggarwal, S., Arent, D., Azevedo, I.L., Benson, S.M., Bradley, T., Brouwer, J., Chiang, Y.-M., Clack, C.T.M., Cohen, A., Doig, S., Edmonds, J., Fennell, P., Field, C.B., Hannegan, B., Hodge, B.-M., Hoffert, M.I., Ingersoll, E., Jaramillo, P., Lackner, K.S., Mach, K.J., Mastrandrea, M., Ogden, J., Peterson, P.F., Sanchez, D.L., Sperling, D., Stagner, J., Trancik, J.E., Yang, C.-J., Caldeira, K., 2018. Net-zero emissions energy systems. *Science* 360, 1419.

Dewald, U., Truffer, B., 2011. Market formation in technological innovation systems—diffusion of photovoltaic applications in Germany. *Industry & Innovation* 18, 285–300.

Dowling, J.A., Rinaldi, K.Z., Ruggles, T.H., Davis, S.J., Yuan, M., Tong, F., Lewis, N.S., Caldeira, K., 2020. Role of Long-duration energy storage in variable renewable electricity systems. *Joule* 4, 1907–1928.

ECIU, 2021. *Taking Stock: A Global Assessment of Net Zero Targets.* Oxford, p. 30.

Foxon, T.J., 2013. Transition pathways for a UK low carbon electricity future. *Energy Policy* 52, 10–24.

Garud, R., Karnøe, P., 2003. Bricolage versus breakthrough: Distributed and embedded agency in technology entrepreneurship. *Research Policy* 32, 277–300.

Geels, F., Kemp, R., Dudley, G., Lyons, G., 2012. *Automobility in transition?: A Socio-Technical Analysis of Sustainable Transport.* Routledge: New York.

Geels, F.W., 2002. Technological transitions as evolutionary reconfiguration processes: A multi-level perspective and a case-study. *Research Policy* 31, 1257–1274.

Geels, F.W., 2005. The dynamics of transitions in socio-technical systems: A multi-level analysis of the transition pathway from horse-drawn carriages to Automobiles (1860–1930). *Technology Analysis & Strategic Management* 17, 445–476.

Geels, F.W., 2006. Co-evolutionary and multi-level dynamics in transitions: The transformation of aviation systems and the shift from propeller to turbojet (1930–1970). *Technovation* 26, 999–1016.

Geels, F.W., 2014. Regime Resistance against Low-Carbon Transitions: Introducing Politics and Power into the Multi-Level Perspective. *Theory, Culture & Society* 31, 21–40.

Geels, F.W., 2018. Low-carbon transition via system reconfiguration? a socio-technical whole system analysis of passenger mobility in great Britain (1990–2016). *Energy Research & Social Science* 46, 86–102.

Geels, F.W., 2019. Socio-Technical transitions to sustainability: A review of criticisms and elaborations of the multi-level perspective. *Current Opinion in Environmental Sustainability* 39, 187–201.

Geels, F.W., Kern, F., Fuchs, G., Hinderer, N., Kungl, G., Mylan, J., Neukirch, M., Wassermann, S., 2016. The enactment of socio-technical transition pathways: A reformulated typology and a comparative multi-level analysis of the German and UK low-carbon electricity transitions (1990–2014). *Research Policy* 45, 896–913.

Geels, F.W., Sovacool, B.K., Schwanen, T., Sorrell, S., 2017. Sociotechnical transitions for deep decarbonization. *Science* 357, 1242–1244.

Gray, N., McDonagh, S., O'Shea, R., Smyth, B., Murphy, J.D., 2021. Decarbonising ships, planes and trucks: An analysis of suitable low-carbon fuels for the maritime, aviation and haulage sectors. *Advances in Applied Energy* 1, 100008.

Grubler, A., Wilson, C., Nemet, G., 2016. Apples, Oranges, and consistent comparisons of the temporal dynamics of energy transitions. *Energy Research & Social Science* 22, 18–25.

Haley, B., 2018. Integrating structural tensions into technological innovation systems analysis: Application to the case of transmission interconnections and renewable electricity in nova scotia, Canada. *Research Policy* 47, 1147–1160.

Haszeldine, R.S., Flude, S., Johnson, G., Scott, V., 2018. Negative emissions technologies and carbon capture and storage to achieve the Paris agreement commitments. *Philosophical Transactions of the Royal Society A: Mathematical, Physical and Engineering Sciences* 376, 20160447.

Hess, D.J., 2014. Sustainability transitions: A political coalition perspective. *Research Policy* 43, 278–283.

Hess, D.J., 2019. Cooler coalitions for a warmer planet: A review of political strategies for accelerating energy transitions. *Energy Research & Social Science* 57, 101246.

Hillman, K.M., Sandén, B.A., 2008. Exploring technology paths: The development of alternative transport fuels in Sweden 2007–2020. *Technological Forecasting and Social Change* 75, 1279–1302.

Hoogma, R., Kemp, R., Schot, J., Truffer, B., 2002. *Experimenting for Sustainable Transport. The approach of Strategic Niche Management.* Spon Press: London/New York.

Hoppmann, J., Peters, M., Schneider, M., Hoffmann, V.H., 2013. The two faces of market support – how deployment policies affect technological exploration and exploitation in the solar photovoltaic industry. *Research Policy* 42, 989–1003.

IEA, 2021. *Net Zero by 2050: A Roadmap for the Global Energy Sector.* International Energy Agency: Paris, p. 224.

IRENA, 2021. *Renewable Power Generation Costs in 2020.* Abu Dhabi.

Isoaho, K., Markard, J., 2020. The politics of technology decline: Discursive struggles over Coal phase-out in the UK. *Review of Policy Research* 37, 342–368.

Jacobsson, S., Bergek, A., 2011. Innovation system analyses and sustainability transitions: Contributions and suggestions for research. *Environmental Innovation and Societal Transitions* 1, 41–57.

Johnstone, P., Hielscher, S., 2017. Phasing out coal, sustaining coal communities? living with technological decline in sustainability pathways. *The Extractive Industries and Society* 4, 457–461.

Kemp, R., Rip, A., Schot, J., 2001. *Constructing Transition Paths through the Management of niches, Path Dependence and Creation.* Lawrence Erlbaum, pp. 269–299.

Kemp, R., Schot, J., Hoogma, R., 1998. Regime shifts to sustainability through processes of niche formation: The approach of strategic niche management. *Technology Analysis and Strategic Management* 10, 175–195.

Köhler, J., Geels, F.W., Kern, F., Markard, J., Wieczorek, A., Alkemade, F., Avelino, F., Bergek, A., Boons, F., Fünfschilling, L., Hess, D., Holtz, G., Hyysalo, S., Jenkins, K., Kivimaa, P., Martiskainen, M., McMeekin, A., Mühlemeier, M.S., Nykvist, B., Onsongo, E., Pel, B., Raven, R., Rohracher, H., Sandén, B., Schot, J., Sovacool, B., Turnheim, B., Welch, D., Wells, P., 2019. An Agenda for sustainability transitions research: State of the art and future directions. *Environmental Innovation and Societal Transitions* 31, 1–32.

Köhler, J., Turnheim, B., Hodson, M., 2020. Low carbon transitions pathways in mobility: Applying the MLP in a combined case study and simulation bridging analysis of passenger transport in the Netherlands. *Technological Forecasting and Social Change* 151, 119314.

Konrad, K., Truffer, B., Voß, J.-P., 2008. Multi-regime dynamics in the analysis of sectoral transformation potentials: Evidence from German utility sectors. *Journal of Cleaner Production* 16, 1190–1202.

Lamb, W.F., Mattioli, G., Levi, S., Roberts, J.T., Capstick, S., Creutzig, F., Minx, J.C., Müller-Hansen, F., Culhane, T., Steinberger, J.K., 2020. Discourses of climate delay. *Global Sustainability* 3.

Lauber, V., Jacobsson, S., 2016. The politics and economics of constructing, contesting and restricting socio-political space for renewables—The German renewable energy act. *Environmental Innovation and Societal Transitions* 18, 147–163.

Levin, K., Cashore, B., Bernstein, S., Auld, G., 2012. Overcoming the tragedy of super wicked problems: Constraining our future selves to Ameliorate global climate change. *Policy Sciences* 45, 123–152.

Mäkitie, T., Hanson, J., Steen, M., Andersen, A.D., 2020. *The Sectoral Interdependencies of Low-Carbon Innovations in Sustainability Transitions.* Trondheim, Norway.

Markard, J., 2018. The next phase of the energy transition and its implications for research and policy. *Nature Energy* 3, 628–633.

Markard, J., 2020. The Life Cycle of Technological Innovation Systems. *Technological Forecasting and Social Change* 153, 119407.

Markard, J., Geels, F.W., Raven, R.P.J.M., 2020. Challenges in the acceleration of sustainability transitions. *Environmental Research Letters* 15, 081001.

Markard, J., Hekkert, M., Jacobsson, S., 2015. The technological innovation systems framework: Response to six criticisms. *Environmental Innovation and Societal Transitions* 16, 76–86.

Markard, J., Hoffmann, V.H., 2016. Analysis of complementarities: Framework and examples from the energy transition. *Technological Forecasting and Social Change* 111, 63–75.

Markard, J., Raven, R., Truffer, B., 2012. Sustainability transitions: An emerging field of research and its prospects. *Research Policy* 41, 955–967.

Markard, J., Rosenbloom, D., 2020. A tale of two crises: COVID-19 and climate. *Sustainability: Science, Practice and Policy* 16, 53–60.

Markard, J., van Lente, H., Wells, P., Yap, X.-S., 2021. Neglected developments undermining sustainability transitions. *Environmental Innovation and Societal Transitions* 41, 39–41.

Martin-Roberts, E., Scott, V., Flude, S., Johnson, G., Haszeldine, R.S., Gilfillan, S., 2021. Carbon capture and storage at the end of a lost decade. *One Earth* 4, 1569–1584.

McMeekin, A., Geels, F.W., Hodson, M., 2019. Mapping the winds of whole system reconfiguration: Analysing low-carbon transformations across production, distribution and consumption in the UK electricity system (1990–2016). *Research Policy* 48, 1216–1231.

Meadowcroft, J., Layzell, D., Mousseau, N., 2019. *The Transition Accelerator: Building Pathways to a Sustainable Future.* The Transition Accelerator, Ottawa, p. 65.

Meckling, J., 2011. The Globalization of Carbon Trading: Transnational Business Coalitions in Climate Politics. *Global Environmental Politics* 11, 26–50.

Meckling, J., Kelsey, N., Biber, E., Zysman, J., 2015. Winning coalitions for climate policy. *Science* 349, 1170–1171.

Meckling, J., Sterner, T., Wagner, G., 2017. Policy sequencing toward decarbonization. *Nature Energy* 2, 918–922.

Miller, S.A., Habert, G., Myers, R.J., Harvey, J.T., 2021. Achieving net zero greenhouse gas emissions in the cement industry via value chain mitigation strategies. *One Earth* 4, 1398–1411.

Moe, E., Hansen, S.T., Kjær, E.H., 2021. Why Norway as a Green Battery for Europe Is Still to Happen, and Probably Will Not, in: Midford, P., Moe, E. (Eds.), *New Challenges and Solutions for Renewable Energy: Japan, East Asia and Northern Europe.* Palgrave Macmillan, Cham, pp. 281–317.

Nemet, G., 2019. *How Solar Energy Became Cheap: A Model for Low Carbon Innovation.* Earthscan, London.

Papachristos, G., Sofianos, A., Adamides, E., 2013. System interactions in socio-technical transitions: Extending the multi-level perspective. *Environmental Innovation and Societal Transitions* 7, 53–69.

Rinscheid, A., Rosenbloom, D., Markard, J., Turnheim, B., 2021. From terminating to transforming: The role of phase-out in sustainability transitions. *Environmental Innovation and Societal Transitions* 41.

Rip, A., Kemp, R., 1998. Technological Change, in: Rayner, S., Malone, E.L. (Eds.), *Human Choice and Climate Change—Resources and Technology.* Battelle Press, Columbus, pp. 327–399.

Roberts, C., Geels, F.W., 2019. Conditions for politically accelerated transitions: Historical institutionalism, the multi-level perspective, and two historical case studies in transport and agriculture. *Technological Forecasting and Social Change* 140, 221–240.

Roberts, D., 2017. *The Key to Tackling Climate Change: Electrify Everything.* Vox, Vox.

Rosenbloom, D., 2018. Framing low-carbon pathways: A discursive analysis of contending storylines surrounding the phase-out of coal-fired power in Ontario. *Environmental Innovation and Societal Transitions* 27, 129–145.

Rosenbloom, D., 2019. A clash of socio-technical systems: Exploring actor interactions around electrification and electricity trade in unfolding low-carbon pathways for Ontario. *Energy Research and Social Science* 49, 219–232.

Rosenbloom, D., 2020. Engaging with multi-system interactions in sustainability transitions: A comment on the transitions research agenda. *Environmental Innovation and Societal Transitions* 34, 336–340.

Rosenbloom, D., Markard, J., Geels, F.W., Fuenfschilling, L., 2020. Why carbon pricing is not sufficient to mitigate climate change – and how "sustainability transition policy" can help. *Proceedings of the National Academy of Sciences* 117, 8664–8668.

Rosenbloom, D., Rinscheid, A., 2020. Deliberate decline: An emerging frontier for the study and practice of decarbonization. *Wiley Interdisciplinary Reviews: Climate Change* 11, e669.

Rotmans, J., Kemp, R., van Asselt, M., 2001. More evolution than revolution. Transition management in public policy. *Foresight* 3, 15–31.

Sandén, B.A., Azar, C., 2005. Near-term technology policies for long-term climate targets – economy wide versus technology specific approaches. *Energy Policy* 33, 1557–1576.

Sandén, B.A., Hillman, K.M., 2011. A framework for analysis of multi-mode interaction among technologies with examples from the history of alternative transport fuels in Sweden. *Research Policy* 40, 403–414.

Saujot, M., Le Gallic, T., Waisman, H., 2020. Lifestyle changes in mitigation pathways: Policy and scientific insights. *Environmental Research Letters* 16, 015005.

Scharlemann, J.P.W., Laurance, W.F., 2008. How green are biofuels? *Science* 319, 43–44.

Schot, J., Geels, F.W., 2008. Strategic niche management and sustainable innovation journeys: Theory, findings, research agenda, and policy. *Technology Analysis & Strategic Management* 20, 537–554.

Schot, J., Kanger, L., 2018. Deep transitions: Emergence, acceleration, stabilization and directionality. *Research Policy* 47, 1045–1059.

Schot, J., Steinmueller, W.E., 2018. Three frames for innovation policy: R&D, systems of innovation and transformative change. *Research Policy* 47, 1554–1567.

Smith, A., Raven, R., 2012. What is protective space? Reconsidering niches in transitions to sustainability. *Research Policy* 41, 1025–1036.

Smith, A., Voß, J.-P., Grin, J., 2010. Innovation studies and sustainability transitions: The allure of the multi-level perspective and its challenges. *Research Policy* 39, 435–448.

Sovacool, B.K., 2016. How long will it take? Conceptualizing the temporal dynamics of energy transitions. *Energy Research & Social Science* 13, 202–215.

Sovacool, B.K., Ali, S.H., Bazilian, M., Radley, B., Nemery, B., Okatz, J., Mulvaney, D., 2020. Sustainable minerals and metals for a low-carbon future. *Science* 367, 30–33.

Spaargaren, G., Cohen, M.J., 2020. Greening Lifecycles and Lifestyles: Sociotechnical Innovations in Consumption and Production as Core Concerns of Ecological Modernisation Theory, in: *The Ecological Modernisation Reader*. Routledge, pp. 257–274.

Staffell, I., Scamman, D., Velazquez Abad, A., Balcombe, P., Dodds, P.E., Ekins, P., Shah, N., Ward, K.R., 2019. The role of hydrogen and fuel cells in the global energy system. *Energy & Environmental Science* 12, 463–491.

Stutzer, R., Rinscheid, A., Oliveira, T.D., Mendes Loureiro, P., Kachi, A., Duygan, M., 2021. Black coal, thin ice: The discursive legitimisation of Australian coal in the age of climate change. *Humanities & Social Sciences Communications* 8, 178.

Swiss Confederation, 2008. Bundesrat lehnt "Offroader-Initiative" ab, Berne.

The Guardian, 2021. *German Green MP Hints at Ban on New Urban Single-Family Houses, by Kate Connolly*, London.

Trencher, G., Rinscheid, A., Duygan, M., Truong, N., Asuka, J., 2020. Revisiting carbon lock-in in energy systems: Explaining the perpetuation of coal power in Japan. *Energy Research & Social Science* 69, 101770.

Turnheim, B., Geels, F.W., 2012. Regime destabilisation as the flipside of energy transitions: Lessons from the history of the British coal industry (1913–1997). *Energy Policy* 50, 35–49.

Unruh, G.C., 2000. Understanding carbon lock-in. *Energy Policy* 28, 817–830.

van den Bergh, J., Folke, C., Polasky, S., Scheffer, M., Steffen, W., 2015. What if solar energy becomes really cheap? A thought experiment on environmental problem shifting. *Current Opinion in Environmental Sustainability* 14, 170–179.

van Renssen, S., 2020. The hydrogen solution? *Nature Climate Change* 10, 799–801.

Verbong, G., Geels, F., 2007. The ongoing energy transition: Lessons from a socio-technical, multi-level analysis of the Dutch electricity system (1960–2004). *Energy Policy* 35, 1025–1037.

Wells, P., Xenias, D., 2015. From 'freedom of the open road' to 'cocooning': Understanding resistance to change in personal private automobility. *Environmental Innovation and Societal Transitions* 16, 106–119.

PART II

Systems and Geographic Dynamics

8

TRANSFORMING OIL REFINING IN A DEEPLY DECARBONIZED WORLD

Madhav Acharya, Deborah Gordon,
Uday Turaga, and Swati Singh

A. Introduction

The electric utilities sector has been in transition for close to two decades now, as fossil-fuel-based power plants are replaced by solar and wind farms supplemented with energy storage capacity (IRENA 2021). The unsubsidized cost of renewable power is now in many cases lower than that provided by fossil fuels (Bloomberg 2021). Future electricity demand is projected to grow as the transportation and industrial sectors are electrified, replacing the combustion of fossil fuels that generate greenhouse gases (GHGs). The impact of this transition on the oil industry, however, has received too little attention beyond recent projections of "peak oil" demand, wherein reduced consumption of select petroleum products are assumed to necessarily result in lower overall oil demand (Sears 2020) (Thune, Engen and Wicken 2019).

Transforming the highly integrated and widely traded petroleum sector, however, will require significant planning and investment. Unlike power, which primarily moves within national borders, refining is part of a highly globalized oil and gas sector, with tens of millions of barrels of various crude oils and petroleum products moving around the world every day from producing regions to refineries, trading hubs, and markets. Furthermore, unlike electricity, oil provides not just energy value but also serves as the feedstock for several building blocks to produce most of the goods we use in our daily lives. Consumers are largely oblivious to the presence of petroleum-derived molecules in their clothes, homes, cars, and medicines; the list is truly exhaustive.

The cost of crude oil, the quantity of proven reserves, and the regional ease of access currently dominate oil industry decision-making. This simplified picture, however, overlooks refining – the most essential link in the oil supply chain, where a barrel of otherwise directly unusable crude oil is converted into the vast array of intermediates and finished products that end up in all sectors of the economy. Petrochemicals make plastics and fertilizers; transportation fuels power cars, planes, and ships; diesel runs industrial equipment; lubricants allow machines to run and improve their longevity; tar prevents roofs from leaking; and asphalt is used worldwide for paving roads, bridges, and airports.

A petroleum refinery enables these applications, as a complex system that is designed to maximize the conversion of the crude oil raw material into finished products. Unlike the production of electricity, which is the direct conversion of an energy source into electrons, refineries have

DOI:10.4324/9781003183020-10

127

to accommodate the individual economic drivers of many different products simultaneously to comprehensively balance inputs and outputs on a mass basis to be profitable. The commodity nature of most of these products means that refiners must accept the prices set by the market, even if it means operating at a loss.

The complex nature of refining along with its large throughput volumes also means that even minor shifts in product demands have the potential to cause disruptions. A typical refinery produces around 60–70% various transportation fuels and 10–20% assorted chemicals, with the rest spread across a wide array of petroleum products, each with differing end-use demand patterns and economics (Gary, Handwerk and Kaiser 2007). Historically, refineries have primarily satisfied the enormous appetite of the global transportation system that relies almost exclusively on petroleum fuels, including gasoline (petrol), diesel, jet, and marine bunker fuel. While there are differences in demand and petroleum product value propositions across countries and regions (the annual trans-Atlantic exchange of diesel and gasoline is well known) (Gagniere, Pucci and Rousseau 2013), continued long-term demand for various oil products is creating an incentive for new refineries to be built.

The prospect of a dramatic shift toward electrified transportation is only just starting to appear on the oil industry's radar. Most projections to date have concluded that oil demand will remain high through 2050. If electric vehicles (EVs) do rapidly globalize, however, the drop in gasoline and diesel demand could introduce new and unusual industry dynamics. A preview of this is being played out as a result of COVID-19, which will be addressed later in this chapter. Over the past year, refiners have had to undertake some unusual steps, such as blending jet fuel into marine fuel or shifting to renewable diesel production (which benefits from government subsidies). Many refineries have also been shut down due to an inability to profitably convert crude oil into finished products and uncertainty around long-term demand (Reuters 2020). And while this kind of rationalization has been a part of the industry for decades, the speed with which this has played out from the start of a global pandemic in early 2020 is potentially a sign of things to come.

B. The Value Proposition of Petroleum

Products of petroleum refining are typically associated with fueling our mobility – cars, trucks, trains, ships, and planes. While it is true that about two-thirds of refined oil ends up flowing to the transportation sector (EIA 2021a), the remaining petroleum products have an invisible presence in just about every other economic sphere – industrial, commercial, agricultural, and residential. Even the flag bearers of renewable power, wind turbines and solar panels, require oil in their manufacturing, installation, and upkeep. And the millions of new EVs envisioned to reduce demand for oil contain many plastic parts (such as seats, dashboards, and bumpers) (Schirber 2009) and rubber tires (Tiremart 2016) that are all derived from oil. So while an EV does not need to be fueled by oil, its existence is still dependent on oil and many petroleum by-products (Bloomberg 2017).

Very little oil in each barrel currently goes to waste; processes have evolved over time to find a use for almost every drop of crude oil fed to a refinery. This high degree of system integration is how petroleum has embedded itself in society over the course of millennia (Forbes 1958). As early as 6,000 years ago, asphalt was quarried and used for mortar between building stones, caulking ships, waterproofing baths, and as an adhesive for weapon handles. In China, by AD 1, bamboo pipes were carrying oil and gas into homes for heat and light (Freudenberg 2020). The invention of the steam engine in the mid-1700s sparked the use of coal in transportation. But coal would soon be replaced as the dominant source of energy by petroleum when the first modern oil well was drilled in 1859 (Solomon and Krishna 2011).

The gush of crude oil that followed, inspired new ways to realize the greatest value of this commodity, which in its raw form is essentially useless to consumers. But once converted to a slew of petroleum products through the refining process, it is invaluable. Today, petroleum is the largest industry in the world, bigger than all other raw material markets combined (Desjardins 2016).

Petroleum Products Substitution

The prospects for replacing petroleum products with low-carbon substitutes is a high priority. Yet studies to date have largely ignored the fact that refining is a large-volume, multi-product business whose profitability critically depends on the ability to derive maximum value from every molecule of hydrocarbon feedstock that is processed. Lower-carbon solutions are easier for some petroleum products but are extremely difficult for others.

This oversight is understandable. The "electrification of everything" is a concept that is simple to grasp. It involves a single end-use product – electrons. Thus, there is really no difference between a coal and gas plant, wind turbine or a solar panel; they all fabricate electrons to produce the exact same thing – electricity – albeit with different environmental and climate impacts.

While the earliest refineries merely distilled oil into kerosene for lighting, today's refineries are very different than power plants and make an electricity generating facility look elementary in comparison. Refineries involve a myriad of chemical processes that separate, catalyze, reconfigure, and treat hydrocarbons into various petroleum products that flow throughout the economy – to the transport, industrial, commercial, residential, and agriculture sectors and even the power sector itself. These highly integrated facilities must transform the diverse molecules in each barrel of heterogeneous crude oils into a large slate of products. The historic success of the oil industry has stemmed as much from its ability to continue to create new markets for each of its products as it has from its ability to constantly find new reserves of oil.

There is, however, the unforeseen twist in the decarbonization debate: the issue of "embedded oil" – namely, the fossil fuels used to manufacture and maintain everyday products and services. This is also true for the production of steel, cement, and aluminum, not to mention mining, processing, and distributing rare metals and minerals that go into various gadgets that have become an indispensable part of our lives. This does not stop at fossil based technologies either. Even a wind turbine requires large amounts of fossil fuels in its construction and operation (Smil 2016). While EVs can eliminate large portions of refined gasoline, the use of renewable energy to charge those vehicles is predicated on fossil fuel availability for the entire life cycle of the wind turbine or solar panel.

Refining Through the Ages

An oil refinery is a gargantuan series of processing equipment that turn every single molecule that it is fed into something of value. The earliest oil refineries were basically kettles for boiling oil into kerosene, a cleaner-burning lamp oil that replaced whale oil and animal fat (Solomon, A.L., Waddams and Carruthers 2018). As new chemical engineering processes were developed using heat, pressure, and reaction-enhancing catalysts, scientists found ways to convert oil into transportation fuels and industrial feedstocks. When Henry Ford launched the world's first mass-produced car, the oil industry responded by producing gasoline and diesel fuels.

The advent of World War II saw a step change in refining innovations with the introduction of high-octane fuels and synthetic rubber that were crucial for an Allied victory. In the West, the growth in refining capacity was spurred on in peace time by the explosive growth in car ownership and the advent of plastic goods. The uptake of petroleum products followed globally, fueled by policies that linked economic growth to energy use (Johnstone and McLeish 2020).

The oil supply shocks of the 1970s arrested global growth in oil consumption due to high prices (Our World in data 2021). In response, public policies were enacted to conserve oil and invest in R&D for new energy sources (National Museum of American History 2021). Refineries followed suit; new refinery construction in the US came to a halt between 1977 (Marathon's Garyville, Louisiana refinery) and 2022 when a new MEG refinery is slated to start construction in North Dakota (Blackmon 2020).

The 1980s and 1990s brought improvements in process efficiency and ongoing compliance with increasingly stringent regulations on transportation fuel compositions. This drove smaller refineries out of business, while larger, more-integrated plants stayed profitable and survived (EIA 1991). More recently, deep cost-cutting and a maturing industry have slowed the pace of new refining process innovations considerably.

The run-up in oil prices in the early 2000s reignited US interest in refining, with billions of dollars invested to enable more complex processing of heavier oils into marketable petroleum products (Oil and Gas Journal 2000). By 2015, refiners were revisiting their decisions so they could also handle the growth in light tight oils and condensates produced from hydraulic fracturing of shales in the US that required different processing parameters (EIA 2019).

Refining is the lynchpin of the oil and gas value chain. Over 600 refineries worldwide – 20% of which are in the US – are the intermediary between unusable crude oil and usable petroleum products (Oil and Gas Journal 2020). Over the past century, a handful of transportation fuels (gasoline, diesel, and jet fuel) have driven refinery output. But highly volatile fuel demand, global trade patterns, and prices have put pressure on refinery profits. By 2017, nearly 90 million barrels of oil were being refined (OPEC 2021a) – roughly double the volume in 1970 and ten times more than in 1950 (US Federal Trade Commission 1975). While the OECD countries used to be the largest consumers of petroleum products, today non-OECD countries are consuming nearly as much gasoline, diesel, jet fuel, naphtha, and petrochemical feedstocks (OPEC 2021a). And the other refined oil products made from a barrel of oil – such as propane, petrochemical feedstocks, petroleum coke, lubricants, waxes, solvents, and sulfur – are in high demand in non-OECD countries (Gordon and Acharya 2018). This is, in part, a reflection of global shifts in petroleum product consumption to Asia and other developing regions. Building a refinery costs billions of dollars in capital (Blackmon 2020); they are also expensive to operate. Fixed costs include manpower, maintenance, insurance, and depreciation. Variable costs include crude oil, chemicals, catalysts, hydrogen, natural gas, and purchased energy, such as electricity.

Over the past several decades, numerous refineries have closed, while others have increased their throughputs. While the global refining sector has withstood short-term market fluctuations in the past, a reliable rebound produced growth even if cyclical over the past several decades. Historically, depressed demand has resulted in lower prices that drove a slow but steady recovery, eventually turning yesterday's recessions into economic booms (S&P Global Platts 2020) (Global Data Energy 2021).

Meanwhile, non-OECD countries have realized the importance of setting up their own processing to drive greater energy security, local employment, and economic returns. India's Reliance Industries has the world's largest single refinery complex with capacity exceeding 1.2 million barrels per day (bpd) (Reliance Industries 2021). With modern equipment and systems, it has a strong competitive advantage relative to older, aging OECD facilities. There are also significant investments being made in refining in OPEC countries from Saudi Arabia to Iran, where a total of 8 million bpd of refining capacity is being planned (OPEC 2021a). It does not appear that non-OECD countries that currently consume 52% of the world's petroleum products are preparing to shift away from oil consumption (S&P Global Platts 2021). So even if the world is warming, the non-OECD refining sector could follow a different path than the OECD in the future.

Demand Disruption due to COVID-19

Through 2020 and into 2021, the world experienced a period of turmoil due to the COVID-19 pandemic, which brought significant changes to the global economy. While the struggle continues to vaccinate the global population and enable societies to return to some degree of normalcy, the disruptions to conventional oil and gas consumption patterns turned the refining industry on its head. This served as a wake-up call for a sector that has long relied on robust demand for all its products to remain profitable. With abrupt shifts in economic activity driven by lockdowns, the industry was left facing huge inventories of products that had no market outlet. The oil industry was also affected by the battle between Russia and the Saudis on oil production levels in the spring of 2020.

For example, with air travel at a virtual standstill, the market for jet fuel evaporated. To offload inventories, refiners priced jet fuel so low that there were reports of it being used as a blend stock for marine fuel. The precipitous drop in diesel demand forced India's refiners to import gasoline (Chakraborty and Pandya 2020); US renewable diesel production was projected to skyrocket 500% by 2024 (EIA 2021a); Shell announced it would sell off a majority of its refineries (Argus Media 2019); and the French oil company, Total, announced it would replace oil refining with clean fuel and plastics plants (Total Energies 2020). Perhaps the most alarming was a headline announcing that the world's "$2-trillion-a-year refining industry was in crisis" (Bloomberg 2020).

In pre-COVID-19 times, these reports might have been dismissed as alarmist fiction, and while there are signs of demand recovery as economies start to reopen after widespread availability of vaccines, the century-old oil refining sector is ripe for a complete overhaul, one that must be reconciled technically, economically, and environmentally.

As refiners take stock of their operations and forecast post-pandemic petroleum demand, they are struggling to determine if there is value in continuing with business as usual. As "price takers", refiners are not free to set product prices because their merchandise is sold on global markets. Moreover, refiners also have little to no influence on the price of their inputs (principally crude oil) which also trades on global markets.

To maintain their profit margins, refiners have historically increased their operating efficiencies, producing more in outputs with the same or lower input. This economic strategy only works if steady demand persists for large volumes of petroleum products. In 2019, global refining capacity was projected to grow significantly in the developing world, with companies like Aramco planning over $100 billion in investments.

Refining Post-Pandemic

The impact from COVID-19 itself immediately resulted in a reduction of some 10 million bpd in global oil consumption (IEA 2020). While this demand destruction is not expected to be permanent, new oil demand projections are increasingly uncertain varying from a marked rise to downward slides. The International Energy Agency (IEA) forecasts that after recovering from COVID-19, base-case world oil demand could rise to over 104 million bpd by 2040 (ibid.). The Organization for Petroleum Exporting Companies (OPEC), on the other hand, sees oil demand seeking higher levels to over 108 million bpd by 2045 (OPEC 2021b). Major oil companies, however, suggest that global oil demand will not recover the cuts seen during the pandemic, and levels could fall by as much as 50% by 2040 in the face of stronger climate action (BP 2020).

Further, McKinsey suggests that even if global oil demand were to level off by 2030, amid certain circumstances, it could accelerate a major energy transition. For this to happen, they suggest that globally: electric vehicles account for 78% of new car sales and 49% of new truck sales; 23% of plastics would be made from recycled feedstock; biofuels, natural gas, and electricity would account for 42% of the fuel mix; and heating, cooking, and industry would be largely electrified (McKinsey and Company 2019). These are tall orders, but if they do not happen in a calculated and carefully coordinated fashion, global refining could be in for a major curveball.

If predicting crude oil demand after a pandemic is challenging, it is even tougher to forecast individual petroleum product consumption trends in an accelerated energy transition of decarbonization. Clearly, future decisions about personal work commutes, air travel, digital commerce, industrial activities, and other economic activities are far from certain. So, too, are EV penetration levels, plastics recycling rates, and the electrification of "everything".

New energy modeling scenarios are recognizing the scale and scope of this challenge. For example, the IEA's new Net-Zero Emissions (NZE) by 2050 scenario concludes that refining throughputs will drop considerably with significant changes in product demand due to rapid electrification of the vehicle fleet (IEA 2021). Even so, demand for other refined products, such as petrochemicals, will increase. The IEA's modeling shows that the traditional 55% of oil demand from gasoline and diesel will drop to less than 15% in 2050 but the share of petrochemical feedstocks (ibid.). This means that ethane, naphtha, and liquefied petroleum gas (LPG) could triple from 20% to almost 60% in 2050. Collectively, IEA is forecasting refinery runs to fall by 85% between 2020 and 2050, leading to refinery closures, especially for those that not able to concentrate primarily on petrochemical operations or the production of biofuels (ibid.).

Even if we could predict petroleum product demand, not all refineries would be equally affected. Shifting overall oil consumption does not necessarily amount to equal or equitable petroleum product increases or decreases across the board. Depending on their location and configuration, refineries can face bottlenecks that affect their throughput, storage capacity, delivery schedules, and future utilization rates. Low utilization, however, can have safety, economic, and operational impacts. This poses a unique challenge for refining, which cannot simply be turned off by pushing a button. Refinery units are complex, and it takes time to close lines, empty vessels, prevent equipment damage, and safeguard against corrosion. Furthermore, restarting units is often responsible for major accidents as failure to follow every single start-up step diligently could be disastrous. The time required to permit and build a new facility can often stretch to a decade, during which time the demand could well be very different from what was originally assumed.

Thus, an industry that has thrived for over a century finds itself confronted by an existential threat, one that will force it to adapt if it wants to maintain its relevance. A handful of high-volume petroleum products derived from oil could eventually be replaced through electrification. However, there are many refined products that will be quite hard to replace. Despite their substantially lower demand by volume compared to transportation fuels, these other petroleum products create significant economic value and form the foundation of future renewable technologies, as will be discussed further. The challenge, then, is to figure out how to pivot the refining sector from its traditional large-volume, multi-product mode to a new smaller-volume, limited-product mode, something which it has not done for over a century. How would that take place?

Given the complexity of conventional refining and ongoing product demands, it is no small task to remake refining. The magnitude of the shift can be seen using the case of lubricants, for example. The global production of lubricant oil base stocks is currently around 1 million bpd – about 1% of global petroleum product supply (Meridian Energy Group 2019). A major portion of lubricants demand is in automotive applications. However, the shift from internal combustion engines to EVs would likely result in a significant permanent reduction in automotive lubricant demand.

Where might lubricants supplied by refineries go? One option is the power sector, to wind turbines. But how do current and future volumes compare? A quick calculation shows that, even with the most optimistic projections for the growth of the wind sector, the total lubricant requirement would be fractions of a percent of current automotive demand. This product shift poses a real problem for the refining industry, which has built up its capacity to supply the automotive sector over decades. Churning out 1 million bpd of lubricants that are no longer in demand is the equivalent of abandoning a portion of each refinery because it no longer serves a marketable purpose. Hence, examples of "stranded assets" in the oil industry will not just be the underground oil reserves. But will likely also be steel aboveground.

This simple example of lubricants underscores the need for refineries to think about marked changes in future demand for a slew of petroleum products. There are two possible future pathways forward, and each must also achieve the overarching goal of decarbonization to meet net-zero GHG targets.

C. Remaking Refining

Refining is a massive global enterprise with many variables at play: economics, technology, geopolitics, demographics, and public policy. Of these, economics is perhaps the most influential, at least in the near term. If refiners cannot fulfill their market mandate to cost-effectively supply nearly 100 million barrels of diverse petroleum products every day to consumers worldwide, this could lead to shortages and price spikes.

It is all too easy to paint an overly simplified climate policy picture where lower gasoline demand reduces oil throughput in lock step. BP projects that if 100 million EVs replace gasoline-powered cars by 2035 (about a 70-fold increase from today), this would result in only a 1.2 million bpd reduction in oil demand worldwide (BP 2019). Taking embedded oil into account, these oil savings may be even smaller. The reason that a rise in EVs does not translate into a commensurate drop in oil consumption is that despite gasoline displacement, global refineries must continue to produce the rest of their product slate for which there are not necessarily low-carbon alternatives currently available at scale.[1]

While financial gains producing and trading oil can be enormous, refining has historically been a far less lucrative business, running on slim margins. When times are good, a single refinery with an optimal crude flow and strategic location can be a cash cow. But an unplanned downtime of even a few days (as happened in the aftermath of Hurricane Harvey in summer 2017) can quickly turn a refinery into a money pit. Just like airlines, which only make money when their planes are in the air, refineries make money only when they are operating at full capacity making a wide range of petroleum products.

But what happens if a growing portion of a refinery's products are no longer required by society? Figure 8.1 shows the major products of refining; in principle, large portions of the product slate including petrochemical feedstock (C1–C5), gasoline, and diesel can be replaced in the short term through alternate routes. However, jet fuel, lubricants, waxes, and asphalt cannot be replaced easily, and developing low-carbon substitutes will be technologically very challenging.

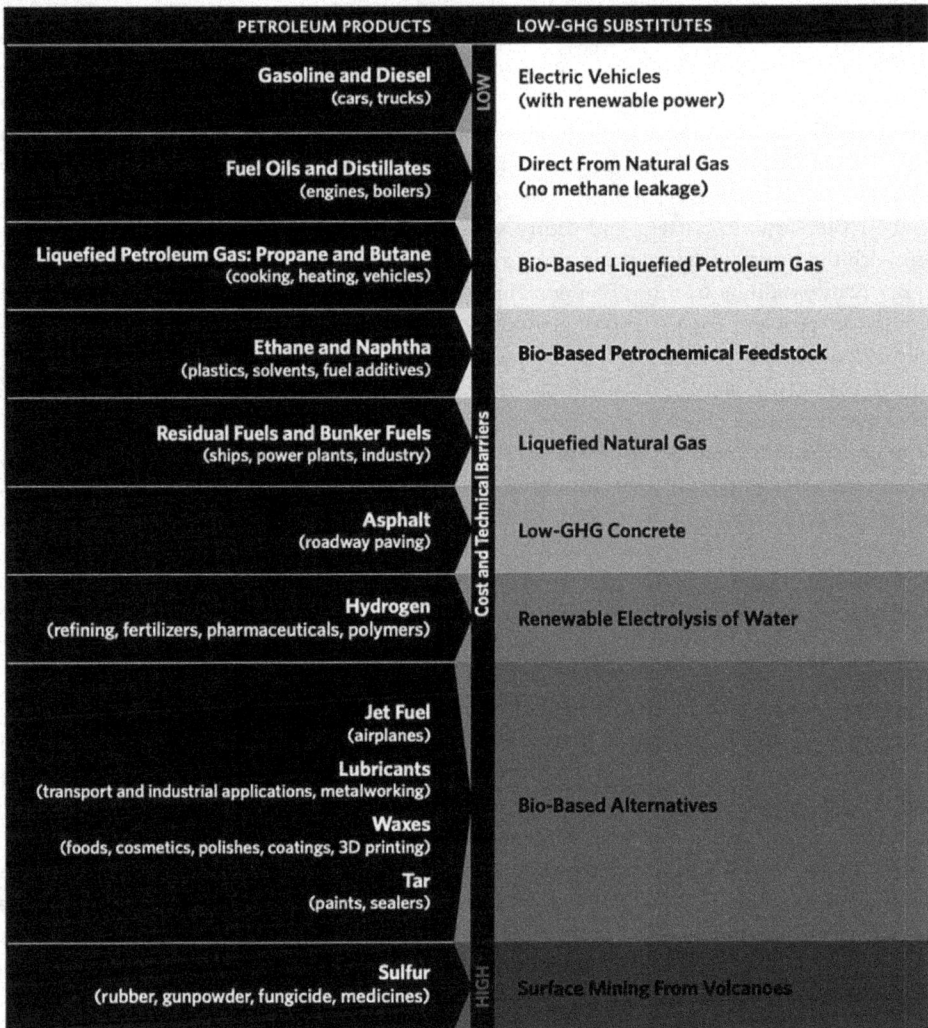

PETROLEUM PRODUCTS	LOW-GHG SUBSTITUTES
Gasoline and Diesel (cars, trucks)	**Electric Vehicles** (with renewable power)
Fuel Oils and Distillates (engines, boilers)	**Direct From Natural Gas** (no methane leakage)
Liquefied Petroleum Gas: Propane and Butane (cooking, heating, vehicles)	**Bio-Based Liquefied Petroleum Gas**
Ethane and Naphtha (plastics, solvents, fuel additives)	**Bio-Based Petrochemical Feedstock**
Residual Fuels and Bunker Fuels (ships, power plants, industry)	**Liquefied Natural Gas**
Asphalt (roadway paving)	**Low-GHG Concrete**
Hydrogen (refining, fertilizers, pharmaceuticals, polymers)	**Renewable Electrolysis of Water**
Jet Fuel (airplanes)	
Lubricants (transport and industrial applications, metalworking)	**Bio-Based Alternatives**
Waxes (foods, cosmetics, polishes, coatings, 3D printing)	
Tar (paints, sealers)	
Sulfur (rubber, gunpowder, fungicide, medicines)	Surface Mining From Volcanoes

Figure 8.1 Refining shifts to low-greenhouse-gas product slates.

Source: Deborah Gordon and Madhav Acharya, "Oil Shake-Up: Refining Transitions in a Low-Carbon Economy", April 2018, Carnegie Endowment for International Peace, https://carnegieendowment.org/files/Gordon_Driving-Change_Article_April2018_final.pdf. Reprinted with permission.

Take jet fuel for instance. While bio-based jet fuel strategies are being developed (World Economic Forum 2020), commercial production is occurring in small amounts, and it will likely be many decades before all airports and aircraft are retrofitted for its universal, safe use. The prospect of liquid fuels obtained directly from renewable energy (so-called solar fuels) is even more distant. Replacing jet fuel will certainly be far slower than the adoption of EVs. The same is true of a product as humble – and cheap – as asphalt. Solar roadways notwithstanding, we will still continue to rely on the bottom of the barrel on which to drive our cleaner cars. The future of a decarbonized product slate is further complicated by the fact that an increasing volume of unconventional oils will be flowing to refineries in the future, lending themselves to be easily turned into different petroleum products.

If a refinery is only able to sell a portion of its product slate, the disposition of the remaining products that no longer have a market will still need to be addressed. In the base case, they have zero value, but every refiner will try and derive some benefit from having produced these molecules. Refiners could seek to use unwanted hydrocarbons in new formulations of existing products or create new petroleum products altogether, provided these options achieve climate goals. However, the easiest course of action would be for refiners to burn these products to generate energy (which would happen to the fuels in any case). While this might help improve energy efficiency of a refinery, it does not cut overall greenhouse gas emissions. Additional carbon capture equipment might be required on site, which would greatly increase cost and overall refining complexity.

Moreover, refiners need to confront changing crude inputs, from ultra-light to extra-heavy variations with different impurities to remove. Crude oil comes out of the ground as a mix of different hydrocarbon molecules. A game changer for the refining industry would be technologies that can selectively "process" an oilfield underground, using nanotechnology, microbial, or other techniques, such that only the desired molecules required in the final products were brought to the surface (Bryant 2016). But while the oil industry has gained a lot of knowledge about field composition, it is still a long way from being able to shift entire production units underground.

Barring such a move, the best a refiner can hope for is to somehow find a way to rework all the processes inside the refinery so that molecules can be converted to only the desired end products. This is easier said than done, as refining processes have been fine-tuned over a hundred years to be extremely efficient at producing the current slate. A recent trend in this direction is the concept of "crude to chemicals", wherein a processing plant converts the entire crude oil stream directly to chemical feedstocks (Hydrocarbon Processing 2016). More of these refineries will likely emerge over time as companies concentrate their efforts on the profitable end of the barrel amid pressure to address climate change. However, it will do little to alleviate the challenge faced by an existing refinery that makes a diverse slate of products. In the long run, we could see large, integrated refinery-petrochemical plants give way to smaller, customized operations that convert different feedstocks into a small number of finished products.

Bespoke Refineries

While the integrated refining mode has been the preferred choice for operators over the past few decades, it is possible that a "remade" refining industry would gravitate toward a model where individual refineries specialize in producing a limited number of products using low-GHG processes, non-fossil feedstocks such as biomass, captured manmade CO_2, or low-methane natural gas. In fact, Reliance Industries, the operator of the world's largest refinery, announced a plan to convert the facility entirely to producing only petrochemicals and jet fuel, which are toward one end of the molecular spectrum (PTI 2019)

One could imagine that another separate set of refineries would produce lubricants and fluids. And a third set of refineries, which is harder to reconcile economically and environmentally, would be left to manufacture lower value products such as asphalt, paving materials, roof shingles, tar, hydrocarbon emulsions, and other solid petroleum by-products grouped as residuum. In today's refineries, these most carbon-laden products come from the heaviest portion of the crude slate, which is also typically produced from the heaviest oils that have some of the highest production and refining GHG emissions with life cycle climate footprints estimated at nearly one metric ton CO_2e per barrel, compared to light, petrochemical-producing condensates with one-half the carbon footprint (Gordon 2021). Therefore, while they might well be economically viable, bespoke refineries reconfigured to produce residuum by-products will require significant rethinking to meet net-zero GHG targets.

The shift to construct new crude-to-chemicals plants is an early sign of the shape of things to come. There are only a few of these plants around today, but they produce a significantly different product slate than a conventional refinery.

Crude-to-chemicals plants are a step in the right direction to move refining in the direction of low-volume, high-value products. But they are far from being a silver bullet, and the refining industry should resist the temptation to think of them as such.

Reintegrated, Decarbonized Refineries

An alternate to the fragmentation of the refining industry is to preserve the integrated refining model by introducing new technologies that eliminate products, like transport fuels, for which its value will decline as energy transition picks pace. An example of this is provided by the Fuels Europe's Vision 2050 study (FuelsEurope 2018). The electrification or gasification of transport would render gasoline, diesel, and marine bunker fuels less profitable for a refiner. Over time, wholesale transport fuel switching away from oil-based sources could render these fuels entirely obsolete. This move would buck conventional refining trends, reversing a century of infrastructure build-out worldwide, whereby the ability to sell gasoline and diesel fuel products made refining the entire barrel of crude oil a profitable venture.

In order to assess the impact of demand shifts on the profitability of an integrated refinery, we created a simple model to assess the economics of refining under different scenarios. Specifically, refining economics were assessed in three different scenarios. The first was a business-as-usual scenario representing the current situation for a large 300,000-bpd integrated refinery on the US Gulf Coast. In addition to producing transportation fuels, such as gasoline, jet fuel, and diesel, such a refinery also produces a number of feedstocks for petrochemical and lubricant plants. It converts more than 85% of the crude oil feedstock into transportation fuels led by gasoline and followed by diesel and jet fuel. Nearly one-tenth of the plant's capacity is used to produce petrochemical feedstocks, while the rest is for production of asphalt, liquefied petroleum gas (LPG), and marine fuels. As shown in Table 8.1 that details these assumptions along with the results of the economic modeling, such a refinery generates a gross margin of a little over $23 per barrel of crude oil processed.

The second economic scenario that was modeled assumes that fuel demand has declined for this refinery by 50% in 2035. As a result, the refinery will cut gasoline and diesel production to ~23% and ~19%, respectively, of the plant's processing capacity. However, the refinery will likely increase jet fuel production to ~23% of capacity since it is unlikely to suffer the same declines as gasoline and diesel due to the challenges of shifting the aviation sector away from petroleum-derived fuels to electric engines or sustainable aviation fuels in the medium term. Similarly, petrochemical, lubricant, and LPG production will also double relative to the business-as-usual case. Although most of these products are priced at a premium over gasoline and diesel, processing volumes fall so drastically in this scenario that overall gross margin declines substantially to nearly $6 per barrel of crude oil processed. In other words, a bleak economic future where margins are cut by nearly 75% awaits in the medium-term for a complex, integrated oil refinery on the US Gulf Coast.

Table 8.1 Economic Modeling Results for Refining Configuration Shifts

	Refining Case 1		*Refining Case 2*		*Refining Case 3*	
Case description	• Business as usual • Integrated, complex US Gulf Coast refinery with 300,000 bpd capacity		• A 50% reduction in fuel demand in comparison to Case 1 by 2035 • Integrated, complex US Gulf Coast refinery with 300,000 bpd capacity		• Crude-to-chemicals type model with no fuels production in 2050 • Integrated, complex US Gulf Coast refinery with 150,000 bpd capacity	
Product	**Yield** (%)	**Price** ($/barrel)	**Yield** (%)	**Price** ($/barrel)	**Yield** (%)	**Price** ($/barrel)
LPG	4.6%	$32.72	9.2%	$32.72	0.0%	N/A
Gasoline	43.4%	$71.44	22.8%	$47.87	0.0%	N/A
Diesel	31.2%	$79.13	18.7%	$59.35	0.0%	N/A
Jet fuel	10.7%	$78.92	22.9%	$78.92	10.0%	$78.92
Lubes	1.2%	$115.50	2.4%	$115.50	10.0%	$115.50
Petchem	9.2%	$68.81	18.4%	$68.81	80.0%	$68.81
Bunker	1.0%	$66.36	1.0%	$56.99	0.0%	N/A
Asphalt + others	6.3%	$95.09	6.3%	$56.99	0.0%	N/A
Margin	**$23.02**		**$5.75**		**$34.60**	
Refinery hydrogen demand, billion cubic feet per day	2.6		1.9		4.6	

Source: Authors' calculations.

The final scenario assumes a significantly reconfigured refinery in 2050 on the US Gulf Coast that converts crude oil entirely to chemicals. Such a refinery would also be much smaller – our case assumes 150,000 bpd of capacity – since petrochemical demand will be a fraction of that for transportation fuels. This refinery manufactures only three products – jet fuel, lubricants, and petrochemical feedstocks with the latter accounting for 80% of output – but delivers nearly $35 in gross margin per barrel of production. Although the refinery enjoys the best margins across all three scenarios, it is operating at 50% the production capacity of the other two scenarios reflecting a significant decline in revenues generated by the plant.

A final observation from this high-level conceptual analysis is around hydrogen demand in each of these scenarios. Table 8.1 shows that the final crude-to-chemicals scenario consumes significantly higher volumes of hydrogen at 4.6 billion cubic feet per day (cfd) due to the larger volumes of lubricants production in comparison to the 2.6 billion and 1.9 billion cfd consumed by the first and second scenarios, respectively. Refineries typically source hydrogen from large-scale steam-methane reformers that are highly CO_2-intensive but will likely have to be displaced with sources of green hydrogen that may likely be more expensive impacting the refinery's gross margins. In addition to the economic implications, this also highlights the highly integrated and interlinked nature of refineries that will face severe disruptions as the world accelerates on its journey toward energy transition and decarbonization.

Biorefineries

Oil, as a source of energy and a building block for materials, is rivaled by only one other material – biomass. (Indeed, oil is simply biomass that existed millions of years ago and is now underground.) Reports such as the US DOE's "Billion Ton Biomass Study" (Perlack 2005) point to the possibility that biomass might potentially be used as a raw material to substitute for the petroleum-based economy. Indeed, because of regulations such as California's Low-Carbon Fuels Standard, growing demand for energy transition, and COVID-19, many refineries are shifting to use bio-based feedstocks instead of crude oil.

Like their petroleum counterparts, however, biorefineries have to contend with a number of issues, some of which are highlighted here:

• Feedstock availability: While some feedstocks, such as palm oil used for biodiesel, are traded globally, most biorefineries tend to use local sources of biomass, such as corn, sugarcane, animal fats, and vegetable oils. This constrains the location of these facilities; the majority of US biorefineries are in the Midwest. The debate over "food versus fuel" has still not been resolved satisfactorily; land use changes to grow raw materials for biorefining have their own climate change impact (Araujo et al. 2017).
• Product mix: Unlike a petroleum refinery, the product mix from a biorefinery tends to be limited by the specific feedstock and processing steps. A biorefinery that mimics the entire range of products from an oil refinery would likely require multiple trains, each using a different feedstock; in effect, it would be multiple small refineries operating in parallel. The bespoke model of refining mentioned earlier would clearly be amenable to the widespread use of different bio-feedstocks.
• Scale: Oil refineries are the ultimate example of economies of scale; except for bespoke small refineries that exist to serve specific geographies, the most profitable sites tend to be the largest. The largest biorefinery in the world currently being planned would be 50,000 bpd, as compared to over 1 million bpd for the largest petroleum refineries (Bioenergy International 2020).

One of the primary drivers for the explosive growth in renewable and biodiesel production in the US is related to the California tax credits that accrue to the producer. Biorefining pits "farm states" versus "oil states" as each try to promote the industry that provides a large source of employment for their residents. The electrification of transportation could, however, provide a shot in the arm for biorefineries that now no longer have to compete against giant oil refineries for the same fuel market. Indeed, smaller biorefineries that are integrated with other operations in an industrially symbiotic model could be highly competitive in producing high-quality products that serve specific sectors.

Climate Impact of Refining

Refining accounts for 10% of global industry GHG emissions, the third largest stationary emissions source after power and petroleum and natural gas systems (Bruckner et al. 2014). There are over 600 operating refineries in just over 100 countries worldwide (Oil and Gas Journal 2020); however, the top 20 countries refine more than 80% of global crude oil (Jing et al. 2020). The average global volume-weighted-average refining carbon intensity (CI) is estimated at 40.5 kilograms of CO_2e per barrel (kg CO_2e/bbl).

Today's refineries can be categorized based on their complexity and the gravity of the crude oil processed. There are four major types of increasing complexity: simple conversion (hydroskimming), medium conversion, and two different deep conversion configurations (coking and hydrocracking), with estimated carbon intensities of 16.3, 34.4, 48.9, and 55.1 kg CO_2e/bbl, respectively (Jing et al. 2020). The US and China handle one-third of global crude oil refining throughput and have higher than average CIs due to their use of deep conversion refining of the heaviest crude oils.

The largest source of refining GHGs results from the production of hydrogen, which is an essential input when processing today's medium-to-heavy oils. Hydrogen is made by steam methane reforming (SMR), an energy-intensive process that generates high CO_2 and fugitive methane emissions. One of the best ways to reduce refinery GHGs is by replacing SMR with green hydrogen generated through renewable electrolysis with very low emissions.

In addition to green hydrogen, numerous additional refining GHG emission reductions, such as cogeneration of heat and power, replacing fuel oil firing, and carbon capture, are technologically feasible for today's integrated refineries. Annual global refining emissions reductions of 31% (low investment) and 70% (high investment) are possible, according to a study of 343 crude oils processed in 478 refineries located in 83 countries representing 93% of global crude oil refining in 2015 (Jing et al. 2020). However, achieving GHG reductions amid shifts in product slates that may necessitate new refining configurations will require further study (ibid.).

The Long Road Ahead

Governments are expanding their decarbonization goals beyond the power sector and into transport and chemicals, where petroleum holds a near-monopoly. In 2015, the G7 nations called for a first-of-its-kind complete decarbonization of their economies by the end of the century (Livingston 2016). In July 2017, the governments of the UK (Castle 2017) and France (Ewing 2017a) – representing the second and third largest car markets in the EU, respectively – announced plans to ban petroleum-fueled vehicles by 2040. The Chinese-owned automaker Volvo (Ewing 2017b) announced that it would design only new hybrid and electric vehicles starting in 2019. In 2016, Germany's Bundesrat passed a non-binding, though more sweeping, resolution, calling for the European Commission to effectively ban internal combustion engine vehicles from EU roads by 2030 (Schmitt 2016). Further, India and China are working on plans to ban cars that run only on fossil fuels (Petroff 2017).

The response of the petroleum industry toward efforts to decarbonize transportation has, not unexpectedly, been quite varied. While European-based companies are already seizing the opportunity to fuel electric vehicles (Egan 2017) or invest in renewable power (Keohane 2017), their US-based counterparts have responded more slowly, citing lack of policy certainty or unfavorable market economics. It is true that, at least in the near term, refineries in the industrialized countries could export surplus gasoline displaced by EVs to non-OECD nations, especially those experiencing economic growth. Depressed gasoline and diesel prices would ignite demand. And regulatory relief that avoids extra refining steps to reformulate gasoline could marginally increase profit margins. Shareholders, however, are getting concerned and demanding refining companies to prepare more aggressively for the global energy transition to deep decarbonization. In 2021, ExxonMobil, Shell, and Chevron were all rebuked by shareholders who voted for deeper and more substantive changes in support of the energy transition (Osaka 2021). In fact, one of the directors that shareholders elected to the company's board in opposition of ExxonMobil management's guidance was a former senior executive at Neste, which has successfully transitioned toward a far more sustainable refining portfolio than its peers in the past decade (Ambrose 2021).

In the long term, there are several ways to remix the underlying components in oil into different product formulations. It is more a matter of the price relative to their substitutes, although safety, availability, ease of use, and other trade-offs merit consideration. Generally, the lower the price, the greater the market potential for new petroleum product configurations. But if automotive decarbonization policies materialize and spread globally, their impacts on refining will not perpetuate the gradual shift that the industry has dealt with over the past half century. Such a major shift away from established petroleum products has never been charted.

To adjust to a world with a vastly different demand profile, the oil and gas industry will need a 21st-century technology makeover, and fast. The industry has done this before, albeit a very long time ago when oil companies devised key processes, such as reforming and fluid catalytic cracking, to develop new high-octane fuel during World War II (Science History Institute 2011).

Making the shift amid a dynamic geoeconomics backdrop will be even more difficult, considering the many market and social pressures on refiners. Refining innovations will have to square with policy mandates to meet decarbonization goals and address local environmental challenges. They will have to be conscious of fickle consumer preferences and changing product demands, as they respond to disruptive innovations in all economic sectors. National security goals will have to be factored in for both foreign and domestic concerns.

The challenge to develop new and retrofit old processes will require mobilizing technical resources on a national and international scale. Even if new manufacturing units, such as biorefineries or renewable energy-powered electrolyzers are technically feasible, entirely new processes and capital will need to be put in place, and the price tag for that will not be cheap. It might require the establishment of joint public-private "blue ribbon panels" to identify key technical gaps in the fuel and chemical manufacturing processes of the future. This could be integrated with creative funding mechanisms, such as the X Prize, to design and implement competition models to solve challenges on a global scale.

D. Conclusion

Energy transitions can often take much longer than policymakers envision. Such slow uptake is likely to be the case with a *clean, global* energy transition. A new approach to oil refining will require major modifications to the long-honed, durable petroleum system. If economy-wide decarbonization is going to be accomplished, it will not be enough to simply advance sales of cleaner substitutes like EVs. It is one thing to craft a policy that steers cars and trucks away from petroleum. It is quite another to assume that these will result in successfully elimination of oil out of the economy.

Still, as momentum grows to adapt the global economy to climate change, it is critical to start thinking deeply about how these changes will impact a century-old industry like oil refining. In order to minimize unintended energy and environmental consequences, a roadmap for the future needs to account for all petroleum shifts, not just the ones that capture the most attention. The better decision-makers understand the workings of the economic activity – petroleum refining – that is poised to be disrupted, the more likely sustainable outcomes will result.

It would be wishful thinking to rely on the development of as-yet-unknown technologies that can place the molecules exactly where we want them, with zero waste. While that has been the history of refining in the 20th century, it was done in a largely unconstrained environment where emissions were left unchecked. Starting in the late 1800s and picking up significantly in the 1930s, refining received a significant amount of academic study, engineering expertise, and capital investment. The same process – research, development, demonstration, and deployment –will be required to remake refining.

What makes this next transition so challenging, however, is that it has to result in demonstrably lower emissions, presumably with net-zero GHGs, or else the goal of limiting global temperatures to well below 2°C and minimizing devastating climate impacts will be overshot by far too much. The oil and gas industry has constantly touted the major role that carbon capture and storage (CCS) will play in decarbonizing the sector. But this does not mean that refining will maintain the status quo amid standalone CCS units. To be sure, the techno-economic challenge of CCS is as great as that involved in reinventing refining; one cannot assume the probability of success will be greater. The "brute force" approach might be to simply combust all the non-essential products from a conventional refinery and capture those emissions to render them carbon-neutral.

The situation is not all doom and gloom, however. The future of clean and secure energy supplies will still depend on large amounts of specific products, both commodities and specialties. Companies that want to capitalize on this will rely on novel technologies and a mastery of logistics that enable them to be the lowest-cost producer. Cooperation, not competition, is warranted to enable an orderly transition and avoid a sudden exodus of suppliers in certain regions. During the World War II, government funding enabled the industry to develop new processes for making high-octane fuels that helped win the war. As the 21st century advances, we may need a similar effort to develop processes that minimize waste and emissions, while supporting the shift toward cleaner technologies. (BP 2020)

Note

1 Globally, 73% of oil consumed is used to make petroleum products other than gasoline. As such, a 25% reduction in gasoline consumption represents only a 7% decline in overall oil demand. See: www.iea.org/media/omrreports/fullissues/2017-08-11.pdf

References

Ambrose, Jillian. 2021. *ExxonMobil and Chevron Suffer Shareholder Rebellions Over Climate.* May 26. www.theguardian.com/business/2021/may/26/exxonmobil-and-chevron-braced-for-showdown-over-climate.

Araujo, K., D. Mahajan, R. Kerr, and M. da Silva. 2017. "Global Biofuels at the Crossroads: An Overview of Technical, Policy and Investment Complexities in the Sustainability of Biofuels Development." *Agriculture* 7(4), 32–54. https://doi.org/10.3390/agriculture7040032.

Argus Media. 2019. *Shell Aims to Keep Only 10 "Top Notch" Refineries.* August 1. www.argusmedia.com/en/news/1950962-shell-aims-to-keep-only-10-top-notch-refineries.

Bioenergy International. 2020. *Phillips 66 Plans to Transform San Francisco Refinery into the World's Largest Biorefinery.* August 15. www.bioenergyinternational.com/biofuels-oils/phillips-66-plans-to-transform-san-francisco-refinery-into-worlds-renewable-fuels-plant.

Blackmon, D. 2020. "First Major New U.S. Oil Refinery Since 1977 Targets BAKKEN Shale Crude." *Forbes*, July 25.

Bloomberg. 2017. *Exxon's Not Afraid of Tesla's Trucks.* www.bloomberg.com/news/articles/2017-10-30/exxon-s-not-afraid-of-tesla-trucks-chemicals-to-carry-demand.

Bloomberg. 2020. *Lost in Oil's Rally: $2 Trillion Refining Industry in Crisis.* July 4. Accessed November 25, 2021. www.bloomberg.com/news/articles/2020-07-05/lost-in-oil-s-rally-2-trillion-a-year-refining-industry-crisis.

Bloomberg. 2021. "Building New Renewables Is Cheaper Than Burning Fossil Fuels." *Bloomberg.com.* June 23. www.bloomberg.com/news/articles/2021-06-23/building-new-renewables-cheaper-than-running-fossil-fuel-plants.

BP. 2019. "Energy Outlook 2019."

BP. 2020. "Energy Outlook 2020." Energy Outlook.

Bruckner T., I.A. Bashmakov, Y. Mulugetta, H. Chum, A. de la Vega Navarro, J. Edmonds, A. Faaij, B. Fungtammasan, A. Garg, E. Hertwich, D. Honnery, D. Infield, M. Kainuma, S. Khennas, S. Kim, H.B. Nimir, K. Riahi, N. Strachan, R. Wiser, and X. Zhang. 2014. Energy Systems. In: *Climate Change*

2014: Mitigation of Climate Change. Contribution of Working Group III to the Fifth Assessment Report of the Intergovernmental Panel on Climate Change. Cambridge: Cambridge University Press.

Bryant, Steven. 2016. *Modern Alchemy: Materials for Green Petroleum.* April 13. www.mcgill.ca/tised/past-events/2016/cerc-dr-steven-bryant-modern-alchemy.

Castle, Stephen. 2017. "Britain to Ban New Diesel and Gas Cars by 2040." *The New York Times.* July 26. Accessed November 25, 2021. www.nytimes.com/2017/07/26/world/europe/uk-diesel-petrol-emissions.html?mcubz=0.

Chakraborty, Debjit, and Dhwani Pandya. 2020. *Diesel Demand Drop Forcing India Refiners to Import Gasoline.* September 28. www.bloombergquint.com/diesel-demand-slump-forcing-indian-refiners-to-import-gasoline.

ClimateTRACE. 2021. www.climatetrace.org/inventory?sector=oil-and-gas&subsector=oil-refining&time=2020&country=all-countries.

Desjardins, Jeff. 2016. *The Oil Market is Bigger Than all Metal Markets Combined.* October 14. www.visual-capitalist.com/size-oil-market.

Egan, Mark. 2017. *Oil Giant Shell Bets on eleCtric Cars.* October 12. Money.cnn.com/2017/10/12/investing/shell-oil-buys-electric-car-charging/index.html.

EIA. 1991. *U.S. Petroleum Refining Industry in the 1980s.* Accessed November 26, 2021. www.osti.gov/servlets/purl/6504843.

EIA. 2019. *Crude Oil used by U.S. Refineries Continues to get Lighter in Most Regions.* October 11. Accessed November 26, 2021. www.eia.gov/todayinenergy/detail.php?id=41653.

EIA. 2021a. *Oil and Petroleum Products Explained.* www.eia.gov/energyexplained/oil-and-petroleum-products/use-of-oil.php.

EIA. 2021b. *U.S. Renewable Diesel Capacity Could Increase Due to Announced and Developing Projects.* July 29. Accessed November 25, 2021. www.eia.gov/todayinenergy/detail.php?id=48916.

Ewing, Jack. 2017a. "France Plans to End Sales of Gas and Diesel Cars by 2040." *The New York Times.* July 6. Accessed November 25, 2021. www.nytimes.com/2017/07/06/business/energy-environment/france-cars-ban-gas-diesel.html.

Ewing, Jack. 2017b. "Volvo, Betting on Electric, Moves to Phase Out Conventional Engines." *The New York Times.* July 5. Accessed November 25, 2021. www.nytimes.com/2017/07/05/business/energy-environment/volvo-hybrid-electric-car.html?mcubz=0.

Forbes, James R. 1958. *Studies in Early Petroleum History.* Brill Archive.

Freudenberg. 2020. "Oil Drilling through the Centuries." www.fogt.com/news-event/oil-drilling-through-the-centuries.

FuelsEurope. 2018. "Vision 2050: A Pathway for the Evolution of the Refinery Industry and Liquid Fuels."

Gagniere, Marielle, Annick Pucci, and Emilie Rousseau. 2013. "Tackling the Gasoline/middle Distillate Imbalance." *Digital Refining.* www.digitalrefining.com/1000726/tackling-the-gasoline-middle-distillate-imbalance.

Gary, James, Glenn E. Handwerk, and Mark J. Kaiser. 2007. *Petroleum Refining: Technology and Economics.* CRC Press.

Global Data Energy. 2021. *Asia Dominates Upcoming Refinery Capacity Additions Globally by 2025.* October 25. Accessed November 26, 2021. www.offshore-technology.com/comment/asia-refinery-capacity-additions.

Gordon, Deborah. 2021. "Oil Climate Index." Dxgordon.github.io/OCIPlus.

Gordon, Deborah, and Madhav Acharya. 2018. *Oil Shake up – Refining Transitions in a Low Carbon Economy.* Washington, DC: Carnegie Endowment for International Peace.

Hydrocarbon Processing. 2016. "ExxonMobil and Saudi Aramco Technologies Produce Ethylene Directly from Crude Oil." *Hydrocarbon Processing*, July 7.

IEA. 2020. *Oil Market Report.* April. www.iea.org/reports/oil-market-report-april-2020.

IEA. 2021. *Net Zero by 2050.* Paris: IEA.

IRENA. 2021. "World Adds Record New Renewable Energy Capacity in 2020." *Irena.org.* April 5. www.irena.org/newsroom/pressreleases/2021/Apr/World-Adds-Record-New-Renewable-Energy-Capacity-in-2020.

Jing, L., H.M. El-Houjeiri, J.C. Monfort, et al. 2020. "Carbon Intensity of Global Crude Oil Refining and Mitigation Potential." *Nature Climate Change* 10, 526–532.

Johnstone, P., and C. McLeish. 2020. "World Wars and the Age of Oil: Exploring Directionality in Deep Energy Transitions." *Energy Research and Social Science* 69.

Keohane, David. 2017. *Total Buys 23% Stake in Renewable Energy Company Eren.* September 19. www.ft.com/content/69b2df00-ba2c-3066-adb8-740f59cf2f11.

Liang, Jing, Hassan M. El-Houjeiri, Jean-Christophe Monfort, Adam R. Brandt, Mohammad S. Masnadi, Deborah Gordon, and Joule A. Bergerson. 2020. "Carbon Intensity of Global Crude Oil Refining and Mitigation Potential." *Nature Climate Change* 526–532.

Livingston, David. 2016. *The G7 Climate Mandate and the Tragedy of Horizons.* Washington DC, February. Accessed November 2021. https://carnegieendowment.org/files/Brief-Livingston-G7.pdf.

McKinsey and Company. 2019. "Global Energy Perspective."

Meridian Energy Group. 2019. *IRPC 19 Keynote Speakers Tackle Shifts in Refining Landscape.* September 27. Accessed November 25, 2021. https://meridianenergygroupinc.com/irpc-19-keynote-speakers-tackle-shifts-in-hpi-landscape.

National Museum of American History. 2021. *Energy Crisis.* Accessed November 26, 2021. https://americanhistory.si.edu/american-enterprise-exhibition/consumer-era/energy-crisis.

Oil and Gas Journal. 2020. *2020 Worldwide Refining Capacity.* www.ogj.com/ogj-survey-downloads/worldwide-refining/document/14196089/2020-worldwide-refining-capacity-summary.

Oil and Gas Journal. 2000. *U.S. Refining Sector Stages Remarkable Recovery in 2000.* April 28. Accessed November 26, 2021. www.ogj.com/refining-processing/article/17254188/us-refining-sector-stages-remarkable-recovery-in-2000.

OPEC. 2021a. *Downstream Capacity Additions and Investments in OPEC Member Countries.* www.opec.org/opec_web/en/650.htm.

OPEC. 2021b. *World Oil Outlook 2045.* Energy Outlook, OPEC.

Osaka, S. 2021. *Why Climate Rage Hit Exxon, Chevron and Shell on the Same Day.* https://grist.org/energy/why-climate-rage-hit-exxon-chevron-and-shell-on-the-same-day/

Our World in Data. 2021. *World Crude Oil Price versus Consumption.* Accessed November 26, 2021. https://ourworldindata.org/grapher/world-crude-oil-price-vs-oil-consumption.

Perlack, Robert D. et. Al. 2005. *Biomass as a Feedstock for a Bioenergy and Bioproducts Industry: The Technical Feasibility of a Billion-Ton Annual Supply.* Washington, DC: US Department of Energy.

Petroff, Alana. 2017. "These Countries Want to Ban Gas and Diesel Cars." *CNN.* September 11. Accessed November 25, 2021. https://money.cnn.com/2017/09/11/autos/countries-banning-diesel-gas-cars/index.html.

PTI. 2019. "Reliance Plans to Produce Only Jet Fuel, Petrochemicals at Jamnagar Refinery." *Bloomberg Quint.* www.bloombergquint.com/business/reliance-plans-to-produce-only-jet-fuel-petrochemicals-at-jamnagar-refinery.

Reliance Industries. 2021. *Dream Big: India on the Global Refining Map.* www.ril.com/ourbusinesses/petroleumrefiningandmarketing.aspx.

Reuters. 2020. *Factbox: Oil Refiners Shut Plants as Demand Losses May Never Return.* November 10. Accessed November 25, 2021. www.reuters.com/article/us-global-oil-refinery-shutdowns-factbox/factbox-oil-refiners-shut-plants-as-demand-losses-may-never-return-idUSKBN27R0AI.

S&P Global Platts. 2020. *Refinery Closures Loom Across the Globe.* November 12. Accessed November 26, 2021. www.spglobal.com/platts/en/market-insights/latest-news/oil/111220-refinery-news-roundup-refinery-closures-loom-across-the-globe.

S&P Global Platts. 2021. *Global Energy Demand to Grow 47% by 2050, with Oil Still Top Source: US EIA.* October 6. Accessed November 25, 2021. www.spglobal.com/platts/en/market-insights/latest-news/oil/100621-global-energy-demand-to-grow-47-by-2050-with-oil-still-top-source-us-eia.

Schirber, Michael. 2009. "The Chemistry of Life: The Plastic in Cars." May 26. www.livescience.com/5449-chemistry-life-plastic-cars.html.

Schmitt, Bertel. 2016. "Germany's Bundesrat Resolves End Of Internal Combustion Engine." *Forbes.* October 8. Accessed November 25, 2021. www.forbes.com/sites/bertelschmitt/2016/10/08/germans-bundesrat-resolves-end-of-internal-combustion-engine/?sh=55ec592260bd.

Science History Institute. 2011. *Cracking Down on Crude Oil.* www.sciencehistory.org/distillations/cracking-down-on-crude-oil.

Sears, R. 2020. "Energy Futures: Oil and the Oil Industry." *The Brown Journal of World Affairs* 26(2).

Smil, Vaclav. 2016. *To Get Wind Power, You Need Oil.* February 29. Spectrum.ieee.org/to-get-wind-power-you-need-oil.

Solomon, Barry D., and Karthik Krishna. 2011. "The Coming Sustainable Energy Transition: History, Strategies and Outlook." *Energy Policy* 39(11), 7422–7431.

Solomon, L.H., A.L., Waddams, and John E. Carruthers. 2018. *Petroleum Refining.* November 9. www.britannica.com/technology/petroleum-refining.

Thune, Taran, Ole Andreas Engen, and Olav Wicken. 2019. *Petroleum Industry Transformations: Lessons from Norway and Beyond.* Routledge.

Tiremart. 2016. "How Much Oil does it Take to Make One Car Tire." April 19. Blog.tiremart.com/how-much-oil-make-one-car-tire.

Total Energies. 2020. *Energy Transition: Total Is Investing More Than €500 Million To Convert Its Grandpuits Refinery Into a Zero-Crude Platform for Biofuels and Bioplastics.* September 24. Accessed November 25, 2021. https://totalenergies.com/media/news/news/energy-transition-total-investing-more-eu500-million-convert-its-grandpuits.

US Federal Trade Commission. 1975. *The International Petroleum Cartel.* Washington, DC. https://play.google.com/store/books/details?id=LvFjyDXzt6EC&rdid=book-LvFjyDXzt6EC&rdot=1.

World Economic Forum. 2020. "Clean Skies for Tomorrow: Sustainable Aviation Fuels as a Pathway to Net-zero Aviation."

9

THE ROLE OF FOOD SYSTEMS IN ENERGY TRANSITIONS

Bruno S.L. Cunha, Joana Portugal-Pereira, Alexandre Szklo, and Roberto Schaeffer

Introduction

The term "energy transition" may at times refer only to the global energy sector's shift from fossil-fuel-based systems of energy production and consumption to renewable energy sources. However, it is possible to expand the concept by highlighting a parallel between the energy sector and food systems. The initial energy transitions occurred mainly in the field of agriculture, from hunter-gatherers discovering and mastering fire for heat and food cooking to the intensification of farming, leading to food surplus, population density, immunity to diseases, specialization, technology, and advanced tools (Diamond 1997; Harari 2014) (see also Zotin et al. in this *Handbook*). The provision of both food and energy are considered basic human needs and are well designated in the United Nations Sustainable Development Goals (SDGs) (United Nations Framework Convention on Climate Change 2015).

Points of connection can be observed in varied ways. First of all, food stores energy. Food energy is defined as the energy released from macronutrients (carbohydrates, fats, and proteins) when consumers eat them. A popular unit for measuring food energy is the kilocalorie (kcal) that represents 4.18 kilojoules (kJ). Humans need food energy to maintain body size, body composition, and a level of necessary and desirable physical activity consistent with long-term good health[1] (Food and Agriculture Organization of United Nations, World Health Organization, and UNU Expert Consultation 2001).

Over the last 50 years, the global human population has more than doubled (United Nations 2019). This factor has inevitably increased the necessity to produce more food to avoid higher numbers of hungry people worldwide. Hence, the Green Revolution, which emphasized higher agricultural productivity of staple grains, increased use of inorganic fertilizers and pesticides, improved irrigation facilities, and genetic component of traditional crops, achieved a certain level of success in reducing undernutrition (Evenson and Gollin 2003). However, food security interventions focused on agricultural production to increase food supplies and meet dietary energy needs have shown to be largely misguided (Food and Agriculture Organization of United Nations et al. 2020). Although global production of food energy has kept pace with population growth, food systems became key drivers of environmental changes. Food systems are responsible for a third of global greenhouse gas (GHG) emissions and half of the world's habitable land is used for agriculture (Intergovernmental Panel on Climate Change 2019). Besides,

DOI:10.4324/9781003183020-11

70% of global freshwater withdrawals are used for agriculture, 78% of global ocean and freshwater eutrophication[2] is caused by agriculture, and 94% of mammal biomass (excluding humans) is livestock (Poore and Nemecek 2018; Bar-On, Phillips, and Milo 2018; Food and Agriculture Organization of the United Nations 2011; Intergovernmental Panel on Climate Change 2019; Crippa et al. 2021; Tubiello et al. 2021).

In addition, it is increasingly evident that food insecurity is no more associated with lack of food resources but rather a matter of inefficient distribution of food supplies. While 821 million people are still undernourished, mainly in sub-Saharan countries, changes in consumption patterns contributed to over two billion people suffering from being overweight and obesity, particularly in upper-middle- and high-income countries (Intergovernmental Panel on Climate Change 2019) (see also Wilson et al. in this *Handbook*). Low-quality diets cause micronutrient deficiencies and contribute to a substantial rise in the incidence of diet-related obesity and disease mortality (Tilman and Clark 2014; Springmann et al. 2018; Willett et al. 2019; Intergovernmental Panel on Climate Change 2019; Food and Agriculture Organization of United Nations et al. 2020).

There is more evidence than ever before that the food system is intimately connected to major contemporary global challenges, from malnutrition to climate change (Swinburn et al. 2019; Rockström et al. 2020; Bodirsky et al. 2020). Our current food system could be regarded as fossil-fuel-intensive food because of its current GHG emissions and have reliance on fossil fuels. In turn, its production causes unacceptable environmental impacts and it is depleting finite resources, which indicates that the global food system we currently rely on is not sustainable.

In addition to the increasing penetration of renewable energy into the energy supply mix in a global energy transition toward greater sustainability, the fostering of electrification-intensive transport, heating, and cooling improvements in energy storage are critical. So, too, are a halt of the agricultural expansion and improvement of crop yields, the changing of diets, and reduction of food loss or waste. Taken together, food intake inextricably links human health, an energy transition and environmental sustainability.

Global Food Systems

According to the High-Level Panel of Experts on Food Security and Nutrition (2014), "a food system gathers all the elements (environment, people, inputs, processes, infrastructures, institutions, etc.) and activities that relate to the production, processing, distribution, preparation and consumption of food, and the output of these activities, including socioeconomic and environmental outcomes". Components of the food system are interdependent and include the entire life cycle of food supply chains, food environments, and individual-level filters that collectively influence consumer behavior and are affected by external drivers.

The food supply chain includes all life cycle steps needed to produce and move foods from field to fork. Food is produced, harvested, gathered, or slaughtered; cleaned, packed, and stored; and typically processed in some way – from cutting and canning to complex manufacturing. The level and type of technology adopted in the food chain play a major role in how the food chain functions. Then food is distributed – transported and traded – and sold and marketed to people in myriad ways. Retail and marketing are the last relevant steps and contribute to shape the food environment in which consumers make purchasing decisions. Food loss and waste are integral parts of the life cycle and reflect the inefficiency of the production and consumption processes. Agriculture and food production affect food availability and affordability as well as dietary quality and diversity. In rural communities, farmers and food producers may eat the food directly, sell it to neighbors in the local market, or engage in more distant markets. In urban areas, food is typically produced farther away, and more people are involved in the supply chain.

Food supply chains can increase the nutritional value of food, by increasing access to macronutrients and micronutrients.

The food environment is the connecting link between supply systems and demand systems. The food environment includes physical places, like stores or markets where people buy food, which is influenced by social, economic, and cultural factors. The key elements of the food environment that influence consumer food choices are food availability (type and diversity of food on offer), food affordability (economic access to food), product properties (safety, quality, appeal, and convenience), vendor proprieties (location and type of retail outlets), and food messaging (food promotion, advertising, and information). In many parts of world, healthy food environments are being converted to ones that are convenient, promoting unhealthy dietary choices for consumers through of a plethora of energy-rich, nutrient-poor foods.

Individual factors influence how people interact with their food environment and what they choose to buy and eat. Cognitive, aspirational, and situational filters, as well as socioeconomic status, shape the basis of the consumer behavior. Income and purchasing power determine what foods are affordable. For some people, nutrition knowledge or environmental awareness affects what they purchase and eat. Others are more driven by desires, values, and preferences. Nowadays, the time people have to shop for and prepare food is affected by work or home environments, as well as mobility. Personal and collective determinants of consumer food choices reflect on what food to acquire, store, prepare, cook, and eat. Consumers face many barriers to healthy eating, and the way that they interact with food is affected not only by their own beliefs and decisions but also by the people in their lives, their community and environment and the culture in which they live (High-Level Panel of Experts on Food Security and Nutrition 2017). Today, more and more countries far exceed the recommended energy intake per capita, yet supplies of micronutrients in the food supply have fallen (Beal et al. 2017).

Food systems are changing rapidly with urbanization, globalization, trade liberalization, increase in incomes, and changes in lifestyles. To better understand the complexity of food systems, the Food Systems Dashboard (Fanzo et al. 2020) divides such systems into five types (Table 9.1). Common patterns in food supply chains and food environments can be seen across

Table 9.1 Food Systems Characteristics

Food System Types	Characteristics		
	Landholders	Agricultural Productivity	Supply Chains
Rural and traditional	Small landholders	Low	Short and fragmented and informal markets.
Informal and expanding	Medium- and some large-scale farms	Higher than in rural and traditional food systems	Modern food supply chains for grains, with large-scale processors and centralized distribution centers.
Emerging and diversifying	Medium- and large-scale commercial farms co-exist alongside large numbers of small-scale farms	Higher than in informal and expanding food systems	Modern supply chains for fresh foods; processed and packaged foods are more available in rural areas; less seasonal fluctuation in the availability and pricing of perishable foods.

(Continued)

Table 9.1 (Continued)

Food System Types	Characteristics		
	Landholders	Agricultural Productivity	Supply Chains
Modernizing and formalizing	Large-scale farms	Higher than in emerging, informal, and traditional systems	Infrastructure is more developed; supermarket chains exist within cities and hold a large share of processed and dry goods sales; food imports enable year-round availability.
Industrialized and consolidated	Small number of large-scale and input-intensive farms	High yields	Well-developed infrastructure; long and integrated supply chains, with national and international sourcing of a wide range of foods.

countries and in different regions of a single country. By comparing these different food system types, some foods could include a mix of traditional and modern characteristics.

Outcomes of the Global Food System

Food systems have become more interconnected, with longer and more complex food supply chains. Globally, agriculture contributes to nearly 40% of the global GDP and agricultural goods represent 43% of all exports (Alston and Pardey 2014; Food and Agriculture Organization of United Nations et al. 2020). Livestock constitutes nearly half of total agricultural revenues by price and provides 17% of global food-caloric consumption and 33% of global protein consumption (Rojas-Downing et al. 2017). Self and wage employment in farming still generates a large share of rural incomes, accounting for about 70% of total employment in low-income countries globally (ibid.). Further, the share in the off-farm segments (food manufacturing and services) is expected to increase in the future (World Bank 2017; International Food Policy Research Institute 2020). In European Union, the food industry provides a larger share of employment than other manufacturing sectors, provides more stable employment during different economic crisis, such as the COVID-19 pandemic and has a higher share of women employed than overall manufacturing (Eurostat 2016).

Urbanization, along with the associated changes in income and lifestyle, expanded consumers' food choices and modified their dietary preferences (Popkin, Corvalan, and Grummer-Strawn 2020). Diets high in grains and tubers and low in animal-sourced foods gave way to more processed and packaged foods, as well as more animal-sourced foods, fruits, and vegetables. However, cooking less and eating away from home can contribute to preferred convenience or unhealthy food choices, which requires little to no preparation and offers far less nutrients than home-cooked food. Poor diets are one of the main risk factors for deaths globally (Afshin et al. 2019). Diets that exceed recommended energy intake can lead to overweight, obesity, and non-communicable diseases, like diabetes and cardiovascular diseases. Yet throughout the world, people still do not have access to adequate calories or a diversity of healthy, nutrient-rich foods.

Nearly one-third of total food produced worldwide does not reach consumers' meals because it is either lost or wasted, which results in cost losses of about $1 trillion (US$ 2012) per year (Mbow et al. 2019). Food waste is mainly generated by consumers and retailers, particularly in developed countries. In 2019, over 900 Mt of food waste was generated by households and retailers, which is equivalent to 17% of total global food production (United Nations Enviroment Programme 2021). Food waste is more frequent for highly perishable foods, such as animal products and fruits and vegetables. Food losses, on the other hand, are due to inefficiency of production and transportation logistics and inadequate storage systems. Generally higher losses are found in fruits and vegetables than for cereals, legumes, and meat.

From a resource depletion point of view, the existing food system can be regarded as unsustainable due to its reliance on fossil energy and mineral resources. Modern agriculture is currently dependent on large-scale land acquisitions, which require fertilizers applications, based on nitrogen, phosphorus, and potassium. Fertilizer potassium (or potash) is important for agriculture because it improves water retention, yield, nutrient value, taste, color, texture, and disease resistance of food crops. Doing so, the peak of potash reserves are of great concern because this can compromise future food availability (Cordell, Drangert, and White 2009; Mohr and Evans 2013; Manning 2015; Rosa et al. 2021). The global food system has converted nonrenewable fossil fuel energy into food by enabling mechanization, amplified fertilizer production, improved food processing, and safe global transportation (Kucukvar and Samadi 2015). Energy is required at every stage in the food system, with crop cultivation and animal rearing being one of the most energy-intensive phases and with limited penetration of renewable energy. In total, food systems consume 15–30% of global primary energy (Schramski, Woodson, and Brown 2020). This causes a low efficiency for food systems, with around 10 kcal of fossil fuel energy to produce 1 kcal of food, on average (Pimentel and Pimentel 2007). Some foods are very efficient in their use of resources to produce a nutritious meal, such as fruits and vegetables, while others are very inefficient, such as animal-based products. In general, animal foods require eight times more energy per calorie than plant-based commodities to produce a nutritious meal (Pelletier et al. 2011; Laso et al. 2018).

Different types of food production have varying environmental impacts. The expansion of agriculture has been one of humanity's largest impacts through deforestation and use of fires, reducing wilderness and threatening biodiversity. Over the past 50 years, the biggest driver of habitat loss has been the conversion of natural ecosystems for crop production or pasture (Intergovernmental Science-Policy Platform on Biodiversity and Ecosystem Services 2019; Benton et al. 2021). The practice of mono-cropping – growing the same crop over an extended area – and cash crop production can result in soil degradation and a food system that is less resilient to droughts or other extreme weather events (Mbow et al. 2019). The land use of foods is largely dependent on the intensity of farming – this is especially true when differentiating crops and animal products. Producing 100 grams of protein from beef requires 104 m², on average, but the range across producers is huge: spanning from 7 m² to 369 m² (Poore and Nemecek 2018). On the other hand, grains, peas, and tofu have a median footprint below 6 m² per 100 grams of protein (ibid.).

In addition to energy, water is an essential input to food production, whether in the form of rainfed sources or pumped irrigation. In recent decades, irrigation has been an important input factor in the observed increase of crop yields across many countries (Food and Agriculture Organization of the United Nations 2012). Globally, food production accounts for approximately 70% of freshwater withdrawals (Gleick 2014). Besides, the runoff of nitrogen and other nutrients from agricultural production systems is a leading contributor to eutrophication. The sum of water requirements across the full supply chain and the quantity of freshwater pollution

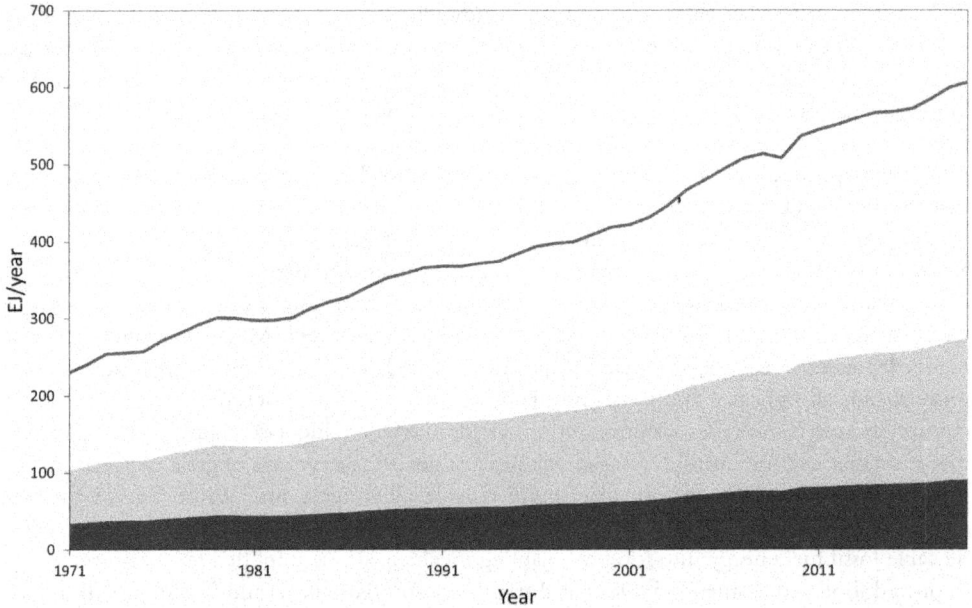

Figure 9.1   Global primary energy consumption and food production by human society. Food systems from farm- or ocean-to-table require from 15% (dark gray) to 30% (light gray) of total primary energy consumed (solid dark line).

Source: Adapted from Schramski, Woodson, and Brown (2020).

as a result of production varies significantly depending on food type. Cheese, nuts, prawns (farmed), and beef production represent the highest water footprints, more than to 1,000 L per 100 grams of protein, while some legumes and tofu use less than 200 L of freshwater (Poore and Nemecek 2018).

In terms of climate change, the entire food system contributes roughly one-third of total anthropogenic GHG emissions – a similar share to that from global electricity generation (Intergovernmental Panel on Climate Change 2019; Rosenzweig et al. 2020; Crippa et al. 2021; Tubiello et al. 2021). Animal-based foods present a higher carbon footprint than plant-based due to the biological inefficiencies inherent to feed conversion. Chicken, eggs, and pork nearly always have a lower footprint than beef and lamb. Beef herd production emits nearly 50 kg CO_2eq per 100 grams of protein, followed by lamb and mutton with 20 kg (Poore and Nemecek 2018). In this sense, animal products contribute with 5–8% of the total annual anthropogenic GHG emissions globally (Herrero et al. 2016; Intergovernmental Panel on Climate Change 2019). The production of plant-based food, such as peas, tofu, and nuts, leads to much lower impact than animal-based foods. For most foods, the majority of GHG emissions result from land use change and from processes at the farm stage. Farm-stage emissions include the use of more fertilizers[3] and pesticides, as well as enteric fermentation.[4] The global food system is becoming more energy-intensive due to a more mechanized agriculture mainly in developing regions. Combined, land-use and farm-stage emissions account for more than 70% of the carbon footprint for most foods (Poore and Nemecek 2018; Crippa et al. 2021). Processing, packaging, transporting, and storage account for around 15–20% of total anthropogenic GHG emissions, while global food loss and waste contribute to 8–10%.

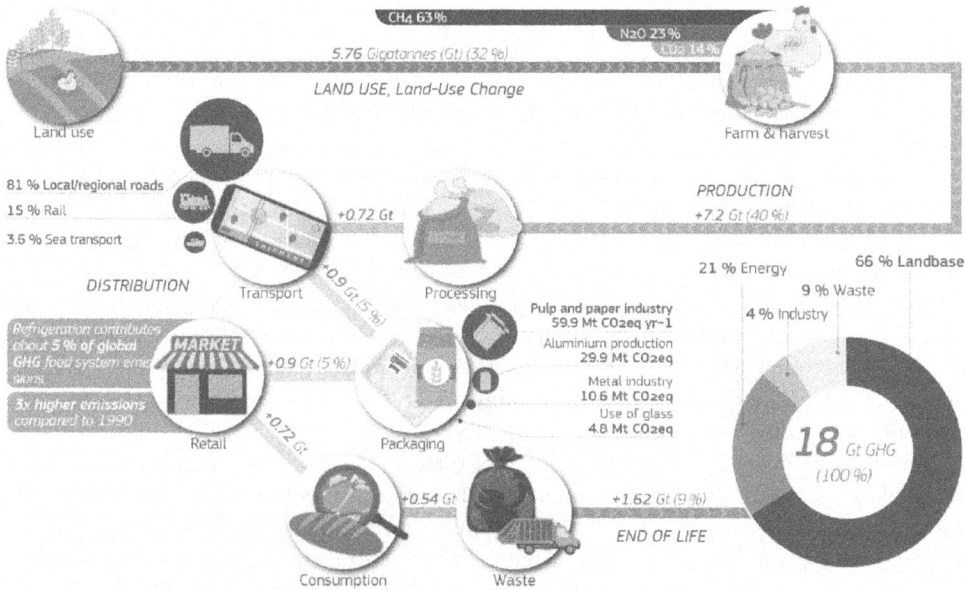

Figure 9.2 GHG emissions of the global food system by activities in 2015.

Source: Adapted from European Union 2021.

The Future of Food-Energy Transitions

As highlighted in previous sections, humanity faces serious challenges to provide enough nutritious food universally to a growing world population. Following a reference baseline, food demand would need to rise by 56% between 2010 and 2050 to feed nearly 10 billion people (World Resources Institute 2019). This would require an unprecedented increase of crop production: from 13,100 trillion kcal in 2010 to 20,500 trillion kcal of crop production in 2050. Furthermore, the level of land expansion is expected to be 3.2 billion hectares, in addition to the 5 billion hectares of farmland in use in 2010, which directly implies losses in native biomes of primary forests and savannas. This reference baseline would also increase by threefold the global GHG emissions from food systems, which would be equivalent to the entire carbon budget to stabilize global temperature increase to *well-below* 2°C, threating the ambitious pledges of Paris Agreement goals (Rogelj et al. 2018).

Given this unstainable reference baseline trend, alternative forecasts suggest that the food gap[5] may be reduced by (1) fostering sustainable intensification of agriculture practices without direct and indirect land-use changes, (2) reducing food losses and waste, and (3) reshaping dietary choices of consumers toward healthier and more sustainable foods (Food and Agriculture Organization of the United Nations 2018; Food and Agriculture Organization of United Nations et al. 2020; World Resources Institute 2019; Mbow et al. 2019). A green transition for agriculture, food, and land-use will transform the food system. Such a transition is an important lever to address simultaneously the complex interactions between multiple food system layers and dimensions of Sustainable Development Goals (SDGs).

To achieve these goals, a diverse menu of solutions is available both at the supply and demand sides, as well as in distribution. On the supply side, improving cropland, livestock, and grazing land management and increasing crop breeding to boost yields through genetic improvements for higher tolerance to heat and drought can reduce the food gap within planetary boundaries. Precision agriculture and practices that promote sustainable manure and residue management also play an important role to increase energy efficiency of food production systems.

Reduction of food loss and waste is another important strategy to increase efficiency of overall food systems, reduce its production costs and improve food security (Food and Agriculture Organization of United Nations 2019). Increasing efficiency in cold storage systems and improving road transport conditions and trade logistics are of key importance to reduce food losses. Furthermore, processing and packaging may also prevent losses, but it may imply potential conflicts with increased consumption of energy and generation of plastic wastes. Food wastes, on the other hand, may be reduced if consumers and retailers adopt better practices, such as improving purchase and meal planning, as well as phasing out aesthetic standards in terms of color, shape, and size.

Behavioral change toward adoption of more sustainable and healthy foods is crucial to close the food gap. Estimates suggest that the technical potential to reduce GHG emissions by shifting to plant-based diets, without undermining nutritional needs, is equivalent to the mitigation potential of supply-side measures (Mbow et al. 2019). This includes reduction of demand for meat by adoption of plant-based foods and the expansion of markets based on cultured meat and insect-based protein sources.

A healthy diet ensures the needed macronutrients (energy) and essential micronutrients to each specific gender, age, physical activity level, and physiological state, comprising diversity within and across food groups, adequacy compared with requirements, moderation, and overall balance (World Health Organization 2018; Food and Agriculture Organization of United Nations and World Health Organization 2019). A growing number of countries have established national food-based dietary guidelines, with dietary recommendations that are appropriate for their unique contexts (Springmann et al. 2020). In 2019, the EAT-Lancet Commission[6] launched what is referred to as the EAT-Lancet reference diet, which quantitatively describes the first attempt to set universal scientific targets for the food system, based on an increase in consumption of nutritious foods (such as vegetables, fruits, whole grains, legumes, and nuts) and a decrease in consumption of energy-dense foods (such as red meat, sugar and refined grains) that would provide major health benefits while also defining sustainable food systems that minimize damage to our planet (Willett et al. 2019). Accordingly, a timeframe for peak livestock (Harwatt 2019; Harwatt et al. 2020) within the next decade, as an analogy to the concept of "peak oil" (Campbell and Laherrère 1998), has emerged as an attractive option for both achieving long-term GHG reduction goals and avoiding near term global temperature rise.

However, producing and distributing sufficient amounts of sustainable and healthy food is costly and unaffordable to the entire population, especially in lower-income countries. Estimates suggest that healthy diets cost 60% more than diets that only meet the requirements for essential nutrients and almost five times more expensive than conventional diets just focused on energy sufficient requirements (Food and Agriculture Organization of United Nations et al. 2020). As part of comprehensive strategies to improve food intake by the human population, prices of healthy diets need to decline to become affordable for all, across and within countries, by regions and country income group. At the same time, food systems transformation should create supportive food environments, encourage people to learn about nutrition and spur behavior change that can lead to healthy food choices for human and for the planet.

Final Remarks

The evolution of food systems, from rural and traditional to industrialized and consolidated, the expansion of international food trade and the ascension of supermarkets and fast-food chains in many places reinforce the interlinkages with a fossil-fuel-based energy system. In an era of climate change and emerging public health issues, it is worth keeping in mind the connections between the consumption of energy and of food for an analysis of transitions in either system. Reaching the Paris Agreement goals is not possible by just decarbonizing the global energy system. *The call for low-carbon or negative emissions technologies in order to improve energy efficiency and reduce GHG emissions in energy sector is as relevant as the more sustainable intensification of agriculture practices in food systems.* Importantly, consumers can and must act as agents of change in their roles as citizens and influence the societal norms and institutional policies for the betterment of society. Healthy diets emerge as a solution both to human and planet health. However, when discussing any sustainability transition, such as that with energy, paying attention to questions of responsibility, accountability, and justice is imperative. There is no healthy planet unless we are all healthy.

Notes

1 Typically, an adult woman needs 2,000 kcal/day and an adult man around 2,500 kcal/day.
2 This represents the pollution of waterways with nutrient- and mineral-rich pollutants.
3 Both based on organic (manure management) and synthetic nitrogen.
4 The production of methane in the stomachs of cattle.
5 Food gap is defined as the difference between the total amount of food produced in 2010 and the required food that needs to be produced to feed to world population in 2050.
6 The EAT-Lancet Commission consists of 37 world-leading scientists from 16 countries and various scientific disciplines. It seeks to reach scientific consensus on targets for healthy diets and sustainable food production.

References

Afshin, Ashkan, Patrick John Sur, Kairsten A. Fay, Leslie Cornaby, Giannina Ferrara, Joseph S. Salama, Erin C. Mullany et al. 2019. "Health Effects of Dietary Risks in 195 Countries, 1990–2017: A Systematic Analysis for the Global Burden of Disease Study 2017." *The Lancet* 393 (10184). https://doi.org/10.1016/S0140-6736(19)30041-8.

Alston, Julian M., and Philip G. Pardey. 2014. "Agriculture in the Global Economy." *Journal of Economic Perspectives* 28(1). https://doi.org/10.1257/jep.28.1.121.

Bar-On, Yinon M., Rob Phillips, and Ron Milo. 2018. "The Biomass Distribution on Earth." *Proceedings of the National Academy of Sciences* 115 (25). https://doi.org/10.1073/pnas.1711842115.

Beal, Ty, Eric Massiot, Joanne E. Arsenault, Matthew R. Smith, and Robert J. Hijmans. 2017. "Global Trends in Dietary Micronutrient Supplies and Estimated Prevalence of Inadequate Intakes." *PLOS ONE* 12 (4). https://doi.org/10.1371/journal.pone.0175554.

Benton, T.G., C. Bieg, H. Harwatt, R. Pudasaini, and L Wellesley. 2021. *Food System Impacts on Biodiversity Loss. Three Levers for Food System Transformation in Support of Nature.* London.

Bodirsky, Benjamin Leon, Jan Philipp Dietrich, Eleonora Martinelli, Antonia Stenstad, Prajal Pradhan, Sabine Gabrysch, Abhijeet Mishra et al. 2020. "The Ongoing Nutrition Transition Thwarts Long-Term Targets for Food Security, Public Health and Environmental Protection." *Scientific Reports* 10 (1). https://doi.org/10.1038/s41598-020-75213-3.

Campbell, C.J., and J.H. Laherrère. 1998. "The End of Cheap Oil." *Scientific American* 278 (3): 78–83.

Cordell, Dana, Jan-Olof Drangert, and Stuart White. 2009. "The Story of Phosphorus: Global Food Security and Food for Thought." *Global Environmental Change* 19 (2). https://doi.org/10.1016/j.gloenvcha.2008.10.009.

Crippa, M., E. Solazzo, D. Guizzardi, F. Monforti-Ferrario, F.N. Tubiello, and A. Leip. 2021. "Food Systems Are Responsible for a Third of Global Anthropogenic GHG Emissions." *Nature Food* 2 (3). https://doi.org/10.1038/s43016-021-00225-9.

Diamond, J.D. 1997. *Guns, Germs, and Steel: The Fates of Human Societies.* New York: WW Norton & Co.

European Union. 2021. "EDGAR – Emissions Database for Global Atmospheric Research." *Energy, Climate Change, Environment. European Comission.* 2021. https://edgar.jrc.ec.europa.eu/edgar_food.

Eurostat. 2016. *Data and Trends: EU Food and Drink Industry 2016.*

Evenson, R.E., and D. Gollin. 2003. "Assessing the Impact of the Green Revolution, 1960 to 2000." *Science.* https://doi.org/10.1126/science.1078710.

Fanzo, Jessica, Lawrence Haddad, Rebecca McLaren, Quinn Marshall, Claire Davis, Anna Herforth, Andrew Jones et al. 2020. "The Food Systems Dashboard Is a New Tool to Inform Better Food Policy." *Nature Food* 1 (5). https://doi.org/10.1038/s43016-020-0077-y.

Food and Agriculture Organization of the United Nations. 2011. "The State of the World's Land and Water Resources for Food and Agriculture (SOLAW) – Managing Systems at Risk." Rome.

Food and Agriculture Organization of the United Nations. 2012. "Crop Yield Response to Water." Rome.

Food and Agriculture Organization of the United Nations. 2018. "The Future of Food and Agriculture – Alternative Pathways to 2050." Rome.

Food and Agriculture Organization of United Nations. 2019. "The State of Food and Agriculture 2019. Moving Forward on Food Loss and Waste Reduction." Rome.

Food and Agriculture Organization of United Nations, International Fund for Agricultural Development, The United Nations Children's Fund, World Food Programme, and World Health Organization. 2020. *The State of Food Security and Nutrition in the World 2020.* Rome: FAO. https://doi.org/10.4060/ca9692en.

Food and Agriculture Organization of United Nations, and World Health Organization. 2019. "Sustainable Healthy Diets: Guiding Principles." Rome.

Food and Agriculture Organization of United Nations, World Health Organization, and UNU Expert Consultation. 2001. "Human Energy Requirements." Rome.

Gleick, Peter H. 2014. *The World's Water.* Edited by Peter H. Gleick. Washington, DC: Island Press/Center for Resource Economics. https://doi.org/10.5822/978-1-61091-483-3.

Harari, Y.N. 2014. *Sapiens: A Brief History of Humankind.* London: Harvill Secker.

Harwatt, Helen. 2019. "Including Animal to Plant Protein Shifts in Climate Change Mitigation Policy: A Proposed Three-Step Strategy." *Climate Policy* 19 (5): 533–541. https://doi.org/10.1080/14693062.2018.1528965.

Harwatt, Helen, William J. Ripple, Abhishek Chaudhary, Matthew G. Betts, and Matthew N. Hayek. 2020. "Scientists Call for Renewed Paris Pledges to Transform Agriculture." *The Lancet Planetary Health* 4 (1). https://doi.org/10.1016/S2542-5196(19)30245-1.

Herrero, Mario, Benjamin Henderson, Petr Havlík, Philip K. Thornton, Richard T. Conant, Pete Smith, Stefan Wirsenius et al. 2016. "Greenhouse Gas Mitigation Potentials in the Livestock Sector." *Nature Climate Change* 6 (5): 452–461. https://doi.org/10.1038/nclimate2925.

High-Level Panel of Experts on Food Security and Nutrition. 2014. "Food Losses and Waste in the Context of Sustainable Food Systems." Rome.

High-Level Panel of Experts on Food Security and Nutrition. 2017. "Nutrition and Food Systems." Rome.

Intergovernmental Panel on Climate Change. 2019. "IPCC Special Report on Climate Change, Desertification, Land Degradation, Sustainable Land Management, Food Security, and Greenhouse Gas Fluxes in Terrestrial Ecosystems."

Intergovernmental Science-Policy Platform on Biodiversity and Ecosystem Services. 2019. "Summary for Policymakers of the Global Assessment Report on Biodiversity and Ecosystem Services of the Intergovernmental Science-Policy Platform on Biodiversity and Ecosystem Services." *Population and Development Review* 45 (3). https://doi.org/10.1111/padr.12283.

International Food Policy Research Institute. 2020. "2020 Global Food Policy Report: Building Inclusive Food Systems." Washington, DC. https://doi.org/10.2499/9780896293670.

Kucukvar, Murat, and Hamidreza Samadi. 2015. "Linking National Food Production to Global Supply Chain Impacts for the Energy-Climate Challenge: The Cases of the EU-27 and Turkey." *Journal of Cleaner Production* 108 (December). https://doi.org/10.1016/j.jclepro.2015.08.117.

Laso, Jara, Daniel Hoehn, María Margallo, Isabel García-Herrero, Laura Batlle-Bayer, Alba Bala, Pere Fullana-i-Palmer, Ian Vázquez-Rowe, Angel Irabien, and Rubén Aldaco. 2018. "Assessing Energy and Environmental Efficiency of the Spanish Agri-Food System Using the LCA/DEA Methodology." *Energies* 11 (12). https://doi.org/10.3390/en11123395.

Manning, David A.C. 2015. "How Will Minerals Feed the World in 2050?" *Proceedings of the Geologists' Association* 126 (1). https://doi.org/10.1016/j.pgeola.2014.12.005.

Mbow, Cheikh, Cynthia Rosenzweig, Luis G. Barioni, Tim G. Benton, Mario Herrero, Murukesan Krishnapillai, Emma Liwenga et al. 2019. "Climate Change and Land. Chapter 5: Food Security." *IPCC SPECIAL REPORT Global Warming of 1.5°C*, 1–200. www.ipcc.ch/site/assets/uploads/2019/08/2f.-Chapter-5_FINAL.pdf.

Mohr, S., and G. Evans. 2013. "Projections of Future Phosphorus Production." *PHILICA.COM*, no. 308. www.resilience.org/wp-content/uploads/articles/General/2013/09_Sep/peak-phosphorus/Phosphorus Projections.pdf.

Pelletier, Nathan, Eric Audsley, Sonja Brodt, Tara Garnett, Patrik Henriksson, Alissa Kendall, Klaas Jan Kramer, David Murphy, Thomas Nemecek, and Max Troell. 2011. "Energy Intensity of Agriculture and Food Systems." *Annual Review of Environment and Resources* 36 (1). https://doi.org/10.1146/annurev-environ-081710-161014.

Pimentel, D., and M. Pimentel. 2007. *Food, Energy, and Society*. Edited by David Pimentel Ph.D. and Marcia H. Pimentel M.S. CRC Press. https://doi.org/10.1201/9781420046687.

Poore, J., and T. Nemecek. 2018. "Reducing Food's Environmental Impacts through Producers and Consumers." *Science* 360 (6392): 987–992. https://doi.org/10.1126/science.aaq0216.

Popkin, Barry M., Camila Corvalan, and Laurence M. Grummer-Strawn. 2020. "Dynamics of the Double Burden of Malnutrition and the Changing Nutrition Reality." *The Lancet* 395 (10217). https://doi.org/10.1016/S0140-6736(19)32497-3.

Rockström, Johan, Ottmar Edenhofer, Juliana Gaertner, and Fabrice DeClerck. 2020. "Planet-Proofing the Global Food System." *Nature Food* 1 (1): 3–5. https://doi.org/10.1038/s43016-019-0010-4.

Rogelj, J., Drew Shindell, Kejun Jiang, Solomone Fifita, Piers Forster, Veronika Ginzburg, Collins Handa et al. 2018. "Mitigation Pathways Compatible with 1.5°C in the Context of Sustainable Development." In *Global Warming of 1.5 °C. An IPCC Special Report on the Impacts of Global Warming of 1.5 °C above Pre-Industrial Levels and Related Global Greenhouse Gas Emission Pathways, in the Context of Strengthening the Global Response to the Threat of Climate Chang.*

Rojas-Downing, M. Melissa, A. Pouyan Nejadhashemi, Timothy Harrigan, and Sean A. Woznicki. 2017. "Climate Change and Livestock: Impacts, Adaptation, and Mitigation." *Climate Risk Management* 16: 145–163. https://doi.org/10.1016/j.crm.2017.02.001.

Rosa, Lorenzo, Maria Cristina Rulli, Saleem Ali, Davide Danilo Chiarelli, Jampel Dell'Angelo, Nathaniel D. Mueller, Arnim Scheidel, Giuseppina Siciliano, and Paolo D'Odorico. 2021. "Energy Implications of the 21st Century Agrarian Transition." *Nature Communications* 12 (1): 2319. https://doi.org/10.1038/s41467-021-22581-7.

Rosenzweig, Cynthia, Cheikh Mbow, Luis G. Barioni, Tim G. Benton, Mario Herrero, Murukesan Krishnapillai, Emma T. Liwenga et al. 2020. "Climate Change Responses Benefit from a Global Food System Approach." *Nature Food* 1 (2). https://doi.org/10.1038/s43016-020-0031-z.

Schramski, John R., C. Brock Woodson, and James H. Brown. 2020. "Energy Use and the Sustainability of Intensifying Food Production." *Nature Sustainability* 3 (4): 257–259. https://doi.org/10.1038/s41893-020-0503-z.

Springmann, Marco, Luke Spajic, Michael A. Clark, Joseph Poore, Anna Herforth, Patrick Webb, Mike Rayner, and Peter Scarborough. 2020. "The Healthiness and Sustainability of National and Global Food Based Dietary Guidelines: Modelling Study." *The BMJ* 370: 1–16. https://doi.org/10.1136/bmj.m2322.

Springmann, Marco, Keith Wiebe, Daniel Mason-D'Croz, Timothy B. Sulser, Mike Rayner, and Peter Scarborough. 2018. "Health and Nutritional Aspects of Sustainable Diet Strategies and Their Association with Environmental Impacts: A Global Modelling Analysis with Country-Level Detail." *The Lancet Planetary Health* 2 (10). https://doi.org/10.1016/S2542-5196(18)30206-7.

Swinburn, Boyd A., Vivica I. Kraak, Steven Allender, Vincent J. Atkins, Phillip I. Baker, Jessica R. Bogard, Hannah Brinsden et al. 2019. "The Global Syndemic of Obesity, Undernutrition, and Climate Change: The Lancet Commission Report." *The Lancet* 393 (10173): 791–846. https://doi.org/10.1016/S0140-6736(18)32822-8.

Tilman, David, and Michael Clark. 2014. "Global Diets Link Environmental Sustainability and Human Health." *Nature* 515 (7528): 518–522. https://doi.org/10.1038/nature13959.

Tubiello, Francesco N., Cynthia Rosenzweig, Giulia Conchedda, Kevin Karl, Johannes Gütschow, Pan Xueyao, Griffiths Obli-Laryea et al. 2021. "Greenhouse Gas Emissions from Food Systems: Building the Evidence Base." *Environmental Research Letters* 16 (6). https://doi.org/10.1088/1748-9326/ac018e.

United Nations. 2019. "World Population Prospects 2019: Highlights."

United Nations Enviroment Programme. 2021. "Food Waste Index Report 2021." Nairobi.

United Nations Framework Convention on Climate Change. 2015. "Transforming Our World: The 2030 Agenda for Sustainable Development." https://sdgs.un.org/sites/default/files/publications/21252030 Agenda for Sustainable Development web.pdf.

Willett, Walter, Johan Rockström, Brent Loken, Marco Springmann, Tim Lang, Sonja Vermeulen, Tara Garnett et al. 2019. "Food in the Anthropocene: The EAT – Lancet Commission on Healthy Diets from Sustainable Food Systems." *The Lancet* 393 (10170): 447–492. https://doi.org/10.1016/ S0140-6736(18)31788-4.

World Bank. 2017. "Shaping the Food System to Deliver Jobs. Future of Food."

World Health Organization. 2018. "Healthy Diet Factsheet." Geneva.

World Resources Institute. 2019. "Creating a Sustainable Food Future: A Menu of Solutions to Feed Nearly 10 Billion People by 2050."

10

ENERGY TRANSITIONS IN LATIN AMERICA THROUGH THE LENS OF VULNERABILITY AND RESILIENCE

Insights from Colombia, Cuba, and Mexico

*A. Mejia-Montero, N.R. Leon-Rodriguez, B. Lorenzo-Yera,
D. Diaz-Florian, H. Thomson, T. Robles-Bonilla,
José Grabiel Luis-Cordova, J.M. Romero-Bravo, K.G. Cedano-
Villavicencio, and Y. Delgado-Triana*

1. Introduction

Although climate change is frequently framed as a global issue, low-carbon transitions can often materialize through the implementation of renewable energy targets at the national level. In this context, reports such as the Global Renewables Outlook: Energy Transformation 2050 (IRENA, 2020a) adopt a regional perspective to assess energy transitions, bridging global aspirations and regional actions. Such regional assessment of energy transitions highlights the strong differences in terms of the deployment of renewable or low-carbon energy technologies across the world, always dependent on available energy resources, as well as economic, institutional, environmental, and political contexts. The IRENA Outlook (ibid.) suggests that Latin American countries are doing remarkably well for low-carbon transitions, presenting the highest share of renewable energy worldwide in terms of the regional total primary energy supply (30%) and in power generation (65%). However, focusing on the country level would present a different panorama, with a diverse mosaic of energy transitions.

To assess the full implications of low-carbon transitions beyond technical solutions or impacts, it is necessary to recognize energy systems as socio-technical systems where social, institutional, economic, political, and environmental processes play a significant role (Sovacool et al., 2020; Adil and Ko, 2016; Jenkins et al., 2016, 2018; Araújo, 2017). Such socio-technical recognition of energy systems then raises additional challenges, as it is not sufficient to only produce low-carbon energy but which also covers the increasing energy needs of a globally growing population, and that is reliably distributed from generation to consumption points. Such balance between clean, sufficient, and accessible or reliable energy is manifested in important instruments like the United Nations Sustainable Development Goal 7, which aims to "Ensure access to affordable, reliable, sustainable and modern energy for all" (UN, 2021).

DOI:10.4324/9781003183020-12

It is within this recognition of energy systems as socio-technical systems that the concepts of energy resilience and energy vulnerability gain relevance. Energy resilience has traditionally been understood as a multidimensional concept (Gatto and Drago, 2020), useful for assessing the ability of an energy system to prevent, absorb, recover, or adapt to disturbances or shocks of different forms (economic, social, environmental, technical, etc.) (Ahmadi et al., 2021; Araújo and Shropshire, 2021). On a related front, energy vulnerability can also be defined as a multidimensional concept, useful this time to assess the risk or potential of an energy system or a household to fall into a state where resulting energy services are inadequate or insufficient due to adverse effects or structural/systemic conditions of marginalization and inequality (Bouzarovski et al., 2015). These two concepts of energy resilience and vulnerability overlap and are key to understanding how low-carbon transitions may enhance our energy systems in terms of making them cleaner and also making them more reliable and robust, ensuring that the general population is able to access the energy services that form the backbone of daily activities in any country.

The rest of the chapter is divided into three sections. Section 2 provides an overview of the concepts of energy vulnerability and energy resilience and why they are important for understanding Latin American energy transitions. Section 3 explores illustrative micro-cases from Colombia, Cuba, and Mexico, analyzing how each one relates differently to energy vulnerability and energy resilience. Finally, Section 4 closes with some final thoughts and priorities for future research.

2. Energy Transitions in Latin America

Considering energy resources, Latin America has considerable amounts of water, biomass, solar, and wind energy distributed across the region, enabling a currently high penetration of renewable energy resources in the power sector, at roughly 65% (IRENA, 2020a). The region also has the potential to eventually reach 100% in the subcontinent (Noura Guimarães, 2020). However, examining trends at the country level reveals significant differences across the region. For example, Colombia produces around 73.2% of the electricity from renewable energy sources such as hydropower, solar, and wind, surpassing the regional average (XM, 2021). Meanwhile, countries like Mexico currently only derive around 21% of electricity from low-carbon sources (SENER, 2021), which is more similar to regions like East Asia or Southeast Asia (IRENA, 2020a), while Cuba has an average share of renewable power similar to countries in the Middle East and North Africa at around 4% (ibid.).

Many countries in Latin America have a high dependency on specific resources and technologies for the generation of electricity. According to Noura Guimarães (2020), one group of Latin American countries relies mainly on low-carbon technologies, such as hydropower, as is evident in Brazil (62.5%), Colombia (79.8%), Costa Rica (74.4%), Panama (61%), Paraguay (99.7%), Uruguay (63.2%), and Venezuela (64%). By contrast, a second group of Latin American nations relies on fossil fuel endowments for electricity generation, as may be seen in Argentina (66%), Bolivia (79.3%), Chile (63.7%), Cuba (96%), Dominican Republic (88%), and Mexico (79%) (ibid.). This diversity of natural resource use translates to differentiated challenges for an energy transition, such as social conflicts around wind power in Mexico (Mejia-Montero et al., 2020), or the integration of the indigenous population's preferences into the production of renewable energy in Chile (Merino et al., 2020). Within this context, stark differences may be expected when considering the work needed to achieve a decarbonized energy sector and successful energy transition. Moreover, successful energy transitions should be understood from the perspective of households and countries and cannot be completed without affordable access and reliable energy services.

2.1 A Quick Glance at Energy Vulnerability and Energy Resilience

The term *resilience*, from the Latin "resilire", refers to a series of physical movements, such as bouncing back and returning. A more modern use of the concept can be traced to France in 1430 where it was used as a legal term to describe the restoration of an original legal situation after the termination of a contract (Gößling-Reisemann et al., 2018). Over time, the concept of resilience has been used under various contexts, adopting different definitions, creating a fertile ground for interdisciplinary work (Thoren, 2014). Within this context, authors such as Holling (1996) reflected on the differences between the concept of resilience from an ecological and an engineering perspective. From an engineering perspective, important characteristics were conceptualized to measure systems resilience, such as its resistance to disturbances and the speed of a system to return to the equilibrium system. On the other hand, characteristics like efficiency, constancy, and predictability can be used to understand resilience as the stability of an ecological system near an equilibrium steady state, allowing us to identify strong resemblances between the engineering and ecological understandings of resilience.

Within this context, recent reviews of the term *resilience* as applied to energy systems (Ahmadi et al., 2021; Araújo and Shropshire, 2021) indicate there is currently no consensus on the terminology to define it. However, even without a shared definition of energy systems resilience, many authors agree that such definition is based on four essential characteristics: planning, absorbing, recovering, and adapting (Roege et al., 2014; Sharifi et al., 2016; Ahmadi et al., 2021).

It is starting from the area of energy research that this chapter seeks to adopt a multidimensional definition of energy resilience, which considers technical, social, institutional, environmental, and economic factors that can be framed and serve to operate within the context of power systems companies in Latin America. In this way, we define the concept of energy resilience as a concept useful for analyzing and evaluating the capacity of an energy system of different scales (from household to national) to anticipate, absorb, recover, or adapt to situations where energy services are insufficient or inadequate.

In a related note, the concept of energy vulnerability has its roots in the 1970s oil crisis, in which the rapid increase in pricing of this energy resource caused a stark increase in household energy bills, causing civil servants in the United Kingdom to coin the term "fuel poverty" (Isherwood and Hancock, 1979), which was later formalized by Brenda Boardman (1991) in her seminal work. This concept spread throughout Europe and other geographical areas, becoming known by the term energy poverty (Thomson et al., 2017), and later "energy vulnerability" (Bouzarovski and Petrova, 2015). To this day, some terminological confusion persists, with some arguing that fuel poverty and energy poverty are distinct concepts, with the latter concerning extreme access issues in low and middle-income countries and the former relating to energy affordability issues in Europe. However, we share the pragmatism of Bouzarovski and Petrova (2015) in seeing that "all forms of household-scale energy deprivation share the same consequence: a lack of adequate energy services in the home, with its associated discomfort and difficulty". Definitions of energy vulnerability often take the household as the unit of analysis, aiming to address social issues of accessibility to energy services and sensitivity to this phenomenon. In connection with this line of thought, Bouzarovski and Petrova (2015) define energy vulnerability as the propensity to experience an inadequate or insufficient amount of energy services. Authors such as Murias et al. (2020) highlight that one of the greatest challenges regarding the study of energy vulnerability from a household perspective is related to the lack of a precise definition of the concept and the persisting confusion that exists between energy and socioeconomic vulnerability. By comparison, some authors such as Gatto and Drago (2020) define energy vulnerability at a larger scale, seeing it as the state in which an energy system is unable to face or deal with adverse events and becomes at risk of falling into economic, social, environmental, and institutional traps. As can be seen in the literature, there is a great diversity of definitions for the concept of energy vulnerability. Gatto and Busato (2020) argue that the lack of a formal definition prevents an effective mainstreaming and measurement of energy vulnerability. However, despite the shortcomings, Murias et al. (2020) argue that a trend toward a commonplace for defining energy vulnerability from a household perspective can be observed in the literature, which coincides in three main points. First, energy vulnerability is considered as something potential or a risk, unlike the concept of energy poverty, which is characterised as an actual situation of need. Second, vulnerability is considered a dynamic phenomenon, dependent on aspects like place and time. Finally, the multidimensional nature of energy vulnerability is recognised as it is conditioned by various factors (social, economic, environmental, etc.). Therefore, energy vulnerability can be understood as a multidimensional concept useful for analysing or evaluating the propensity, risk, or likelihood of an energy system or a household to experience a state where energy services are inadequate or insufficient due to conditions of structural inequality or marginalisation over time or due to specific and punctual adverse events or disturbances.

It can be argued that the concepts of resilience and vulnerability share a strong connection across the different levels of the energy system, and in general it is possible to conceptualise them as different sides of the same coin that represents energy systems. While energy vulnerability represents the elements that define the propensity of an energy system or a home to fall into a deficient state, causing inadequate or insufficient energy services, energy resilience comprises all the existing elements in that same system to anticipate, absorb, recover, or adapt avoiding such a deficient state.

Due to the primary importance that energy services have for human development, the concepts of energy resilience and energy vulnerability are of great importance for the governance of the energy systems that provide them, especially within the context of climate change, where extreme natural events and energy transitions are becoming increasingly important (Murias et al., 2020). Within this context, the development of resilience represents one of the greatest challenges to improve the quality of life and well-being of people, especially in low- and middle-income countries (Gatto and Drago, 2020), like those in Latin America. Issues of energy vulnerability and resilience can also hold a special significance for vulnerable groups, such as rural populations, women, elderly people, or people with different health needs. Therefore, in the context of energy transitions, the concepts of energy vulnerability and resilience can be understood through frameworks, such as UNESCO's four dimensions of sustainable development: society, environment, culture, and economy (2021), in which a technical dimension could be added to represent the sociotechnical nature of energy systems (Jenkins et al., 2016; Sareen and Haarstad, 2018).

3. Illustrative Case Studies: The Diversity of Energy Transitions in Latin America

The following illustrative case studies provide the reader with a snapshot of how environmental or climate aspects (e.g., freezing temperatures, lack of rain due to climate events, and strong winds from hurricanes) constitute relevant vulnerabilities at the systems and household level for different Latin American countries. At the same time, such case studies provide a good entry point for discussing how these countries are building or could build more resilient power systems within their own processes of energy transition, to provide energy to their population that is not only clean but also reliable.

3.1 Colombia and the Influence of El Niño on a Vulnerable Hydropower

The Colombian case study highlights the importance of energy resilience and diversification of energy resources in the context of low-carbon transitions by illustrating how an initially "clean" power system highly reliant on hydropower can be vulnerable to environmental variables linked to climate change. Electric power generation in Colombia is mostly derived from renewable sources (73.2%), with only 26.8% of generation coming from nonrenewable energy (XM, 2021). As can be seen by Table 10.1, hydropower is a key element of installed capacity in Colombia, accounting for nearly 70% in 2018. Therefore, it is extremely important for the country to maintain permanent monitoring of water reservoirs since most hydropower plants are large capacity. However, coal (9.75%) and gas (9.61%) still remain very relevant players in non-variable electricity generation. By comparison, solar and wind power account for less than 1% of total power generation capacity (UPME, 2018).

In combination with the previously mentioned additions to the Colombian power generation matrix, it is expected that 11 solar projects accounting for 800 MW of solar photovoltaic will be added following a third energy auction held at the end of 2021 (Minergia, 2021). Such capacity is in addition to the 1,365 MW of combined wind and solar capacity awarded in the second auction in 2019, which are currently under construction or entering operation (ibid.). However, from 2019 to 2020, Colombia also increased the total percentage of power generation based on nonrenewable energy sources, going from 21% to 27%, in order to counteract seasonal low water inputs that represent a serious threat to the Colombian power system (XM, 2021).

The problem of climate change may be observed in Colombia with the El Niño southern oscillation (ENSO) phenomenon, which is the primary driver of the hydroclimate and has a strong influence on the power sector (Henao et al., 2020). This phenomenon occurs with greater intensity, frequency, and duration (ibid.). Since the beginning of the nineties, the episodes of the El Niño phenomenon, aggravated by climate change, have affected the economy and well-being of the Colombian population, reaching $564 million in damages from 1997 to 1998, with losses due to electricity supply issues representing around $308 million alone (CEPAL, 1999). This is due to the broad predominance of hydroelectricity in the country's energy matrix and the confidence of governments in hydroelectricity as a source of energy security that the country would enjoy in the near future (ibid.). Indeed, during the last third

Table 10.1 Installed Capacity for Electricity Generation by Technology in Colombia (MW) Across Years

Technology	2015	2016	2017	2018	Increase (%)
Hydro	10,945.80	11,532	11,682	12,258.40	12
Wind	18.4	18.4	18.4	18.4	0
Solar PV	0	0	9.8	9.8	–
Bioenergy	77.2	83.1	132.7	146.7	90
Coal	1,016	1,369	1,352	1,727	70
Oil	297	299	187	309	4
Gas	1,848	1,698	2,095	1,703.30	−7.8
Diesel	1,023	1,247	931	1,240	21.2
Jet A1	46	46	44	44	−4.3
Mix Gas–Jet A1	276	264	264	264	−4.3
Total	**15,547.40**	**16,556.50**	**16,715.90**	**17,720.60**	

Source: SIEL, 2018.

of the 20th century, the expansion strategy of Colombian energy policy was to replace the then precarious thermoelectric plants with hydropower plants (Ministerio de Minas y Energía, 2021). In doing so, investments also accelerated the build-out of the nascent National Interconnection System (SIN), with notable anticipation of the installation of the respective distribution networks and interconnection in the five most populated regions of the country (Ministerio de Minas y Energía, 2021; XM, 2021).

The predominance of hydroelectric generation in the power supply, in combination with the extreme drought of 1992–1993, positioned Colombia for a situation in which there was a decline in potential of the reservoirs without thermoelectric plant readiness to compensate for the power demand of the interconnected Andean and Caribbean areas (Planas-Martí et al., 2019). The programmed rationing of electricity for the Colombian population was not long in coming but stayed for over six months (Gobierno de Colombia, 2015). The recessive effects on the economy and the welfare of the Colombian population proved to be a hard lesson that prompted rapid institutional and regulatory changes in 1994 to diversify the energy mix. Such diversification was mainly based on incentives to increase the number and capacity of thermal plants to generate firm backup power in the event of reduced hydroelectric availability caused by the El Niño phenomenon (Colombia, 1994; Ministerio de Minas y Energía, 2021).

Within this context, the El Niño phenomenon between 1994 and 2014 tested the resilience acquired by the construction and commissioning of suitable thermoelectric plants to serve as backup during critical junctures in hydropower generation. In fact, the 1997–1998 El Niño was as intense as that of 1992–1993, but its economic impact was more mild to moderate (CEPAL, 1999). Moreover, the phenomenon did not affect the energy sector either, which at that juncture was more flexible and resilient, allowing it to adapt to the climatic phenomenon thanks to advances in interconnection and the support of non-variable or non-intermittent thermal power. Therefore, in these occasions, the dramatic reduction in reservoir levels did not translate into forced energy rationing.

The economic incentive that drove private investment in thermal and hydropower generation based on the complementary strategy between hydropower and thermal plants was known as the Reliability Charge, implemented in 2006 (Botero Duque et al., 2016). This mechanism represents an additional income to generators, financed in part by the collection of tariffs from consumers. It is distributed, through an auction process, in proportion to the offer of firm power and must always be available, especially at critical junctures when El Niño causes severe droughts, for which impacts are intensified by global warming.

This novel mechanism operates as a short-term incentive for the water system, and long-term as support for the financial sustainability of generators, and control of energy price volatility (Juvinao, 2021). The antecedents of the Reliability Charge go back to Laws 142 and 143 in 1994 that led to the expansion of the power system, and the interconnection based on incentives for private investment and the organization of a regulated energy market. These laws respectively established the provision of home public services and the power system, with a market regulated by a commission that is independent of public powers and the private sector.

The litmus test to the Colombian system's resilience after expansion and strengthening came roughly two decades following 1992–1993, during El Niño phenomenon in 2015–2016. However, despite expansion of the system's infrastructure and capacity, a succession of shortcomings cast doubt on the effectiveness of short-term management to such an extent that the specter of electricity rationing from the early '90s once again haunted the country's interconnected region (Mateus Valencia, 2016).

Due to this, generating companies and Public Ministry entities have shown concern since mid-2014 for the delay in readjusting the scarcity price, a key mechanism/incentive used to stimulate the supply of backup power. Delays by the regulatory commission resulted in losses from supporting generation technologies (Ahumada, 2017), and there were concerns about projected fluctuations in reservoir levels being out of date by the time the climatic phenomenon was confirmed (ibid.). Such predictions were finally updated in 2016 at the height of the climate crisis experienced by the Colombian power system, which according to Mateus Valencia (2016) would only need to lose an additional 130 MW to trigger the imposition of generalized energy rationing across the country.

A voluntary plan to subsidize energy savings and penalize excessive consumption took three months to apply, resulting in achieved savings remaining below the target of 5% per month, established to avoid stressing the power system (Mateus Valencia, 2016). Additionally, during these two years of drought, natural gas supply to thermal generation stations also presented some problems, being unable to keep up with the crisis. Under this context, Colombia had no alternative but to replace electricity generation based on hydropower or natural gas with higher-priced liquid fuels (Mateus Valencia, 2016; Morcillo et al., 2020).

At the peak of the drought experienced in 2016, water contributions to the reservoirs dropped to an average of 31% – lower than that of 1992 (Mateus Valencia, 2016). During this year, Colombia's power system generated 52% of its electricity by thermal power and 48% by hydropower plants (ibid.). However, despite many shutdowns of hydroelectric and thermal power plants due to technical and logistical difficulties, Colombia's power system was able to avoid blackouts and scheduled rationing by operating at full capacity across the remaining power plants. Overcoming the crisis caused by El Niño, amid so many signs of failures in the financial and technical management of the interconnected system, provides two insights: (1) the interconnected system demonstrated sufficient resilience amid all the technical, climatic, administrative, and logistical adverse events; (2) without ignoring the seriousness of the technical and administrative failures determining power losses in the midst of the drought, the delay in readjusting the scarcity price stands out as the most serious deficiency suffered by the Colombian power system since it is a key parameter for a reliable backup thermal generation. Therefore, it is imperative that the allocation of economic benefits from regulatory mechanisms are able to outweigh the potential challenges of climatic phenomena to guarantee a resilient Colombian power system.

3.2 Cuba and the Development of Resilient Wind Power

The Cuban case study exemplifies the vulnerabilities faced by a power system that heavily relies on costly foreign fossil fuel for electricity production. This case study also illustrates the quest of Cuba to build a more resilient and clean energy matrix based on a recent increase in the installation of wind power and its adaptation process to the Cuban environmental and sociopolitical landscape. As Table 10.2 illustrates, electric power in Cuba is produced with a generation capacity of over 6,500 MW. The main component of the National Electric Power System (SEN) is thermal generation, which produces approximately 60% of the country's electricity (ONEI, 2021). Nevertheless, Cuba also has about 2,700 MW in both diesel and other fuel engines distributed throughout the country (ibid.), providing electricity access to the general population, and stability to the Cuban grid. Currently, only around 4.5% of the total generation is produced from renewable energy sources. At the moment, the consumption of fossil fuels at the national level is close to five million tons, of which about half is imported fuel, representing a considerable expense for the country (ibid.). Within this context, Table 10.2 illustrates Cuba's dependence on electricity from fossil fuel sources. It also shows how Cuban power production has been diversifying since 2016 with the entry of wind and solar photovoltaic generation.

Table 10.2 Electricity Generation Capacity Installed in Cuba from 2016 to 2020 in MW

Year	Total	Thermoelectric	Combined-Cycle Gas Turbines (CCGT)	Diesel	Public Service Power Plants				
					Isolated Diesel	*Other Fossil (a)*	*Hydro*	*Wind and Solar PV*	*Other Thermal (b)*
2016	6,453.6	2,525	580	117.8	32.3	2,592.0	65.9	46.6	494.0
2017	6,475.9	2,528	580	113.0	75.2	2,497.6	65.9	85.1	531.0
2018	6,661.0	2,498	580	114.1	95.8	2,617.2	64.0	139.0	553.0
2019	6,507.8	2,498	580	114.1	106.1	2,527.5	64.0	159.2	458.9
2020	6,660.5	2,498	580	111.2	105.2	2,515.0	64.6	221.5	565.0

Note:
(a) Includes different scales of electricity generation using fossil fuels.
(b) Includes self-supply schemes from the Energy and Mines Ministry and from sugarcane producers (AZCUBA).
Source: ONEI, 2021.

Despite wind power's novel application in the Cuban power system, it reflects the strongest annual growth in terms of installed capacity, among renewables and other energy technologies. Such growth is driven both by large investments and by the large size of the turbines being built or tested today in Cuba.

It is estimated that the development of eight new wind projects in the east and central region of Cuba would prevent the island from emitting around 184,000 tons of carbon dioxide and other greenhouse gases per year (Reyes Tamayo and Rodríguez Córdova, 2018). The Engineering and Projects Company for Electricity (INEL), from the Cuban government, estimated the potential wind resources in the Greater Antilles could reach 1,200 MW and is thus a promising environmental solution for Cuban energy transition (ibid.). Nevertheless, the elongated and narrow shape of the island makes it vulnerable throughout its geography to extreme meteorological events such as hurricanes and tropical storms, these two natural elements being the main external agents causing damage to wind technologies (Medrano Hernández et al., 2019).

Future investment in wind power is planned for the north coast of the island, in the easternmost region (Ministerio de Minas y Energía, 2021a). However, its coastal position presents a double risk to investment costs in the presence of extreme winds and saline spray of seawater, which present problematic conditions in terms of the structure, electronics, and control of the turbines. Facing such technical and operating vulnerabilities, the Cuban Wind Program has considered strict compliance with policies defined through the investment process of wind farms in Cuba (MEP, 2021). These policies aim to guarantee a resilient wind energy sector by adopting a multidimensional approach, including the assimilation of this energy technology at the social level, as well as the inclusion and stabilization of wind energy in the national energy matrix. The most relevant characteristics taken into account to ensure a resilient development of wind power in Cuba are listed here:

- National ownership of wind power data and training of Cuban personnel: Studies of wind potential resources at the national level must be in collaboration with national companies and endorsed by international institutions (Reyes Tamayo and Rodríguez Córdova, 2018).

This decision allowed the wind resource database to be owned by the country and not by foreign companies, providing also an organic process of training and preparation for Cuban personnel throughout the measurement of wind resource and evaluation campaign (Ministerio de Minas y Energía, 2021a).

- Wind power matching the expected energy consumption in the country: Investments in the wind power sector must be in line with the energy demand policies coming from different sectors and with the future development plans at the national and local level (Martín Barroso et al., 2021).
- Linking national companies, institutes, and universities: Knowledge and know-how must be part of the investment made so the development of this energy source remains a transcendent public good for the country over time (Moreno Figueredo, 2012).
- Selection of appropriate technologies according to geographic location: The use of completely folding turbines in areas where the passage of hurricanes is frequent, or the selection of specific wind turbines for certain wind class sites reflects relevant examples to illustrate this point (Torres-Durán and Moreno-Figueredo, 2018).
- Social appropriation of technologies: Realizing that any wind power project proposal modifies how social groups interact with their vital spaces leads to the need for such projects to positively impact the cultural, economic, organizational, and consumer spheres. In this way, wind farms can be awarded new meanings, uses, and purposes, allowing different groups to control their own narrative of social transformation within the context of a technological boom and energy transition (Reyes Tamayo and Rodríguez Córdova, 2018).
- Development of own technologies: The Cuban wind power program requires the development of its own technologies based on the scientific and technical capacity available in the country. Within this context, national electro-mechanical workshops, specialized in the construction of large technological equipment, start with the development of medium and small machines that provide the basis for greater efforts. Such workshops also provide technical assistance and spare parts for technologies already installed by imports.

(Ibid.)

All these factors have allowed the Cuban Wind Program to mitigate some of the existing vulnerabilities of a nascent wind power industry using a multidimensional approach that includes technical, social, environmental, and economic approaches. Total independence is discernible in technological decision-making, access to technologies, availability of data for analysis, development of new investments, and preparation of highly qualified personnel in the different areas related to wind power generation. Such an approach has fostered the resilience of the entire Cuban Wind Program, both from a technological point of view, as well as its linkage with the social and economic programs planned by the country.

However, the wind program has not been free of technical, organizational, and human mistakes, and setbacks, the most notable of which are related to the investor process, given limited experience in the installation and execution of wind projects. According to Reyes Tamayo and Rodríguez Córdova (2018), the nonexistence of a program from the Cuban government specialized in wind energy in combination with the struggles of wind farm power operators has translated into difficulties for the effective integration of wind power in the national electrical system.

Even if these past experiences allow us to identify where there is still room for improving the energy resilience of Cuban wind power, the experience so far has shown the success of Cuban wind power development at the national and local level (MEP, 2021). Within this context, investment in the Cuban wind power sector has also generated wider benefits for the population, such as the creation of new jobs in areas removed from urban centers and the renovation of highways (Reyes Tamayo and Rodríguez Córdova, 2018), which was essential for connecting remote areas where the wind parks are located. Such measures aim for the appropriation of technology by residents and the contribution of wind power to the local development, becoming a heritage asset of the communities and not only a company operating in its territory.

3.3 Mexico and Rolling Blackouts due to Low Temperatures and Texan Natural Gas Dependence

The Mexican case study portrays the technical and geopolitical vulnerabilities faced by the Mexican power system, which is heavily reliant on natural gas imported from the neighboring United States, by analyzing the domino effect that the Texan power outages in February 2021 had at the other side of the border. At the same time the case study portrays how the aforementioned energy vulnerabilities at the system level between Texas and Mexico cascade down to the household level impacting millions of Mexican families. As can be seen in Table 10.3, the highest installed capacity of power generation in Mexico is represented by combined-cycle gas turbines, with 39.2% of the national total installed capacity.

Table 10.3 Installed Capacity by Technology in MW

Technology	2017	2018	2019	2020	2021	MAGR (2017–2021)*
Hydro	12,612	12,612	12,612	12,612	12,614	0.0%
Geothermal	899	899	899	951	976	2.1%
Wind	3,898	4,866	6,050	6,504	7,691	18.5%
Solar PV	171	1,878	3,646	5,149	7,026	153.2%
Bioenergy	374	375	375	378	408	2.2%
Nuclear	1,608	1,608	1,608	1,608	1,608	0.0%
Efficient cogeneration	1,322	1,709	1,710	2,305	2,309	15.0%
Clean total percentage	**30.7%**	**32.8%**	**34.3%**	**35.5%**	**36.5%**	
CCGT	25,340	27,393	30,402	31,948	35,060	8.5%
Conventional thermoelectric	12,665	12,315	11,831	11,809	11,809	−1.7%
Turbogas	2,960	2,960	2,960	3,545	3,781	6.3%
Internal combustion	739	880	891	850	734	−0.2%
Coal	5,463	5,463	5,463	5,463	5,463	0.0%
Conventional percentage	**69.3%**	**67.2%**	**65.7%**	**64.5%**	**63.5%**	
Total installed capacity	**68,051**	**72,958**	**78,447**	**83,122**	**89,479**	**7.1%**

Note: *MAGR = mean annual grow rate. The change in the value of measurement over the period between 2017 and 2020.
Source: CENACE, 2021.

Hydroelectric energy represents the highest share of renewables at 14.1% and all the clean renewable technologies contribute 32.1% of the total in the country (CENACE, 2021). However, it should be noted that there has been sustained growth in solar photovoltaic power, which has grown annually at an average rate of 153%, rising from 171 MW of capacity in 2017 to 7,026 MW in 2021 (ibid.).

A large share of gas used in the combined-cycle gas turbines originates from the US, which as the following section outlines, left Mexico exposed to winter blackouts in 2021, and highlighted the need for power systems that are not only cleaner but also more resilient and less vulnerable.

In February 2021, an uncharacteristically harsh winter weather in Texas (recording temperatures of $-14°C$) resulted in the loss of electricity for about ten million people in the state during peak demand and, in the following days, for the less fortunate (Busby et al., 2021). According to preliminary reports from the Electrical Reliability Council of Texas (ERCOT), climate-related problems, fuel limitations, and equipment failures led to the loss of 51 GW of electricity generation capacity, causing blackouts in the state (ERCOT, 2021). Climate-related problems were one of the most obvious causes of the blackouts related to the inability of the Texan system to react to the low temperatures. These causes are related but not limited to the freezing of gas pipes and water used in nuclear power, accumulation of ice on the blades of wind turbines, snow or ice covering solar panels, and the flooding of equipment due to melting of ice and/or snow (ibid.). From an energy resilience point of view, the aforementioned causes are related to environmental variables such as the management of natural resources necessary for power generation (such as water), effects due to extreme weather and the effects of climate change, and the lack of institutional capacities to face extreme weather events. Other reports, such as that of the University of Texas at Austin (2021), indicate that there is a great diversity of causes of a multidimensional nature, such as the lack of regulations related to winterization, a spike in energy demand, generation units experiencing outage, and so on (King et al., 2021). Finally, from the perspective of energy vulnerability in households, King et al. emphasize the effect of electricity price escalation in the market, which impacted energy affordability for households with variable price contracts.

The events leading to failures in the Texan power system had a domino effect in Mexico, with almost five million clients of the state electricity company Comisión Federal de Electricidad (CFE) experiencing blackouts in states bordering Texas due to a "generation deficit" (BBC, 2021a). These first disruptions in turn led to rotating load outages in 12 other states across the country (BBC, 2021b). On the Mexican side, the problems of vulnerability and resilience were closely related to dependency on the United States for the supply of natural gas and specifically due to the lack of supply from the pipelines that import the fuel from Texas.

According to information reported by the CFE (2021), the price of natural gas increased 5,000%, going from $3 per unit of volume to more than $200, reaching $600 in some parts of the United States. This strong reaction to the lack of natural gas made evident a double dependence that puts the Mexican power system in a strong state of vulnerability: the heavy reliance on imported natural gas as a primary source of electricity generation and the reliance on the United States as the main supplier.

When looking at these problems from the point of view of energy resilience, it is possible to name a number of variables that influenced this domino effect that heavily affected the Mexican system. From a technical point of view, this phenomenon is a wake-up call to increase the resilience of the Mexican power system by diversifying the energy supply, making it possible to balance the reliance on natural gas in the power generation at the national level, as well as a call to improve the reliability of the energy infrastructure by increasing the capacity of the natural gas reserves when a similar event happens in the future. From an environmental point of view, the crisis caused by dependence on gas pipelines, specifically in Texas, demonstrated that the geographic distribution of energy infrastructure must be considered to support a power system that is resilient to extreme weather events. At the same time, this incident highlights the need to build or strengthen institutional capacities to deal with natural disasters or extreme weather events. It is necessary to find strategies that allow Mexico to reduce its dependence on imported natural gas, specifically from Texas, or, failing that, to generate agreements that assure Mexico greater energy security, reducing the risk of events such as those of February 2021.

Analyzing this case study from an energy vulnerability perspective using an economic and market dimension, variables such as uncertainty in energy prices and energy affordability are represented very differently at the system or domestic level in Mexico. This is mainly because electricity prices are controlled by state agencies to be kept at a fixed price for domestic users, despite the sudden rise in natural gas prices for electricity generators. In the context of energy resilience, one of the most important dimensions to note is organizational due to the strong energy dependence of the Mexican electricity system on natural gas from the United States. In this sense, thinking about vulnerability from an institutional and regulatory point of view can lead one to surmise the need to develop and implement public policies and regulatory frameworks that analyze the weaknesses of the energy system and the necessary measures to reduce vulnerability (Gnansounou, 2008), as well as the promotion of programs for energy efficiency or energy saving (Michalec et al., 2019).

An analysis from an energy vulnerability perspective runs parallel to the lack of resilience of the Mexican power system but places emphasis on how different populations in Mexico may be prone to facing a lack of energy services in different ways when departing from initial circumstances of energy poverty (Cedano et al., 2021). The ways in which Mexican households experienced and coped with the power outages of February 2021 depended on a range of variables linked to the social dimension of vulnerability. In the first instance, the physical conditions of domestic properties influence the efficiency, risk and other characteristics of the home and play a great role in the way in which people experience or cope with energy vulnerability (Llera-Sastresa et al., 2017; Thomson et al., 2017; Willand et al., 2021). Likewise, energy vulnerability largely depends on the demographic characteristics of the people in the household, such as health conditions (Murias et al., 2020), stability of wages (Middlemiss et al., 2015), and gender (Robinson, 2019).

3.4 Case Studies Synthesis and Discussion

An initial look at energy transition processes in Latin America as a block provides a picture of fairly advanced attainment of low-carbon power systems, with 65% renewable energy share in power generation (IRENA, 2020a), yet national realities of energy transition processes are quite diverse. Within this context, this chapter analyzed the struggle experienced by countries like Cuba or Mexico, to overcome a power system that relies heavily on fossil fuels for the generation of electricity for domestic consumers and businesses. For these countries, the transition to low-carbon electricity generation and the diversification of their energy matrices also represents a path to reduce energy vulnerability. Even though both countries rely on fossil fuels for power generation, the environmental, economic, institutional, and geopolitical contexts in which they are both embedded represent a very different set of challenges toward a low-carbon and reliable power system.

The Mexican case study provides a multidimensional analysis of the general failure of the electrical system in February 2021, evidencing the geopolitical, environmental, and technical vulnerabilities on both sides of the US-Mexico border, stressing an exacerbated dependence on natural gas (Araújo and Shropshire, 2021; Busby et al., 2021). Within the Mexican context, the failure of the power systems in that cold month of February 2021 highlights the geopolitical dimension of energy vulnerability through the devastating effects that high dependence on natural gas from Texas produced. Aside from being a reminder of the need to promote the diversification of power generation to endow power systems with resilience, the Mexican case study highlights the need to develop means of collaboration between neighboring countries to overcome such energy vulnerability dilemmas and their effects on the household and systems levels.

However, there are some positive signs emerging in Mexico related to the recent acceleration in the amount of renewable capacity within the energy matrix (especially in terms of wind and solar power). By comparison, while Cuba experiences similar challenges to those faced by Mexico in terms of an energy matrix heavily reliant on fossil fuels, the island faces specific economic, environmental, and social challenges that play an important role in the relative energy vulnerability of its power system as a member of the Small Island Developing States (SIDS) (Genave et al., 2020). Nevertheless, the case study of Cuba's wind power illustrates some of the measures to improve the resilience of Cuban wind power, such as national ownership of data, training of Cuban personnel, and selection of appropriate technologies according to geographic location. Overall, the case study highlights the need for specific economic, geopolitical, and institutional considerations for Cuba and other SIDS that allow for the successful inclusion of renewable energy technologies within an energy matrix that is currently vulnerable to a heavy dependence on expensive foreign oil imports (ONEI, 2021).

Contrasting with the experiences of Mexico and Cuba to strive for greater low-carbon power generation, Colombia experiences a particular conundrum when dealing with issues of resilience and vulnerability, having a mostly clean or decarbonized electricity generation matrix based on a large share of hydropower generation. The case of Colombian hydropower and the influence of El Niño provides an opportunity to reflect on how issues of energy vulnerability can hamper processes of energy transition and decarbonization in countries with an already deep penetration of low-carbon or renewable energy generation technologies. The case of Colombia also demonstrates how a country can appeal to multidimensional solutions, including the development of financial mechanisms in combination with firm generation technologies to successfully mitigate the risks posed by extreme weather events to the power system.

Based on the analysis that was carried out using the different multidimensional variables of energy resilience and vulnerability for the three case studies in Colombia, Cuba, and Mexico, we call for further interdisciplinary research to deepen our understanding of these concepts. A more nuanced understanding of energy resilience and vulnerability could facilitate the development of power systems capable of dealing with the great diversity of technical, economic, social, institutional, environmental, and geopolitical challenges that the future holds. In particular, research is needed to explore the connections of energy vulnerability at the systems and the household level and to develop conceptualizations of these terms that better encapsulate the diversity of contexts where they are experienced.

Table 10.4 provides a summary of the most relevant features related to energy vulnerability and resilience present in the case studies from Colombia, Cuba, and Mexico, analyzed according to environmental, technical, economic, social, and institutional dimensions.

Table 10.4 Summary of Relevant Features Related to Energy Vulnerability and Energy Resilience Across the Three Case Studies

	Colombia	*Cuba*	*Mexico*
Environmental	Vulnerability to changes in precipitation due to El Niño southern oscillation (ENSO)	Vulnerability of wind power to tropical storms and strong winds	Vulnerability to extremely low temperatures
Technical	Vulnerability to lack of water reserves for hydropower to function properly. Resilience through polluting and expensive, but non-variable thermal power	Vulnerability to salinity and extreme winds. Resilience through the selection of technologies like wind turbines with folding blades	Vulnerable to failing infrastructure due to low temperatures and to the dependence on natural gas from the United States
Economic	Vulnerability toward increased fuel costs of thermal power. Resilience through Reliability Charge mechanisms	Vulnerable to limited experience in the installation and execution of wind projects. Resilience through nationally owned data and training of Cuban personnel	Vulnerability to a sharp increase in natural gas prices
Social	Vulnerability to forced energy rationing. Resilience through a voluntary plan to subsidize energy savings	Resilience through social appropriation and assimilation of wind power	Vulnerability to rolling blackouts. Resilience through a vast interconnected electricity network and subsidized domestic tariffs
Institutional	Vulnerability related to delayed institutional decision making to discourage increases in energy consumption	Vulnerabilities due to a lack of specialized national programs on wind power. Resilience through national ownership of energy operation	Vulnerability to geopolitical dependence on energy resources from the United States

4. Conclusions and Priorities for Future Research

The concepts of energy resilience and energy vulnerability so far lack definitions upon which the academic community unanimously agrees. Nevertheless, within this chapter, the authors provide a working definition for the concept of energy resilience as a multidimensional concept useful for analyzing and evaluating the capacity of a power system, at different scales (from household to national), to anticipate, absorb, recover, or adapt to situations where energy services are insufficient or inadequate. On a related front, energy vulnerability can be understood as a multidimensional concept useful for analyzing or evaluating the propensity, risk, or likelihood of a system or a household, to experience a state where energy services are inadequate or insufficient due to conditions of structural inequality or marginalization over time or due to specific and punctual adverse events or disturbances.

The findings delineated across this chapter highlight the need to generate research dedicated to reviewing the conceptualization of energy resilience and energy vulnerability, pulling from the different strings of literature to develop comprehensive definitions of both concepts. Such comprehensive research looking to deepen our conceptual understanding of energy resilience and energy vulnerability would also allow us to better identify the similarities and differences between both concepts in a much more nuanced way. Therefore, this chapter makes a call for a comprehensive review of the various existing definitions that circulate in the academic literature. The authors of this chapter believe that such a contribution would facilitate the embedding of the energy resilience and vulnerability concepts in policy- and decision-making, influencing the development of more comprehensive and more integrated policies and regulatory frameworks.

Finally, this chapter also identifies the need to promote a pan–Latin American energy vulnerability and resilience network, where members of the public and private sectors, universities, and civil society can collaborate in an inclusive way and become better informed in order to generate effective and efficient solutions to address the major issues that pertain to energy resilience and vulnerability.

References

Adil, A.M. and Ko, Y. (2016) 'Socio-Technical Evolution of Decentralized Energy Systems: A Critical Review and Implications for Urban Planning and Policy', *Renewable and Sustainable Energy Reviews* 57: 1025–1037. doi: 10.1016/j.rser.2015.12.079.

Ahmadi, S., Saboohi, Y. and Vakili, A. (2021) 'Frameworks, Quantitative Indicators, Characters, and Modelling Approaches to Analysis of Energy System Resilience: A Review', *Renewable and Sustainable Energy Reviews* 144, 1–17. doi: 10.1016/j.rser.2021.110988.

Ahumada Rojas, Ómar (2017) 'Usuarios quedarían blindados contra el riesgo de apagones en sequías', www.eltiempo.com/economia/sectores/nuevo-modelo-de-precio-de-escasez-protegeria-a-usuarios-de-apagones-144452.

Araujo, Kathleen (2017) *Low Carbon Energy Transitions: Turning Points in National Policy and Innovation.* New York: Oxford University Press.

Araújo, Kathleen, and Shropshire, David. (2021) 'A Meta-Level Framework for Evaluating Resilience in Net-Zero Carbon Power Systems with Extreme Weather Events in the United States', *Energies* 14(14).

BBC (2021a) 'Apagones en México: la Enorme Dependencia Mexicana del gas de EE.UU. Que Dejó la Descubierto la Tormenta Invernal en Texas', *BBC.* www.bbc.com/mundo/noticias-america-latina-56106262 (Accessed: 20 September 2021).

BBC (2021b) 'Apagones en México: La Histórica Tormenta Invernal en TEXAS que ha Causado Cortes Eléctricos en la Mitad del País Latinoamérica', *BBC.* www.bbc.com/mundo/noticias-internacional-56078326 (Accessed: 20 September 2021).

Boardman, B. (1991) *Fuel Poverty: From Cold Homes to Affordable Warmth.* London: Belhaven Press.

Botero, Duque, Pablo, Juan, García, John J. and Velásquez, Hermilson. (2016) 'Efectos Del Cargo Por Confiabilidad Sobre El Precio Spot de La Energía Eléctrica En Colombia', *Cuadernos de Economía* 35(68): 491–519.

Bouzarovski, S. and Petrova, S. (2015) 'A Global Perspective on Domestic Energy Deprivation: Overcoming the Energy Poverty-Fuel Poverty Binary', *Energy Research and Social Science* 10: 31–40. doi: 10.1016/j.erss.2015.06.007.

Busby, J.W. et al. (2021) 'Cascading Risks: Understanding the 2021 Winter Blackout in Texas', *Energy Research and Social Science* 77: 102106. doi: 10.1016/j.erss.2021.102106.

Cedano, K.G., Robles-Bonilla, T., Santillan, O.S. and Martinez, M. (2021) 'Assessing Energy Poverty in Urban Regions of Mexico: The Role of Thermal Comfort and Bioclimatic Context', *Sustainability* 13(19): 10646. https://doi.org/10.3390/su131910646.

CENACE (2021) 'Programa de Ampliación y Modernización de la Red Nacional de Transmisión y Redes Generales de Distribución del Mercado Eléctrico Mayorista PAMRTNT 2021–2035', *Centro Nacional de Control de Energía*, p. 896.

CEPAL (1999) 'Efectos Macroeconómicos Del Fenómeno El Niño de 1997–1998. Su Impacto En Las Economías Andinas ENSO 101.'

CFE (2021) 'Ante Bajas Temperaturas en EU, Texas Suspende Suministro de Gas Natural a la CF, Comunicación', https://app.cfe.mx/Aplicaciones/OTROS/Boletines/boletin?i=2104 (Accessed: 21 December 2021).

Congreso de Colombia (1994) Law 143.

ERCOT (2021) 'February 2021 Extreme Cold Weather Event: Preliminary Report on Causes of Generator Outages and Derates', Texas, www.ercot.com/content/wcm/lists/226521/51878_ERCOT_Letter_re_Preliminary_Report_on_Outage_Causes.pdf.

Gatto, A. and Busato, F. (2020) 'Energy Vulnerability Around the World: The Global Energy Vulnerability Index (GEVI)', *Journal of Cleaner Production* 253. doi: 10.1016/j.jclepro.2019.118691.

Gatto, A. and Drago, C. (2020) 'Measuring and Modelling Energy Resilience', *Ecological Economics* 172(November 2019): 106527. doi: 10.1016/j.ecolecon.2019.106527.

Genave, A., Blancard, S. and Garabediana, S. (2020) 'An Assessment of Energy Vulnerability in Small Island Developing States', *Ecological Economics* 171. doi: 10.1016/j.ecolecon.2020.106595.

Gnansounou, E. (2008) 'Assessing the Energy Vulnerability: Case of Industrialised Countries', *Energy Policy* 36(10): 3734–3744. doi: 10.1016/j.enpol.2008.07.004.

Gobierno de Colombia (2015) 'El Fenómeno de El Niño. Riesgos de La Suspensión de Operaciones de El Quimbo', Bogota, Colombia.

Gößling-Reisemann, S., Hellige, H.D. and Thier, P. (2018) 'The Resilience Concept: From Its Historical Roots To Theoretical Framework for Critical Infrastructure Design', *Universität Bremen*, Nr. 217: 81. www.uni-bremen.de/artec.

Henao, F. et al. (2020) 'Annual and Interannual Complementarities of Renewable Energy Sources in Colombia', *Renewable and Sustainable Energy Reviews* 134(September). doi: 10.1016/j.rser.2020.110318.

Holling, C.S. (1996) 'Engineering Resilience Versus Ecological Resilience', in Schulze, P.C. (ed.) *Engineering within Ecological Constraints*. Washington, DC: National Academy of Sciences, pp. 1–222. http://books.google.com/books?hl=en%7B&%7Dlr=%7B&%7Did=lv2cAgAAQBAJ%7B&%7Doi=fnd%7B&%7Dpg=PT39%7B&%7Ddq=Engineering+Resilience+versus+Ecological+Resilience%7B&%7Dots=38gzTibxnS%7B&%7Dsig=XJGD-4ZsCr0H0-AsOOu4k6zJQqY.

IRENA (2020a) *Global Renewables Outlook: Energy Transformation 2050, International Renewable Energy Agency*. Abu Dhabi. www.irena.org/publications/2020/Apr/Global-Renewables-Outlook-2020.

Isherwood, B.C. and Hancock, R.M. (1979) *Household Expenditure on Fuel: Distributional Aspects*. London: Economic Adviser's Office, DHSS.

Jenkins, K., Sovacool, B.K. and McCauley, D. (2018) 'Humanizing Sociotechnical Transitions Through Energy Justice : An Ethical Framework for Global Transformative Change', *Energy Policy* 117(February): 66–74. doi: 10.1016/j.enpol.2018.02.036.

Jenkins, K. et al. (2016) 'Energy Justice : A Conceptual Review', *Energy Research & Social Science* 11: 174–182.

Juvinao, J.C. (2021) 'Lecciones Del Cargo Por Confiabilidad En Colombia Como Un Mecanismo de Incentivo a La Generación de Energía Eléctrica', *Desarrollo y Sociedad* 1(87): 113–148.

King, C.W. et al. (2021) *The Timeline and Events of the February 2021 Texas Electric Grid Blackouts*. Aust. Available at: https://energy.utexas.edu/sites/default/files/UTAustin %282021%29 EventsFebruary-2021TexasBlackout 20210714.pdf.

Llera-Sastresa, E. et al. (2017) 'Energy Vulnerability Composite Index in Social Housing, from a Household Energy Poverty Perspective', *Sustainability (Switzerland)* 9(5). doi: 10.3390/su9050691.

Martín, Barroso, Manuel, Ariel, Leyva Ferreiro, Grisell, and Cantero García, Mariela Francisca. (2021) 'La Sostenibilidad Económica de La Inversión Renovable En Cuba.' Perspectivas Pasadas, Presentes y Futuras Desde El Marco Regulatorio Nacional', *Cofin* 15.

Mateus Valencia, A.C. (2016) 'Crisis Energética en Colombia', *Tecnología Investigación y Academia* 4(2): 74–81.

Medrano Hernández, J.A., Moreno Figueredo, C. and Vaillant Rebollar, J.E. (2019) 'Estudio de Prefactibilidad Técnica del Aprovechamiento del Viento Como Recurso Energético en zonas Pre-Montañosas', *Ingeniería Energética* 40(3). www.redalyc.org/journal/3291/329160723006/html.

Mejia-Montero, A., Alonso-Serna, L. and Altamirano-Allende, C. (2020) 'The Role of Social Resistance in Shaping Energy Transition Policy in Mexico: The Case of Wind Power in Oaxaca,' in Noura Guimaraes, L. (ed.) *The Regulation and Policy of Latin American Energy Transitions*. 1st edn. Sao Paulo: Elsevier, pp. 397–415.

MEP (2021) Informe Nacional Voluntario: Cuba 2021. La Habana. https://sustainabledevelopment. un.org/content/documents/280872021_VNR_Report_Cuba.pdf.

Merino, F., Mejia-Montero, A. and Dastres, C. (2020) 'An Inclusive and Participative Model for Energy Transition in Latin America: the Case of Chilean Generación Comunitaria', in Noura Guimaraes, L. (ed.) *The Regulation and Policy of Latin American Energy Transitions*. 1st edn. Sao Paulo: Elsevier, pp. 392–412.

Michalec, A. (Ola), Hayes, E. and Longhurst, J. (2019) 'Building Smart Cities, the Just Way. A Critical Review of "Smart" and "Just" Initiatives in Bristol, UK', *Sustainable Cities and Society* 47(July 2017): 101510. doi: 10.1016/j.scs.2019.101510.

Middlemiss, L. and Gillard, R. (2015) 'Fuel Poverty from the Bottom-Up: Characterising Household Energy Vulnerability through the Lived Experience of the Fuel Poor', *Energy Research and Social Science* 6: 146–154. doi: 10.1016/j.erss.2015.02.001.

Minergia (2021) 'Nuevo Hito En La Transición Energética: Colombia Multiplicará Por Más de 100 Veces Su Capacidad En Energías Renovables', Ministerio de Minas y Energía de Colombia. www.minenergia.gov.co/en/web/10180/1332?idNoticia=24314285.

Ministerio de Minas y Energía (2021) 'Transición Energética: Un Legado Para El Presente y El Futuro de Colombia', Bogota, Colombia.

Ministerio de Minas y Energía (2021a) 'Eólica', www.minem.gob.cu/es/actividades/energias-renovables-y-eficiencia-energetica/eolica.

Morcillo, J.D., Angulo, F. and Franco, C.J. (2020) 'Analyzing the Hydroelectricity Variability on Power Markets from a System Dynamics and Dynamic Systems Perspective: Seasonality and ENSO Phenomenon', *Energies* 13(9). doi: 10.3390/en13092381.

Moreno Figueredo, C. (2012) 'Estado Actual y Desarrollo de La Energía Eólica En Cuba', ISPJAE. La Habana.

Murias, P., Valcárcel-Aguiar, B. and María, R. (2020) 'A Territorial Estimate for Household Energy Vulnerability: An Application for Spain', *Sustainability (Switzerland)* 12(15). doi: 10.3390/SU12155904.

Noura Guimarães, L. (2020) 'Is there a Latin American Electricity Transition? A Snapshot of Intraregional Differences', in Noura Guimarães, L. (ed.) *The Regulation and Policy of Latin American Energy Transitions*. 1st edn. Elsevier Inc., pp. 3–20. doi: 10.1016/b978-0-12-819521-5.00001-2.

ONEI (2021) 'Oficina Nacional de Estadistica e Informacion', www.onei.gob.cu.

Planas-Martí, María Alejandra, and Juan Carlos Cárdenas (2019) 'La Matriz Energética de Colombia Se Renueva', https://blogs.iadb.org/energia/es/la-matriz-energetica-de-colombia-se-renueva.

Reyes Tamayo, J.R. and Rodríguez Córdova, C.R.R. (2018) 'Necesidad de la apropiación social de la tecnología eólica en Cuba', *Universidad y Sociedad* 10(5): 113–120. Available at: http://scielo.sld.cu/pdf/rus/v10n1/2218-3620-rus-10-01-336.pdf.

Robinson, C. (2019) 'Energy Poverty and Gender in England: A Spatial Perspective', *Geoforum* 104: 222–233. doi: 10.1016/j.geoforum.2019.05.001.

Roege, P.E. et al. (2014) 'Metrics for Energy Resilience', *Energy Policy* 72: 249–256.

Sareen, S. and Haarstad, H. (2018) 'Bridging Socio-Technical and Justice Aspects of Sustainable Energy Transitions', *Applied Energy* 228(July): 624–632. doi: 10.1016/j.apenergy.2018.06.104.

SENER (2021) 'Programa de Desarrollo del Sistema Eléctrico Nacional. PRODESEN 2021–2035', *Secretaria de Energia* 148: 222.

Sharifi, A. and Yamagata, Y. (2016) 'Principles and Criteria for Assessing Urban Energy Resilience: A Literature Review', *Renewable and Sustainable Energy Reviews* 60: 1654–1677. doi: 10.1016/j.rser.2016.03.028.

SIEL (2018) 'Estadísticas y Variables de Generación', Sistema de Informacion Electrico Colombiano. www.siel.gov.co/Inicio/Generación/Estadísticasyvariablesdegeneración/tabid/115/Default.aspx.

Sovacool, B.K., Turnheim, B., Martinskainen, M., Brown, D. and Kivimaa, P. (2020) 'Guides or Gatekeepers? Incumbent-Oriented Transition Intermediaries in a Low-Carbon Era', *Energy Research and Social Science* 66(August 2019): 101490. doi: 10.1016/j.erss.2020.101490.90.

Thomson, H., Bouzarovski, S. and Snell, C. (2017) 'Rethinking the Measurement of Energy Poverty in Europe: A Critical Analysis of Indicators and Data', *Indoor and Built Environment* 26(7): 879–901. doi: 10.1177/1420326X17699260.

Thoren, H. (2014) 'Resilience as a Unifying Concept', *International Studies in the Philosophy of Science* 28(3): 303–324. doi: 10.1080/02698595.2014.953343.

Torres-Durán, Armando, and Conrado Moreno-Figueredo (2018) 'Evaluation of the Wind Potential in the Popular Council of Cojímar', *Revista Cubana de Meteorologia* 24(3). http://rcm.insmet.cu/index.php/rcm/article/view/432/554.

UNESCO (2021) *Sustainable Development*, https://en.unesco.org/themes/education-sustainable-development/what-is-esd/sd (Accessed: 22 December 2021).

UPME (2018) 'Informe Mensual de Variables de Generación y Del Mercado Eléctrico Colombiano – Agosto de 2018', *Unidad de Planeación Minero Energética* (69): 1–14. www.siel.gov.co/portals/0/generacion/2018/Informe_de_variables_Ago_2018.pdf.

Willand, N., Middha, B. and Walker, G. (2021) 'Using the Capability Approach to Evaluate Energy Vulnerability Policies and Initiatives in Victoria, Australia', *Local Environment* 26(9): 1109–1127. doi: 10.1080/13549839.2021.1962830.

XM (2021) 'Generación del SIN', *Reporte Integral de Sostenibilidad, Operación y Mercado 2020*. https://informeanual.xm.com.co/2020/informe/pages/xm/24-generacion-del-sin.html.

11

TRANSFORMATION OF THE DANISH WIND TURBINE INDUSTRY THROUGH PATH CREATION[1]*

Peter Karnøe and Raghu Garud

Introduction

In an op-ed piece in the *New York Times*, Paul Krugman, a Nobel laureate in economics, noted the remarkable decrease in the cost of wind energy over the past decade. The decline in cost is such that wind energy is more economical than energy from coal as of 2019. Asked if this decline in cost was because of luck, Krugman opined that it was "a consequence of good policy decisions" (Krugman, 2021).

While agreeing with Mr. Krugman, we argue that it is more than policy that shapes and nurtures technologies such as wind turbines, which are disadvantaged in the present but offer benefits in the future. It is through the interactions between the actors of an emergent cluster of users, producers, evaluators, concerned citizen groups, and regulators that a technological trajectory emerges and takes shape. We offer the notion of path creation to understand the dynamics involved (Garud and Karnøe, 2001; Garud, Kumaraswamy and Karnøe, 2010; Karnøe and Garud, 2012).

Our path creation perspective is based on a relational-temporal ontology, which conceptualizes agency as emergent in unfolding action-nets (Czarniawska, 2004, 2008). What distinguishes path creation from path dependence is the notion of situated action-nets, wherein actors have the capacity to pursue options based on their specific socio-material entanglements (see also sympathetic arguments by Sharp et al., this volume). Entanglements in action-nets is an interactive endeavor, with actors modulating their spheres of interactions with other actors and artefacts, all the while realizing that they can only attempt to influence but not determine the processes that unfold. Based on situated actions, actors try and mobilize existing resources to generate new options for the future based on generative memories of the past through active experimentation, discussions, and dialogue with others (Garud, Gehman, and Karnøe, 2019).

The setting for the path creation of wind energy in Denmark was an industrial district comprising diverse small- and medium-sized enterprises with a culture that valued hands-on machine design engineering besides collaborative learning between producers, suppliers, national regulators, and evaluators (Araújo, 2017; Karnøe, 1999a, Karnøe, 1999b, Garud and Karnøe, 2003, Lundvall, 1992). At the beginning, wind power was seen by incumbents as an irrelevant relic (Meyer, 2000). Instead, nuclear power was the promise of the future, serving as

DOI:10.4324/9781003183020-13

a vehicle for the upgradation of industrial competences by industry and labor unions, a new advanced energy source by electricity companies, and as the source of future energy by policy-makers and media.

We describe here the process underlying the transformation of heterogeneous resources in the Danish wind turbine district. While departing from existing tools, skills, and components, entrepreneurs confronted several challenges in their efforts at mobilizing support for funding and for developing their competencies (Karnøe, 1991, 1999b). It was not easy to activate and enroll suppliers to embark on exploratory learning processes. Moreover, a political plan for R&D support and the innovative model of market subsidies for small-scale wind turbines were not a part of the first official Energy Plan 1976.

The transition to wind energy required the transformation of artifacts, skills, and the identities of actors. Transformation was involved when (1) the electrical utilities agreed to grid-connect and pay for wind power due to political pressure; (2) the energy policy shifted to prioritize domestic import substituting energy sources, such as natural gas and wind power, support small-scale wind turbines with a test and research center, and institute an innovative model of market subsidies; (3) national and municipal planning transformed the natural landscape into categories that were used for shaping the location of wind turbines; and (4) the Wind Atlas from meteorological science became a calculative device that transformed the wind speed in different landscape types into an estimated energy production.

The generative process was not an orderly linear one, as emergent contingencies influenced the non-linear learning processes. An implication is that public policy to guide transitions cannot assume that linear learning dynamics will unfold. The non-linear learning dynamics included the scale-efficiency effects of wind power, which gradually transformed expectations and interests of both supporting and opposing stakeholders with the overall effect of increased commitments to wind power. Such a view is consistent with the one offered by Sharp et al. (this volume) where transitions are not based on radical changes brought about to overcome strong path dependencies. Overall, the emergence of wind energy in Denmark is an accomplishment based on a long journey involving multiple transformative acts.

The Emergence of the Danish Wind Turbine Cluster

In the 1950s, Juul (a utilities engineer) experimented with various wind turbine designs to develop a 200 kW wind turbine. This wind turbine had three blades with an asynchronous generator, double-brake system (blade tip brake and shaft brake), and active electromechanical yaw that operated on the grid (Karnøe, 1991:168–173).[2] Despite this accomplishment, a special utility commission decided in 1962 that the technical efficiency and cost economics associated with Juul's design were unattractive (Nielsen, 2001:11–20). However, Juul's wind turbine design is called the mother of the modern Danish wind turbine design, as the it informed the subsequent wind turbine entrepreneurs in the late 1970s. We describe the efforts of these entrepreneurs and other dynamics of activation and co-creation of heterogeneous resources (please see Table 11.1 for details of key activities).

Emerging Action-Nets in Design and Production

The Riisager Design

In the early 1970s, Riisager (a carpenter by trade) began experimenting with renewable energy after he was called in by a local resident to repair an old windmill (Jensen, 2003:19–20). Using

Table 11.1 Summary of Critical Activities in the Emergence of the Danish Wind Turbine Cluster

	Design and Production	Suppliers	Users and Market	Policy and Regulation	Test and Research Center
1940	FL Smidth,	KK Electronic	Industry, utility	Nuclear power is	Helge Pedersen
1962	Juul	Components	Riisager unlocks	the future	design aircraft
1974	Riisager	Økaer Blades	grid	Energy Plan '76	wings
1976	NIVE	Alternegy	DEF Guidelines	Large-scale wind	TRC begins,
1977	TVIND	LM Blades	Møller reports on	turbine program	Helge Pedersen
1978	Herborg		grid connected	Small-scale wind	director
1979	Vestas		turbine	turbine program	Reset TRC
1980	Nordtank		Farmers and	Subsidy scheme	interaction with
	Bonus		cooperatives buy	(30%) of	industry
			wind turbines	investment,	
			DWO established	approval	
				procedure	
1981	DWT established	Suppliers	Slow sales in	Enforcement of	Load paradigm and
1982	Search for new	continue	Denmark	policies for wind	approval criteria
1985	markets	New suppliers	Sales to California	power	Wind Atlas
1986	Number of firms	due to	Export boom	Extra subsidies for	method
1987	increase	sourcing	DK wind farms	wind farms	
1989		strategy	Export slows	New payment	
		and export	Utilities direct	scheme	
		growth	buyers	100 MW	
			New export markets	agreement	
			begin	Invest. Subsidy	
				removed	

a transformed gearbox from an old military tank, Riisager developed a 15 kW wind turbine. For blades, he used ship sails, which he mounted on wooden frames. As these blades did not function effectively, Riisager invited his son, a pilot with knowledge of aerodynamics, to help design a pear-shaped wooden blade. In an experiment with a hairdryer, Riisager discovered the now famous lift and drag mechanism, wherein the aerodynamic shape of the blade creates a low pressure as it rotates, thereby increasing the force of energy on the blade, leading to its acceleration (ibid.).

Riisager's wind turbine, which he installed in his garden, caught the attention of Medelbye, an engineer whose father had worked with Juul (Wistoft et al., 1992). Upon Medelbye's suggestion, Riisager tried an asynchronous motor (as in the Juul design) to connect his wind turbine to the grid. After operating the wind turbine for four days, which he connected to the grid using a washing machine plug, Riisager approach the director of the local electric utility to seek his permission. The director refused for technical and legal reasons.

However, a journalist who talked with Riisager after a meeting mistakenly reported in a local newspaper that the utility had granted its permission. Upon reading this article, other journalists in bigger cities started visiting Riisager's house to see his wind turbine and then wrote about it. Because of all the publicity that Riisager's wind turbine received, the utility eventually granted Riisager permission to connect the wind turbine to the grid (Møller, 1978).

Following this incident, a customer who bought Riisager's first turbine also applied for grid connection in another district. Along with the political publicity, this customer's action led to the emergence of the first set of guidelines by the Danish Electrical Utility Association. These guidelines essentially granted users the rights to connect wind power to the grid, thus activating several action-nets around wind energy.

Tvind Design

One early Danish action-net emerged in 1974 in the small town of Ulfborg at the local Tvind Schools, an educational institution based on socialist values (Jensen, 2003:60–63). The leader of the wind turbine project at Tvind contacted Copenhagen Airport to see if they could help with the design of the blade as Tvind teachers did not possess the necessary skills. The project leader was redirected to Helge Pedersen, an aeronautic engineer at Risoe, the Danish nuclear power research center (Jensen, 2003:62).

Helge Pedersen had designed blades for a series of F.L. Smidth wind turbines during World War II (Karnøe, 1991). When approached by Tvind, Helge and two of his colleagues negotiated from Risoe time to work on the Tvind project. The Tvind project led to important principles for blade design and manufacturing including a mold-based technique for holding the shape of the wrapped fiberglass film on the steel structure of the blade (ibid.). Tvind's efforts also led to a realization that three-bladed designs would be able to balance loads better than two-bladed design.

The Herborg Design Action-Net

Another noteworthy contribution was from Jørgensen, a skilled mechanic who lived in Herborg (located near Tvind). Jørgensen wanted to build a wind turbine to produce electricity for his workshop to mitigate against the high cost of energy (Jensen, 2003). In contrast to Riisager, Jørgensen used his preference for metal to design a multi-bladed wind-rose design based on historical drawings and pictures, as well as his own sense of component dimensions.

In February 1978, engineering student Henrik Stiesdal from the alternative energy movement visited Jørgensen for assistance with mechanical components for his own wind turbine design. The young engineering student and the mechanic began collaborating on an upgraded design of the wind turbine. The multi-bladed wind-rose rotor was replaced with three aerodynamically shaped blades and an active yaw system like Juul's design. Following a newspaper article about public funding of "new ideas for products" for renewable energy, the two filed an application for funding for the new Jørgensen-Stiesdal design in the spring of 1978. With modest funding, they were able to design a three-bladed wind turbine with an asynchronous generator. Grove-Nielsen, who had bought the Tvind molds, designed the fiberglass blades (Jensen, 2003).

The wind turbine they assembled operated without problems for several months. But when it encountered stormy winds, the wind turbine accelerated out of control. Rather than give up, Stiesdal and Jørgensen used this incident to learn about the weak points of their design. Grove-Nielsen even published a paper titled "We can always use a storm" in *Natural Energy*, February 1982. By November 1978, the wind turbine had new blades and a tip-brake system. This time, the wind turbine worked successfully.

The action-nets around Riisager, Tvind, and Herborg designs were all located close to one another and near Vestas (now the world-leading wind turbine manufacturer), one of the entrepreneurial firms at that time. There were many other entrepreneurial firms, do-it-yourself builders, and institutional actors in the region who experimented with wind turbines. By 1980, at least 16 producers of wind turbines were registered (Karnøe, 1991: Table 2.2, p. 331). North

Jutland Institute for Renewable Energy (NIVE), an important driver of the process inspired by Juul's design, was instrumental in activating local labor by publishing technical reports to support do-it-yourself builders and wind turbine start-ups (Maegaard, 2000:67–69).

Action-Nets of Users Constituting an Emergent Market

Møller, an award-winning journalist for a Copenhagen newspaper, was the first to buy a Riisager wind turbine in early 1976 for his new countryside home in Jutland. His motivation to buy a wind turbine was to safeguard against future oil prices. He sought permission to grid-connect his wind turbine; however, there were no guidelines at that time, and only Riisager had obtained permission, as described before.

In the meantime, "haunted by the press" and forced by the "growing private interest in wind energy", the Association of Danish Electric Utilities (DEF) pressed the national price and tariff committee to come up with rules for connecting wind turbines to the electricity grid (van Est, 1999:78). Guidelines granting independent owners the right to connect wind turbines to the local low-voltage electricity distribution grid emerged in 1976 (ibid.). Per these guidelines, owners received a modest price from the utilities who now had to buy the electricity generated by the wind turbine; both had to share the costs associated with grid connections (ibid.).

After Møller's 22 kW wind turbine was grid-connected in November 1976, a series of widely read articles began appearing in the media promoting the "new recipe for how users could make money on private investments in wind power" (Møller, 1978:43–46). These articles generated a snowball effect, as they contributed greatly to the interest in wind power (ibid.). Early owners were mostly farmers who came together to form cooperatives. These initiatives confronted resistance from local utilities and municipalities despite the DEF '76 guidelines. In addition, many users continued with their concerns about their investments because of failures. For all these reasons, the market for wind turbine between 1976 and 1978 remained modest at best – that is, less than 150 wind turbines (see Table 11.2)

Table 11.2 Industrial Activity in the Danish Wind Turbine Cluster (1974–1989)

	1974–1979	*1980*	*1982*	*1985*	*1987*	*1989*
Number of Firms	1–6	10	10	8	11	14
Sales in Denmark (1$/6 dkr)	200 wind turbines	3	10	30	30	70
Export sales (1$/6 dkr)*			5	350	65	65
Wind turbine kW and rotor**	15–30 kW 4–10 m	45–55 kW 12–16 m	55 kW 14–16 m	75–100 kW 17–20 m	100–180 kW 20–23 m	180–450 kW 23–35 m
Employees in cluster		50	200	3300	900	1200

Source: Various tables in Karnøe, 1991.

Note: *$1 = €0.8.
**Dominant wind turbine size in the Danish market by rotor diameter measured in meters, a key measure for wind turbine size.

In 1978, the Danish Windmill Owners Association (DWO) was established at Møller's house (Møller, 1978). Comprising about 40 people, the mandate for the DWO was to secure the investments of members with the new and unknown technology (Tranaes, interview 2010). Six domains of interactions were targeted: (1) with politicians to create a regulatory framework, (2) with wind turbine producers to make design improvements, (3) with electrical utilities association to establish payments and conditions for grid connection, (4) with insurance companies to cover component failures, (5) with consultants to seek assistance for individuals and cooperatives in making investments in wind turbines, and (6) with the government to provide national planning tools and regulations for local siting of wind power, such as landscape classification (without a classified landscape, municipalities could not give permission to install wind turbines) (ibid.).[3]

This user action-net complemented the already existing "wind meetings", an important collective forum for mutual knowledge sharing among producers, users, and self-builders. These wind meetings, held four to eight times a year, offered a forum for the exchange of ideas between users and entrepreneur-producers but eventually also involved regulators and utilities. In addition, interactions between early wind turbine entrepreneurs and users created a strong environment for the coupling of learning-by-using and learning-by-doing, resulting in innovations, such as a double-brake system. A monthly publication called *Natural Energy* detailed the performance of different brands of wind turbines. The location, wind speed, operational hours, and component failures for each wind turbine were all reported, stimulating mutual learning in the wind turbine cluster. Moreover, the data that emerged became important later in negotiations on payments and subsidies with utilities and politicians (Tranaes, interview 2010). Users' ability to reduce the uncertainty of their investment was linked to the invention of a calculative devices, the so-called Wind Atlas method, developed by meteorology researchers for calculating the wind speed given the landscape-type ruggedness (Nielsen, 2001).

Action-Nets for Policies and Regulation

Since the mid-1950s, nuclear power was promoted nationally in Denmark as part of the US-driven Nuclear for Peace program (Nielsen et al., 1998, Garud, Gehman, and Karnøe, 2010) to become the preferred energy technology of the future in most energy policy discourses (Karnøe and Jensen, 2016). The oil crisis of 1973–1974 brought energy and environmental consciousness to the forefront (Jamison et al., 1990:90, van Est, 1999:70). For instance, concerned groups emerged to problematize the then existing discourse on nuclear energy.

However, the Danish Electrical Utilities (DEF) claimed that "nuclear power was the only realistic alternative to fossil fuels" (Hansen, 1985). The director of the Association of Danish Electric Utilities even commented, "you may discuss as much as you want, but nuclear power you will get" (Karnøe and Buchhorn, 2008). In response, a new organization – Organization for Information about Nuclear Power (OOA) – emerged. This organization played a key role later in challenging nuclear power in the energy policy discourse over time.

New initiatives for alternative energy sources were funded by the 1974 Energy Information Campaign (Meyer, 2000:87, Nielsen, 2001:70). Shortly thereafter, in 1975, the social democratic minister found that the campaign was too successful and so formally closed this initiative to support nuclear power. However, the momentum of the action-nets built up from the many distributed initiatives was difficult to stop. The movement had already yielded many artifacts, such as books and movies depicting different realities, and solutions that fueled more debates and experiments.

Nybroe's practical guide to building *Sun and Wind* (1975) is illustrative. Because of this effort, American scientist Amory Lovins presented his groundbreaking book *Soft Energy Paths*

(1975) (Meyer, 2000:77), which he presented to the Danish Energy Agency. Further, the Network of Energy Offices, an initiative organized via folk high schools, spread around Denmark with the goal to formalize systematic education in building wind power plants, biogas, solar heating, and so on (Beuse, 2000:62).

An important initiative emerged around experts inspired by the Club of Rome and Limits to Growth report from 1972, when the president of the Academy of Technical Sciences (ATV), physics professor Niels Meyer, established two wind power committees (in October 1974 and September 1975). President Meyer hoped that the reports would spur wind power. Two wind power reports were published – the first in May 1975, and the second in January 1976. The second ATV report recommended two programs – a large-scale wind turbine (more than 30 m rotor diameter), and a small-scale wind turbine (less than 15 m rotor diameter) (van Est, 1999:76, 310, fn. 21).

Despite these reports, wind power was not a part of the first official energy plan. Instead, the first Danish Energy Plan, EP '76, depicted a future with power plants based on coal that replaced imported oil and four to six nuclear power plants by 1995. However, the role for domestic energy sources that could reduce import dependency was also prioritized in Denmark's first energy policy plan (Handelsministeriet (Ministry of Trade), 1976). In addition, the reliance on nuclear power was criticized by OOA. The parliamentary Energy Policy Committee asked the minister to draft a plan without nuclear power. The minister refused (van Est, 1999:74).

This refusal motivated the formation of a group of scientists (including Professor Meyer from ATV) to draft an Alternative Energy Plan (AE '76) in the summer 1976 (Meyer, 2000). Utilizing assumptions on economic growth like the ones in the official EP '76 plan, new techno-economic calculations generated a socio-technical imaginary with a much stronger role for renewable energy. These proposals served as the basis of strong media discussions, parliamentary debates, and hundreds of other debates arranged by OOA and the Organization for Renewable Energy (OVE) (Meyer, 2000:80).

The action-nets involving OOA, ATV, and Professor Meyer produced technical reports. These reports served as discursive devices that intervened in existing realities, by proposing alternative realities (Callon et al., 2007) that were able to change the discourse. As a result, large-scale wind power was officially incorporated into the Danish Energy research program in the fall of 1976. Building on new knowledge performed by the ATV report "for large-scale wind power", the Wind Power Program was run by the Ministry of Trade and DEFU (Danish Electrical Utilities Research Unit, which took over Juul's Gedser mill) to investigate the possibilities associated with large-scale wind power (Nielsen, 2001:108–109). However, this initiative was seen by many as a tactic to delay the introduction of wind power while nuclear power advanced (Møller, 1978:38, van Est, 1999).

At this point in time in 1976, there were no plans to support small wind turbines. Pedersen from Risoe was the main driver behind small-scale wind power (Jensen, 2003). Inspired by wind power policy efforts in the US, Pedersen advocated the creation of a Test and Research Center (TRC) for small wind turbines. In 1978, his efforts resulted in a $1 million grant to initiate a three-year program for small-scale wind turbines (Meyer, 2000:97, Nielsen, 2001:114). Only few believed that small wind turbines could contribute significantly to Danish electricity generation. And no one imagined that Juul's design could be scaled up to MW-size.

The Action-Net Around the Test and Research Center

Given the temporary nature of the grant, the TRC's approach was to serve as a resource for firms with the hope that the TRC would grow if the industry grew. Petersen, the designer of the blades for the Tvind machine, became its first director. He hired three engineers who had worked on the Tvind and Juul wind turbines as students. Comprising theoretical thinkers and hands-on engineers, TRC researchers specialized in aerodynamics, structural dynamics, and other practical engineering bases of knowledge important for designing the wind turbine (Karnøe, 1991, 1999b). They had to learn about the more holistic properties of wind turbines since existing engineering knowledge was poor (Garud and Karnøe, 2003). However, despite the general openness within the network of wind turbine practitioners, many firms did not want the TRC looking into their specific products and design rules (Karnøe, 1991). In this context, TRC worked very hard to cultivate trust with practitioners in the field. For instance, to attract designs to evaluate, they offered an award for the best wind turbine design installed at Risoe's site.

In the late 1970s, political negotiations emerged around subsidies for wind power. Several actors, including the Ministry of Trade, the media, and industry associations, contested this subsidy scheme. However, the scheme was passed in June 1979, with the minister of housing arguing that "the 30% subsidy was not meant to support basic development work" but instead for "creating production opportunities for the Danish industry in such a way that series production could be achieved" (van Est, 1999:79).

In its passage, the new government regulation required firms to seek the TRC's approval for their wind turbine designs so that buyers could get subsidies. This requirement forced the industry to open its doors to the TRC. It is important to note that the institution of the TRC did not stem from a Danish industrial or technology policy tradition. Indeed, it was a major institutional innovation that not only stimulated the market but also induced actors in industry and those in institutional bodies such as the TRC to collaborate (Karnøe, 1991). Over time, TRC researchers were gradually accepted by the industry as TRC's positive influence in the creation of the cluster became evident.

As interactions with industry progressed, TRC researchers found that they did not possess all the knowledge required to create robust criteria for the evaluation of wind turbines. Consequently, TRC researchers began interacting heavily with early wind turbine users and manufacturers. Criteria were borrowed from the DWO whose members demanded safety; these were supplemented with other structural criteria used in buildings. Because of the uncertainties involved, these criteria were on the conservative side, encouraging heavy-weight constructions in the tradition of machine shops. Although not anticipated, such heavy-weight constructions reduced the structural vulnerability of the Danish wind turbines when they encountered unexpected high dynamic wind loads (Stoddard, 1986, 1993, Garud and Karnøe, 2003).

The TRC's rules were flexible and continued evolving as wind turbine manufacturers' suggestions were incorporated into the overall framework (Lundsager and Hjuler Jensen, 1982). The TRC did not think of itself as conducting ivory-tower research; instead, it tried to bridge different knowledge communities. Sharing knowledge about critical parameters in designing wind turbines through early reports put out by the TRC occurred as much through "qualitative stories" and "metaphors" as through the precise language of mathematical formulations.[4] These reports and the interactions between TRC researchers and manufacturers resulted in forming a path of learning that stabilized wind turbine designs and greatly enhanced their performance. For instance, the performance of the 55 kW design improved more than 50% from 1981 to 1984 (Karnøe, 1991:39).

Renewal and Continued Upgradation of Action-Nets in Design and Production

The 1979 subsidy-approval scheme was an industrial policy innovation in Denmark that completely reset the economics of the wind power industry (Karnøe and Jensen, 2016). Before the approval-subsidy scheme, subsidies as an industrial policy instrument had only been used for emergent industries such as textile and shipyards that in 1970s faced competition from Asian producers (Karnøe, 1991). Subsequently, Vestas, Nordtank, Bonus, and Micon all decided to diversify into wind turbines by licensing and copying elements from Riisager's and Herborg's designs. Over time, these firms became key players in the global wind turbine field – Vestas was a world leader by 2019, and Bonus was in 2004 taken over by the German multinational company Siemens and is now a world leader in offshore wind power. By 1980, about ten firms were actively selling wind turbines in Denmark (Karnøe, 1991:331).

Around 1979, knowledge about who was doing what with wind turbines was "in the air" in the region. For instance, word that Herborg wind power could not fund the production of 50 wind turbines became common knowledge. In the meantime, Stiesdal (of Herborg) heard that Vestas had not been successful with its vertical Darrieus turbine design. Discussions ensued between the two companies, resulting in a license agreement that transferred Herborg's design to Vestas (Jensen, 2003).

Similarly, Nordtank suffered declines in sales of their traditional products – that is, oil tanks and watering equipment for farmers. A production engineer convinced Nordtank's owner to go into wind power (Jensen, 2003:78). The project team used design solutions from designs pioneered by Riisager and Herborg, including blade designs from Grove-Nielsen and Økær Energy, which were Herborg's blade suppliers (ibid.). Nordtank's newly reinvented and redesigned wind turbine was presented at Farmers Fair in May 1980 (ibid.).

Bonus, too, was suffering declines in the sales of its traditional products. Bonus' owner did not want to spend money on "loose ideas" proposed by his son, a skilled mechanic. But after seeing Nordtank's wind turbine at the Farmers fair, the son was able to convince his father to provide funding for this new venture. Subsequently, Bonus re-engineered a Nordtank wind turbine (Jensen, 2003:82–83) and obtained design knowledge from interacting with Risoe Test Center. Bonus made the metal frame for the nacelle of the wind turbine (the housing for the generating components) and activated local suppliers to supply components such as bearings, blades, and main shaft (Karnøe, 1991).

All these companies became convinced about wind power as a future source of energy. In the pursuit of that opportunity, they began building on and transforming their existing skills and routines for product designs and manufacturing. They had all the typical hands-on approach to design and development with a base of skilled mechanics, Teknikum engineers, and a few engineers from technical universities. Knowledge for the design and construction of wind turbines gradually became embedded in new shop floor routines and practices as competencies and design rules began accumulating through practical experimentation and learning-by-using from operational experience (Karnøe, 1999b). To the people involved, the wind turbine presented itself as a set of problems similar to those they had confronted in the construction of machines that demand the integration of different parts, and the Danish wind turbines became known as "heavy-weight" (Karnøe, 1991:184). During the 1980s, the work method went from an informal process of design and experimentation with a low degree of knowledge explication to a more formalized process involving formalized R&D interacting with manufacturing and use.

Suppliers' Action-Nets

Supplier competencies did not pre-exist and had to be built over time in collaborative network settings that encouraged and created mutual knowledge sharing. Pioneers such as Riisager and Herborg Wind Power worked to translate the interests of key suppliers for electronics, blades, and bearings from the local industrial district to see an opportunity in the wind turbine business. Local suppliers in the industrial district were specialized, who then reoriented their skills to address new problems and issues to make "just another component for a new customer" (Andersen and Kristensen, 1999). For example, Riisager activated and enrolled KK Electronics to his action-net. KK Electronics had 20 employees in 1976 and supplied various electronic control board systems. Upon receiving Riisager's order, KK developed a simple relay control and improved the hardware and critical measurements points and control loops. This was critical for all new Danish wind turbine manufacturers. KK Electronics eventually became the national and international market leader.

For blades, Riisager approached LM Fiber Glass, a supplier to boat manufacturers, to see if they could offer their skills to design blades for wind turbines. But LM was skeptical about wind power and so did not pursue this opportunity. Another boat company, Coronate Boats (renamed to Alternegy), became involved (Jensen, 2003). Alternegy took over the finished product that Grove-Nielsen and his company Økær Blades had developed between 1975 and 1980 in interaction with Herborg and others. LM Fiber Glass overcame their skepticism already in 1980, only to become a national and international leader of blade design and production. In 2017 LM was bought by American GE Wind to have this critical knowledge and component in-house.

Components like gears and generators were not available in Denmark and had to be sourced from suppliers in Sweden, Finland, and Germany. During the 1980s, standard components became more specialized (Karnøe, 1991). All the major wind turbine manufacturers collaborated with their own suppliers to design their own special components. For each new and bigger wind turbine design, wind turbine firms had their own "gear engineer" located for months at the supplier's premises. Overall, the wind turbine cluster emerged as local firms located in the Mid Jutland industrial district responded entrepreneurially to their declining business in farming related equipment by exploring new opportunities in wind power.

There were overlaps in the suppliers for Vestas, Micon, Nordtank, and Bonus. However, to avoid dependency on any one supplier and to ensure competitive upgrading in quality and prices of the parts, each firm had a second source. In contrast to Micon and Bonus, Vestas began building up competences for internal design and manufacturing of blades and electronics in 1985 to control critical component knowledge in-house for certification purposes (Karnøe, 1991) – that is, a generative process of repurposing, experimentation, and collective learning (Garud, Gehman and Karnøe, 2019).

Such supplier relations followed the institutionalized practices of the industrial region – a mix of competition, collaboration, and sharing risk (Maskell et al., 1998; Lorenzen, 2003). The industrial district was also typical for the Danish industry structure composed of collaborative small- and medium-sized companies (Karnøe, 1999a). Such a structure offered wind turbine manufacturers an opportunity to exploit external economies such as supplier competences rather than creating all specific technological competencies in-house. This "external economy" both maintained specialization of competences and reduced the initial investments in new competences for the focal company designing and developing wind turbines. In addition, the long-term interactive learning between suppliers and the wind turbine entrepreneurs let to a "systemic effect" in the distributed action-nets in industrial district in the sense that suppliers and wind turbine companies transformed and upgraded their skills (Andersen and Drejer, 2008).

International Action-Nets: Developers

From 1980 to 1982, the home market grew slowly with total revenues of $10 million for the ten manufacturers. However, the learning experiences from 1974–1981 prepared the emerging cluster well for a new growth opportunity that emerged in California. The Danish Wind Turbine manufacturers association funded the creation of a report on the California market in spring 1982. This report highlighted a new market in California – the "1980 program" aimed at obtaining 10% of the energy from wind power before 2000. There were federal and state tax credits for investors and electric utilities would pay a ten-year fixed price for wind power (via the PURPA)[5]. Noticing this opportunity, all big Danish manufacturers went to California to establish export opportunities.

The number of Danish wind turbines exported annually increased from 30 to 2,000 per annum between 1982 and 1985 (Karnøe, 1991:25). Industry employment grew from 300 to 2,500, making the "ugly duckling" (i.e., the wind turbine) famous in Denmark. The Danish market share in California increased from 0% in 1981 to 65% in 1985, while the market expanded from 71 MW to 1,250 MW in California. Although the large weight of Danish winds turbines created a legitimacy problem in the home market, the designs outperformed their lightweight American counterparts (Stoddard, 1986). Also, the strong dollar favored Danish export and profits.

This dramatic California growth to serve the developers generated tremendous organizational challenges even as it offered growth opportunities for the SMEs (Karnøe, 1991). California growth jump-started the creation of new industrial routines, such as (1) "large batch production", (2) sales of wind farms to developers, (3) three-to-five-year product guarantees, (4) sourcing, and (5) shipping logistics. Developers came with "big money and big stakes", and this led insurance firms and certification companies to demand explicit criteria and the explication of knowledge used in the design of turbines before granting any funds and before extending insurance coverage. The load paradigm and Risoe researchers helped to legitimize the highly tacit knowledge (Lundsager and Hjuler Jensen, 1982, from Karnøe 1991:200). Also, in servicing the export market, wind turbines took on a new meaning when wind turbine manufacturers understood the magnitude of the efforts required to sell and service a capital good with a life span of 15–20 years with the three-to-five-year product guarantees. New competences and procedures for service, repair, and guarantees had to be instituted through a costly process of trial-and-error learning. Moreover, today such service packages have become a large business unit for world-leading Vestas, and knowledge about need for repair and wear and tear of components is critical for both new product design and competitive bidding.

Regulations and Domestic User Action-Nets in the 1980s

As mentioned, the booming export market in California had a positive effect on the domestic market as it shaped expectations and market policies at a point in time when the existence of wind power in Denmark was fragile and contested. The national Energy Plan of 1981 mentioned that wind power could contribute 10% electricity by 2000, but it was still a nuclear power energy plan. However, the 10% number was seen more optimistically by chairman Madsen of Wind Turbine Manufacturers Association (Nielsen, 2001). Optimism was important, as critics and skeptics contested wind power strongly to negatively influence adaptive expectations about this source of energy.

In one debate centered on the size of the 22–30–55 kW wind turbines, opponents claimed that it was "too small to matter energy production wise" in comparison to the large-scale wind power program. Intense criticisms came from a member of the Nuclear Power Promoting Association in Denmark and a contributor to the overall Danish energy policy. Madsen, the chairman of responded and offered counterevidence from recent performance data (Karnøe, 1991). Despite this evidence, though, many had doubts about the potential of this seemingly low-tech system and the small-scale program.

Reports on the California wind power market offered the red-green political coalition arguments to save wind power (Mortensen, 2018). Wind power policies could now be justified with reference to export and new jobs. During the 1980s, the activities to stabilize wind power in the energy policy discourse were intensified by concerned groups. OOA's activities continued after the Parliament voted nuclear power out of the Danish energy policy in 1985.

The California wind power program also legitimized Danish wind farms, promoted by a new direct investment subsidy (25% plus an additional 15% for large-scale, privately owned wind farm installations). In addition, the economics of wind power improved by a regulation in 1984 that phased out the investment subsidy and replaced it with a fixed subsidy per kilowatt-hour produced (a forerunner of the feed-in tariff model), and with a 1984–1994 agreement among electrical utilities. The private users who were also the investors now had a predictable investment horizon. However, the growth of wind farm installations and profits delegitimized the subsidy program because it allowed rich investors to operate wind farms without involving local inhabitants. Strong public debates for and against subsidies dominated media (Mortensen, 2018). Public funding of wind farms that would yield large profits to the rich was not considered to be a legitimate use of public funds (Worsaae, 1996). Further, speculative investors were not wanted by WTO as they wanted to maintain political and public legitimacy in the subsidies for locally owned wind turbines (Tranaes, interview, 2010, Kirkegaard, Cronin, Nyborg and Karnøe, 2021).

Consequently, a new governmental act from December 1985 imposed restrictions on private ownership of wind turbines (Karnøe, 1991). This act placed geographical boundaries on ownership (the geographical distance between an owner's residence and the wind turbine and the two locations had to be in the same municipality). The second restriction limited the extent to which individuals could invest in wind power relative to their use of electric power. Specifically, the act stipulated that the share of wind power must not exceed one's electricity consumption by more than 35% (ibid.).

This act led to losses of sales in the private market and resulted in a crisis in the industry as many contracts were cancelled. But all was not lost. As part of this act, energy authorities sought an agreement from the action-net of the electric utilities as they accepted to install 100 MW of wind power energy from 1986 to 1990. This regulation that made electrical utilities a direct buyer of wind turbines represented a substantial increase in the home market as private installations from 1978 to 1985 totaled only 75 MW. Moreover, this agreement had the effect of establishing the action-net of the electric utilities as a new group of direct users and a resource for wind power development, as they had to invest in wind power (ibid.).

The Danish governance regime for organizing and monitoring the regulations was an important part of the success. The governance regime included the innovative model with subsidies for wind power production (now known as feed-in tariffs), which stimulated an emergent technology and industry that Riisager and Herborg already had started. Linking the subsidies to the approval from the Test and Research Center fostered two-way interaction and learning. Moreover, this experimentation combined both hands-on and abstracted knowledge by those involved with the emerging technological knowledge of design of wind turbines and components of the wind turbine cluster.

Indeed, in 2021, the wind turbine approach was promoted by the Danish National Council for Innovation as a possible "model" for mission oriented Danish research program for Climate Change.[6] In addition, the 100 MW Act forced the electrical utilities to become an action-net, and build up their own competences in installing and operating wind turbines already in the late 1980s. These competences became the seeds for the global leadership in offshore wind farming by Danish electricity company Oersted.

Another example was the Renewable Energy Committee (within the Danish Board of Technology) sponsored small-scale entrepreneurial activities from 1982 (a total of $45 million in smaller grants were allocated in 1982–1991) (Meyer, 2000). It became a forum for generating legitimacy for renewable energy and wind energy, as the chairman was Professor Niels Meyer, former chairman of ATV, had a network of contacts to ministries and politicians. However, the regulations linked to the subsidy schemes were, as mentioned, heavily contested by the right-wing sociopolitical coalitions as the subsidies were seen as violating the supposedly self-regulating market forces.

It is in this context that the regulations supporting the wind turbine cluster throughout the 1980s were dependent upon a stable red-green political coalition in the Parliament, which annually had to secure a majority to renew the subsidies on the state budget. Further, despite national landscape planning and allocation of sites for wind turbine installation, resistance from local municipalities often stalled the process, and implementation of existing regulations were dependent upon political enforcement to be implemented. Such delicate regulations were not based upon a strict bureaucratic approach but were flexible and involved compromises that were reached in committees and forums where involved actors had legitimate seats. Overall, this governance regime of the 1980s was full of contestation from the right-wing sociopolitical coalition, media, and industrial actors (Mortensen, 2018). The red-green sociopolitical coalition had to navigate conflicts, pressure, and seek compromises to maintain and transform the governance arrangement enabling wind power to exist in the Danish energy system in these formative years (Karnøe and Jensen, 2016).

A New Beginning and Lucky Timing

The decline in the California market was not offset by growth in the Danish and other international markets. One can argue that this late 1980s slump was detrimental for the wind turbine companies in Denmark. However, the wind turbine companies took this slowdown as an opportunity to invest in product development and design of the next generations of larger and more efficient 180–450 kW wind turbine designs by early 1990s (Karnøe, 1991). At the same time, the electric utilities were delayed in their implementation of the 100 MW agreement because they now had to develop organizational competencies besides having to identify relevant sites. By the time these issues had been sorted out, it was already 1988.

This delay became a "lucky" timing as the sales to utilities boosted the home market sales at a time when it was most needed, and the utilities installed the new and improved 180–450 kW wind turbine designs. One of the utility installations was the world's first offshore wind farm at Vindeby, which was opened in 1991 with a capacity of 5 MW (11 wind turbines of 450 kW). Moreover, because of the technical performance of the newly designed wind turbines, the electricity utilities shifted their attitudes from skepticism to acceptance. With these efforts, the transformed Danish wind turbine industry became leading through market domination in the new European markets like Germany, UK, and Spain (Karnøe, 1991). Supplier relations were also upgraded and a central strategic concern both in Denmark and when Danish wind turbine manufacturers set up production abroad (Andersen and Drejer, 2008).

However, this world-leading position was only possible due to regulatory continuity that almost was lost when the supportive red-green majority lost power in the 1991 election. The right-wing coalition acted on new concerns from wind turbines, such as noise and visual landscape pollution (Mortensen, 2018). These concerns were mobilized against wind turbines and, waiting for new guidelines for classification of the landscape, fostered great uncertainty among producers and users of wind turbines in Denmark. As a result, private Danish installations of wind turbines almost stopped in the years 1992–1993; only utility installations remained.

However, an unexpected political scandal brought the red-green coalition back into government with a majority in 1993. So before conclusions could be made from the studies of concerns with noise and visual pollution, the new minister of energy and environmental affairs, Svend Auken, who participated in the first IPCC Rio meeting in 1992, declared his enthusiastic support to combat climate change with renewable energy. The new ambitions were accompanied by a new Energy Plan (E-21) from 1996 that made wind power a central technology in reducing CO_2 emissions from electricity generation. Also, a new feed-in payment regime was in 1994 negotiated with the state, the wind turbine owners, and the electrical utilities. This led to increased private installations, and with some utility owned offshore wind farms, wind power reached in 2000 a world record of 10% share of Danish electricity generation. Deliberations by the new policy commitment chaired by Svend Auken of the minister of energy and environmental affairs led in late 1990s to energy policy regulation that mandated electric utilities install four offshore wind farms after 2000. While the utilities, right-wing policymakers, and stakeholders resisted these steps (Mortensen, 2018), the technical and business competencies that utilities developed paved the way for the subsequent leading role for offshore wind farming (see the next section).

Situation Today and the Future of Wind Energy

From 2000 to 2019, the wind turbine cluster continued its transformative upgradation of skills, the size of wind turbines, and its global leadership role. The Danish wind turbine cluster had about 33,000 employees by 2019, and the export of wind turbine technology increased from €1.5 billion in 2002 to about €20 billion in 2019. With a 19% market share, Vestas was world-leading among the global top 15 wind turbine manufacturers.[7] An additional sign of the strong competences in the Danish wind turbine cluster is that 10 of the top 15 global wind turbine companies has R&D and/or production facilities in Denmark.

The typical wind turbine in Denmark increased gradually from 55 kWh to 250–450 kWh in 1990 and then to 600 kW in 1995. From around 2000, wind turbine sizes were measured in MW (2 MW in 2002). Now, Vestas has launched a 15 MW wind turbine with a 236-meter rotor diameter to be ready for the market in 2024.[8] Vestas claimed that its V236 15 MW turbine would be able to produce 80 GWh/year, enough to power around 20,000 European households and save up to 38,000 tons of CO_2.[9]

In 2019, Danish wind power set a record by offering 46.9% of the total Danish electricity production.[10] However, new dynamics of contestation emerged. For example, the size of MW wind turbines is such that these turbines require massive investment and planning. Their introduction has marginalized local cooperative ownership and has generated head winds from the public who offer resistance toward onshore installations (Kirkegaard, Cronin, Nyborg, and Karnøe, 2021). In this way, the rise of MW-size wind turbines has paradoxically outgrown local wind communities that served as the driving for the creation of a market because they believed in an alternative energy future. This Danish case shows that values and identities are dynamically constituted and transformed during the transition process itself. This observation complements Brinkman and Hirsh's findings (this volume) that values and identities of technological users change and influence the success or failure of energy transitions.

In February 2021, a new red-green Danish government passed a new Energy Act depicting the construction of two "energy islands" that can host from 3 GW to 10 GW of wind power capacity (see endnote 11). This is the costliest infrastructural investments for the Danish society ever. The 3 GW island is expected to generate electricity for more than 3 million households, whereas the 10 GW island can generate to 9–12 million households or be used for the new power-to-X liquid fuels based on hydrogen.[11] Wind power is defining the basic infrastructure of the future of low- or zero-carbon energy system just as fossil fuels defined the infrastructure for the electricity and transportation sectors.

Due to the wider acceptance of climate change and the dramatic cost reductions for wind power as also discussed by Krugman, the global growth in investments is unprecedented. The International Renewable Energy Agency wrote in 2020 "that renewable energy's share of all new generating capacity rose considerably for the second year in a row. More than 80 per cent of all new electricity capacity added last year was renewable, with solar and wind accounting for 91 per cent of new renewables".[12]

Discussion and Conclusion

The innovation journey for the Danish wind turbine cluster involving the co-production of skills, technology, and critical policy regulations described earlier is one that very few could have planned. The most recent energy policy agreement on energy islands with gigawatt-scale wind power demonstrates the ongoing co-productive transformation of policy goals and the interests of many action-nets and stakeholders that originally were against wind power (Karnøe and Jensen, 2016). Wind powering Denmark is not the result of strategic planning by policymakers or entrepreneurs as claimed by reductionist single factor explanations. Rather, wind power in Denmark has emerged because of the activation of multiple distributed action-nets that gradually became entangled with one another over time, bringing about a series of transformations of artifacts/technologies, discourses, tools, competences, policy regulations, and stakeholder interests and frames. Today, wind power has been transformed from a fragile micro-actor to a robust macro-actor (Callon and Latour, 1981) composed of multiple interacting action-nets with different roles and frames, which provides momentum (Hughes, 1983), thereby making it easier for the action-nets to continue the path of transformative development (Callon, 1991, Unruh, 2000).

Employing path creation as a processual lens highlights important facets of this developmental process. Following the action-nets involved, we can see that the wind turbine cluster did not emerge because of a central regulatory planning effort or because of innovations from one university or a company. In the mid-1970s, when the path had been prepared for nuclear energy as the future energy source by incumbent actors, other actors engaged in path creation by envisioning a different future with wind power as an alternative energy source.

Understanding action as situated and distributed, path creation makes it possible to understand that agency is not only constrained but also enabled by structures that have emerged from the past (see also Sharp et al., this volume). Path creation views all action as being situated, with multiple forms of inclusion-exclusion of actors and their artifacts in their relational entanglements with dominant paths. In contrast, path dependence perspective theorizes an all-inclusive inertia generated by existing arrangements, which provides an impoverished notion of agency. Path creation is about paths in-the-making, which entails maintenance, continuity, and possible transformation dependent upon the temporal frames implicated in situated action-nets.

From the path creation perspective, the Danish wind turbine cluster originated across multiple localized action-nets from which embedded actors framed new worldviews as they experimented with wind power. Multiple participants understood these issues differently depending upon their frames and their entanglements in their local action-nets. Unlike the path dependence perspective of lock-in, actors were not only constrained by their entanglements in action-nets but also enabled. We saw how their specific experiences of actors and firms such as Riisager, Jørgensen and Stiesdal, and Vestas induced them to develop new identities and frames of reference, which made it possible for them to imagine new future states and to reorganize their action-nets by translating and enrolling other actors and artifacts into these new projects.

As the process unfolded, non-linear learning by actors was subject to the timing and sequencing of events associated with increasing returns. For instance, (1) storms showcased weaknesses in components; (2) unemployment became a justification for market subsidies; (3) the Californian market for wind power allowed for volume-based learning, profits, and competence upgrading; (4) the collapse of the California market not only was a problem but also offered actors more time (i.e., a pause) to engage in R&D to design even bigger and more efficient wind turbines; and (5) delays in Danish utility installations of wind power let to installations of better designs, which impacted performance and expectations positively.

Our account of the emergence of the Danish wind turbine cluster shows it was not a result of policymakers picking a winner. Nor was it shaped by pure market forces. Clusters based on new technologies do not offer clear-cut valuations *ex ante*, which can qualify the economic viability of novel innovations. Yet Danish regulators stepped in with subsidies and grants to nurture and shape the emerging technology. In a similar fashion, cluster-forming initiatives, such as defense contracts in Silicon Valley, Korean industrial development, Chinese development of wind power, and German solar power, suggest the importance of the active role of the state in investing in technological missions that can foster distributed and emergent actions in new knowledge ecosystems (Fagerberg, 2019). Such investments set the stage for the interweaving of resources and knowledge from market use, production, testing, and basic research, thereby making it possible for the gradual transformation of existing resources over time.

The characteristics of the complex and non-linear processes involved may be like those theorized by Cohen, March, and Olsen (1972). They used the metaphor of a garbage can to offer an alternative to linear rational decision-making. In their model, different solutions and problems become connected in non-linear ways.[13] In the wind turbine case, precipitating these non-linear processes were multiple events, such as Stiesdal visiting Jørgensen, Tvind schools activating Risoe and Helge Pedersen, the Three-Mile accident generating nuclear concerns, high Danish unemployment used to justify the subsidy scheme, and the 100 MW agreement from 1985 that mandates electrical utilities to install wind power. The central idea here is that effects of many links between actors, problems, and solutions depend on the simultaneity of their arrivals. The interweaving of resources is an emergent process that appears disorderly and impossible to foresee. Such non-linear learning results in changes in the trajectory of paths dynamically, a process that is different from the more linear and predictable learning dynamics suggested by Martin and Sunley (2003).

In this regard, Mazzocato (2013) suggests that public policymakers can play an important role as partners in the entrepreneurial process by fostering a fertile ground through R&D and market subsidy programs for novel path creation processes to unfold (see also Araújo, 2017). It can be through the creation of "hybrid forums" (Callon, Lascumes, Barthe, 2009) that foster interactions and dialogues between actors, artifacts, and other resources from which clusters can emerge. In these forums, concerns and issues related to the emergent cluster are constantly raised and addressed to create trust, legitimacy, and commitment for all the involved, such that they can co-create resources in a transformative fashion.

Notes

1* Chapter written for *Routledge Handbook of Energy Transitions*, edited by Kathleen M. Araújo. Sizable portions of the text, reproduced with permission, are from an earlier version of the paper "Path Creation: Co-creation of Heterogeneous Resources in the Emergence of the Danish Wind Turbine

Cluster", *European Planning Studies* 2012, 20(5): 733–752. We thank Kathy Araújo for her inputs on an earlier version of this chapter.

2 The account we offer is based on our synthesis of events from multiple sources including historical accounts, interviews, observations, questionnaires, and site visits (Araújo, 2017; Karnøe, 1991; Karnøe, 1995; Karnøe, 1996, Karnøe 1999b; Garud and Karnøe, 2003; Karnøe and Garud, 2012; Karnøe and Jensen, 2016; Kirkegaard, Cronin, Nyborg and Karnøe, 2021). Besides, we have used several non-academic sources that have informed us about the efforts by multiple situated actors entangled in their local action-nets. For example, "Mænd i Modvind" (Men in headwind) by Ib Konrad Jensen (2003) is a book by an investigative journalist who followed the wind turbine industry over many years. In addition, we refer to the book *Vinden Vender* (*The Wind Turns*), which is an account by Møller who bought the second wind turbine from the wind turbine producer Riisager. Møller became so fascinated with his story and struggles that he detailed his account almost as it was happening. The book *OVE 2000* (cited as Beuse et al. (2000), *Vedvarende Energi I Danmark – 25 år 1975–2000*; in English, *Renewable Energy in Denmark – 25 Years of Becoming 1975–2000*) is based on accounts written in the many chapters contained in the book by people directly involved with NGOs, policymaking, and technical demonstrations.

3 Several other devices framed the economics of wind power. For instance, wind turbines were classified as a building. Consequently, wind entrepreneurs could get a 20-year loan rather than a loan for four to six years if it had been classified as a machine.

4 Stoddard (1993) noted that, of the 25 most important research reports of the 1980s on wind turbine designs, 12 came from the Danish TRC.

5 The Public Utility Regulatory Policies Act (PURPA), part of the National Energy Act of 1978, attempted to eliminate barriers for wider adoption of renewable energy sources in the US. Utilities were compelled to buy electricity from independent power producers at fair prices. Jointly with federal and California tax credits amounting to some 40–45%, these incentives encouraged the use of alternative energy (Gipe, 1995). In fact, tax credits were so high that profits were made even from wind turbines that generated hardly any electricity (ibid.).

6 https://ufm.dk/forskning-og-innovation/rad-og-udvalg/danmarks-forsknings-og-innovationspolitiske-rad/projekter/klimamaal-og-midler.

7 https://winddenmark.dk/nyheder/vestas-fastholder-position-verdens-stoerste-moelleproducent, retrieved September 2, 2021.

8 www.offshorewind.biz/2021/02/10/vestas-launches-15-mw-offshore-wind-turbine/, retrieved September 2, 2021.

9 www.windpowermonthly.com/article/1706915/vestas-launches-new-15mw-offshore-wind-turbine-236-metre-rotor, retrieved September 2, 2021.

10 https://energiwatch.dk/Energinyt/Renewables/article11852237.ece, retrieved September 3, 2021.

11 www.altinget.dk/artikel/bredt-flertal-enige-om-danmarkshistoriens-stoerste-anlaegsprojekt, retrieved September 2, 2021.

12 www.irena.org/newsroom/pressreleases/2021/Apr/World-Adds-Record-New-Renewable-Energy-Capacity-in-2020, retrieved September 2, 2021.

13 Interview with James March by Diane Coutu, *Harvard Business Review*, October 2006, p. 88.

References

Andersen, Poul H. and Drejer, I. (2008). Systemic Innovation in a Distributed Network: The Case of Danish Wind Turbines 1972–2007, *Strategic Organization* 6: 13, DOI: 10.1177/1476127007087152.

Andersen, Poul H. and Kristensen, P.H. (1999). The Systemic Qualities of Danish Industrialism, in Karnøe, P., Kristensen, and Andersen (eds), *Mobilizing Resources and Generating Competencies: The Remarkable Success of Small and Medium-sized Enterprises in the Danish Business System*, Copenhagen Business School Press.

Araújo, K. M. (2017). *Low Carbon Energy Transitions: Turning Points in National Policy and Innovation*. Oxford University Press: New York.

Beuse, E. (2000). Grundtvig køber vindmølleanlæg, in Beuse, E. et al. (eds), *Vedvarende Energi i Danmark 1975–2000 (Renewable Energy in Denmark 1975–2000)*, OVE Forlag (Press).

Callon, M., Millo, Y. and Muniesa, F. (Eds.) (2007). *Market Devices – A Set of Novel Contributions Exploring the Sociology of 'Market Devices' – the Analysis of the Various Sorts of Technical Instruments That Intervene in the Shaping and Reshaping of Markets*, Wiley.

Callon, M. (1991). Techno-Economic Networks and Irreversibility, in J. Law (ed.), *A Sociology of Monsters: Essays on Power, Technology and Domination*, Routledge, London.

Callon, M., Lascumes, P., and Barthe, Y. (eds) (2009). *Acting in an Uncertain World – An Essay on Technological Democracy*, MIT Press.

Callon, M., and Latour, B. (1981). Unscrewing the Big Leviathan; or How Actors Macrostructure Reality, and How Sociologists Help Them To Do So, in K. Knorr et A. Cicourel (eds), *Advances in Social Theory and Methodology*, Routledge and Kegan Paul, London, pp. 277–303.

Cohen, M., March, J., and Olsen, J. (1972). A Garbage Can Model of Organization Choice. *Administrative Science Quarterly* 17: 1–25.

Czarniawska, B. (2004). On Time, Space, and Action Nets. *Organization* 11: 773–791.

Czarniawska, B. (2008). *A Theory of Organizing*. Elgar, Northampton, MA.

Fagerberg, Jan (2019). Mission (im)possible? Mobilizing Innovation – and Policies Supporting it – in the Transition to Sustainability, in *Working Papers on Innovation Studies 20190923, Centre for Technology, Innovation and Culture*, University of Oslo.

Garud, R., Gehman, J., and Karnøe, P. (2021). Winds of Change: A Neo-Design Approach to the Regeneration of Regions. *Organization & Environment* 34(4): 634–643.

Garud, R., and Karnøe, P. (2001). Path Creation as a Process of Mindful Deviation, in Garud, R. and Karnøe, P. (eds), *Path Dependence and Creation*, Lawrence, Earlbaum Associates

Garud, R., and Karnøe, P. (2003). Bricolage versus Breakthrough: Distributed and Embedded agency in Technological Entrepreneurship, *Research Policy* 32: 277–300.

Garud, R., Kumaraswamy, A., and Karnøe, P. (2010). Path Dependence or Path Creation. *Journal of Management Studies* 47(4).

Gipe, P. (1995). *Wind Energy Comes of Age*, Wiley.

Handelsministeriet (Ministry of Trade) (1976). Dansk Energipolitik.

Hansen, Jacob L. (1985). Samarbejde med Folketinget på sidelinjen, *Naturlig Energi* No. 9, p. 7.

Hughes, Thomas (1983). *Networks of Power: Electrification in Western Society*, 1880–1930. Johns Hopkins University Press, Baltimore. ISBN 0-8018-4614-5.

Jamison, A., Cramer, J., Eyerman, R., and Læssøe, J. (1990). *The Making of the New Environmental Consciousness: A Comparative Study of the Environmental Movements in Sweden*, Edinburgh University Press, Denmark and the Netherlands

Jensen, Ib Konrad (2003). Mænd i Modvind, (Men in Headwind), Børsens Forlag

Karnøe, P. (1991). *Danish Wind Turbine Industry – A Surprising International Success: On Innovations, Industrial Development and Technology Policy* (in Danish), Samfundslitteratur: Copenhagen.

Karnøe, P. (1995). Institutional Interpretations and Explanations of Differences in American and Danish Approaches to Innovation, in Richard Scott, W. and Christensen, Søren (eds), The *Institutional Construction of Organizations*, Sage Publications: Thousand Oaks.

Karnøe, P. (1996). The Social Process of Competence Building. *International Journal of Technology Management* 11(7/8): 770–790.

Karnøe, P. (1999a). The Business Systems Framework and the Danish SME's, in Karnøe, Peter, Kristensen, Peer Hull and Andersen, Poul Houman (eds), *Mobilizing Resources and Generating Competencies: The Remarkable Success of Small and Medium Sized Enterprises in the Danish Business System*. Copenhagen Business School Press.

Karnøe, P. (1999b) When Low-Tech Becomes High-Tech: The Social Construction of Technological Learning Processes in the Danish and American Wind Turbine Industry, in Karnøe, Peter, Kristensen, Peer Hull, Andersen, Poul Houman (eds), *Mobilizing Resources and Generating Competencies: The Remarkable Success of Small and Medium Sized Enterprises in the Danish Business System*, Copenhagen Business School Press.

Karnøe, P. and Buchhorn, Adam (2008). Path Creation Dynamics and Winds of Change in Denmark, in Lafferty et al. (eds), *Promoting Sustainable Electricity in Europe*. Edgar A.

Karnøe, P. and Garud, R. (2012). Path Creation: Co-creation of Heterogeneous Resources in the Emergence of the Danish Wind Turbine Cluster. *European Planning Studies* 20(5): 733–752.

Karnøe, P. and Stissing Jensen, J. (2016). Struggles in the Danish Path Towards a Low Carbon Future: Fractures and New Momentum in Shifting Towards a Low-Carbon Energy Technology Assemblage, in: *Handbook on Energy Policy*. Routledge

Kirkegaard, J. K., Cronin, T., Nyborg, S., and Karnøe, P. (2021). Paradigm Shift in Danish wind Power – the (un)Sustainable Transformation of a Sector. *Journal of Environmental Policy and Planning* 23(1).

Krugman, P. (2021). Who Created the Renewable-Energy Miracle? *New York Times*. August 17, 2021. www.nytimes.com/2021/08/17/opinion/us-obama-renewable-energy.html

Lorenzen, M. (2003). Low-Tech Localized Learning: The Regional Innovation System of Salling, Denmark, in: Asheim, B. T., Coenen, L., and Svensson-Henning, M. (eds), *Nordic SMEs and Regional Innovation Systems* – Final Report. Nordic Industrial Fund, Oslo (www.nordicinnovation.net).

Lundsager, P., and Jensen, P. H. (1982). Licensing of Windmills in Denmark, in: *Proceedings of the Fourth International Symposium on Wind Energy Systems*, Stockholm, 21–24.

Lundvall, B.-A (ed.) (1992). *National Systems of Innovation: Towards a Theory of Innovation and Interactive Learning*. Pinter: London.

Maegaard, P. (2000). Nordvestjysk Folkecenter for Vedvarende Energi, in: Beuse, E. et al. (eds), *Vedvarende Energi i Danmark 1975–2000 (Renewable Energy in Denmark1975–2000)*. OVE Forlag (Press)

Martin, R. and Sunley, P. (2003). Deconstructing Clusters: Chaotic Concept or Policy Panacea? *Journal of Economic Geography* 3: 5–35.

Maskell, P., Eskelinen, H., Hannibalsson, L., Malmberg, A., Vatne, E., (1998). *Competitiveness, Localised Learning and Regional Development*. Routeledge: London.

Mazzocato, M. (2013). *The Entrepreneurial State – Debunking public vs. Private Sector Myths*. Penguin Books.

Meyer, N. I. (2000). Politik og Vedvarende Energi, in: Beuse, E. et al. (eds), *Vedvarende Energi i Danmark 1975–2000 (Renewable Energy in Denmark1975–2000)*. OVE Forlag (Press)

Møller, T. (1978). Vinden Vender (The Wind Turns), Knebel

Mortensen, H. B. (2018). The Valuation History of Danish Wind Power: The Ongoing Struggle of a Challenger Technology to Prove its Worth to Society. Aalborg Universitetsforlag. Ph.D. Dissertation, Det Tekniske Fakultet for IT og Design, Aalborg Universitet https://doi.org/10.5278/vbn.phd.tech.00040.

Nielsen, Hvidtfelt, K. (2001). Tilting at Windmills. 2nd Edition. Phd Dissertation. History of Science Department, Aarhus University, and Department of Organization, Copenhagen Business School.

Nielsen, H., Nielsen, K., Petersen, F. and Jensen, H. S. (1998). Til Samfundets Tarv – Forskningscenter Risøs Historie. Danmarks Tekniske Universitet, Risø Nationallaboratoriet for Bæredygtig Energi, Risø.

Stoddard, F. S. (1986). The California Experience, in *Proceedings of the Danish Wind Energy Association Wind Energy, Conference*, Herning, Denmark.

Stoddard, F. S. (1993). Detailed Comments on the Development of Wind Turbines in a Letter to Peter Karnøe, 5 January.

Tranaes, F. (2010). First Chairman of Danish Wind Turbine Owners Association. Interview.

van Est, R. (1999). *Winds of Change: A Comparative Study of the Politics of Wind Energy Innovation in California and Denmark*, Amsterdam: International Books.

Unruh, G. (2000). Understanding Carbon Lock-In. *Energy Policy* 28(12): 817–830.

Wistoft, B. (1992). *Elektricitetens Århundrede. Dansk Elforsynings Historie*, København: Danske Elværkers Forening, bd. II (1992).

Worsaae, K. (1996). Director in Danish Energy Agency, Personal Conversation.

12

THE INTERNATIONAL POLITICAL ECONOMY OF CROSS-BORDER ELECTRICITY TRADE IN EAST ASIA

A Case Study Analysis of the Brunei/ Indonesia/Malaysia/Philippines–East Asia Growth Area

Clare G. Richardson-Barlow

1. Introduction: Global Energy Transitions and Regional Climate Solutions

The East Asian region includes many of the dominant actors in the organized, global responses to climate change and economic development challenges, such as the UN's Sustainable Development Goals (SDGs) (UN et al., 2017). East Asian economies have contributed significantly to the root causes of global climate change concerns, including increased air pollution and carbon dioxide (CO_2) emissions, but they are also major global players in clean energy technology development, investment, and policy mechanisms aimed at green growth and sustainable development solutions (Renewable Energy Network for the 21st Century [REN21], 2019). The economic development story that made East Asia one of the largest contributors to global emissions also made the region one of the dominant global contributors to policy responses and solutions.

The significance of addressing decarbonization and renewable energy integration challenges and of incorporating an evaluation of East Asian governance and economic policies as they relate to regional electricity markets lies in the pressing need for coordinated, multi-level (global, sub-regional, regional, national) responses to climate change and the global energy transition from fossil fuel energy to low-carbon and/or renewable energy. One way to enable this transition is by strengthening renewable energy integration and electricity capacity through sub-regional and regional electricity trade and the buying and selling of excess capacity depending on energy needs. However, problems arise in policy coordination and coherence, particularly related to national policy measures that impact sub-regional and regional clean energy utilization, and in turn have an impact on global markets, policy, and trade. Understanding the national and sub-regional linkages and addressing any arising challenges can positively impact East Asian clean energy development strategies and policy responses to the global energy transition.

DOI:10.4324/9781003183020-14

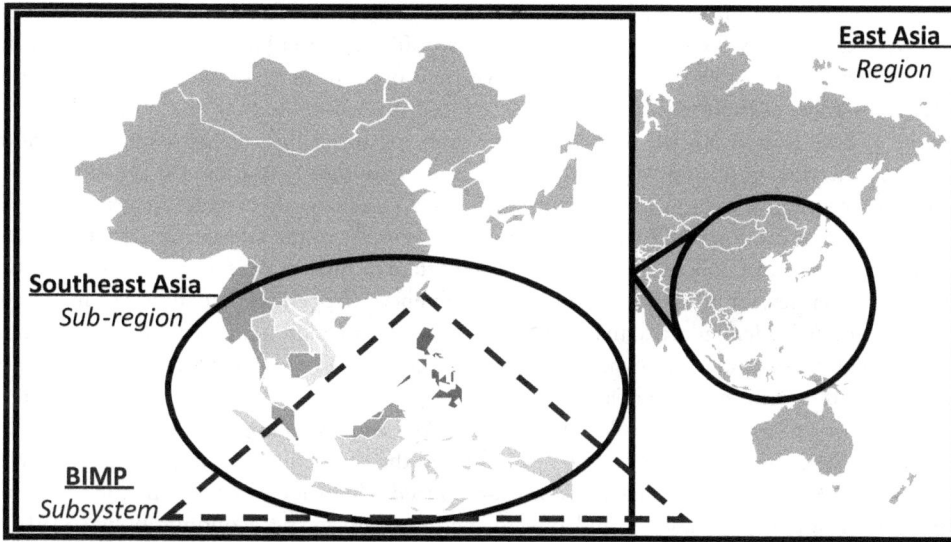

Figure 12.1   Geographical context.
Source: Author.

This analysis seeks to explain how East Asian nations, particularly those in the sub-region of Southeast Asia, view the imperative to construct and operate new power sector integration projects and how the tapestry of competing interests (state, quasi-state, private, and corporate utilities) that form individual energy markets position themselves with regard to interconnection between markets. Policies are utilized that make the best case for optimizing clean energy's contribution in addressing global climate goals, as well as examples from the East Asian sub-region of Southeast Asia and the Brunei/Indonesia/Malaysia/Philippines–East Asia Growth Area (BIMP-EAGA).

The geographical focus of this research lies in the East Asian[1] sub-region of Southeast Asia due to Southeast Asia's major cross-border electricity trade initiatives and sub-region-wide goals of deeper energy market integration (EMI).[2] The collection of ten Association of Southeast Asian Nations (ASEAN) member states makes up one of the fastest growing sub- region in world (IMF, 2021), with the associated economic and environmental issues that come with rapid growth, making it a highly relevant place to EMI and the IPE of cross-border electricity trade. The focus of the case study selection on one of the identified subsystems and its relevant EMI initiative, BIMP-EAGA, and this geographical context can be visualized as follows.

A number of factors play a part in how this research has taken shape: the symbiotic relationship between economic development and climate change, the role of East Asia in global responses to climate change, and the significance of electricity markets in any coordinated policy response to the economic and political challenges associated with energy transitions.

1.1 *Methodology and Theoretical Framework*

This research generates a detailed overview of the international political economy (IPE)[3] of cross-border electricity trade in Southeast Asia in the context of clean energy transitions. The variability of electricity produced from clean energy resources provides an opportunity for regional and international electricity market expansion and the efficient disposal and purchase of power capacity. Based on the potential for cross-border electricity trade, this research examines the expansion of these markets in East Asia and opportunities for their further development. This political economy analysis includes a case study examination of a potential growth market in Southeast Asia's BIMP-EAGA subsystem following informative expert interviews that shape the view of this subsystem, the sub-region of Southeast Asia, and the broader region. A mix of quantitative and qualitative methods is used to address these issues in relation to the regional political economy of EMI and result in an overview of the relevant electricity markets in the subsystem and a cohesive view of the current state of EMI in the sub-region. In addition, 28 expert interviews and 14 workshops and conferences (Appendix 1) conducted in Southeast Asia among leaders in academia, business, and policy (including multilateral organizations, development banks, and current and past government officials) provide added insight in terms of market and reform recommendations.

The history of economic development in East Asia and its continued presence within the global free-trade-driven economic system has shaped the theoretical framing of this case study analysis. East Asian economies have historically utilized a high-level of government economic intervention, foreign technology, and inward- and outward-facing economic policies (Wade, 1988; White & Wade, 1988) while simultaneously engaging in the global system of free trade as a means of increasing economic development (Cai, 2008). This combination of economic policymaking is commonly identified as developmental statism (DS) and represents a blend of state directed policies that includes a focus on industrialization, industry targeting, and mercantilist trade policies common among East Asian countries (Leftwich, 1995).

The prevalence of East Asian developmental statism (DS) within a capitalist values-driven economic system is interesting in light of the competing economic values within the global economic system. Being a member of this global economic system requires a commitment to market values, trade liberalization, openness to foreign direct investment, privatization, fiscal discipline, and belief in market fundamentalism (Serra & Stiglitz, 2008). Yet the East Asian DS model persists nationally and regionally. These two seemingly contradictory theoretical frameworks are combined to explain the political economy of current electricity markets in Southeast Asia, forming what the author refers to as *neo-developmental statism*. Neo-developmental statism is rooted in the concepts of liberal developmentalism and neo-developmentalism – the combination of both engagement with free market and liberal economic practices while utilizing state-directed economic policymaking common in East Asian countries. Neo-developmental statism, while an evolution of traditional DS paradigms, represents a theoretical approach to explaining cross-border electricity trade and barriers to EMI in ASEAN that have not been explored in other theoretical iterations of DS paradigms. Ultimately, the overarching aim of this analysis is framed by the concept of neo-developmental statism and critical approaches to energy transitions in East Asia, Southeast Asia, and BIMP-EAGA. Neo-developmental statism offers an alternative viewpoint to traditional ideas of economic activity, providing a multidisciplinary approach to power sector development and renewable energy transitions in ASEAN.

1.2 Aims and Highlights

This case study analysis focuses on making a number of contributions to energy transition research. Broadly, this analysis contributes a unique way of approaching cross-border electricity trade by combining IPE, Asian studies, and sustainability research to examine sub-regional EMI. Overall empirical, methodological, and theoretical contributions have revealed the difficulty in addressing EMI in systems in which national and sub-regional interests prioritize neoliberal market factors. In addition, this approach offers alternatives to traditional East Asian studies development paradigms that struggle to account for the political and economic dynamics at play in power sector integration and cross-border interconnections

While there are some academic studies of the BIMP-EAGA sub-region and its potential for EMI, this study examines the IPE of individual electricity markets in the BIMP-EAGA subsystem. Here, the case study analysis of individual BIMP-EAGA markets, a synthesis of common reform recommendations, and the contrast of these recommendations against the backdrop of developmental state practices in the region add a unique viewpoint to regional and sub-regional energy transition literature. The analysis conducted includes not only mapping existing electricity interconnectors but also identifying shortcomings in current markets for increased EMI and electricity trade. The energy and economic data analyzed is publicly available, however, here it is uniquely collated, synthesized, and compared and contrasted with research into the market structure of the subsystem, resulting in a more detailed image of sub-regional interconnections and the national limits of these markets.

This analysis also results in two distinct themes within BIMP-EAGA EMI: sub-regional market factors and national market factors. In the following sections, analysis of the market structures in BIMP-EAGA frames cross-border electricity trade and EMI more fully within the subsystem and the wider sub-region. This case study analysis finds that sub-region and national interests complicate the ability of EMI to address renewable energy integration challenges. It also finds that the ASEAN Power Grid (APG) progress is complicated by uncertainty at both sub-regional and national levels and various power dynamics at play within regional and sub-regional energy transitions. Ultimately the overarching thread described through this research is that energy transitions can be messy, policy has the potential to be flawed in its incentives, and there is a delicate balance between national, sub-regional, and global goals and the necessary means to achieve them (if at all).

2. Clean Energy Transitions in East Asia

East Asia has a complicated relationship with climate change – the region is one of the world's ascendant actors in the global fight against climate change. East Asia has significantly contributed to the deterioration of the global environment, climate, and air quality (REN21 2019), but it is also the site of increased alternative energy investment and development, as well as innovative clean energy policy mechanisms (ASEAN Centre for Energy [ACE] 2017). The same phenomenon that contributed to the global challenge of climate change – fossil-fuel-powered economic development and industrialization – has also made East Asia an important geographical locus of the mitigation of global emissions for preventing further degradation of the global climate.

Clean energy, energy with neutral emissions, often (but not always) from a renewable energy source, holds the potential to address two critical challenges facing East Asia (and the world) – (1) the dual eradication of energy poverty and quest for energy security and (2) environmental and climate repercussions that occur because of traditional, fossil-fuel-based energy consumption (REN21, 2019). In the case of some renewables, integrating clean energy sources into global power grids and markets can be challenging for three reasons: variability, uncertainty, and flexibility (Jones, 2014). These reasons apply to the resource supply, financial support, physical infrastructure, and policy mechanisms (IRENA, 2018a). Variability is the hour-by-hour, minute-by-minute availability of intermittent resources (Impram et al., 2020); uncertainty refers to the longer-term questions regarding intermittency and difficulty predicting future weather and solar and wind power. The existing solution to this is to build more flexibility into current systems (National Renewable Energy Laboratory [NREL], 2014). Variability and uncertainty create the need for flexibility and the need for management through flexibility (NREL, 2014).

At the region-wide level, cross-border electricity trading aids in managing diversity in loads and resources, can increase supply security and efficiency, can reduce average electricity prices, and can encourage further deployment of clean energy technologies (UN, 2006; Kunstýř & Mano, 2013; Chang & Li, 2015). In the case of traditional renewable energy, such as solar and wind power, intermittency in supply poses challenges for larger increases in a country's energy mix. For example, wind power, which is created via the movement of wind turbines, is only available during times when the wind is blowing. Cross-border electricity trade would allow countries to trade excess power across interconnectors to countries that are experiencing power deficits due to intermittent resources, like, in this example, those that have surplus solar energy resources. EMI through power connectivity is one-way electricity trading could be encouraged.

2.1 Cross-Border Electricity Trade in ASEAN

The global challenge of climate change has forced governments around the world to reconsider their energy mix and approach to electricity production – transitioning from carbon-intensive energy resources toward more renewable sources to reduce global CO_2 emissions. This is illustrative of part of the energy transition and climate imperatives for increased cross-border electricity trade – higher shares of clean energy in a country's energy mix have a wide variety of advantages.

In East Asia, where geographical, policy, and financial limitations differ from country to country, cross-border electricity trade is viewed with growing interest and hope for contributing to the attainment of future renewable energy and climate targets (Chang & Li 2013, 2015). In Southeast Asia, under the sub-regional organizational umbrella of ASEAN, member states have set shared targets associated with expanding cross-border electricity trade and growing renewable energy's share in the sub-regional and national power sectors (ACE, 2013). This includes a 23% renewables target and a 20% reduction in energy intensity by 2020 and a 30% renewables target by 2025 (APAEC, 2015).

The need for increased cross-border electricity trade and increased renewable energy integration for the ASEAN stems from its own poverty alleviation, sub-regional economic development, and the success of the ASEAN Economic Community (AEC).[4] Linkages between electricity consumption and economic growth demonstrate the positive relationship between electricity access and the development process (Apergis & Payne, 2011). Additional benefits when energy poverty alleviation intersects with climate change mitigation can be realized via policy efforts on efficiency and energy mix diversification (Ürge-Vorsatz & Herrero, 2012). Rapid economic development requires increased energy access, which puts pressures on the climate and environment and requires energy solutions that respond to global greenhouse gas emissions while also incorporating responses to climate change concerns (Gulagi et al., 2017). ASEAN's energy poverty reductions, economic growth success, and environmental challenges are all interlinked in the development and growth story of the sub-region. The 1997 ASEAN Power Grid Project (APG) is based on recognition that integrating electricity systems across the region would benefit ASEAN Vision 2020 goals, including greater energy access, reduced energy poverty, and increased sub-regional connectivity (ASEAN, 1997).

Reaching the ASEAN goal of 23% renewables by 2025 requires national commitments in addition to the sub-regional goal, aimed at helping each member economy make the transition based upon their own particular economic and policy mechanisms (Huang et al., 2019). In the case of Brunei Darussalam, for example, this means increasing renewables to 10% share of energy by 2035; for Indonesia, that number goes up to 23% by 2025 (ACE, 2017). According to the ASEAN Energy Outlook ASEAN member economies will not reach this goal unless significant policy and infrastructure changes are made (ACE, 2017). In addition, while power capacity is expected to increase threefold to accommodate sub-regional electricity demand, coal is estimated to provide 42% of new capacity by 2040; renewables maintain a lower share of 29% if all proposed policy and structural adjustments are made to accommodate increased variable capacity (ACE, 2017).

2.2 ASEAN Power Grid Interconnections

Three primary models of electricity market integration are employed globally – (1) consolidation (of markets and system operators) and (2) coordination (of system operators) or (3) a hybrid of both 1 and 2 (IEA, 2014). Consolidation is more feasible in already integrated or similar markets, coordination ideal for different market structures and areas with geographical variability (ibid.). In the case of ASEAN, the IEA (2014) recommends a hybrid approach – coordination and consolidation at the ASEAN and sub-ASEAN level, utilizing ASEAN's own incremental approach.

This ASEAN approach to incremental integration is being done via the progression of ASEAN subsystems, which are made up of smaller groupings of ASEAN member countries based on geographical location and power systems (ASEAN, 2004). This progression begins with cooperation via individual markets, progression to multilateral markets, and eventually the joining of multilateral markets into a sub-regional electricity market, where electricity can be seamlessly traded among partners and their synchronized and standardized national electricity markets (APAEC, 2010).

Development of the APG has followed recommendations outlined in the ASEAN Interconnection Master Plan Studies (ASEAN, 2004 and 2010). These studies have recommended an incremental approach to the APG via subsystems first, then linkages developed between and among subsystems, which would eventually form the wider APG. These subsystems are the upper west system, consisting of the Greater Mekong Sub-Region (GMSR, established in 1992), including Cambodia, Lao PDR, Myanmar, Thailand, and Vietnam; the lower west system, consisting of the Indonesia, Malaysia, and Thailand Growth Triangle (referred to as IMT-GT, established in 1994) and Indonesia, Malaysia, and Singapore Growth Triangle (referred to as IMS-GT); and the east system (established in 1994), or BIMP-EAGA (ASEAN 2010). To date, the hybrid approach (IEA, 2014) still works for ASEAN economies and is complemented by the variety of market structures represented in the sub-region.

Via the APG, ASEAN member economies primarily trade electricity between transmission system operators (TSO), and interconnections have been implemented primarily on a bilateral basis (ACE, 2018b). Based on data available as of January 2019, projected and planned interconnectors are 30 and 9, respectively.[5] Updates are expected when new sub-regional assessments become available; however, this data is standardized and compiled by the author, as many data sets do not include updated interconnectors. Generating capacity is expected to quadruple and, through sub-regional interconnectors ASEAN, aims to reach 100% electrification.

The status of the APG is fraught with uncertain progress and moving slowly. Multilateral trading is virtually nonexistent (Figure 12.2). A few bilateral interconnectors have been put in place, but multilateral trading is yet to be realized on the scale APG set out to achieve originally (APAEC, 2010; ASEAN, 2017a,b). Forecast interconnections are set out in each subsystem, with the least advanced interconnections in the BIMP-EAGA subsystem (Pacudan, 2016; APAEC, 2004; APAEC, 2010). Forecast and current ASEAN interconnections are depicted in

Existing	Ongoing & proposed	Future outlook
9	21	30
Operating interconnections	Ongoing and proposed new interconnections	Total regional interconnections
4k	30k	34k
Megawatt capacity	Megawatt forecast new capacity	Total regional megawatt capacity

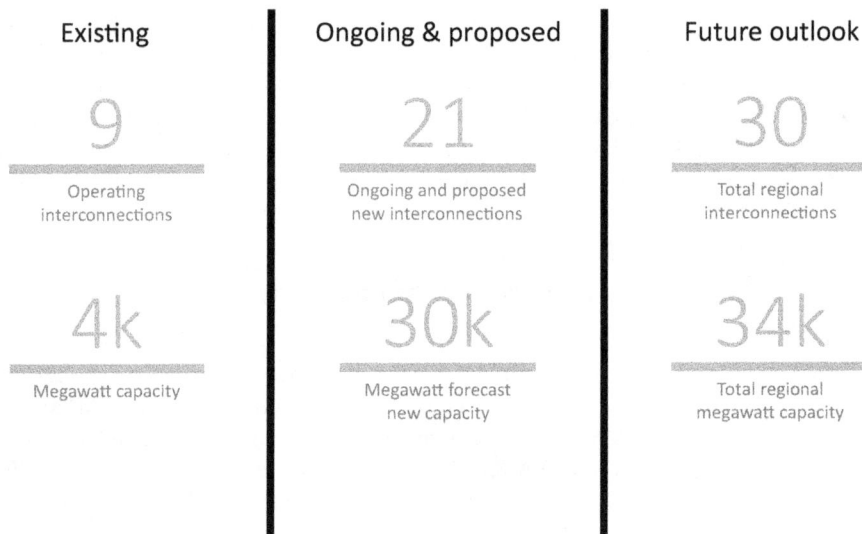

Figure 12.2   ASEAN interconnections summary.

Source: Compiled by author using APAEC (2004, 2010); Srisuping (2013); Andrews-Speed & Hezri (2013); Ibrahim (2014); Shi (2014, 2016); IEA (2015, 2019b); Hermawanto (2016); Li & Kimura (2016); ASEAN (2017a,b); Halawa et al. (2018); IRENA (2018b); ACE (2018b). The author is also grateful to interview subjects for their feedback and suggestions.

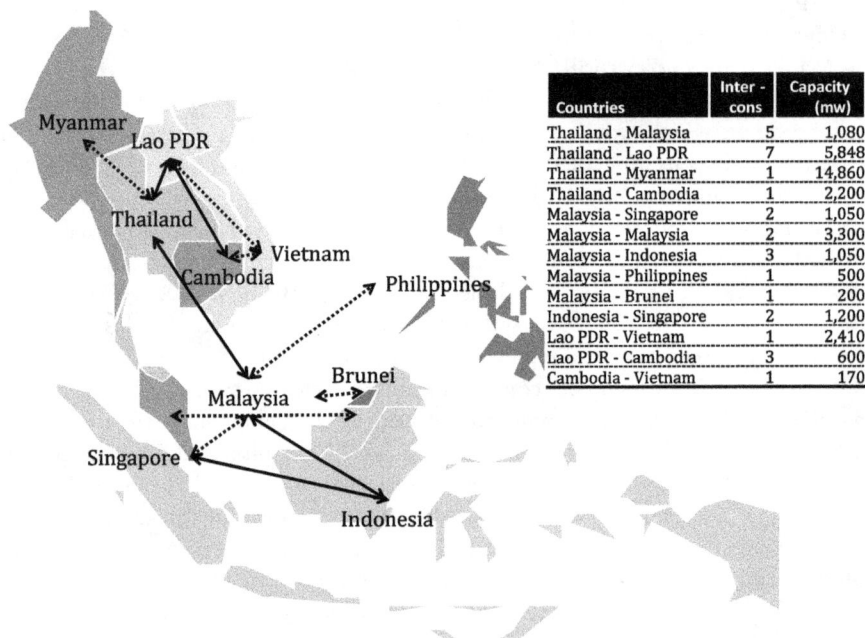

Countries	Inter-cons	Capacity (mw)
Thailand - Malaysia	5	1,080
Thailand - Lao PDR	7	5,848
Thailand - Myanmar	1	14,860
Thailand - Cambodia	1	2,200
Malaysia - Singapore	2	1,050
Malaysia - Malaysia	2	3,300
Malaysia - Indonesia	3	1,050
Malaysia - Philippines	1	500
Malaysia - Brunei	1	200
Indonesia - Singapore	2	1,200
Lao PDR - Vietnam	1	2,410
Lao PDR - Cambodia	3	600
Cambodia - Vietnam	1	170

Figure 12.3   Forecast and current ASEAN interconnections.

Source: Compiled by author using APAEC (2004, 2010); Srisuping (2013); Andrews-Speed & Hezri (2013); Ibrahim (2014); Shi (2014, 2016); IEA (2015, 2019b); Hermawanto (2016); Li & Kimura (2016); ASEAN (2017a,b); Halawa et al. (2018); IRENA (2018b); ACE (2018b). The author is also grateful to interview subjects for their feedback and suggestions.

the following figure, with forecasted ones in dotted lines and current interconnections depicted in solid lines.

2.3 Subsystem Interconnections: BIMP-EAGA

The BIMP-EAGA subsystem, also known as the east system, is the least advanced of the three subsystems identified by ASEAN (Pacudan, 2016). This subsystem has only one interconnection to date, established as recently as 2016 (ibid.). However, the potential for continued expansion of interconnections, integration of renewable energies, development of national electricity markets, and growth in electricity access is quite high in this subsystem. Central to the opportunity of BIMP-EAGA is the island of Borneo, made up of three different BIMP-EAGA countries (Brunei Darussalam, Indonesia, and Malaysia) (ADB, 2014). Forecast and current interconnections are as follows, with forecasted ones depicted in dotted lines and current interconnections in solid lines.

Borneo (the largest island in Asia and split between Malaysia, Indonesia, and Brunei) has quite a large imbalance of electricity supply, with varied infrastructure across the island. Depending on the Borneo country, this geographical variation and accompanying infrastructure deficit results in roughly 70–90% electrification ratio, apart from Brunei Darussalam, which has 99% electrification (ADB, 2014). Indonesia and the Philippines have the largest to gain, as they have total populations without electricity reaching 23 million and 11 million, respectively (IEA,

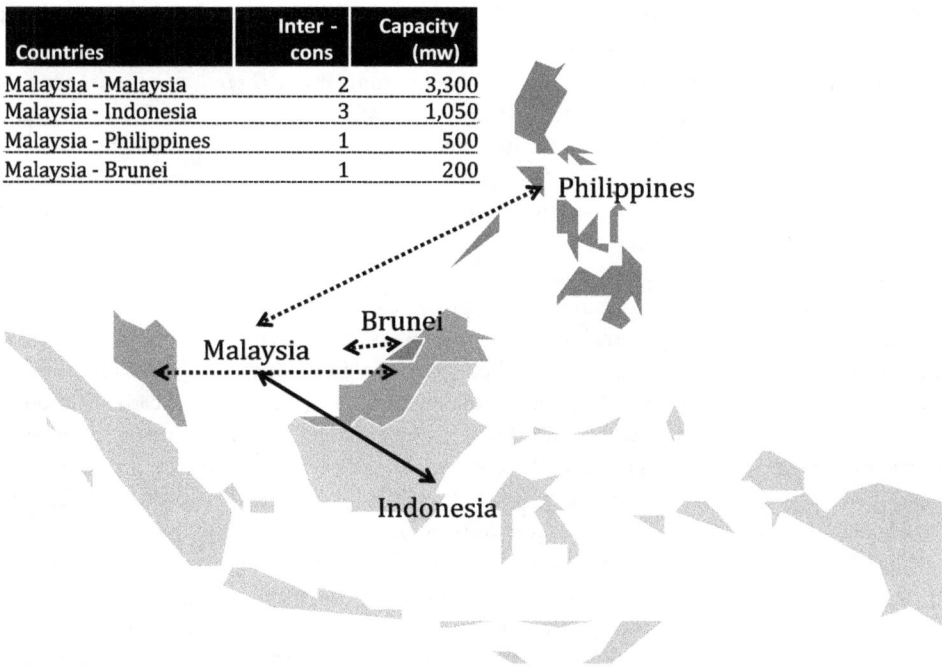

Countries	Inter - cons	Capacity (mw)
Malaysia - Malaysia	2	3,300
Malaysia - Indonesia	3	1,050
Malaysia - Philippines	1	500
Malaysia - Brunei	1	200

Figure 12.4   BIMP-EAGA forecast and current interconnections.

Source: Compiled by author using APAEC (2004, 2010); Srisuping (2013); Andrews-Speed & Hezri (2013); Ibrahim (2014); Shi (2014, 2016); IEA (2015, 2019b); Hermawanto (2016); Li & Kimura (2016); ASEAN (2017a,b); Halawa et al. (2018); IRENA (2018b); ACE (2018b). The author is also grateful to interview subjects for their feedback and suggestions.

2017b,c). Each country also has national targets for renewable energy integration and potential for growth in renewable sectors. However, each nation's fuel dependency is currently dominated by fossil fuels, with the majority dependent on coal and oil (IEA, 2017a,c).

In addition to reliance on fossil fuels and fossil fuel imports, the BIMP-EAGA subsystem is made up of a variety of market structures. Brunei Darussalam's power system is managed by both a department and a private management company and Malaysia by independent companies (Pacudan, 2016). Indonesia's five separate power systems are managed by individual state-owned branch offices, whereas both Malaysia and the Philippines have a combination of state-run corporations and national grid offices (Yokota & Kutani, 2017). Some of the power systems are vertically integrated with liberalization efforts under initial development, while others have transmission systems and distribution networks that are closed to outside power producers. This variety of structure is not uncommon and is reflective of the tendency toward a combination of state-run and private-sector management.

The Philippines offers interesting insight into power sector development in the region, as its structure is quite different from the rest of BIMP-EAGA. It is the only country in the subsystem with market competition and has allowed third-party involvement in the transmission of power, as well as an independent regulatory body and independent electricity market operator (Pacudan, 2016). This models global and sub-regional recommendations for development of power systems (IEA, 2014) and offers a unique example to the remainder of BIMP-EAGA. The Philippine system bodes well for the future coordination of neighboring BIMP-EAGA systems, as it more feasible to coordinate in a highly diverse international power system than to consolidate (IEA, 2014; Pacudan 2016).

To date, a single interconnection exists between Malaysia and Indonesia and is the second smallest current international interconnection in the region, with a capacity of roughly 250 MW (Pacudan 2016; author compilation). This interconnection is seen as a model for future BIMP-EAGA development, with plans for expansion of further interconnections leading out to 2030 (Sarawak Energy, 2017). These are forecast as Malaysia (2 interconnections), Malaysia–Brunei Darussalam (1), Malaysia-Philippines (1), and Malaysia-Indonesia (3), resulting in roughly 5,000 MW of additional capacity (author compilation). Based on the outlined Borneo interconnection plan, the harmonization of the BIMP-EAGA region will be minimal compared to other subsystems in ASEAN (Pacudan, 2016).

Progress for the BIMP-EAGA subsystem hinges on a few key similarities with other APG subsystems. First and foremost, this includes encouragement of national incentives. Where ASEAN has made efforts to present and engage member economies in understanding broad incentives for interconnection (e.g., increased energy access, the buying and selling of excess capacity, modernization of power systems, financial support, knowledge exchange, and higher shares of renewables to reach renewable integration targets), these incentives are not providing enough movement or engagement among national-level policymakers (See: Kimura & Shi, 2011; Kimura et al., 2013; Pacudan, 2016; and the breadth of literature from ACE and ASEAN, respectively).

Once national-level policy adjustments have occurred, economic exchanges can be developed, third-party access can be provided, and multi-buyer, multi-seller market systems can develop. National incentives for these power system changes and increased cross-border interconnections are currently tied closely to the sub-regional imperatives for these power system changes in the first place – greater energy access and reductions in energy poverty, associated climate and energy gains, and potential economic gains via regional and global economic market liberalization.

In addition to climate incentives, ASEAN appears to understand the IPE of cross-border electricity trade as part and parcel to general market liberalization practices and expectations of emerging economies. In fact, the majority of literature (see Kimura & Shi, 2011; Kimura et al., 2013; Pacudan, 2016; Wu, 2016) on EMI in ASEAN all advocate for market liberalization and increased competition within systems dominated by state-run utilities and transmission operators, which most of the region's systems are. BIMP-EAGA economies similarly understand the IPE of cross-border electricity trade; however, integration remains minimal currently due to market factors not meeting capacity and infrastructure requirements and national-level policies not pushing these changes quickly enough (Kimura et al., 2013; Pacudan, 2016). Analysis of current interconnections and market factors did not, unfortunately, yield any major surprises in BIMP-EAGA, further supporting expert insight that BIMP-EAGA is the weakest of ASEAN's EMI subsystems among academic and policy experts interviewed. The u--nderlying issue recurring here, national versus sub-regional incentives and requirements, repeatedly emerges as a hindrance to EMI and resurfaces again and again across analysis.

3. Development of National Market Structures

Development of national electricity markets has been identified widely across the literature[6] as a necessary prerequisite of increased multilateral cross-border interconnection, regional power sector development, and EMI. This is because national electricity markets require the physical capacity and regulatory frameworks to manage not just domestic energy but imported energy as well (Li & Kimura, 2016) where multilateral interconnection develops and system capability, efficiency, flexibility, and reliability become even more important than when bilateral connections occurred. National markets have to be physically developed – have the interconnection and the capability to transport – in order for exchange to occur. Development of national markets includes the implementation of regulatory standards, transmission processes, legal requirements, physical requirements, and even local government support. Finenko et al. (2017) argue that while physical bilateral or very basic multilateral interconnection can take place in undeveloped markets, any deep integration and long-term trade requires alignment of individual markets and their standards, policies, and practices.

If the benefits from EMI and cross-border electricity trade are going to be available to all parts of the market (both rural and urban populations), then modernization and integration are necessary to provide low-cost energy access in an intermittent environment that crosses borders (IEA 2015). Analysis of individual power markets in ASEAN shows sector development is varied. Power markets in the sub-region are a mix of public, private, and hybrid forms, with varying levels of state regulation and oversight. Many power markets are not unbundled. Generation, distribution, transmission, and/or retail all take place within one or multiple vertically integrated companies who oversee most processes and function very close to a monopoly (Jones, 2014). In fact, development of power markets, including transmission and generation infrastructure, are at varying levels depending on the nation (Wu, 2016).

Brunei Darussalam's power market is dominated by fossil fuel interests. The abundance of hydrocarbon resources and heavily subsidized energy and electricity sectors has contributed to a reliance on fossil fuels (Energy and Industry Department of Prime Minister's Office, 2016). This overreliance on fossil fuels makes Brunei's economy closely tied to fossil industries, with fossil fuels making up 90% of the country's exports and 44% of its GDP (OECD, 2018). Brunei Darussalam's power sector is regulated by a government agency, with separate and unconnected networks for generation, transmission, and distribution (Ahmed & Othman, 2014) that operate in a single-buyer market that is not open to independent power producers (Pacudan, 2016). Brunei's power sector is the most heavily regulated of the BIMP-EAGA economies (Navarro & Sambodo, 2013) and data, the most limited. However, Brunei is also the economy for which the electricity industry is most dominated by fossil fuel industries. Brunei produces a surplus of power domestically, roughly 100–106% of annual consumption being covered by domestic production in 2016 (World Meters, 2020).

Indonesia is one of the national markets with discrepancies between policy and action, in part due to the intersection of business and policy interests in energy sectors (Navarro & Sambodo, 2013). Government control over Indonesia's energy industries (particularly oil and gas), lack of government clarity over national energy regulations, close relationships between illegal energy ventures and government leaders, and tight restrictions on foreign investment top the list of conflicting interests and policy messaging (Dutu, 2016). Multiple periods of reform have influenced the continued removal of fossil fuel subsidies; however, the use of subsidies still exists and is closely tied to the prevalence of fossil fuels in the country's energy mix (Chelminksi, 2018). Overall Indonesia's power sector is dominated by the national utility, PLN (Perusahaan Listrik Negara); Indonesia has a state-owned, vertically integrated electricity market with little private sector participation and limited independent power producer involvement at just over 20% of total generation (PWC, 2016). Indonesia is one half of the existing cross-border electricity interconnection with Malaysia, with plans for expansion and assessments underway (Wu, 2019).

Malaysia makes up the second half of the existing cross-border electricity interconnection, with plans for expansion. Similar to the electricity markets of Brunei and Indonesia, Malaysia's market is not fully liberalized, is vertically integrated, and allows limited independent power producer involvement (Yokota & Kutani, 2017). Electricity market regulation is overseen by an independent regulatory body, unlike Brunei and Indonesia. Malaysia also has vertically integrated transmission and distribution, operated by either Tenaga Nasional Berhard (TNB) or Sarawak Energy Berhard (SEB), depending on the region within the country (Samsudin et al., 2016). Malaysia has significant government-funded petroleum subsidies in place and also produces more than 100% of its annual consumption needs, with 108% annual consumption generated domestically in 2016 (World Meters, 2020).

The Philippine power sector is perhaps the most advanced in the subsystem, as it is farthest along in the liberalization process. The Philippine market is regulated by an independent body, as in Malaysia, but the Philippines has also introduced some competition into the market, with a hybrid public-private model and the only of its kind in the BIMP-EAGA subsystem (Pacudan, 2016). However, the Philippine market still is only partially competitive – it was one of the first Southeast Asian economies to introduce independent power producers into the market, but the majority of power is still produced by NAPOCOR, the National Power Corporation (Hall & Nguyen, 2017). In addition to initial liberalization practices, the Philippine market also has the most advanced forecasting capabilities among BIMP-EAGA economies (Huang et al., 2019). Domestically, the Philippines generated 111% of its total electricity consumption needs in 2018 (DOE, 2019). Among interview subjects (Multilateral Organization – Informant 2), the Philippines has been held up as the most advanced of the BIMP-EAGA economies.

Overall, ASEAN electricity markets are structured around similar factors: state-driven markets with state oversight even in liberalized markets, limited competition generally, a predominance of subsidies, a range of development stages, and varying levels of electrification across the region (Wu, 2019). There is no one-size-fits-all recommendation for EMI and electricity market development in ASEAN given the variety across markets; however, there are similar solutions to these limitations across the sub-region. Overall, the economies of BIMP-EAGA are making physical progress in developing their power sectors, yet market structure is a dominant factor hindering advancement (Matsuo & Tsunoda, 2016).

5. Conclusions

The inclusion of higher shares of clean energy in a county's energy mix places pressure on undeveloped and minimally integrated regional electrical power systems as they transition from historically fossil-fuel-based electricity production. Where previously it was possible to build power sectors within national boundaries, increased pressure to source a substantial and growing percentage of a nation's load from renewable resources means national systems become more variable as more intermittent resources are introduced. The prospect of electricity interconnections where nations can buy surplus power during low generation hours from a neighboring market that may not be experiencing scarcity (and vice versa) becomes a more pressing national concern and more attractive international investment than has previously been the case. Put simply, electricity is good for development, but old models of centralized fossil-fuel-based systems are not compatible with global climate commitments and the global energy transition. The opportunities for cross-border electricity trade in East Asia broadly and Southeast Asia and BIMP-EAGA specifically are abundant and well justified. However, it appears that economic gains via trade in surplus energy (renewable or not) may be behind in meeting the regional and sub-regional interest as well as national measures.

Unfortunately, APG progress to date is quite limited. This is in part attributed to, among financing challenges, difficulty implementing the necessary national power sector reforms in a timely manner. As discussed previously, progress to date is largely bilateral in nature, with no region-wide interconnections, though ASEAN has attributed this to be the first step to wider connection (APAEC, 2004). Complicated power relations make bilateral agreements more realistic, as governments must work within complicated systems of interests, including business, consumer, and public interest groups (Jones, 2016).

Power grids are, largely, confined by national-level boundaries; therefore, reinforcement or growth of power grids needs to first happen at the national level. However, political will and governance strength are required for this buttressing to occur. Where state-owned utilities dominate the power sector (Indonesia and Brunei, for example), there may be difficulty in expanding renewable energy options within the national energy mix; for example, the prevalence of fossil fuel subsidies and the nationalized companies that benefit from their continued use is one potential barrier in a system (Shi, 2016) where fossil fuel companies are linked with political leadership and vice versa. In addition, power dynamics, in the form of dominant national economic actors, influence political will. National governments are in a complicate position; there are multiple vested interests informing policymakers and multiple factors pushing policy many different directions. It is no wonder that connectivity is not fully clarified and imagined yet in ASEAN.

Analysis of market factors and how they assist in understanding the IPE of cross-border electricity trade in the sub-region points predominantly to liberalization and national-level limitations in achieving liberalization. EMI requires liberalization of domestic energy markets, increased competition, and standardization across markets – traditionally solutions found in Western, developed countries for already advanced power sectors that are prescribed as an ideal, without being moderated for an East Asian context. Commitments to free trade (Oseni & Pollitt, 2014), amid centrally planned economies that utilize state-directed economy policy, reflects this author's characterization of neo-developmental statism and its emerging role in understanding the political economy of cross-border electricity trade. In examining the sub-regional market factors the IPE of cross-border electricity trade emerges as a link between developmental state practices and liberalization identified for EMI to advance. However, it appears that states are "committed" to free markets and liberalization in so far as free markets and liberalization benefit state goals, where those goals fall in the hierarchy of needs is debatable and, based on market needs identified so far, not always in line with the market factors. The political economy of energy policy development in the sub-region or subsystem can influence the development of policies that respond to broader resource concerns – such as commitments to free trade, market competition, and even energy transitions.

5.1 Policy Implications

The BIMP-EAGA case study demonstrates that market realities do not necessarily reflect a current positive outcome for increased EMI and cross-border electricity trade in Southeast Asia. This is for two reasons in particular: (1) Decarbonization of the power sectors, via increased shares of renewables for electricity generation, is not evident via the current sub-regional and subsystem energy mix; coal is still the dominant resource in individual markets and projected to continue as the dominant fuel source. (2) Because, ultimately, there is very little (if any) gap between electricity needs and domestic supply, each country can supply its own electricity needs based on 2016 rates; however, this could change depending on future added electricity demand (a).

These two points, coal dominance and electricity needs, ultimately signal that ASEAN's BIMP-EAGA subsystem is not, system-wide, committed to immediately decarbonizing electricity sectors and energy mixes. This research, while not focused solely on energy diversity, but the usefulness of cross-border electricity trade for regional energy transitions, exposes an uncomfortable truth: cross-border electricity for renewable energy integration is not the same priority as cross-border electricity trade for economic gain. The selling of excess electricity and continued use of hydrocarbon resources in said electricity generation is a very real possibility for EMI in Southeast Asia. Combining climate-driven choices for sub-regional, regional, and global energy transformation is a necessary addition to sub-regional and regional EMI initiatives.

Several policy implications have emerged. Some of these are standard energy policy recommendations, such as better coordination of policy goals and initiatives with sub-regional and national competencies, a consolidation of needs across sectors, and prioritization of deregulation in order to encourage other power sector needs, such as liberalization and increased FDI. These recommendations are reflected in the literature on electricity market integration, they are reflected among the opinions of experts interviewed, and they are reflected in the work of organizations and multinational institutions on the ground in the region pushing for EMI. What is clear from the consistency in responses is that these needs are not being adequately addressed or at a pace in line with sub-regional and national energy goals. The regional and sub-regional energy transition could be improved. The primary policy implications of these improvements are as follows:

1. National: Physical infrastructure development, market liberalization, and refinement of market design are all necessary at national levels. Without these three overarching adjustments EMI progress will not be realized among ASEAN and BIMP-EAGA countries. The decarbonization of national power sectors should also be prioritized. Current national interconnection efforts appear to be driven by economics, not decarbonization goals and climate targets.
2. Sub-regional alignment: National priorities and sub-regional targets are not aligned in policy responses at either level. Better coordination and development of join initiatives/support policies (political and economic/financial) could help in this area, but ultimately, national needs may not align with sub-regional targets for many years.

3. Technical concerns: More technical targets, including enhanced capacity, sub-regional coordination, including grid codes, sub-regional transmission systems, and other sub-regional jurisdiction considerations are secondary to national power sector reform. These needs are years away among BIMP-EAGA countries.

4. Power dynamics and political will: Strong political and economic relationships need to be present in tandem with national power sector reform. These trusting relationships are a necessary by-product for national and sub-regional reform measures to be realized. Political and economic power dynamics will be challenged if EMI measures are realized; this will be difficult in the short term but have benefits for the sub-regional and global energy transition.

5. Global climate targets: sub-regional, regional, and global climate targets are not fully integrated into national policymaking and local-level concerns. Engagement in global and sub-regional climate commitments is performative and limit the regional and global energy transition.

This research has combined the barriers to EMI to illustrate the competing interests at both sub-regional and national levels of EMI within the BIMP-EAGA subsystem. A variety of different reforms have been proposed to address these barriers. These political and economic reforms were observed via interviews and data collection to be realistic but contradictory in how they are carried out at different levels. In addition to the contradiction between sub-regional versus national-level policy implementation and priority, there is the possibility of certain populations being left behind in the benefits to be gained from increased renewable energy integration and electricity market expansion and integration. As such, neo-developmental statism can combine power and equity in explaining energy and climate priorities; perhaps, ultimately, this is an unrealistic call for action considering the dominant sub-regional, regional, and global neoliberal economic system, and the climate change mitigation linkages discussed.

APPENDIX 1

Anonymous Research Interview Subjects (adopted from 36 original interviews reduced to 28 relevant one-to-one interviews and 14 events based on relevant information gathered)
Former Government – Informant 1
Multilateral Organization – Informant 2
Business 3
Multilateral Organization 4
Think Tank 5
Academic 6
Think Tank 7
Development Bank 8
Business 9
Academic 10
Think Tank 11
Think Tank 12
Academic 13
Academic 14
Think Tank 15
Former Government 16
Former Government 17

Multilateral Organization 18
Business 19
Academic 20
Academic 21
Business 22
Multilateral Organization 23
Academic 24
Development Bank 25
Academic 26
Former Government 27
Former Government 28
Think Tank 29
Journalist 30
Business 31
Academic 32
Academic Conference 1
Academic Conference 2
Academic Conference 3
Development Bank Conference 4
Academic Conference 5
Academic Workshop 6
Policy Conference 7
Policy Conference 8
Policy Conference 9
Academic Workshop 10
Academic Conference 11
Government Conference 12
Government Conference 13
Government Workshop 14

Notes

1 The definition of East Asia used in this context encompasses the two sub-regions of Northeast Asia (which is made up of the economies of Japan, the Democratic People's Republic of Korea [or DPRK], the People's Republic of China [hereby referred to as China], the Republic of Korea [ROK], and Mongolia) and Southeast Asia (Brunei Darussalam, Cambodia, Indonesia, Lao PDR, Malaysia, Myanmar, Philippines, Singapore, Thailand, and Vietnam). This definition was chosen in line with definitions used in East Asian studies at the University of Leeds.
2 In this research, "region" is used to refer to the area of East Asia, and "sub-region" is used to refer to the sub-region of Southeast Asia. In addition, "subsystem" is used to refer to the smaller, sub-regional groupings of countries in Southeast Asia, such as BIMP-EAGA, as defined and used by the Association of Southeast Asian Nations (ASEAN), the sub-regional governing body of Southeast Asia (ASEAN, 2004).
3 International political economy (IPE) is, broadly, the combination of political science, international relations, and economics to explain global events and economic interactions, often among nation-states and relating to the management of political and economic affairs (Strange, 1988).
4 The AEC is one of ASEAN's flagship initiatives, focused on increasing economic prosperity, developing a "rules-based, competitive, resilient" and globally integrated regional market and stronger regional economic community by a 2015 due date (Shi, 2014). The AEC was largely achieved, in part due to the broad and generalized goals of improving development and governance across the sub-region. Many of ASEAN's subsequent initiatives involve aspects of the AEC, which remains an important cornerstone of ASEAN identity and achievement (APAEC, 2015) and is often referenced as a successful ASEAN initiative.

5 All interconnection data in Figures 13.2–13.4 are estimated based on a variety of sources: APAEC (2004; 2010); Srisuping (2013); Andrews-Speed & Hezri (2013); Ibrahim (2014); Shi (2014; 2016); IEA (2015; 2019b); Hermawanto (2016); Li & Kimura (2016); ASEAN (2017a,b); Halawa et al. (2018); IRENA (2018b); ACE (2018b). The author is also grateful to interview subjects for their feedback and suggestions.
6 See, for example, Andrews-Speed & Hezri (2013); Oseni & Pollitt (2014); Li & Kimura (2016); Pacudan (2016); Shi et al. (2019).

Bibliography

ACE. 2013. *ASEAN Energy Market Integration (AEMI): From Coordination to Integration.* AEMI Group. Bangkok: ASEAN Studies Center.

ACE. 2017. *The 5th ASEAN Energy Outlook (AEO) 2015–2040.* Jakarta: ACE.

ACE. 2018b. *Provision of the ASEAN Interconnection Masterplan Study (AIMS) III.* Jakarta: ACE.

ADB. 2014. An *Evaluation of the Prospects for Interconnections Among the Borneo and Mindanao Power Systems.* Manila: ADB.

Ahmed, Azman and Othman, Hilda Maya. 2014. Electricity Consumption in Brunei Darussalam: Challenges in Energy Conservation. *International Energy Journal.* 14: 155–166.

Andrews-Speed, Philip and Adnan Hezri. 2013. Institutional and Governance Dimension of ASEAN Energy Market Integration. In *ASEAN Energy Market Integration (AEMI): From Coordination to Political Integration,* 149–174. Bangkok, Thailand: ASEAN Studies Centre.

APAEC. 2004. *ASEAN Plan of Actions for Energy Cooperation 2004–2009.* Jakarta: ASEAN Secretariat.

APAEC. 2010. *ASEAN Plan of Actions for Energy Cooperation 2010–2015.* Jakarta: ASEAN Secretariat.

APAEC. 2015. *ASEAN Plan of Actions for Energy Cooperation 2016–2025.* Jakarta: ASEAN Secretariat.

Apergis, Nicholas and James E. Payne. 2011. A Dynamic Panel Study of Economic Development and the Electricity Consumption-Growth Nexus. *Energy Economics.* 33: 770–781.

ASEAN. 1997. ASEAN Vision 2020. https://asean.org/?static_post=asean-vision-2020.

ASEAN. 2004. *MPAC. (AIMS I).* Jakarta: ASEAN Secretariat.

ASEAN. 2010. *MPAC (AIMS II).* Jakarta: ASEAN Secretariat.

ASEAN. 2017a. *MPAC 2025.* Jakarta: ASEAN.

ASEAN. 2017b. *The ASEAN Electricity Exchange (AEE): An International Perspective.* Jakarta: HAPUA-UNESCAP.

Cai, Kevin G. 2008. *The Political Economy of East Asia.* New York: Palgrave Macmillan.

Chang, Y. and Li, Y. 2013. Power Generation and Cross-Border Grid Planning for the Integrated ASEAN Electricity Market: A Dynamic Linear Programming Model. *Energy Strategy Reviews.* 2: 153–160.

Chang, Y. and Li, Y. 2015. Renewable Energy and Policy Options in an Integrated ASEAN Electricity Market: Quantitative Assessments and Policy Implications. *Energy Policy.* 85: 39–49.

Dutu, R. 2016. Challenges and Policies in Indonesia's Energy Sector. *Energy Policy.* 98: 513–519.

Energy and Industry Department of Prime Minister's Office. 2016. Brunei Darussalam Country Report. In Kimura, Shigeru and Han Phoumin (eds), *Energy Outlook and Energy Saving Potential in East Asia 2016,* pp. 79–90. Jakarta: ERIA.

Finenko, Anton, Owen, Anthony D., and Tao, Jacqueline. 2017. *Power Interconnection in ASEAN Region: Lessons Learned from International Experiences.* Singapore: Energy Studies Institute.

Gulagi, A., Bogdaanov, D., Fasihi, M., and Breyer, C. 2017. Can Australia Power the Energy-Hungry Asia with Renewable Energy? *Sustainability.* 9(233): 1–26.

Halawa, Edward, Geoffrey James, Xunpeng Shi, Novieta H. Sari, and Rabrinda Nepal. 2018. "The Prospect for an Australian-Asian Power Grid: Critical Appraisal." *Energies.* 11(20): 2–23.

Hermawanto, Bambang. 2016. "ASEAN Power Grid." 4th North East Asia Energy Security Forum, December 15, 2016, Seoul, Republic of Korea. www.unescap.org/sites/default/files/Session%20 4.1.%20Bambang%20Hermawanto_ASEAN.pdf

Huang, Yu Wen, Noah Kittner, and Daniel M. Kammen. 2019. ASEAN Grid Flexibility: Preparedness for Grid Integration of Renewable Energy. *Energy Policy.* 128C: 711–726. DOI: 10.1016/j. enpol.2019.01.025.

Ibrahim, Syaiful B. 2014. Barriers and Opportunities for Electricity Interconnection the Southeast Asian Experience. *APEC Conference,* March 27, 2014, Tokyo, Japan. https://aperc.or.jp/file/2014/4/4/ S2-2-2_IBRAHIM.pdf

IEA. 2014. *Seamless Power Markets: Regional Integration of Electricity Markets in IEA Member Countries.* Paris: IEA.

IEA. 2015. *Development Prospects of the ASEAN Power Sector*. Paris: IEA.

IEA. 2017a. Southeast Asian Energy Outlook 2017. *World Energy Outlook Special Report*. Paris: OECD/ IEA.

IEA. 2017b. Country Profile: Indonesia. https://www-pub.iaea.org/MTCD/Publications/PDF/cnpp2017/ countryprofiles/Indonesia/Indonesia.htm

IEA. 2017c. Electricity Database 2017. https://webstore.iea.org/weo-2017-special-report-energy-access-outlook

IEA. 2019a. *Renewables 2019*. Paris: IEA.

IEA. 2019b. *Establishing Multilateral Power Trade in ASEAN*. Paris: IEA.

IMF. 2021. Regional Economic Outlook: Asia and the Pacific. https://data.imf.org/?sk=abff6c02-73a8-475c-89cc-ad515033e662.

Impram, Semich, Secil Varbak Nese and Bülent Oral. 2020. Challenges of Renewable Energy Penetration on Power System Flexibility: A Survey. *Energy Strategy Reviews*. 31: 1–12.

IRENA. 2018a. *Power System Flexibility for the Energy Transition*. Abu Dhabi: IRENA.

IRENA. 2018b. *Renewable Energy Market Analysis: Southeast Asia*. Abu Dhabi: IRENA.

Jones, Lawrence E. 2014. *Renewable Energy Integration: Practical Management of Variability, Uncertainty and Flexibility in Power Grids*. London: Academic Press.

Jones, Lee. 2016. Explaining the Failure of the ASEAN Economic Community: The Primacy of Domestic Political Economy. *The Pacific Review*. 29(5): 647–670. https://doi.org/10.1080/09512748.2015.1022 593.

Kunstýř, Jan and Shuta Mano. 2013. "Energy Security and Cross-Border Electricity Trade: Can the Asian Super Grid Project Pose Security Risks for Japan? Can the Electricity Imports be Used as an Extortion Weapon?" Renewable Energy Institute. www.renewable-ei.org/en/images/pdf/20140130/Energy_weapon_final.pdf

Leftwich, Adrian. 1995. Bringing Politics Back in: Towards a Model of the Development State. *The Journal of Development Studies*. 31(3): 400–427.

Li, Yanfei, and Shigeru Kimura, ed. 2016. *Achieving an Integrated Electricity Market in Southeast Asia: Addressing the Economic, Technical, Institutional, and Geo-political Barriers*. ERIA Research Project Report FY2015, Vol. 16. Jakarta: ERIA.

Matsuo, Y., and Tsunoda, M. 2016. Integrated, Trans-Boundary Energy and Electricity Markets in the BIMP Region – A Quantitative Assessment. In Li, Yanfei and Shigeru Kimura (eds), *Achieving an Integrated Electricity Market in Southeast Asia: Addressing the Economic, Technical, Institutional, and Geo-political Barriers*, 41–58. Jakarta: ERIA.

Navarro, Adoracion and Maxensius Tri Sambodo. 2013. The pathway to ASEAN Energy Market integration. *Philippines Institute for Development Studies Discussion Paper Series*. 40: 1–21.

NREL. 2014. Flexibility in 21st Century Power Systems. *21st Century Power Partnership*. Colorado: NREL.

OECD. 2018. *Economic Outlook for Southeast Asia, China and India 2019*. Paris: OECD.

Oseni, Musiliu O., and Michael G. Pollitt. 2014. Institutional Arrangements for the Promotion of Regional Integration of Electricity Markets: International Experience. *Policy Research Working Paper*, No. WPS 6947. Washington, DC: World Bank Group.

Pacudan, Romeo. 2016. "Road Map for Power Market Integration in the Brunei-Indonesia-Malaysia-Philippines (BIMP) Region." In Li, Yanfei and Shigeru Kimura (eds), *Achieving an Integrated Electricity Market in Southeast Asia: Addressing the Economic, Technical, Institutional, and Geo-Political Barriers*, 41–58. Jakarta: ERIA.

PWC. 2016. *Power in Indonesia*. Jakarta: PWC.

REN21. 2019. *Asia and the Pacific Renewable Status Report 2019*. Paris: REN21 Secretariat.

Samsudin, Mohamed Syahril Nazreen, Rahman, MD Mizanur and Wahid, Mazlan Abdul 2016. Power generation sources in Malaysia: Status and prospects for sustainable development. *Journal of Advance Review on Scientific Research*. 25(1): 11–28.

Serra, N. and Stiglitz, J.E. 2008. *The Washington Consensus Reconsidered*. London: Oxford University Press.

Shi, X. 2014. ASEAN Power Grid, Trans-ASEAN Gas Pipelines and ASEAN Economic Community: Vision, Plan and the Reality. *Global Review*. Fall: 116–131.

Shi, X. 2016. The Future of ASEAN Energy Mix: A SWOT Analysis. *Renewable and Sustainable Energy Reviews*. 53: 672–680.

Shi, Xunpeng, Lixia Yao, and Han Jiang. 2019. Regional Power Connectivity in Southeast Asia: The Role of Regional Cooperation. *Global Energy Interconnection*. 2(5): 444–456. https://doi.org/10.1016/j.gloei.2019.11.020.

Srisuping, Kornphat. 2013. *ASEAN Power Grid*. Expert Group Meeting on Conceptualizing the Asian Energy Highway, September 3–5, 2013, Urumqi, China. www.unescap.org/sites/default/files/EGAT_ASEANPowerGrid.PDF

Strange, Susan. 1988. *States and Markets*. London: Pinter Publishers.

UN. 2006. Multi-*Dimensional Issues in International Electric Power Grid Interconnections*. New York: UN.

UN, ADB, UNDP. 2017. *Asia-Pacific Sustainable Development Goals Outlook*. Thailand: UN.

Ürge-Vorsatz, Diana and Sergio Tirado Herrero. 2012. Building Synergies Between Climate Change Mitigation and Energy Poverty Alleviation. *Energy Policy*. 49: 83–90.

Wade, Robert. 1988. State Intervention: Taiwanese Practice. In Gordon White (ed.), *Developmental States in East Asia*, 30–67. London: The Macmillan Press.

White, Gordon and Robert Wade. 1988. Developmental States and Markets in East Asia: An Introduction. In Gordon White (ed.). *Developmental States in East Asia*, 1–29. London: The Macmillan Press.

World Meters. 2020. Electricity Generation. [Online]. [Accessed 12 December 2019]. www.worldometers.info.

Wu, Yanrui. 2016. Electricity Market Integration in ASEAN: Institutional and Political Barriers and Opportunities. In Li, Y. and S. Kimura (eds). *Achieving an Integrated Electricity Market in Southeast Asia: Addressing the Economic, Technical, Institutional, and Geo-political Barriers*, 109–125. Jakarta: ERIA.

Wu, Yanrui. 2019. Electricity Market Integration: Global Trends and Implications for the EAS Region. *Energy Strategies Review*. 2(19): 138–145. https://doi.org/10.1016/j.esr.2012.12.002.

Yokota, Emiri and Ichiro Kutani. 2017. *Study on Electricity Supply Mix and Role of Policy in ASEAN*. Tokyo: ERIA.

13

REACTIVE DECARBONIZATION IN CUBA AND VENEZUELA

D. Kingsbury

1. Reactive Decarbonization in an Unequal World

Cuba's Special Period in Times of Peace – as the decade that followed the Soviet Union's collapse was known – and the implosion of Venezuela's oil industry provide glimpses of the unintended consequences of global energy transitions. They are illustrations of *reactive decarbonization*: fundamental reconfigurations of carbon-intensive sociopolitical energy regimes, which include some move away from carbon, triggered by external shocks to prevailing ways of organizing state, society, and nature. During the Special Period, Cuba's carbon-intensive electricity, agriculture, mobility, and industrial capacities were reduced to near zero when the disintegration of the Soviet Bloc cut off the island's access to oil and petrochemical imports. Venezuela's petroleum export-dependent economy was decimated when oil prices fell to historic lows in the 2010s (Bull and Rosales, 2020), severely limiting the government's ability to provide basic services from electricity provision to health care and education. Crises in both countries, furthermore, occurred in the context of United States and international sanctions, further intensifying the consequences of reactive decarbonization by limiting room for maneuver.

Reactive decarbonization also suggests the need to think of transitions as occurring within larger sociopolitical energy regimes and beyond purely technical matters of replacing "dirty" inputs like fossil fuels for "clean" ones. By *sociopolitical energy regimes*, I refer here to the power relations and decision-making processes by which societies organize the production, distribution, and consumption of energy in its various forms. Sociopolitical energy regimes, furthermore, operate at multiple scales. They shape whether or not a family has the capacity to feed and care for themselves in their households; the ways in which we relate to our environments via infrastructure (e.g., generation facilities, powerlines, and home construction), extraction (e.g., oil wells and mines), and waste (the "externalities" of pollution and interrupted landscapes); and the geopolitical relations between states that provide, regulate, and depend upon sources of energy (see also Boyer, 2014; Cederlöf and Kingsbury, 2019; Mitchell, 2013).

Energy transitions entail more than energy, and while reactive decarbonization could be seen as a worst-case scenario the world over, the experiences examined here suggest their consequences will be most dire in those places with the least options and alternatives. Responses to the climate crisis are unfolding in a global order long characterized by inequality and maldevelopment.[1] Transitions offer opportunities to address these structural conditions as they revise

DOI:10.4324/9781003183020-15

and reconfigure sociopolitical energy regimes across multiple scales. Failure to do so, as the cases of Cuba and Venezuela illustrate, risk undermining the success of present and future decarbonization.

The chapter proceeds in three sections. In the first section, I explore the disruption to Cuba's carbon-intensive development model that took place during the Special Period (roughly 1991–2001) and its consequences for politics and society on the socialist island nation. The second considers Venezuela and the mutually amplifying crises that have characterized the social and political situation there since the death of the former president Hugo Chávez (1999–2013) and the subsequent crash in global oil prices. The concluding section draws lessons from Cuba and Venezuela for global energy transitions in the context of historically persistent global inequalities.

2. Cuba's Post–Cold War Transitions

The Special Period in Times of Peace was characterized by the upheaval of ecological, political, and economic systems constituted since the Cuban Revolution's victory in 1959. With the dissolution of the Soviet Union, Cuba lost a key ally and shield against the United States. It also lost its primary source for oil. The ensuing decade of reactive decarbonization entailed a creative process pursued under duress of finding alternatives to petroleum while also revising the state-society-nature dynamics of carbon-intensive socialism. Understood as an energy transition, looking back on the Special Period allows us to consider human and natural consequences of forced decarbonization, to ask who drives transitions, and ultimately, to acknowledge the reversibility of decarbonization.

Shortly after its victory in January 1959, the Cuban Revolution aligned itself with the Soviet Union. Fidel Castro's initial hopes that ties to the USSR would accelerate Cuba's exit from neocolonial dependency, monocrop agriculture, and the influence of the United States were quickly dashed. Even though trade with the Soviet Bloc did not diversify Cuba's productive base, it did accelerate the industrialization of Cuba's sugar industry (Pollitt, 2004). For this, petroleum was key. Throughout the 1970s and 1980s, sugarcane cultivation was increasingly mechanized and became progressively capital and chemical intensive (Febles-González et al., 2011). Following Soviet approaches to economic development, electrification was also vigorously pursued as the revolutionary government moved to deliver on promises of modernization (Cederlöf, 2017). The good life of a socialist middle class was promoted throughout the 1970s and into the 1980s, marked by an increased availability of CEMA-made (Council for Economic Mutual Assistance) consumer goods, many of which relied on petroleum-generated electricity from the newly expanded grid (Cederlöf, 2017). Electrification and increased access to consumer goods marked progress, even as the island remained tethered to a sugarcane economy that had been dominant for centuries. Rather than the result of endogenous or organic shifts within the Cuban economy, improvements in the quality of life were dependent on economic planning, the export of sugar and the import of oil, petrochemicals, and industrial and consumer goods (Cabrera Abrús, 2019).[2] It was a carbon-intensive and socialist sociopolitical energy regime incubated under siege conditions.

During this era, "anthropocentric and social issues, such as education, health care, economy and security ranked high on the country's priority list", while "environmental issues were of no importance" (Maal-Bared, 2006: 350). Soviet-Cuban trade and development encouraged a Green Revolution model of carbon-intensive agriculture, utilizing "higher levels of fertilizers, mechanization, and irrigation than the USA, and certainly more than other Latin American countries" (Wright, 2009: 58). Industrialized agriculture and mining polluted waterways; intensified soil salinization, degradation, and erosion; and contaminated the air (Díaz-Briquets and Pérez-López, 1995). By the 1970s and 1980s, the situation had deteriorated such that the United Nations Development Program provided technical and financial assistance to deal with pollution (Díaz-Briquets and Pérez-López, 1995).

Carbon-intensive socialism ended with the disappearance of the USSR and the increased exposure of the island to the US embargo.[3] As Louis Pérez Jr. (2015) notes, prior to the Special Period, CEMA imports amounted to over 85% of Cuban trade across all sectors; oil imports decreased by 90% within a year. Cuba's agriculture system was also thrust into a dangerously precarious state: fertilizer production declined by 80%, dairy products by half, meat by two-thirds, poultry by 80%, pork production declined by 70% (ibid.). All told, "between 1990 and 1993, the Cuban economy shrank by more than 40%" (ibid.). The impact was severe and constant as average caloric intake fell by as much as 30% (Wright, 2009). Population outflows also increased during the Special Period, as loosening emigration controls were used to lessen sociopolitical tensions. Pérez concludes that

> the years of the *período especial* evoked signs of an apocalyptic premonition, especially in the cities . . . an eerie silence descended upon urban neighborhoods in the evenings as the sights and sounds of the city so much associated with gasoline and electricity ceased.
>
> (Ibid.)

Emily Morris (2014) argues that while outsiders saw the Special Period as a grim euphemism, "within Cuba it was immediately understood as a reference to established civil-defence procedures, in case of natural disasters or US attack" (17). It activated an anticipated but not yet realized shift in the norms and practices of life on the island at every scale. Morris expands that

> the Economic Defence Exercise of 1990 – in which electricity and water supplies were cut off for short periods, to rehearse emergency collective responses involving factories, offices, households, schools and hospitals – used methods of collective organization and multi-agency coordination similar to those of hurricane-preparedness or military-defence exercises. The same types of mobilization were evident in the early 1991 Food Programme, in which farmers and city dwellers were called upon to contribute to food production; the December 1991 Spare Parts Forum, on ideas for recycling machinery and substituting for imports; and the January 1992 Energy Plan, in which households, enterprises, and local authorities identified ways to cut fuel consumption.
>
> (Ibid.)

The energy transition that was reflected in the Special Period called for ingenuity, discipline, and sacrifice to navigate extreme circumstances. It forced a shift in Cuba's productive capacities and in so doing altered the daily lives of Cubans across the island – a transvaluation of the means by which Cubans judged progress and identity on a scale similar to the 1959 revolution (Pérez, 2015).

The revolution had disrupted the urban-rural divide, class systems, and neocolonial status that defined Cuba since independence from Spain at the end of the nineteenth century. The Special Period, in turn, reorganized the state-society-nature dynamics of the post-revolutionary period. Among other reforms, the Cuban government called for a revision of the "ideological rigidity" of the preceding years toward "pragmatic innovations" that included decentralization of the state's most important social welfare and economic practices, the introduction of new forms of collective ownership and decision-making, and limited forms of private property and legalized entrepreneurship while pivoting toward organic and biodynamic agriculture in smaller-scale and urban farms (Pérez, 2015; see also Cederlöf, 2017; Fernández et al., 2018; Kapcia, 2018). Decentralized and urban smallholder agriculture began as key survival strategies as massive state-owned industrial farms became less viable – "by 1998, Havana had more than 26,000 urban gardens, producing 540,000 tons of fresh fruits and vegetables" (Wright, 2009: 6) – but also became a base for reimagining the place of the collective in the state-society-nature relationship (Premat, 2003).

As an energy transition, the shift from a carbon-intensive sociopolitical energy regime to a decarbonizing one poses a few troubling questions for decarbonizations more generally. First and foremost, to highlight the Special Period's successes makes a virtue of necessity if it fails to register the human toll and geopolitical context of imposed transition. As Bonneuil and Fressoz (2016) summarize, Cuba comprehensively overhauled its social and technological energy matrices, implementing the rapid degrowth of the economy, decentralization of political life, and shift to biodynamic and organic agriculture (ibid.). Despite this comprehensive response, however, and even with the aid of a consolidated authoritarian state and established command economy model, reductions in carbon dioxide emissions during the Special Period were only ever relatively "modest, falling in ten years from 10 million to 6.5 million tonnes, much less than the 40 to 70 percent reduction of world emissions by 2050 called for by the IPCC that could cap global warming to a two-degree Celsius increase" (104). The Cuban energy transition is perhaps the most comprehensive case of reactive decarbonization in recent years, barring perhaps the case of North Korea in the same period. Even with this in mind, the net reduction of the island's carbon footprint – a footprint that was miniscule to begin with in comparison with the United States, Europe, and China and, thanks to the prevalence of a small set of industries on the island, relatively straightforward – was minimal.

Reactive decarbonization of this sort is also particularly vulnerable to reversals. In response to the disappearance of Soviet oil, for example, the Castro government "went all out" to increase domestic production of petroleum, which is "of very low quality and has a very high sulphur content, thus generating a great deal of air pollution" (Díaz-Briquets and Pérez-López, 1995: 287). Cuba did not transition in the name of conservation; it had to do so and did so with the perfectly rational, understandable, and humane desire to normalize – that is, recarbonize – the state-society-nature matrix as soon as possible. This is why the Special Period ended when oil returned to Cuba. Starting with the election of Hugo Chávez in Venezuela in 1998, Latin America was swept by a "Pink Tide" of left-of-center presidencies (Ellner, 2019; Goodale and Postero, 2013). While this left turn was always ideologically diverse, Cuba's reintegration into the regional community was a shared norm across administrations (Kapcia, 2008). China's rise in the region provided a new source of foreign direct investment and trade, and by 2008, China was the island's second largest trading partner, after Venezuela (Pérez, 2015). Finally, while the blockade persisted, the United States' focus after 2001 on a "global war on terror" in the Middle East at least moderately eased pressure on the island (Pineo and Birns, 2013). This pressure was further released with the diplomatic thaw between Cuba and the United States during the late years of the Obama administration (Hirschfeld Davis, 2017; Roberts, 2016) before the reversals of the Trump administration and the resurgence of the right throughout Latin America (Romano and Tirado, 2018; TeleSur, 2019).

In 2017, after over a decade of piecemeal responses to the climate crisis and in an increasingly unfavorable regional political context, the Council of Ministers approved a comprehensive plan under the Ministry of Science, Engineering, and the Environment (CITMA, for its initials in Spanish) to identify, prioritize, and execute progressive investments in climate change adaptation for the short (2020), medium (2030), and long term (2100) (Pérez Cabrera 2019). The century-long vision of this plan, *Tarea Vida* (Project Life), gained urgency later that year as Hurricane Irma glanced off of Cuba's northern coastline, inundating Havana's picturesque Malecón with ten-meter waves and leaving a wide swath of destruction in its wake (Stone, 2018). The government requested technical and material aid from partners in the European Union (Dutch Risk Response Team 2018, 9), and adaptations have prioritized shoreline defense and relocating potentially vulnerable populations and farms – with particularly at-risk villages being moved immediately after the adoption of the plan in October 2017 (Dutch Risk Response Team, 2018). Among other tasks, *Tarea Vida* also prioritizes relocating agricultural production inland and adapting farming methods – including the replanting of its coastal mangrove swamplands, which constitutes the island's first line of defense to sea level rise – to changing climactic conditions (ibid.). The radical speed of these adaptations matches the scale of the climate crisis faced by Cuba. However, even here, competing demands blunt the impact of necessary reforms.

Cuba's controlled economy, strong central state, and small population all potentially lend themselves to rapid action in ways not found elsewhere. Ecological and geopolitical limits have remained persistent to what Cuba has been able to accomplish. The government's 2030 Strategy (drafted in 2011) frames needed changes to the energy matrix, for example, by noting how dependence of imported oil has both "drained the state economy of valuable hard currency and left it dependent on an unpredictable geopolitical environment" (Hornborg, Cederlöf, and Roos et al., 2019: 995). Despite this, "in 2014, 95% of Cuba's electricity was still generated from petroleum-based resources, both imported and domestically produced" (ibid.). The transition from fossil fuels, Hornborg, Cederlöf, and Roos (2019) conclude, is hamstrung by "socio-metabolic constraints" imposed by the combination of human labor, physical space, capital, and infrastructure required to successfully utilize post-carbon energy technologies that Cuba lacks (1002). Existing infrastructures, lifestyles, and spatial organization have all locked Cuba into a carbon-intensive energy regime. Breaking with oil, even on this comparatively small scale, would necessarily entail nothing short of a complete reorganization of space and social relations. Even an authoritarian state like Cuba has been unable to do so.

While Cuba has not fully returned to its carbon-intensive and mono-agricultural past, the end of the Special Period in 2001 has meant its energy mix is no longer of necessity as green and lean as in the 1990s. On the whole, petroleum consumption remains much lower than during the Soviet era (Cederlöf, 2017). However, "oil for doctors" agreements with Venezuela in the early 2000s and other bi- and multilateral agreements allowed Cuba to once again utilize and extend fossil fuel infrastructures (Cederlöf, 2017). What is more, Cuba has opened its territorial waters to foreign firms seeking to tap into the estimated 15 billion barrels located in deep water (Fabricant and Gustafson, 2017). The immediate causes of Cuba's transition during the Special Period were exogenous in nature. When those conditions shifted, so too did the scope and trajectory of the transition.

In Cuba, the sudden absence of oil exposed the limits of its existing sociopolitical energy regime as seemingly unrelated systems – food production and the university sector, for example – as being wholly integrated into and dependent upon the networks of fossil fuel production and consumption (Cunha et al., this edition). That Cuba was able to navigate the seismic shift of losing its primary energy source at all – even at the tremendous human cost of the Special Period – is a remarkable feat. However, this does not mean the entire island has now successfully transitioned to a sustainable model. Its transition entailed necessary and difficult adaptations rather than long-term commitments. The Cuban government navigated the transition that was the Special Period while maintaining its focus on human health and welfare (Morris, 2014). While infant malnutrition rates spiked to their highest levels in decades during the Special Period, life expectancy and other welfare indicators in Cuba remained favorable in relation to many of its more developed neighbors (Pérez, 2015). Public health crises in other disaster zones – Haiti after the 2010 earthquake or Puerto Rico after 2017's hurricane season, for example – offer glimpses of how much worse the Special Period could have been, how disastrous the coming global transitions could be, and how unevenly their impacts may be distributed (Bonilla and LeBrón, 2019; Dupuy, 2010).

3. Venezuela: Boom, Bust, Transition?

For many, Venezuela was the first "petro state". According to Terry Lynn Karl (1999) and others (e.g., Ross 2012), the boom-and-bust cycles of petroleum markets malform institutions, encourage corruption, and sow the seeds for political and social unrest. However, while many versions of the petro state argument are limited to events within states, focusing on domestic stakeholders, institutions, and policies, petro states are better understood as nodes in a transnational web of energy, power relations, and ecology than as dysfunctional islands in an otherwise prosperous free market sea (Kramarz and Kingsbury, 2021). The consequences are significant. In 2018, petroleum exports accounted for 99% of Venezuela's national export earnings (OPEC, 2018), despite the fact that Venezuela's production declined by over half between 2015 and 2018 due to sanctions and deteriorating infrastructure (Monaldi, 2018; Rousseau, 2015; Vázquez, 2015). Governments across regime types have depended on oil production, as the self-reinforcing cycles of extractivism – an ideology linking extraction of natural resources to national development (Gudynas, 2015; Terán-Mantovani, 2014) – has become an imperative spanning civilian, military, democratic, and Bolivarian governments for a century (see also Keahey et al., this volume). The resulting sociopolitical energy regime is one in which life across multiple scales – from state power to household access to basic foodstuffs and consumer goods – is wholly dependent on petroleum.

It makes some sense, then, that rather than decarbonization, Venezuela consistently worked to increase oil and natural resource exports (Grisanti, 2011; Vázquez, 2015). During boom years, this facilitated state formation and has radically transformed physical and infrastructural space in pursuit of modernization (Blackmore, 2017). Only very recently has Venezuela attempted to merge extractivist development and discourses of environmental sustainability, with cognitively dissonant results. The 2007–2013 Plan de la Patria, for example, recognizes climate change as a pressing national concern and calls for the construction of "ecosocialism" while aiming to cement Venezuela's role as a "global energy power" and suggesting "the production and use of petroleum and energy resources ought to contribute to the preservation of the environment" (Terán-Mantovani, 2014: 194).

Different forms of petroleum ambivalence have characterized Venezuela's relationship to oil from the early days of the industry. As early as 1936, Arturo Uslar Pietri called for planners to "sow the oil" to make extraction productive rather than parasitical, and policy and opinion makers cautioned against the destabilizing effects of the oil economy (D'Angelo, 2017). The industry and the anxieties that surround it continue to define national identity and political fault lines into the 21st century (Coronil, 1997; Kingsbury, 2016; Salas, 2009). Oil, and extractivism more generally, must thus be understood not merely as elements in economic production but rather as the foundation for state-society-nature relations in Venezuela. The specter of decarbonization thus overshadows more than economics. Decarbonization threatens to rupture the state from within, undermining development models and sending shockwaves through nearly every aspect of the national fiber vision that could prove too dire to navigate.

The Venezuelan petro state formed as a result of global and domestic pressures. The Great Depression and World War II weakened demand for its chief agricultural exports (mostly cacao and coffee), and the postwar reconstruction and Cold War economic boom in Europe required more and more fossil fuels (Mommer, 2010). Real GDP growth in Venezuela was the highest in Latin America of the era (Coronil, 1997; DiJohn, 2009), and the military government of Marcos Pérez Jiménez (1948–1958) enjoyed tremendous material capacities during the boom: "in the period between 1950 and 1957, ordinary oil revenue grew an average of 11.6 percent and government income an average of 13.9 percent annually, while the total value of oil exports increased 250 percent and treasury reserves 400 percent" (Coronil, 1997: 187). During this period, the dictator reconstructed the country, investing in modern and monumental infrastructure as a means of securing modernization (Blackmore, 2017). With the oil boom of the mid-1970s, it was the turn of civilian governments now organized under the exclusionary two-party Puntofijo pact to use expanded revenue to underwrite a class compromise between masses and elites under the banner of what President Carlos Andrés Pérez (1974–1979) described as *La Gran Venezuela* – Venezuela the Great (Neuhouser, 1992; Terán-Mantovani, 2014). However, the political logic of the Puntofijo system also meant Venezuela's foreign debt ballooned during these years as well, as funds earmarked for development disappeared in capital flight and corruption (DiJohn, 2009).

Dwindling revenues, skyrocketing debt, and the neoliberalization of the global economy reduced the capacity of the Venezuelan developmentalist state to fund a social peace and petroleum-financed progress in the 1980s and 1990s.[4] The era of austerity that began with Venezuela's "Black Friday" currency devaluation in 1983 became inescapable as oil prices foundered (Lander, 1996: 50). These difficulties boiled over in the *Caracazo* of late February 1989, a spontaneous uprising against austerity measures and subsequent government crackdown that murdered scores of Venezuelans (Wilpert, 2007). By 1992, a year which saw of two attempted military coups and the impeachment of the president on corruption charges, the three decades' old political order imploded amid freefalling social and economic conditions (Terán-Mantovani, 2014: 148).

In the ensuing decade, successive administrations implemented piecemeal structural adjustment measures, including the opening of the oil sector and privatization of other public industries. As in other cases throughout the region, the results of neoliberalization in Venezuela were "ambiguous at best" (Escobar, 2010: 2). As in Cuba during the Special Period, the crash was affective and socioeconomic, as it followed decades of the oil-inspired confidence that *La Gran Venezuela* was on the horizon. This "lost decade" was characterized by downward social mobility and deteriorating public services (Ellner and Tinker Salas, 2006). These were also years of social and spatial fragmentation as crime rates spiked and the fear of crime spawned a generalized securitization of daily life, and the distrust of public institutions, the police, and the political class deepened (Sanjuán, 2002, 88; Zubillaga, 2013). In two decades, Venezuela plummeted from the richest country in Latin America to the one with the highest rates of per capita debt (DiJohn 2009). According to CORDIPLAN,[5] by 1999 the poverty rate reached 80% and extreme poverty 39%, and malnutrition rate among children stood at 37%. Unemployment reached 15%, with over half of the workforce employed in the informal sector (cited in Terán-Mantovani, 2014). This drastic deterioration of social, political, and economic conditions opened the door for radical outsiders like Hugo Chávez.

Upon taking office, Chávez dedicated his new government's diplomatic energies to rejuvenating OPEC and to increase global oil prices (Lander, 2007). By 2003, when the government took control of the national oil industry, the state-society-nature complex had fully returned to the dynamics of the petro state. For the next ten years, despite echoing long-held concerns about oil's unpredictability and negative side effects, the Bolivarian Revolution intensified the state's reliance on the industry. Using historically high returns that came as a result of rising oil prices and reforms that saw the state capture increasing shares of oil rents, Chávez favored policies to reverse plummeting living conditions and increased investment in social and civic infrastructures. The resulting sociopolitical energy regime was one in which revenues from the sale of one energy source, petroleum, underpinned an expansion in social rights across the country, including access to energy sources like food and electricity for marginalized sectors of the population. Like the formation of Cuba's carbon-intensive socialism after 1959, for many in Venezuela the early 21st century was an aspirational moment characterized by enhanced (or reclaimed) sovereignty and increased access to basic necessities, and upward mobility (Escobar, 2010; Kramarz and Kingsbury, 2021). Gains across indicators were significant:[6] poverty was reduced and extreme poverty nearly eradicated; Venezuela was certified illiteracy-free by the United Nations in 2005; access to health care, education, safe and dignified housing, and food security were extended, if imperfectly, to the country's poor majority through an expanding array of government-sponsored social programs, the *Misiones Bolivarianas* (Bolivarian Missions) as the rhetoric of the government increasingly centered on the drive to build a "socialism for the twenty-first century; Chávez also asserted Venezuela's sovereignty in international and regional forums, insisting on the need for south-south solidarity and the construction of a multipolar world freed from the overbearing influence of the United States (Buxton, 2016; Ellner, 2007; 2019; Ellner and Tinker Salas, 2006; Wilpert, 2007).

While Chávez used an oil boom in the early 2000s toward social ends, his successor, Nicolás Maduro, has instead ruled in the context of reactive decarbonization. The collapse of oil prices after the great recession of 2008 deepened significantly in 2014. Structural problems compounded in the Venezuelan economy. State resources were depleted, further encouraging corruption and triggering staggering rates of inflation (Cohen, 2018). Economic difficulties and Maduro's failure to forge a charismatic tie to a majority of Venezuelans in the same manner as Chávez has emboldened the perennially intransigent opposition, who have used violent protests, electoral contests, and appeals for international sanctions to isolate, destabilize, and they hope, overthrow Maduro and the Bolivarian Revolution (Buxton, 2016). What is more, the negative consequences of difficult global markets in petroleum have been exacerbated by problems inside the state-owned oil company, Petróleos de Venezuela, SA (PDVSA), itself, which, after "opening" and increased internationalization in the 1990s, has been faltering due to mismanagement, lack of capital upkeep and investment, and foreign pressures in the 2000s (Monaldi, 2018; Rousseau, 2015; Vázquez, 2015).

Without contributing to the voyeuristic tone of much media coverage of Venezuela's troubles since 2014, the human cost of this transition should not be understated. The Maduro government attempted – and failed – to address the economic situation when it announced a *reconversión monetaria* (monetary reform) in August of 2018 that, among other measures, devalued the *Bolívar Fuerte* (Strong Bolívar) – itself rolled out only in 2007 – and replaced it with the *Bolívar Soberano* (Sovereign Bolívar), removing six zeros from the hyperinflated currency (Moleiro, 2021a, b). Meanwhile, shortages in basic foodstuffs and medicine were exacerbated by sanctions, estimated to have caused the death of 40,000 people between 2017 and 2019 (Weisbrot and Sachs, 2019).

Blackouts have also become increasingly common. As in the case of the Special Period in Cuba, blackouts permeate the fabric of life itself, forcing populations to revise their daily schedules and reweigh the balance between work, leisure, and reproductive labor. In both cases, blackouts added to the general anxieties of societies navigating uncertain political realities. While in Cuba the power outages were the result of rationing as the island ran out of fuel for its diesel electric generators, in Venezuela the story is slightly more complex. Rolling blackouts were triggered throughout 2016 as three years of persistent droughts brought water levels in the El Guri dam – which produces 60% of national electricity supply – to historic lows (Cawthorne, 2016).[7] As one commentator noted, "contrary to what we might assume, in today's sustainability-minded world, the choice [to invest in hydroelectric generation] did not arise from a noble commitment to renewable energy" but rather "to preserve as much of its oil as possible for export" (Bakke, 2016). However, multiple administrations have failed to diversify generation capacity, and the proper maintenance of existing electricity infrastructure has gone neglected. As a result, when climate change intensifies and extends droughts and other weather cycles, as in 2016, shocks to existing energy systems are deeper than they might have been. In sum, electricity crises of 2016 (repeated in 2018 and 2019) provided jolts to daily life Venezuelans had to normalize as part of larger crises of the sociopolitical energy regime established in the early 21st century.

The deteriorating situation triggered a mass outflow of Venezuelans. The United Nations High Commissioner for Refugees estimates over 4 million – of a population of roughly 32 million – Venezuelans fled by 2019 (UNHCR, 2019). As opposed to previous waves, the current migrants come from the popular sectors, further casting doubt on the medium- and long-term viability of the Bolivarian project under Maduro's leadership. Venezuela's ongoing energy transition in other words much more closely resembles a collapse as the Maduro administration has proven lacking in the political resources and conditions, let alone skill, to effectively steer the country beyond the implosion of carbon-intensive socialism for the 21st century.

4. Energy Transitions and Reactive Decarbonization

This chapter explored the Special Period in Cuba's and the collapse of Venezuela's oil industry to ask questions of ongoing energy transitions and their effects on sociopolitical energy regimes. Energy transitions may be disruptive, but not everyone experiences disruption in the same way. Cuba's and Venezuela's experiences are both clear and striking, but they are also familiar in a global order long marked by inequality and maldevelopment. If they make for grim reading, it is because they are projections of the world as it currently exists. In the words of the Petrocultures Research Group, all too often thinking around "energy futures tell us more about the present than they do about the future. The energy transition [*sic*] characterizes the global present, but the lived experience of that transition is not the same the world over and is characterized by inequalities on varying scales . . . this is no apocalyptic vision. This is the energy present" (Petrocultures Research Group, 2016: 63). If transitions too often look to the same mechanisms that produced the current climate crisis to address it and if they reproduce through omission the inequalities of the present, then understanding existing transitions in the Global South can help produce not only more holistic responses to present crises but more just ones as well.

Countries facing greater challenges in infrastructure, vulnerability, development, and state capacity will likely continue to pursue carbon-reliant fiscal strategies as they become available. A striking example of this can be seen in Ecuador. Ecuador has been a member of OPEC since 1973 and an agricultural extractivist state since the early 20th century, Despite this, its constitution of 2008 was widely celebrated as leading the charge in sustainability by endowing nature with rights and elevating ecological conservation as a core mandate of the country. However, this window of environmental progress was closed by the failure of the Yasuní-ITT supply-side decarbonization initiative in 2013,[8] after which point Ecuador's reliance on oil and mineral extraction, regardless of the environmental or human consequences, intensified (Kingsbury, Kramarz, and Jacques, 2018; Kramarz and Kingsbury, 2021; Riofrancos, 2017).

Ecuador's course of action after the failed Yasuní-ITT Initiative was similar to that in Cuba and Venezuela in the periods examined here. In each case, a crisis in one extractive sector was dealt with by enhanced extraction elsewhere, and both countries have also attempted to extend the lives of their respective oil economies. Cuba has increasingly relied on tourism, mining, and later, offshore drilling to fill the economic hole left by the Soviet Union. Similarly, Maduro has proposed a shift to mega-mining projects for gold, diamonds, and coltan – the last a necessary component for the so-called green technologies of the post-carbon economy – to take oil's place. These immediate responses to crises seek alternative revenue streams to maintain existing sociopolitical energy regimes: new inputs for established realities that maintain both countries' subordinate positions in the global division of power, labor, and nature. Arguably, they have little other choice.

The Castro regime did not pass through the difficulties of the Special Period unchanged, nor did the lives of Cubans. Even if it was eventually brought back into the Latin American community of nations without regime change with the emergence of the Pink Tide, decentralizing and decarbonizing reforms altered the shape and substance of Cuban socialism. While the Castro government insisted on upholding sovereignty and emphasized its commitments to health, education, and equality despite diminished capacities, it could not pursue these ends with what had come to be conventional means (Morris, 2014). The Special Period was, in short, a reconfiguration of Cuba's sociopolitical energy regime in which power relations and the organization of work, life, and future aspirations were forced to adapt under the exogenous constraints of reactive decarbonization, the US embargo, and increasingly, the climate crisis.

Similarly, the energy transition underway since 2014 in Venezuela has seen social conditions deteriorate, but it has yet to culminate into the regime change sought by the opposition and its backers in Washington. Much less have the present troubles enabled the construction of the "communal state" pursued by social movements at the grassroots. The ongoing energy transition has been occasioned by a seismic shift in the country's sociopolitical energy regime that has relied on petroleum exports for a century. In the more immediate term, reactive decarbonization has undermined President Maduro's authority in the eyes of current and former Chavistas, as increasingly precarious access to basic necessities from electricity to medications renders daily life more tenuous in communities and workplaces, contributing to the growing numbers of Venezuelans leaving the country in the early 2020s. Disappearing oil revenues have reduced the caloric intake, health, and security of Venezuelans as the extractive state fumbles to reinvent itself in the face of geopolitical and market-imposed decarbonization (Hodal, 2019; Human Rights Watch, 2019; Kurmanaev, 2019; UNHROHC, 2019).

This type of reactive decarbonization attempts to limit the energy transitions' negative social consequences. Moves to decentralize food production and to collectivize distribution in both Cuba and Venezuela through urban agriculture, for example, are necessary responses to systemic shocks. These reactions have the added virtue of mobilizing people to directly participate in responding to transitions' challenges. Such attempts are also, however, intensely local in nature, and at present supplement rather than replace the internationalized, industrial, and carbon-intensive agricultural networks on which both countries rely.

Looking toward future research, we might ask how the experiences of Cuba and Venezuela inform current approaches to energy transitions and how they can help us understand the evolution of sociopolitical energy regimes from global to local scales. The first question lends itself to growing calls for a global and more socially attentive approach to net-zero carbon systems (e.g., Aronoff, Battistoni, Cohen, and Riofrancos, 2019). Failure to approach energy transitions globally, in an inclusive and egalitarian fashion, leaves little other likelihood than the sort of crash-landing reactive decarbonizations of Cuba and Venezuela. However, at the same time, the continuation of top-down, paternalistic, and otherwise exclusionary developmentalist approaches to transitions, global or otherwise, are prone to reversibility and offer little incentive for uncoerced buy-in from the Global South.

The Cuban and Venezuelan cases also illustrate how transitions span the social, political, and ecological at multiple scales. Beyond replacing dirty fuels for clean ones, transitions shift state-society-nature relations. In the process, they put into question how risk and benefit are distributed in present and future sociopolitical energy regimes, how daily life is negotiated, how power is negotiated among states, and how future aspirations are engendered, promised, and disappointed.

Reactive decarbonization in Cuba and Venezuela illustrates consequences of unfolding energy transitions beyond the electrification of actually existing supply chains. While the sort of scaling up and catalyzing effects of local responses to climate change are of vital importance and perhaps inevitable, and while academics, activists, and policymakers should do all in their power to amplify them, one should also avoid making a virtue of necessity. Piecemeal transitions or Green New Deals confined to the existing national boundaries and geopolitical alliances of late carbon capitalism displace and intensify the trauma of transition. What is worse, they risk undermining the necessarily global reach of decarbonization by encouraging – and rationally so – countries to avoid the new energy regimes. Unless energy transitions entail system transitions, and unless the systemic inequalities that produced these crises are rectified, any future *after* the transition will look much like the unsurvivable present.

Acknowledgments

This article has benefited greatly from ongoing conversations with Gustav Cederlöf, Teresa Kramarz, and Victor Rivas. Parts of this argument were considered previously in "Combined and Uneven Energy Transitions: Reactive Decarbonization in Cuba and Venezuela", *Journal of Political Ecology* 27(1, 2020): 558–579, and are reproduced here with the permission of the journal. Many thanks to Carmen Bezner Kerr for her very capable research assistance. All mistakes or oversights are purely the fault of the author.

Notes

1 *Maldesarrollo*, or maldevelopment, has been developed as a concept to counter the teleological and often Eurocentric notions of development in circulation in the Western world since the Cold War. For critics like Maristella Svampa and Enrique Viale (2014), maldevelopment highlights not only the failed promises of industrialization and progress, but the consequences Western modes of development have always entailed: the disruption of ecosystems and pollution; the subordination of human rights to economic profit – especially for women, indigenous, and racialized peoples; undemocratic forms of governance and accountability; and in the case of Latin America's primary-product exporting economies, socioeconomic and political volatility that comes with highly unstable global commodity markets (28).

2 While Cabrera Abrús notes real shifts in the standard of living of Cubans in terms of access to consumer goods, she also cautions that the representation of Soviet-style economic stability and abundance often outstripped the actual experience of Cubans themselves (190).

3 The United States imposed an increasingly strict set of economic and military sanctions on Cuba, starting with the Revolution's victory in 1959, in response to the Castro government's nationalization of the assets of US-based companies on the island. Beginning with limits on all but non-humanitarian aid exports to the island and bans on military exports, the embargo was internationalized with Cuba's expulsion from the Organization of American States in 1962, the same year the US banned nearly all trade with the country. Further intensifications of the sanctions took various forms at crucial junctures, including the Helms-Burton Act of 1996 which aimed to penalize non-US companies for trading with the socialist island (Pérez, 2015a).

4 Starting with a military coup lead by Augusto Pinochet in Chile in 1973 until the Pink Tide of the early 2000s, governments throughout Latin America implemented a series of structural adjustment reforms in response to the economic and social crises of the late 20th century. Seeking to "return" national economies to liberal market principles, such as comparative advantage in the face of post–Cold War globalization, the neoliberalization of Latin American and global economies emphasized privatizing state-owned enterprises, deregulated businesses, slashing social welfare capacities, opening of domestic markets to foreign direct investment, and strengthening the police and judiciary to enforce enhanced property rights and stifle resistance (among other reforms). Throughout the region, neoliberalism resulted in an upward redistribution of wealth, increased inequality and poverty rates, and heightened vulnerability to volatile global prices in commodities and basic consumer manufactured goods. They also contributed to the popular mandate for the socializing reforms of the Pink Tide governments that responded to neoliberalism's overlapping and compounding social, political, environmental, and ecological crises (Escobar 2010; Harvey 2005; Svampa and Viale 2014).

5 Ministry of Coordination and Planning.

6 All but the most recalcitrant sectors of the Venezuelan opposition have begrudgingly admitted as much, though they often spin such gains as elements of a petro-populist model constructed by (and for) Chávez. See, for example, López Maya 2014.

7 The Guri dam and power plant, one of the largest such complexes in the world, is itself vulnerable to climate change, as seen in the 2014–2016 El Niño cycle (Cawthorne 2016). By the spring of 2016, prolonged drought reduced capacity at the dam to such an extent that rolling blackouts spread throughout the country power outages throughout the country, prompting the government to shorten the official workweek to two days and exacerbating already strained social and political tensions.

8 Shortly after taking office in 2007, the former president Rafael Correa took on an initiative proposed by social movements since the 1990s to protect fragile ecosystems and indigenous territories in the Ecuadorian Amazon by leaving oil in the ground in exchange for compensation from the international community (Acosta et al. 2000). The Yasuní-ITT Initiative – named for the Ishpingo, Tambococho, and Tiputini blocks of the Yasuní National Reserve – proposed to forego extraction in exchange for half of the estimated value of the oil located therein. Over the ensuing five years less than 10% of the requested $3.6 billion was pledged to the United Nations' trust created for the human development projects to which the monies would be dedicated. Correa withdrew the proposed initiative in 2013, and extraction began shortly thereafter (Kingsbury, Kramarz, and Jacques 2018).

Bibliography

Acosta, Alberto. 2000. El Ecuador post-petrolero. Quito: Acción Ecológica.

Aronoff, Kate, Alyssa Battistoni, Daniel Cohen, and Thea Riofrancos. 2019. A planet to win: why we need a Green New Deal. New York: Verso, 2019.

Bakke, Gretchen. 2016. The electricity crisis in Venezuela: A cautionary tale. The New Yorker. Retrieved May 17 2019 from www.newyorker.com/tech/annals-of-technology/the-electricity-crisis-in-venezuela-a-cautionary-tale.

Blackmore, Lisa. 2017. *Spectacular modernity: Dictatorship, space, and visuality in Venezuela*. Pittsburgh: University of Pittsburgh Press.

Bonilla, Yarimar and Marisol LeBrón, eds. 2019. *Aftershocks of disaster: Puerto Rico before and after the storm.* Chicago, IL: Haymarket Books.

Bonneuil, Christophe and Jean-Baptiste Fressoz. 2016. *The shock of the Anthropocene: The earth, history, and us.* New York: Verso.

Boyer, Dominic. 2014. Energopower: an introduction. Anthropological Quarterly 87(2),: 309–333.

Bull, Benedicte, and Antulio Rosales (2020) The crisis in Venezuela: drivers, transitions, and pathways. European Review of Latin American and Caribbean Studies 109: 1–20.

Buxton, Julia. 2016. Venezuela after chávez. New Left Review 99 (2): 5–25.

Cabrera Abrús, María (2019) The material promise of socialist modernity: fashion and domestic space in the 1970s. In M. Bustamante and J. Lambe (eds.). *The Revolution from Within: Cuba, 1959–1980*, Durham: Duke University Press, pp. 189–218.

Cawthorne, Andrew. 2016. 'Drought hit Venezuela awaits rain at crucial Guri dam', *Reuters* April 13, 2016. Retrieved February 11, 2020 from www.reuters.com/article/us-venezuela-energy/drought-hit-venezuela-awaits-rain-at-crucial-guri-dam-idUSKCN0XA1WL

Cederlöf, Gustav. 2017. *Energy Revolution: Oil Dependence and the Political Ecology of Energy use in Socialist Cuba*. PhD Dissertation. London: King's College.

Cederlöf, Gustav and Donald Kingsbury. 2019. On PetroCaribe: Petropolitics, energopower, and post-neoliberal development in the Caribbean Energy region. *Political Geography* 72: 124–133.

Cohen, Luc. 2018. IMF sees Venezuela inflation at 10 million percent in 2019. *Reuters Online*, October 9. Retrieved October February 7, 2020 from https://in.reuters.com/article/venezuela-economy/imf-sees-venezuela-inflation-at-10-million-percent-in-2019-idINKCN1MJ1YX.

Coronil, Fernando. 1997. *The Magical State: Nature, Money, and Modernity in Venezuela*. Chicago: University of Chicago Press.

D'Angelo, Oriette. 2017. Sembrar el petróleo, por Arturo Ulsar Pietri (Caracas, 1906–2001). Retrieved February 7, 2020 from https://digopalabratxt.com/2017/05/16/sembrar-el-petroleo-por-arturo-uslar-pietri-caracas-1906-2001.

Díaz-Briquets, Sergio and Jorge Pérez-López. 1995. The special period and the environment. *Annals of the Association for the Study of the Cuban Economy* 5.

Dupuy, Alex. 2010. Disaster capitalism to the rescue: The international community and Haiti after the earthquake. *NACLA Report on the Americas* 43(4): 14–19.

Dutch Risk Response Team. 2018. *DRR Team Mission Report: Cuba – Havana Province and City*. The Netherlands: DRR Team. Retrieved February 7 2020 from www.drrteam-dsswater.nl/wp-content/uploads/2018/11/Report-DRR-Mission-Cuba-final-20181011.pdf.

Ellner, Steve. 2007. Toward a 'multipolar world': Using oil diplomacy to sever Venezuela's dependence. *NACLA Report on the Americas* 40(5): 15–22.

Ellner, Steve, ed. 2019. *Latin America's Pink Tide: Breakthroughs and Shortcomings*. Bouder: Lynne Reinner.

Ellner, Steve, and Miguel Tinker Salas, eds. 2006. *Venezuela: Hugo Chávez and the decline of an "Exceptional Democracy"*. Lanham, Boulder and New York: Rowman & Littlefield Publishers.

Escobar, Arturo. 2010. Latin America at a crossroads: Alternative modernities, post-liberalism, or post-development? *Cultural Studies* 24(1): 1–65.

Fabricant, Nicole and Bret Gustafson. 2017. Revolutionary oil?: Offshore drilling in Cuba. *NACLA Report on the Americas* 49(4): 441–443.

Febles-González, J., A. Tolón-Becerra, X. Lastra-Bravo, and X. Acosta-Valdés. 2011. Cuban agricultural policy in the last 25 years. From convenetional to organic agriculture. *Land Use Policy* 28: 723–735.

Fernandez, M., J. Williams, G. Figueroa, G. Graddy Lovelace, M. Machado, L. Vasquez, N. Perez, L. Casimiro, G. Romero, and F. Funes Aguilar. 2018. New opportunities, new challenges: Harnessing Cuba's advances in agroecology and sustainable agriculture in the context of changing relations with the United States. *Elementa Science Anthology* 6(1): 76.

Goodale, Mark, and Nancy Postero, eds. 2013. *Neoliberalism, Interrupted: Social Change and Contested Governance in Contemporary Latin America*. Palo Alto: Stanford University Press.

Granma: Official Voice of the Communist Party of Cuba Central Committee. 2019. *Consitución de la República de Cuba*. Havana, Cuba: Granma. Retrieved February 7 2020 from www.granma.cu/file/pdf/gaceta/Nueva%20Constituci%C3%B3n%20240%20KB-1.pdf.

Grisanti, Alejandro. 2011. Venezuela's oil tale. *Americas Quarterly* 2(5): 39–42.

Gudynas, Eduardo. 2015. *Extractivismos: Ecología, Economía y Política de un Modo de Entender el Desarrollo y la Naturaleza*. Cochabamba: Centro de Documentacíon e Informacíon Bolivia.

Harnecker, Camila 2014. Nonstate enterprises in Cuba: Building socialism? *Latin America Perspectives* 41(4): 113–128.

Harvey, David. 2005. *A Brief History of Neoliberalism*. New York: Verso.

Hirschfeld Davis, Julie. 2017. Moving to scuttle Obama legacy, Donald Trump to crack down on Cuba. *The New York Times*, June 15. Retrieved February 7 2020 from www.nytimes.com/2017/06/15/us/politics/cuba-trump-obama.html

Hodal, Kate. 2019. UN urged to declare full-scale crisis in Venezuela as health system 'collapses'. *The Guardian*, April 5. Retrieved February 7 2020 from www.theguardian.com/global-development/2019/apr/05/un-urged-to-declare-full-scale-crisis-in-venezuela-as-health-system-collapses

Hornborg, Alf, Gustav Cederlöf and Andreas Roos. 2019. Has Cuba exposed the myth of 'free' solar power? Energy, space, and justice. *Environment and Planning E: Nature and Space* 2(4): 989–1008.

Human Rights Watch. 2019. *Venezuela's humanitarian emergency: Large-scale UN response needed to address health and food crisis*. New York: Human Rights Watch, April 4. Retrieved February 7 2020 from www.hrw.org/report/2019/04/04/venezuelas-humanitarian-emergency/large-scale-un-response-needed-address-health

International Crisis Group. 2004. *Venezuela: Headed Toward Civil War?* Quito/Brussels: International Crisis Group. Retrieved February 7, 2020 from https://d2071andvip0wj.cloudfront.net/venezuela-headed-toward-civil-war.pdf

Intergovernmental Panel on Climate Change (IPCC). 2014. Summary for policymakers. In O. Edenhofer, R. Pichs-Madruga, Y. Sokona, E. Farahani, S. Kadner, K. Seyboth, A. Adler, I. Baum, S. Brunner, P. Eickemeier, B. Kriemann, J. Savolainen, S. Schlömer, C. von Stechow, T. Zwickel and J. C. Minx (eds.). *Climate change 2014: Mitigation of climate change. Contribution of Working Group III to the Fifth assessment report of the Intergovernmental Panel on Climate Change.* Cambridge and New York: Cambridge University Press.

Kapcia, Antoni. 2008. *Cuba in Revolution: A History since the Fifties.* London: Reacktion Books.

Karl, Terry Lynn. 1999. The perils of the petro-state: Reflections on the paradox of plenty. *Journal of International Affairs* 53(1): 31–48.

Kingsbury, Donald. 2016. Oil's colonial residues: Geopolitics, identity, and resistance in Venezuela. *Bulletin of Latin American Research* 35(4): 423–436.

Kingsbury, Donald, Teresa Kramarz and Kyle Jacques. 2018. Populism or petrostate?: The afterlives of Ecuador's Yasuní-ITT initiative. *Society and Natural Resources* 32(5): 530–547.

Kramarz, Teresa and Donald Kingsbury. 2021. *Populist Moments and Extractivist States in Venezuela and Ecuador: The People's Oil?* New York: Palgrave.

Kurmanaev, Anatoly. 2019. Venezuela's collapse is the worst outside of war in decades, economists say. *The New York Times*, May 17. Retrieved February 7 2020 from www.nytimes.com/2019/05/17/world/americas/venezuela-economy.html

Lander, Luis. 1996. Insurrección de la tecnocracia petrolera en Venezuela. In H. Ochoa Henríquez and A. M. Estévez (eds.). *El Poder de los Expertos: Para Comprender la Tecnocracia.* Maracaiba: Centro de Estudios de la Empresa, Facultad de Ciencias Económicas y Sociales.

Lander, Luis. 2007. Venezuela's balancing act: Big oil, OPEC and national development. *NACLA Report on the Americas* 34(4): 25–31.

Maal-Bared, Rasha. 2006. Comparing environmental issues in Cuba before and after the Special Period: Balancing sustainable development and survival. *Environmental International* 32: 349–358.

Maya, Margarita. 2014. Venezuela: The political crisis of post-Chavismo. *Social Justice* 40(4): 68–87.

Mitchell, Timothy. 2013. *Carbon Democracy: Political Power in the Age of Oil.* New York: Verso.

Moleiro, Alonso. 2018a. Venezuela resta seis ceros a su moneda ante la hiperinflación y lanza el 'bolívar digital'. *El País.* August 5, 2021. Retrieved https://elpais.com/economia/2021-08-05/venezuela-resta-seis-ceros-a-la-moneda-ante-la-hiperinflacion-y-lanza-el-bolivar-digital.html

Moleiro, Alonso. 2018b. Maduro anuncia una reconversión monetaria y le quite tres ceros al bolívar. *El País* March 23, 2021. Retrieved https://elpais.com/internacional/2018/03/23/america/1521767129_975583.html#?rel=mas

Monaldi, Francisco. 2018. La implosión de la industria petrolera venezolana. *ProDavinci*, August 15. Retrieved November 25, 2019 from https://prodavinci.com/la-implosion-de-la-industria-petrolera-venezolana.

Morris, Emily. 2014. Unexpected Cuba. *New Left Review* 88: 5–45.

OPEC. 2019. Venezuela facts and figures. Retrieved November 27, 2019 from www.opec.org/opec_web/en/about_us/171.htm

Pérez Cabrera, Freddy. 2019. Tarea Vida señala el camino frente al cambio climático. Granma: Órgano oficial del Comité *central del partido comunista de Cuba,* June 4. Retrieved Feburary 7 202 from www.granma.cu/cuba/2019-06-04/tarea-vida-senala-el-camino-frente-al-cambio-climatico-04-06-2019-20-06-01.

Pérez, Louis. 2015. *Cuba: Between Reform and Revolution.* New York: Cambridge University Press.

Pérez, Louis. 2015a. Change through impoverishment: A half century of cuba-US relations. *NACLA Report on the Americas* December 14, 2015. Retrieved https://nacla.org/news/2015/12/13/change-through-impoverishment-half-century-cuba-us-relations

Petrocultures Research Group. 2016. *After Oil.* Edmonton: Petrocultures Research Group.

Pineo, Ronn and Larry Birns. 2013. Latin America's backyard. *Council on Hemispheric Affairs*, August 23, 2013. Retrieved February 7 2020 from www.coha.org/latin-americas-backyard.

Pollitt, Brian 2004. The rise and fall of the cuban sugar economy. *Journal of Latin American Studies.* 36(2): 319–348.

Premat, Adriana. 2003. Small-scale urban agriculture in Havana and the reproduction of the 'new man' in contemporary Cuba. *Revista Europea de Estudios Latinoamericanos y del Caribe* 75: 85–99.

Riofrancos, Thea. 2017. Extractivismo unearthed: A genealogy of a radical discourse. *Cultural Studies.* 31(2–3): 277–306.

Roberts, Dan. 2016. Obama lands in Cuba as first US president to visit in nearly a century. *The Guardian*, March 21. Retrieved December 10 2019 from www.theguardian.com/world/2016/mar/20/barack-obama-cuba-visit-us-politics-shift-public-opinion-diplomacy

Romano, Silvina and Arantxa Tirado. 2018. Trump, el 'influencer' de la derecha latinoamericana. *Centro Estratético Latinoamericano de Geopolítica*, October 28. Retrieved December 1 2019 from www.celag.org/trump-influencer-derecha-latinoamericana.

Rousseau, Isabelle. 2015. The dynamic of Latin American national oil companies' evolution case studies: Pemex and PdVSA. In A. V. Belyi and K. Talus (eds.). *States and Markets in Hydrocarbon Sectors*. London: Palgrave Macmillan.

Stone, Richard. 2018. Cuba embarks on a 100-year plan to protect itself from climate change. *Science Mag of the American Association for the Advancement of Science*, January 10. Retrieved November 29 2019 from www.sciencemag.org/news/2018/01/cuba-embarks-100-year-plan-protect-itself-climate-change.

Svampa, Maristella and Enrique Viale. 2014. *Maldesarrollo: La Argentina del Extracitivsmo y el Despojo*. Buenos Aires: Katz Editores.

TeleSur. 2019. Relaciones Cuba-EE.UU. viven franco retroceso con Trump. *TeleSur Net*, July 20. Retrieved February 7 2020 from www.telesurtv.net/news/cuba-eeuu-relaciones-diplomaticas-retroceso-gobierno-donald-trump-20190720-0006.html

Terán-Mantovani, Emiliano 2014. *El fantasma de la Gran Venezuela: un estudio del mito y los dilemas del petro-estado en la revolución bolivariana*. Caracas: Centro de Estudios Latinoamericanos Rómulo Gallegos.

Tinker Salas, Miguel. 2009. *The Enduring Legacy: Oil, Culture, and Society in Venezuela*. Durham: Duke University Press.

UNHCR. 2019. Refugees and migrants from Venezuela top 4 million: UNHCR and IOM. Retrieved June 15 2019 from www.unhcr.org/news/press/2019/6/5cfa2a4a4/refugees-migrants-venezuela-top-4-million-unhcr-iom.html.

UNHR Office of the High Commissioner. 2019. UN Human Rights report on Venezuela urges immediate measures to halt and remedy grave rights violations. Retrieved February 7 2020 from www.ohchr.org/EN/NewsEvents/Pages/DisplayNews.aspx?NewsID=24788&LangID=E

Vázquez, Gonzalo. 2015. 25 Years of contemporary history of PDVSA. In B. M. Bagley, H. S. Kassab and S. Moulioukova (eds.). *The Impact of Emerging Economies ON Global Energy and the Environment: Challenges Ahead*. London: Lexington Books.

Weisbrot, Mark and Jeffery Sachs. 2019. *Economic Sanctions as Collective Punishment: The Case of Venezuela*. Washington, DC: Center for Economic and Policy Research.

Wilpert, Gregory. 2007. *Changing Venezuela by Taking Power: The History and Policies of the CHÁVEZ Government*. New York: Verso.

Wright, Julia. 2009. *Sustainable Agriculture and Food Security in an era of Oil Scarcity: Lessons from Cuba*. New York: Routledge.

14

SOUTH AFRICA'S NEW GOLD

Building a Global Leader in Green Hydrogen

Tobias Bischof-Niemz and Terence Creamer

Introduction: New Gold

South Africa has the potential to become a global leader in green hydrogen, similar to its role in gold production for large parts of the 20th century.[1] Based on an abundance of renewable energy potential and export-oriented industry focus, the country is almost uniquely positioned to produce green hydrogen, as well as carbon-neutral fuels, chemicals, and steel from low-cost electricity. Such products could be exported to countries that require these inputs to meet net-zero commitments. Additionally, South Africa has the potential to competitively decarbonize its transport, heating, cooling, and industrial sectors and processes.

By producing new and highly sought-after commodities, this chapter outlines how South Africa could transition its coal-heavy and environmentally unsustainable energy system to one based on renewable energy. What if the new commodities allowed South Africa to decarbonize its own hard-to-abate sectors of land and marine freight, aviation, steel, cement, fertilizers, and chemicals, as well as those of its export partners? In doing so, what if millions of well-paying and decent jobs were created in the process, allowing the country to move away from an industrial playing field with legacy forms of racism?

The chapter highlights how green hydrogen and other products derived from green hydrogen, such as carbon-neutral kerosene and green ammonia, could become South Africa's "new gold". Such a transition could be accomplished by South Africa leveraging surplus electricity to produce new, high-demand green products – a power sector pathway known as power-to-X.[2]

Resource Advantage

Similar to South Africa's gold history, the foundation for a potentially game-changing "green hydrogen" industry lies in South Africa's natural resource advantages. In this case, the resource advantages include strong solar energy, high-quality wind, and abundant and relatively low-cost land. This combination is not unique to South Africa, as similar conditions exist in North Africa and the Middle East and parts of South and North America and Australia. Such favorable conditions, however, do not exist in many of the countries in Europe and Asia where plans have been announced to adopt green hydrogen to decarbonize hard-to-abate transport and manu-facturing sectors. To meet such decarbonization objectives, a significant portion of global green

DOI:10.4324/9781003183020-16

hydrogen demand will need to be supplied through hydrogen and hydrogen-based product exports that are produced in countries where the conditions exist for cost-competitive production and export-oriented infrastructure. To date, the European Union and Japan have already committed to the bulk import of hydrogen produced elsewhere (Roos and Wright, 2021). These commitments provide a ready-made market for green hydrogen and tradeable hydrogen derivatives, which is an opportunity for South Africa and for businesses in South Africa (ibid.).

South Africa's resource advantage positions it strongly to produce the main input required for unlocking the competitive manufacture of green hydrogen: cheap renewable electricity. A joint study conducted in 2016 by the Council for Scientific and Industrial Research (CSIR), state-owned electricity utility Eskom, the South African National Energy Development Institute, Fraunhofer IEE of Germany, and the Danish Technical University highlights South Africa's natural advantage. South Africa has, for example, almost two times more solar energy on average annually reaching each square meter of land compared to Middle and Northern European countries (Fraunhofer IWES, CSIR, SANEDI, Eskom, 2016). As a result, the electricity generated from solar photovoltaics (PV) in South Africa is roughly 50% cheaper (ibid.). Likewise, the country's wind resource is significantly better, and more widely spread, than initially assumed. An analysis of five years of historical wind data from satellite observations concluded that a 35% capacity factor or higher could be achieved on almost 70% of all suitable land in South Africa (ibid.). In fact, using less than 1% of its land area, South Africa would be able to generate more electricity from wind and solar today than it currently needs and will also be sufficient to meet the country's future demand (ibid.).

Over the past ten years, a rapid decline in technology costs for solar PV and onshore wind has resulted in a more accelerated global rollout: between 2014 and 2019, an average of 55 GW of wind and 79 GW of solar PV was installed worldwide yearly, compared with an annual average of 22 GW of wind and 10 GW of solar PV between 2000 and 2013 (Roos and Wright, 2021). The fall in costs has also meant that electricity from wind and solar, blended with small amounts of more expensive but flexible electricity, represents the cheapest way to produce new electricity in South Africa (Bischof-Niemz and Creamer, 2018).

Analysis of actual solar PV and wind tariff bids into the country's Renewable Energy Independent Power Producer Procurement Program (REIPPPP), reverse energy auctions from 2011 to 2015, highlighted the growing competitiveness of the two technologies (ibid.). Under the reverse auction scheme, independent power producer (IPP) bids are adjudicated based on the lowest submitted prices or tariffs while also meeting socioeconomic criteria. To account for multiple aims, a 70% weighting was initially attributed to the bid price, with 30% for socioeconomic development (ibid.). The weighting has since transitioned to one whereby a 90% weighting is given to the price, with a several of the socioeconomic criteria embedded as minimum thresholds for bid compliance (ibid.). During a bidding round of the REIPPPP in November 2015, bid prices of R0.62/kWh ($0.042/kWh) were obtained for both solar PV and wind, reflecting a tariff reduction of 80% and 60% respectively from prices contracted in the first bidding round four years earlier (Ngobeni, 2016). This auction outcome was 40% lower than bid prices of R1.03/kWh ($0.07/kWh) for new-build coal (Bischof-Niemz and Fourie, 2016). It was also considerably lower than the cost assumptions of R1.09/kWh ($0.074/kWh) for new-build nuclear and R1.15/kWh ($0.078/kWh) from new-build combined-cycle power stations powered by natural gas (all prices adjusted to 2016) reported in South Africa's Integrated Resource Plan (IRP) at the time (Roos and Wright, 2021).

In 2021 which represents the most recent REIPPPP competitive auction round, the favorable tariff trajectory continued. Another 45% reduction was evident in Rand-terms for solar PV (2021 results in 2021 prices: R0.429/kWh or $0.03/kWh) and 36% reduction for wind (2021 results in 2021 prices: R0.495/kWh or $0.035/kWh) when compared with the last round six years ago (Mantashe, 2021).

The continued cost decline of wind and solar technology underscores how South Africa's cost-optimal and new-build electricity mix should consist primarily of solar PV and onshore wind generators. Such a mix can then be supported by flexible solutions, such as pumped hydro, gas, and battery energy storage. This emerging reality is largely reflected in the country's latest IRP, which allocates most of the new capacity to be added between 2020 and 2030 to onshore wind and solar PV, the installed base of which would rise to 17.7 GW and 8.8 GW, respectively, by 2030 (Department of Energy, 2019).

The near-term composition of the country's electricity mix is expected to shift, although not entirely. By 2030, coal is to decrease from about 86% to roughly 55% if progress aligns with the IRP (ibid.). The combination of demand growth and coal-plant decommissioning will result in the energy contribution from renewables rising to about 39% from less than 10% (ibid.). Dedicated infrastructure may arguably be necessary to enable hydrogen production from renewables, but renewable allocations in the IRP 2019 in the form of solar PV and wind enable significant new-build capacity for this to happen (Roos and Wright, 2021). The precise scale of the renewables fleet that is needed to unlock the country's true green hydrogen potential is covered later in this chapter.

Another factor in South Africa's potential to produce green hydrogen will be the trajectory of electrolyzer technology costs and whether these can be reduced to make green hydrogen competitive with fossil-fuel-derived gray or blue hydrogen. In fact, this remains the key hurdle, as green hydrogen costs are, on average, between two and three times higher currently than gray hydrogen (ibid.). However, research indicates that, with larger production facilities, design standardization, and insights from early adopters, the proposed strategies could cut costs by 40% in the short term and up to 80% in the long term (IRENA, 2020). "In price terms, the resulting green hydrogen could fall below the $2-per-kilogram mark – low enough to compete – within a decade" (ibid.).

The cost structure of green hydrogen is also significant. From cost information available in the public domain, the cost breakdown for bulk green hydrogen production can be roughly extrapolated as 60% are related to electricity input, 30% to the electrolyzer itself, and the remaining 10% to the balance of plant, for which no further cost reductions are expected (Roos and Wright, 2021). Hence, while cost reductions for electrolyzers will contribute to the reduction of green hydrogen costs, the exploiting of the best solar and wind resources globally will still be the biggest cost-reduction driver. This will be the case at least until the initial costs of early adoption and learning for integrated solar-wind-electrolyzer plants have been paid and projects become more standardized (IEA, 2021).

Should electrolyzer costs fall as anticipated, South Africa's valuable combination of wind, sun, and land could be paired with another national advantage: a long-established and sophisticated heritage of exporting bulk commodities such as coal, iron ore, chrome, manganese, and processed commodities, such as aluminum, chemicals, ferrochrome, ferromanganese, and steel. Relatively sophisticated and efficient logistics infrastructure has been developed (National Development Plan 2030, 2012). This includes dedicated commodity rail corridors that are linked primarily to three deep-level ports at Richards Bay, Ngqura, and Saldanha Bay, which are able to handle larger vessels required for low-cost seaborne trade. South Africa also exports chemicals and fertilizer products and imports large volumes of liquid fuels and natural gas. On balance, the country has the skills and infrastructure foundations that could quite feasibly be repurposed for large-scale exportation of hydrogen-based products. In line with this, the country's emerging green-hydrogen strategy now includes the possible creation of a green hydrogen special economic zone (SEZ) at a new port known as Boegoebaai, in the Northern Cape (Ramaphosa, 2021).

Looking more fully at South Africa's resource profile, a relative disadvantage at first glance is apparent in the area of water, with the country ranked as the 30th driest in the world (National Development Plan 2030, 2012). Yet the country has an extensive coastline and access to sea water. Its ability to produce low-cost renewable electricity means there is also an opportunity to progressively link an emerging hydrogen infrastructure with desalination of sea water. Moreover, as South Africa's existing coal fleet is decommissioned, water that has hitherto been used for the production of electricity could, instead, be redirected for the production of green hydrogen. In fact, the least-cost electricity scenario for South Africa developed in 2017 shows that a renewables-led electricity system will be substantially less water-intensive (CSIR, 2017). South Africa's coal-fired power stations currently consume more than 300 million tons every year (ibid.). That consumption would fall by 97% to about 10 million tons a year by 2050 in a scenario where solar PV and wind supply 82%, or 388 TWh/a, of South Africa's primary electricity production of 417 TWh/a (ibid.).

The 300 million tons of water consumption by the existing coal fleet can, conservatively estimated, produce between 15 and 20 million tons of electrolysis-based hydrogen per year or roughly 20% to 25% of the global pure-hydrogen demand today (Roos and Wright, 2021) Even under a scenario where South Africa's electricity fleet is scaled to cater for green hydrogen production, the consumption of water would not grow considerably, as solar PV and wind do not require much water (Bischof-Niemz, 2021).

Globally, water and land are also not expected to represent barriers to scaling up water electrolysis. Nevertheless, in places with water stress, the source of water for hydrogen production should be explicitly considered (IRENA, 2020). Where access to sea water is available, desalination can be used with limited impact on cost and efficiency, potentially deploying multipurpose desalination facilities to provide local benefits (ibid.). A 1 GW plant, for example, could occupy about 0.17 km² of land, which translates to 1,000 GW of electrolysis occupying an area equivalent to Manhattan, New York (ibid.).

For green hydrogen to qualify as truly sustainable, however, the water source must also be sustainable. Put differently, using water for fuel production must not negatively affect communities, agriculture, or the environment (Roos and Wright, 2021). The use of potable water for hydrogen generation will, therefore, come under intense scrutiny, including by importers of South African commodities. Costs of desalination or non-potable water are not a barrier to development in South Africa, given that the costs will represent a near-negligible fraction of overall hydrogen production costs (ibid.). The Japanese cost targets for hydrogen of $3/kg in 2025 and $2/kg in 2050 will, thus, be more than accommodative, given that the desalination cost components are calculated to vary between $0.005 and $0.020 per kilogram of hydrogen produced, or less than 1% of the 2025 target price (ibid.).

Looking still further in the value chain, additional gains could be made by using the financial platform created by the export of green hydrogen to shore up water resources in South Africa for communities and other productive activities, such as agriculture (Bischof-Niemz, 2021). As part of the "social license to operate", desalination plants supplying inland electrolysis plants could be scaled to at least 300% of the capacity required for the electrolysis plant alone, with the desalination capital expenditure built into the economics of the hydrogen plant (ibid.). For export-oriented coastal hydrogen production, the desalination plants should operate at reduced capacity in times of good rains and full dams, supplying only the electrolyzer plants (ibid.). In times of drought, by comparison, the desalination plants should operate at full capacity. The local water utility then can buy the excess water, paying only for the electricity component: the capital repayment costs are paid for by the hydrogen business (Roos and Wright, 2021). Overall, water supply is not expected to be a binding constraint to the development of a large-scale green hydrogen manufacturing complex in South Africa.

Energy World Turning on Its Head

What is emerging globally and in South Africa is that the backbone of the future energy system will be a combination of solar, wind, and hydrogen (Bischof-Niemz, 2021). Solar and wind may become the main source of primary energy. Energy flowcharts would then be turned on their head. Electricity produced using wind and solar will be utilized either directly or, following electrolysis, in the form of hydrogen. Where possible, wind and solar electricity should be used directly, as this represents the most efficient outcome. However, there will be times when that solar- and wind-based electricity may be recycled, as the variable renewable electricity can be required at a different time of the day, month, or year. To achieve this, electricity will be recycled through batteries for short-term storage tasks of intra-day or intra-week storing of energy. It will also be used directly for industrial heat, plus household space and water heating (combined with heat pumps wherever possible) and for cooking. In the transportation sector, meanwhile, renewable electricity will charge battery-electric vehicles, especially passenger cars, and drive electric trains, trams, and off-road mining vehicles.

For those energy services that cannot use electricity directly, solar- and wind-derived energy will be used indirectly to produce green hydrogen, by using electrolyzers to split water into hydrogen and oxygen. This green hydrogen could then be used directly or indirectly. Direct applications include a limited share of heating (during longer wind doldrums in sunless winter) and re-electrification, as well as in fuel-cell electric cars, trucks, and trains. Green hydrogen would also replace grey hydrogen in today's fertilizer-production processes. It could similarly replace coking coal in steelmaking and be used to reduce the heat-related (i.e., non-process-related) emissions of cement making.

Should the entire global energy system evolve in this direction (Bischof-Niemz, 2021), the world will require between 500 and 600 million tons a year of green hydrogen (conservatively estimated, as it is based on today's final energy consumption levels). This would imply a massive

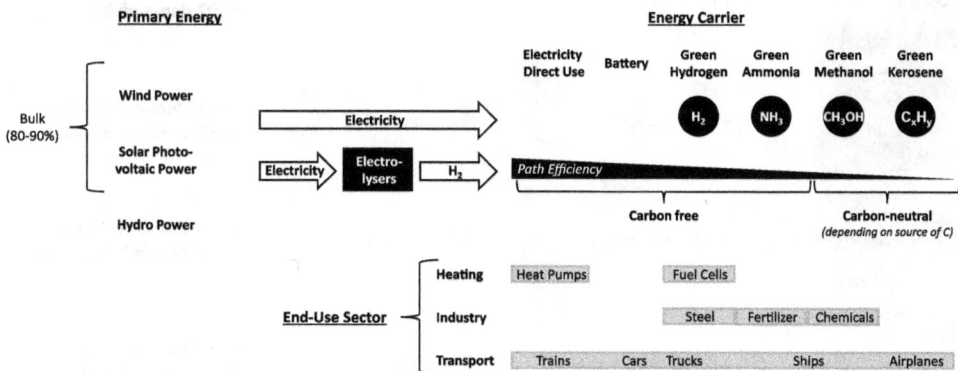

Figure 14.1 Structure of the future energy system based on solar and wind electricity as the primary sources of energy.

Source: Own graphic. Reprinted with permission from Creamer Media.

expansion of hydrogen production given that only about 120 million tons of hydrogen is currently produced yearly, with 95% of that derived from gas or coal (Bischof-Niemz, 2021).

From a transport logistics perspective, not all hydrogen would be used in pure form by the end user, because various end-use sectors require different energy carriers. Hydrogen, combined with nitrogen derived from air, is used today to produce ammonia for fertilizers and for explosives. Ammonia can also be utilized as a carbon-free fuel for the global shipping industry, for example in marine combustion engines. Alternatively, hydrogen combined with carbon atoms from a sustainable carbon source can produce a carbon-neutral hydrocarbon, such as methanol for shipping and as feedstock for the chemicals industry or kerosene for use as an aviation fuel. All those green-hydrogen-derived fuels (ammonia, methanol, kerosene) can easily, inexpensively, and with minimal emissions be transported across the globe. In fact, they are already today (Bischof-Niemz, 2021). Steel derived from green hydrogen, though not a fuel, is another end product that would have little to no difference in the ease and expense to transport.

Because of the very low temperatures that are required to liquefy hydrogen and the associated costs, the global trading of green hydrogen will probably be in the form of derived fuels or products such as green ammonia, methanol, and steel rather than hydrogen itself. The green hydrogen feedstock can be produced anywhere where sun, wind, land, and access to water (seawater or sufficient river/groundwater) are in abundance. Until such time that direct air capture becomes necessary, locations with a sustainable carbon source will be advantageous. As for green steel, locations that, in addition to abundant natural resources, also possess iron ore will be well positioned to begin production.

Based on these conditions, large volumes of green hydrogen that are produced in areas with a natural-resource advantage can be converted into more easily tradeable, hydrogen-rich products. Once the end-uses are assessed, estimates point to about 320 million tons of annual green hydrogen output (500–600 million tons) will most likely be converted into four broad categories of tradeable products: green ammonia, green methanol, green kerosene, and green steel (Figure 14.2).

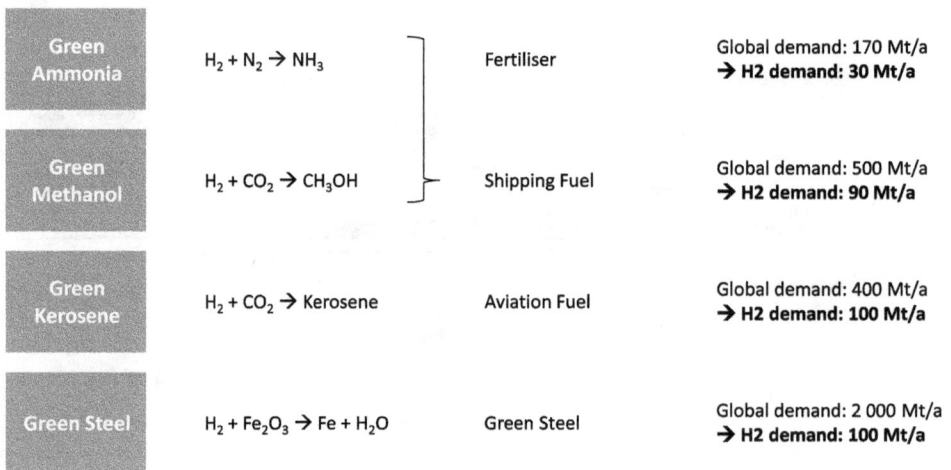

Green Ammonia	$H_2 + N_2 \rightarrow NH_3$	Fertiliser	Global demand: 170 Mt/a → H2 demand: 30 Mt/a
Green Methanol	$H_2 + CO_2 \rightarrow CH_3OH$	Shipping Fuel	Global demand: 500 Mt/a → H2 demand: 90 Mt/a
Green Kerosene	$H_2 + CO_2 \rightarrow Kerosene$	Aviation Fuel	Global demand: 400 Mt/a → H2 demand: 100 Mt/a
Green Steel	$H_2 + Fe_2O_3 \rightarrow Fe + H_2O$	Green Steel	Global demand: 2 000 Mt/a → H2 demand: 100 Mt/a

Figure 14.2 The four broad categories of tradeable products: green ammonia, green methanol, green kerosene, and green steel (Bischof-Niemz, 2021).

Annual global demand for ammonia as a fertilizer is roughly 170 million tons today and the respective demand for green hydrogen to produce that ammonia is ~30 million tons (IEA, 2021). Today, that is produced by gray hydrogen, but it could be green with the right adaptations. The market for green ammonia and green methanol as a shipping fuel will be far larger at about 500 million tons by 2050 (Bischof-Niemz, 2021), implying a yearly green hydrogen demand of some 90 million tons. Likewise, to meet today's global yearly steel demand of 2 billion tons using green hydrogen instead of coking coal to reduce the iron will translate into yearly demand of 100 million tons of hydrogen (ibid.). In addition, another 100 million tons of green hydrogen may be required annually to produce a carbon-neutral fuel for long-haul aviation (ibid.).

Export Platform

Green hydrogen offers South Africa an opportunity to effectively "beneficiate" its wind and solar resources. This term may sound unusual, as the word beneficiation in South Africa has come to refer to "adding value to mined products such as gold, platinum, chrome and iron ore ahead of export". The same logic could also be applied to the country's wind and solar resources if they were to be "harvested" to produce value-added products primarily for export into international markets.

The export earnings potential for South Africa, should it succeed in capturing even a modest share of the emerging market for tradeable hydrogen-based products, is enormous. Based on the assumption that the annual global market for green hydrogen to produce green ammonia, methanol, steel and aviation fuel could be 320 million tons by 2050 (Bischof-Niemz, 2021), the total market value of the end products by that date may then be €2.3 trillion. This assumes a €80 billion market for green fertilizers (at a €450–500/t price tag assumption for green ammonia), a €250 billion market for green shipping fuel (assumed €500/t price tag), a €500 billion yearly market for carbon-neutral kerosene (assumed €1,250/t price tag), and a €1,500 billion market for green steel (assumed €750/t price tag) (ibid.). If South Africa were to position itself to capture 10% of the green fertilizer (€8 billion), shipping fuel (€25 billion), and sustainable aviation fuel (€50 billion) markets, together with 5% of the global green steel (€75 billion) market, its annual export earnings from hydrogen-based products could be approximately €160 billion by 2050 (ibid.). To put the scale of that opportunity into perspective, South Africa's annual gross domestic product was about €300 billion in 2020.

As the global energy system undergoes change, the South African society will likewise need to adjust its view of the role of renewable energy, if it is to fully capture the green hydrogen and the power-to-X opportunity that is emerging. Instead of solar and wind being seen primarily as a replacement for the country's aging coal-fired power station fleet, with hydrogen as a by-product, the solar and wind fleet would become a green-hydrogen input to ultimately produce exportable, high-value products, and electricity for domestic use as the by-product. Given South Africa's current electricity deficit that has persisted for more than a decade, largely because of a chronic decline in the performance of the coal stations, the current focus of policymakers is simply to address power instability. Such an approach will definitely, if accelerated and upscaled, restore security of supply and help to sustain and grow existing industries, including South Africa's energy-intensive mining and smelting industries.

Given the demand pull being created in regions such as Europe and Asia for carbon-free or carbon-neutral alternative energy sources, South Africa should adopt a more ambitious and entrepreneurial strategy with green hydrogen, opting for an energy-wide rather than an electricity-only approach to solving its power crisis. In other words, policymakers should be aiming for a large renewables fleet that fully caters to current and future direct electricity demand in lighting, heating, industrial production and electric mobility, but also for indirect electricity demand in the form of green hydrogen and green-hydrogen derivatives.

To match that new level of ambition, South Africa would need to facilitate considerable investment in electrolyzers and prepare for a far larger investment in solar and wind than would be the case if the renewables fleet were to be scaled purely for electricity supply. To meet South Africa's current average weekly load profile using only wind and solar, roughly 50 GW of wind and another 50 GW of solar PV capacity would be required versus an existing coal-heavy fleet of some 40 GW, much of which is not reliable currently (ibid.). The upshot would be a system in which solar and wind supply the bulk of the system load, complemented by flexible generators, such as the pumped-storage plants, battery energy storage, or gas-fired generation. There would be times when the renewables component of this fleet would produce too much electricity, leading to periods of curtailment, whereby that energy is wasted. This curtailment notwithstanding, such a fleet would be considered cost-optimal, as it makes sense to ensure that the cheap renewable electricity is used as the system workhorse, with the expensive flexible generators deployed only to balance the system.

If South Africa were to scale its fleet to produce green hydrogen as a by-product of its cheap renewable electricity, the solar fleet could be expanded to 80 GW and the wind fleet to 100 GW. These plants would be coupled to a 45 GW electrolyzer fleet. The excess energy in such a system would be consumed by the electrolyzers first, to ensure as high a load factor as possible for the electrolyzers, and the balance would be absorbed by direct-electricity heat production, with a small share being curtailed. In this hypothetical configuration, South Africa's 45 GW of electrolyzer capacity would produce about 5 million tons of green hydrogen per year.

Another option exists – one that seeks to capitalize more fully on South Africa's advantage in the production of cheap renewable electricity to capture a share of an expected, fast-growing market for green hydrogen and hydrogen-based fuels or chemicals. Under such a scenario, South Africa would need to invest in 300 GW each for solar PV and wind power, alongside an electrolyzer fleet of 250 GW. In such a scenario, South Africa's entire electricity demand would become a solid and stable by-product of the production of green-hydrogen products for export – much as the domestic oil-demand market in Saudi Arabia is a modest side business of the country's main export business (ibid.). With this projection, South Africa would produce 25 to 30 million tons of green hydrogen per year.

Land of H$_2$ Opportunity

Looking beyond the potential to earn much-needed foreign exchange, the electricity scenario would enable South Africa to combat one of its most pressing social problems: unemployment. A renewables fleet comprising 300 GW of solar PV and 300 GW of wind by 2050 would meet not only the country's power needs (with 50 GW of wind and solar PV capacity, respectively, required for electricity alone at current consumption levels) but also create the platform for the annual export of green-hydrogen-based products, containing between 25 and 30 million tons of hydrogen. To build such a fleet, 12 GW to 14 GW of new wind and new solar PV installed capacity would need to be built every year in perpetuity to sustain an installed base of 600 GW as older plants are decommissioned. This alone would yield more than 500,000 permanent jobs, even before adding the jobs associated with the production and exportation of green hydrogen and the downstream tradeable derivatives (ibid.).

At such a scale, questions turn to whether South Africa has sufficient land to accommodate such a large renewables fleet. The availability of land is undoubtedly a binding constraint on

Total capacity
in GW

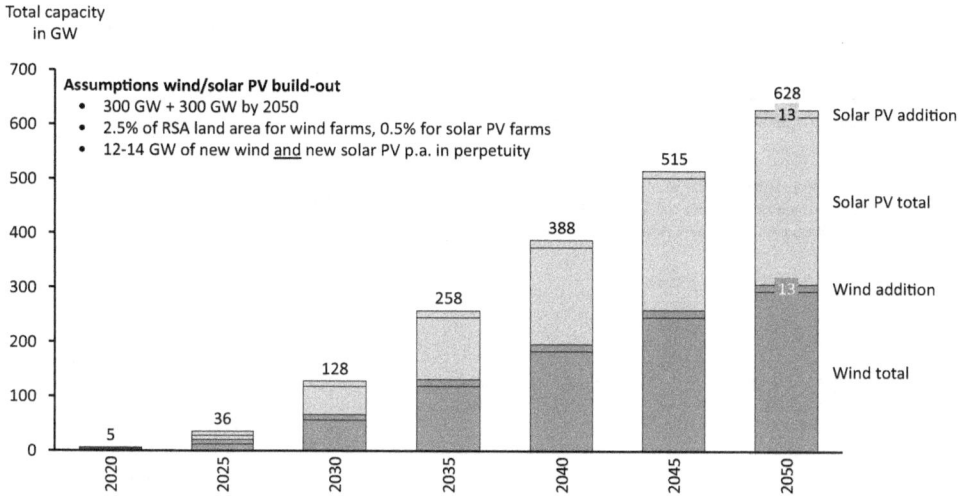

Figure 14.3 The buildup of solar and wind capacity under the hypothetical scenario whereby South Africa would produce 25 to 30 million tons of green hydrogen per year.

Source: Own estimates, 2021.

renewables–led systems in several countries but definitely not in South Africa. The country has a surface area of 1,220,000 km², substantially larger than the UK's 240,000 km² or Germany's 360,000 km² (Bischof-Niemz, 2020). As a result, South Africa's population density is substantially lower at 48 people per square kilometer (versus 274 and 232 people per square kilometer for the UK and Germany, respectively). Its electricity demand per square kilometer is also much lower at 190 MWh per year per square kilometer (versus 1,460 and 1,680 MWh per year per square kilometer for the UK and Germany, respectively) (ibid.). South African power demand relative to its land use reflects an order of magnitude lower than in the UK or Germany, meaning there is far more land available for electricity and green-hydrogen production (ibid.).

Once South Africa's solar and wind resource advantage is included, a quite startling picture of abundance emerges. A wind and solar aggregation study conducted by Fraunhofer and the Council for Scientific and Industrial Research in partnership with Eskom and the South African National Energy Development Institute shows that large parts of South Africa have average wind speeds of more than 6 m/s at 100 m aboveground (Bischof-Niemz and Creamer, 2018), which are superior to the wind resource in many other countries. Likewise, South Africa's solar resource is one of the best in the world, with 50% to 100% more irradiation reaching 1 m² in South Africa on an annual basis than in most parts of Europe (ibid.). If 1% of South Africa's land area, equaling a square with 110 km edge length, was dedicated to harvesting electricity from solar and wind farms, the present value of the potential electricity generation over the coal equivalent 300 years, using a discount rate of 8%, would be more than 18,000 TWh. By comparison, if South Africa were to convert all its recoverable coal into electricity over 300 years, the present value of the potential generation, at the same discount rate, would be only 2,750 TWh (Bischof-Niemz, 2021).

In other words, the electricity generation potential of wind and solar for just 1% of South Africa's surface area is 6.5 times larger than the country's currently known reserves of unmined coal. Meeting South Africa's present annual electricity demand of about 250 TWh, using renewable energy, meanwhile would require only 0.3% of South Africa's surface area to be

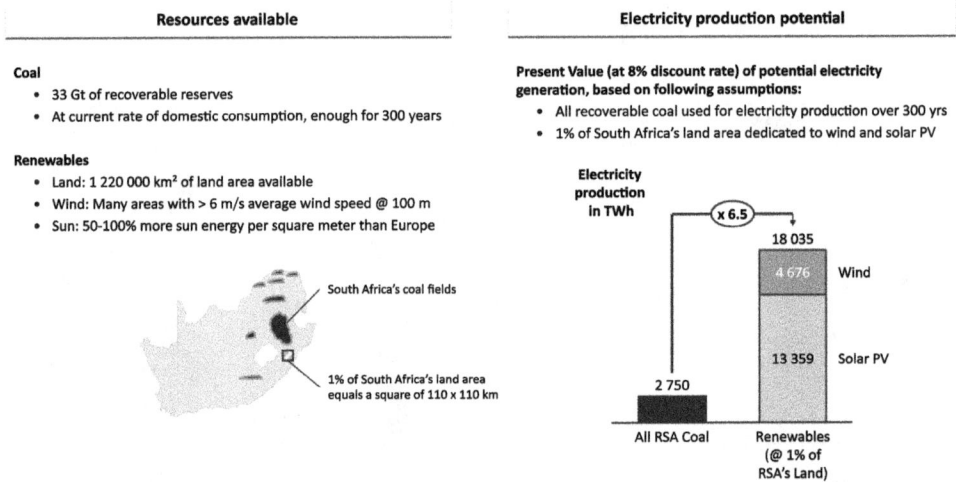

Figure 14.4 A comparison of the present value of potential electricity production from South Africa's coal resources versus its combined solar and wind resources; own calculations based on basic coal and area availability data (EIA, 2021).

dedicated to wind farms and 0.15% to the deployment of solar PV plants. For the full export-focused build-out of 300 GW of wind and solar each, 2.5% of South Africa's land area would have to be dedicated to wind farms and 0.5% to solar plants.

Naturally, if all of South Africa's wind and solar plants were concentrated into a single square comprising 1% of the country's surface area, it would represent a massive industrial complex in which no other economic activities could take place. In reality, though, these facilities would be spread across the country and, in many cases, co-exist with other productive activities, such as farming, manufacturing, mining, and retail, as well as social activities, such as human settlements, schools, and hospitals. In addition, South Africa has large tracts of land that would otherwise be considered unproductive for agriculture or mining. This land could well be deployed in the interest of electricity and hydrogen "harvesting", the reason being that, while solar and wind resources may be stronger in certain regions, the generic renewable-energy resource base across all nine of South Africa's provinces is typically superior to those found in other countries (Bischof-Niemz and Creamer, 2018).

The water consumption of a corresponding and fully scaled-up 250 GW electrolyzer fleet would be 400–500 million tons per year – more than what the existing coal fleet consumes today in fresh water (300 million tons per year). Hence, if 50% of the electrolyzer capacity was built inland and 50% along the thousands of kilometers of South African seashore, the water could be supplied half by the reduced water consumption of the electricity sector and half by seawater desalination (Bischof-Niemz, 2021).

Carbon-Neutral Jet Fuel Super Major

Besides South Africa's resource advantages, the country has one other unique asset base that, if repurposed, could position it as a "super major" in the production of carbon-neutral power fuels, specifically jet fuel.

While most ground-based transport can be decarbonized with the help of battery- and fuel-cell-powered electric vehicles, the aviation industry will for the foreseeable future continue to require hydrocarbons, based on the high energy density and the ease of handling. This is expected at least for long-distance flights, which consume the majority of all aviation fuel today.

At the same time, there is growing pressure on the airline industry to reduce its carbon footprint. The implication is that airlines will be willing to pay a premium for a fuel source that is carbon-neutral. In addition, mandatory blending requirements for electricity-derived jet fuels are already getting implemented (European Commission, 2021). Naturally, the cost burden will be paid by travelers, but a growing number of travelers are already voluntarily seeking to neutralize their air-travel carbon footprints by paying offset fees. Hence, the production of synthetic, carbon-neutral aviation fuel is increasing in focus.

This is another opportunity for South Africa. The country is a world leader in the production of synthetic hydrocarbons. Sasol's Secunda plant produces diesel, petrol, kerosene (aviation fuel), and other hydrocarbons from coal. Roughly, 40 million tons of coal is consumed today in order to produce the 8 million tons of product output yearly by Sasol (Bischof-Niemz and Creamer, 2018). As a result, Secunda is one of the largest point sources of carbon dioxide (CO_2) emissions in the world (ibid.). In a decarbonizing world, this is a serious liability, but there may well be an emerging opportunity to draw on this synthetic-fuels asset and skills base, combining it with green hydrogen to meet the needs of the aviation sector.

Currently, Sasol uses coal gasification to produce carbon monoxide and hydrogen from coal, water, and air (oxygen). This gas is carbon-monoxide-rich and does not include enough hydrogen yet to be fed into a Fischer-Tropsch reactor, which makes the synthetic hydrocarbons. Therefore, additional water and some of the carbon monoxide is converted in a water-gas-shift reactor into extra hydrogen – and CO_2. The now stoichiometrically correct mix of hydrogen and carbon monoxide, so-called syngas, is then converted into hydrocarbons in the Fischer-Tropsch reactors. To produce a ton of aviation fuel, more than 4 tons of CO_2 is emitted in the Sasol process from making hydrogen alone (in addition to the 3 tons of CO_2 emissions when burning the fuel in the airplane turbines). This amount of 4 tons can be reduced by gradually replacing the coal-based hydrogen from the water-gas-shift reactor with green hydrogen from renewables-based electrolysis (Bischof-Niemz, 2019).

In this way, Sasol's hydrogen-production-related carbon emissions can be reduced down to zero. Sasol would then essentially convert fossil carbon (coal) into fossil hydrocarbons (kerosene) but without any CO_2 emissions during that process. The airline that burns the aviation fuel, of course, still emits CO_2 into the atmosphere that ultimately stems from a fossil carbon source, in this case South African coal.

The carbon in the synthetically produced hydrocarbon still needs to be replaced from its current fossil origin. In the short run, sustainably sourced biomass can provide the carbon, as well as unavoidable emissions from industrial processes (e.g., cement-making). In the long run, coal as the source of carbon can be replaced entirely with the direct capturing of CO_2 from the air. The water-gas-shift reactor now runs in reverse mode, absorbing CO_2 and hydrogen and producing carbon monoxide and water. This would make Sasol a carbon sink, or carbon negative. The amount of CO_2 that Sasol sucks out of the air exactly equals the amount of CO_2 burned by the aviation industry supplied by Sasol – a closed carbon cycle with no net carbon emissions. Sasol's aviation fuel would be 100% carbon-neutral. Sasol would "beneficiate" or convert South African wind and sun into an exportable, high-value good (Bischof-Niemz, 2019). To put things into perspective: one of the international flight routes from Europe to South Africa would require 100 MW of electrolyzer capacity, plus an associated 100 MW to 120 MW of wind and another 100 MW to 120 MW of solar PV capacity to be completely supplied with such green aviation fuel.

Related to the mentioned repurposing, South Africa's state-owned synthetic fuels producer, PetroSA, could likewise be adapted (Roos and Wright, 2021). Located at the coast at Mossel Bay, in the Western Cape, PetroSA has a gas-to-liquids Fischer-Tropsch synthesis plant, which uses fast-depleting offshore natural gas as a feedstock. It is a 45,000 bbl/day facility (ca. 2-million tons of product per year) with annual emissions of about 2 million tons of CO_2.

> Converting PetroSA operations to sell green powerfuels to Europe on a long-term offtake contract, using the existing refinery and South African renewable resources should prove cheaper than a European business building a new refinery or repurposing an existing refinery, with European renewable resources.
>
> (Ibid.)

In summary, South Africa has two distinct advantages in the production of green aviation fuel. First, the country's renewables electricity is cheaper than in most other countries, owing to its intrinsic resource advantages of more wind, sun, and land. Second, South Africa has vast experience in and substantial existing infrastructure for the production of synthetic liquid fuels, using water-gas shift and Fischer-Tropsch reactors. Of course, other countries could invest to catch up, but South Africa has a substantial head start. This is a first-mover advantage that naturally only has value if the holder moves first. In a world hungry for carbon-neutral aviation fuels, there is considerable opportunity to turn South Africa into a substantial exporter of green aviation fuels. In the process, the country could also convert Secunda from the largest global single-site producer of CO_2 into a national and international green-energy champion.

To conclude, South Africa is well placed to decarbonize its electricity and energy systems over the coming two to three decades. The country has the solar, wind, and land resources not only to meet its current and future electricity needs but to leverage this relatively low-cost power to competitively decarbonize its transport, heating, cooling, and industrial sectors and processes, as well as those of its trade partners. The relatively mature nature of South Africa's coal assets, which have hitherto been South Africa's electricity workhorses, has materially reduced any stranded-asset risk, while the past trade-off between clean and cheap electricity has been entirely eliminated by the steep fall in the technology costs associated with solar photovoltaic and onshore wind. South Africa is almost uniquely positioned internationally to take a critical lead in producing green hydrogen and carbon-neutral fuels and chemicals that could be exported to countries or sectors that require such inputs to meet their own net-zero commitments. It is no overstatement to indicate that South Africa could become a green hydrogen superpower and the Saudi Arabia of carbon-neutral aviation fuels. A new golden age beckons!

Notes

1 South Africa's Witwatersrand Basin is the largest known gold anomaly on Earth and has produced more gold, about 50,000 tonnes than any other gold province in the world. These goldfields have been responsible for the creation of the metropolis of Johannesburg and have had a major influence in the shaping of South Africa's society and its economy. The country's industrial development has been heavily influenced by gold mining, as has its fractious political and social trajectory. The unique natural endowment of the Witwatersrand Basin catapulted South Africa to the top of the gold rankings, producing as much as one-third of world output as recently as 1993 (Frimmel et al., 2005). Output has since declined sharply, from a peak of around 1,000 tonnes in 1970, the nation's gold output fell to 130 tonnes in 2018 (www.gold.org). Gold mining also crowded in foreign and domestic capital and skills, which supported the rapid industrialization of the country and spawned a minerals and energy complex that dominated the country's financial, business, and economic landscape for decades.

2 Power-to-X is a concept for when power generated from solar and wind sources is converted to different types of energy carriers for use across multiple sectors or to be reconverted back into power.

References

Bischof-Niemz, T, 2019. Green Hydrogen Export Opportunity for South Africa, Webinar Presentation, EE Business Intelligence and EU Delegation to South Africa, 9 February.

Bischof-Niemz, T., 2020, Engineering News, Is Land a Constraint to a Renewables-Led Energy System in South Africa? www.engineeringnews.co.za/article/is-land-a-constraint-to-a-renewables-led-energy-system-in-south-africa-2020-01-24-1, 24 January 2020.

Bischof-Niemz, T. and T. Creamer, 2018, *South Africa's Energy Transition: A Roadmap to a Decarbonised, Low-Cost and Job-Rich Future*. London: Routledge.

Bischof-Niemz, T., Engineering News, 2019, Could South Africa be the Saudi Arabia of Green Aviation Fuel?, www.engineeringnews.co.za/article/could-south-africa-be-the-saudi-arabia-of-green-aviation-fuel-2019-10-25, 25 October 2019.

Bischof-Niemz, T. and R. Fourie, 2016, Cost of New Power Generators in South Africa: Comparative Analysis based on Recent IPP Announcements, https://cisp.cachefly.net/assets/articles/attachments/65080_new_power_generators_rsa_-_csir_-_14oct2016.pdf, 14 October 2016.

Council for Scientific and Industrial Research Energy Centre, Presentation on Future Wind Developments for South Africa, www.csir.co.za/sites/default/files/Documents/20171109-WindAc-Future_Wind_Scenarios-RSA.pdf, 14 November 2017.

European Commission, 2021, Directive of the European Parliament and of the Council as Regards the Promotion of Energy from Renewable Sources, and Repealing Council Directive (EU) 2015/652, https://ec.europa.eu/info/sites/default/files/amendment-renewable-energy-directive-2030-climate-target-with-annexes_en.pdf, 14 July 2021.

Fraunhofer IEWS, CSIR, SANEDI, Eskom, 2016. Wind and Solar PV Resource Aggregation Study for South Africa, www.csir.co.za/sites/default/files/Documents/Wind%20and%20Solar%20PV%20Resource%20Aggregation%20Study%20for%20South%20Africa_Final%20report.pdf. 1 November, 2016.

Frimmel H. E., D. I. Groves, J. Kirk, J. Ruiz, J. Chesley, W. E. L. Minter, 2005, Economic Geology 100th Anniversary Volume The Formation and Preservation of the Witwatersrand Goldfields, the World's Largest Gold Province, *Society of Economic Geologists*, p. 769, www.gold.org/goldhub/gold-focus/2019/06/south-african-production-important-no-longer-globally-significant

International Energy Agency (IEA), 2021, Ammonia Technology Roadmap, www.iea.org/reports/ammonia-technology-roadmap/executive-summary, accessed 20 December 2021.

IRENA, 2020, *Green Hydrogen Cost Reduction: Scaling up Electrolysers to Meet the 1.5°C Climate Goal*. Abu Dhabi: International Renewable Energy Agency.

Mantashe, G., 2021, Announcement of Preferred Bidders for Sound Five of the Renewable Energy Independent Power Producer Procurement Programme, 28 October 2021.

National Planning Commission of South Africa, August 2012, National Development Plan: Vision 2030.

Ngobeni, M., 2016, Presentation of the State of Renewable Energy Report: Market Overview and Current Levels of Renewable Energy Deployment, www.energy.gov.za/files/renewable-energy-status-report/Market-Overview-and-Current-Levels-of-Renewable-Energy-Deployment-NERSA.pdf, 25 November 2016.

Ramaphosa, C., 2021, Keynote Address by President Cyril Ramaphosa at the Sustainable Infrastructure Development Symposium of South Africa, 7 October 2021.

Roos, T. and J. Wright, 2021, Powerfuels and Green Hydrogen, Pretoria, EU–South Africa Partners for Growth, Council for Scientific and Industrial Research.

South African Government Gazette 42784, 2019, Integrated Resource Plan 2019, Pretoria, Department of Energy, October 2019.

US Energy Information Administration (EIA). www.eia.gov/international/overview/country/ZAF, accessed 20 December 2021.

15

GEOTHERMAL ENERGY

Decarbonization of the Heating Sector Through Direct Use and Related Applications

Jack Kiruja and Fabian Barrera

1. Introduction

As countries implement measures to decarbonize their energy systems to achieve climate targets and sustainable development goals, geothermal energy is uniquely positioned to play a significant role in the energy transition. Currently, only a small fraction of the global energy consumed today is derived from geothermal resources, despite enormous potential. This chapter examines direct heating end-use examples from China, Africa, Europe, and Latin America to highlight geothermal energy's potential and forms to contribute more prominently in the decarbonization of energy systems. Diverse heat applications that give geothermal a competitive advantage over other renewable energy resources are considered.

Subsequent sections of the chapter are structured as follows. An overview of key characteristics of geothermal energy is provided as a basis to better understand the opportunity. Direct geothermal energy uses are detailed next with country-specific examples. Recommendations are subsequently provided to move geothermal utilization forward, including the identification of resources, demonstration, policy and planning, and knowledge sharing.

1.1 Understanding the Resource

Geothermal energy is a clean and renewable resource that can be used to generate electricity or for direct heating and cooling applications. It is energy that is stored in the form of heat beneath the surface of the earth. This form of energy originates from the magma in the core of the earth and migrates naturally to the surface. Geothermal energy can be found in different geological and geographical settings across the world. Its heat can be used for a range of applications across multiple end-use sectors.

Though geothermal utilization technologies are mature, their applications are limited by various challenges including high upfront costs, a high-risk profile in the early stages of development, inadequate policies and licensing procedures, and limited technical capacity and awareness (IRENA, 2020a). Nonetheless, a number of countries have developed geothermal resources and used them to generate electricity, to heat space such as in buildings and greenhouses, and to provide heat for industrial heating.

This chapter explores the uses of geothermal energy in the heating of buildings and in the agri-food sector to drive energy transition beyond the power sector. It highlights examples

DOI:10.4324/9781003183020-17

from China, Africa, Europe, and Latin America in which the resource has been successfully developed in the two sectors.

1.1.1 Occurrence and Types of Geothermal Energy Resources

Geothermal resources may be found in various forms, largely characterized by location and temperature (IRENA and Aalborg University, 2021).

High-temperature geothermal resources tend to be situated on top of the tectonic plate boundaries, such as the Pacific Ring of Fire (Araujo, 2017), the East Africa Rift, and the Mid-Atlantic Ridge, among others. Such resources can develop during volcanic eruptions, where magma may be trapped at shallow depth, usually <10 km, creating an anomalous[1] thermal gradient. This phenomenon is common along the tectonic plate boundaries, with geothermal resource temperature which can be in excess of 200°C. Such resources can easily be accessible by drilling down to 2–3 km. However, these types of geothermal systems are found in a few places where the heat source underlays permeable rock, and subsurface water circulates within the permeable rocks creating very good geothermal reservoirs. Regions with high temperature geothermal resources include Iceland, Western United States, Indonesia, East Africa Rift, Central America, and so on (World Energy Council, 2013).

Lower-temperature geothermal resources are located away from the plate boundaries (e.g., in the sedimentary basins). Good-quality geothermal reservoirs can be found and harnessed by deep drilling to around 3–5 km. Such forms of geothermal resources are usually less than 150°C and can be used to generate electricity with binary technologies.

In most places, hot rocks also exist without water circulating underground. In these conditions, the geothermal energy in the rocks cannot be accessed using conventional technologies. Research is currently underway to create pathways in the subsurface for water to circulate in the subsurface around such hot rocks to extract the energy for utilization, either through fracturing the rocks (enhanced geothermal systems) or drilling vertical wells that are then interconnected by a series of horizontal wells across the hot rocks (IDB and IRENA, 2021). The success of these research projects has the potential to make geothermal accessible almost everywhere.

Beyond the geothermal energy forms that are harnessed with deep drilling, the geothermal industry is witnessing an increase in the utilization of geothermal resources found in shallow ground (usually <400 m), mainly for space heating (IRENA and Aalborg University, 2021). The energy in this form is obtained from soil and rocks or from the groundwater. The temperature of these resources is usually low and, in most cases, ranges below 60°C. These resources can be accessed at significantly lower costs than those requiring deep drilling. However, heat pumps may be necessary to boost the temperature of the extracted fluids for utilization.

Looking still further, geothermal energy is also being harnessed from abandoned coal mines. Mine waters benefit from the earth's thermal gradient, maintaining a constant temperature, usually 18–34°C, throughout the year. In Spain, the 2 MWt Barredo Colliery district heating project uses water at 23°C from an abandoned coal mine to heat 245 residential and two public buildings (HUNOSA, 2020). In addition, oil and gas wells located close to population centers may provide geothermal energy through co-production or by repurposing the wells.

Based on the wide variety and occurrence of geothermal resources in diverse geological and geographical settings, a large portion of the world may benefit from tapping geothermal energy.

1.1.2 Classification of Geothermal Resources

Several classifications have been adopted for geothermal resources depending on the characteristics of the geothermal system from which the energy is extracted. The main classifications are

Table 15.1 Classification of Geothermal Resources[3]

Classification	A	B	C	D	E
Low-temperature resources	<90	<125	<100	<150	<190
Medium-temperature resources	90–150	125–225	100–200		
High-temperature resources	>150	>225	>200	>150	>190

Source: Dickson and Fanelli, 2004.

based on temperature, enthalpy,[2] physical state of the fluids and geological setting. Temperature and enthalpy are mainly used for classifying geothermal resources for utilization purposes.

Classification of geothermal resources according to temperature is based on three categories: low, medium, and high temperature. Temperature ranges can vary by author or report, as shown in Table 15.1.

2. Direct Uses of Geothermal Heat

As global efforts are underway to limit the rise in global temperature well below 2°C relative to the pre-industrial average global temperature, the heating sector represents an important area to consider. This sector accounts for about 50% of the total energy usage today (IRENA and Aalborg University 2021) and roughly 40% of energy-related carbon dioxide emissions (Cole, 2020). In this context, geothermal energy can advance a decarbonization agenda with resources for direct heating distributed around the world.

Focusing on the heating and cooling sector, geothermal water can be used for a large number of applications. Such use has been done since ancient times. Particular attention is expected to be on lower-temperature resources, which are more widely available than the high-temperature resources (IRENA, 2019).

The global direct use of geothermal energy was recently estimated to equal 107,727 MWt (Lund and Toth, 2020). Direct use applications are reported in 88 countries worldwide with China (40,610 MWt), United States (20,713 MWt), Sweden (6,680 MWt), Germany (4,806 MWt), and Turkey (3,488 MWt), reflecting the top five countries for direct-use applications, accounting for 71.1% of the world capacity (ibid.). It is worth highlighting that the countries with the largest percentage increase in geothermal installed capacity (MWt) in the period between 2015–2020 were Iceland, Hungary, France, Egypt, and Australia. These demonstrate a clear example of the regional diversity of direct use applications worldwide (ibid.).

Figure 15.1 shows the notable increase in the installed capacity of geothermal direct use worldwide in the past 25 years. In the period between 2015 and 2020, the average annual growth rate was around 9%. This increase in installed capacity is likely due in part to an improvement in the reporting and data system for this type of applications, which has been a constant challenge in the promotion of widespread development of geothermal direct use. Additionally, there is more awareness in the direct utilization of geothermal energy, which has promoted the development of new projects focused on decarbonizing heating and cooling applications of the energy sector.

Many international agencies, multilateral and international organizations, and development banks, among other important stakeholders are focusing their global and regional efforts on promoting the direct use of geothermal energy as a key component for the decarbonization of countries around the globe. Organizations engaged in this area include the International Renewable Energy Agency (IRENA),[4] International Energy Agency (IEA),[5] United Nations Environment Program (UNEP) through the Africa Rift Geothermal Facility (ARGeo),[6] and the

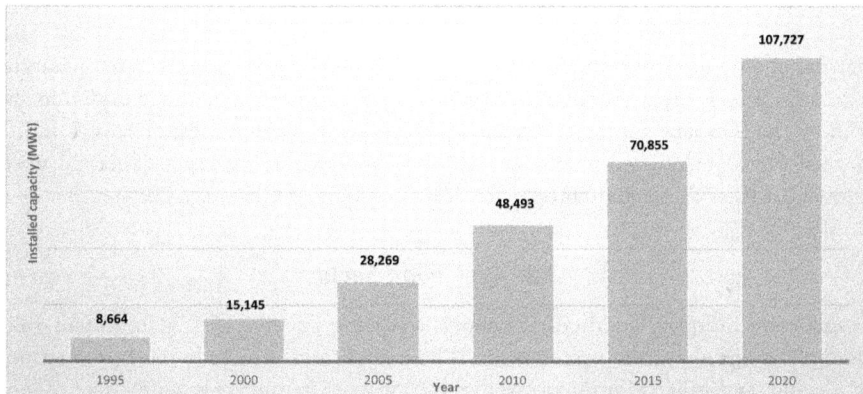

Figure 15.1   Global installed capacity (MWt) of geothermal direct use in the last 25 years.

Source: Adapted from Lund and Toth (2020).

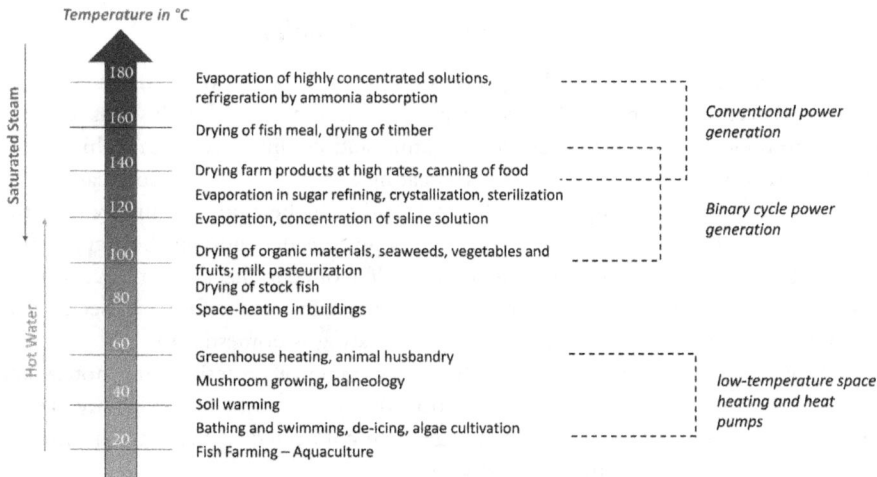

Figure 15.2   Direct applications of geothermal energy.

Source: Adapted from Lindal, 1973.

World Bank through the Energy Sector Management Assistance Program (ESMAP),[7] among others. As a result of such work, countries are increasingly including direct use of geothermal energy as part of their energy policies, decarbonization strategies, and low-carbon development. Some examples can be found in Chile, China, El Salvador, Kenya, and the Netherlands.

As noted earlier, applications for tapping geothermal heat vary depending on the temperature of the resource. Major areas for the direct utilization of geothermal energy are (1) space heating and cooling, including district heating; (2) the agri-food sector, such as with aquaculture and greenhouse applications; (3) industrial processes; (4) swimming, bathing, and balneology; and (5) heat pumps (Popovski and Vasilevska, n.d.). Importantly, geothermal heat pumps are the most common application of geothermal heat to heat and cool buildings (Van Nguyen et al., 2015). A Lindal diagram provides a common representation of the different uses of geothermal direct applications and their correspondent temperature (Figure 15.2). The following section details more specific applications.

2.1 Swimming, Bathing, and Balneology

Geothermal energy use in swimming, bathing and balneology requires low-temperature geothermal fluids. This is the most common application of geothermal energy worldwide, used for centuries by the Romans, Chinese, Ottomans, Japanese, and Central Europeans (Lund, 1996). Geothermal fluids are utilized in hot springs, spas, and swimming pools. Thermal waters are well known for their therapeutic properties (Jóhannesson and Chatenay, 2014).

2.2 Agri-Food Sector

In the agri-food industry, geothermal energy is playing an increasingly important role with various applications along the value chains. These applications include food production, processing, drying, and value addition, as well as cold storage, requiring a wide range of temperatures. Direct use of geothermal heat in the agri-food sector can contribute to enhanced food security and economic activities for local communities. An expanded discussion can be found in Section 3.1.

2.3 Space Heating and Cooling

The temperature characteristics of the geothermal fluids allow for wide use in space heating and cooling. Recently, centralized low-temperature heating and cooling systems have become a popular component of decarbonization in heating and cooling due to their high efficiency (IRENA and Aalborg University, 2021). Space heating and cooling applications can be found in the building sector, directly using the geothermal fluid or by heating a secondary fluid through heat convectors in the room. The type of technology use depends on the temperature and attributes of the fluid and can be applied to an individual or centralized system. A more detailed discussion on geothermal district heating and cooling systems can be found in Section 3.2.

Another application commonly linked to space heating is domestic water heating for sanitation. This utilization of hot or warm water is very common in residencies, hotels, and restaurants. Geothermal heat can be a useful option due to its constant availability throughout the year. Usually, domestic water heat exchangers are connected to the space heating system, achieving the required temperature for the water.

2.4 Industrial Processes

The industrial sector also utilizes geothermal fluids at a wide range of temperatures. Geothermal fluids can be used for process heating applications including evaporation, drying, distillation, refrigeration (absorption cycle), sterilization, washing, and deicing (mining operations), plus salt and chemical extraction (Lund, 1996). Despite a large number of possible industrial uses, geothermal resources are yet not widely employed in the industrial sector.

Through cascaded use, geothermal fluid progressively yields its heat when it is harnessed in a series of thermal processes, resulting in a staged decrease of temperature. If the fluid at the outlet of an initial process still contains sufficient energy, it can be utilized further in a subsequent process (Rubio-Maya et al., 2015). The cascading use of geothermal fluids can be commonly found linked to a power generation process, where a high-temperature geothermal resource is used to produce electricity and then used for one or more heating processes such as milk pasteurization, drying of food or wood, and even low-temperature applications, such as aquaculture and balneology. The cascade utilization of geothermal energy can improve the economics of small and large projects, including electricity generation projects. In addition, this approach increases resource efficiency and reduces the need to develop new energy resources to meet increasing demand. Cascaded use for geothermal energy has been deployed in a variety of ways around the world (ibid.). The most common configurations include electricity production and thermal uses, as well as systems consisting only varied thermal uses. Europe is the region with the most diverse uses for cascade systems, tapping medium- and low-enthalpy geothermal energy, with most of the applications related to heating networks, greenhouses, and balneology. The Americas represent the regions using most geothermal cascade systems linked to electricity generation (ibid.).

3. Geothermal Heat Utilization in the End-Use Sectors

3.1 Direct Uses in the Agri-Food Sector

Energy is a key input in the food value chain, including production, harvesting, processing, storage, transportation, retail, and cooking (IRENA, 2019, 2022; see also chapter by Cunha et al. in this *Handbook*). The use of sustainable and indigenous energy sources, such as geothermal energy, becomes very relevant to enhancing energy and food security, as well as addressing energy poverty and climate change.

The 2030 Agenda for sustainable development addresses the nexus between energy and food, aiming to end hunger and ensure access to affordable, reliable, sustainable, and modern energy for all (IRENA, 2019). To achieve these objectives, challenges need to be addressed relating to access to sustainable energy sources, modern technology, mitigating food waste and promoting the development of the agricultural sector and rural economies.

The potential direct use of geothermal heat along many stages of the food chain (production, processing, and storage) makes geothermal a reliable energy option to decarbonize such systems. Geothermal energy also can serve as a vehicle to promote the development of the agricultural sector. Many countries around the world currently utilize geothermal energy for agri-food applications (IRENA, 2022). The following section describes some examples of the uses of geothermal energy in the agri-food sector, based on the temperature of the available geothermal fluids.

3.1.1 Aquaculture

Aquaculture is the production of aquatic organisms including fish and plants, such as seaweed and algae. With temperatures from 20°C to 40°C, geothermal water can be used to heat fresh water through heat exchangers or be directly mixed to obtain suitable temperatures for aquaculture. Cultivating fish in aquaculture units that are heated to the desirable temperature for a species of fish to mature enables enhanced growth rates of fish due to an improved metabolism-feed conversion ratio. Geothermal energy has also been widely used in the cultivation of algae at optimum temperatures of 35°C to 37°C (Van Nguyen et al., 2015).

3.1.2 Soil Heating

Geothermal energy can be used to heat soil for cultivating crops by attaining two main objectives: (1) achieving an optimal temperature for growing a given crop, such as carrots or cabbages (IRENA, 2019), and (2) pasteurizing the soil by killing insects, fungus, worms, and some bacteria that may be harmful to crops (Lund, 1996). For both processes, the use of geothermal heat results in better production of the crop and requires temperatures of around 20°C to 60°C.

3.1.3 Irrigation

Cooled geothermal water can be used for irrigation of crops in areas with water scarcity if its chemical composition is suitable. In Tunisia, geothermal water is first piped to the greenhouses for heating purposes, then collected in cooling ponds and used in the irrigation of oases (Box 15.1).

3.1.4 Greenhouse Heating

The use of greenhouses has been a common practice for cultivating crops that provides a controlled environment to achieve the optimal conditions for a specific crop, especially in cold climates. Greenhouses enable the production of crops during off-season, as well as a protecting the crops from harsh environment and infestation of pests and diseases. Greenhouses may utilize locally sourced geothermal energy at a temperature range between 25°C and 80°C to control the temperature and humidity levels that are suitable for crop production. The heating of greenhouses with geothermal energy results in improved crop production, minimizes the chance of diseases, as noted earlier, and reduces the use of chemicals to control diseases and pests. Geothermal energy can be supplied to the greenhouses directly if the temperature is sufficient or heat pumps can be employed to boost the temperature of the geothermal fluid to the required level.

Box 15.1: Greenhouse Heating – Producing Tomatoes in Tunisia

The use of geothermal energy in Tunisia is limited to direct use (versus electricity generation) based on the availability of low-temperature resources, which are mainly in the southern part of the country. In 2010, 90% of the geothermal energy utilized in Tunisia was used for agricultural purposes, the main use being oasis irrigation, followed by heating of greenhouses (Mouldi Ben, 2015).

Greenhouses in Tunisia are mostly utilized for cultivation of tomatoes (57%), followed by cucumbers (23%), melons (12%), and watermelons (3%) (Mouldi Ben, 2015). The south of Tunisia is prone to high temperature variations, which can damage crops. To solve this problem, the use of geothermal water has been a good approach, as it controls the environment inside greenhouses, especially during the night. To achieve heating, geothermal water at 45°C to 80°C is passed through pipes laid on the ground between the crops (Mouldi Ben, 2015).

Data from Tunisia show that exports of geothermally heated, early crops in the governorate of Gabes reached 14,000 tons by June 2018, compared to 9,700 tons in the same period the previous year (Richter, 2018). Among these products, tomatoes stand out due to their very good quality and, therefore, are exported to the European markets and the Gulf countries. The prices for Tunisian tomatoes are high in the European market due to their high quality and taste (ibid.).

3.1.5 Sterilization

In the case of food processing such as in the fish and meat industry, geothermal heat is used for sterilization. The process is undertaken to eliminate bacteria and other microorganisms responsible for the spoilage of the products. Usually, sufficient sterilization can be achieved by exposing the food to a temperature of 121°C for three minutes, as this is adequate to kill the most common bacteria in meat. Geothermal hot water or steam at a temperature of 105–120°C can be used to sterilize equipment in the food processing, canning, and bottling industries (Van Nguyen et al., 2015).

3.1.6 Evaporation and Distillation

Geothermal heat can also be used in evaporation and distillation processes that result in increased concentrations of products and separation of mixtures. Common products that require evaporation and distillation processes include sugar, milk concentrate, mint, and liquor products. Geothermal fluids in the range between 80°C and 120°C may be used for such processes, depending on the product being processed. Usually, heat exchangers are employed to separate the product from the geothermal fluids providing the energy (Van Nguyen et al., 2015).

Due to the wide range of temperature that may be used for this process, evaporation and distillation are sometimes developed in a cascade arrangement, where the food processing is done close to the geothermal field that is used for power generation.

3.1.7 Milk Pasteurization

Pasteurization is yet another process that can be carried out with geothermal heat. It is the thermal treatment of milk to stop enzymatic reactions and kill microorganisms thereby extending milk's shelf life. The most common pasteurization techniques are low-temperature, long-duration pasteurization (60–65°C for 30 minutes) and high-temperature, short-duration pasteurization (72–75°C for 15–20 seconds). Ultra-high-temperature treatment of milk requires 135–140°C for 1–3 seconds. Such temperature ranges of 60–140°C can be obtained from geothermal hot water or steam (Kiruja, 2011).

3.1.8 Food Drying and Dehydration

In addition to the agriculture and food treatments with geothermal heat, drying and dehydration can also utilize geothermal energy. Drying and dehydration are important processes that are performed to preserve food by reducing moisture content. Common foods that require drying and dehydration include grains, fruits, vegetables, onions, meat, and fish. In drying, the moisture content is reduced to less than 13%, while in dehydration, moisture content of less than 20% may be desirable depending on the type of food. Food drying and dehydration facilitate storage of food for off-season consumption and minimize the chances of spoilage. To enhance the efficiency of the drying and dehydration processes, thermal energy in the range of 60–120°C may be used in a food dryer (Kiruja, 2017).

Energy for such processes can be obtained from geothermal fluids in a number of ways. Usually, a heat exchanger is used in the drying processes to transfer the heat from the fluids to the air circulated in the drying chamber of the food dryer. Its design depends on the nature of the geothermal fluid, and some countries have successfully developed their own food dryer designs, such as in Mexico (Box 15.2), where a fruit dehydrator was developed in collaboration with academia. Geothermal food/crop drying and dehydration has been successfully applied in drying cereals, coffee, fish, shrimps, tobacco, tomatoes, and mangos, among others. For El Salvador's application, which includes food drying, see Box 15.3.

Box 15.2:   Geothermal Innovation for Food Drying in Mexico

A group of researchers from the Autonomous University of Mexico (UNAM) developed a food geothermal drying system, as an initiative to address the growing food waste problem in Mexico (UNAM, 2020). This local innovation requires geothermal fluids with temperatures below 150°C.

The project built a first pilot demonstration in the geothermal field of San Pedro Lagunillas in Nayarit with a capacity to process 200 kg per day of dried food, which can be easily stored and transported. The equipment has a continuous operation, based on the availability of the geothermal resource, and has been utilized to dry shrimp, pineapple, carrots, mango, papaya, and so on (Romero, 2018).

The geothermal food dryer was expanded to reach an industrial capacity of around 900 kg per day of dried food. With this growth, the facility employs 50 people directly and 60 indirectly, besides mitigating the food waste from local farmers (Romero, 2018).

The success of the geothermal food dryer has brought together private companies to create a new food branding, GeoFood, which commercializes food products resulting from clean energy and technologies, such as the dried products from the geothermal field in Nayarit (Geofood, 2019).

Box 15.3:   Harvesting Geothermal in El Salvador

El Salvador is one of the top producers of electricity coming from geothermal energy. By 2020, the installed capacity of geothermal energy in the country for power generation accounted for 204 MW (IRENA, 2020b).

El Salvador's agricultural sector has low production rates and is heavily dependent on biomass and fossil fuels, which signify a very low income for local farmers, especially women. The non-sustainable use of biomass promotes deforestation and fossil fuels contribute to the growth of greenhouse gas emissions in the country (United Nations Framework Convention on Climate Change [UNFCCC], n.d.)

The Salvadorian electricity company LaGeo has initiated a program focusing on empowering women living next to the geothermal field of Berlin. Using waste heat from the geothermal power plant, women from the local communities grow and sell plants watered with geothermal condensates and dehydrate fruits with geothermal steam. In this way, this initiative aims to promote local economies and provide skills and new abilities to local women, enhancing their leadership in the communities (UNFCCC, n.d.)

El Salvador has recently launched a new National Energy Policy 2020–2050, which looks to promote renewable energy to mitigate, among others, the strong dependence of the country on fossil fuels, enhancing its energy security. Additionally, the new Energy Policy highlights the role of direct use of renewable energy such as solar thermal and geothermal, as key components for the country's energy transition, especially for the decarbonization of the agri-industrial sector.

3.2 Direct Use for Space Heating in the Building Sector

As with the diversity in temperature and location, geothermal resources can be found at a variety of depths, ranging from a few meters to several kilometers below the ground. Shallow geothermal resources, usually <400 m, consist of heat stored in the upper layers of soil/

rocks and the heat contained in the groundwater resources. These resources are mainly utilized through the use of geothermal heat pumps – both ground source and water source types. Deep geothermal resources, including hydrothermal systems and hot dry rock can also be accessed through drilling.

The global utilization of geothermal energy to supply heat to buildings through the use of heat pumps was estimated at about 77.5 GWt in 2020. The number of heat pumps installed around the world totaled roughly 6.46 million units, mostly in China, United States, Sweden, Germany, and Finland. The installed heat pump capacity ranges from 5.5 to 150 kW (Lund and Toth, 2020).

Apart from heat pumps, geothermal energy is also supplied to buildings through district heating systems and individual heating systems. District heating refers to a system of insulated pipes and auxiliary equipment for distributing heat to meet the space heating and hot water needs of residential and commercial building located in close proximity to each other (IRENA and Aalborg University, 2021). The heat may be generated centrally or from several locations using a single source or distributed energy sources, including waste heat from power genera- tion, industrial processing, or cooling operations. One way of scaling up the use of geothermal energy for space heating is through the use of district heating systems, with a distributed heat supply, including in combination with large-scale heat pumps.

3.2.1 Geothermal District Heating and Cooling

In 2020, the global installed capacity of geothermal district heating was about 11.6 GWt while individual heating with geothermal energy was about 1.1 GWt (Lund and Toth, 2020). The largest installed capacity for geothermal district heating is found in China, Iceland, Turkey, France, and Germany, which have a combined share of about 90% of the world's total district heating capacity (Lund and Toth, 2020).

Given the widespread availability of geothermal resources noted earlier (Figure 15.2), their application in heating and cooling of buildings has the potential for further growth. As a result, geothermal resources can play a key role in supplying the heating requirements of buildings through district heating systems, thereby reducing the emissions and pollution from the building sector. A key requirement for utilizing geothermal energy in district heating is the geographical co-location of the geothermal resource and the heat demand centers such as cities or munici- palities (DOE/EE, 2019). Some of the advanced examples of geothermal energy for heating of buildings can be found in China, Europe, and the United States, as detailed in the following sections.

CHINA – GEOTHERMAL DISTRICT HEATING CASE 1

China is endowed with vast geothermal resources that (similar to many countries around the world) remain unexplored and unutilized to a large degree. The distribution of geothermal resources in China is mainly concentrated to the northern and northeastern parts of the country (Zhu et al., 2015).

Efforts have been made in China to estimate the potential of the geothermal resources. Table 15.2 shows that the largest potential for recoverable geothermal resources is the low- to medium-temperature hydrothermal resources. Shallow geothermal resources also present a sig- nificant option while the potential for high-temperature resources is much lower. While hot dry rock presents significant potential, the technology for extracting the heat is still under research and development.

Table 15.2 Estimates of China's Geothermal Potential

Type	Reserves		Regions
Shallow geothermal	0.7 billion tce (recoverable reserves)		Huanghuaihai Plain, middle and lower plains of the Yangtze River
Hydrothermal	Medium to low temperature	1.865 billion tce (recoverable reserves)	Plain areas: North China basin, Songliao basin, Subei basin, Jianghan basin, Ordos basin
			Hilly areas: southeastern coast, Jiaodong Peninsula, and Liaodong Peninsula
	High temperature	18 million tce (recoverable reserves)	South of Tibet, west of Yunnan, and west of Sichuan
Hot-dry-rock	856 trillion tce (depth 3 km to 10 km, total reserves)		Tibet, Yunnan, Guangdong, and Fujian

Source: China Geological Survey, 2018.

Note: tce: tons of coal equivalent.

According to the 13th Five-Year Plan for Geothermal Energy Development, the total space heating and cooling area for geothermal in 2020 in China was expected to be 1.6 billion m^2 (NDRC, 2017). In 2019, around 841 million m^2 of floor area was heated using geothermal from shallow wells in China (China Geological Survey, 2018). This was achieved following a growth rate of 28% since 2010 (ibid.). Geothermal heating is common in the cities of Northern China, including the Beijing-Tianjin-Hebei region, Liaoning, Shandong, Hubei, Jiangsu, Shanghai, and so on (ibid.).

Besides shallow geothermal resources, China is increasingly using its hydrothermal resources to provide energy for space heating. In 2018, in excess of 150 million m^2 of floor area was heated using this form of geothermal resources, mainly in Shandong, Hebei, and Henan provinces, following a growth rate of about 10% since 2018 (China Geological Survey, 2018).

Geothermal heating still has broad development prospects for China. Geothermal resources in China are mainly found in Northern China, which coincidentally has a large heating demand during the cold months. Presently, the heating demand is extensively met using coal, resulting in heavy pollution (IRENA and Aalborg University, 2021). The switch to geothermal heating in Northern China would result in clean heating, hence less pollution in the cities.

In the Beijing-Tianjin-Hebei region, the area with the most geothermal potential in China, such heating is expected to be equivalent to about 20 million tce (NDRC, 2017). By 2021, the heating area for geothermal was estimated to reach 1 billion m^2 – 0.5 billion m^2 for hydrothermal and 0.5 billion m^2 for shallow geothermal energy (ibid.; see also Box 15.4).

Box 15.4:   Geothermal District Heating in China

The city of Xiongxian in China's Hebei Province is a city of about 390,000 people in Northern China that derives the energy for heating of buildings entirely from geothermal energy (Richter, 2017). This represents 4.5 million m^2 of floor area heated with geothermal energy, a contrast to

---

the neighboring Beijing where most of the heating is provided using coal, resulting in massive pollution (ibid.). The Chinese authorities in 2017 put in place plans to establish the Xiong'an special economic zone, an area that will include Xiongxian and the neighboring counties of Rongcheng and Anxin (ibid.).

Clean energy is a key priority for the economic zone, where natural gas, biomass, geothermal, and solar energy are expected to play a major role. The region is endowed with abundant geothermal resources as indicated by preliminary studies carried by China Geological Survey in 2017, which showed the presence of significant shallow geothermal resources down to 200 m (ibid.).

EUROPE – GEOTHERMAL DISTRICT HEATING CASE 2

According to the European Geothermal Energy Council (EGEC), the installed capacity of geothermal district heat and cooling projects in Europe in 2019 was 5.5 GWt from 327 projects in 25 countries. The installed geothermal heat pumps in Europe reached two million units in 2019 (EGEC, 2021).

Most of the installed district heating capacity in Europe is found in Iceland, Turkey, France, Germany, Hungary, the Netherlands, and Italy, accounting for more than 85% of the total European capacity (ibid.; see also Box 15.5). Studies in Europe indicate that in excess of 25% of the EU's population dwells in areas where their space heating needs can be met using geothermal district heating systems. This refers to areas with temperature of at least 60°C at a depth of less than 3,000 m and an urban density that can support a district heating project. Unfortunately, most of the resources remain largely untapped today (GeoDH, 2015).

On the other hand, several geothermal district heating projects are planned to be implemented in the coming years in Europe. The realization of these projects will result in significant decarbonization of the heating sector in Europe.

Box 15.5:   Danube Region District Heating Potential

The Danube region, consisting of Bosnia and Herzegovina, Croatia, Hungary, Romania, Serbia, and Slovenia, has a large share of fossil fuels in the heating and cooling sector and great geothermal potential that remains untapped (Danube Region Strategy, 2017). The main users of heat in the region include buildings, industries, and agriculture. In addition, the district heating as well as the individual heating systems used to supply energy operate at low efficiency (ibid.).

With the support of the Danube Transnational Program, Darlinge Project was established to promote the utilization of the region's untapped geothermal potential in the heating sector, while promoting sustainability of the resource. The project resulted in the mapping of the region's geothermal resources based on the data shared by the member countries. The geothermal potential was matched with the demand in the region and an interactive web portal (Danube Region Geothermal Information Platform) was developed.

This elaborated the Transnational Geothermal Strategy and the Danube Region Action Plans.

Hungary, one of the countries within the assessed geothermal resource area in the Danube region had an installed geothermal spacing heating capacity of 223 MWt in 2017. Most of the geothermal based systems supply a portion of the thermal energy needed, mainly for heating in public buildings.

Geothermal energy is also used in district heating in Hungary. The two largest geothermal-based heating systems in Hungary that are connected to the local district heating network are the Miskolc project, with an installed capacity of 55 MWt, and the Győr project, with a capacity of 52 MWt. Both projects displaced the use of gas to supply heat to residential and commercial buildings (Nádor et al., 2019).

Source: Danube Region Strategy, 2017.

4. Considerations to Move Geothermal Heat Use Forward

Geothermal resources, especially the ones at low- and medium-temperature, are widely available, making this form of renewable energy an important component of a global transition to low-carbon energy. As presented in this chapter, geothermal energy is a clean source of energy that has a high potential to decarbonize the heating sector through its utilization in buildings, industries, and agriculture.

Geothermal energy can mitigate greenhouse gas emissions, provides a stable energy source that is not dependent on the weather conditions and at the same time supports the realization of some of the objectives of the 2030 UN Agenda for Sustainable Development, such as climate change, food security, energy access and security, and gender empowerment, among others. Direct use of geothermal energy has the potential to transform the socioeconomic well-being of the local communities through creation of new jobs, development of infrastructure, and improving standards of living.

The following recommendations are outlined to facilitate the strategic development and utilization of geothermal heat as a means to achieve higher levels of decarbonization (IRENA, 2020a)

- *Identify geothermal potential in the urban and rural areas:* The identification of geothermal resources should not only focus on high-temperature resources suitable for electricity generation but also low- and medium-temperature resources in order to promote direct use. Several techniques can be applied for the identification of geothermal resources and geological, geophysical, and geochemical measurements. Encouraging the sharing of data about the underground conditions, including on temperature and rock formation, as is the case in some countries, where companies in the oil and gas and mining are required to make such data publicly available, can support the identification of geothermal resources. Abandoned coal mines and repurposed oil and gas wells could also be sources of geothermal energy for direct use applications, especially those located close to areas with high demand for heat.

- *Develop dedicated policy, licensing procedures, and regulatory framework for geothermal direct use applications:* Geothermal resources of low and medium temperature do not always require deep exploration and can be found in shallow aquifers and even superficial waters. Therefore, the use and exploitation of these geothermal resources can benefit from more simplified licensing and regulatory processes. A simplified licensing procedure for specific applications can attract investors and develop geothermal direct-use projects in shorter time periods. Likewise, clear rules on the regulatory aspects of the use of geothermal heat, classification of the geothermal resources, and remuneration schemes, among others, can provide a clear environment for the development of direct-use projects, attracting investors and project developers. A more streamlined licensing procedure and dedicated regulatory framework should be supported by a clear policy framework for the promotion of geothermal direct use. A tailored policy for geothermal direct use can include clear targets for the adoption of the energy sources, highlighting its linkages with the sustainable development goals and the benefits to the population, especially the local communities surrounding the geothermal resources. A clear policy framework can serve as the basis for rapid deployment of direct use applications by also promoting financial incentives and the development of pilot projects in the country.

- *Include geothermal energy in the national, regional, and local development plans:* The promotion of direct use applications should be a common effort resulting from the coordination of several institutions beyond the energy sector. For instance, environmental authorities should recognize the benefits of geothermal energy in the mitigation of greenhouse emissions, highlighting geothermal energy's role in the national determined contributions (NDCs) and the achievement of the objectives from the Paris Agreement. Furthermore, the application of geothermal energy in the buildings sector requires close coordination with regional and local authorities for the utilization of geothermal resources in the urban setting, especially when promoting district heating and cooling systems. In the case of agri-food applications, coordination is recommended between the energy, agricultural, and industrial sectors on the development of applications nearby the available geothermal resources while also considering the local economies to enhance the impact of such activities to the local population. These coordination efforts can include the development of plans or roadmaps with clear targets and responsibilities and establishment of multi-stakeholder implementation groups. The aim is to have a coordinated action that enlarges the impact of such projects and guarantees their sustainability in the long term.

- *Promote pilot projects and innovation:* The development of pilot or demonstration projects for geothermal direct use applications is always a good start to identify technical, financial, and technology barriers that need to be addressed before scaling up a project to the industrial or district level. Pilot projects are learning experiences especially in economies where the application of geothermal heat is not yet widely deployed. For instance, pilot projects can contribute to the creation of capacities for technicians on the operation of equipment part of the agri-food processes and contribute to the identification of best options for heating systems in buildings, such as the need to refurbish existing ones or the creation of new ones. Working together with the private sector and the academia, demonstration projects can lead to the creation of local technology and promote cooperation and technology transfer among countries willing to promote geothermal direct use applications.

- *Raise awareness on the benefits of direct use of geothermal resources:* Many people are not very aware of geothermal energy and especially the potential application of such resources for food production or heating/cooling of buildings. In addition, negative public perception of geothermal energy should be addressed, which in some countries includes policymakers and the general public. Information campaigns on the benefits of geothermal energy may facilitate the implementation of projects, as the inhabitants may better understand and be willing to adopt such projects. The development of direct-use projects can boost local economies, provides stable income, and has the potential to strengthen capacities for local women to play a more leading role in their communities.

- *Encourage cooperation, networking, and knowledge sharing:* The development of direct-use projects is occurring in different stages of development within various countries, a situation that can facilitate international cooperation, sharing of experiences, lessons learned and best practices on the application of geothermal resources in the agri-food industry and the buildings sector. Cooperation between countries can entail existing platforms such as south-south and triangular cooperation and promote innovation and technology transfer from economies with a more developed geothermal market. Likewise, the development of networks of experts can facilitate the address of barriers for the development of direct-use projects, create peer-to-peer platforms for technical cooperation, and in some cases, explore options for financial cooperation of pilot projects or technology development

Notes

1 The change in temperature with depth that is higher than the average for a given location due to tec-tonic/volcanic activity or other geological phenomena.
2 This refers to the total heat content of a system – the internal energy of the system plus the product of pressure and volume.
3 (A) Muffler and Cataldi (1978); (B) Hochstein (1990); (C) Benderitter and Cormy (1990); (D) Nicholson (1993); (E) Axelsson and Gunnlaugsson (2000).
4 www.globalgeothermalalliance.org.
5 https://iea-gia.org.
6 https://theargeo.org.
7 https://esmap.org/GGDP.

References

Araujo, K. 2017. *Low Carbon Energy Transitions: Turning Points in National Policy and Innovation*. New York: Oxford University Press.

Axelsson, G. and E. Gunnlaugsson. 2000. *Geothermal Utilization, Management and Monitoring*. Morioka: World Geothermal Congress Short Course, 3–10.

Benderitter, Y. and G. Cormy. 1990. "Possible Approach to Geothermal Research and Relative Cost Estimate." In: Dickson M. H. and Fanelli M. (eds) *Small Geothermal Resources*. Rome, Italy: UNITAR-RJNDP Centre for Small Energy Resources, 61–71.

China Geological Survey. 2018. *China Geothermal Energy Development Report (2018)*. China Petrochemical Press.

Cole, L. 2020. *Climate Change How to Cut the Carbon Emissions from Heating*. London: British Broadcasting Corporation. www.bbc.com/future/article/20201116-climate-change-how-to-cut-the-carbon-emissions-from-heating.

Danube Region Strategy. 2017. *Darlinge Project—New Ways of Thermal Water Utilization*. April. Accessed June 2021. https://energy.danube-region.eu/darlinge-project-new-ways-of-thermal-water-utilization.

Dickson, H. M., and M Fanelli. 2004. *What Is Geothermal Energy?* Pisa: Istituto di Geoscienze e Georisorse.

DOE/EE. 2019. *GeoVision: Harnessing the Heat Beneath our Feet*. U.S. Department of Energy.

EGEC. 2021. *2019 EGEC Geothermal Market Report: Key Findings*. Brussels: European Geothermal Energy Council.

GeoDH. 2015. *Developing Geothermal District Heating in Europe*. Brussels: GeoDH Project.

Geofood. 2019. *GeoFood- Nutrición sostenible*. Accessed June 2021. https://geofood.mx/quienes-somos.

Hochstein, M. P. 1990. "Classification and Assessment of Geothermal Resources." In: Dickson M. H. and Fanelli M. (eds) *Small Geothermal Resources*. Rome, Italy: UNITAEWNDP Centre for Small Energy Resources, 31–59.

HUNOSA. 2020. "Barredo Colliery District Heating."

IDB and IRENA. 2021. *Geothermal: The Solution Underneath – The Value of Geothermal for a Clean Energy Transition*. Abu Dhabi, Washington DC: International Renewable Energy Agency, Inter-American Development Bank.

IRENA. 2019. *Accelerating Geothermal Heat Adoption in the Agri-Food Sector: Key Lessons and Recommendations*. Abu Dhabi: International Renewable Energy Agency.

IRENA. 2020a. *Geothermal Development in Eastern Africa: Recommendations for Power and Direct Use*. Abu Dhabi: International Renewable Energy Agency.

IRENA. 2020b. *Renewable Readiness Assessment: El Salvador*. Abu Dhabi: International Renewable Energy Agency.

IRENA. 2022. *Powering Agri-Food Value Chain with Geothermal Heat: A Guidebook for Policy Makers*. Abu Dhabi: International Renewable Energy Agency.

IRENA and Aalborg University. 2021. *Integrating Low-Temperature Renewables in District Energy Systems*. Abu Dhabi, Aalborg: IRENA, Aalborg University.

Jóhannesson, T., and C. Chatenay. 2014. *Direct Use of Geothermal Resources*. Reykjavik: UNU-GTP.

Kiruja, J. 2011. *Use of Geothermal Energy in Dairy Processing*. Reykjavik: UNU-GTP.

Kiruja, J. 2017. *The Viability of Supplying an Industrial Park with Thermal Energy from Menengai Geothermal Field, Kenya*. Reykjavik: UNU-GTP.

Lund, J. W. and A. N. Toth. 2020. "Direct Utilization of Geothermal Energy 2020 Worldwide Review." *Proceedings World Geothermal Congress 2020*. Reykjavik: World Geothermal Congress.

Mouldi Ben, Mohamed. 2015. "Geothermal Energy Development: the Tunisian Experience." *Proceedings World Geothermal Congress 2015*.

Muffler, P., and R. Cataldi. 1978. "Methods for Regional Assessment of Geothermal Resources." *Geothermics* 7, 53–89.

Nádor et al. 2019. "Geothermal Energy Use, Country Update for Hungary." *European Geothermal Congress 2019*. Den Haag, The Netherlands: European Geothermal Congress.

NDRC. 2017. *13th Five Year Plan for the Development and Utilization of Geothermal Energy*. Beijing: National Development and Reform Commission.

Nicholson, K. N. 1993. *Geothermal Fluids. Chemistry and Exploration Techniques*. Berlin, Heidelberg, New York, London, Paris, Tokyo, Hong Kong, Cambridge University Press xv + 263 pp.

Popovski, K. and S. P. Vasilevska. n.d. "Direct Application of Geothermal Energy."

Richter, A. 2017. *Think Geoenergy – Chinese City of Xiongxian in Hebei Province Deriving all Heating from Geothermal*. April. Accessed June 2021. www.thinkgeoenergy.com/chinese-city-of-xiongxian-in-hebei-province-deriving-all-heating-from-geothermal.

Richter, A. 2018. *Think Geoenergy – Tunisia Increasing Output and Export of Geothermally Grown Tomatoes to European and Gulf Countries*. June. Accessed June 2021. www.thinkgeoenergy.com/tunisia-increasing-output-and-export-of-geothermally-grown-tomatoes-to-european-and-gulf-countries.

Romero, Laura. 2018. *Gaceta UNAM – Primer Deshidratador Geotérmico de Alimentos*. December. Accessed June 2021. www.gaceta.unam.mx/primer-deshidratador-geotermico-de-alimentos.

Rubio-Maya, C. et al. 2015. "Cascade Utilization of Low and Medium Enthalpy Geothermal Resources—A Review." *Renewable and Sustainable Energy Reviews* 52(C), 689–716.

UNAM, Energy Institute. 2020. *Deshidratador Geotérmico de Alimentos (DGA) IIDEA Frutos del Vapor*. April. Accessed June 2021. www.ii.unam.mx/es-mx/AlmacenDigital/Gaceta/GacetaMarzo-Abril2020/Paginas/dga-iidea.aspx.

United Nations Framework Convention On Climate Change (UNFCCC). n.d. *Harvesting Geothermal Energy | El Salvador*. Accessed June 2021. https://unfccc.int/climate-action/momentum-for-change/women-for-results/harvesting-geothermal-energy.

Van Nguyen, M., S. Arason, M. Gissurarson and P. G. Pálsson. 2015. *Uses of Geothermal Energy in Food*. Rome: FAO.

World Energy Council. 2013. *World Energy Resources: Geothermal*. London: World Energy Council. www.worldenergy.org/assets/images/imported/2013/10/WER_2013_9_Geothermal.pdf

Zhu, J. et al. 2015. "A Review of Geothermal Energy Resources, Development, and Applications in China: Current Status and Prospects." *Energy Journal* 93(1), 466–483. https://www.sciencedirect.com/science/article/abs/pii/S0360544215011755.

PART III

Policy, Politics, and Behavior

16

THE ROLE OF CARBON PRICING IN ENERGY-TRANSITIONS POLICY AND RESEARCH

Jeroen van den Bergh and Wouter Botzen

1. Introduction

There is currently much attention for energy transitions, interpreted as major changes in the way of producing, transporting, and utilizing energy, aimed at contributing to an environmentally sustainable and socially equitable economy. Many scientific disciplines contribute insights about policies and institutions needed to foster such transitions, based on theoretical arguments and empirical lessons drawn from ongoing transitions. These insights tend to reflect distinct assumptions and methodological traditions, resulting in considerable disagreement about both the ideal policy package and the core instrument required to enforce major changes in energy-relevant behaviors, practices, organizations, and technologies. We focus here on decarbonization, notably low-carbon behaviors and technologies, while connecting with the broader literature on sustainability transitions, notably its suggestions for policy design. Our concern is that this literature ignores or, at best, downplays carbon pricing. Instead, we forcefully argue in favor of giving this instrument a key role in transition policy, without denying an important complementary role for other instruments. Assessing the latter requires – as we will argue – a careful assessment of the positive and negative synergies of instrument combinations, to which we also devote attention. We evaluate studies that arrive at a negative judgment of carbon pricing to counter the neglect of and aversion against carbon pricing that still exists in many social sciences and that has spilled over to sustainability-transition studies. In addition, we will give attention to the many advantages of the instrument, including overlooked ones such as its unique transparency offering an exclusive opportunity for international or even global harmonization of national climate policies. We will not hide that, like any effective instrument of climate policy, carbon pricing faces serious challenges, but suggest ways to overcome these.

But before we enter into arguments pro and contra carbon pricing, it is important to note that many policymakers have already embraced it. One can even say it is on the rise. Almost 80 jurisdictions in the world have implemented, or are planning to implement, a carbon tax or a cap-and-trade scheme (World Bank, 2020; Haites, 2018). Taken together, they cover some 20% of global greenhouse gas emissions, while about 20 schemes already price emissions above $20/ tCO_2. Internal carbon pricing by private companies and institutions is growing as well (Gillingham et al., 2017; CDP, 2017). In 2018, a carbon pricing proposal was even put on the table of

DOI:10.4324/9781003183020-19

US legislators in the form of the Energy Innovation and Carbon Dividend Act, which enjoyed some bipartisan support across the aisle (HR7173, 2018). Early January 2019, a group of 3,554 economists, including 27 Nobel laureates, four former chairs of the US Federal Reserve, and 15 former chairs of the US Council of Economic Advisers, expressed strong support for carbon pricing and dividends (ESOCD, 2019). Importantly, even prominent conservatives advocate for it (Baker III et al., 2017).

In a recent article with many co-authors (van den Bergh et al., 2020), we argued that international, post–Paris Agreement climate negotiations should capitalize on the recent expansion of carbon pricing to harmonize and strengthen national climate policies. Without harmonization, one cannot expect countries to implement sufficiently stringent and effective policies, as witnessed by the current situation in which actual emission reductions and pledges for the Paris Agreement are grossly insufficient (Roelfsema et al., 2020). As the Paris Agreement did not harmonize policies but instead focused on voluntary and ad hoc national emission targets, countries in effect are faced by the challenge of unilaterally implementing climate policies. This invites for free-riding, which explains the extreme variety in country pledges (nationally determined contributions – NDCs). These can be broadly categorized in four types (King and van den Bergh, 2019).[1] This is likely to translate into huge differences in the stringency of policies among countries, which in turn is likely to generate considerable carbon leakage (King and van den Bergh, 2021). The Paris Agreement did not opt for any joint policy approach, which at the time was the easy way out (i.e., politically expedient), given strong resistance from several main emitters (USA, Russia, China, Australia) along with others (fossil-fuel suppliers like Saudi Arabia) to seriously participate in the agreement. We claim in our study (van den Bergh et al., 2020) that initiatives outside the agreement, notably through a coalition or club of likeminded countries worldwide in terms of climate policy ambition, are urgently needed to overcome this extreme weakness of the Paris Agreement. This seems to be the only option to avoid an impasse characterized by insufficient global emissions reduction from which it is near impossible to escape. To overcome this impasse, we proposed a two-track, five-phase transition approach. This involves the UNFCCC climate negotiations being supplemented by an expanding coalition or club of national and possibly subnational jurisdictions, implementing a uniform or gradually converging carbon price. Formation of a carbon-pricing coalition would enable the coalition to speak with a single, powerful voice at UN climate conferences. In addition, the coalition would put economic and moral pressure on non-members to join the coalition and adopt a constructive attitude in UN negotiations. A stick and carrot consisting of carbon border tax adjustments on imports and access to redistribution of tax revenues would incentivize joining the coalition. Such a growing coalition of ambitious countries would serve as the next stage of the Paris Agreement, necessary to overcome its fundamental flaws.

The focus on carbon pricing in achieving global policy harmonization for the purpose of a feasible path to strengthen national policies is not only logical as carbon prices can be easily compared and harmonized among countries. They will also moderate free-riding and fear of competitiveness losses. Additional advantages are that a global price can be gradually strengthened over time, that it limits national energy and carbon rebound and international carbon leakage, that it automatically generates revenues to compensate low-income households and countries, and that it enables reaching emission reduction objectives at low if not the lowest societal costs.[2] The point that other instruments lack such advantages is insufficiently recognized. For a more complete summary of unique advantages of carbon pricing, see Baranzini et al. (2017).

Incidentally, the effectiveness of carbon pricing is often questioned by critics (Rosenbloom et al. 2020a; Markard and Rosenbloom, 2020) but is clearly shown to be robust by rigorous empirical studies. Sen and Vollebergh (2018) estimate the long-run effect of a broad-based carbon tax on energy consumption for OECD countries. They find that a €1 increase in energy taxes is associated with a reduction of carbon emissions from fossil fuel consumption by 0.73% in the long run. A recent study by Best et al. (2020) uses data for 142 countries over a period of two decades, 43 of which had a carbon price in place at the national or subnational level by the end of the study period. The authors find that the average annual growth rate of CO_2 emissions from fuel combustion has been around 2 percentage points lower in countries with a carbon price. In addition, they estimate that an additional euro per tonne of CO_2 in carbon price is associated with a reduction in the subsequent annual emissions growth rate of approximately 0.3 percentage points. These results are for rather weak carbon prices. Imagine what can be achieved with high carbon prices as proposed by economists and expert panels (HLCCP, 2017; IMF, 2019; van den Bergh and Botzen, 2014), such as above €100 according to these studies, one might expect emissions reduction on the order of 30–70%.[3] Under such a scenario, investments, technologies, firm routines, consumer habits and social influence will all be directed toward low-carbon options, making a full shift to a low-carbon economy realistic. Additional policy instruments, such as information provision, physical and social nudges, and innovation support, can overcome any remaining gaps.

In view of the foregoing, there can be no doubt about the effectiveness of carbon pricing. As a result, it should play a key role in a wider policy package if we want to stand a chance to solve climate change effectively and with minimal economic and social costs. Policymakers have realized this for a while – witness the many carbon prices implemented in countries and subnational states worldwide, with the EU-ETS already harmonizing carbon pricing in 30 countries (Appunn, 2021). In spite of all this, the importance of the instrument does not seem to be embraced by many transition researchers. Here, we offer a self-contained discussion of relevant issues related to carbon pricing and wider policy for a low-carbon transition, motivated by a number of recent developments and discussion in both policy domains. We argue that more integration of insights about carbon pricing is needed to arrive at effective transition policy and clarify that many critics show insufficient understanding of the intricacies of carbon pricing. In addition, we argue that the complementarity and synergy of distinct policy instruments, including carbon pricing, deserves a more rigorous and systemic analysis. We will end by providing a set of concrete suggestions for integrating carbon pricing in ongoing research within the field of sustainability-transition studies.

2. Skepticism About Carbon Pricing

Despite broad-spectrum support by most economists, many climate policy studies by social scientists ignore the unmatched effectiveness of reducing emissions through carbon pricing (Kallbekken et al. 2011; Sorrell, 2015; van den Bergh, 2008; IMF, 2000). This could be called carbon pricing denial. Instead, they tend to suggest some form of bottom–up solution through voluntary and local action (Seddon and Ramanathan, 2013), or a rigid scheme of person carbon limits intended to promote global equity (Fawcett, 2010). As the latter would limit consumption by affluent households, regardless of income potential, it would face considerable political resistance, casting doubt on its political feasibility.

To illustrate, note that many travel organizations, flight companies, and environmental NGOs recommend that air travelers offset their flight emissions voluntarily. However, evidence for countries with relatively environmentally conscious citizens, such as in Sweden, shows that the relative amount of air traffic emissions offset remains negligible, and what is offset usually involves an unrealistically low implicit cost of carbon emissions (Gössling et al., 2009). On the other hand, air travelers are willing to pay more if they know that others pay as well, through a tax imposed on flight emissions (Brouwer et al., 2008). This illustrates a general problem: voluntary action happens as long as it is not too expensive, but even then, it remains rare. It suggests we should put our hopes not on voluntary action to solve climate change but on policies that equally incentivize everyone to reduce carbon emissions. Moreover, voluntary action, or energy sufficiency, tends to result in considerable rebound that reduces its already small impact. In this context, Sorrell et al. (2020) find that sufficiency hardly affects aggregate energy use. One might think of bans and quotas on consumers (e.g., limits to flight frequency or kilometers) as a good alternative option, but these are overly costly in terms of foregone welfare, while there is no reason for optimism about their politically feasibility. An intermediate incentive solution, like carbon pricing, restricts firms and travelers while leaving them freedom on how exactly to reduce their emissions. This contributes to minimizing welfare costs, which together with leaving basic freedoms intact, contributes to political feasibility in a democracy – indeed, people will not easily vote for bans that interfere with their personal consumption.

If we consider other instruments, such as information provision or nudges, one can see broad support and little opposition, arguably because people consider their personal and administrative costs to be limited, as they leave people complete freedom regarding consumption and pollution and as they affect everyone equally (Drews and van den Bergh, 2016). However, this is matched by a low effectiveness in terms of emissions reduction – namely, on the order of 5–10% of prevailing emissions, according to various meta-analyses and reviews (Delmas et al. 2013; Andor and Fels, 2018; Wynes et al., 2018). Fortunately, such instruments can serve a complementary role by creating positive synergy with pricing or standards. One cannot count on this, though, as empirical and experimental studies offer mixed evidence on synergy – which can be as well absent or negative, for example, in case nudges backfire (Drews et al., 2020).

Although it is not uncommon to hear condescending remarks about carbon pricing in meetings and personal communications, few have written them down in an argumentative way. An exception is Ball (2018a), who seems to think that carbon pricing is about regulating industries. But this represents a clear misinterpretation of the literature on and proposals about carbon pricing. It is about a systemic change, a fundamental correction of all prices in the economy, by charging all fossil fuels for carbon content. This will seep through all prices of intermediate goods, capital goods, and final goods and services, affecting choices by all types of agents in the economy – consumers, producers, investors, and innovators. In another article, the author criticizes implemented carbon prices for being low and not covering all emissions (Ball, 2018b). We agree, but this is no reason to set aside carbon pricing as an instrument. That many implementations have not followed the textbook advice is a reason to improve, not to give up. Moreover, the author should then be equally critical of other instruments that are also imperfect, weak, and ad hoc in terms of implementation, explaining why climate policies worldwide have not achieved serious emissions reduction. The absence of significant climate-change mitigation efforts is more a reflection of a strong political resistance against stringent climate policy actions in general, instead of being a problem specific to carbon pricing. One needs also to take into account that lobbying is easier for ad hoc climate policies focused on one sector, such as through negotiated technology-specific standards for which many exceptions and alternative technological specifications are possible, than for economy-wide policies, such as a single carbon price. The only way so far we have been successful in integrating and harmonizing national policies has been through carbon pricing, notably carbon markets – with the EU as a prime example.

A widespread idea among social scientists is that ecolabels, supported by life cycle analysis, will allow consumers to voluntarily reduce their carbon footprint (Baldo et al., 2009). However, limited human capacities of information processing and altruism means this approach cannot be counted on to achieve considerable emissions reduction (Waechter et al., 2015). In achieving local, bottom-up climate solutions, cities are also frequently mentioned (Watts, 2017). Not denying their potential contribution, one must realize that cities only exert direct control over a small portion of total emissions generated by industry, electricity production and consumption, and transport, while their control over emissions caused by agricultural production and land use change (deforestation) is largely absent (Satterthwaite, 2008). Moreover, the implementation of uncoordinated policies at the city level may generate carbon leakage. Complementing city strategies with carbon pricing will reduce their detrimental systemic effects, which will improve their effectiveness (van den Bergh, 2020). These lessons about the effectiveness of carbon pricing are also relevant for a vibrant branch of social science research that examined local, often urban-scale, experimental initiatives for sustainability transitions (Bernstein and Hoffmann, 2018; Fuenfschilling et al., 2018; Grandin et al., 2018). However, these initiatives have been criticized for often being temporary (Grandin and Sareen, 2020), non-binding (Biermann et al., 2017), and challenging to diffuse and upscale (Naber et al., 2017). Carbon pricing could overcome these shortcomings by acting as a long-term incentive for sustaining, diffusing and upscaling of niche experiments that have proven locally successful in reducing carbon emissions.

Subsidies for research and deployment of new technologies, such as electric vehicles[4], are often assumed to contribute to reducing emissions. However, without carbon pricing we cannot ensure that the full life cycle of new innovations will actually use less carbon (Popp, 2006). For instance, the production cycles of batteries for electric vehicles or solar PV panels might be unnecessarily intensive in carbon dioxide emissions, often relying on cheap coal power for manufacture, delaying a low-carbon transition. More generally, production of cleaner technologies generates emissions in an economy that is still running mainly on fossil fuel energy. To limit the carbon intensity of such production, subsidies fall short – we need to penalize the dirty next to rewarding the clean if we aim for a quick low-carbon transition. One can see this by considering the following Kaya identity (Kaya and Yokoburi, 1997): CO_2 *emissions = carbon intensity of energy (CO_2/energy)* × *energy intensity of economy (energy/GDP)* × *income level (GDP/population)* × *population*.

Subsidizing renewables will only affect the first factor (i.e., the *carbon intensity of energy*), while a carbon price will simultaneously influence the first and second factors (i.e., also the *energy intensity of economy*). Subsidizing R&D of clean technologies can still be desirable for various reasons: capturing positive externalities of knowledge spillovers, keeping trajectories of promising but still expensive options open, and accelerating learning curves. But subsidies should be complemented with carbon pricing as otherwise emissions reduction will go far too slow.

In a very contentious article, Spash (2010) offers a very radical and ideological critique. He argues that claims by economists about the cost-effectiveness of carbon emission trading are not substantiated based on arguments that focus on uncertain marginal emission abatement costs, lobbying affecting the design of emission markets, and markets crowding out voluntary actions. Uncertain abatement costs are generally seen as a reason for favoring carbon taxes as a carbon pricing instrument over emission trading (as the latter can lead to high costs for firms), but not as a reason for preferring regulatory instruments over carbon pricing (Weitzman, 1974). Moreover, arguments by Spash that oppose emission trading because of lobbying are not convincing. We agree that lobbying can result in suboptimal emission control, but lobbying is likely to be worse with alternative climate policy instruments that are more sector-tailored as well as more costly for firms than carbon pricing. No convincing empirical evidence is given by Spash showing that carbon markets have a net effect of crowding out voluntary carbon reductions. A meaningful crowding-out effect would contradict the aforementioned empirical evidence of emission reductions achieved because of carbon prices. Moreover, this argument does not recognize the limited potential of voluntary action in reducing emissions in the first place. Spash argues that direct regulations of emissions are much easier to implement. This reasoning neglects that almost all consumption and production decisions in reality involve carbon emissions, and regulating all involved technologies would be a huge, if not impossible, task. Spash concludes that emission markets distract from the need to change human behavior, institutions, and infrastructure, but he does not detail the policies that should be put in place to trigger such a change. Carbon pricing that alters all relative prices of goods and services in the economy based on their carbon content is the only instrument that is able to have a widespread simultaneous influence on decisions by consumers, producers, and investors in an economic system. Such widespread control is a good basis for setting in motion the required changes to move toward a low-carbon economy.

3. Responding to Criticism of Carbon Pricing from a Sustainability-Transitions Perspective

Recently, Rosenbloom et al. (2020a) wrote a critical article on carbon pricing[5] in relation to transition policy in which they downplay the role of carbon pricing in mitigating climate change and transitioning to a low-carbon economy. Despite the sympathetic title of their contribution, the authors say nothing positive about carbon pricing, instead emphasizing five supposed shortcomings. As this is the first and only published criticism, it is worthwhile to devote some attention to it here. This elaborates several points only touched upon in a necessarily brief response by van den Bergh and Botzen (2020), defending carbon pricing as having an essential and irreplaceable role in a wider policy package aimed at fostering a low-carbon economy. Moreover, in our argument we react to Rosenbloom et al. (2020b), which to a large degree repeats Rosenbloom et (2020a) but in addition raises some new arguments as a response to van den Bergh and Botzen (2020). Our hope is to contribute to a more nuanced, theoretically informed, and evidence-based perspective on carbon pricing.

3.1 Market Failure Versus System Problem

According to Rosenbloom et al. (2020a), framing climate change as a market failure fails to seriously appreciate its scope and depth. They suggest that it is better understood as a system problem. Unfortunately, they never enter into any details, and hence overlook that "market failures" is a broad category that includes, among others, negative externalities of carbon emissions, positive externalities of innovation and knowledge generation, the public-bad nature of climate change, and the public-good nature of international policy coordination and agreement-formation. All of these are systemic issues central to the economic theory underlying carbon pricing (Perman et al., 2011). This underpins what economists well recognize as the systemic global nature of the climate-change externality that originates from consumption, production, and investment decisions in a large diversity of markets around the world (Stern, 2007; Aldy et al., 2010; Cramton et al., 2017). Hence, its complexity is not downplayed by framing it as market failure, as Rosenbloom et al. (2020a) argue.

Carbon pricing is moreover a systemic policy that matches well a systemic problem like climate change. Indeed, carbon emissions are generated by all kinds of productive, consumptive, investment, and transport activities. Moreover, who causes carbon emissions or where this occurs does not matter for the global warming effect. This feature strongly supports charging a uniform price on emissions.

An advantage of carbon pricing is that one can implement it in relatively few sectors – namely, exploration and imports of fossil fuels – which then affects all other prices of goods, services, intermediate products, materials, and electricity to signal societal costs of direct and indirect emissions over their respective product life cycles. No other instrument is capable of achieving such consistent and precise system-wide control. It would shift choices by consumers, producers, investors, and innovators in all sectors to low-carbon inputs, outputs, and processes. This means it reaches everyone and every decision in the economy, without discrimination.

We agree with Rosenbloom (2020a) that lock-in of high-carbon technologies and practices is a serious system challenge. There are distinct types of lock-in, such as related to demand or supply sides and to networks or complementary technologies or infrastructures. Incidentally, network lock-in is often conceptualized as an externality. Each lock-in type requires a particular policy, as recognized in both economics and innovation studies (Seto et al., 2016). A historical absence of high carbon prices has contributed to this lock-in by unintentionally steering investments toward high-carbon production and consumption, which, in contrast to arguments in Rosenbloom et al. (2020a,b), suggests that carbon pricing should be part of a policy mix to unlock these investments.

We disagree with the argument by Rosenbloom (2020b) that infrastructure, technological capacity, and routinized consumption practices are difficult to change with carbon pricing. One should realize that a sufficiently high carbon price could in principle unlock any high-carbon technology or practice. In fact, behavioral changes triggered throughout the economy by a carbon price can amplify through social interactions such as conformism, imitation and status-seeking (Konc et al., 2020), creating increasing returns that counteract the increasing returns underlying the existing lock-in of high-carbon options. By combining carbon pricing and unlocking policies, such as innovation support and information provision, one can escape lock-in using lower carbon prices. However, it is improbable that these alternative policies alone (i.e., without price signals) can achieve the large-scale reallocation of investments needed to escape lock-in of high-carbon infrastructure and technologies and achieve the sociotechnical transition Rosenbloom (2020b) aims for.

3.2 Efficiency Versus Effectiveness

Rosenbloom et al. (2020a) suggest that carbon pricing means that efficiency is an overriding priority of climate policy. It is good to realize that a policy can only be cost-effective or efficient if it is effective in the first place, meaning that efficiency and effectiveness are not necessarily a trade-off as Rosenbloom et al. (2020a) argue. An ineffective policy will always be overly expensive, as it does not achieve much for a given effort. Probably, economic writings have not stressed this sufficiently and we should better clarify that carbon pricing is also among, if not *the* most effective instrument – on its own and especially if well combined with other instruments.

The effectiveness of carbon pricing is due to the fact that no decision in the economy escapes it influence, as already clarified in Section 3.1. It will steer both purchase and use decisions and affect strategies by investors and innovators. It will work like an instrument that fills all the holes where carbon emissions could escape. A very important reason that carbon pricing is so effective, in comparison with other instruments, is that it can control energy and carbon rebound (Baranzini et al., 2017). This involves direct rebound due to more intense use of energy-efficient technologies, by pricing energy proportionally to carbon content. Regarding the challenge of avoiding indirect rebound, carbon pricing has the unique advantage of discouraging money savings due to energy conservation being spent on high-carbon goods and services. The reason is that it makes these goods and services relatively expensive compared with low-carbon alternatives. Regulatory measures, like emission standards proposed by Rosenbloom et al. (2020a) do not limit rebound effects, unless all production technologies and consumption goods that contribute to carbon emissions are regulated. This would require a huge set of standards – which likely will be inconsistent in terms of implicit carbon prices, hence resulting in much higher costs for society for the same emissions reduction. This said, it seems Rosenbloom et al. (2020a) do not value efficiency much. Inefficient policies contribute, however, to less emissions reduction than is feasible with carbon pricing. To limit rebound these regulations would also require continuous updating over time to account for any changes in technologies and consumer preferences, which is practically impossible.

This said, it is our impression that efficiency and cost-effectiveness are regarded as fairly unimportant criteria by Rosenbloom et al. (2020a). This is unfortunate, as it ignores that efficiency translates to higher employment and household income (Rengs et al., 2020), more emissions reduction for the same money, a higher government budget for distributional compensation, and – as a result – also more public support, while inefficient policies imply the opposite, in turn hampering stable political support. Efficiency is important in all times, but especially in the coming decade when rapid cuts in carbon emissions are needed and at the same time government policies will be focused on limiting impacts from the economic recession that is predicted to follow the COVID-19 pandemic (IMF, 2020).

3.3 *Optimizing Versus Transforming*

Rosenbloom et al. (2020a) suggest that carbon pricing cannot transform systems. Their statement is void of any proof. This is not surprising: if all purchase and use decisions by consumers and firms, investors and innovators in all production sectors are affected by a serious carbon price, transformation is likely to come about. Additional instruments – notably to support innovation and escape of lock-in, will help. In this respect, we do not disagree with Rosenbloom et al. (2020b) that technology-specific deployment policies and green industrial policies have contributed to low-carbon solutions, such as photovoltaics, wind energy, and electric vehicles. However, the historical absence of high carbon prices has implied that these innovations in low-carbon solutions and their uptake have been too slow for solving the climate problem. The systemic nature of carbon pricing provides more certainty that demand, supply, adoption, and innovation decisions are altered in concert, which seems to offer a pretty good starting point for a major transition.

Rosenbloom et al. (2020a, b) further overlook the critical role of carbon pricing in realizing low-carbon innovations. In fact, carbon pricing contributes to steering low-carbon innovations of all kinds. Such innovation patterns will not unfold as quickly as is possible and neither in the right direction unless one implements uniform and serious carbon pricing. The reason is that many private innovators and investors are driven by expectations about prices as these co-determine future profit opportunities. In view of this, carbon pricing would help steer the direction of innovations toward energy-efficient and low-carbon production life cycles (Aghion et al., 2016; Calel and Dechezleprêtre, 2016). Also the older evidence on energy prices clearly affecting the direction and speed of energy innovations is highly relevant given the close association between energy use and carbon emissions and between energy and carbon prices (e.g., Popp, 2002). For a recent evaluation of the broader literature on this topic of low-carbon innovation, see van den Bergh and Savin (2021).

The opinion that a carbon price only affects incremental innovations is debatable. Model studies indicate that a sufficiently high carbon price is able to enforce large changes in the economy (Jorgenson et al., 2009; Rengs et al., 2020). Carbon prices have the effect that low-cost solutions for reducing carbon are taken first as Rosenbloom et al. (2020a) rightly argue, and many would see as an attractive feature instead of a disadvantage. However, the authors fail to recognize that more expensive transformations become attractive when sufficiently high carbon prices are implemented. Anyway, most radical changes, if analyzed well, turn out to be composed of many incremental changes that happened in a relatively short period of time or take the form of combining preselected technological modules. Nothing prohibits carbon pricing from triggering either type of innovation (van den Bergh, 2013a).

In addition, both economics and innovation studies recognize that it is important to support promising but still expensive technologies. This may be done through R&D subsidies to account for knowledge spillover effects, which avoid foreclosing technological trajectories too early by a high carbon price (Jaffe et al., 2005).

Unlike other instruments, carbon pricing would be able to highlight carbon differences between the more and less "clean" technologies: for example, solar PV panels produced with different processes or electricity using distinct energy sources. It is difficult to know which products or technologies are more low-carbon over their life cycle as production processes are complex and roundabout, involving many intermediate deliveries between firms and sectors. Through a cumulative carbon price signal, the high-carbon options would be effectively discouraged, which would be very challenging, if not impossible, to accomplish with regulatory instruments only (Liu and van den Bergh, 2020).

Finally, Rosenbloom et al. (2020a) suggest that we need policies that discourage carbon-intensive technologies and policies that encourage low-carbon innovations, suggesting that the latter are innovation policies and the first so-called "decline policies" (note that, here, the authors surprisingly suggest an entirely new and unclear term, the difference of which with the traditional term "regulatory policies" remains unclear), without assigning a concrete and specific role to carbon pricing. Moreover, this focus on technologies overlooks the link with demand. More importantly, the authors overlook that carbon pricing has both effects: it reduces use of high-carbon technologies and goods/services, and it promotes the use of low-carbon ones. Again, we agree support of low-carbon technologies is warranted under certain conditions, mainly through investment in public R&D, subsidizing cleverly relevant private R&D (when it clearly falls short), and (with moderation) subsidizing adoption of low-carbon technologies by firms and consumers. However, pricing the high-carbon options is generally better as it equally closes the gap between prices of low-carbon and high-carbon alternatives, but in a correct way — namely, by punishing the polluter and not rewarding the adopter. Subsidies easily lead to expansion of energy use rather than substitution of high- by low-carbon options as they lower the cost of energy production and use.

3.4 Universal Versus Context-Sensitive Policy

Rosenbloom et al. (2020a) prefer a context-sensitive over a universal approach. Two comments are relevant here. First, while not denying that attention for specific sector context can be relevant, such as agriculture policies that stimulate sustainable food production, sector-specific policies or strategies run the risk of being inconsistent and overly costly. This will translate in leakage among sectors and in high costs and possibly unemployment, respectively. The costly nature of this approach is due to sector-specific policies not guaranteeing that the cheapest emission-reduction options in the economy are selected. Instead, arbitrary goals (considered fair somehow) are set for sectors, which will result in distinct marginal and average costs of emissions reduction between them.

Second, approaches tend to be ad hoc, costly, and susceptible to lobbying, while causing inter-sectoral carbon leakage. Moreover, climate policy is bound to remain weak whenever implemented in a unilateral manner, that is, without coordination between jurisdictions, from cities through provinces to countries. The Paris Agreement was focused on voluntary pledges rather than coordinated policies. We need policy harmonization to overcome free-riding and concerns about competition and exports by national governments. Carbon pricing is our single hope to achieve policy harmonization. Its universal approach of putting a monetary price on CO_2 makes it relatively easy to harmonize, as shown by the various carbon markets around the world that cover multiple countries or provinces/states (World Bank, 2020). Also, carbon taxation has been convincingly argued to satisfy these advantages (Weitzman, 2014). Other instruments are less easily harmonized: for instance, technical standards for millions of products and technologies are difficult to coordinate, while countries with high stakes in certain industries will resist associated standards (e.g., countries with an important car industry will fight ambitious fuel-efficiency standards). Harmonization will not only allow for a gradual increase in the stringency of national policies but also discourage carbon leakage. This means that emissions move from countries with strong to ones with weak climate policies due to shifts in international trade and relocation of emitting firms. It is surprising that Rosenbloom et al. (2020a) highlight that carbon leakage is a problematic aspect of non-uniform carbon pricing but fail to recognize that this policy coordination problem is likely to be worse with alternative regulatory policies. This also applies to other related arguments, like the absence of high levels of regulatory competences and monitoring systems in some countries, which would hamper implementing any type of climate policy. More generally, despite Rosenbloom et al. (2020a)'s first criticism (Section 3.1), the previous discussion suggests that the field of sustainability transitions lacks a genuine systemic approach that accounts for shifts and leakages between sectors, regions and countries.

It is good to add that especially fossil-fuel supplying countries will not come on board easily and, so far, have resisted a good climate deal. Hence, the way forward is not a full participatory agreement immediately, as Rosenbloom (2020b) interpreted our argument in van den Bergh and Botzen (2020) by saying our assumptions require all countries implement a unified carbon-pricing framework. Instead, we propose a climate coalition or club of ambitious countries with a uniform carbon price and border carbon tariffs, which can put economic and moral pressure on non-members to join, leading to club expansion over time (Nordhaus, 2015; Victor, 2015). This is elaborated, extended, and generalized for a carbon tax and carbon market (cap-and-trade or emissions trading) in van den Bergh et al. (2020).

3.5 Political Realities

Regarding political realities, Rosenbloom et al. (2020a) suggest that carbon pricing faces a lot of resistance. Several other authors express the same view (Cullenward and Victor, 2020). Three considerations nuance this. First, all serious climate policies meet strong political resistance. In line with this, voters and politicians are attracted to less effective strategies and policies, such as information provision to trigger voluntary action or subsidies for solar PV and electric vehicles. Second, Rosenbloom et al. (2020a,b) do not provide any evidence that policy instruments other than carbon pricing, with similar effectiveness, can count on more political support. This is not surprising since even if other policy instruments would be as effective, they will be more limiting, costly and economically harmful, which is unlikely to appeal to lobbyists and voters (Baranzini et al., 2017). Third, carbon pricing is in fact already quite popular among policy instruments. Almost 60 jurisdictions have implemented some form of it (Haites, 2018). The 27 countries of the European Union (along with additional countries) even have a joint carbon emissions trading system (ETS)[6] that had a price between €50 and €60 per ton of CO_2 for most of 2021 and reached a value above €80 in 2022. Of course, the prices of current carbon taxes and ETS are still low due to climate policy lacking ambitions and being still unilateral in nature. An exception is the EU-ETS due to harmonizing carbon prices among 30 countries. That is exactly why we need more integration and harmonization worldwide. A carbon price has a clear advantage over other instruments in this respect (van den Bergh et al., 2020).

3.6 Other Issues

It is not true, as Rosenbloom et al. (2020a) suggest, that carbon pricing is only supported by neoclassical economics and rational-agent assumptions. Many different types of empirical studies provide evidence for its capacity and effectiveness to enforce major changes that translate in significant emissions reduction (see references in Narassimhan et al., 2017). In addition, various agent-based models that allow for heterogeneous boundedly rational and socially sensitive behavior have studied carbon pricing, notably carbon markets, and provide further support for its effectiveness (Castro et al., 2020). However, regarding the effectiveness of transition policies, the jury is still out, as evidence here is thin, conceptual, and anecdotal. This suggests a need for modesty in policy advice and application of research methods that provide stronger evidence.

So far, most applications of renewable energy do not replace fossil fuels but just add to growing energy demand. As long as we subsidize renewables but fail to implement a serious carbon price, this demand will continue to grow, and fossil/renewable substitution will remain disappointing. So we need not just a carrot for low-carbon options but also a stick for high-carbon ones.

Finally, many of the examples by Rosenbloom et al. (2020a) refer to changes in production or household sectors (e.g., "restoring peatlands", "mobility-as-a-service", "biobased materials") without clarifying how these are triggered by concrete policies. The same critique applies to the "green industrial policy" Rosenbloom et al. (2020b) call for. Such an approach falls short of arguing that carbon pricing is insufficient. Instead of talking about policy consequences, as Rosenbloom et al. (2020a) do, they would do better to directly compare the alternative policy instruments or mixes and their performance.

4. Positive Synergy of Carbon Pricing and Other Instruments in the Policy Mix

Rosenbloom et al. (2020a) never clarify if and how carbon pricing can complement other instruments in a broader transition policy. Neither we nor most economists claim that carbon pricing should be the only instrument of climate policy or low-carbon transition policy. However, there is a tendency in transition studies to prefer many policy instruments, not always with a very solid analysis of synergies underpinning it (Kivimaa and Kern, 2016; Rogge and Reichardt, 2016; Howlett et al., 2017; Rogge et al., 2017). These studies often, ironically, ignore or give little weight to pricing. The same preference for many policy instruments can be seen in policy practice. To illustrate, the German government recently agreed on a climate policy package covering dozens of sectoral measures (Edenhofer et al., 2019).

Reasons for and against using multiple instruments are multifold. Important arguments in favor are (Lehmann, 2012; van den Bergh, 2013b): accomplishing multiple criteria, such as effectiveness, efficiency and equity; complementariness or even positive synergy in terms of the set goal (e.g., emissions reduction); or addressing distinct market failures, such as negative environmental and positive innovation externalities (Jaffe et al., 2005). In more abstract terms, a policy mix often reflects a second-best (non-optimal) response to a first-best (theoretically optimal) single instrument not being feasible – due to political constraints or imperfect monitoring and hence compliance with policy (Bennear and Stavins, 2007). A policy mix can also result from political compromises between stakeholders with distinct policy preferences, or from a political strategy to camouflage insufficiency of core policy. Indeed, complex transition policy mixes run a serious risk that they focus on soft policies and lack destabilizing policies. This is, in fact, a finding of various policy assessments, including for Finland and the UK (Kivimaa and Kern, 2016).

Arguments to limit the number of instruments in a policy mix include (van den Bergh et al., 2021): instruments overlap or create negative synergies; the risk of introducing potentially multiple distortions into the economy; each policy instrument generates a cost for the government in terms of human resources, transaction costs of political and policy processes until implementation, costs of monitoring and control, and sometimes serious budgetary sacrifices such as with subsidies. In the context of a global challenge, like climate policy, an additional concern is that policy stringency is comparable among regions and countries, which is not trivial (Schmidt and Sewerin, 2019) and which, in turn, may limit international policy harmonization (Howlett et al., 2017).

A variety of disciplines offer insights on reasons for and composition of policy mixes: economics, psychology, policy sciences, innovation studies, and more recently transition studies (Jaffe et al., 2005; Bulkeley and Kern, 2006; Howlett and Rayner, 2007; Rogge and Reichardt, 2016; Rogge et al., 2017; Mundaca et al., 2019). However, they do not agree and tend to emphasize distinct criteria and motivations.

According to a recent survey of policy mixes by van den Bergh et al. (2021), the clearest evidence for positive synergy is based on theoretical modeling and experimental studies. The reason is that these, unlike empirical studies, allow clearly separating between effects caused by each particular instrument. The literature suggests that combining a carbon tax with other regulatory instruments, such as sector-level targets or technical standards, has an advantage over doing this with carbon markets. The latter weaken a good functioning (i.e., a high price) and thus the effectiveness of such markets and may give rise to carbon leakage between sectors (Fankhauser et al., 2010). To understand this, note that (through other regulatory instruments) particular sectors reduce more than they would do with a carbon market only and thus demand fewer permits, causing the permit price to drop, in turn allowing other sectors to pollute more. Likewise, combining carbon markets with stringent renewable-energy targets or quota can generate carbon leakage. Adoption subsidies are generally better combined with a carbon tax than market as they negatively affect the carbon price and hence reduce the effectiveness of the market. With carbon taxes, this problem does not appear as the carbon price is exogenous. Two other important instrument types – innovation support and information provision – tend to be complementary to other instruments. Apart from what was said about information provision in Section 2, certain types of information provision can create positive synergy by reinforcing the effectiveness of regulation by prices through social network functioning, raising the social multiplier of carbon pricing (Konc et al., 2020). Innovation support, such as with R&D subsidies, can counter the short-term selection pressure against promising but still immature and thus expensive technologies created by regulatory instruments, like standards, targets, or carbon pricing, while also serving to enhance or speed up escape from lock-in of high-carbon technologies or practices. Other subtle issues of innovation support are well recognized in transition studies (Rosenbloom et al., 2020a).

A climate policy package that combines a carbon tax, adoption subsidies, innovation support, and information provision scores well on the criteria of effectiveness, efficiency, and potential for international harmonization, and also on equity if revenues are recycled inversely proportional with income. Another relevant policy mix is a carbon market with innovation support and information provision. This is possibly a better approach in the long run as international harmonization is easier given that this mix is simpler in structure: namely, it omits adoption subsidies since these interfere with the intended functioning of the carbon market. The great advantage is, then, that it in turn allows policy to be strengthened more easily over time.

Regarding the specific role of other instruments in this policy mix, information provision can garner understanding of, and support for, carbon pricing. Behavioral nudges can address informational failures and bounded rationality, for example, by presenting a low-carbon product as the default option for consumers (Ebeling and Lotz, 2015). So far, the discussion implicitly focuses on energy-related CO_2 or other greenhouse gas (CO_2-equivalent) emissions. In addition, non-price regulatory instruments are needed to control certain non-energy greenhouse gas emissions as from land conversion, deforestation, and landfills. Innovation policies are required as well, to ensure further development of promising low-carbon technologies that are still too expensive to compete in markets. The main justification of public sector support is well-known – namely, that R&D has positive externalities and knowledge spillovers. But it cannot address negative environmental externalities – for this externality (i.e., carbon), pricing is the most effective climate policy. As a corollary, subsidies for technological innovation and adoption cannot stand alone.

Whereas carbon pricing was in the past often criticized for being inequitable, there is now wide agreement that it should be complemented by equitable revenue recycling, for both ethical reasons and political feasibility (Klenert et al., 2018). In fact, no other instrument allows for such automatic revenue compensation: technical standards, performance targets, and quota impose costs on economic agents and have distributional effects but do not generate revenues, while subsidies use up revenues. Moreover, adoption subsidies for renewable energy and electric vehicle are rather inequitable as they tend to fall on relatively well-off households. The revenue-generating capacity of carbon pricing is increasingly relevant now that many governments around the world see their already high debt burdens rise rapidly. Moreover, pressures on employment and incomes strengthen the need to compensate households and companies for costs from additional climate policy measures. The equity issue has seen more conceptual and empirical elaboration for carbon pricing than for other instruments (Klenert et al., 2018; Maestre-Andrés et al., 2019). This is somewhat surprising, given the expressed concern for equity by many social and policy scientists who seem more charmed about other instruments than carbon pricing.

Ironically, many writings on transition policy are not consistent with the basic idea of a gradual, multi-phase transition covering the stages of pre-development, take-off, acceleration, and stabilization (Rotmans et al., 2001). The ambitious, complex policy mixes suggested are approached more from a static design perspective than from a dynamic and multi-phase transition angle. Instead, the literature on carbon pricing spends a great deal of attention on dynamics, adaptation and strengthening of prices of time (or tightening caps in emission trading systems), and even improving feasibility through a transition path. Regarding the latter, van den Bergh et al. (2020) propose a dual-track, five-phase transition approach to progressively harmonize and strengthen national climate-mitigation policies.

5. Conclusions and Research Suggestions

Sustainability-transition studies tend to ignore carbon pricing altogether (Savin and van den Bergh, 2021). As we hope to have shown, this is unwarranted. Carbon pricing has unmatched advantages, including consistent regulation of direct and indirect emissions (covering also adoption and use behavior), limiting rebound and life-cycle effects, selecting for low-cost abatement options, generating revenues for compensating energy-poverty and other inequity effects, contributing to transparency and comparability between policies in distinct jurisdictions, and facilitating international harmonization as witnessed by the EU-ETS (i.e., it harmonizes carbon prices in 30 countries). This clearly indicates that carbon pricing has a unique and powerful capacity to effectively reduce emissions with minimal economic harm – in terms of costs, macroeconomic effects like employment, and income or purchase equity. The implication is that the function of carbon pricing cannot be taken over by any other instrument – something that is insufficiently valued in transitions research. Soft instruments like information provision or networking, aimed at encouraging voluntary action or fostering community initiatives, will only create significant effects where alternative sustainable options for behavior and technology are not burdensome or expensive but will fall short in achieving deep decarbonization. In addition, large-scale bans and quotas on energy use not only will severely harm production and welfare but are politically infeasible in a democracy. Carbon pricing represents an intermediate solution that restricts firms and consumers in a gentle but effective way through clear price incentives, leaving them sufficient freedom to decide on how to reduce their emissions. In addition, carbon pricing of fossil fuels will seep through the economy to affect all prices of intermediate and final goods and services and, hence, all decisions by consumers, producers, investors and service providers – and hence, nothing will escape its influence, which also means that social and economic harm will be maximally spread rather than concentrate in a few sectors or social groups.

Summarizing our assessment of the studies that are negative about carbon pricing (where we have not tried to be exhaustive since many studies/authors repeat the same arguments), a first observation is that most of these are qualitative in nature instead of based on methods like modeling and statistical data analysis, which arguably provide more definite and thorough insights. In addition, critics tend to offer unbalanced accounts where positive aspects of carbon pricing are ignored or downplayed and shortcomings are exaggerated. They also often overlook that certain shortcomings – notably, political resistance, inequity effects, and barriers to strengthening policy over time – apply equally or more strongly to other instruments. Reading into these studies, we get a strong feeling that authors arrive at their conclusions based on an incomplete assessment of instruments and criteria rather than by comparing systematically pros and cons of all instruments. Worse even, we find that some critics of carbon pricing show a superficial or erroneous knowledge of it. This all evidently contributes to unfounded and unfair resistance against carbon pricing, in turn hampering rather than promoting high-quality debate about how to solve climate change. We hope with this chapter has clarified many issues surrounding carbon pricing and its connection with sustainability transitions and transition policy and has provided an entry into the broader and rich theoretical and empirical literature on carbon pricing.

All these considerations do not deny that we need a climate-policy package. Van den Bergh et al. (2021) examine how this can be best achieved, by employing positive and avoiding negative synergies between instruments. Innovation support on its own will not change consumer and firm behavior quickly and sufficiently; technical standards or adoption subsidies for energy efficiency alone may, due to a focus on purchase decisions, lead to much energy and carbon rebound, which can only be limited by carbon pricing as it controls the use phase; and information provision alone will according to meta-analyses and surveys reduce energy use and emissions at most with 5–10% (Delmas et al., 2013; Andor and Fels, 2018; Wynes et al., 2018). Therefore, carbon pricing should be a key element of energy-transition policy. The reason is that the fundamental transformation needed is unlikely to be achieved through niche experimentation in the absence of carbon pricing. So-called "new business models for new technologies", which Rosenbloom et al. (2020a) mention, without detailing how they should be achieved, are unlikely to scale up toward a large-scale transition to a low-carbon economy without serious financial incentives that make these business models and low-carbon technologies economically attractive for investors and clients.

Our countercriticism does not deny the creative ideas one finds in the literature on transitions thinking about policy. It is surprising, though, that it has not undertaken a serious effort to integrate important and broadly supported insights from more established policy sciences such as environmental economics. This presents a challenge for future research. Based on the previous information, one can identify various research suggestions for transitions studies:

- We should give attention to a transition of policies themselves – that is, how to achieve a more stringent and effective policy mix over time. Some ideas are provided in van den Bergh et al. (2020) about how to do this for carbon pricing in a multistage and dual-track stetting with interactions between an expanding climate club and international (UNFCCC) climate negotiations. This could receive attention from different perspectives, using particular transition theories, such as innovations systems, multi-level perspective, strategic niche management, complex systems, and evolutionary systems (Nill and Kemp, 2009; Rotmans and Loorbach, 2009; Jacobsson and Bergek, 2011; Geels, 2011; Safarzynska et al., 2012).
- Regarding the mix of policies, a widely accepted division into four main modes of urban climate governance is as follows (Bulkeley and Kern, 2006): (1) self-governance of urban public sector activities; (2) provision of public services, such as public transport; (3) enabling emissions reduction by firms and households, such as through information or adoption subsidies; and (4) regulation of firms and households, such as through zoning, standards, or carbon pricing. Transition policy could be clarified along these lines, and one could test if there is sufficient balance between the four modes at different points over the transition path. This should take into account the lessons from the literature as sketched in Section 4.
- A systemic approach to assess effectiveness of transition policy is missing yet. It was recently proposed to decompose effectiveness of emissions reduction into reach, ability, and stringency (van den Bergh, 2020). Assessment of the effectiveness of transition policy could build on this decomposition approach and possibly extend it with dynamic elements to capture the role of essential environmental technological innovations and slow societal transition processes.
- One could connect specific instruments in the transition policy mix to the different levels of action, systemic processes, and governance, such as specified in the multi-level perspective (MLP) framework (niche, regime, and landscape), economics (micro, meso, and macro), or other fields (local, regional, and global). Here, the role of carbon pricing should be clarified – for example, would it be relevant to multiple levels or not?
- A challenge of local niche experimentation with mitigation projects is that scalability is limited but essential to effectively reduce greenhouse gas emissions sufficiently across the economy and space. Future studies should give serious attention to whether and how carbon pricing can provide adequate financial incentives for the upscaling of niche projects.
- Finally, more attention is needed for the feasibility and international harmonization of effective transition policy, to assure that it is comparable between regions and countries so that it can be consistent in terms of regulatory strength. If this is not achieved, it is very likely that transition policies will remain too weak to solve climate change.

Acknowledgment

The authors are grateful to Stefan Drews, Ivan Savin, and the editor Kathy Araujo for their helpful comments. Jeroen van den Bergh received support through ERC Grant 741087 in EU's Horizon 2020.

Notes

1 These four categories of pledges/NDCs (nationally determined contributions) are absolute emission reduction targets versus (distinct) base years, reduction relative to future emissions growth under BAU scenario, reduction of emission intensity of national income (GDP), and projects without identifying associated emissions reduction. For more details, see King and van den Bergh (2019).

2 Newell and Stavins (2003) provide a good entrance to the theoretical and empirical literature supporting cost-effectiveness of pricing instruments in environmental policy. Classic contributions are Pigou (1920), Dales (1968), Baumol (1972) and Baumol and Oates (1975). Other influential contributions are Russell (1979), Tietenberg (1985), Mendelsohn (1986), and Hahn and Hester (1989).

3 For instance, the World Bank (2020) expects that carbon prices should be at least $40–80/tCO$_2$ by 2020 and $50–100/tCO$_2$ by 2030 to reduce emissions in line with the objectives of the Paris Agreement.

4 For example, www.bmwi.de/Redaktion/EN/Artikel/Industry/electric-mobility-r-d-funding.html.

5 According to one of the co-authors the article was inspired by the interview of Jochen Markard with Jeroen van den Bergh, "Carbon Pricing", in the STRN newsletter no. 31, March 2019, pp. 8–11 (https://transitionsnetwork.org/wp-content/uploads/2016/09/31st-STRN-Newsletter.pdf). Interestingly, Markard concluded from the interview, "It has become clear that there are many subtle aspects to carbon pricing. Transitions research could pay more attention to, and learn from, the debate on carbon pricing. Given 'our' experience with the innovation dimension of policies, the relevance of industry and market creation, as well as politics and the formation of coalitions, there is certainly much we can contribute. Also, the interaction of e.g. carbon pricing and innovation policies, together with ongoing changes in policies and policy mixes as transitions progress, deserve our attention". Unfortunately, a similar message cannot be found in Rosenbloom et al. (2020a).

6 Although Rosenbloom (2020b) argue that carbon pricing is difficult to implement for heavy industry, the EU-ETS covers energy-intensive industrial sectors.

References

Aghion, P., A. Dechezleprêtre, D. Hémous, R. Martin, and J. Van Reenen (2016). Carbon taxes, path dependency, and directed technical change: Evidence from the auto industry. *Journal of Political Economy* 124(1): 1–51.

Aldy J., A. Krupnick, R. Newell, I. Parry, and W. Pizer (2010). Designing climate mitigation policy. *Journal of Economic Literature* 48(4): 903–934.

Andor, M.A., and K.M. Fels (2018). Behavioral economics and energy conservation—A systematic review of non-price interventions and their causal effects. *Ecological Economics* 148: 178–210.

Appunn, K. (2021). Understanding the European Union's Emissions Trading System (EU ETS). *Clean Energy Wire*, 20 Jul 2021. www.cleanenergywire.org/factsheets/understanding-european-unions-emissions-trading-system

Baker III, J.A., M. Feldstein, T. Halstead, N.G. Mankiw, H.M. Paulson Jr, G.P. Schulz, T. Stephenson, and R. Walton, 2017. The conservative case for carbon dividends. *The Climate Leadership Council*, February 2017, www.clcouncil.org.

Baldo, G.L., M. Marino, M. Montani, and S.-O. Ryding (2009). The carbon footprint measurement toolkit for the EU Ecolabel. *The International Journal of Life Cycle Assessment* 14(7): 591–596.

Ball, J. (2018a). Hot air won't fly: The new climate consensus that carbon pricing isn't cutting it. *Joule* 2: 2491–2494.

Ball, J. (2018b). Why carbon pricing isn't working: Good idea in theory, failing in practice. *Foreign Affairs*, July/August 2018.

Baranzini, A., J. van den Bergh, S. Carattini, R. Howard, E. Padilla and J. Roca (2017). Carbon pricing in climate policy: Seven reasons, complementary instruments, and political-economy considerations. *WIREs Climate Change* 8(4): e462.

Baumol, W.J. (1972). On taxation and the control of externalities, *American Economic Review* 62(3): 307–322.

Baumol, W.J. and W.E. Oates (1975). *The Theory of Environmental Policy* (second edition 1988). Cambridge University Press, Cambridge.

Bennear, L.S., and R.N. Stavins (2007). Second-best theory and the use of multiple policy instruments. *Environmental and Resource Economics* 37(1): 111–129.

Bernstein, S. and M. Hoffmann (2018). The politics of decarbonization and the catalytic impact of subnational climate experiments. *Policy Sciences* 51: 189–211.

Best, R., P.J. Burke and F. Jotzo (2020). Carbon pricing efficacy: Cross-country evidence. *Environmental and Resource Economics* 77: 69–94.

Biermann, F., N. Kanie and R.E. Kim (2017). Global governance by goal-setting: The novel approach of the UN sustainable development goals. *Current Opinion in Environmental Sustainability* 26–27: 26–31.

Brouwer, R., L. Brander and P. Van Beukering (2008). A convenient truth": Air travel passengers' willingness to pay to offset their CO_2 emissions. *Climatic Change* 90: 299–313.

Bulkeley, H. and K. Kern (2006). Local government and the governing of climate change in Germany and the UK. *Urban Studies* 43(12): 2237–2259.

Calel, R. and A. Dechezleprêtre (2016). Environmental policy and directed technological change: Evidence from the European carbon market. *Review of Economics and Statistics* 98: 173–191.

Castro, J., S. Drews, F. Exadaktylos, J. Foramitti, F. Klein, T. Konc, I. Savin, J. van den Bergh (2020). A review of agent-based modelling of climate-energy policy. *WIREs Climate Change*, e647.

CDP (2017). *Putting a Price ON Carbon: Integrating Climate Risk into Business Planning*. Carbon Disclosure Project, London.

Cramton, P., D.J.C. MacKay, A. Ockenfels and S. Stoft (2017). *Global Carbon Pricing: The Path to Climate Cooperation*. The MIT Press, Cambridge, MA.

Cullenward, D., and D.G. Victor (2020). *Making Climate Policy Work*. Polity Press, Cambridge.

Dales, J. (1968). *Pollution, Property and Prices*. University Press, Toronto.

Delmas, M., M. Fischlein and O. Asensio (2013). Information strategies and energy conservation behavior: A meta-analysis of experimental studies from 1975–2011. *Energy Policy* 61(C): 729–739.

Drews, S., F. Exadaktylos and J. van den Bergh (2020). Assessing synergy of incentives and nudges in the energy policy mix. *Energy Policy* 144: 111605.

Drews, S., and J. van den Bergh (2016). What explains public support for climate policies? A review of empirical studies. *Climate Policy* 16(7): 855–876.

Ebeling, F., and S. Lotz (2015). Domestic uptake of green energy promoted by opt-out tariffs. *Nature Climate Change* 5(9): 868–871.

Edenhofer, O., C. Flachsland, M. Kalkuhl, B. Knopf, and M. Pahle (2019). *Assessment of the German Climate Package and Next Steps: Carbon Pricing, Social Balance, Europe, Monitoring*. Mercator Research Institute on Global Commons and Climate Change, Berlin.

ESOCD (2019). Economists' statement on carbon dividends: The largest public statement of economists in history. *The Wall Street Journal*, 17 January 2019, www.clcouncil.org/economists-statement.

Fankhauser, S., C. Hepburn and J. Park (2010). Combining multiple climate policy instruments: How not to do it. *Climate Change Economics* 1(3): 209–225.

Fawcett, T. (2010). Personal carbon trading: A policy ahead of its time? *Energy Policy* 38(11): 6868–6876.

Fuenfschilling, L., N. Frantzeskaki, L. Coenen (2018). Urban experimentation & sustainability transitions. *European Planning Studies* 27: 219–228.

Geels, F.W. (2011). The multi-level perspective on sustainability transitions: Responses to seven criticisms. *Environmental Innovation and Societal Transitions* 1: 24–40.

Gillingham, K., S. Carattini, and D. Esty (2017). Lessons from first campus carbon-pricing scheme. *Nature* 551: 27–29. https://doi.org/10.1038/551027a

Gössling, S., L. Haglund, H. Kallgren, M. Revahl and J. Hultman (2009). Swedish air travellers and voluntary carbon offsets: Towards the co-creation of environmental value? *Current Issues in Tourism* 12(1): 1–19.

Grandin, J., H. Haarstad, K. Kjærås, and S. Bouzarovski (2018). The politics of rapid urban transformation. *Current Opinion in Environmental Sustainability*, 31: 16–22.

Grandin, J., and S. Sareen (2020). What sticks? Ephemerality, permanence and local transition pathways. *Environmental Innovation and Societal Transitions* 36: 72–82.

Hahn, R.W., and G.L. Hester (1989). Marketable permits: Lessons for theory and practice. *Ecology Law Quarterly* 16(2): 361–406.

Haites, E. (2018). Carbon taxes and greenhouse gas emissions trading systems: What have we learned? *Climate Policy* 18(8): 955–966.

HLCCP (2017). *Report of the High-Level Commission on Carbon Prices*. World Bank, Washington, DC.

Howlett, M., and J. Rayner (2007). Design principles for policy mixes: Cohesion and coherence in 'new governance arrangements'. *Policy and Society* 26(4): 1–18.

Howlett, M., J. Vince and P. del Rio (2017). Policy integration and multi-level governance: Dealing with the vertical dimension of policy mix designs, *Politics and Governance* 5(2): 69–78.

HR7173 (2018). Energy Innovation and Carbon Dividend Act of 2018.115th Congress (2017–2018) of the USA. www.congress.gov/bill/115th-congress/house-bill/7173.

IMF (2000). The impact of higher oil prices on the global economy. IMF Research Department, December 8, 2000. www.imf.org/external/pubs/ft/oil/2000.

IMF (2019). *Fiscal Policies for Paris Climate Strategies—from Principle to Practice*. Fiscal Affairs Department, International Monetary Fund, Washington, DC.

IMF (2020). World Economic Outlook, April 2020, www.imf.org/en/Publications/WEO/Issues/2020/04/14/weo-april-2020.

Jacobsson, S.B., and A. Bergek (2011). Innovation system analyses and sustainability transitions: Contributions and suggestions for research. *Environmental Innovation and Societal Transitions* 1: 41–57.

Jaffe, A.B., R.G. Newell, and R.N. Stavins (2005). A tale of two market failures: Technology and environmental policy. *Ecological Economics* 54: 164–174.

Jorgenson, D., R. Goettle, M. Sing Hoc and P. Wilcoxen (2009). Cap and trade climate policy and U.S. economic adjustments. *Journal of Policy Modeling* 31: 362–381.

Kallbekken, S., S. Kroll and T.L. Cherry (2011). Do you not like Pigou, or do you not understand him? Tax aversion and revenue recycling in the lab. *Journal of Environmental Economics and Management* 62(1): 53–64.

Kaya, Y. and K. Yokoburi (1997). *Environment, Energy and Economy: Strategies for Sustainability*. United Nations University Press, Tokyo.

King, L., and J. van den Bergh (2019). Normalisation of paris agreement NDCs to enhance transparency and ambition. *Environmental Research Letters* 14: 084008.

King, L., and J. van den Bergh (2021). Potential carbon leakage under the Paris Agreement. *Climatic Change* 165: 52.

Kivimaa, P., and F. Kern (2016). Creative destruction or mere niche support? Innovation policy mixes for sustainability transitions. *Research Policy* 45(1): 205–217.

Klenert, D., L. Mattauch, E. Combet, O. Edenhofer, C. Hepburn, R. Rafaty and N. Stern (2018). Making carbon pricing work for citizens. *Nature Climate Change* 8: 669–677.

Konc, T., I. Savin and J. van den Bergh (2020). The social multiplier of environmental policy: Application to carbon taxation. EVOCLIM Working Paper, ICTA-UAB, revised submission.

Lehmann, P. (2012). Justifying a policy mix for pollution control: A review of economic literature. *Journal of Economic Surveys* 26(1): 71–97.

Liu, F., and J. van den Bergh (2020). Differences in CO2 emissions of solar PV production among technologies and regions: Application to China, EU and USA. *Energy Policy* 138: 111234.

Maestre-Andrés, S., S. Drews, and J. van den Bergh (2019). Perceived fairness and public acceptability of carbon pricing: A review of the literature. *Climate Policy* 19(9): 1186–1204.

Markard, J., and D. Rosenbloom. (2020). Political conflict and climate policy: The European emissions trading system as a Trojan Horse for the low-carbon transition? *Climate Policy* 20(9): 1092–1111.

Mendelsohn, R. (1986). Regulating heterogeneous emissions. *Journal of Environmental Economics and Management* 13: 301–312.

Mundaca, L., J. Sonnenschein, L. Steg, N. Höhne, and D. Ürge-Vorsatz (2019). The global expansion of climate mitigation policy interventions, the talanoa dialogue and the role of behavioural insights. *Environmental Research Communications* 1: 061001.

Naber, R., R. Raven, M. Kouw, T. Dassen (2017). Scaling up sustainable energy innovations. *Energy Policy* 110: 342–354.

Narassimhan, E., K.S. Gallagher, S. Koester and J. Rivera Alejo (2017). Carbon Pricing in Practice: A Review of the Evidence. Center for International Environment & Resource Policy, The Fletcher School, Tufts University, Medford, Mass.

Newell, R.N. and R.N. Stavins (2003). Cost heterogeneity and the potential savings from market-based policies. *Journal of Regulatory Economics* 23(1): 43–59.

Nill, J., and R. Kemp (2009). Evolutionary approaches for sustainable innovation policies: From niche to paradigm? *Research Policy* 38(4): 668–680.

Nordhaus, William (2015). Climate clubs: Overcoming free-riding in international climate policy. *American Economic Review* 105(4): 1339–1370.

Perman, R., Y. Ma, M. Common, D. Maddison and J. McGilvray (2011). *Natural Resource and Environmental Economics*. 3rd ed. Pearson Education Limited, Harlow.

Pigou, A.C. (1920). *The Economics of Welfare*. MacMillan, London.

Popp, D. (2002). Induced innovation and energy prices. *The American Economic Review* 92(1): 160–180.

Popp, D. (2006). R&D subsidies and climate policy: Is there a "free lunch"? *Climate Change* 77: 311–341.

Rengs, B., M. Scholz-Wäckerle and J. van den Bergh (2020). Evolutionary macroeconomic assessment of employment and innovation impacts of climate policy packages. *Journal of Economic Behavior and Organization* 169: 332–368.

Roelfsema, M., van Soest, H.L., Harmsen, M. et al. (2020). Taking stock of national climate policies to evaluate implementation of the Paris Agreement. *Nature Communications* 11: article 2096.

Rogge, K.S., F. Kern and M. Howlett (2017). Conceptual and empirical advances in analysing policy mixes for energy transitions. *Energy Research & Social Science* 33: 1–10.

Rogge, K.S. and K. Reichardt (2016). Policy mixes for sustainability transitions: An extended concept and framework for analysis. *Research Policy* 45(8): 1620–1635.

Rosenbloom, D., J. Markard, F.W. Geels and L. Fuenfschilling (2020a). Opinion: Why carbon pricing is not sufficient to mitigate climate change – and how "sustainability transition policy" can help. *Proceedings of the National Academy of Sciences of the U.S.A. (PNAS)* 117(16): 8664–8668.

Rosenbloom, D., J. Markard, F.W. Geels, and L. Fuenfschilling (2020b). Reply to van den Bergh and Botzen: A clash of paradigms over the role of carbon pricing. *Proceedings of the National Academy of Sciences of the U.S.A. (PNAS)*, 117(38): 23221–23222.

Rotmans, J., R. Kemp and M. van Asselt (2001). More evolution than revolution: Transition management in public policy. *Foresight* 3(1): 15–31.

Rotmans, J., and D. Loorbach (2009). Complexity and transition management. *Journal of Industrial Ecology* 13(2): 184–196.

Russell, C.S. (1979). What can we get from effluent charges? *Policy Analysis* 5(2): 155–180.

Safarzynska, K., K. Frenken and J.C.J.M. van den Bergh (2012). Evolutionary theorizing and modelling of sustainability transitions. *Research Policy* 41: 1011–1024.

Satterthwaite, D. (2008). Cities' contribution to global warming: Notes on the allocation of greenhouse gas emissions. *Environment & Urbanization* 20(2): 539–549.

Savin, I. and J. van den Bergh (2021). Main topics in EIST during its first decade: A computational-linguistic analysis. *Environmental Innovation and Societal Transitions* 41: 10–17.

Schmidt, T.S. and S. Sewerin (2019). Measuring the temporal dynamics of policy mixes—An empirical analysis of renewable energy policy mixes' balance and design features in nine countries. *Research Policy* 48(10): 103557.

Seddon, J. and V. Ramanathan (2013). Bottom-up solutions to mitigating climate change. *Stanford Social Innovation Review* 11(3): 48–53.

Sen, S. and H. Vollebergh (2018). The effectiveness of taxing the carbon content of energy consumption. *Journal of Environmental Economics and Management* 92: 74–99.

Seto, K.C., S.J. Davis, R.B. Mitchell, E.C. Stokes, G. Unruh and D. Ürge-Vorsatz (2016). Carbon lock-in: Types, causes, and policy implications. *Annual Review of Environment and Resources* 41: 425–452.

Sorrell, S. (2015). Reducing energy demand: A review of issues, challenges and approaches. *Renewable and Sustainable Energy Reviews* 47: 74–82.

Sorrell, S., B. Gatersleben and A. Druckman (2020). The limits of energy sufficiency: A review of the evidence for rebound effects and negative spillovers from behavioural change. *Energy Research & Social Science* 64: 101439.

Spash, C.L. (2010). The brave new world of carbon trading. *New Political Economy* 15(2): 169–195.

Stern, N. (2007). *The Economics of Climate Change: The Stern Review*. Cambridge University Press, Cambridge.

Tietenberg, T.H. (1985). *Emissions Trading: An Exercise in Reforming Pollution Policy*. Resources for the Future, Washington D.C.

van den Bergh, J. (2008). Environmental regulation of households? An empirical review of economic and psychological factors. *Ecological Economics* 66: 559–574.

van den Bergh, J.C.J.M. (2013a). Environmental and climate innovation: Limitations, policies and prices. *Technological Forecasting and Social Change* 80(1): 11–23.

van den Bergh, J.C.J.M. (2013b). Policies to enhance economic feasibility of a sustainable energy transition. *Proceedings of the National Academy of Sciences of the U.S.A. (PNAS)* 110(7): 2436–2437.

van den Bergh, J.C.J.M. (2020). Systemic assessment of urban climate policies worldwide: Decomposing effectiveness into 3 factors. *Environmental Science and Policy* 114: 35–42.

van den Bergh, J.C.J.M. and W. Botzen (2014). A lower bound to the social cost of CO_2 emissions. *Nature Climate Change* 4(April): 253–258.

van den Bergh, J. and W. Botzen (2020). Low-carbon transition is improbable without carbon pricing. *Proceedings of the National Academy of Sciences of the U.S.A. (PNAS)*, forthcoming.

van den Bergh, J.C.J.M., A. Angelsen, A. Baranzini, W.J.W. Botzen, S. Carattini, S. Drews, T. Dunlop, E. Galbraith, E. Gsottbauer, R.B. Howarth, E. Padilla, J. Roca and R.C. Schmidt (2020). A dual-track transition to global carbon pricing. *Climate Policy* 20(9): 1057–1069.

van den Bergh, J., J. Castro, S. Drews, F. Exadaktylos, J. Foramitti, F. Klein, T. Konc and I. Savin (2021). Designing an effective climate-policy mix: Accounting for instrument synergy. *Climate Policy* 21(6): 745–764.

van den Bergh, J. and I. Savin (2021). Impact of carbon pricing on low-carbon innovation and deep decarbonisation: Controversies and path forward. *Environmental and Resource Economics* 80: 705–715.

Victor, D. (2015). The Case for Climate Clubs. E15 Expert Group on Measures to Address Climate Change and the Trade System, January 2015, International Centre for Trade and Sustainable Development (ICTSD) and the World Economic Forum, Geneva, Switzerland.

Waechter, S., B. Sütterlin and M. Siegrist (2015). The misleading effect of energy efficiency information on perceived energy friendliness of electric goods. *Journal of Cleaner Production* 93: 193–202.

Watts, M. (2017). Cities spearhead climate action. *Nature Climate Change* 7: 537–538.

Weitzman, M.L. (1974). Prices vs. quantities. *Review of Economic Studies* 41(4): 477–491.

Weitzman, M.L. (2014). Can negotiating a uniform carbon price help to internalize the global warming externality? *Journal of the Association of Environmental and Resource Economists* 1: 29–49.

World Bank (2020). *State and Trends of Carbon Pricing 2020*. World Bank, Washington, DC. https://openknowledge.worldbank.org/handle/10986/33809.

Wynes, S., K.A. Nicholas, J. Zhao and S.D. Donner (2018). Measuring what works: Quantifying greenhouse gas emission reductions of behavioural interventions to reduce driving, meat consumption, and household energy use. *Environmental Research Letters* 13: 113002.

17

PLANNING FOR THE FUTURE OF EXISTING ONSHORE RENEWABLE ENERGY INFRASTRUCTURE

Rebecca Windemer

Introduction

Central to a global transition to low-carbon energy is its temporal dynamics due to the need for the energy transition to occur quickly in response to global challenges (IPCC 2021). However, despite this urgency of action, there is expected to be a limit to the speed at which new energy technologies can be implemented due to the requirements on human and industrial capacity and the time required to scale up technologies, as well as the need for land and enabling technologies to support developing energy systems (Kramer and Haigh 2009). Meanwhile, the development of new sites for established renewable energy technologies such as onshore wind and solar face a host of challenges, including complicated planning systems, public opposition to new sites (Roddis et al. 2018), and increasing competition with land for other uses such as food production (Scheidel and Sorman 2012).

In the context of these challenges, it is remarkable that little attention has been given to the temporal dynamics of renewable energy rollout beyond initial development – that is, how we can ensure the long-term future or increase energy generation at existing renewable energy generation sites. Considerations of a global transition in energy sources usually focus on the development of new technologies and sites, considering existing sites and infrastructure as in place, with little or no consideration to their future. However, our existing energy infrastructure is not infinite, parts will wear out, the output will decrease over time, and in some countries, planning consents will run out. Thus, a key element of a global transition to low-carbon energy will involve considering the future of existing sites, including potentially increasing output through repowering. This is particularly significant in many countries where the older infrastructure occupies some of the best locations in terms of renewable energy resources, such as wind.

In this chapter, in the context of considering the global transition to low-carbon energy (i.e., the move away from fossil fuels to low-carbon options), I propose that we need to look beyond the implementation of new infrastructure to consider the future of our existing renewable energy sites. This is a particularly timely issue as, globally, our existing renewable energy infrastructure is starting to reach the end of its operational life (through the infrastructure wearing out) or consented life (through time-limited planning consents or land leases running out),

DOI:10.4324/9781003183020-20

raising the question of what happens next. The decisions that are made in the next decade will likely have a significant impact on the global transition in energy sources both in terms of impacting the speed of the transition to renewables and in terms of achieving the aims of a "just transition" – the ability for the global transition in energy sources to occur in a way that does not create distributional or procedural impacts on particular groups (Newell and Mulvaney 2013).

The end-of-life point provides a potentially important opportunity to increase the amount of energy that we are generating from existing sites through repowering a site with new, more efficient technology. This can also create further benefits such as for the environment and communities. However, end-of-life decision-making is not straightforward because, as I explain in this chapter, renewable energy sites and the communities living close to these sites do not remain static over time, raising questions regarding perceptions of original siting decisions and how decisions should be made regarding their future. There will thus be wider consequences to the decisions made, beyond whether to change energy output. A particularly important element of this, if we are to secure a just transition, is the experience and the opinions of those who have been living close to existing renewable energy infrastructure. Meanwhile, if the operation of existing sites is not continued or repowered with new, more energy-efficient infrastructure, then there is a potential that our overall renewable energy output could decrease, negatively impacting the speed of the global transition to low-carbon energy.

The global transition in energy sources may also bring new, more efficient technologies that may replace existing renewable energy technologies, thus understanding the decommissioning process and the wider effects of developments on places and ecosystems may also be necessary. The extent to which project owners are made liable for the total effects of their infrastructure could affect the economics of renewable energy, as it does with all energy systems, in addition to impacting the overall reputation and social acceptance of renewables.

In the context of these challenges, understanding the potential future of existing renewable energy generation capacity is imperative. The key argument of this chapter is thus that we need to take a broader temporal approach to exploring the global transition to low-carbon energy by thinking about the long-term future of our existing renewable energy assets. To demonstrate the importance of this, I draw upon empirical research undertaken in the UK, which investigated how the oldest wind UK farms changed over time and how this influenced end-of-life decision-making. However, while the focus of the chapter is on onshore wind, the lessons learned are applicable to other forms of energy infrastructure.

This chapter begins by providing a more detailed discussion of the options for onshore wind farm sites that reach the end of their operational or consent life and the significance of this decision-making process. It then explores how sites in the UK have changed over time, impacting the context in which end-of-life decisions are made. The challenges and opportunities of repowering and decommissioning are then discussed before the conclusion is drawn, providing insights for the wider global transition to low-carbon energy alongside policy recommendations.

Why We Need to Consider the Future of Existing Onshore Renewable Energy Sites

The expansion of renewable energy appears caught between several temporal dilemmas. Low-carbon-transition thinking often assumes that renewable energy sites are permanent; however, infrastructure is finite as it is limited by its operational or consent life. The longevity of infrastructure may also be impacted by several factors, including the development and popularity of other renewable energy technologies and policy changes – the future of our infrastructure is neither permanent nor a given. Increasingly, a growing proportion of the international wind energy capacity (as well as other forms of energy infrastructure) will have to negotiate end-of-life decisions, with broader implications for the energy mix and potentially wider decarbonization policy. It is therefore necessary to understand how decisions are being made and the challenges that have been faced so far. Exploring onshore wind farms in the UK provides an illustration of the key challenges and opportunities facing aging renewable energy infrastructure.

The oldest wind farms in the UK were developed in 1991 and most UK wind farms have been subject to 25-year conditions on their planning consents, requiring complete removal at the end of the set period. Wind farms have already started to reach that end-of-life point (see Windemer 2019a). Meanwhile, internationally, some onshore wind farms are also starting to reach the end their operational life, creating the need to consider their future. While the primary focus of this chapter is onshore wind; similar patterns can be expected to occur for other forms of onshore infrastructure. For example, field-scale solar farms will soon need to be upgraded with new infrastructure, as first-generation modules have an expected lifespan of 20–25 years (Malandrino et al. 2017). Furthermore, irrespective of the technology in question, the oldest infrastructure is expected to be occupying some (but not all) of the best located sites.

Investigating end-of-life decision making for onshore wind farms in the UK identified that at the end of their consented or operational life, developers have four main options: (1) to decommission the site, removing the energy infrastructure (and usually returning the land to its previous use); (2) to increase the existing consent life of the infrastructure without making any material changes to the site through asset life extension; (3) to repower the site, involving replacing the existing infrastructure with new infrastructure (in the case of wind, this is often involves a different number and size of turbines), (4) partial repowering (changing certain parts of the infrastructure egg the blades but keeping the existing foundations and layout).

These decisions are just starting to occur in the UK. Data from the UK government renewable energy planning database reveals that 23 sites have been granted permission to repower; one site was granted permission for a partial repowering, and two were refused repowering permission. Eighteen of these sites have now completed the repowering process. Two sites have been decommissioned, approximately ten have their life extended, and one site has been partially repowered. However, there is expected to be a significant increase in repowering of the UK fleet after 2027 as the financial subsidies that were attached to a lot of the older wind farms will end. To date, in the UK, repowering has, on average, occurred after 18 years of operation, it has not occurred when the infrastructure has worn out, but rather repowering at this earlier stage enables developers to sell the existing turbines and also benefit from larger turbines with the latest technology. We can thus see that this decision-making point may soon be arising for wind farms internationally.

Continuing generation on existing sites through repowering or life extension can provide many benefits, notably sustained or increased energy generation and thus income to owners, landowners and possibly communities. In development terms, such sites can benefit from existing access and grid connectivity and existing performance data, which helps to inform design. Land is a finite resource and through installing more efficient infrastructure repowering provides a way of enabling greater efficiency from the same or less land. This is particularly significant given the pace of technological change in recent decades. Technology has developed rapidly since the first turbines were developed, leading to modern turbines being much larger and capable of producing significantly more energy. Evidencing this, as of 2021, in the UK, repowering has increased the installed capacity (in megawatts) of a site by 143%. In doing so, this has on average involved a 41% decrease in the overall number of turbines on the site, but the average turbine height has increased by 98.8%. It is worth considering that despite the significant increases noted here, in some locations land restrictions may create potential barriers to repowering due to the increased space requirements of larger, more efficient turbines with greater rotor diameters. Moreover, the greatest increases in installed capacity are likely to occur from upgrading the earliest sites due to the substantial improvements in turbine technology. Nevertheless, we can see wind farms potentially becoming larger and more economically and energetically efficient during repowering.

However, as will be discussed, decision-making for repowering is not as straightforward as simply deciding to increase the energy output from a site. The continuation of renewable energy generation on a site will mean the continuation (or potential increase) of visual impacts and other externalities. Keeping existing sites as areas for renewable energy generation raises questions regarding the duration of impacts on particular communities, but continuing genera-tion also has ramifications for the duration of decarbonization benefits. The decision-making process is thus multilayered. This raises questions regarding how decisions are made, the weight given to the different actors involved, and to what extent anyone is considering and reflecting the interests of future generations.

Renewable Energy Sites Are Not Static

Existing literature on wind energy often appears to consider wind farm development and decision-making process in simple binary terms (i.e., the infrastructure was not there and now it is, or in the case of decommissioning, it was there and now it is not), ignoring the complex reality and scope for projects to evolve and change over time. There are a small number of studies exploring issues surrounding decommissioning, repowering, and life extension of onshore wind (for exam-ple, Möller 2010; Himpler and Madlener 2012; Ziegler et al. 2018). However, end-of-life stud-ies tend to be focused on specific aspects, such as economic benefits, and fail to provide details regarding what factors are being considered and controlled within regulatory systems. Looking beyond wind infrastructure, other studies have adopted a broader perspective by considering the social-ecological-technical aspects of energy project planning. In this book, such considerations are particularly evident in Sharp et al., Brinkman and Hirsh, and Ferry and Monoian. These considerations are important as it is problematic to treat renewable energy infrastructure as an object simply occupying space rather than as a facility that evolves over time with diverse relations to the surrounding environment and society. A renewable energy site is not simply a fixed entity. While it is fixed in location, often for a certain duration, other elements of the site and its context are likely to evolve and change. Such changes may include changes in policy, social changes, and physical changes to the equipment, the site, and the surrounding area. Each of these changes are likely to occur over a different timespan, potentially impacting decisions regarding the duration of the infrastructure in situ and the future use of the site. Consequently, these changes impact the ability of a site to continue to contribute to the global transition to low-carbon energy.

Exploring the oldest onshore wind farms in the UK illustrates the range of changes that have occurred over a 20-to–25-year period, impacting the setting of the sites, the continued suitabil-ity of the sites for renewable energy generation, and the context in which end-of-life decisions are being made. Where physical landscapes (in terms of features of the land and landscapes) and social landscapes (involving peoples' relationship with and perceptions of landscapes) change over the lifespan of infrastructure, this can impact end-of-life decision making. Landscapes are not fixed or stable, characteristics of landscapes can alter significantly over the lifespan of renew-able energy developments. The landscapes of energy infrastructure can be seen to be in flux, changing the context in which decisions are made. The scale of changes will be different glob-ally, particularly in areas that are becoming increasingly impacted by extreme weather events. For example, an increase in the frequency and scale of wildfires over the longer term is requiring planners to rethink how infrastructure, such as the power grid, is designed and used.

In the case of onshore wind in the UK, further development, particularly of other wind farms but also of residential dwellings or other built development, has occurred in proximity to sites. Such physical changes in the surrounding area can influence the development in a number of ways. Physical changes that have occurred over time can impact the developable area of the site, impacting the size of potential future turbines as there are minimum distances that are required between wind turbines and surrounding development, if the turbines are larger so too are the distances required between the turbines and other built development. New development may also impact site access, in particular the ability to bring new, larger turbines onto a site. Further development around a wind farm may also change how communities view the site. For example, they may consider it as less suitable for a wind farm if there has been a lot of residential development nearby and thus may not support an application to repower. The development of additional renewable energy infrastructure may also impact community perceptions of the repowering of an existing site. This has been seen in cases in the UK, where in responding to repowering applications, members of communities that have seen many turbines developed in their surrounding area may oppose the repowering in the hope that a site will be removed. Elsewhere, there appears to be an appetite from community energy organizations to become involved in co-owning a wind farm during repowering (Philpott and Windemer, 2022).

Landscape designations are another aspect of renewable energy sites that can change over time, providing constraints on their ability to provide a long-term contribution to the global transition to low-carbon energy. This is expected to be a particular concern in a number of European countries where many existing sites are now designated as Natra2000 sites or have been subject to other forms of spatial exclusion criteria (see, for example, the case of Germany, Grau et al. 2021; Unnewehr et al. 2021). Even when landscape designations are not impacting a site directly, a change to protect nearby land can impact people's perceptions of the suitability of a wind farm in that particular landscape. For example, in the UK, the Lake District National Park became designated as a World Heritage Site, leading people to object to the repowering of an existing wind farm on the edge of the national park, despite the change in designation occurring while the wind farm was in situ. Current experience in the UK shows this trend flowing one way in terms of increasing constraints over time rather than in terms of some landscape changes being a loosening of constraints and restrictions (e.g., de-designation of landscapes or other wind farms being decommissioned).

The variability of economic changes can also influence strategies regarding the future of existing wind farm sites. If subsidies change over time this can have a significant impact on the viability of repowering. In the UK, the ending of the onshore wind subsidy scheme in 2017 has led to a situation in which continuing to operate the existing older turbines that have subsidies attached until 2027 is often considered more viable than developing new turbines. In this context, if a site was to be repowered, then significantly taller turbines are likely to be deemed necessary in order to make the repowering viable. Developers in the UK regularly test end-of-life options in order to assess if it is more viable to run existing sites with subsidies (potentially involving life extension) or to repower. Linked to this, investment or business decisions are often shaped by the (un)certainty of the planning horizon. In the UK, policy changes have impacted the rate of onshore wind repowering. For example, changes in subsidy regimes and an uncertain government planning policy context in England can be seen to have impacted the end-of-life decision-making context for developers, creating difficulties for both deciding whether to submit repowering applications and deciding if it is viable to implement existing repowering consents.

Social changes are important, the changes that occur over time are not just background landscape changes but also made by the actors involved. The communities living close to energy infrastructure are likely to change over the lifespan of the infrastructure both in terms of the composition and scale of the community and potentially how they consider the site and its various impacts. A range of factors can influence community considerations of the future of the infrastructure, including the history of how the land was used before the wind farm (e.g., the history of using land as a resource in some rural areas) and how they consider the impacts of ancillary equipment that is often located closer to the community. Public attitudes can also be influenced by how people conceptualize the baseline against which to judge end-of-life applications – that is, if they consider the long-term use of the site to be for wind energy generation or if they consider the wind energy infrastructure as a temporary use of the land. As a result, community responses to end-of-life applications in the UK have varied.

There is a widespread perception among industry and local authority planners that acceptance is likely to occur among communities living with wind farm infrastructure. Such a perspective suggests that wind farms will become a familiar part of the landscape. This perspective reflects existing literature suggesting that those living close to wind turbines become more supportive or at least less active in their objection or opposition to them after installation (Warren et al. 2005; Damborg and Krohn 1999; Wolsink 2007). Such perceptions of familiarity are reflected in developer confidence that people will not object to life-extending applications. There is also widespread expectation that local communities will prefer repowering of an existing site rather than a wind farm in a completely new site. However, cases in the UK have shown that familiarity and acceptance is not always engendered over time, particularly in cases where people perceive the wind farm as not working or not creating any local benefits (Windemer 2019b). Changes in perceptions over time have also been identified as occurring for other forms of energy infrastructure due to changes in information provision. For example, in recent decades, hydropower went from being the prime choice for new energy projects in Iceland to one that was increasingly contested or seen as needing protection. In this example, green nongovernmental organizations were identified as an important actors in the debate with the rise of awareness, organizational capacity, and means of engagement through the internet and social media (see Araujo 2017).

UK experience suggests that the social history developed over time between communities and developers/operators can influence community views of end-of-life options. Relations can improve over time through communities recognizing the value of community benefits and through good communication and the establishment of trust with the developer. This reflects existing literature that identified the importance of trust on responses to renewable energy schemes (see Wolsink 2007; Aitken 2010a; Ricci et al. 2010; Walker et al. 2010; Friedl and Reichl 2016). In other cases, relations worsen over time, negatively impacting community perceptions of the site and its long-term suitability. In cases where poor relations are formed between the community and developer, end-of-life applications provide an opportunity for the community to oppose or renegotiate, raising issues that may have long been of concern.

Additionally, communities may attribute less value to a wind farm over time, particularly in cases where more wind farms have been developed in the local area. Such diminishing value can be seen to have occurred in cases in the UK, such as Windy Standard in Scotland, where, due to the volume of wind farms that have been developed in the area, the community is receiving a number of community benefit funds (financial payments paid to the local community from the wind farm developers). As community benefit funds need to be spent on community projects, the small rural community is running out of possible projects to spend the money on and thus attribute less value to their benefit (Windemer 2019b). This reflects existing literature identifying communities' varying capacities to make use of community benefit funds and the challenges and opportunities associated with such payments (see Aitken 2010b; Cowell et al. 2011; Munday et al. 2011; Bristow et al. 2012).

The Challenges and Opportunities of Repowering

As renewable energy sites change over time, challenges and opportunities emerge, impacting the viability of different end-of-life options. Understanding the types of challenges and opportunities that are faced provides useful insights regarding the future of the global transition in energy sources, particularly regarding the potential for increased energy generation from existing sites through repowering. In considering this, it is firstly important to recognize that end-of-life decisions, such as repowering, life extension, and decommissioning are not always taken at the end of the operational or consent life of the infrastructure. As noted earlier in this chapter, the average age for repowering applications to be submitted in the UK so far is after 18 years of operation. Multiple factors can be seen to come into play to influence timing (e.g., economics, policy context, condition of the infrastructure). Developers may undertake applications early due to financial reasons, such as selling the existing turbines or to achieve greater efficiency benefits. Developers may also undertake these applications when it is considered least risky to do so, reflecting the power of developers to trigger end-of-life considerations at a chosen point. Likewise, if a renewable energy site is not working efficiently, then it may be decommissioned early.

The most obvious benefit of repowering is the opportunity to significantly increase energy generation from the same area of land. As noted earlier, repowering for the earliest UK sites has been significant, creating an average increase of 143% in installed capacity (in megawatts). The repowering of existing sites can thus clearly provide an important contribution to the global transition to low-carbon energy. Existing sites also have a number of benefits that are likely to make the process quicker than developing on a new site; this includes aspects such as grid connection, site access, and data about the site performance. The repowering of an existing wind farm also provides an opportunity to respond to community experiences and preferences. By involving communities in the repowering of a site, there is an opportunity to address any challenges that they may have faced with the existing site and to design the site in a way that responds to their preferences. Linked to this, repowering provides the opportunity to renegotiate community benefits. As discussed earlier, communities will not always value a community benefit fund and may be looking for another form of benefit. Repowering thus creates an opportunity to implement a new form of community benefit, such as community co-ownership whereby communities can own part of the wind farm or opportunities for local communities to receive reduced energy bills. Alongside this, repowering provides opportunities to implement other benefits, such as enhanced environmental improvements that reflect (or go beyond) updated environmental standards, science, and priorities.

However, while improved community benefits and community involvement are some of the biggest opportunities of repowering, in other cases a lack of community support can act as a barrier, preventing a site getting consent. As discussed earlier, not all communities will have had positive experiences of living with wind farms. However, developers often fail to consider this, assuming that if they increase the value of a community benefit fund, communities will be supportive.

Current industry understanding of the process, impacts, and perceptions of repowering can be seen to foreground certain elements such as measurable dimensions of landscape improvement, and place others, particularly potential challenges of acceptance, outside of consideration. However, in some instances, end-of-life applications provide an opportunity for opposition to (re) surface, demonstrating that even where there may have been some acceptance of facilities while the publics were unable to exert impact, an end-of-life application reopens the opportunity, creating new potentials for influence. As discussed earlier, opposition to life-extending applications is particularly likely where publics perceive few benefits from the existing site, have bad relations with the developer, or if elements have changed over the life of the site (such as designated landscapes). Moreover, for those who have accepted them, the changing impacts of repowered schemes, such as larger turbines, may lead to additional opposition. There are no guarantees.

In the context of the global transition to low-carbon energy, there is a need to consider questions of justice, particularly how decisions are made, and impacts distributed over time and space. Decisions regarding the future of sites (particularly concerning the location and longevity of infrastructure) raise questions regarding the possible impacts on future generations and the current generation. Of particular importance is how the future is represented and considered, especially given the well-known and seemingly deeply entrenched short-term bias in policy-making (Boston 2016). However, challenges will be likely to emerge if there is widespread local opposition to repowering, particularly in cases such as the UK where communities were originally ensured that infrastructure would be removed after 25 years. There is a difficult decision to be made in such circumstances regarding the preferences of local communities compared to the need to rapidly transition to renewable energy sources.

A further potential challenge for repowering lies in the change in visual impact. Visual impacts have long been identified as a critical factor shaping wind energy consent decisions (Wolsink 2007), and this remains the case for repowering. The embedded nature of technologies with landscapes can be seen to impact end-of-life decisions in numerous ways. Due to wider landscape shifts, it is not as simple as judging whether the application, in its narrow dimensional terms, creates an acceptable net change in visibility. Local planning authority decision-makers in the UK revealed that assessing the impacts of the visual change upon the landscape was particularly difficult in terms of deciding if a smaller number of larger turbines or a larger number of smaller turbines was visually preferable. Existing industry research asserted that a smaller number of larger turbines is usually preferred (Sustainable Energy Ireland 2003). However, in practice this assumption is not always born out. The context of the site, including perceptions of the suitability of the turbines on the landscape, is also significant (see also Hirsh and Sovacool 2013; Otto and Leibenath 2014). Peoples' considerations of the impact of the infrastructure on the landscape relates to wider concerns such as (but not limited to) a person's assessment of the quality and value of landscape, the change in the character of the landscape over time, and a person's attachment to a place or landscape (see Wolsink 2018).

Site-specific factors, such as the ability to extend the land lease, the condition of the turbines, site access, and calculations of the possibility of obtaining planning permission, all provide further challenges. Meanwhile, the lack of existing policy, industry, and research focus on end-of-life decision-making also creates a potential challenge. In the UK, through focusing on time-limited consents, there has been a lack of consideration regarding what happens to sites when their consent period ends, including how applications to replace the turbines with more efficient turbines through repowering should be considered. A lack of detailed end-of-life policy in the UK has been seen to lead to developers delaying submitting repowering applications due to uncertainty. It is likely that similar challenges will occur elsewhere in the absence of a supportive end-of-life policy. Similarly, the lack of financial support has, in a number of UK sites, made it more viable to continue operating aging turbines with decreasing energy output rather than replace them with new more efficient infrastructure. Meanwhile, for the UK local planning authorities, a lack of strategic consideration regarding the long-term future of sites has led to significant challenges in decision-making, with local authorities largely being reactive to developers' applications and the limited government-level guidance steering decisions.

The Need to Consider Decommissioning of Existing Renewable Energy Sites

While the repowering and continuation of existing onshore renewable energy sites provides an important opportunity for the global transition to low-carbon energy, it is important to consider that this may not be possible for all sites. As discussed earlier, sites change over time and therefore there will likely be some sites that will require decommissioning. Moreover, as the transition continues, innovative technology may emerge, creating the need to remove existing infrastructure and replace with new technology. New, better-located sites may also emerge, creating the need for decommissioning of sites that have not performed as well as expected. Decommissioning evidently forms a part of the global transition in energy sources, and we need to consider the opportunities and challenges it may bring. This is particularly significant as inadequate decommissioning on existing renewable energy sites could negatively impact the reputation of the renewables industry, potentially hampering the transition.

Existing literature has provided limited attention to decommissioning. The greatest consideration of onshore wind farm decommissioning can be seen in life cycle assessment (LCA) studies that explore the total impacts over the lifetime of renewable energy developments; however, these studies often only give limited consideration to decommissioning, failing to consider the complexities of how it may occur in practice (Windemer and Cowell 2021). Ferrell and DeVuyst (2013) identified that while decommissioning is a policy concern for wind energy, there is little experience of decommissioning or public information on the costs. Writing from experiences in the USA, they identified a lack of standard decommissioning procedures and in some cases (particularly for the oldest wind farms) a lack of regulatory obligation for decommissioning, raising questions as to whether the wind industry will learn from the abandonment issues experienced by other sectors such as oil and gas.

The term reversible has often been used within planning practice and particularly by developers, to highlight that renewable infrastructure can easily be decommissioned, leaving limited or minimal impact on a site with no off-site effects (Pasqualetti, Gipe, and Righter 2002). Particularly in the case of wind, the lack of permanency of the infrastructure has been argued to enable sites to easily return to their previous condition following decommissioning (Jaber 2013). There appears to be an assumption among UK developers that decommissioning will not present a challenge, reflecting market logic which itself makes assumptions about the future. Considerations of decommissioning by many developers and local planning authorities in the UK appear to assume that it will be relatively straightforward and will, at a minimum, involve removal of all visible aboveground impacts. Reflecting this, there is a lack of UK planning policy or guidance regarding how the decommissioning process should be carried out, leaving this domain highly open, for developers, local planning authorities, and maybe others to argue about questions such as "How much decommissioning?" "To what end?" and "How secured?" In the face of future uncertainty, this may not be unreasonable. Furthermore, no one would claim that the physical legacy of renewable energy developments equates in scale and severity to that of nuclear power (Blowers 2017) or open-pit mining (Ibarra and De las Heras 2005) which have left immensely costly remediation challenges. Nonetheless, challenges may occur in the future.

There is also an expectation that longer-term legacy issues will not present an issue for decommissioned sites. Decommissioning bonds are typically used and decommissioning method statements (often required in conditions attached to planning consents), which usually cover how the site will be reinstated and any monitoring of the landscape, which may be carried out following decommissioning. In the past, long-term legacy issues were given very little consideration from local planning authorities, with some of the oldest sites having been granted permission with inadequate decommissioning requirements or without financial decommissioning bonds in place. While decommissioning requirements in the UK and elsewhere have improved over time, the lack of requirements on the oldest sides does not appear to be of a concern to UK planning authorities. This is due to the assumption that developers will either repower or sell the infrastructure, either way securing decommissioning. The focus of local planning authorities thus remains largely on granting consent and reacting to applications as they arise, reflecting wider literature identifying the short-term focus of the planning system (Couclelis 2005; Myers and Kitsuse 2000; Van Der Knapp and Davidse 2010).

To date, the UK has limited experience of decommissioning with only two sites having been decommissioned. In both cases the site was restored to agricultural use without issue. However, there is some concern from publics (reflected in discussions with UK government officers) that in some cases, inadequate decommissioning requirements could lead to legacy issues like long-term dereliction. Indeed, there may be more considerable challenges in the future in cases where the relationship between regulation and the range of interests are not as straightforward or where the goals are more complex than returning to agricultural use. In the two cases of successful decommissioning in the UK, the wind farms had planning conditions requiring the removal of the infrastructure after 25 years of operation; however, such time-limited planning consents are not commonplace internationally. Many sites across the globe, as well as some of the earliest wind farms in the UK, have planning consents that require removal of the turbines when the infrastructure stops working for a specified period of time (often 6–12 months). Such consents rely on enforcement action from the local authority to ensure turbines are removed unless the developer decides it is in their interest to do so.

There is potential for greater challenges to occur for some of the earliest wind farms where planning permission failed to legally require full decommissioning of the site. Such situations create the potential for infrastructure abandonment. This is a concern that has been raised in the UK and exemplified in the case of the Kirkby Moor wind farm, which reveals that where decommissioning planning conditions are lacking, there is a potential for large visible elements to be left in perpetuity (see the following box). Potential site abandonment and dereliction risk marring the "green" connotations of wind energy, whatever claims might be made about the potential reversible nature of the impacts of such technologies. Concerns regarding abandonment and dereliction already form a dimension of wind farm opposition, evident in the narratives of campaign groups.

Box 17.1:   The Potential for Infrastructure Abandonment in Kirkby Moor

The original planning consent for the Kirkby Moor wind farm in England (granted permission in 1992) imposed legal decommissioning conditions requiring removal of the turbines but not of any of the associated infrastructure, such as transformer substations, cabling, and access tracks, after 25 years of operation. There was no legal onus on the developer or any other party to remove these items, creating the potential for their abandonment on the moorland. This created a situation for both the repowering application (in 2015) and life-extension application (in 2017) where the developer was able to offer significantly improved decommissioning and site restoration as part of the applications (see Windemer and Cowell 2021). Members of the local community felt that improved decommissioning provisions were used as a bribe during the applications (Windemer 2019b). Both the repowering application and the life-extension application were refused by the local planning authority, with the life-extension application (to keep the existing infrastructure in place for an additional eight years) granted at the national decision-making level through a planning appeal. The choice between partial infrastructure removal after 25 years of operation versus more complete removal of the infrastructure at a future point proved to be a highly disputed and difficult decision, with decision makers ultimately deciding to grant permission for the lifetime extension (Windemer and Cowell 2021).

Figure 17.1   Photograph (author's own) showing the scale of the transformer substation.

Challenges of potential abandonment will be particularly acute in sites that lack both adequate legal decommissioning conditions or financial decommissioning bonds. Even for some sites that have decommissioning bonds in place, it has been recognized that 25 years ago the expected costs of decommissioning would have been an estimate, and thus, the bonds used are likely to be far lower than the actual costs required for decommissioning. UK developers consider that decommissioning will be self-funding through the value of the materials in cases where the turbines are in a good condition and can be resold. However, while there are some elements for which market incentives may be sufficient to encourage removal, such as turbines, for which there is a recognized secondhand market (Andersen et al. 2014), it is more difficult to envisage how markets could develop for others, such as removal of foundations, access roads, and grid connections. Moreover, if the turbines have no life left in them, then selling the turbines and parts is not expected to cover decommissioning costs.

The challenge of infrastructure removal and potential abandonment is significant for the global transition to low-carbon energy. If existing sites are left abandoned, then this could create a significant reputational risk for the sector, potentially impacting public attitudes toward future renewable energy applications. Repowering or life extension could provide an opportunity to mediate potential issues of abandonment through providing an opportunity to construct a new legal agreement that includes elements not included the first time around – improved decommissioning requirements. However, while repowering and life extension can provide the benefit of improving decommissioning requirements, this raises difficult questions regarding the trade-off between the duration of turbines and the duration of other elements of the infrastructure. This creates situations in which site restoration and the difficulties of securing this without regulatory support may be used to try to gain support for the new application, similar to cases where opencast coal mining has been justified in relation to the restoration of sites previously used for deep mining (see Milbourne and Mason 2017). In such cases, a longer life for a wind farm – and potentially an increase in energy capacity – is the public "price" of greater assurance of the eventual end-of-life outcome.

Wind energy does have material impacts, and the spatially extensive nature of wind farms in rural areas gives the legacy issue significance. While there have been cases of traditional windmills becoming positive, nostalgic elements of rural landscapes (Pasqualetti 2000), modern turbines are much larger and different in appearance. A further aspect of decommissioning with particular relevance to the image and reputation of renewables and thus with the potential to impact the global transition in energy sources is how the land is restored on sites that are decommissioned and if future land uses are facilitated. Wind farms reaching end-of-life may leave concrete foundations or other equipment, and projects may shape landscape meanings and perceptions if left in situ. The use of legal planning agreements can hold developers to land remediation and habitat creation actions; however, such agreements can be hampered by difficulties in valuing the natural environment (see Boucher and Whatmore 1993), and there appear not to be any studies assessing the efficacy of such planning agreements in the renewable energy sector. The works on end-of-life of minerals provide potentially useful insights regarding longer-term considerations of the land, particularly concerning decommissioning, facilitating a future land use, and how social attitudes may change over time, influencing future uses of the land. McHaina (2001) identified that the subsequent use of mining sites is decided based on several factors, including the current surrounding land use at the time of decommissioning, environmental impacts, and the possibility of reusing site infrastructure. This links to ongoing debates regarding the possibility of creating environmental, social, and economic benefits through developing new land uses on mineral sites (Zhang et al. 2011).

Conclusion

Planning for the future of existing renewable energy sites is vital if renewables are to make a long-term contribution to a global decarbonized energy system and command social legitimacy. Shifts in the move to low-carbon energy systems can be seen to occur quickly when happening alongside policy learning, innovations, and industrial and societal advancements (Araujo 2017). Thus, there is need to consider the wider context of how the future of existing energy sites is being considered. Increasing competition for land that is suitable for renewable energy infrastructure adds further importance to this issue. As this chapter has identified, the challenges that are arising as existing onshore renewable energy infrastructure reaches the end of its operational or consent life have the potential to decrease energy output or create reputational risks, impacting the speed of the global transition to low-carbon energy. On the other hand, upgrading existing renewable energy infrastructure through repowering provides a key opportunity to increase energy output from existing sites while also potentially creating additional social and environmental benefits.

Existing considerations of the global transition to low-carbon energy have, understandably, focused largely on the development dynamics of new renewable energy sites or the transition away from fossil fuels. Meanwhile, considerations of wind energy have concentrated overwhelmingly on initial consenting decisions, tacitly assuming that this is the critical decision-point shaping the evolution of wind energy capacity into the future. Such research often appears to consider the development and decision-making process in simple binary terms (i.e., the infrastructure was not there and now it is), ignoring the scope for projects to evolve and change over time. Where end-of-life options have been considered, this has largely been from an economic or technical perspective, lacking consideration of how end-of-life decision making occurs in practice and how it is considered by the various actors involved.

The example of aging onshore wind assets in the UK has demonstrated the importance of thinking about the future of existing renewable energy sites when considering the global transition to low-carbon energy. By doing so, this chapter has revealed the challenges that may arise at the end-of-life point. End-of-life decision-making will not necessarily be straightforward as sites change over time. Such changes may include physical changes such as accumulating developments and land designations and changes in the social landscape that can influence the decision-making context. This chapter has also revealed the significance of changes in economic or policy support. Periods of policy absence and turbulence and challenging economic conditions, including the removal of subsidies, have led to delays in submitting repowering applications or in developers pursuing a lower-risk strategy of life extension rather than replacing existing turbines with new, more efficient turbines that would increase overall energy output. These challenges have the potential to impact the long-term ability of existing energy generation sites to future energy production, particularly if a longer-term approach to sites is not considered.

Maintaining consented renewable energy capacity in place over time involves a range of opportunities, issues, and complexities that ought to be seen dynamically. This chapter has revealed the potential challenges surrounding end-of-life decision-making. It has shown that end-of-life is a bundle of concerns, affecting (1) the renewable energy equipment, with developer assessments of viable physical or commercial life and the benefits of replacement; (2) the temporal (and other) terms of any planning consent; and (3) the ongoing presence of a renewable energy-generating facility and its relationship to the changing site, landscape, and public. Each has its own temporality, which require coordination but also create the possibility for challenges. There are also a host of factors influencing how local attitudes to renewable infrastructure may change over time, impacting the possibility for future repowering. The most significant factors influencing public responses include community relations with the developer, existing benefits from the site, and changes in the surrounding landscape. Public attitudes can also be influenced by how people conceptualize the baseline against which to judge end-of-life applications – that is, if they consider the site to have become suitable for use as a wind farm in the long term or if they are waiting for the land to return to its previous use.

End-of-life considerations are increasingly important as the global stock of renewable energy capacity is not fixed; rather, it is a fluid entity subject to ongoing needs for repair, adaptation, and re-consent. Although most UK end-of-life applications so far have been consented, there are no guarantees that renewable energy sites will continue in the future as subsidies end and existing infrastructure wears out (see also Kooij et al. 2018). UK research unearthed examples of refusals (Windemer 2019a), and there is undoubtedly potential for the multiple factors pervading end-of-life decisions to slip into new alignments that make repowering or life extension less certain. Some sites may not have performed as well as originally expected, or the changes that occur over time, impacting sites and the wider setting of sites, may impact the suitability of sites for repowering. Meanwhile, the global transition to low-carbon energy may bring new, more efficient technologies that may replace existing technology. Decommissioning of existing renewable energy thus also forms an important but often little considered element of the global transition in energy sources. To date, studies have paid limited attention to decommissioning with an assumption that it will not cause challenges. While two cases of decommissioning have occurred in the UK without challenge, it may not always be so simple, particularly in circumstances where adequate financial support is not in place. As this chapter has revealed, cases of inadequate decommissioning or infrastructure abandonment could create a significant reputational risk for the renewables sector; thus, there is a need for future work to consider decommissioning of existing renewable energy sites as a part of the global transition to low-carbon energy.

This chapter has demonstrated how adopting longer-term considerations of existing energy sites is an important, but currently under considered, aspect of the global transition to low-carbon energy. While this chapter has focused on the experience of wind energy, it can be seen that the same challenges will occur for other forms of renewable energy infrastructure and thus lessons to be learned remain the same. The object at stake – in this case, renewable energy infrastructure – makes a difference, whether it is treated as an entity like a product occupying space or as a more complex assemblage of machinery, social, and ecological relations that evolve over time. There is a need for future research to consider how renewable energy sites change over time, how this will impact future decision making, and how this should be regulated in the context of considering the ability of these sites to continue to contribute to the low-carbon energy system of the future.

This chapter provides a number of recommendations for policymakers. First, there is a need for planning policy to guide the future of existing renewable energy sites. Replacing existing infrastructure with new, more efficient technology has the potential to significantly increase energy generation; however, such decision-making is not straightforward as sites will have changed over time. Policymakers need to ensure that decisions on the future of existing renewable energy sites will be made in a way that ensures that communities and the environment also benefit. Potential considerations here could involve opportunities for communities to co-own part of repowered sites. Given the speed of technological change, and the often-slow speed of the permitting system, it is also suggested that permitting or siting licenses enable some leeway in the project technology specifications (e.g., a percentage change in the size of turbines) rather than requiring exact specifications that cannot then be altered. Speeding up the permitting system for renewables also has an important role to play here. One element of this could involve a quicker permitting system for smaller modifications, such as changing the size of blades on existing turbines. Additionally, as this chapter has shown, there is also a need to ensure that adequate policy attention is given to decommissioning.

References

Aitken, Mhairi. 2010a. "Why We Still Don't Understand the Social Aspects of Wind Power : A Critique of Key Assumptions within the Literature." *Energy Policy* 38 (4): 1834–1841.

———. 2010b. "Wind Power and Community Benefits: Challenges and Opportunities." *Energy Policy* 38 (10): 6066–6075.

Andersen, P.D., A. Bonou, J. Beauson, and P. Brøndsted. 2014. "Recycling of Wind Turbines." In *DTU International Energy Report 2014: Wind Energy – Drivers and Barriers for Higher Shares of Wind in the Global Power Generation Mix.*, 91–97. Technical University of Denmark (DTU).

Araujo, K. 2017. *Low Carbon Energy Transitions: Turning Points in National Policy and Innovation.* New York: Oxford University Press.

Blowers, Andrew. 2017. "Nuclear's Wastelands Part 1–Landscapes of the Legacy of Nuclear Power." *Town and Country Planning*, 2017.

Boston, J. 2016. *Governing for the Future: Designing Democratic Institutions for a Better Tomorrow.* Emerald Group Publishing.

Boucher, Susan, and Sarah Whatmore. 1993. "Green Gains ? Planning by Agreement and Nature Conservation." *Journal of Environmental Planning and Management* 36 (1): 33–49.

Bristow, Gillian, Richard Cowell, and Max Munday. 2012. "Windfalls for Whom ? The Evolving Notion of 'Community' in Community Benefit Provisions from Wind Farms." *Geoforum* 43 (6): 1108–1120.

Couclelis, Helen. 2005. "'Where Has the Future Gone?' Rethinking the Role of Integrated Land-Use Models in Spatial Planning." *Environment and Planning A* 37 (8): 1353–1371.

Cowell, Richard, Gill Bristow, and Max Munday. 2011. "Acceptance, Acceptability and Environmental Justice: The Role of Community Benefits in Wind Energy Development." *Journal of Environmental Planning and Management* 54 (4): 539–557.

Damborg, S. and S. Krohn. 1999. "On Public Attitudes towards Wind Power." *Renewable Energy* 16 (1–4): 954–960.

Ferrell, Shannon L., and Eric A. DeVuyst. 2013. "Decommissioning Wind Energy Projects: An Economic and Political Analysis." *Energy Policy* 53: 105–113.

Friedl, Christina, and Johannes Reichl. 2016. "Realizing Energy Infrastructure Projects – A Qualitative Empirical Analysis of Local Practices to Address Social Acceptance." *Energy Policy* 89: 184–193.

Grau, Leonie, Christopher Jung, and Dirk Schindler. 2021. "Sounding out the Repowering Potential of Wind Energy—A Scenario-Based Assessment from Germany." *Journal of Cleaner Production* 293: 126094. https://doi.org/10.1016/j.jclepro.2021.126094.

Himpler, Sebastian, and Reinhard Madlener. 2012. "Repowering of Wind Turbines : Economics and Optimal Timing." FCN Working Paper No. 19. https://papers.ssrn.com/sol3/papers.cfm?abstract_id=2236265.

Hirsh, Richard F., and Benjamin K. Sovacool. 2013. "Wind Turbines and Invisible Technology: Unarticulated Reasons for Local Opposition to Wind Energy." *Technology and Culture* 54 (4): 705–734.

Ibarra, José Manuel Nicolau, and Mariano Moreno De las Heras. 2005. "Opencast Mining Reclamation." In *Forest Restoration in Landscapes*, 370–378. New York: Springer.

IPCC. 2021. "Summary for Policymakers." In Masson-Delmotte, V., P. Zhai, A. Pirani, S.L. Connors, C. Péan, S. Berger, N. Caud, Y. Chen, L. Goldfarb, M.I. Gomis, M. Huang, K. Leitzell, E. Lonnoy, J.B.R. Matthews, T.K. Maycock, T. Waterfield, O. Yelekçi, R. Yu, and B. Zhou (eds.), *Climate Change 2021: The Physical Science Basis. Contribution of Working Group I to the Sixth Assessment Report of the Intergovernmental Panel on Climate Change*, 3–32. Cambridge, United Kingdom and New York, NY, USA: Cambridge University Press. doi:10.1017/9781009157896.001.

Jaber, Suaad. 2013. "Environmental Impacts of Wind Energy." *Journal of Clean Energy Teachnologies* 1 (3): 251–254.

Kooij, Henk-jan, Marieke Oteman, Sietske Veenman, Karl Sperling, Dick Magnusson, Jenny Palm, and Frede Hvelplund. 2018. "Between Grassroots and Treetops : Community Power and Institutional Dependence in the Renewable Energy Sector in Denmark, Sweden and the Netherlands." *Energy Research & Social Science* 37: 52–64. https://doi.org/10.1016/j.erss.2017.09.019.

Kramer, Gert Jan, and M Haigh. 2009. "No Quick Switch to Low-Carbon Energy." *Nature* 462: 568–569.

Malandrino, Ornella, Daniela Sica, Mario Testa, and Stefania Supino. 2017. "Policies and Measures for Sustainable Management of Solar Panel End-of-Life in Italy." *Sustainability (Switzerland)* 9 (4): 1–15. https://doi.org/10.3390/su9040481.

McHaina, D.M. 2001. "Environmental Planning Considerations for the Decommissioning, Closure and Reclamation of a Mine Site." *International Journal of Surface Mining, Reclamation and Environment* 15 (3): 163–176.

Milbourne, Paul, and Kelvin Mason. 2017. "Environmental Injustice and Post-Colonial Environmentalism: Opencast Coal Mining, Landscape and Place." *Environment and Planning A* 49 (1): 29–46.

Möller, Bernd. 2010. "Spatial Analyses of Emerging and Fading Wind Energy Landscapes in Denmark." *Land Use Policy* 27 (2): 233–241.

Munday, Max, Gill Bristow, and Richard Cowell. 2011. "Wind Farms in Rural Areas : How Far Do Community Benefits from Wind Farms Represent a Local Economic Development Opportunity ?" *Journal of Rural Studies* 27 (1): 1–12.

Myers, Dowell, and Alicia Kitsuse. 2000. "Constructing the Future in Planning: A Survey of Theories and Tools." *Journal of Planning Education and Research* 19 (3): 221–231.

Newell, Peter, and Dustin Mulvaney. 2013. "The Political Economy of the 'Just Transition.'" *Geographical Journal* 179 (2): 132–140. https://doi.org/10.1111/geoj.12008.

Otto, Antje, and Markus Leibenath. 2014. "The Interrelation between Collective Identities and Place Concepts in Local Wind Energy Conflicts." *Local Environment* 19 (6): 660–676.

Pasqualetti, Martin. 2000. "Morality, Space, and the Power of Wind-Energy Landscapes." *The Geographical Review* 90 (3): 381–394.

Pasqualetti, Martin, Paul Gipe, and Robert Righter. 2002. *Wind Power in View: Energy Landscapes in a Crowded World. Windpower in View.* Academic Press.

Philpott, A., and R. Windemer. 2022. "Repower to the People: The Scope for Repowering to Increase the Scale of Community Shareholding in Commercial Onshore Wind Assets in Great Britain". *Energy Research & Social Science* 92: 102763.

Ricci, Miriam, Paul Bellaby Å, and Rob Flynn. 2010. "Engaging the Public on Paths to Sustainable Energy : Who Has to Trust Whom ?" *Energy Policy* 38 (6): 2633–2640.

Roddis, Philippa, Stephen Carver, Martin Dallimer, Paul Norman, and Guy Ziv. 2018. "The Role of Community Acceptance in Planning Outcomes for Onshore Wind and Solar Farms: An Energy Justice Analysis." *Applied Energy* 226 (May): 353–364. https://doi.org/10.1016/j.apenergy.2018.05.087.

Scheidel, Arnim, and Alevgul H Sorman. 2012. "Energy Transitions and the Global Land Rush : Ultimate Drivers and Persistent Consequences." *Global Environmental Change* 22 (3): 588–595.

Sustainable Energy Ireland. 2003. "Attitudes towards the Development of Wind Farms in Ireland." 2003. www.sei.ie/uploads/documents/upload/publications/Attitudes_towards_wind_pdf.

Unnewehr, Jan Frederick, Eddy Jalbout, Christopher Jung, Dirk Schindler, and Anke Weidlich. 2021. "Getting More with Less? Why Repowering Onshore Wind Farms Does Not Always Lead to More Wind Power Generation—A German Case Study." *Renewable Energy* 180: 245–257. https://doi.org/10.1016/j.renene.2021.08.056.

Van Der Knapp, W.G.M and Bart Jan Davidse. 2010. "While Time Goes by ; Dealing with Time and Multi-Dynamics in Spatial Planning and Design." *Resource Management, Energy and Planning* 438–439.

Walker, Gordon, Patrick Devine-Wright, Sue Hunter, Helen High, and Bob Evans. 2010. "Trust and Community: Exploring the Meanings, Contexts and Dynamics of Community Renewable Energy." *Energy Policy* 38 (6): 2655–2663.

Warren, Charles R., Carolyn Lumsden, Simone O'Dowd, and Richard V. Birnie. 2005. "'Green On Green': Public Perceptions of Wind Power in Scotland and Ireland." *Journal of Environmental Planning and Management* 48 (6): 853–875.

Windemer, Rebecca. 2019a. "Considering Time in Land Use Planning: An Assessment of End-of-Life Decision Making for Commercially Managed Onshore Wind Schemes." *Land Use Policy* 87 (June): 104024. https://doi.org/10.1016/j.landusepol.2019.104024.

———. 2019b. "Managing (Im)Permanence: End-of-Life Challenges for the Wind and Solar Energy Sectors," Unpublished Thesis.

Windemer, Rebecca, and Richard Cowell. 2021. "Are the Impacts of Wind Energy Reversible? Critically Reviewing the Research Literature, the Governance Challenges and Presenting an Agenda for Social Science." *Energy Research and Social Science* 79: 1–26.

Wolsink, Maarten. 2007. "Wind Power Implementation: The Nature of Public Attitudes: Equity and Fairness Instead of 'Backyard Motives.'" *Renewable and Sustainable Energy Reviews* 11 (6): 1188–1207.

———. 2018. "Co-Production in Distributed Generation: Renewable Energy and Creating Space for Fitting Infrastructure within Landscapes." *Landscape Research* 43 (4): 542–561.

Zhang, Jianjun, Meichen Fu, Ferri P. Hassani, Hui Zeng, Yuhuan Geng, and Zhongke Bai. 2011. "Land Use-Based Landscape Planning and Restoration in Mine Closure Areas." *Environmental Management* 47 (5): 739–750.

Ziegler, Lisa, Elena Gonzalez, Tim Rubert, Ursula Smolka, and Julio J Melero. 2018. "Lifetime Extension of Onshore Wind Turbines: A Review Covering Germany, Spain, Denmark, and the UK." *Renewable and Sustainable Energy Reviews* 82: 1261–1271.

18

THE POLICIES AND POLITICS OF JAPAN'S MISSED OPPORTUNITY FOR A POST-FUKUSHIMA ENERGY TRANSITION

Masahiro Matsuura

Introduction: An Opportunity to Transition to Renewables Created by a Disaster

Natural disasters and other external shocks often trigger large-scale transitions in socioeconomic systems. A disaster can abruptly terminate all traditional activities and operations, force citizens to recover from the impacts, and spur rethinking of more resilient policies and systems (Kingdon, 1995; Birkland, 1998). While natural disasters and other kinds of crises are often tragic, they can also catalyze transitions toward a more resilient and sustainable society by bringing about rethinking and new action (Boin, Hart and McConnell, 2009; Loorbach and Huffenreuter, 2013; Patterson et al., 2021).

The Fukushima accident of 2011 provides important insight into how a catastrophe could but might not be a trigger for transition. The disaster began with a magnitude 9 earthquake that was followed by a tsunami and nuclear plant accident (Fire and Disaster Management Agency, 2013). It was the most severe earthquake in Japan since records began. More than 18,000 individuals were confirmed dead and 2,800 remain missing (Fire and Disaster Management Agency, 2013). The meltdown at the Fukushima Daiichi power plant forced the immediate evacuation of 1,600 km^2 of land, containing more than 100,000 individuals (Fukushima Prefecture, 2021).

Immediately before the disaster, nuclear power plants in Japan produced 288.2 TWh, which corresponds to 25.1% of the total generation in 2010 (Figure 18.1) (Ministry of Economy, Trade and Industry, 2021a). By 2014, none of Japan's commercial nuclear reactors were operating. In 2019, nuclear power plants in Japan produced 63.8 TWh, equaling 6.2% of total electricity generation. Meanwhile, the share of fossil-fuel-based (oil, gas, and coal) generation increased from 65.4% of total generation in 2010 to 75.7% in 2019 (ibid.).

Japan could have accelerated the transition to renewable sources during the post-earthquake decade by leveraging the crisis. On the contrary, it was the decade characterized by a return to fossil fuel sources, as is evident in Figure 18.1. Why did Japan miss the opportunity to transition to renewables after the Fukushima disaster?

DOI:10.4324/9781003183020-21

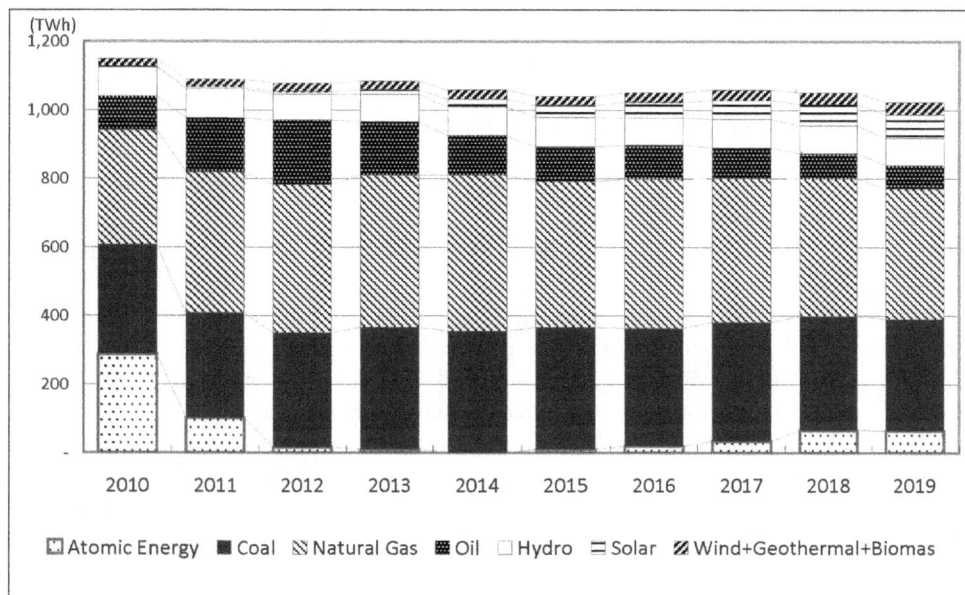

Figure 18.1 Japan's electricity generation portfolio by fuel type (2010–2019).

Source: Ministry of Economy, Trade and Industry 2021a.

While multiple explanations of the phenomena are plausible, one possible answer is agenda setting by the Liberal Democratic Party's (LDP) administration. This chapter will explore the political maneuvers that brought Japan to a standstill with respect to an energy transition to renewable sources.

The Beginning: Great East Japan Earthquake

Meltdown at the Fukushima Daiichi Nuclear Power Plant

At 2:46 p.m. on March 11, 2011, a magnitude 9 earthquake occurred 80 miles off the coast of the Miyagi Prefecture in the northern part of Japan (Fire and Disaster Management Agency, 2013, p. 12). No earthquake with such force had ever been recorded in the modern history of Japan (Cabinet Office, 2011, p. 2). Damage to buildings and direct casualties from the quake's vibration were relatively contained compared to other major Japanese earthquakes in the past that had epicenters beneath the densely inhabited region (Fire and Disaster Management Agency, 2013, pp. 82–84). The leading cause of subsequent damage was tied to the tsunami that eliminated most towns and villages along the Pacific coast. As much as 90.6% of the total death toll was associated with drowning (Fire and Disaster Management Agency, 2013, p. 84).

In addition to the natural consequences of the disaster, the March 11 earthquake was unique owing to the meltdown at the Fukushima Daiichi Nuclear Power Plant operated by the Tokyo Electric Power Company (TEPCO). The initial earthquake hampered the electricity supply to the plant from the local grid; however, the plant had multiple auxiliary diesel-powered generators, so the reactor could be monitored and managed. Greater operational damage ensued when a 14–15 m high tsunami struck the plant approximately 50 minutes after the earthquake (TEPCO, 2011). The plant's protective barriers were designed for a 5.7 m tsunami, so the plant was literally engulfed by the tsunami (National Diet Independent Investigation Commission, 2012). All auxiliary power supplies, both AC and DC, were completely damaged. The control room suffered power outage. The reactor lost its active cooling function and became almost uncontrollable. Although the passive water supply remained operational for some time and the field operators made several attempts to vent the steam inside the containment vessel, the three reactors that were fully operating at the time of the earthquake all melted down within a few days of the earthquake (NHK, 2021).

In the days following the earthquake, the reactors released radioactive substances into the air. The contaminated air traveled to different parts of Japan. On March 15, contamination levels were recorded 140 miles away in downtown Tokyo (Shinjuku), measuring 0.496 micro Gy/h, which was more than ten times higher than the typical readings (Tokyo Metropolitan Institute of Public Health, n.d.). Multiple embassies in Tokyo encouraged their citizens to evacuate Tokyo based on concerns over heightened radioactivity (Szep and Jones, 2011).

The area that was most impacted by radiation was to the northwest of the Daiichi plant due to rainfall that contaminated the soil with radioactive substances (Japan Atomic Energy Agency, 2011). The residents who lived within the 20 km radius from the plant were forced to evacuate the area immediately and could not return for several years (Fukushima Prefecture, 2013).

Nationwide Reactor Shutdowns

The Tokyo Electric Power Company, as well as other regional utilities in Japan, suffered from the forced shutdown of their nuclear power plants. At the time of the earthquake, 54 commercial reactors were operational, with 34 of them providing electricity to the grid (Japan Atomic Industrial Forum, 2012). Four months after the earthquake, the Nuclear and Industrial Safety Agency required operators "stress tests" for reevaluating the safety of nuclear reactors (Nuclear and Industrial Safety Agency, 2011).

By May 5, 2012, all reactors in Japan, even those at a great distance from Fukushima, had been shut down for reevaluation. In particular, the Kansai Electric Power Company, which provided electricity to the areas around Osaka and Kyoto, demanded the immediate restart of their reactors that passed a new round of tests. The national government, as well as local governments, albeit reluctantly, agreed to restart these reactors. This triggered a massive opposition movement against the administration by civil society groups (Hasegawa, 2014). The National Diet building in Tokyo was surrounded every Friday evening by more than 100,000 individuals who chanted "stop the nuclear" (ibid.). It was the first instance of mass public protest in front of the Diet building since the anti-US-Japan Security Treaty movement in 1960 (ibid.).

In September 2012, the Nuclear Regulation Agency was established in order to ensure the independence of regulators from the Ministry of Economy, Trade, and Industry, which promoted nuclear power. A new set of regulations for disaster prevention was issued immediately after its inception (Nuclear Regulation Agency, 2019). The process revision indicated that municipal governments within a radius of 30 km from the plants, which is defined as Urgent Protective Action Planning Zones, should prepare emergency evacuation plans (Cabinet Office, n.d.). The process instituted a new round of negotiations for operators with local governments. Unlike the pre-accident period in which local authorities of the municipalities adjacent to those where a plant is located did not have veto power, it is now required that plant operators secure their consent to operate the reactors (Nakazawa and Tatsumi, 2022).

Dealing with the host communities is a major challenge for operating nuclear power plants in Japan (Aldrich, 2010). As of February 2022, ten reactors in the western part of the country have restarted, but six other reactors have not been able to restart despite obtaining the national government's approval (Federation of Electric Power Companies of Japan, n.d.). Two reactors at TEPCO's Kashiwazaki-Kariwa Atomic Energy power plants, for example, passed the national government's renewal inspections in 2017 but have not been granted approval by the local government (Nagahashi, 2021).

The Short-Lived DPJ Administration

Tainted Image of Prime Minister Kan's Responses

When the magnitude 9 earthquake occurred off the coast of Miyagi on March 11, 2011, the national government was led by Prime Minister (PM) Naoto Kan with the Democratic Party of Japan (DPJ) being the majority in the lower house of the Diet. After a landslide victory in the 2009 election, the DPJ had lost the upper house election in 2010, and its prominence as an alternative to the long-dominating LDP had weakened.

The response to the earthquake was a major challenge to the administration for its survival. PM Kan, who has an engineering degree from the Tokyo Institute of Technology, made several attempts to take decisive leadership in the aftermath of the meltdown in Fukushima. According to investigations by independent committees organized by both the Cabinet and Diet, a lack of streamlined communication between the PM's office, the TEPCO, and staff members at the Fukushima Daiichi site was a confounding factor (National Diet of Japan Fukushima Nuclear Accident Independent Investigation Commission, 2012). Within a few days of the earthquake, a number of reactors eventually experienced meltdowns, emitting substantial radiation.

PM Kan became a symbol of the post-accident chaos. For instance, he visited an evacuation shelter in Fukushima on April 21 (ANN News, 2011). As he left the room, one of the evacuees shouted to him desperately, "Are you leaving us so soon?" He returned to the person and apologized for the insensitivity. The image was captured by the media's cameras and repeatedly broadcast on news programs.

In the nationwide local government elections on April 24, a number of DPJ-affiliated councilpersons lost their seats. By then, the PM had almost lost his leadership within the party. After weathering a motion of no confidence by the opposition party in the Diet, PM Kan announced on June 27 that he would resign only after enacting the law for establishing the feed-in-tariff (FIT) system for promoting renewable energy (Asahi Shinbun, 2011). On August 26, 2011, PM Kan resigned after the FIT legislation was passed (Mainichi Shinbun, 2011).

FIT Legislation

The previously noted FIT legislation was a means for facilitating an energy transition after Fukushima. Prior to that, the Japanese government had employed the renewable portfolio standard system since 2002, which required the ten regional monopolies to source a certain proportion of renewable energy in their electricity generation (Ministry of Economy, Trade and Industry, n.d.-b). The system was insufficient for increasing the share of renewables: from 0.7% in 2002 to 1.1% in 2009 (Ministry of Economy, Trade and Industry, 2020). Renewable energy advocates argued for the FIT system to assure the investors a rate of return (Iida, 2002; Asahi Shinbun, 2008). The FIT system was enacted in 2011, only seven months after the earthquake, as the Act on Special Measures Concerning Procurement of Electricity from Renewable Energy Sources by Electricity Utilities. This system began in April 2012.

What may be seen as a rapid introduction of the FIT could be considered, in policy terms, a case of a natural disaster serving as a trigger for a transition. Citizens were shocked by the meltdown, and advocates for nuclear power who had argued for its safety lost their political power, albeit temporarily. The sudden shortage in electricity supply and the pressure to reduce the use of fossil fuels under the Kyoto Protocol supported the rapid introduction of renewables. These factors, along with PM Kan's introduction of the FIT system, acted as three streams of agenda setting that enabled a policy window to open, enabling the legislation to pass the Diet in a relatively short timeframe (Kingdon, 1995; Tanaka et al., 2020).

The Japanese FIT system guarantees payment of ¥40 (approximately $0.50) per kilowatt-hour for 20 years to a solar photovoltaic power plant with a capacity exceeding 10 kW (Ministry of Economy, Trade and Industry, n.d.-a). This rate was very favorable to investors, producing a number of projects in 2012 (Ministry of Economy, Trade and Industry 2021a). One might argue that this was a moment of significant energy transition in Japan. However, the rate for newly approved plants was soon reduced after the DPJ was ousted from national politics in 2012 (ibid.). The momentum for the rapid introduction of solar photovoltaic plants was not sustained to realize a major energy transition in Japan; this is discussed in subsequent sections.

Energy and Environment Council

During Kan's administration (2010 to 2011), the Energy and Environment Council was established on June 7, 2011 under the authority of the Cabinet Secretariat. Consisting mostly of cabinet members from the administration in power, the council revised the national Strategic Energy Plan. Guided by the Basic Act on Energy Policy of 2002, these plans must be periodically revised every three years. While the most recent plan predating the Fukushima accident was established in 2010, the circumstances warranted an update. By June 29, 2011, the council had published three goals: (1) less dependence on nuclear power, (2) a transition to consumer-driven distributed systems, and (3) expanded citizen's dialogue (Energy and Environment Council, 2012).

Amid the controversy over the restart, the Energy and Environment Council released three scenarios of Japan's energy mix in June 2012 (ibid.). The options primarily focused on the share of nuclear power in 2030 being 0%, 15%, or 20–25%. The Council organized two venues for soliciting public input to choose an option: what was termed (1) "town hall meetings" that allowed a few selected citizens to speak out in conference halls in 11 different cities in Japan and (2) a "deliberative polling" event that invited 285 randomly selected individuals from all over Japan to a venue in Tokyo for two days in August (Mikami, 2015; Mah et al., 2021). This process was supervised directly by James Fishkin, the inventor of the participatory method (Ackerman and Fishkin, 2008; Fishkin, 2011). The invited citizens were asked to learn about the options from experts (both pro- and anti-nuclear), deliberate on these options, and express their choice in polling. While the results were not decisive, these events revealed that citizens, in general, were most supportive of the 0% option (Mikami, 2015; Mah et al., 2021).

Subsequently, the Council released the "Innovative Strategy for Energy and the Environment" on September 14, 2012. It sets out three goals: (1) realization of a society that is not dependent on nuclear power in the earliest possible future, (2) realization of a green energy revolution, and (3) ensuring a stable supply of energy. The first goal of denuclearization became the principal foci of the controversy.

The participatory processes and public opinion polls (Asahi Shinbun, 2012a) indicated public support for the denuclearization option. On the other hand, the privately owned electric companies that had invested in the reactors (Federation of Electric Power Companies of Japan, 2012) and manufacturing industries (Japan Association of Corporate Executives, 2012) feared rate hikes after the introduction of renewables and publicly denounced the strategy. One of the main supporters of the DPJ, the labor unions in the electric industry, also voiced their concerns regarding the strategy in relation to maintaining their jobs at nuclear power plants (Asahi Shinbun, 2012b). In fact, the institutional structures for nuclear power remained mostly the same even after the Fukushima incident (Hughes, 2021).

Another area of concern was the use of plutonium that would occur in reprocessing plants. Under the US-Japan cooperation agreement on civilian nuclear energy, the Japanese government was allowed to reprocess the spent nuclear fuel to produce plutonium, which would be mixed in fuel rods for purposes unrelated to the military (Watanabe, 2017). The Japanese government's strategy appeared contradictory in that it promised to continue the operation of the reprocessing plant while discontinuing the use of nuclear fuels. This nonproliferation concern of the US government was conveyed to PM Noda directly by the then secretary of state Hillary Clinton in the APEC summit meeting on September 8 (Asahi Shinbun, 2012c).

Opposition to the proposed strategy from many influential stakeholders was a significant blow to the DPJ government. The initial plan was to adopt the council's "strategy" document as the Cabinet's decision and then to translate this into a more formal national Strategic Energy Plan in a very short timeframe. In the end, however, the Cabinet could not adopt the strategy as such. The Cabinet's decision on September 19 read as follows:

> The energy and environmental policy from now on will follow up on the "Innovative Strategy for Energy and the Environment" (the Energy and Environment Council decision on September 14, 2012) and be implemented after responsible discussions with local governments and international communities, with understanding of the citizens, and with flexible and continuous evaluations and revisions.
>
> (Cabinet Secretariat, 2012, translation by the author)

Despite the ambiguity of the statement, the DPJ executives insisted that the strategy was adopted as the Cabinet's decision and ready for implementation (Asahi Shinbun, 2012d). Following the decision, the Ministry of Economy, Trade, and Industry convened an Advisory Committee for Natural Resources and Energy on September 18, 2012, to reformulate the formal Strategic Energy Plan. In the meeting, some of its members, including industry representatives who openly supported the pro-nuclear option, indicated their concerns about deciding the plan based on the published strategy (Ministry of Economy, Trade and Industry, n.d.-c). The national government could not move forward as it had planned.

Demise of the DPJ Administration

Repeated blunders by the administration raised questions for the public regarding the competence of the DPJ administration. By 2012, public support for the DPJ had nearly disappeared. Opinion polling by the *Asahi* newspaper in October 2012 indicated that only 18% of the citizens supported the Noda administration, the lowest ever (Asahi Shinbun, 2012e). The LDP repeatedly attacked the DPJ's poor performance in the Diet (Yomiuri Shinbun, 2012a). In a debate session of the lower house on November 14 between PM Noda and Shinzo Abe, who was elected as LDP's head in September, PM Noda suddenly announced that he would disband the lower house on November 16 (Asahi Shinbun, 2012f; Yomiuri Shinbun, 2012b). This decision led to a general election for the lower house, which was held on December 16.

The result was a disastrous one for the DPJ. Abe's LDP had a landslide victory, winning 294 of the 480 seats (Asahi Shinbun, 2012g; Yomiuri Shinbun, 2012c). The DPJ, which had won 308 seats in the previous election in 2009, won only 57 seats in the 2012 election (ibid.). In fact, a number of DPJ Diet members left the party before the election and formed small independent parties, but many of them also lost the election (primarily based on the single-member districts) (ibid.).

On December 26, 2012, Shinzo Abe assumed the role of PM. The new PM won the election with the slogan "Get Back Japan" (*Nihon Wo Torimodosu*) (Liberal Democratic Party, 2012). It was obvious to everyone that the words were primed to be followed by "from the DPJ" without mentioning it. The new PM would establish an entirely different agenda that would nullify the legacy of the DPJ administrations.

The Abe Administration's Plans and Realities of Revitalizing Nuclear Energy

Renewed Energy Plans

In the 2012 election, the LDP's manifesto promised to introduce as much renewable energy as possible in the following three years (Liberal Democratic Party, 2012). It also promised to determine the restarting of the nuclear power plants in three years and to establish "the best mix" for the electricity generation portfolio within ten years (ibid.).

The formal national Strategic Energy Plan, which the previous DPJ administration failed to finalize, had to be prepared immediately after the inception of the Abe administration. A completely new plan was adopted by the Cabinet on April 16, 2014 (Ministry of Economy, Trade and Industry, 2014).

The new Plan of 2014 did not acknowledge the efforts of the DPJ administration (ibid.). It ignored the existence of the Innovative Strategy for Energy and the Environment, determined by the DPJ government, and the public participation efforts for preparing the strategy. The plan stated,

> Nuclear power is an important base-load power source as a low carbon and quasi-domestic energy source . . . Dependency on nuclear power generation will be lowered to the extent possible by energy-saving and introducing renewable energy as well as improving the efficiency of thermal power generation, etc.
>
> (Ministry of Economy, Trade and Industry, 2014, p. 24)

While the language is unclear, which is typical of Japanese policy documents, the new policy indicated the Abe administration's intention of continuing the nuclear program as the "base-load power source" (ibid.). This was a major shift from the previous administration's strategy document that suggested its phasing out.

The Strategic Energy Plan of 2014 did not set any targets for the electricity generation portfolio (ibid.). A subsequent study at the Ministry of Economy, Trade, and Industry by the Electricity Supply-Demand Verification Subcommittee explored this in 2015. Its report, published in July 2015, expected that in 2030, the electricity generation sources would be oil (3%), coal (26%), natural gas (27%), nuclear (20–22%), hydro (8.8–9.2%), solar (7%), wind (1.7%), biomass (3.7–4.6%), and geothermal (1–1.1%) (Ministry of Economy, Trade and Industry, 2015). From this, one may surmise that the Abe administration had the intent of reviving nuclear power to a level that was similar to the pre-Fukushima era. Although the policy stated that it would expand renewables as much as possible, wind energy was predicted to constitute only 1.7% of total energy generation.

The Strategic Energy Plan was revised again in 2018. In this round, the Paris Accords was an additional mandate in preparing the new plan. The new policy, which was adopted by the Cabinet in July 2018, stated that renewable energy sources would be the "major power source" (Ministry of Economy, Trade and Industry, 2018). However, the position of nuclear power as a baseload power source did not change substantially in the new version. Moreover, the new policy document promoted the use of reprocessed fuel containing plutonium in commercial reactors. Taking these policy measures together, the Abe administration consistently supported the use of nuclear power.

Political Success Amid Public Opposition to the Nuclear Option

In contrast to the policy developments of the time, public perception remained opposed to nuclear energy. For instance, a survey by the Japanese national newspaper *Asahi Shinbun* in September 2014 indicated that 57% of the respondents were *against* the restarting of the nuclear power plants (25% *for* the restart) (Asahi Shinbun, 2014). Another survey in February 2018, indicated that 61% of the respondents were against the restarts while 27% were in favor (Asahi Shinbun, 2018). As citizen perception remained largely unchanged during the Abe administration, only a quarter of existing reactors could be restarted by the electric companies during the period (Ministry of Economy, Trade and Industry, 2022).

Meanwhile, the Abe administration had been successful in winning local and national elections. The PM disbanded the lower house twice (in 2014 and 2017), but the LDP maintained the majority in both elections. The LDP-Komei alliance also maintained its majority in the upper house through 2013, 2016, and 2019 elections. It is somewhat stunning to observe the contradiction between the citizen's animosity toward nuclear power and the strength of the Abe administration in the national elections.

In brief, the nuclear energy issue was not a key issue affecting voter choice of candidates, even after the Fukushima incident. From its inception, the Abe administration started a number of aggressive public spending programs and interventions into the market, dubbed *Abe-nomics* by the PM, which had some positive effects on the economy (Hausman and Wieland, 2014; Fukuda, 2015). Surveys consistently indicated that people were concerned about social security and economic revitalization. In a June 2016 survey that asked about priority issues for choosing a candidate in the 2016 upper house election, 35% of the respondents chose "pension and social security", 24% chose "economy and employment", while only 2% chose "energy policy" (Yomiuri Shinbun, 2016).

The administration was also able to distance itself from the controversial nuclear issues at the local level. Although the 20–22% nuclear energy in the 2030 target was presented by the METI (Ministry of Economy, Trade and Industry, 2015), the administration did not pursue the goal aggressively. The national government did not intervene, for example, in the debates over the restarting of reactors at the local level. In an August 2015 press conference, for instance, the then chief Cabinet secretary Yoshihide Suga reiterated that the decision to restart the reactors was a matter for private electric companies when he was asked about the controversy over the restarting of the Sendai Nuclear Power Plant (Umekawa, 2015). The national government could distance itself from negotiations between local communities and operators. In fact, the regulatory structure of the Japanese electricity generation industry allowed such non-involvement of the national government.

Transition to Renewables Slowing Down During the Abe Administration

Introduction of Solar Slowing Down

Based on what has been outlined thus far, one may see a lack of clear direction in Japanese energy policy during the Abe administration and the likely effect on the renewable energy industry. Despite the Strategic Energy Plan of 2014 promoting the introduction of renewables as much as possible for three years, implementation did not reflect these shifts.

Local controversies involving solar power development emerged in the mid-2010s. The favorable FIT rate of ¥40/kWh that was adopted by the previous DPJ administration triggered a sudden wave of investments in solar power plants in Japan (Li, Xu and Shiroyama, 2019). Along with grassroots initiatives for developing solar power, entrepreneurial businesses rushed to develop solar power plants in order to benefit from the favorable rate. The explosion of developments in different parts of Japan, based on the secured return from the FIT system, raised concerns from environmentalists and the general public regarding their negative effects on local environments. In September 2015, for instance, the riverbanks of the Kinu River in the Ibaraki Prefecture broke after an extraordinarily heavy rainfall. One of these riverbanks, which existed on private land, had been lowered in 2014 to develop a solar power plant (Ministry of Land, Infrastructure, Transport and Tourism, 2017), and the popular media attacked the solar plant by indicating it as one of the causes of the flood (Asahi Shinbun, 2015).

The electric companies, which were required to allow solar plant connection to the grid, were concerned about the natural fluctuations of supply. In early 2014, an extraordinary number of applications was made for grid connection by independent solar plants to grid operators before the announced lowering of the national FIT rate from ¥36/kWh to ¥32/kWh in April 2014 (Ministry of Economy, Trade and Industry, n.d.-a). In September 2014, the Kyushu Electric Power Company indicated that it would temporarily stop responding to requests from independent producers for grid connection (Kyushu Electric Power Company, 2014). Although the regional monopolies, as distributors, were required to purchase electricity from independent renewable generators, they could effectively decline the grid connection in order to balance the demand and supply. Here, the electricity industry was impeding the rapid expansion of solar power in Japan.

By 2017, roughly a few years after the Fukushima accident and the introduction of the FIT, the annual growth of solar power generation had slowed in Japan (Figure 18.2). The FIT rate

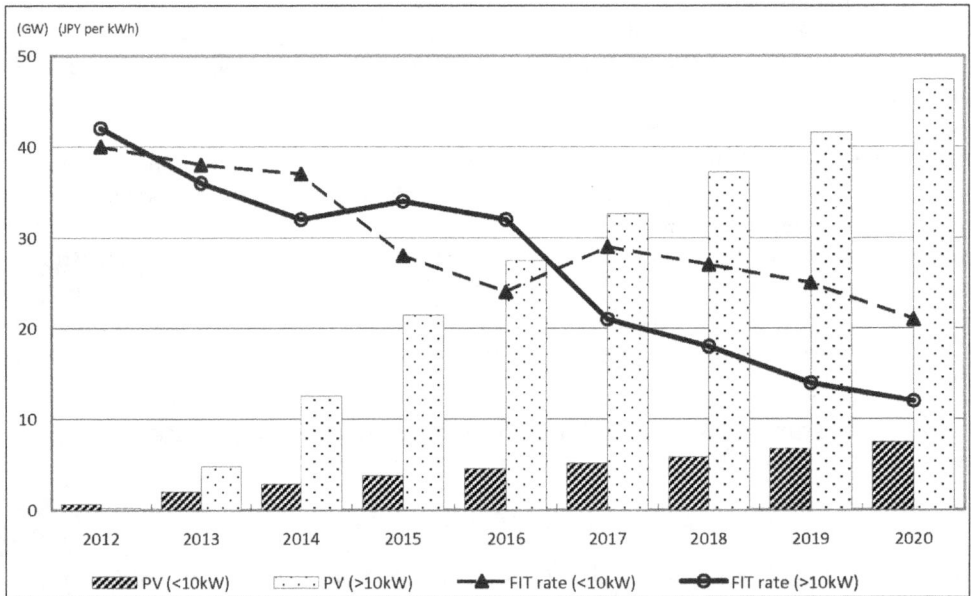

Figure 18.2 Cumulative installed solar power generation capacity and the feed-in tariff rate.

Sources: Ministry of Economy, Trade and Industry 2021a and n.d-a.

for solar power was reduced annually, making the new investments less profitable (Ministry of Economy, Trade and Industry, n.d.a). The additional capacity from newly approved solar projects each year crumbled from 25.4 GW in 2013 to 3.3 GW in 2017 (Ministry of Economy, Trade and Industry, 2021b, p. 9). The expansion of solar energy in Japan may have slowed since it was nearing the policy target. According to the Ministry of Economy, Trade, and Industry, solar power was expected to take a 7% share of power sources in 2030, equaling approximately 64 million kWh (Ministry of Economy, Trade and Industry, 2015). At the end of 2020, the installed capacity totaled 54.9 million kWh or 86% of the 2030 target (Ministry of Economy, Trade and Industry, 2021a). Here, some may argue that it was reasonable for the government to slow down the pace in order to avoid an oversupply from solar.

In fact, power fluctuation caused by weather changes is a major concern for connecting solar power to the grid. On the other hand, it can be alleviated with storage. The use of electric vehicles as a form of auxiliary battery, such as with a V2H (vehicle-to-home) system has been experimented with by Toyota as early as 2012 (Toyota Motor Corporation, 2012). Recently developed large-scale solar power plants, particularly in the Hokkaido region, where land is readily available, are now equipped with batteries for moderating natural fluctuations (Morais, 2020).

While the Abe administration provided financial support for such R&D projects, it was not aggressive enough to expand the use of solar power beyond the "outlook" stipulated in 2015. Drastic measures, such as the prospective ban of gas-powered vehicles for promoting EVs in 2035, would be taken after the PM resigned in 2020.

Forgotten Wind Options

In order to expand renewables, the administration could have focused on other renewable sources, such as wind power. In fact, to compensate for the failures of restarting reactors for meeting the 2030 target, the administration could have invested more in the development of wind energy much earlier.

Onshore wind power had already been explored in many parts of Japan, even before the 2011 earthquake (Maeda and Kamada, 2009). Of the 144 GW of onshore and 608 GW of offshore wind potential, up to roughly 2.55 GW of capacity was developed both onshore and offshore (Watanabe, 2014; New Energy and Industrial Technology Development Organization, 2018). Its expansion started around the turn of the century and continued to increase until 2011 (Figure 18.3). While the DPJ government, particularly during the Kan administration, had demonstrated its intention of expanding renewables by introducing the FIT system, the introduction of wind power slowed down significantly from 2012, even before the LDP assumed power.

Offshore projects had been explored in Japan before the 2011 earthquake. The New Energy and Industrial Technology Development Organization (NEDO), a subsidiary of the METI, had been undertaking offshore wind farm experiments since as early as 2008. Multiple offshore turbines, seven nearshore turbines by Wind Power Kashima in 2010, a 2 MW turbine by the TEPCO in 2012, and a 2 MW floating turbine by the Toda Company in 2014 had already been deployed at an experimental scale (Aoki, 2009; Fukumoto, 2013; Utsunomiya et al., 2014). These did not, however, expand to the commercial scale until the 2020s. No construction of offshore wind farms involving multiple turbines was initiated during the Abe administration.

While a complex set of factors may explain the wind power introduction in Japan (Mizuno, 2014), one possible reason for the tapering effect was the FIT rate for wind energy. Initially,

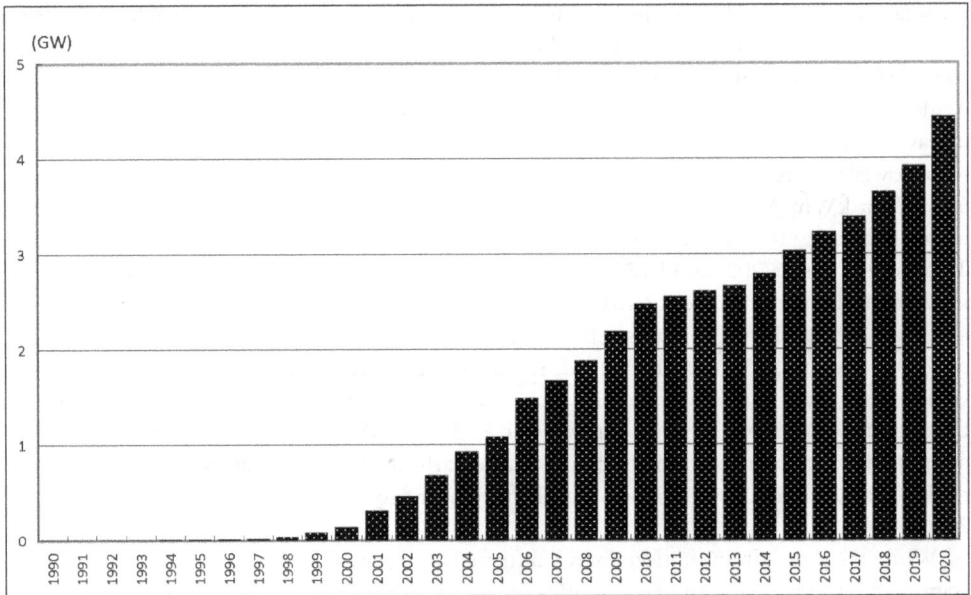

Figure 18.3 Cumulative installed wind power generation capacity in Japan.

Sources: New Energy and Industrial Technology Development Organization 2018 and Japan Wind Power Association 2021.

in 2012, the rate was set as ¥22/kWh for industrial-scale wind with a capacity of 20 kW or more (Ministry of Economy, Trade and Industry, n.d.a.). This was significantly lower than that for solar (¥40/kWh) and was also the lowest rate among all renewable sources, including geo-thermal and hydropower. Moe (2012) and Li, Xu, and Shiroyama (2019) suggest that the wind industry in Japan was not integrated in the vested interest structure, while the solar was. This unfavorable rate-setting for wind options could have discouraged possible investors.

Looking further, the same rate was applied for offshore projects even though only a few experimental projects existed in 2012. The rate for offshore was determined in 2014 as ¥36/kWh, which was almost equivalent to that for solar projects (¥32). While the rates for other sources reduced every year, the rate for offshore remained the same until 2020 but did not produce uptake. This suggests that the initial rate in 2014 for offshore projects was too low to induce market expansion and cost reduction.

Another reason for the sluggish growth after 2012 was the introduction of the environmental impact assessment (EIA) for almost all industrial-scale wind projects exceeding 10 MW in capacity in 2012. The Environmental Impact Assessment Act requires certain procedures to assess and mitigate adverse impacts by collecting and analyzing data and to offer multiple opportunities for public meetings. This created an additional burden on the project developers and required ongoing projects to reformulate their project plan according to the law.

Interestingly, both of these measures were implemented during the DPJ administration, and one might suggest that it was the DPJ's, not the later LDP's, fault that the introduction of wind power in Japan was slowed down. However, the Abe administration did not take drastic measures to repeal the policies that slowed the growth of wind power until its final years. The FIT rate for wind power remained low even after 2013, and the EIA requirement remained the same. In fact, the 2015 outlook assumed that wind would represent only a 1.7% share of electricity generation in 2030, which was approximately 10 GW. At the end of 2020, less than half of the 2030 target was achieved (Figure 18.3).

The slow pace of the Abe administration in enabling wind farm development crushed the wind turbine industry in Japan. The manufacturers – Mitsubishi Heavy Industry, Japan Steel Works, and Hitachi – all discontinued the domestic production of wind turbines in the 2010s (Asada and Shibata, 2019).

A Swift Shift to Decarbonization Transition by Abe's Successor

The Abe administration has long been criticized not so much for its energy policy but for alleged favorable treatment of organizations that had personal relationships with the PM (Reuters, 2018; Johnston, 2019, 2020). While the PM denied any wrongdoing, favoritism seemed to prevail in the government, at least in reports by the popular media (Asahi Shinbun, 2021a; Mainichi Shinbun, 2020). The initial dealings with the COVID-19 spread following a cluster development on a cruise ship harbored in Yokohama and the postponement of the Olympic Games in 2020 after PM Abe's intensive involvement in bringing the game to the city affected popular sentiment toward the Abe administration. Public support for the Abe administration dwindled throughout 2020, from 44% in January to 34% in August (NHK, n.d.). In fact, in June, as much as 49% of those who responded to the NHK survey disapproved of the Abe administration. PM Abe started to suffer from his long-sustaining ulcerative colitis in the summer, and he had to resign from his position on August 28, 2020 (Asahi Shinbun, 2020; Yomiuri Shinbun, 2020).

At his inaugural speech in the Diet on October 26, the LDP's new leader, PM Yoshihide Suga, surprised the energy and climate change circles in Japan by announcing that he would aim for carbon neutrality in 2050 (Prime Minister's Office of Japan, 2020). In his speech, he argued for a transition in industrial and socioeconomic structure for dealing with climate change. While acknowledging the continuation of nuclear policy, he announced that he would completely change the long-standing policy for coal-fired plants (ibid.). This announcement appears to have been accepted favorably among citizens. In November, the NHK's poll indicated that 56% of respondents supported the Suga administration (NHK, n.d.). The majority of its respondents (61%) positively evaluated the 2050 carbon neutrality ambition.

Following the announcement, the Suga administration started to pursue the decarbonization agenda quite aggressively with the help of the minister of environment, Shinjiro Koizumi, and the minister for administrative and regulatory reform, Taro Kono. In preparing the nationally determined contributions (NDCs) to be submitted to the UNFCCC secretariat in April 2021, for instance, Japan's greenhouse gas reduction target for 2030 was upgraded to 46% or higher, relative to the emission base year in 2013, from the 26% target that had been announced as the interim NDC in 2015.

PM Suga also announced the discontinuation of the government's support for exporting coal-fired plants without CO_2 emission reduction technologies to developing nations in the G7 meeting in 2021 (Nihon Keizai Shinbun, 2021). While the limitation to "those without CO_2 emission reduction technologies" still draws criticism from environmentalists, the termination of the export promotion program is a major shift from the Abe administration's policy. In fact, one of the key economic policy agendas of the Abe administration was the export of Japanese energy technologies, including nuclear power plants, to other nations. PM Abe was personally involved by participating in the promotion events held in these nations (Nihon Keizai Shinbun, 2013; Asahi Shinbun, 2019).

Abe administration's focus on promoting the export of coal-fired and nuclear power plants to developing nations would be contradictory to transitioning to renewable sources in Japan.

His resignation was, in fact, the best timing to terminate the dysfunctional policy agenda that adhered to nuclear options, including their exports, but failed in restarting the reactors. The pressure to accelerate the introduction of renewable energy in the face of the paradox of increasing reliance on fossil fuels in Japan was extremely high particularly after the Paris agreement. The new Suga administration, as Abe's successor, was in an ideal position to fill the gap widened by the lost eight years of the Abe administration.

Discussion and Conclusion

The Suga administration's robust acceleration toward a renewable energy transition highlights a clear contrast with the reluctance of PM Suga's predecessor, PM Abe, in reducing Japan's reliance on fossil fuels and introducing renewable energy sources. PM Shinzo Abe was not a climate denier and occasionally expressed his concerns about climate change (Ministry of Foreign Affairs, 2014). He is well-known for his commitment to constitutional reforms, however, and the energy and climate issues were not among his personal priorities. Why did PM Abe's administration not pursue the popular denuclearization and renewable energy agenda during his tenancy? Multiple hypotheses may be considered.

First the Abe administration wanted to maintain an arms-length distance from the nuclear power subject in order to maintain public support. The administration maintained a delicate distance from the efforts made by the private electric companies to restart each reactor. While the government indicated its support for restarting the reactors by framing nuclear as the base-load power source in its policy documents, it also maintained that the safety measures must be approved through the quasi-independent review by the Nuclear Regulation Agency, and the operator must negotiate with the local governments. The highly private nature of electricity generation in Japan since its emergence in the 19th century and privatization in 1951 (Kikkawa, 2012) allowed the government to stay away from local controversy over the reactor restarts. Implementation of the national pro-nuclear energy policy was, in fact, thwarted at the local level (Koppenborg, 2021).

In the end, while the energy mix target for 2030 that was determined in 2015 was that 20–22% of total electricity should be produced by nuclear power plants, only 6.2% (63.8 TWh) was achieved in 2019 (Ministry of Economy, Trade and Industry, 2021a). Considering that the figure was 25.1% (288.2 TWh) in 2010 before the Fukushima incident, the Abe administration failed to restart the reactors as it had committed to in its energy policy documents.

The existence of non-operating nuclear power plants remains a bottleneck for a swift energy transition in Japan because the decision-makers seemed compelled to take an ambivalent stance toward their restart since the Fukushima incident. The majority of citizens continue to indicate their disapproval of restarting the reactors, as discussed in a previous section. Decision-makers, on the contrary, are bound by their key supporters – the manufacturing industry for the LDP and labor unions for the opposition parties – that prefer the nuclear option. Terminating the nuclear power program would also require a renegotiation of the US-Japan cooperation agreement on nuclear power, which allows Japan to produce plutonium. The ideal solution for the decision-makers, who must worry about their next election, would be to maintain a low profile on the nuclear issue. It is probably one of the reasons for Shinzo Abe's success as the longest-serving PM in Japan.

Next, PM Abe's commitment to exporting Japanese energy technologies to developing nations, as one of the countermeasures against China and Russia's aggressive efforts of exporting their technologies, could have discouraged a transition in domestic energy policy. Pursuing the radical development of renewable sources in the domestic market would contradict the effort of exporting nuclear and fossil-fuel-based power plants to other nations. While the export of nonrenewable plants would have boosted the Japanese economy, on which Abe's popular support was mostly based, the plan provided additional legitimacy to the incumbent energy sources. In the end, all projects by the Japanese companies to build commercial reactors abroad failed by 2020. The plans to facilitate the export of conventional coal-fired plants are also due to be canceled, as discussed in the previous section, in the face of pressures from the international community (Nihon Keizai Shinbun, 2021). The export agenda of the Abe administration not only failed but also slowed the transition to renewable sources, which would be pursued by its successor.

Other possible reasons for the missed transition include the Abe administration's worries about the United States' climate policy under former President Donald Trump, the influence of the nuclear industry through the METI's close involvement in the management of the Cabinet (Mainichi Shinbun, 2018; Asahi Shinbun, 2021b), and the geographical features of the Japanese archipelago. In fact, Samuels (2013) had already noted the Japanese government's tendency to return to the status quo in examples of security, energy, and local governance, even after the Fukushima disaster. Whatever the reasons, the fact that the energy transition toward more renewable sources did not take place in Japan after the meltdown of the Fukushima Daiichi Nuclear Power Plant remains the same.

Japan could have reflected on the disaster and used the moment for transition. The DPJ administration immediately after the quake attempted to terminate nuclear power by around 2030 by issuing the Innovative Strategy for Energy and the Environment in September 2012. Due to the unpopularity in other policy areas, however, the DPJ almost wholly lost its power through the general election in December 2012. The LDP government, led by PM Shinzo Abe, reversed the course and aimed for 20–22% nuclear energy by 2030. Given public resentment toward the reactors, the administration maintained a distance from the energy agenda, particularly at the local level, including an accelerated introduction of renewables. In the end, nothing much has changed in the portfolio of electricity generation since 2012. A few reactors were restarted, and solar power has risen to a 7% share, thanks to the FIT system introduced by the DPJ administration. The share of fossil-fuel-based generation, however, has risen from 65.4% in 2010 to 75.7% in 2019 (Figure 18.1).

Like other chapters that have focused on the trajectory of energy transition in particular regions, like Southeast Asia, India, South Africa, and Latin America, this chapter explored a case of an energy transition in a particular country, in this case one that was missed. The Japanese case is unique, however, in that the nation missed the opportunity to transition to renewable sources for almost a decade after the Fukushima incident. Unlike developing nations, Japan had enough resources and technologies that would have allowed such a transition to succeed. Political leaders at the national level distanced themselves from local controversies regarding reactor restarts while promoting nuclear option in energy policy. In the end, neither renewables nor the nuclear options that would have reduced GHG emissions were implemented, leading to the return to fossil fuels.

The missed transition following the Fukushima meltdown incident might be a case of technological lock-in (Trencher et al., 2020). It will not be an easy course for the following administrations to catch up with the global trend for transitioning to a carbon-free society. Japanese leaders will not be able to stay away from the energy agenda, like their predecessors. They will have to discuss the role of nuclear in 2030 in the face of public resentment over reactor restarts. Only after recognizing the complex nature of the sociotechnical systems that impeded the transition to renewable energy sources can the new administration initiate a transition toward more sustainable energy sources ten years after the meltdown.

Bibliography

Ackerman, Bruce, and James Fishkin. 2008. *Deliberation Day*. New Haven: Yale University Press.

Aldrich, Daniel. 2010. *Site Fights: Divisive Facilities and Civil Society in Japan and the West*. Ithaca: Cornell University Press.

ANN News. 2011. "Kan-souri-ni Juumin-kara Ikari-no Koe." [Angry voice from local residents to PM Kan] Apri 21, 2011. www.youtube.com/watch?v=FuXW2pq014I.

Aoki, Hiroaki. 2009. "Wind Power Kamisu Fuuryoku Hatsudensho-no Gaiyou" [Outline of Kamisu wind power plant] *Journal of JWEA*. 33, no. 4: 56–61.

Asada, Kenji and Shibata Nana. "Yojo Fuuryoku Kyodaika Kisou." [Competing for larger offshore wind turbines] *Nihon Keizai Shinbun*, January 29, 2019.

Asahi Shinbun. 2008. "Shin-ene Kaitori Seido, RPS-Ho-ga Gyakufyuu?" [Feed-in-tarrif obstructred by the RPS Act?] July 24, 2008.

Asahi Shinbun. 2011. "Kan-shusho, Jinin 3-Joken-wo Meigen." [PM Kan reveals 3 conditions for resignation] June 28, 2011.

Asahi Shinbun. 2012a. "Genpatsu Zero, 30-nendai Yori Maeni 36%." [36% supports nuclear phaseout before 2030] October 3, 2012.

Asahi Shinbun. 2012b. "Genpatsu-Zero Giron, Denryoku Soren-ga Hihan." [Federation of electric power union criticizes nuclear phaseout] September 5, 2012.

Asahi Shinbun. 2012c. "Datsu Genpatsu Yoron-ni Osare." [Nuclear phaseout supported by the public opinion] September 15, 2012.

Asahi Shinbun. 2012d. "Ene Senryaku-no Kyokusetu, Furukawa Zen-Kokka Senryaku-sho ni Kiku." [Controversies involving energy strategy, according to the former minister Furukawa] November 16, 2012.

Asahi Shinbun. 2012e. "Noda Naikaku Shiji Saitei 18%." [Noda administration's worst approval rate at 18%] October 22, 2012.

Asahi Shinbun. 2012f. "Ippatsu-de Taosu, Noda Shusho, Zenya-no Ketsui." [Resolution of PM Noda to defeat the opposition at once] November 15, 2012.

Asahi Shinbun. 2012g. "Jiko 320 Cho, Abe Seiken-e Minshu, Kaimetsuteki Haiboku." [LDP/Komeito 320 seats for Abe administration, DPJ catastrophic defeat] December 17.

Asahi Shinbun. 2014. "Shohizei 10%-ni Hantai 69%," [69% oppose rasing consumption tax to 10%] September 8, 2014.

Asahi Shinbun. 2015. "Kekkai-no Machi, Doro-no Umi" [Breached town, sea of mud] September 12, 2015.

Asahi Shinbun. 2018. "9-jo Kaiken Wareru Sanpi." [Divided over constitutional reform article 9] February 20, 2018.

Asahi Shinbun. 2019. "Genpatsu Yushutsu, Koshu-shita Seiken." [Administration adhered to exporting nuclear plants] January 18, 2019.

Asahi Shinbun. 2020. "Abe Shusho, Jinin Hyomei." [PM Abe resigns] August 29, 2020.

Asahi Shinbun. 2021a. "(Shasetsu) Jiminto Sosaisen Abe-shi Sontaku, Yoron-to Kairi." [(Editorial) LDP president election and Abe's cronyism, not matching the public opinion] September 15, 2021. p. 10.

Asahi Shinbun. 2021b. "Kanbann Korokoro Keisansho Naikaku." [Everchanging headlines, the METI-driven cabinet] January 29, 2021.

Birkland, Thomas A. 1998. "Focusing events, mobilization, and agenda setting." *Journal of Public Policy* 18, no. 1: 53–74. https://doi.org/10.1017/S0143814X98000038.

Boin, Arjen, Paul 't Hart and Allan McConnell. 2009. "Crisis exploitation: political and policy impacts of framing contests." *Journal of European Public Policy* 16, no. 1: 81–106. https://doi.org/10.1080/13501760802453221.

Cabinet Office. 2011. *Bosai-Hakusho*. [Disaster prevenation white paper] February 2, 2022. www.bousai.go.jp/kaigirep/hakusho/pdf/H23_zenbun.pdf

Cabinet Office. n.d. "Genshiryoku Bosai. [Nuclear disaster prevention]" February 2, 2022. https://www8.cao.go.jp/genshiryoku_bousai.

Cabinet Secretariat. 2012. "Kongo-no Energuri Kankyo Seisaku-ni Tsuite [On future energy and environmental policy] (Cabinet Decision of September 19, 2012)." November 30, 2021. www.cas.go.jp/jp/seisaku/npu/policy09/pdf/20120919/20120919_1.pdf

Energy and Environment Council. 2012. "Enerugi, Kankyo-ni-kansuru Sentakushi. [Options for energy and the environment]" Accessed February 6, 2022. www.cas.go.jp/jp/seisaku/npu/policy09/pdf/20120629/20120629_1.pdf

Federation of Electric Power Companies of Japan. 2012. "Kakushinteki Enerugi Kankyo Senryaku-no Kettei-ni Tsuite. [On the decision of innovative strategy for energy and the environment]" December 8, 2021. www.fepc.or.jp/smp/about_us/pr/oshirase/__icsFiles/afieldfile/2012/09/14/press_20120914.pdf

Federation of Electric Power Companies of Japan. n.d. "Kokunai-no Genshi-ryoku Hatsudensho-no Saikado-ni Muketa Taiou Joukyou. [Current state of restarting nuclear power plants in Japan]" Accessed February 4, 2022. www.fepc.or.jp/theme/re-operation.

Fire and Disaster Management Agency. 2013. *Higashi Nihon Dai-Shinsai Kiroku-Shu*. [Record of Great East Japan Earthquake] December 8, 2021. www.fdma.go.jp/disaster/higashinihon/item/higashinihon001_12_03-01_02.pdf

Fishkin, James. 2011. *When the People Speak: Deliberative Democracy and Public Consultation*. Oxford: Oxford University Press.

Fukuda, Shin-ichi. 2015. "Abenomics: Why was it so successful in changing market expectations?" *Journal of the Japanese and International Economies* 37: 1–20. https://doi.org/10.1016/j.jjie.2015.05.006.

Fukushima Prefecture. 2013. "Fukkou Saisei no Ayumi [Progress of recovery] (5th edition)" November 29, 2021. www.pref.fukushima.lg.jp/uploaded/attachment/476293.pdf

Fukumoto, Yukinari. 2013. "Choshi-Oki-ni-okeru Chakusho-Shiki Yojo Furyoku Hatsuden-no Jissho Kenkyu [Experimentation of fixed offshore wind power at Choshi]" *Journal of JWEA*. 37, no. 2: 146–150.

Hasegawa, Koichi. 2014. "The Fukushima nuclear accident and Japan's civil society: Context, reactions, and policy impacts." *International Sociology* 29, no. 4: 283–301. https://doi.org/10.1177/0268580914536413.

Hausman, Joshua K. and Johannes F. Wieland. 2014. "Abenomics: Preliminary analysis and outlook." *Brookings Papers on Economic Activity* 1: 1–63.

Hughes, Llewelyn. 2021."Energy Policy in Japan: Revisiting Radical Incrementalism." In *The Oxford Handbook of Japanese Politics*, edited by Robert J. Pekkanen, Saadia Pekkanen, 377–394. Oxford: Oxford University Press.

Iida, Tetsunari. 2002. Shizen Enerugi Suishin-no-tameno Hou-Seibi-ni-tsuite. [Developing legal systems for promoting natural energy] *Journal of Environmental Conservation Engineering*. 31, no. 5: 356–362.

Investigation Committee on the Accident at the Fukushima Nuclear Power Stations of Tokyo Electric Power Company. 2012. *Final Report*. December 8, 2021. www.cas.go.jp/jp/seisaku/icanps/eng.

Japan Association of Corporate Executives. 2012. "Enerugi Seisaku-ni-kansuru Keizai Dantai Kyodo Kisha Kaiken Hatsugen Yoshi. [Summary of joint press conference by business associations on energy policy]" Accessed December 8, 2021. "www.doyukai.or.jp/chairmansmsg/pressconf/2012/120918a.html

Japan Atomic Energy Agency. 2011. *Tokyo-denryoku Fukushima Daiichi Genshiryoku Hatsudensho Jiko-niyoru Plant Hoku-sei-tiiki-no Senryo Josho Process-wo Kaiseki*. [Analyzing radiation-level increase on the northeastern side of the TEPCO Fukushima Daiichi Nuclear Plant] (Press release). February 2, 2022. www.jaea.go.jp/02/press2011/p11061302/index.html

Japan Atomic Industrial Forum. 2012. "Nihon-no Genshi-ryoku Hatsudensho-no Unten Jokyo. [Current operation of nuclear power plants in Japan]" Accessed December 8, 2021. www.jaif.or.jp/ja/news/2012/jp-npps-operation120326.pdf

Japan Wind Power Association. 2021. "Nihon-no Fuuryoku Hatsuden Donyuryo. [Introduction of wind power in Japan]" December 8, 2021. http://jwpa.jp/pdf/dounyuujisseki2020graph.pdf

Johnston, Eric. 2019. "Cherry blossom-viewing party: Breaking down Abe's latest cronyism scandal" *The Japan Times*. November 27, 2019.

Johnston, Eric. 2020. "A Tokyo prosecutor's delayed retirement spurs more allegations of Abe cronyis" *The Japan Times*. March 1, 2020.

Kikkawa, Takeo. 2012. "The history of Japan's electric power industry before World War II." *Hitotsubashi Journal of Commerce and Management* 46, no. 1: 1–16.

Kingdon, John. 1995. *Agendas, Alternatives, and Public Policies* (2nd Ed.). New York: Addison-Wesley.

Koppenborg, Florentine. 2021. "Nuclear restart politics: How the 'nuclear village' lost policy implementation power." *Social Science Japan Journal* 24, no. 1: 115–135. https://doi.org/10.1093/ssjj/jyaa046.

Koshizawa, Akira. 2001. *Tokyo Toshi Keikaku Monogatari*. [Story of Tokyo's city planning] Tokyo: Chikuma Shobo.

Kyushu Electric Power Company. 2014. *Kyushu Hondo-no Saisei Kanou Enerugi Hatsuden Setzubi-ni Taisuru Setsuzoku Moushikomi-no Kaito Horyu-ni Tsuite*. [On reserving responses to grid connection requests from renewable plants on Kyushu Island] December 8, 2021. www.kyuden.co.jp/press_h140924-1.html

Li, Aitong, Yuan Xu, and Hideaki Shiroyama. 2019. "Solar lobby and energy transition in Japan." *Energy Policy* 134: 110950. https://doi.org/10.1016/j.enpol.2019.110950.

Liberal Democratic Party. 2012. *Jimin-to Sogo Seisaku-shu*. [General policy document of LDP] December 8, 2021. https://jimin.jp-east-2.storage.api.nifcloud.com/pdf/j_file2012.pdf

Loorbach, Derk and Lijnis Huffenreuter,. 2013. "Exploring the economic crisis from a transition management perspective" *Environmental Innovation and Societal Transitions* 6: 35–46. https://doi.org/10.1016/j.eist.2013.01.003.

Mainichi Shinbun. 2011. "Kan-shusho: Taijin Hyoumei." [PM Kan announces resignation] August 27, 2011.

Mainichi Shinbun. 2018. "Abe Naikaku-de Seisaku Shudo-no Keisansho Wagayo-no Haru-ni Kageri." [The glory of METI, leading the Abe administration, withering] June 15, 2018.

Mainichi Shinbun. 2020. "Shasetsu: Abe Seiji-no Heigai Minshu Shugi Yugameta Fukai Tsumi." [Editorial: Abe administration's harms, distorting the democracy] August 30, 2020.

Maeda, Takao, and Yasunari Kamada. 2009. "A review of wind energy activities in Japan." *Wind Energy* 12, no. 7: 621–639.

Mah, Ngar-yin Daphne, Alice Siu, Ka-yan Li, Yasunori Sone, and Victor Wai Yin Lam. 2021. "Evaluating deliberative participation from a social learning perspective: A case study of the 2012 National Energy Deliberative Polling in post-Fukushima Japan." *Environmental Policy and Governance* 31, no. 2: 125–141.

Mikami, Naoyuki. 2015. "Public Participation in Decision-Making on Energy Policy." In *Lessons from Fukushima*, edited by Yuko Fujigaki, 87–122. Berlin: Springer.

Ministry of Economy, Trade and Industry. 2014. "Strategic energy plan." December 8, 2021. www.enecho.meti.go.jp/en/category/others/basic_plan.

Ministry of Economy, Trade and Industry. 2015. "Choki Enerugi Jukyuu Mitooshi. [Long-term energy outlook]" December 8, 2021. www.enecho.meti.go.jp/committee/council/basic_policy_subcommittee/mitoshi/pdf/report_01.pdf

Ministry of Economy, Trade and Industry. 2018. "Strategic energy plan." December 8, 2021. www.enecho.meti.go.jp/en/category/others/basic_plan/5th/pdf/strategic_energy_plan.pdf

Ministry of Economy, Trade and Industry. 2020. *Enerugi Hakusho 2020*. [Energy white paper 2020] December 2, 2021. www.enecho.meti.go.jp/about/whitepaper/2020html/2-1-4.html

Ministry of Economy, Trade and Industry. 2021a. "Sogo Enerugi Toukei (Jikeiretsu-hyo). [General energy statistics (time series)]" December 2, 2021. www.enecho.meti.go.jp/statistics/total_energy/xls/stte_jikeiretu2019b_kakuhour.xlsx

Ministry of Economy, Trade and Industry. 2021b. "2030-Nen-ni-okeru Saisei Kanou Enerugi-ni-tsuite. [Renewable energy in 2030]" February 6, 2022. www.meti.go.jp/shingikai/enecho/denryoku_gas/saisei_kano/pdf/031_02_00.pdf

Ministry of Economy, Trade and Industry. 2022. "Genshiryoku Hatsudensho-no Genjo. [Current state of nuclear plants]" February 6, 2022. www.enecho.meti.go.jp/category/electricity_and_gas/nuclear/001/pdf/001_02_001.pdf

Ministry of Economy, Trade and Industry. n.d-a. "Kaitori-kakaku Kikan-tou (2012–2020 nendo). [Purchasing rate and period (fiscal years 2012–2020)]" December 2, 2021. www.enecho.meti.go.jp/category/saving_and_new/saiene/kaitori/kakaku.html

Ministry of Economy, Trade and Industry. n.d.-b "What's RPS System in Japan?" Accessed February 6, 2022. www.rps.go.jp/RPS/new-contents/top/toplink-english.html

Ministry of Economy, Trade and Industry. n.d.-c "Sogo Shigen Enerugi Chosa-kai Kihon Mondai Iinkai Dai-32-kai Kaigo. [32nd meeting of advisory committee for natural resources and energy]" February 6, 2022. https://warp.da.ndl.go.jp/info:ndljp/pid/8556468/www.enecho.meti.go.jp/info/committee/kihonmondai/32th/gijiroku32th.pdf

Ministry of Foreign Affairs. 2014. "Statement by Prime Minister Shinzo Abe at the Plenary Session of the UN Climate Summit 2014." February 7, 2022. www.mofa.go.jp/ic/ch/page3e_000235.html

Ministry of Land, Infrastructure, Transport and Tourism. 2017. "Heisei 27-nen 9-gatsu Kanto-Tohoku Gouu-ni Kakawaru Kozui Higai oyobi Fukkyu Jokyo-ni Tsuite. [On damages and recovery from the flood from heavy rains in Kanto and Tohoku in September 2015]" www.ktr.mlit.go.jp/ktr_content/content/000687586.pdf

Mizuno, Emi. 2014. "Overview of wind energy policy and development in Japan." *Renewable and Sustainable Energy Reviews* 40: 999–1018. https://doi.org/10.1016/j.rser.2014.07.184.

Moe, Espen. 2012. "Vested interests, energy efficiency and renewables in Japan." *Energy Policy* 40: 260–273. https://doi.org/10.1016/j.enpol.2011.09.070.

Morais, Lucas. 2020. "SB Energy, Mitsubishi complete 102-MW solar park in Hokkaido" *Renewables Now.* Accessed December 2, 2021. https://renewablesnow.com/news/sb-energy-mitsubishi-complete-102-mw-solar-park-in-hokkaido-717619.

Nagahashi, Akifumi. 2021. "Mittsu-no Kensho, Kento Ichibu-iin Tairitsu Kashiwazaki Kariwa, Giron Honkakuka [Debate over Kashiwazaki Kariwa getting upbeat, some members disagree on three points]" *Asahi Shinbun (Niigata)*, March 10, 2021.

Nakazawa, Takashi, and Tomoyuki Tatsumi. 2022. "Disagreeing with 'local agreement': A survey of public attitudes toward restarting the Hamaoka nuclear power plant." *Journal of Environmental Planning and Management* 65, no. 1: 150–167.

National Diet of Japan Fukushima Nuclear Accident Independent Investigation Commission. 2012. "The Official Report." December 8, 2021. https://warp.da.ndl.go.jp/info:ndljp/pid/3856371/naiic.go.jp/en.

New Energy and Industrial Technology Development Organization. 2018. "Nihon-ni Okeru Fuuryoku Hatsuden Donyuryo-no Suii. [Introduction of wind power in Japan]" December 8, 2021. www.nedo.go.jp/library/fuuryoku/pdf/02_dounyuu_suii.pdf

NHK. 2021. "Toden Fukushima Daiichi Genpatsu Jiko-towa. [What was TEPCO Fukushima Daiichi accident?]" February 2, 2022. https://www3.nhk.or.jp/news/special/nuclear-power-plant_fukushima/feature/article/article_08.html

NHK. n.d. "Naikaku Shiji Ritsu. [Approval rate of cabinet]" Accessed June 15, 2021. www.nhk.or.jp/senkyo/shijiritsu.

Nihon Keizai Shinbun. 2013. "Shusyo, Sekkyoku Gaikou-no 5-Kagetsu. [PM's 5-month agressive diplomacy]" May 30, 2013.

Nihon Keizai Shinbun (online). 2021. "Sekitan Karyoku-no Yushutsu Shien, Shusho-ga Nennai Shuryo-wo Hyomei. [PM announce ending the exporting coal-fired plants]" June 13, 2021. February 7, 2022. www.nikkei.com/article/DGXZQOUA132V40T10C21A6000000.

Nuclear and Industrial Safety Agency. 2011. "Tokyo Denryoku Kabushiki-gaisha Fukushima Daiichi Genshiryoku Hatsudensho-ni-okeru Jiko-wo Fumaeta Kisetsu-no Hatsuden-yo Genshiro Shisetsu-no Anzen-sei-ni-kansuru Sogo-Hyoka-no Jissi-ni Tsuite. [Evaluation of safety of existing nuclear plants after Fukushima Daiichi plant accident] (press release)." February 4, 2022. https://web.archive.org/web/20111019215454/www.meti.go.jp/press/2011/07/20110722010/20110722010.html

Nuclear Regulation Agency. 2019. "Genshiryoku Saigai Taisaku Shishin. [Guidelines for preparing for nuclear disaster]" February 4, 2022. www.nsr.go.jp/data/000024441.pdf

Patterson, James, Carina Wyborn, Linda Westman, Marie Claire Brisbois, Manjana Milkoreit and Dhanasree Jayaram. 2021. "The political effects of emergency frames in sustainability." *Nature Sustainability* 4: 841–850. https://doi.org/10.1038/s41893-021-00749-9.

Prime Minister's Office of Japan. 2020 "Dai 203-kai Kokkain-ni Okeru Suga Naikaku Souri Daijin Shoshin Hyoumei Enzetsu. [Policy speach by PM Suga at the 203rd diet session]" December 8, 2021. www.kantei.go.jp/jp/99_suga/statement/2020/1026shoshinhyomei.html

Reuters. 2018 "Japan's Abe apologizes amid cronyism scandal, vows to revise constitution" March 25, 2018. February 6, 2022. www.reuters.com/article/us-japan-politics-idUSKBN1H104T

Samuels, Richard. 2013. *3.11: Disaster and Change in Japan.* Cornell University Press.

Statistic Bureau of Japan. 2010. "Population census." Accessed December 8, 2021. www.stat.go.jp/english/data/kokusei.

Szep, Jason and Terril Yue Jones. 2011. "Radiation fears spark panic, evacuations in Tokyo," *Reuters.* March 15, 2011. www.reuters.com/article/us-japan-quake-tokyo-idUSTRE72E0ZR20110315.

Tanaka, Yugo, Andrew Chapman, Tetsuo Tezuka and Shigeki Sakurai. 2020. "Multiple streams and power sector policy change: Evidence from the feed-in tariff policy process in Japan." *Politics & Policy* 48, no. 3: 464–489. https://doi.org/10.1111/polp.12357.

Tokyo Electric Power Company. 2011. "To-sha Fukushima Daiichi Genshiryoku Hatsudensho, Fukushima Daini Gensiryoku Hatsudensho-ni-okeru Tsunami-no Chosa Kekka-ni-tsuite [Investigation into tsunami at Fukushima Daiichi and Daini nuclear power plants] (press release, April 9). February 2, 2022 www.tepco.co.jp/cc/press/11040904-j.html

Tokyo Metropolitan Government. n.d. "To Shiko-ni Yoru Tochi Kukaku Seiri Jigyo. [Land readjustments by Tokyo Metropolitan Government]" June 15, 2021 www.toshiseibi.metro.tokyo.lg.jp/bosai/tk_seiri_02.htm

Tokyo Metropolitan Institute of Public Health. n.d. "Environmental radiation levels in Shinjuku-weekly summary." June 15, 2021 http://monitoring.tokyo-eiken.go.jp/report/shinjuku/mon_air_week_201112w.html

Toyota Motor Corporation. 2012. "Toyota Jidosha Denki Riyo Syaryo Juutaku Kan-no Sogo Denryoku Kyokyu Sisutemu-wo Kaihatsu. [Toyota MOTOR develops mutual power supply system between electric vehicle and house]" Accessed December 8, 2021. https://global.toyota/jp/detail/1695335.

Trencher, Gregory, Adrian Rinscheid, Mert Duygan, Nhi Truong and Jusen Asuka. 2020. "Revisiting carbon lock-in in energy systems: Explaining the perpetuation of coal power in Japan" *Energy Research and Social Science* 69: 101770. https://doi.org/10.1016/j.erss.2020.101770.

Umekawa, Takashi. 2015. "Sendai Gempatsu Saikado, Suga Kanbo-Chokan 'Handan Suruno-wa Jigyo-sya' ['It's up to operators,' Chief cabinet secretary Suga on Sendai nuclear reactor restart]" *Reuters.* August 11, 2015. https://jp.reuters.com/article/suga-sendai-nuclear-idJPKCN0QG08U20150811.

Utsunomiya, Tomoaki, Takashi Shiraishi, Iku Sato, Etsuro Inui and Shigesuke Ishida. "Floating offshore wind turbine demonstration project at Goto Islands, Japan," *OCEANS 2014:* 1–7.

Watanabe, Chisaki. 2014. "GE says japan has more potential to harness wind power," *Bloomberg*, February 27, 2014.

Watanabe, Taro. 2017. "Nichibei Genshiryoku Kyotei-to Sono Yukue. [Future of US-Japan nuclear cooperation agreement]" *NDL Issue Brief.* 980.

Yomiuri Shinbun. 2012a. "Daihyo Shitsumon Abe-shi Sekinin Seitou-wo Kyocho. [Abe stress responsible party at Diet questioning]" November 1, 2012. p. 4.

Yomiuri Shinbun. 2012b. "Shu-in Sen Raigetsu 16-nichi. Asu Kaisan. Shusho Meigen. [General election 16th of next month, disband tomorrow, PM announced]" November 15, 2012. p. 1.

Yomiuri Shinbun. 2012c. "Dai 46-kai Shuuin-Sen Jikou 320-cho Abe Seiken-e. [46th general election, LDP/Komei 320+ seats, Abe Administration starts]" December 17, 2012. p. 1.

Yomiuri Shinbun. 2016. "San-in Hirei-sen Tohyosaki Jimin 35% Minshin 12%. [Upper house election, 35% to LDP and 12% to DP]" June 20, 2016.

Yomiuri Shinbun. 2020. "Abe Shusho Jinin Hyomei. Korona-ka Jibyo Akka. [Abe resigns, illness worsened during the coronavirus pandemic]" August 29, 2020.

19

PEOPLE, POLITICS, AND PLACE

An Interdisciplinary Agenda for the Governance of Urban Energy Transitions

Darren Sharp, Sarah Pink, Rob Raven, and Megan Farrelly

1. Introduction

Energy Transitions is a topical field in scholarly research that has been empirically and methodologically explored and theorized through many different framings. Sparked by socioeconomic and political interests in the discovery of new forms of energy fuels, such as nuclear energy, natural gas, and shale gas, as well as geopolitical concerns and national oil dependencies, market liberalization, concerns over climate change, innovation in digital and renewable energy systems, and social acceptance and energy justice issues have been generative to a substantial and interdisciplinary exploration of topics across engineering, economic, social sciences, and humanities research. The interest in this topic is evidenced by an increasing number of dedicated interdisciplinary research forums, such as the *Routledge Studies in Energy Transitions* book series (Marquardt, 2016), the *Routledge Studies Sustainability Transitions* book series (Verbong and Loorbach, 2012), and the increase in the number of publications on energy transitions in journals, such as *Energy Research and Social Sciences* (e.g., Araujo, 2014), *Energy Policy* (e.g., Grubler, 2012), and *Environmental Innovation and Societal Transitions* (e.g., Markard et al., 2016).

In this chapter, we draw on debates in sustainability transitions literature specifically, which has been home to substantial exploration of energy transitions topics. Sustainability transitions is an interdisciplinary field of scholarship that posits that sociotechnical systems like energy, transport, and the built environment exhibit strong path dependencies. Radical change is argued to be necessary because incremental change is insufficient to address sustainability challenges. Indeed, so the argument goes, a systems-level transition is required, referring to far-reaching changes in institutions, infrastructures, markets, policies, behaviors, and culture (Markard et al., 2012).

In this chapter we bring together a particular approach to the governance of transition known as transition management (TM) with design anthropology (DA) to outline a new interdisciplinary agenda on generating change toward sustainability, which can also be applied to transitions in energy systems. TM is a governance framework that suggests how various actors can be mobilized for sustainability to overcome path dependencies and enable system-level transitions (Kemp et al., 2007). A key aspect of transition management is that it uses future visioning and backcasting techniques to identify short-term opportunities for interventions that can inform

DOI:10.4324/9781003183020-22

reflexive learning and coalition building. The operational model for TM can be characterized as a multi-stakeholder governance approach that is process-oriented and shaped by experiences from practice through a process of "searching, learning, and experimenting" (Loorbach, 2007).

DA is a theoretical, ethnographic, experimental, and interventional practice. It is rooted in the ethics of the everyday, while being attentive to the politics that situate life as lived and is both critical and practical. Design anthropologists critique the predictive narratives and assumptions regarding the societal impacts and benefits of (for instance) emerging technologies and energy systems proffered by industry, policy, and government organizations (Pink et al., 2019, Dahlgren et al. 2020, Pink, 2021). Design anthropologists seek to create future visions that are realistic and plausible because they attend to knowledge about, and imagined futures derived from the experience, ethics, and sites of the everyday. That is, they offer insights and foresights (Lindgren et al., 2021) that are rooted in knowledge from those sites where change is often incremental. In doing so, DA develops theoretical and situated empirical accounts of how incremental change already happens as well as how radical changes are experienced and engaged with in everyday contexts, which are place- and time-sensitive. These accounts are used to produce understandings of the characteristics of change processes in the present and the possible futures these invoke or imply by harnessing uncertainty as a generative site for learning and change (Pink, Akama et al., 2015, Akama et al., 2018).

DA shares with TM an ambition to co-create futures with stakeholders by creating new sociotechnical human-centered knowledge about current and possible near and far future alternative transition scenarios. However, in contrast with the system-level focus in TM, it does so through in-depth theoretical-ethnographic research, innovative futures ethnography experiments and participatory design techniques (Smith and Otto, 2016, Pink and Salazar, 2017, Salazar et al., 2017). When undertaken from a human-centered design perspective, the approach directly responds to how humans are conceptualized in technical and planning disciplines. In doing so, it focuses on how the complex and contingent lives of people are entangled with wider sociotechnical systems and how this enables and constrains possible energy transition futures.

Hence, our proposition is that theoretically and ethnographically informed DA future visions can potentially complement the systems-focused approach in TM by bringing into view the implications, potential, and limitations of such transitions for the everyday experience of citizens and end users in two key ways: first, by revising the one-dimensional notions of users and citizens, which define people vis a vis just one facet of their being to one that understands people as acting within more complex and contingent circumstances, and second, by using these understandings to rethink how the relationship between radical and incremental change might underpin transition processes (Garud and Karnøe, 2003, Araujo, 2017, Karnøe and Garud, forthcoming).

Our chapter is also situated in a wider research ambition to improve the governance of urban energy transitions at the precinct scale (Marvin et al., 2018; Sharp and Raven, 2021; Sharp and Salter, 2017). Experimentation, such as in living labs, in particular is more widely mobilized as a mode of governance in urban transitions (Bulkeley et al., 2016). In the context of net-zero energy transitions, we observe nascent interest from urban actors to move beyond the scale of individual buildings or sector-bounded infrastructures to articulate precinct-scale ambitions for net-zero futures. Precincts are also natural sites of everyday experiences associated with commuting, working, and living – places where meaningful urbanism at the human scale is situated (Hajer et al., 2020). While precincts are potentially an appropriate scale for energy transitions, applying TM at this scale invigorates three challenges that, to some extent, have already been raised in earlier critiques of TM. These challenges relate to (1) people (i.e., challenges of TM to connect with the everyday experiential reality of people); (2) politics (i.e., challenges of TM to accommodate complex issues of community participation and empowerment and navigating the risks of capture by vested interests and incumbency); and (3) place (i.e., challenges of TM to embed transition processes in particular places and their transformation). Against this background, this chapter examines how the conceptual apparatus and practice of DA can shed new light on these challenges.

Methodologically, the chapter builds on general principles of a critical review of literature (Grant and Booth, 2009) on DA and TM with examples from existing DA research into urban energy and mobilities. We use the conceptual lenses of people, politics, and place to review, compare and contrast insights across both literatures. Then, we discuss a number of implications for future research. We end with summarizing conclusions.

2. Transition Management: A Brief Review of Claims, Critiques, and Advancements

TM is a process-oriented mode of sustainability science that uses action research to drive transformation in sociotechnical systems (Wittmayer and Schäpke, 2014). TM has arguably gone through different phases of application and development, each of which with distinctive conceptual, spatial, and methodological orientations and emphases, although we note that these phases are not as clear-cut as suggested here.

Initially, TM was proposed in the early 2000s as a new governance perspective for public policy (Rotmans et al., 2001) in the particular context of low-carbon energy transitions. Rotmans et al. (2001) provided five key elements of TM: (1) long-term thinking as a framework for shaping short-term policy, (2) thinking in terms of multiple domains and multiple actors at multiple levels, (3) a focus on learning and in particular learning-by-doing and learning-by-trying, (4) bringing about system innovation alongside system improvement, and (5) keeping a large number of options open. Conceptually, the approach relied strongly on the multi-level perspective of sociotechnical transitions (Rip and Kemp, 1998; Geels, 2002), distinguishing between sociotechnical regimes, innovation niches, and exogenous landscape trends and events as key conceptual apparatus to understand and explain major changes in sociotechnical systems. An S-curve with different transition phases was proposed to distinguish between pre-development, take-off, acceleration, and stabilization of sociotechnical transitions (Rotmans et al., 2001). Participatory future visioning and backcasting techniques to inform short-term experimentation and a multitude of possible transition pathways was seen as core to the practice of TM. Subsequent work in this first phase of TM evolution aimed to ground the approach into complex systems thinking (Rotmans and Loorbach, 2010; Loorbach, 2010) and introduced the idea of a TM Cycle of four types of activities (Figure 19.1): (1) strategic (problem structuring, envisioning, and establishing of a transition arena); (2) tactical (developing coalitions, images, and transition-agendas); (3) operational (mobilizing actors and executing projects and experiments); and 4) reflexive (monitoring, evaluation, and learning).

TM found a policy home when the approach was integrated in the Fourth National Environmental Policy Plan in the Netherlands in 2002 and benefited from long-term funding in the context of the Dutch Knowledge Network on System Innovations (KSI). Most prominently, TM was tested, refined, evaluated, and critiqued in the context of the Dutch Energy Transition Policy developments (Kemp et al., 2007; Kern and Smith, 2008; Shove and Walker, 2007; Voß et al., 2009; Meadowcroft, 2009). These highlighted various challenges and dilemmas for applying and implementing TM, such as how to navigate ambivalence and controversies about goals, dealing with challenges of reflexive governance and prevention of lock-in, being more

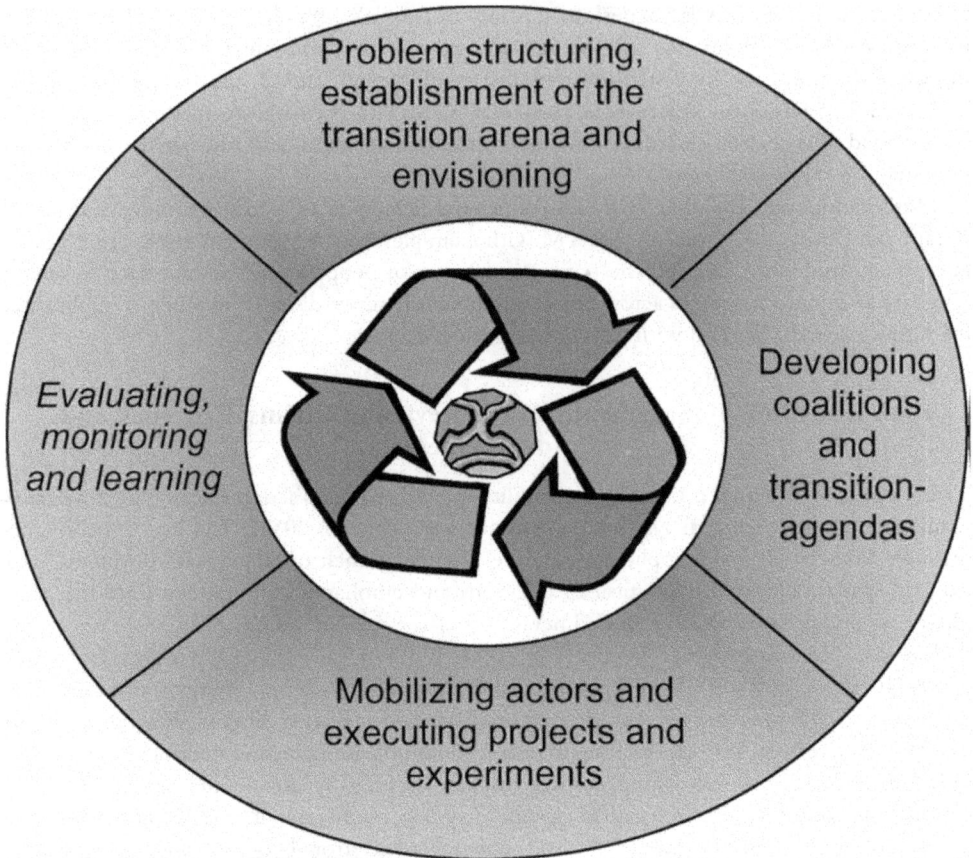

Figure 19.1 The transition management cycle. Reprinted from Loorbach, Derk, and Jan Rotmans. "The practice of transition management: Examples and lessons from four distinct cases". Futures 42, no. 3 (2010): 237–246), with permission from Elsevier.

explicit about the politics of TM, and the risk of capture and control by incumbent interests (see Ford and Newell, 2021).

A second phase of TM can be labelled as "an urban turn", mostly with a focus on European cities. Enabled by a range of successfully funded European research projects such as MUSIC (Mitigation in Urban Areas: Solutions for Innovative Cities) (Roorda and Wittmayer, 2014), ARTS (Accelerating and Rescaling Transitions to Sustainability) (Enhert et al., 2018), GuST (Governance of Urban Sustainability Transitions) (Voytenko et al., 2016), IMPRESSIONS (Frantzeskaki et al., 2018), SUSTAIN (Frantzeskaki et al., 2018), and Urbana (Urban Arenas for Sustainable and Just Cities), the evolution of TM turned to the application of the approach to the cities scale. Initially the urban turn in TM drew strongly on the TM cycle (Nevens et al., 2013). However, as the applications of TM increased, so did the research focus, widening beyond what was prescribed by the initial model.

First, increasingly the TM approach has engaged with debates on geography and in particular the practice around urban living labs. As urban actors increasingly mobilized living labs as a way to explore, navigate and govern transitions locally (Evans et al., 2016), so did TM scholarship (and urban transition scholarship more generally) increasingly engage with the notion of urban experimentation (Bulkeley et al., 2016; Voytenko et al., 2016; Von Wirth et al., 2019). Second, aspects of co-creation, citizen engagement, social innovation, and the challenges of action research received more attention (Frantzeskaki and Kabisch, 2016; Wittmayer and Schäpke, 2014; Hölscher et al., 2019; Frantzeskaki and Rok, 2018). Third, the topic of acceleration and trans-local diffusion of urban initiatives received more attention in an attempt to connect place-based experimentation with multi-level governance and cross-scale diffusion processes (Loorbach et al. 2020; Ehnert et al., 2018).

More recently, a third and recent phase of TM can be identified. In line with an increasing trend in transition studies more generally to turn its gaze to processes of decline, exnovation, disempowerment and destabilization, an emerging research agenda for TM is its engagement with and transformation of the broader sociopolitical institutions that enable or prevent the decline of unsustainable systems of production and consumption (Loorbach, 2014; Loorbach et al., 2017). Frantzeskaki et al. (2018: 9) also note the need for "new process orientations including destabilization and disempowerment".

The next three sections will use our conceptual lenses of people, politics, and place to further discuss critiques and challenges to TM and opportunities for DA for advancing the debates on and practices of TM.

2.1 People

A major concern of transition scholarship is how people can be engaged and empowered to influence the speed and direction of desired transitions toward sustainability (Loorbach et al., 2017). The role of people in transition studies is frequently understood through sociological concepts of "actors" and "agency" yet remains largely neglected in the literature from a theoretical and empirical perspective (Fischer and Newig, 2016). When people do appear, there is "conceptual ambiguity" about actors in transition research, including the problematic overuse of "civil society" to denote anything outside of the market and state and a "vagueness" in how actors are aggregated from the level of specific individuals to more collective formations like organizations or social movements (Avelino and Wittmayer, 2016: 630). It has been suggested that actors are poorly represented in transition studies due to a bias in conceptualizing transitions as a fight between competing systems and technologies resulting in a gap so that "the actions of people are somewhat of an afterthought in the explanations we are used to in the field" (de Haan and Rotmans, 2018: 275).

As an operational approach, TM is concerned with the participatory engagement of change agents known as frontrunners in transition arenas (living labs for e.g.) for problem framing and the co-creative design of future visions and pathways and to undertake transition experiments in real-world settings (Nevens et al., 2013). The question of who gets to govern is an ongoing concern of transition scholars who have criticized TM's use of frontrunners as problematic for its elitism and lack of accountability (Smith and Stirling, 2010). It has also been observed that TM has the potential to "sustain unequal power relations and support those already empowered" (Wittmayer et al., 2015: 947). Other actors, like incumbent participants in ongoing projects, have experienced disempowerment by not feeling "innovative or risk-taking enough for TM" relative to frontrunners (Avelino, 2009: 382). We adopt the frontrunner terminology here as it denotes an approach that works with individuals who are currently ahead in sustainable development, while acknowledging that sustainable development is an ongoing process. Hence, who (should) participate and who does not, and who wins and who loses is understood to be continuously in flux. This is different from for instance the terminology of a champion, which suggests the "race" of sustainable development is already finished, with only winners being involved and prioritized over non-champions – the losers of sustainable development. Front-runners are selected by members of the transition team, as individuals who may be non-experts from different backgrounds but share an appetite for using radical innovation to solve complex societal problems (Loorbach, 2007: 117–118).

TM has also faced criticism from practice-based scholars who have raised questions about the approach's democratic legitimacy, indifference to the role of malevolent actors outside of the "consensus vision", and the perceived bias of socio-technical change, which, so it is argued, downplays "patterns of demand" and the role of consumers in everyday life (Shove and Walker, 2007). Policy scholars have used discursive analysis of participant experiences in the Dutch Energy Transition Program to suggest that TM's "dominant narrative" is built on underlying assumptions that technocratic knowledge should be privileged over democracy, that actor competency provides the necessary creativity for innovation, that expert-led policy will garner public acceptance, and that public engagement in transitions can be delayed (Hendriks, 2009: 356).

People, in contrast, are often at the center of DA, and indeed, there has been an explicit focus on the concept of people (Podjed et al., 2021; Pink, 2021) among some anthropologists whose work intersects with public and applied questions and practice. We note that this point should be contextualized with the need to understand people as part of a multispecies environment, or as the anthropologist Tim Ingold (2000) long since pointed out, we are organisms within an environment rather than as separate from and acting on the environment. However, the emphasis on people is important as a response to the tendency in technology, energy, and urban studies disciplines – as well as their corresponding industry and policy domains – to respectively use the one-dimensional categories of users, consumers, and citizens. People, instead, are complex and lead complicated lives in which their material and social circumstances, bodies, feelings, and decisions are never static but always changing and contingent, often in ways that are barely perceptible to either them or researchers and that are not and cannot be articulated verbally.

The DA perspective we advance does not necessarily use the concept of agency since its commitments to a non-representational and processual approach are rooted in an understanding that people do not "act on" things and have impact on them in order to have outcomes. While acknowledging a "social constructivist" perspective (Bijker et al., 2012) in the sense that agency is not an objective asset grounded in some combination of (financial, material) resources controlled by an individual or organization but rather is understood from the eye of the beholder, the DA approach we use is firmly grounded in a relational perspective. This means that outcomes of encounters between people and things are emergent and contingent and always changing. They become perceptible when held still, which entails momentarily objectifying and contemplating a situation. In this sense, it is inevitable that human activity will participate in shaping the incremental everyday processes through which (measurable) transition happens or through which it is curtailed. However, people are not understood as participating in a "fight" between competing systems and technologies but in improvising in order to be able to move forward in the complex, contingent, and continually changing circumstances in which they find themselves.

DA research shows how people improvise to make their lives work for them in everyday worlds by getting under the surface through intensive engagement with them in the social, material, and technological circumstances of their lives. What people actually do and their reasons often complicate the ambitions associated with transition narratives. For example, the recent AHA2 Design Ethnographic Living Labs for Future Urban Mobility in Sweden (https://aha2.hh.se/) has shown that dominant predictions made by industry and consultancy reporting relating to future urban mobilities, and the desirability and benefits offered by transitioning to future mobility-as-a-service (MaaS) systems are challenged by the ways in which people already conduct their everyday mobilities. It is assumed that first and last mile mobilities – the short distances people travel between home or work to the railway or bus station or other mobility hub – are inconveniences for people that need to be rectified by new technological mobility solutions such as self-driving cars or micro-mobility services. Such new devices are also seen as part of a sustainable transition toward new MaaS systems. However ethnographic research in a diverse neighborhood of a small Swedish city showed that, to the contrary, local people enjoyed and treasured their first and last mile mobilities in their neighborhood – for instance, either to walk, exercise, and encounter friends or neighbors on the way or to drive a status symbol car through the local streets (Pink et al. forthcoming).

This example prompts us to ask how we might better account for people in TM and how proposed sociotechnical interventions might be created in such ways to support the development of existing practices of local people toward transitions that work for them. DA starts from the basis of people's lived experiences that are contingent and change over time. TM involves working with a select group of frontrunners to frame challenges and develop transition visions, pathways, and experiments from a systems perspective (Roorda et al., 2012). Yet this approach inherently excludes other community participants who are not considered as "change agents" or for whom step change innovations proposed externally do not provide viable ways forward. DA could enrich the TM process by problematizing who the key actors in transition arenas are and questioning their roles, worldviews, and underlying assumptions. Ethnographic research could reveal how diverse people experience, adapt to, or resist energy transitions in the sites where the everyday is lived. The insights could inform more inclusive engagement opportunities. DA could also introduce people-engaged processes throughout the TM cycle, especially in relation to the analysis and selection of frontrunners, through attention to the experiences and imagined futures of diverse community members and by accounting for how alternative frontrunner trajectories might also lead to transition.

2.2 Politics

The emergence of transition experiments for sustainability in cities around the world has created new political spaces for urban climate governance between municipal, NGO, and community actors (Bulkeley and Castán Broto, 2013). Sustainability transitions are inherently political modes of reflexive knowledge generation that attempt to institutionalize transformative sustainable development practices and in doing so remain susceptible to capture unless they are able to transcend "the existing constellation of actors and develop more participatory agendas" (Karvonen et al., 2014: 113). The politics of participation remains central to debates over the legitimacy of sustainability transitions, especially in relation to who participates in and who has power over transition processes (Markard, Raven and Truffer, 2012). The role of power in transitions has been conceptualized as two-sided through the notion of "(dis)empowerment", drawn from empirical observations that governance processes can lead to "unintended consequences" that result in actors sensing a lack of impact, competence, and choice (Avelino, 2009; Avelino, 2017).

Critics have suggested that TM is blind to the politics of transition processes and point toward the risks of disempowerment through the reinforcement of existing power structures and role patterns that can create relationships of dependence (Hölscher et al., 2018). It has also been suggested that the open-ended nature of TM and the asymmetries of political power leaves it vulnerable to capture by incumbents interested in maintaining the status quo (Voß, Smith and Grin, 2009). The politics of incumbency relates to power imbalances between incumbent actors and niche innovators and between formal actors and grassroots actors. TM is concerned with shifting power from regimes to niches and empowering niche innovators that are able to counter resistance from incumbent actors (Rotmans, Loorbach and Kemp, 2007). TM interventions in four European cities where local governments played a leading role revealed the challenges of empowering civil society actors due to the "uneven distribution of responsibilities" and the difficulties in realizing multi-actor co-creation through "mutual deliberation and agreement" (Hölscher et al., 2019).

The politics of participation in TM is aligned with DA principles of researching, learning, and intervening with participants rather than studying about and intervening for them. However, underpinning anthropological research is a rigorous dialogue between theory and ethnography, which alerts us to the need to not simply involve people, collect their imagined futures, and enable these to be put into the mix with those of other stakeholders. Rather than advancing participatory agendas, which bring stakeholder future visions into design processes unfiltered by the expertise and frameworks developed through sustained critical debate about the nature of human experience and anticipation, DA involves conceptual work that enables new understandings of possible futures and of how participation in such futures might be envisaged. An analysis of the usually invisible and unspoken elements of life and of the taken for granted narratives and discourses that are part of our representational worlds needs to underpin how we seek realistic, plausible, and ethical everyday and societal futures. Otherwise, we risk centering popular rather than everyday imagined futures and inadvertently supporting the continuation of ill-advised techno-solutionist societal narratives.

Therefore, from a DA perspective the concepts of the experiment and participation both need to be problematized and defined but simultaneously offer a starting point for interdisciplinary collaboration. TM offers sites in which DA research can be undertaken in order to understand how people actually engage, improvise with, and incorporate the (for instance) technologies, infrastructures, and systems that they involve. Yet they open up critical questions concerning the theories, assumptions, choices, and stakeholders who shape such experiments, and the extent to which these are coherent with change as it happens in everyday life contexts. For example, what are the problem-solution narratives that inform transition experiments? In what ways do they see societal changes and benefits coming about?

DA accounts for what people say and do in everyday worlds but understands their experiences, knowledge, and practices as situated both in the present and in possible futures. DA could address some of the challenges related to the politics of participation and incumbency identified in critiques of TM processes. The transition arena is a social learning environment that brings together frontrunners and the transition team to co-create a shared vision, discourse, and agenda with the intention for radical sustainability innovations to be translated across contexts (Loorbach et al., 2015). The transition arena is also the key site where conflicts can arise between niche and regime, business, government, and civil society actors. DA could augment social learning within the transition arena by revealing underlying tensions, alternative framings, and resistance to transition processes rather than aiming to reach consensus through deliberation (Hölscher et al., 2019, Frantzeskaki et al., 2012).

2.3 Place

While there is growth in energy transitions studies examining the importance of place (Coutard and Rutherford, 2010; Bridge et al., 2013; Coenen et al., 2021), for a long time it had been overlooked within transition studies (Murphy, 2015). However, there is growing recognition within the broader umbrella of sustainability transitions scholarship that transition experiments are deeply embedded in place (Coenen and Truffer, 2012), respond to and influence particular geographies of transitions (Truffer et al., 2015), and carry "pre-configured" actor networks, power relations, and institutional arrangements (Raven et al., 2019: 3). Cities and precincts are increasingly regarded as places whereby multiple domains (energy, buildings, mobility, and water, among others) co-exist, co-evolve, and interact to shape (and be shaped by) the functionality, governance, understanding, and experiences of place (and place-making) by a variety of actors (e.g., Nevens et al., 2013; Nielsen and Farrelly, 2019). Transition scholars have noted the utility and importance of place in TM, acknowledging the precinct or city as a material, physical location for urban transition laboratories (Nevens et al., 2013), and urban living labs (Marvin et al., 2018) and as a space/context to explore the relational elements of experimental processes (Broto and Bulkeley, 2013; Pickerill, 2011; Späth and Rohracher, 2010). Yet at the same time, TM is less explicit around its engagement with a city or a precinct as the object of transition and the need to institutionalize long-term transformation agendas in urban planning and policy practice (Hölscher and Frantzeskaki, 2021).

Place-specific critiques of TM and, more broadly, transitions studies relate to its exclusive focus on the transformation of sociotechnical systems over other concerns like sociospatial relations (Murphy, 2015), which is central to the field of, among others, collaborative and strategic planning (Wolfram, 2018). Indeed, "relational place-making" (cf. Pierce et al., 2011) provides a concept and construct of place that could benefit transition studies, by focusing on how "spatial relations, scale, embeddedness, place and power relations" can influence transition pathways (Murphy, 2015: 88). Similar criticism relates to examples of TM at the urban community scale where, in the case of Carnisse in Rotterdam, challenges arose in identifying "ideal-type" frontrunners from the local community who were able to think in "abstract terms" and from a larger "systems perspective" (Wittmayer et al., 2018). This created opportunities to adapt TM processes and incorporate place-specific concerns of community participants in the formulation of transition visions at the neighborhood scale that went beyond systems and strengthened local agency (Frantzeskaki et al., 2018: 12). Other researchers have developed bottom-up foresight processes for community-led agenda formation to overcome TM's perceived elitism and propensity to negate differences in people's lived experiences and "alternative problem framings" (Eames and Egmose, 2011). Overall, these critiques and insights signal the need for new methodological steps to integrate sociospatial relational perspectives into the transformative agenda building phase of TM.

Concepts of place and practices of place design are also integral to anthropology and design respectively. Place, in anthropology, is a theoretical concept rather than a physical locality. This does not mean that the concept of place cannot be applied to describe localities, but rather that place encompasses more than the locality itself, and that place is never circumscribed, static, or given but rather always emergent. For example, the anthropologist Tim Ingold has argued, "Places, then, do not so much exist as occur – they are topics rather than objects, stations along ways of life. Instead of saying that living beings exist in places, I would thus prefer to say that places occur along the life paths of beings" (Ingold, 2008: 1801). As such, places can be seen as intensities or, as Ingold would put it, "entanglements" of things and processes (Pink, 2012), which, drawing on the geographer Doreen Massey's conceptualization of place as a "constellation of processes" (2005, p. 41), have different qualities and move at different speeds. In this sense, while designers suppose that places can be "made" or "designed", anthropologists are more likely to consider places as emerging through these encounters, some of which can be planned, but others of which are part of the serendipity of life. Developing this framework Sarah Pink and Lisa Servon have proposed a theory of place "requires us to understand place not only as constituted through intersections between global and local flows, power relations and the like but one where human experience and perception are equally important" (2013: 253). This, they argue, offers a way in which to understand how sensory and experiential engagements with sustainability at the local everyday life level can become (or fail to be) bound up with globally transferable frameworks for sustainable urban development.

Through the example of the Slow City movement – which could be seen as a model for indirectly managed transition – they have suggested that it is specifically the "intertwining of the experience of the local with the global agenda and framework of Cittaslow that engages town leaders with the movement's model and explains its cross-cultural appeal and continuing global growth" (Pink and Servon, 2013). To explain, Cittaslow (the Slow City offshoot of the better-known Slow Food movement) has been successful in creating a transferable model of the sustainable development of small towns (under 50,000 population), which has been adopted by many towns internationally. The key to the success of this model is that rather than directing sustainability initiatives, the movement's application process involves city leaders and sustainability groups and advocates working together to match the framework and its over 60 criteria to what their town is already doing and (importantly) to its ambitions and hopes for its future. As Pink's own and collaborative ethnographic research in UK, Italy, Spain, and Australia (Pink, 2008, Pink and Lewis, 2014; Pink and Servon, 2013) has shown, the attention to bringing together existing social relationships, practices, and sensory and affective ways of knowing that the Cittaslow model offers, means that transitions to, for instance, sustainable packaging, produce, or energy systems always invest the local into the model rather than seeking to transition local practices and relationships through radical urban experimentation.

The example of the Cittaslow movement is significant in its emphasis on how, when already existing local practices, values, and future hopes can be invested in sustainability processes, they can become successfully entangled with the ambitions of wider change or transition frameworks. Design anthropologically speaking, this is where a theory of place that encompasses a physical locality but serves as its meeting point with national and global flows and processes of things of different kinds and qualities comes into play. Here, place becomes the encounter between these different things, and the intervention becomes the process of shifting the terms and constituents of the encounter in such a way that the towns involved will continue to move forward dynamically in response to the framework. People, as discussed earlier, are integral to these processes; any design for transition must account for how localities are sensed, experienced, and known within the constellations that constitute place and should ask how possible local futures are similarly sensed.

As outlined, it appears DA can offer some core conceptual and methodological insights for TM, to assist in going beyond the physical, technical materiality of constructing and understanding place. By introducing processes of engagement with a different set of actors from the conventional TM "frontrunners", DA's approach to embracing and drawing upon the "entanglements" of the various (multi-scalar) sociospatial relations can perhaps chart new forms of urban transition pathways.

3. Discussion: Future Research Directions

In the previous sections we started to examine how DA can critique and complement TM theory and practice at the precinct scale. This is only the start of a journey to find new ways of advancing transitions to sustainable development in a way that connects with the lived experiences of people in particular circumstances. Here, we identify and discuss three areas that require future research and unpacking: 1) incremental versus radical change; 2) the role of theory and concepts; and 3) the contribution of TM to DA.

3.1 Radical Versus Incremental Change

One of the first publications on TM was entitled "More Evolution than Revolution" (Rotmans et al., 2001), arguing that long-term transformative change comes about through short-term evolutionary diversification. This echoed early thinking in evolutionary accounts of technological change and transformation and how, much like biological evolution, new technological species emerge in specific niches, eventually evolving into distinctive species through processes of niche branching (Schot et al., 1994, Schot and Geels, 2007). Arguably, and perhaps because of an increased sense of urgency to tackle climate change and other grand societal challenges, such nuances on evolutionary and incremental change in the contact of radical transformation have submerged to the background in more recent discourse on sustainability transitions, although this does not necessarily apply to the energy transitions field more generally (Araujo, 2017). Here, we argue that the relationship between radical and incremental change is easy to view retrospectively, something routinely done in historical accounts of sustainability transitions. However, bringing together TM and DA opens up the challenge of how to effectively and realistically view this relationship as a future-focused and forward-moving process. How can we fruitfully construct a relationship between practical gestures toward change at these different levels that is, on the one hand, change with a projected end point or ambition and, on the other change, something incremental and uncertain that weaves its way through the contingent circumstances of life as lived? How might we see incremental change (which is always indeterminate) as part of the forward moving process through which radical change becomes sedimented in everyday life experience and routine? Or is this even possible? Does it require us to rethink our understanding of radical change in favor of a more flexible model? Is there an opportunity to start to think about such questions at the precinct scale, where global challenges such as climate change and the need to decarbonize cities must be confronted with the everyday politics and lived experiences of people within particular places? How can arguments on the necessity to accelerate transitions (Markard et al., 2020) into directions that are consistent with planetary boundaries be married to normative and instrumental arguments to situate these aspirations in, perhaps more evolutionary and incrementally changing, place-based transitions?

3.2 Theory and Concepts

The theoretical basis of DA, as outlined, frames DA practice with a set of assumptions about how change happens, as people, politics, and place are always in progress or in a state of emergence. However, as a theoretical framing, such an approach also remains open to the concepts that might be generated as prisms through which to understand ethnographic realities and imagined futures. Concepts engaged might vary in relation to the questions being asked; for instance, the study of Slow Cities discussed earlier was attentive to categories of sensory experience, since the Slow City movement emphasizes the sensory aesthetics of its cities; research into self-driving cars has focused on the concept of trust because this concept recurs across fields of research in this area. However, other concepts emerge from the fieldwork context; thus, for example, in an earlier study of domestic work in homes, the concept of "freshness" emerged as a participant category (Pink, 2004) and became a key analytical concept in the study. More broadly concepts – what the anthropologist Henrietta Moore (2004) has called "concept metaphors" are useful as containers through which to speak about and debate things or ideas that are similar across different contexts of disciplines. They open up spaces for debate and consensus. In this chapter, we have identified people, politics, and place as areas of work that both unite our approaches and create new and inspiring challenges for us going forward. For example, as a concept metaphor, place offers us a container through which to consider questions relating to the materialities, phenomenology, politics, technologies, and relationships of particular physical, digital, and social sites, as interpreted and theorized by different disciplines. Through this particular concept metaphor, we would expect to generate new understandings and collaborations focused around, for instance, the experiential and measurable dimensions of the processes of place. More broadly, the three concepts of people, politics, and place themselves open up the spaces through which we might evolve our understandings across TM and DA. Here, we note that there is always a need to consider the underpinning fundamental assumptions about what constitutes social worlds (ontologies), how we can develop knowledge about these worlds (epistemologies), and why that is important to researchers and subjects (axiologies) (Geels, 2010; Zolfagharian et al., 2019). To what extent are TM and DA frameworks, which evolved in different epistemic communities, complementary, contradictory, or even incompatible? What are the opportunities but also challenges, tensions, or limitations of combining both frameworks? How could possible ontological tensions become productive in research and practice of precinct-scale transitions? We might also ask these questions of the energy transitions field more generally to develop a more comprehensive approach of TM. Can these frameworks be combined and even integrated, or is it more productive to offer them as multiple lenses exploring people, politics, and place and mobilizing the strength of each of them independently but in parallel? For instance, co-creation is a recurring theme for both TM and DA, which acknowledges a shared commitment to collaboration and engagement through processes of change. For TM, co-creation involves bringing actors together in transition arenas through social learning processes to frame problems, develop long-term visions, undertake transition experiments, and perform reflexive evaluation in an iterative cycle (Loorbach, 2010). For DA co-creation is closer to the design end of DA, while the idea of creating shared ways of knowledge or knowing with participants in research is fundamental to anthropological ethnography.

3.3 Transition Management's Contribution to a New Design Anthropology?

This chapter has considered the contributions that DA can make to TM at precinct scale but has remained mostly silent about the potential of TM to contribute to a DA agenda. DA is itself an evolving field of practice. It is fundamentally an interdisciplinary space that brings together anthropology and design, but that does not have a unique or single ambition or domain of application. DA has been engaged or blended with various fields in existing projects, including construction management and workers safety (Pink et al., 2015), energy demand reduction (Pink et al., 2013), and transport mobilities (Pink, Fors and Glöss, 2018). TM brings new and relevant challenges and inspiration to DA theory and practice. For instance, it invites consideration of how to understand the relationship between radical change and incremental change from a futures perspective as discussed earlier. Radical change is an underpinning requirement for transition and indeed for the survival of our planet. This creates a set of key and fruitful challenges to design anthropologists concerning how to theorize and practice in relation to this ambition, what we must learn and reveal from the sites of the everyday in which we research, and the ethics required to realize this in ways that are responsible and accountable. Nevertheless, such questions may only be the beginning of an exciting journey ahead on cross-fertilization between DA approaches and TM and their related approaches and communities. Similarly, future research on the interplay between TM and DA can bring new insights on epistemological and ontological challenges and complementarities between different approaches to the wider field of energy transitions research.

4. Summarizing Conclusion

Cities are at the forefront of energy transitions as they are both places of energy carbon lock-in and of innovation and experimentation. TM is a process-based approach that uses transition arenas like living labs that bring together different actors to imagine alternative futures and undertake real-world experiments to accelerate sustainability transitions. This raises questions about how individuals and groups participate in transitions and how we might understand these processes through new interdisciplinary concepts. In this chapter, we performed a critical review of literature to discuss the challenges to TM at the urban precinct scale in relation to incorporating the everyday experiences of people, accommodating complex issues of politics and community empowerment, and embedding transition processes in particular places. In line with Grin et al.'s (2017) suggestion that urban sustainability transitions call for a degree of "theoretical promiscuity" to enable broad engagement with diverse perspectives, we called on DA to advance debates on and practices of TM. We considered how DA might complement TM's systems-based approach through an expanded understanding of people, politics, and place, especially at the urban precinct scale. We find that there is considerable promise in bringing the two approaches together. We suggest future directions for research toward an interdisciplinary agenda between TM and DA on the interplay between incremental versus radical change, the role of theory and concepts, and the contribution of TM to DA. Many of these insights also have the potential to resonate with the burgeoning field of energy transitions studies more widely.

References

Akama, Yoko, Sarah Pink, and Shanti Sumartojo. *Uncertainty and Possibility: New Approaches to Future Making*. Bloomsbury, London, 2018.

Araújo, Kathleen. "The emerging field of energy transitions: Progress, challenges, and opportunities." *Energy Research & Social Science* 1 (2014): 112–121.

Araújo, Kathleen M. *Low Carbon Energy Transitions: Turning Points in National Policy and Innovation*. Oxford University Press, 2017.

Avelino, Flor. "Empowerment and the challenge of applying transition management to ongoing projects." *Policy Sciences* 42, no. 4 (2009): 369–390.

Avelino, Flor. "Power in sustainability transitions: Analysing power and (dis) empowerment in transformative change towards sustainability." *Environmental Policy and Governance* 27, no. 6 (2017): 505–520.

Avelino, Flor, and Julia M. Wittmayer. "Shifting power relations in sustainability transitions: A mutli-actor perspective." *Journal of Environmental Policy & Planning* 18, no. 5 (2016): 628–649.

Bijker, Wiebe, Thomas P. Hughes, and Trevor Pinch. (eds.). *The Social Construction of Technological Systems. New Directions in the Sociology and History of Technology*. MIT Press, Cambridge/London, 2012.

Bridge, Gavin, Stefan Bouzarovski, Michael Bradshaw, and Nick Eyre. "Geographies of energy transition: Space, place and the low-carbon economy." *Energy Policy* 53 (2013): 331–340.

Broto, Vanesa Castán, and Harriet Bulkeley. "Maintaining climate change experiments: Urban political ecology and the everyday reconfiguration of urban infrastructure." *International Journal of Urban and Regional Research* 37, no. 6 (2013): 1934–948.

Bulkeley, Harriet, and Vanesa Castán Broto. "Government by experiment? Global cities and the governing of climate change." *Transactions of the Institute of British Geographers* 38, no. 3 (2013): 361–375.

Bulkeley, Harriet, Lars Coenen, Niki Frantzeskaki, Christian Hartmann, Annica Kronsell, Lindsay Mai, Simon Marvin, Kes McCormick, Frank van Steenbergen, and Yuliya Voytenko Palgan. "Urban living labs: Governing urban sustainability transitions." *Current Opinion in Environmental Sustainability* 22 (2016): 13–17.

Coenen, Lars, Teis Hansen, Amy Glasmeier, and Robert Hassink. "Regional foundations of energy transitions." *Cambridge Journal of Regions, Economy and Society* 14, no. 2 (2021): 219–233.

Coenen, Lars, and Bernhard Truffer. "Places and spaces of sustainability transitions: Geographical contributions to an emerging research and policy field." *European Planning Studies* 20, no. 3 (2012): 367–374.

Coutard, Olivier, and Jonathan Rutherford. "Energy transition and city—region planning: Understanding the spatial politics of systemic change." *Technology Analysis & Strategic Management* 22, no. 6 (2010): 711–727.

Dahlgren, Kari, Yolande Strengers, Sarah Pink, Larissa Nicholls, and Jathan Sadowski. *Digital Energy Futures: Review of Industry Trends, Visions and Scenarios for the Home.* Emerging Technologies Research Lab, Monash University, Caulfield East, 2020.

De Haan, Fjalar J., and Jan Rotmans. "A proposed theoretical frameworks for actors in transformative change." *Technological Forecasting and Social Change* 128 (2018): 275–286.

Eames, Malcolm, and Jonas Egmose. "Community foresight for urban sustainability: Insights from the Citizens Science for Sustainability (SuScit) project." *Technological Forecasting and Social Change* 78, no. 5 (2011): 769–784.

Ehnert, Franziska, Niki Frantzeskaki, Jake Barnes, Sara Borgström, Leen Gorissen, Florian Kern, Logan Strenchock, and Markus Egermann. "The acceleration of urban sustainability transitions: A comparison of Brighton, Budapest, Dresden, Genk, and Stockholm." *Sustainability* 10, no. 3 (2018): 612.

Evans, James, Andrew Karvonen, and Rob Raven (eds.). *The Experimental City.* Routledge, 2016.

Fischer, Lisa-Britt, and Jens Newig. "Importance of actors and agency in sustainability transitions: A systematic exploration of the literature." *Sustainability* 8, no. 5 (2016): 476.

Ford, Adrian, and Peter Newell. "Regime resistance and accommodation: Toward a neo-Gramscian perspective on energy transitions." *Energy Research & Social Science* 79 (2021): 102163.

Frantzeskaki, Niki, Katharina Hölscher, Julia M. Wittmayer, Flor Avelino, and Matthew Bach. "Transition management in and for cities: Introducing a new governance approach to address urban challenges." In *Co-Creating Sustainable Urban Futures,* pp. 1–40. Springer, Cham, 2018.

Frantzeskaki, Niki, and Nadja Kabisch. "Designing a knowledge co-production operating space for urban environmental governance – Lessons from Rotterdam, Netherlands and Berlin, Germany." *Environmental Science & Policy* 62 (2016): 90–98.

Frantzeskaki, Niki, Derk Loorbach, and James Meadowcroft. "Governing societal transitions to sustainability." *International Journal of Sustainable Development* 15, no. 1–2 (2012): 19–36.

Frantzeskaki, Niki, and Ania Rok. "Co-producing urban sustainability transitions knowledge with community, policy and science." *Environmental Innovation and Societal Transitions* 29 (2018): 47–51.

Garud, Raghu, and Peter Karnøe. "Bricolage versus breakthrough: Distributed and embedded agency in technology entrepreneurship." *Research Policy* 32, no. 2 (2003): 277–300.

Geels, Frank W. "Ontologies, socio-technical transitions (to sustainability), and the multi-level perspective." *Research Policy* 39, no. 4 (2010): 495–510.

Geels, Frank W. "Technological transitions as evolutionary reconfiguration processes: A multi-level perspective and a case-study." *Research Policy* 31, no. 8–9 (2002): 1257–1274.

Grant, Maria J., and Andrew Booth. "A typology of reviews: An analysis of 14 review types and associated methodologies." *Health Information & Libraries Journal* 26, no. 2 (2009): 91–108.

Grin, John, Niki Frantzeskaki, Vanesa Castán Broto, and Lars Coenen. "Sustainability transitions and the city: Linking to transition studies and looking forward." In *Urban Sustainability Transitions,* pp. 359–367. Routledge, 2017.

Grubler, Arnulf. "Energy transitions research: Insights and cautionary tales." *Energy Policy* 50 (2012): 8–16.

Hajer, Maarten A., Peter Pelzer, Martijn Van Den Hurk, Chris ten Dam, and Edwin Buitelaar. *Neighbourhoods for the Future: A Plea for a Social and Ecological Urbanism.* Trancity Valiz, 2020.

Hendriks, Carolyn M. "Policy design without democracy? Making democratic sense of transition management." *Policy Sciences* 42, no. 4 (2009): 341–368.

Hölscher, Katharina, Flor Avelino, and Julia M. Wittmayer. "Empowering actors in transition management in and for cities." In *Co-creating Sustainable Urban Futures,* pp. 131–158. Springer, Cham, 2018.

Hölscher, Katharina, and Niki Frantzeskaki. "Perspectives on urban transformation research: Transformations in, of, and by cities." *Urban Transformations* 3, no. 1 (2021): 1–14.

Hölscher, Katharina, Julia M. Wittmayer, Flor Avelino, and Mendel Giezen. "Opening up the transition arena: An analysis of (dis) empowerment of civil society actors in transition management in cities." *Technological Forecasting and Social Change* 145 (2019): 176–185.

Ingold, Tim. *The Perception of the Environment: Essays on Livelihood, Dwelling and Skill.* Routledge, London, 2000.

Ingold, T. "Bindings against boundaries: Entanglements of life in an open world." *Environment and Planning A: Economy and Space* 40, no. 8 (2008): 1796–1810. https://doi.org/10.1068/a40156

Karnøe, Peter, and Raghu Garud (2023). Transformation of the Danish wind turbine industry through path creation. In K. Araujo (ed.). *The Routledge Handbook of Energy Transitions,* Routledge, Abingdon.

Karvonen, Andrew, James Evans, and Bas van Heur. "The politics of urban experiments: Radical change or business as usual?." In *After Sustainable Cities?* pp. 104–115. Routledge, 2014.

Kemp, René, Jan Rotmans, and Derk Loorbach. "Assessing the Dutch energy transition policy: How does it deal with dilemmas of managing transitions?." *Journal of Environmental Policy & Planning* 9, no. 3–4 (2007): 315–331.

Kern, Florian, and Adrian Smith. "Restructuring energy systems for sustainability? Energy transition policy in the Netherlands." *Energy Policy* 36, no. 11 (2008): 4093–4103.

Lindgren, Thomas, Sarah Pink, and Vaike Fors. "Fore-sighting autonomous driving – An Ethnographic approach." *Technological Forecasting and Social Change* 173 (2021).

Loorbach, Derk. "To transition! Governance panarchy in the new transformation." Inaugural Address, DRIFT/Erasmus University Rotterdam, 2014.

Loorbach, Derk. "Transition management for sustainable development: A prescriptive, complexity-based governance framework." *Governance* 23, no. 1 (2010): 161–183.

Loorbach, Derk. "Transition management." New Mode of Governance for Sustainable Development (PhD Diss.). Utrecht, Erasmus University, 2007.

Loorbach, Derk, Niki Frantzeskaki, and Roebin Lijnis Huffenreuter. "Transition management: Taking stock from governance experimentation." *Journal of Corporate Citizenship* 58 (2015): 48–66.

Loorbach, Derk, Niki Frantzeskaki, and Flor Avelino. "Sustainability transitions research: Transforming science and practice for societal change." *Annual Review of Environment and Resources* 42 (2017): 599–626.

Loorbach, Derk, Julia Wittmayer, Flor Avelino, Timo von Wirth, and Niki Frantzeskaki. "Transformative innovation and translocal diffusion." *Environmental Innovation and Societal Transitions* 35 (2020): 251–260.

Markard, Jochen, Frank W. Geels, and Rob Raven. "Challenges in the acceleration of sustainability transitions." *Environmental Research Letters* 15, no. 8 (2020): 081001.

Markard, Jochen, Rob Raven, and Bernhard Truffer. "Sustainability transitions: An emerging field of research and its prospects." *Research Policy* 41, no. 6 (2012): 955–967.

Markard, Jochen, Marco Suter, and Karin Ingold. "Socio-technical transitions and policy change – Advocacy coalitions in Swiss energy policy." *Environmental Innovation and Societal Transitions* 18 (2016): 215–237.

Marquardt, Jens. *How Power Shapes Energy Transitions in Southeast Asia: A Complex Governance Challenge.* Taylor & Francis, 2016.

Marvin, Simon, Harriet Bulkeley, Lindsay Mai, Kes McCormick, and Yuliya Voytenko Palgan, (eds.). *Urban Living Labs: Experimenting with City Futures.* Routledge, 2018.

Massey, D. *For Space.* Sage, London, 2005.

Meadowcroft, James. "What about the politics? Sustainable development, transition management, and long term energy transitions." *Policy Sciences* 42, no. 4 (2009): 323–340.

Moore, Henrietta L. "Global anxieties: Concept-metaphors and pre-theoretical commitments in anthropology." *Anthropological Theory* 4, no. 1 (2004): 71–88.

Murphy, James T. "Human geography and socio-technical transition studies: Promising intersections." *Environmental Innovation and Societal Transitions* 17 (2015): 73–91.

Nielsen, Joshua, and Megan A. Farrelly. "Conceptualising the built environment to inform sustainable urban transitions." *Environmental Innovation and Societal Transitions* 33 (2019): 231–248.

Nevens, Frank, Niki Frantzeskaki, Leen Gorissen, and Derk Loorbach. "Urban transition labs: Co-creating transformative action for sustainable cities." *Journal of Cleaner Production* 50 (2013): 111–122.

Pickerill, Jenny. "Building liveable cities: Urban Low Impact Developments as low carbon solutions?." In: H. Bulkeley, V. Castán Broto, M. Hodson, and S. Marvin. (eds.). *Cities and Low Carbon Transitions*, pp. 194–213. Routledge, 2011.

Pierce, Joseph, Deborah G. Martin, and James T. Murphy. "Relational place-making: The networked politics of place." *Transactions of the Institute of British Geographers* 36, no. 1 (2011): 54–70.

Pink, Sarah. *Home Truths.* Berg, Oxford, 2004.

Pink, Sarah. "An urban tour: The sensory sociality of ethnographic place-making." *Ethnography* 9, no. 2 (2008): 175–196.

Pink, Sarah. *Situating Everyday Life: Practices and Places.* Sage, London, 2012.

Pink, Sarah. "Anthropology in an Uncertain World." In Podjed, Dan, Meta Gorup, Pavel Borecky and Carla Guerron Montero (eds.). *Why the World Needs Anthropologists.* Routledge, Abingdon & New York, 2021.

Pink, Sarah, Yoko Akama, and contributors. Un/Certainty. iBook, RMIT, Melbourne, 2015. http://d-e-futures.com/wp-content/uploads/2015/01/Un_certainty_smllowres.pdf

Pink, Sarah, Michelle Catanzaro, Katrina Sandbach, Alison Barnes, Joanne McNeill, M. Gushesh, and Enrico Scotece. "Making and sharing the commons: Reimagining 'the West' as Riverlands, Sydney through a dialogue between design and ethnography." *Global Media Journal Australian Edition* 9, no. 2 (2015).

Pink, Sarah, Vaike Fors and Mareike Glöss. "The contingent futures of the mobile present: Automation as possibility." *Mobilities* 13, no. 5 (2018): 615–631.

Pink, Sarah, Vaike Fors, and Mareike Glöss. "Automated futures and the mobile present: In-car video ethnographies." *Ethnography* 20, no. 1 (2019): 88–107.

Pink, Sarah, Vaike Fors, Katalin Osz, Peter Lutz. "Future Mobility Solutions?: Design ethnography as an interventional device." In D. Lanzeni, K. Waltorp, S. Pink and R.C. Smith (eds.). *Anthropology of Futures and Technologies*. Oxford, Routledge. Forthcoming.

Pink, Sarah, and Tania Lewis. "Making resilience: Everyday affect and global affiliation in Australian slow cities." *Cultural Geographies* 21, no. 4 (2014): 695–710.

Pink, Sarah, Kerstin Leder Mackley, Val Mitchell, Marcus Hanratty, Carolina Escobar-Tello, Tracy Bhamra, and Roxana Morosanu. "Applying the lens of sensory ethnography to sustainable HCI." *ACM Transactions on Computer-Human Interaction* (TOCHI) 20, no. 4 (2013): 1–18.

Pink, Sarah, and Juan Francisco Salazar. "Anthropologies and Futures: Setting the Agenda." In Salazar, Juan Francisco, Sarah Pink, Andrew Irving, and Johannes Sjöberg (eds.). *Anthropologies and Futures*. Bloomsbury, Oxford, 2017.

Pink, Sarah, and Lisa J. Servon. "Sensory global towns: An experiential approach to the growth of the Slow City movement." *Environment and Planning A* 45, no. 2 (2013): 451–466.

Podjed, Dan, Meta Gorup, Pavel Borecký, and Carla Guerrón Montero (eds.). *Why the World Needs Anthropologists*. Routledge, Abingdon & New York, 2021.

Raven, Rob, Frans Sengers, Philipp Spaeth, Linjun Xie, Ali Cheshmehzangi and Martin De Jong. "Urban experimentation and institutional arrangements." *European Planning Studies* 27, no. 2 (2019): 258–281.

Rip, Arie, and René Kemp. "Technological change." *Human Choice and Climate Change* 2, no. 2 (1998): 327–399.

Roorda, Chris, Niki Frantzeskaki, Derk Loorbach, Frank Van Steenbergen, and Julia Wittmayer. *Transition Management in the Urban Context: Guidance Manual*. DRIFT, Erasmus University Rotterdam, Rotterdam, 2014.

Roorda, Chris, and Julia Wittmayer. *Transition Management in Five European Cities – an Evaluation*. DRIFT, Erasmus University Rotterdam, Rotterdam, 2014.

Rotmans, Jan, René Kemp, and Marjolein Van Asselt. "More evolution than revolution: Transition management in public Policy." *Foresight* (Cambridge) 3, no. 1 (2001): 15–31.

Rotmans, Jan, Derk Loorbach, and René Kemp. "Transition Management: Origin, Evolution, Critique." In *Workshop on Politics and Governance in Sustainable Socio-Technical Transitions*, 19–21 September 2007, Schloss Blankensee, Berlin, Germany. Dutch Research Institute for Transitions, 2007.

Salazar, Juan Francisco, Sarah Pink, Andrew Irving, and Johannes Sjöberg (eds.). *Anthropologies and Futures: Techniques for Researching an Uncertain World*. Bloomsbury, Oxford, 2017.

Schot, Johan, and Frank W. Geels. "Niches in evolutionary theories of technical change." *Journal of Evolutionary Economics* 17, no. 5 (2007): 605–622.

Schot, Johan, Remco Hoogma, and Boelie Elzen. "Strategies for shifting technological systems: The case of the automobile system." *Futures* 26, no. 10 (1994): 1060–1076.

Sharp, Darren, and Robert Salter. "Direct impacts of an urban living lab from the participants' perspective: Livewell Yarra." *Sustainability* 9, no. 10 (2017): 1699.

Sharp, Darren, and Rob Raven. "Urban planning by experiment at precinct scale: Embracing complexity, ambiguity, and multiplicity." *Urban Planning* 6, no. 1 (2021): 195–207.

Shove, Elizabeth, and Gordon Walker. "CAUTION! Transitions ahead: Politics, practice, and sustainable transition management." *Environment and planning A* 39, no. 4 (2007): 763–770.

Smith, Adrian, and Andy Stirling. "The politics of social-ecological resilience and sustainable sociotechnical transitions." *Ecology and Society* 15, no. 1 (2010): 11.

Smith, Rachel Charlotte, and Ton Otto. "Cultures of the Future: Emergence and Intervention in Design Anthropology." In Smith, Rachel Charlotte, Kasper Tang Vangkilde, Ton Otto, Mette Gislev Kjaersgaard, Joachim Halse, and Thomas Binder (eds.). In *Design Anthropological Futures*, pp. 19–36. Bloomsbury Academic, London, 2016.

Späth, Philipp, and Harald Rohracher. "The 'eco-cities' Freiburg and Graz: The social dynamics of pioneering urban energy and climate governance." In: Bulkeley, Harriet, Vanesa Castán Broto, Mike Hodson, and Simon Marvin (eds.). *Cities and Low Carbon Transitions*, pp. 104–122. Routledge, London, 2010.

Truffer, Bernhard, James T. Murphy, and Rob Raven. "The geography of sustainability transitions: Contours of an emerging theme." *Environmental Innovation and Societal Transitions* 17 (2015): 63–72.

Verbong, Geert, and Derk Loorbach (eds.). *Governing the Energy Transition: Reality, Illusion or Necessity?* Routledge, 2012.

Von Wirth, Timo, Lea Fuenfschilling, Niki Frantzeskaki, and Lars Coenen. "Impacts of urban living labs on sustainability transitions: Mechanisms and strategies for systemic change through experimentation." *European Planning Studies* 27, no. 2 (2019): 229–257.

Voß, Jan-Peter, Adrian Smith, and John Grin. "Designing long-term policy: rethinking transition management." *Policy sciences* 42, no. 4 (2009): 275–302.

Voytenko, Yuliya, Kes McCormick, James Evans, and Gabriele Schliwa. "Urban living labs for sustainability and low carbon cities in Europe: Towards a research agenda." *Journal of Cleaner Production* 123 (2016): 45–54.

Wittmayer, Julia M., and Niko Schäpke. 2014. "Action, research and participation: Roles of researchers in sustainability transitions." *Sustainability Science* 9 (2014): 483–496.

Wittmayer, Julia. M., Frank van Steenbergen, Ania Rok, and Chris Roorda. "Governing sustainability: a dialogue between Local Agenda 21 and transition management." *Local Environment* 21, no. 8 (2016): 939–955.

Wittmayer, Julia M., Frank van Steenbergen, and Matthew Bach. "Transition Management in Urban Neighbourhoods: The Case of Carnisse, Rotterdam, the Netherlands." In Frantzeskaki, Niki, Katharina Hölscher, Matthew Bach, and Flor Avelino (eds.). *Co-creating Sustainable Urban Futures. A Primer on Applying Transition Management in Cities*, Future City, vol. 11. Springer, Cham, 2018.

Wolfram, Marc. "Urban planning and transition management: Rationalities, instruments and dialectics." In Frantzeskaki, Niki, Katharina Hölscher, Matthew Bach, and Flor Avelino (eds.). *Co-creating Sustainable Urban Futures. A Primer on Applying Transition Management in Cities*, Future City, vol. 11, pp. 103–125. Springer, Cham, 2018.

Zolfagharian, Mohammadreza, Bob Walrave, Rob Raven, and A. Georges L. Romme. "Studying transitions: Past, present, and future." *Research Policy* 48, no. 9 (2019): 103788.

20

MAPPING THE USE OF PUBLIC POLICY THEORIES IN ENERGY TRANSITIONS RESEARCH

A Bibliometric Review and Computational Text Analysis

Nihit Goyal, Araz Taeihagh, and Michael Howlett

Introduction

There is broad consensus in the literature that public policy plays a key role in initiating, accelerating, or supporting how energy is sourced, delivered, or utilized and, thereby, in any future energy transition (Loorbach 2010; Kittner, Lill, and Kammen 2017; Kern and Smith 2008). However, how public policy influences and is influenced by energy transitions appears to be hardly examined empirically (Marquardt 2016; Moore 2018).

To address this gap, some scholars have appealed for closer synthesis among the literature on energy research, public policy, and sustainability transitions. The avenues proposed for this include the use of concepts from policy studies in energy research (Hoppe, Coenen, and van den Berg 2016; Goyal 2021b; Goyal and Howlett 2021), the adaptation of policy process theories to analyze the politics of transitions (Kern and Rogge 2018; Goyal and Howlett 2020), engagement with the notion of policy feedback for co-evolutionary assessment of policy mixes and sociotechnical transitions (Edmondson, Kern, and Rogge 2019), and the application of the research on policy transfer to study in the internationalization of sociotechnical transitions (Heyen et al. 2021).

One (related) concept for such an integration that is under-explored is that of policy innovation. In a polycentric context, policy innovation refers to a multifaceted process consisting of policy invention, policy diffusion, and policy effects (Jordan and Huitema 2014; Goyal 2019). While policy invention entails a radical change in policy objectives or instruments in a jurisdiction (Howlett 2014), policy diffusion denotes the (potential) spread of policy from one jurisdiction to other interdependent jurisdictions (Walker 1969). Meanwhile, the notion of policy effects recognizes the need to analyze the influence of public policy on a system (Vedung 2006; McConnell 2010). Therefore, the study of policy innovation can shed light on the co-evolution of public policy and technological development and generate knowledge for accelerating energy transitions (Goyal and Howlett 2018; Goyal 2019; Goyal, Howlett, and Taeihagh 2021). Whether, to what extent, and how the literature on energy transitions has engaged with policy studies, more broadly, and policy innovation, more specifically, is unclear.

DOI:10.4324/9781003183020-23

This chapter clarifies the relationship between energy transitions and policy presented in the literature that self-identifies as being about energy transitions or energy in sustainability transitions. It presents an overview of the literature and identifies the key themes in this research area, including the discussion about public policy therein, through a bibliometric review and computational text analysis.

Research Methods

The bibliometric data for the analysis was obtained through the Web of Science (WOS) database. We searched the titles, abstracts, and keywords of the publications in the Social Sciences Citation Index (SSCI) and the Book Citation Index – Social Sciences and Humanities (BKCI-SSH) to obtain the bibliometric data. Our final search query – executed on July 5, 2021 – was as follows: *(electricity OR energy OR "power generation" OR "power system★" OR renewable OR smartgrid OR "smart grid★") AND (MLP OR "multi-level perspective" OR SNM OR "strategic niche management" OR "technological innovation system★" OR TIS OR transition) NOT ("demographic transition★" OR "energy intake" OR "land?cover transition★" OR "land?use transition★" OR "nutrition transition★" OR "phase transition★").*

We employed structural topic modeling (Roberts, Stewart, and Tingley 2014) to cluster the publications based on their titles and abstracts, and identify themes not pertaining to the energy transitions. Topic modeling is an unsupervised machine learning technique for "discovering" latent themes (or *topics*) in a document collection based on the distribution of terms (i.e., words or phrases) in the text (Blei 2012). After removing publications from unrelated themes, our final dataset consisted of 8,442 publications on the energy transitions. We then used the *bibliometrix* package (Aria and Cuccurullo 2017) to conduct the bibliometric analysis. Subsequently, the main themes in the final dataset were once again identified using structural topic modeling. Finally, we used count and correlation statistics of terms mentioning "policy" OR "polici" (the word stem of "policy" or "policies") to identify the key analytical lenses from policy studies that have been employed in the dataset.

An Overview of the Literature

As mentioned earlier, the final dataset consists of 8,442 publications on energy transitions. Although some studies on energy transitions – for example, on energy futures or energy scenarios by businesses – preceded the 1970s, those are not captured in our dataset as they did not use the term "energy transition". The earliest publications in our dataset – written in the aftermath of the oil crisis – discussed the impending energy transition, resource scarcity, sustainable energy, and the role of public policy (Aliaattiga 1978; Berry 1975; Hayes 1977; Perelman 1980; Yergin 1979). Since 2015, scientific production has grown exponentially (Figure 20.1). During this period, the annual publications in the research area have increased from less than 500 in 2015 to over 1,500 by 2020.

Over 18,000 scholars have authored publications on energy transitions. While they represent over 4,000 research institutions, the set of core institutions is also relatively small, albeit significant. Approximately 130 institutions worldwide are mentioned on 25 or more occasions (counting multiple authors from an institution or multiple publications by an author as distinct occurrences). A look at the top institutions in this area – the University of Sussex, Utrecht University, the University of Leeds, the Delft University of Technology, and the University of Oxford – reveals a large European (specifically, British and Dutch) presence. In fact, among the top 25 institutions in terms of the number of occurrences, only four are from outside Europe:

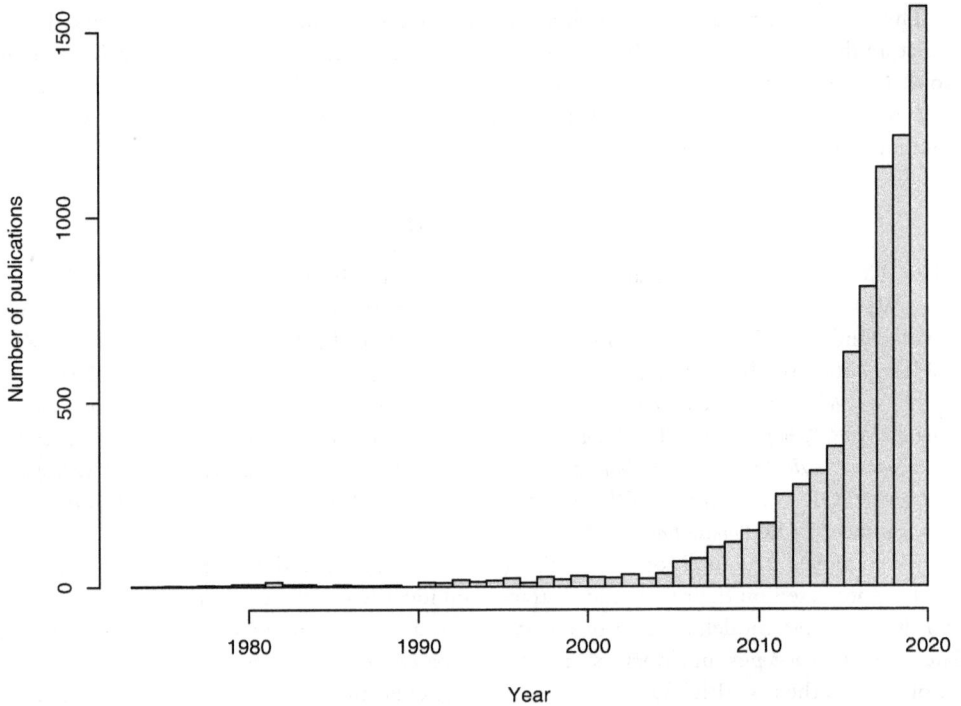

Figure 20.1   The number of publications on energy transitions over time. Publications for the year 2021 are not shown in the figure for consistency.

Source: Data source: Web of Science. Authors' analysis.

Tsinghua University, the University of California Berkeley, the University of Queensland, and the Arizona State University. These institutions fall into one of three clusters (see Figure 20.2).

Based on the institutional affiliation of the corresponding author, the countries with the most publications are the United States, the United Kingdom, China, Germany, the Netherlands, Australia, Sweden, Canada, Spain, and Italy. The presence of the United States and Germany high on this list – but less so on the list of institutions – indicates that research on the energy transitions is diffused across several institutions in these countries. Further, except China, the only non-OECD countries among the top 30 are India (19), Brazil (22), and South Africa (27). This suggests that the Global South has played a limited direct role in shaping the research area.

Key Themes in the Research on the Energy Transitions

The topic model reveals the presence of various analytical approaches and empirical issues that have been covered in the literature on energy transitions (Figure 20.3).

The most prevalent theme in the dataset is Theme 1 on "sociotechnical transition". This theme's focus is the conceptual advancement of the transitions and its empirical application to the climate change or energy system. Illustratively, Geels (2020) proposes a model of agency within the multi-level perspective, while Kivimaa et al. (2019) develop a typology of inter-mediaries, and Chilvers, Pallett, and Hargreaves (2018) conceptualize ecologies of participa-tion in sustainability transitions. In more empirical work, scholars have examined the role of

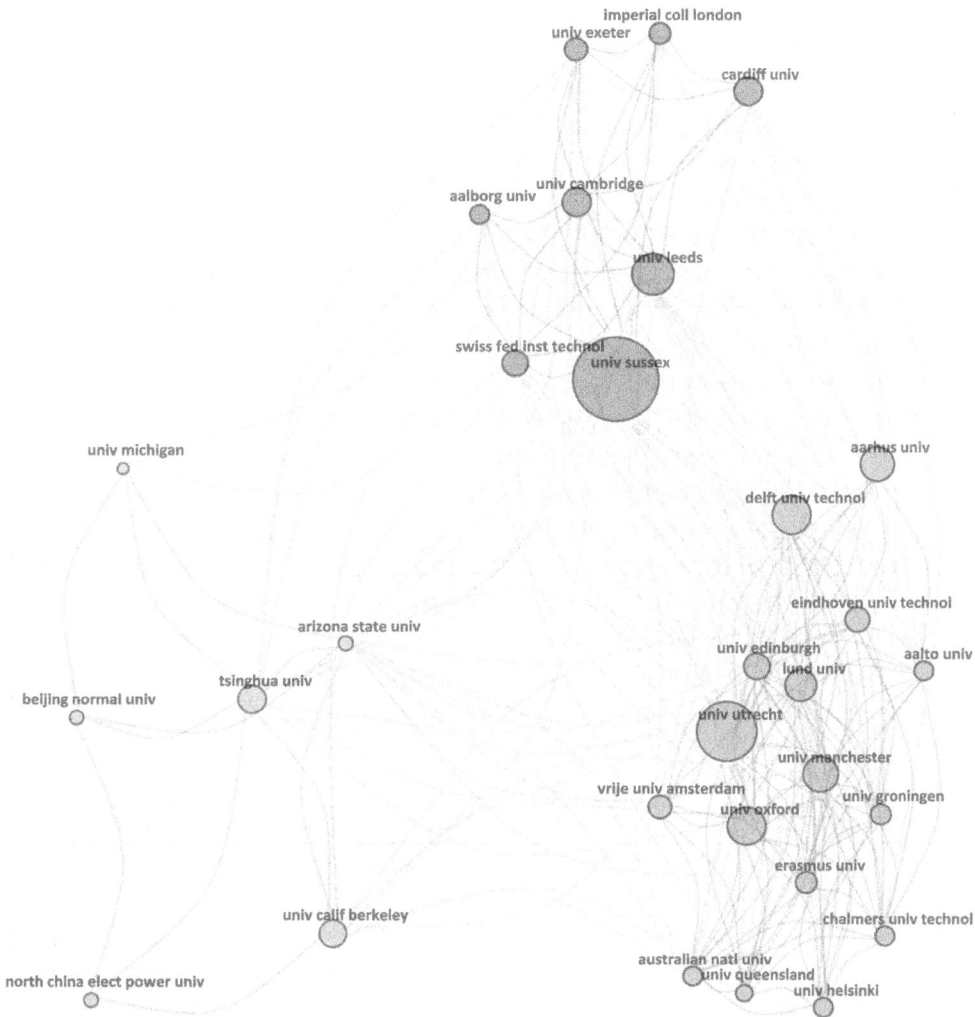

Figure 20.2   Inter-institutional collaboration in the research on energy transitions. Here, collaboration is defined as a co-authorship relationship. A link between two nodes indicates a co-authorship relationship. Node size indicates the number of co-authorship relationships identified in the dataset. The nodes are clustered based on the edges connecting them.

Source: Data from Web of Science. Analysis by the authors.

non-traditional actors in multi-level climate governance (Schroeder, Burch, and Rayner 2013), studied the effect of community leadership on grassroots innovation in energy (Martiskainen 2017), and used the energy justice lens to assess local ownership in community energy (Forman 2017). Scholarship in this theme has also witnessed an emphasis on place, space, and scale (Faller 2016; Hansen and Coenen 2015). A key area for future research in this theme includes the conceptualization and examination of multi-sector transitions, including the role of policy integration therein. "Climate change mitigation" – a key motivation for focusing on the energy transitions – is discussed predominantly in Theme 5. Here, studies have evaluated the nationally determined contributions (NDCs) submitted as part of the Paris Agreement (van Soest et al.

Theme 1: Sociotechnical transition	Theme 2: Fossil fuel dependency	Theme 3: Economy and energy

sustainability transit

conceptu
sociotechn sociotechnical transit
energy justic
actor
practic transit
social commun

global

miner transit
world
coal
energy secur
energi

oil

countri china
effect
energy intens
economic growth
provinc
gdp

region

Theme 4: Resource flows	Theme 5: Climate change mitigation	Theme 6: Politics and power

circular economi citi
system
sustain

cc water
wast urban build

carbon
low carbon
emiss
cc scenario
decarbonis reduc
ghg low

govern
power
discours
democrat
state
conflict polit
contest

Theme 7: Industry and innovation	Theme 8: Governance and policy	Theme 9: Renewable energy integration

compani

industri
technologi
green develop
innov
firm sector
business model

instrument
reform market

regulatori

polici
eu
regul
implement govern

power system

increas
solar
cost
power pl electr
plant gener
electricity gener
renew

Theme 10: Energy modelling	Theme 11: Wind energy	Theme 12: Energy future

predict
base
comput price model
simul time

forecast
optim

diffus support
germani
ti
cooper region

german

develop
local

decis
sustainable energi
cluster
expert energy system futur
energy transit
energi

Theme 13: Behavior and consumption	Theme 14: Energy access	Theme 15: Financing and investment

consum
smart grid
electric vehicl
prefer
user
particip
inform
car ev

lpg

stove
rural
india
access
energi
cook household

invest
risk
renewable energi
financ
investor
project
asset financi

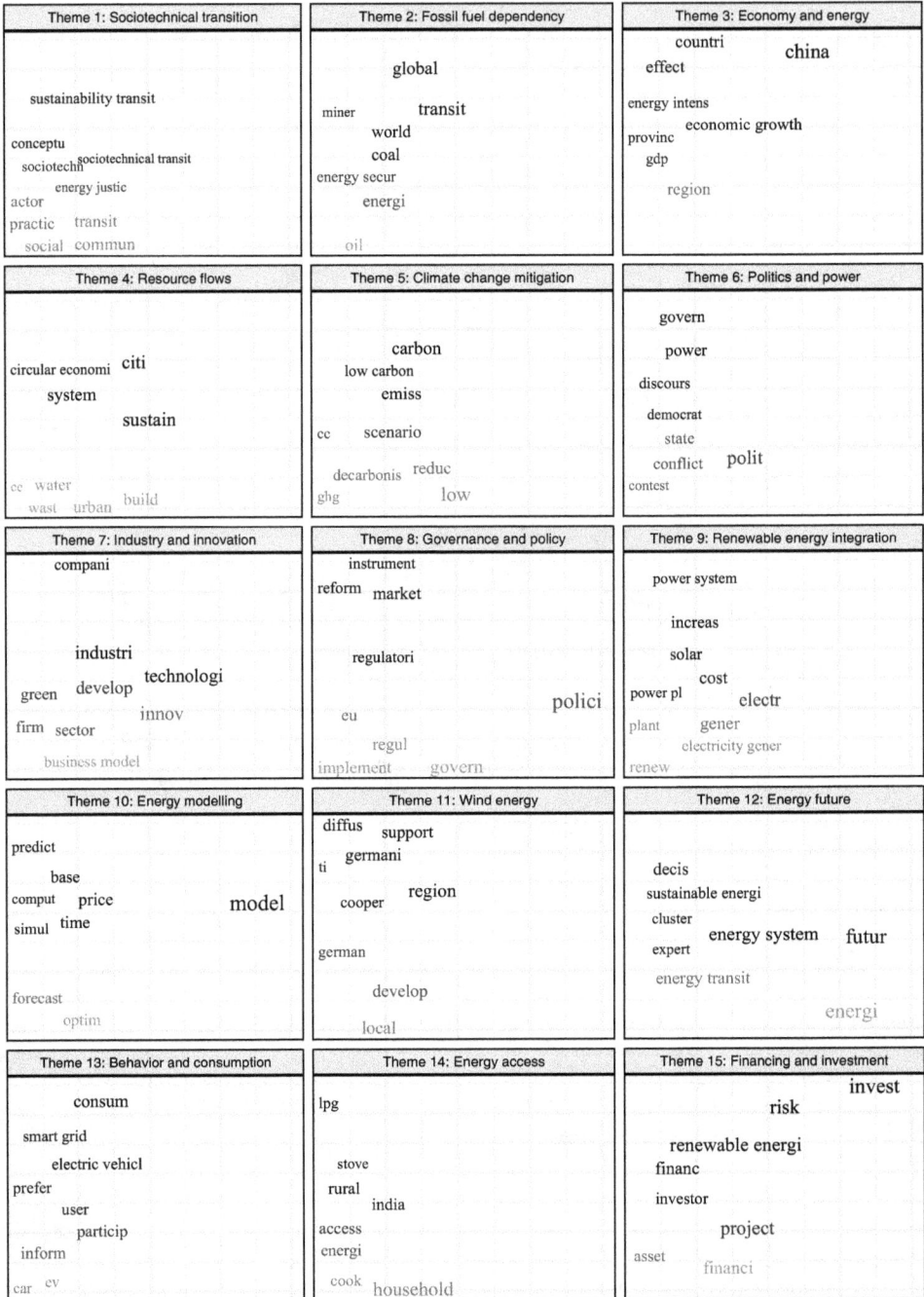

Figure 20.3 The themes in energy transitions research. The themes are arranged in descending order of prevalence, from left to right and top to bottom. The key terms associated with each theme are arranged based on probability of occurrence within the theme (x-axis) and exclusivity to that theme (y-axis).

Source: Data from Web of Science. Analysis by the authors.

2017), proposed short-term milestones to achieve a net-zero energy system (Kuramochi et al. 2018), and examined various alternatives for decarbonization (Oshiro et al. 2021; van Vuuren et al. 2018). Bertram et al. (2015), however, emphasize the need for strong policy signals to avoid carbon lock-in. Future research in this theme will need to pay more attention to pathways to a net-zero world, including the role of urban climate change mitigation, greening industries, and negative emissions technologies in these pathways.

The challenge of transitioning away from high-carbon energy is further deliberated in more detail in Theme 2 on "fossil fuel dependency". Studies on this theme for example, on drawing lessons from historical transitions to fossil fuels (Hölsgens 2019), discuss issues surrounding peak oil (Greene, Hopson, and Li 2006) and shale gas (Le et al. 2021), and question the energy security of a coal-based economy (Sovacool, Cooper, and Parenteau 2011). Outstanding issues in this theme include analyzing and facilitating the phaseout of coal energy from the perspective of energy justice.

Theme 15 on "financing and investment" is concerned with realizing renewable energy on the ground. Several scholars examine the characteristics that affect renewable energy investment (Steffen 2018), including the risk of stranded assets (Curtin et al. 2019), investor preferences concerning rewards and risks (Salm, Hille, and Wüstenhagen 2016), individual behavior in crowdfunding (Bento, Gianfrate, and Groppo 2019), and the role of electricity utilities (Patala et al. 2021). Emphasizing the importance of public policy, Yanosek (2012) proposes policies for financing the transition to renewable energy. The role of new types of financing, such as blended finance, and the effect of cryptocurrencies on the energy transition are examples of topics that will require attention within this theme.

Closely related to the previous theme, Theme 9 focuses on "renewable energy integration" in the energy system. Here, scholars have focused on potential of renewable energy in replacing nuclear energy (Kung and McCarl 2020), assessed the viability of a zero-emissions electricity grid (Martinez-Jaramillo, van Ackere, and Larsen 2020; Lugovoy et al. 2021), discussed challenges involved in high renewable energy penetration (Trondle 2020; Quirapas and Taeihagh 2021), and investigated specific alternatives that facilitate grid integration (Caldera et al. 2018; Kopiske, Spieker, and Tsatsaronis 2017; Hahn et al. 2017). Future work in this theme is likely to focus increasingly on issues such as the influence of green hydrogen and vehicle-to-grid technologies on renewable energy integration.

Predominantly focusing on a "wind energy" perspective, Theme 11 delves largely into the opportunities and challenges posed by clean energy technologies. Studies in this theme focus often on the local level but also on the national or supranational level. The technological innovation systems perspective has been frequently applied as a diagnostic tool in this theme to study the diffusion of clean energy (Negro, Hekkert, and Smits 2007; Edsand 2017; Kebede and Mitsufuji 2017; Negro, Alkemade, and Hekkert 2012). In other works, scholars have analyzed the factors influencing wind energy implementation and its public acceptance – for example, in Germany (Lutz et al. 2017; Langer et al. 2016). Public policy is a focal issue in this theme, as reflected by studies analyzing the relationship between high wind turbine density and wind energy policy (Mauro 2019), identifying government interventions that support community energy (Meister et al. 2020), and proposing a GIS-based approach for municipal renewable energy planning (Wang et al. 2016).

Theme 7 on "industry and innovation" discusses the role of businesses in the energy transitions. A prominent strand of research within this theme delves into the roles of industries in innovation in the clean technology industry and promoting technology diffusion (Bergek 2020; Marra et al. 2015). In other research, scholars have investigated the characteristics of enterprises, ecosystem capabilities, and system-wide (rather than enterprise-wide) business models that accelerate energy transitions (Mlecnik 2013; Hellstrom et al. 2015; Lutjen et al. 2019). Future research should also examine the relationship between private policy and public policy, the role of intermediaries in transitions, and the drivers of nonstate climate action to present a more complete understanding of the role of industry in energy transitions.

Theme 13 on "behavior and consumption" engages with the demand for clean energy. Much recent research on this theme is concerned with the adoption of electric vehicles (Peters, van der Werff, and Steg 2018; Du et al. 2018). Public perception of smart grids – considered an enabler of demand response – is another sub-theme that has received attention in this theme (Li et al. 2017; Lazowski, Parker, and Rowlands 2018). Scholars have also analyzed attitudes and behavior in the context of green energy (Hojnik et al. 2021; Sloot, Jans, and Steg 2019). Future research in this theme should pay more attention to the influence of behavioral assumptions of energy policy on outcomes in energy transitions and the emerging notion of energy citizenship.

Similar to Theme 13 in its focus on energy use, Theme 14 delves into household energy, primarily in the context of "energy access". The topics covered here include energy inequity and energy poverty (Nguyen et al. 2019), fuelwood use (Win et al. 2018), household preferences for energy (Rahut et al. 2017; Chindarkar, Jain, and Mani 2021), and the gendered effects of the incomplete transition to clean cooking fuels (Maji, Mehrabi, and Kandlikar 2021), respectively. In an instance of policy analysis, Aung et al. (2021) study the effect of conditional cash transfer on enhancing energy access in Malawi. Topics for future work within this theme include granular and multidimensional measurement of energy access, the relationship between energy access and renewable energy integration, and a more holistic understanding of the effect of energy access on well-being.

Theme 3 on "economy and energy" highlights the complex relationship among the economy, energy production and use, and the environment (often in the case of China). Issues explored in this theme include the interdependence between economic growth and carbon dioxide emissions (Kim, Lee, and Nam 2010); the link between hydroelectricity consumption and economic growth (Apergis et al. 2016); the effect of technological progress on energy intensity (Wang et al. 2021); the impact of economic growth, foreign direct investment, and energy intensity on carbon dioxide emissions from industry (Ma et al. 2019); and the influence of energy prices on carbon productivity (Tian and Yang 2020). Theme 4 on "resource flows" also adopts a nexus approach to the energy system, delving into life cycle analysis Banias et al. (2020), food-energy-water nexus Zhang et al. (2019), and circular economy (Joensuu, Edelman, and Saari 2020). Scholars will increasingly need to shed light on the relationship among circular economy, resource flows, and sustainable development.

Theme 10 on "energy modeling" is concerned with trend analysis and short-term forecasting (Gutierrez, Gutierrez-Sanchez, and Nafidi 2009). A range of variables related to the energy system have been examined through various modeling approaches in this theme. Some scholars have used regression analysis, for example, to forecast residential electricity consumption (Gabreyohannes 2010) and volatility in the electricity market (Qu et al. 2016). In contrast to Theme 10, Theme 12 on "energy futures" focuses on long-term scenario building and analysis, including scenario construction (Pregger et al. 2020; Ernst et al. 2018), decision-making under uncertainty (Tavana et al. 2020), and multi-criteria approaches for scenario evaluation (Schar and Geldermann 2021). Integration of short-term, medium-term, and long-term energy modeling will be critical for understanding and anticipating the effect of energy transitions on energy, environment, and society.

The remaining themes focus on the governance of energy transitions. Theme 6 on "politics and power" emphasizes the contested nature of the energy transitions, for example, by studying conflicts among the actors involved (Villo, Halme, and Ritvala 2020), analyzing discursive struggles (Hess 2019), examining the concept of energy democracy (Angel 2017), and urging the public to engage in substantive collective action (Stirling 2019). Finally, Theme 8 on "governance and policy" delves into various aspects of policymaking, such as historical policy development (Zhang 2015), inefficiencies in existing governance arrangements (Ibarra-Yunez 2015;

Adom 2016), governance capacities (Cabeca et al. 2021), policy implementation, (Jakob et al. 2019), and policy impact (Zhou et al. 2021). The topics for future research within these themes include, among others, furthering the convergence between innovation studies and policy studies and, relatedly, investigating the role of political entrepreneurship in energy transitions.

This analysis shows that several themes acknowledge the importance of policy. Most prominently, Theme 8 on "governance and policy" focuses directly and explicitly on public policy. In addition, the themes on renewable energy financing, innovation system, industry and innovation, and – to a lesser extent – climate change also emphasize the role of public policy in the energy transitions. However, the extent to which the literature has engaged with the research in policy studies analytically – rather than descriptively – is not entirely clear. In the next section, we analyze the text mentioning policy and review the pertinent literature in more detail.

The Use of Lenses from Public Policy

As mentioned, much of the literature on energy transitions has acknowledged the role of public policy in this process. The word "policy" has in fact been mentioned over 9,000 times in this dataset, either by itself or as part of a phrase. The commonly used phrases containing the word policy in this dataset include the following: "policy mak[ing]" (*count* >800), "energy polici" (*count* >600), "climate polici" (*count* >400), "policy mix" (*count* >200), "policy impl[ication]" (*count* >150), "policy instru[ment]" (*count* >150), "environmental polici" (*count* >150), "public polici" (*count* >100), "renewable energy polici" (*count* >100), "policy design" (*count* >75), "policy recommend[ation]" (*count* >75), and "policy chang[e]" (*count* >75). More analytical or conceptual terms, however, are used less frequently: "policy process" (*count* ~50), "policy implement[ation]" (*count* ~30), "policy innov[ation]" (*count* ~25), "policy feedback" (*count* ~25), "policy outcom[es]" (*count* ~20), and "policy evalu[ation]" (*count* ~20).

A correlation analysis of terms including the word "policy" (or its stemmed form, "polici") sheds light on the context in which such terms have been used (Figure 20.4).

We observe that domain-specific reference to policy – such as climate policy, energy policy, environmental policy, and foreign policy – typically has a low correlation with analytical concepts in policy studies. In contrast, the term innovation policy has a high correlation with the notion of the policy mix. Kern and Howlett (2009), for example, argue that characteristics of policy mixes – such as consistency, coherence, and congruence – influence sociotechnical outcomes in transitions (see also, Costantini, Crespi, and Palma 2017). Meanwhile, Kivimaa and Kern (2016) argue that policy mixes for sustainability transitions should combine instruments that facilitate creating the new with those that "destabilize the old". The typology of policy instruments and the conceptualization of policy process in this strand of literature, however, offer significant scope for further synthesis with the research in policy studies. One example of such a synthesis of policy change, policy mixes, and energy transitions research from a policy design perspective is the examination of policy change and the policy evolution in China between 1981 and 2020 for low-carbon energy transition (Li and Taeihagh 2020).

In general, while many uses of the term are descriptive, the literature on sociotechnical transitions does engage with the concept of policy change – especially using the advocacy coalition framework (Sabatier 1988) – while examining regime dynamics. Markard, Suter, and Ingold (2016), for instance, use the advocacy coalition framework to explain regime stability in the Swiss energy system. In a synthesis with the literature on policy feedback, Schmid, Sewerin, and Schmidt (2019) use the latter concept to explain policy outcomes and advocacy coalition change in the German energy transition. Relatedly, the notion of policy feedback has been invoked in the context of energy transitions (Lockwood et al. 2017; Strauch 2020).

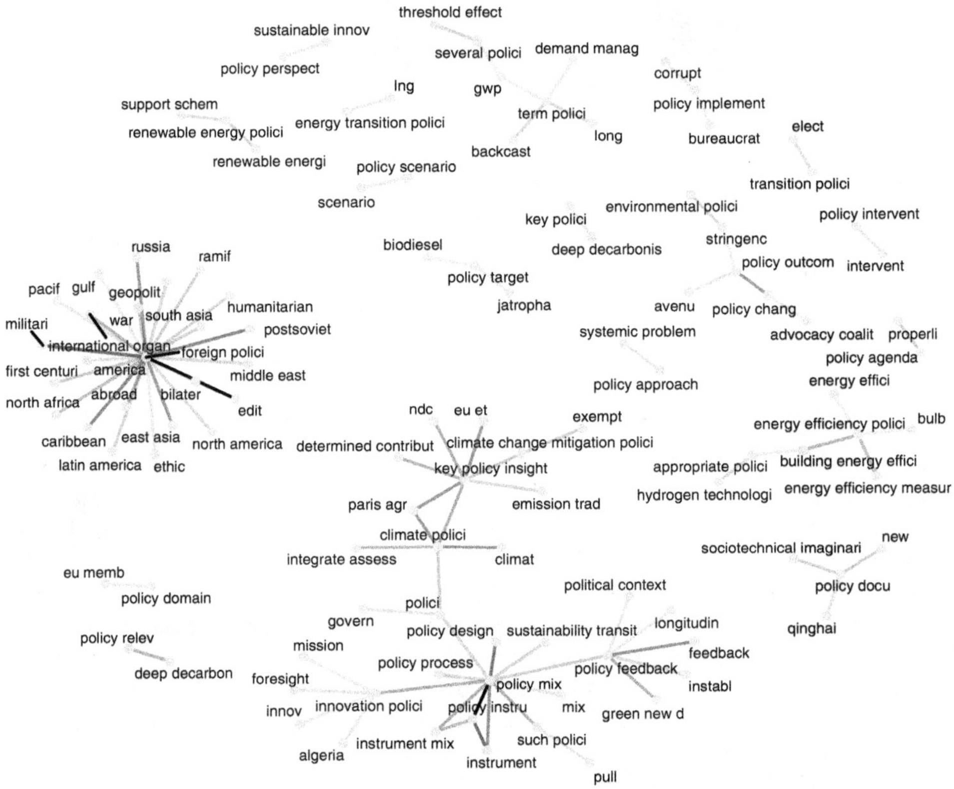

Figure 20.4 Correlation network of terms associated with policy. The nodes represent the phrases used in the literature. A link between two nodes indicates a reasonable correlation between the terms. The color intensity of the link indicates the strength of the correlation.

Source: Data from Web of Science. Analysis by the authors.

Edmondson, Kern, and Rogge (2019), illustratively, propose a framework to examine the co-evolution of policy mixes and sustainability transitions by synthesizing the concept of policy feedback with a framework of policy mixes.

While some uses of policy change and policy feedback overlap with the notion of policy invention, the concepts of policy entrepreneurship, policy invention, or policy innovation are rarely explored (Goyal, Howlett, and Chindarkar 2020). In an exception, Llamosas, Upham, and Blanco (2018) use the multiple streams framework to show that "regime resistance" has thwarted efforts to introduce policy innovation in the energy system in Paraguay (see also Karapin 2019). In assessing a "successful" case, Argyriou (2020) demonstrates that a combination of socioeconomic characteristics, political orientations, and third-sector entities drive policy innovation in commercial energy efficiency in Philadelphia. In a recent article, Derwort, Jager, and Newig (2021) have complemented the use of the multi-level perspective with that of the multiple streams framework to show that, in the case of the German energy transition, policy innovation is the result of an interplay between sociotechnical and political dynamics (see also Goyal and Howlett 2018).

Policy diffusion or transfer can contribute to the spread of policy innovation and, thereby, help accelerate the energy transition. However, few studies have applied these concepts in the context of energy transitions. As an exception, Zimm (2021) analyzes the global diffusion of policies concerning electric vehicles to find the socioeconomic characteristics, political factors, and international mechanisms that can accelerate the transition to electric vehicles. Relatedly, Goyal (2021c) conceptualizes policy diffusion and policy transfer using the multiple streams framework to explain the slow adoption of building energy codes in India. At the intersection of research on policy transfer and sustainability transitions, Pitt and Jones (2016) introduce "scaling up and scaling out" as a new mechanism of transfer and identify the conditions under which it can lead to success. Bhamidipati, Haselip, and Hansen (2019) document the process of policy "translation" through which a coherent policy outcome was achieved in the case of renewable energy in Uganda. Recently, Heyen et al. (2021) have argued for closer integration between the research on policy transfer and sustainability transitions to catalyze transformative change.

Policy effects are another aspect of policy innovation and have hardly been analyzed empirically in the literature. However, some studies have undertaken this task, often using some of the concepts from policy studies discussed earlier (Goyal 2021a). Illustratively, Fontaine, Fuentes, and Narváez (2019) use the policy design framework to show that the (intended) lack of congruence within the policy mix can help actors resisting change and undermine policy outcomes. Meanwhile, studies on policy evaluation in China have examined the effectiveness of the new electric vehicles industry policy (Liu et al. 2020), the energy conservation policies of buildings (Han, Yao, and Li 2021), and the consolidation policy in the coal mining industry (Liu et al. 2016). In the case of the transition from solid fuels in South Africa, Matinga, Clancy, and Annegarn (2014) find that symbolic use of policy explained the non-implementation of the pro-poor energy policy of the South African government. Also, through a comparison of coal phase-out in Germany and the United Kingdom, Brauers, Oei, and Walk (2020) demonstrate that policy outcomes are affected by several actors, such as industries, environmental groups, and the government.

Thus, this analysis reveals that despite the frequent reference to public policy, use of the term in the energy transitions research is typically descriptive (or prescriptive) rather than analytical. Although the literature has engaged in some depth with the concept of policy mixes, its use of other public policy theories is still rather limited. Specifically, few scholars have engaged with the three dimensions of policy innovation: (radical) policy change or invention, policy diffusion, and policy outcomes.

Conclusion

This study examined over 8,000 publications on energy transitions through a bibliometric review and computational text analysis. A bibliometric overview of this dataset indicated that scientific production in this research area is driven by Europe and a few institutions – in the Netherlands and the United Kingdom, particularly – play a dominant role in advancing the scholarship, while limited research is coming out of low- and middle-income countries. Further, institutional collaboration is limited, and scholars typically co-author with others in the same country.

The limited engagement of the scholarly community in the Global South reduces the diversity of perspectives in the research and hinders conceptual and theoretical development of the field. It also skews the research agenda and leaves issues such as energy access and energy efficiency at the margins of the discussion. In addition, it limits the generalizability of the knowledge created in this field and undermine the policy relevance of the research.

The limited cross-fertilization between policy studies and energy transitions research (or more broadly sustainability transitions research) is also an issue that causes fragmentation in knowledge on policy innovation and technological innovation. Even though policy innovation and technological innovation are co-evolutionary, the literature in policy studies has underplayed the role of technological development in the policy process even as the research on energy and sustainability transitions has largely treated public policy as a black box. Adapting and applying public policy theories for research on the energy system is one way of bridging this gap and creating policy-relevant knowledge for accelerating the energy transitions.

References

Adom, P. K. 2016. "The transition between energy efficient and energy inefficient states in Cameroon." *Energy Economics* 54: 248–262. https://doi.org/10.1016/j.eneco.2015.11.025. <Go to ISI>:// WOS:000371942000023.

Aliaattiga, H. E. 1978. "Impact of energy transition on the oil-exporting countries." *Journal of Energy and Development* 4 (1): 41–48. <Go to ISI>://WOS:A1978GF89400003.

Angel, James. 2017. "Towards an energy politics in-against-and-beyond the state: Berlin's struggle for energy democracy." *Antipode* 49 (3): 557–576. https://doi.org/10.1111/anti.12289. https://onlinelibrary.wiley.com/doi/abs/10.1111/anti.12289.

Apergis, Nicholas, Tsangyao Chang, Rangan Gupta, and Emmanuel Ziramba. 2016. "Hydroelectricity consumption and economic growth nexus: Evidence from a panel of ten largest hydroelectricity consumers." *Renewable and Sustainable Energy Reviews* 62: 318–325. https://doi.org/https://doi. org/10.1016/j.rser.2016.04.075. www.sciencedirect.com/science/article/pii/S1364032116301083.

Argyriou, Iraklis. 2020. "Urban energy transitions in ordinary cities: Philadelphia's place-based policy innovations for socio-technical energy change in the commercial sector." *Urban Research & Practice* 13 (3): 243–275. https://doi.org/10.1080/17535069.2018.1540654. https://doi.org/10.1080/17535069 .2018.1540654.

Aria, Massimo, and Corrado Cuccurullo. 2017. "Bibliometrix: An R-tool for comprehensive science mapping analysis." *Journal of Informetrics* 11 (4): 959–975.

Aung, T., R. Bailis, T. Chilongo, A. Ghilardi, C. Jumbe, and P. Jagger. 2021. "Energy access and the ultra-poor: Do unconditional social cash transfers close the energy access gap in Malawi?" *Energy for Sustainable Development* 60: 11. https://doi.org/10.1016/j.esd.2020.12.003. <Go to ISI>:// WOS:000610364600010.

Banias, G., M. Batsioula, C. Achillas, S. I. Patsios, K. N. Kontogiannopoulos, D. Bochtis, and N. Moussiopoulos. 2020. "A life cycle analysis approach for the evaluation of municipal solid waste management practices: The case study of the region of central Macedonia, Greece." *Sustainability* 12 (19). https:// doi.org/10.3390/su12198221.

Bento, Nuno, Gianfranco Gianfrate, and Sara Virginia Groppo. 2019. "Do crowdfunding returns reward risk? Evidences from clean-tech projects." *Technological Forecasting and Social Change* 141: 107–116. https://doi.org/https://doi.org/10.1016/j.techfore.2018.07.007. www.sciencedirect.com/science/ article/pii/S0040162517310260.

Bergek, A. 2020. "Diffusion intermediaries: A taxonomy based on renewable electricity technology in Sweden." *Environmental Innovation and Societal Transitions* 36: 378–392. https://doi.org/10.1016/j. eist.2019.11.004. <Go to ISI>://WOS:000569713900003.

Berry, R. S. 1975. "Crisis of resource scarcity–transition to an energy-limited economy." *Bulletin of the Atomic Scientists* 31 (1): 31–36. https://doi.org/10.1080/00963402.1975.11458186. <Go to ISI>:// WOS:A1975V189400014.

Bertram, Christoph, Nils Johnson, Gunnar Luderer, Keywan Riahi, Morna Isaac, and Jiyong Eom. 2015. "Carbon lock-in through capital stock inertia associated with weak near-term climate policies." *Technological Forecasting and Social Change* 90: 62–72. https://doi.org/https://doi.org/10.1016/j.techfore.2013.10.001. www.sciencedirect.com/science/article/pii/S004016251300259X.

Bhamidipati, Padmasai Lakshmi, James Haselip, and Ulrich Elmer Hansen. 2019. "How do energy policies accelerate sustainable transitions? Unpacking the policy transfer process in the case of GETFiT Uganda." *Energy Policy* 132: 1320–1332. https://doi.org/https://doi.org/10.1016/j.enpol.2019.05.053. www. sciencedirect.com/science/article/pii/S030142151930357X.

Blei, David M. 2012. "Probabilistic topic models." *Communications ACM* 55 (4): 77–84. https://doi. org/10.1145/2133806.2133826.

Brauers, Hanna, Pao-Yu Oei, and Paula Walk. 2020. "Comparing coal phase-out pathways: The United Kingdom's and Germany's diverging transitions." *Environmental Innovation and Societal Transitions* 37: 238–253. https://doi.org/https://doi.org/10.1016/j.eist.2020.09.001. www.sciencedirect.com/science/ article/pii/S2210422420301143.

Cabeca, A. S., C. O. Henriques, J. R. Figueira, and C. S. Silva. 2021. "A multicriteria classification approach for assessing the current governance capacities on energy efficiency in the European Union." *Energy Policy* 148: 19. https://doi.org/10.1016/j.enpol.2020.111946. <Go to ISI>://WOS:000600551000028.

Caldera, U., D. Bogdanov, S. Afanasyeva, and C. Breyer. 2018. "Role of seawater desalination in the management of an integrated water and 100% renewable energy based power sector in Saudi Arabia." *Water* 10 (1): 32. https://doi.org/10.3390/w10010003. <Go to ISI>://WOS:000424397400002.

Chilvers, J., H. Pallett, and T. Hargreaves. 2018. "Ecologies of participation in socio-technical change: The case of energy system transitions." *Energy Research & Social Science* 42: 199–210. https://doi.org/10.1016/j.erss.2018.03.020. <Go to ISI>://WOS:000439444400027.

Chindarkar, N., A. Jain, and S. Mani. 2021. "Examining the willingness-to-pay for exclusive use of LPG for cooking among rural households in India." *Energy Policy* 150: 10. https://doi.org/10.1016/j.enpol.2020.112107. <Go to ISI>://WOS:000636259200006.

Costantini, V., F. Crespi, and A. Palma. 2017. "Characterizing the policy mix and its impact on eco-innovation: A patent analysis of energy-efficient technologies." *Research Policy* 46(4): 799–819. https://doi.org/10.1016/j.respol.2017.02.004. <Go to ISI>://WOS:000399627600007.

Curtin, J., C. McInerney, B. Ó Gallachóir, C. Hickey, P. Deane, and P. Deeney. 2019. "Quantifying stranding risk for fossil fuel assets and implications for renewable energy investment: A review of the literature." *Renewable and Sustainable Energy Reviews* 116: 109402. https://doi.org/https://doi.org/10.1016/j.rser.2019.109402. www.sciencedirect.com/science/article/pii/S1364032119306100.

Derwort, P., Jager, N., & Newig, J. (2022). How to explain major policy change towards sustainability? Bringing together the multiple streams framework and the multilevel perspective on socio-technical transitions to explore the German "Energiewende". *Policy Studies Journal*, 50(3): 671–699.

Du, H. B., D.Y. Liu, B.K. Sovacool, Y.R. Wang, S. F. Ma, and R.Y.M. Li. 2018. "Who buys New Energy Vehicles in China? Assessing social-psychological predictors of purchasing awareness, intention, and policy." *Transportation Research Part F-Traffic Psychology and Behaviour* 58: 56–69. https://doi.org/10.1016/j.trf.2018.05.008. <Go to ISI>://WOS:000447357900006.

Edmondson, D. L., F. Kern, and K.S. Rogge. 2019. "The co-evolution of policy mixes and socio-technical systems: Towards a conceptual framework of policy mix feedback in sustainability transitions." *Research Policy* 48(10). https://doi.org/10.1016/j.respol.2018.03.010. <Go to ISI>://WOS:000503091200009.

Edsand, H.E. 2017. "Identifying barriers to wind energy diffusion in Colombia: A function analysis of the technological innovation system and the wider context." *Technology in Society* 49: 1–15. https://doi.org/10.1016/j.techsoc.2017.01.002. <Go to ISI>://WOS:000404517100002.

Ernst, A., K.H. Biss, H. Shamon, D. Schumann, and H.U. Heinrichs. 2018. "Benefits and challenges of participatory methods in qualitative energy scenario development." *Technological Forecasting and Social Change* 127: 245–257. https://doi.org/10.1016/j.techfore.2017.09.026. <Go to ISI>://WOS:000423643700020.

Faller, F. 2016. "A practice approach to study the spatial dimensions of the energy transition." *Environmental Innovation and Societal Transitions* 19: 85–95. https://doi.org/10.1016/j.eist.2015.09.004. <Go to ISI>://WOS:000379410600007.

Fontaine, G., J. L. Fuentes, and I. Narváez. 2019. "Policy mixes against oil dependence: Resource nationalism, layering and contradictions in Ecuador's energy transition." *Energy Research and Social Science* 47: 56–68. https://doi.org/10.1016/j.erss.2018.08.013. www.scopus.com/inward/record.uri?eid=2-s2.0-85053015989&doi=10.1016%2fj.erss.2018.08.013&partnerID=40&md5=24b75567741d745bc9b3b944160df67d.

Forman, A. 2017. "Energy justice at the end of the wire: Enacting community energy and equity in Wales." *Energy Policy* 107: 649–657. https://doi.org/10.1016/j.enpol.2017.05.006. <Go to ISI>://WOS:000405158200064.

Gabreyohannes, E. 2010. "A nonlinear approach to modelling the residential electricity consumption in Ethiopia." *Energy Economics* 32(3): 515–523. https://doi.org/10.1016/j.eneco.2009.08.008. <Go to ISI>://WOS:000276326200002.

Geels, F. W. 2020. "Micro-foundations of the multi-level perspective on socio-technical transitions: Developing a multi-dimensional model of agency through crossovers between social constructivism, evolutionary economics and neo-institutional theory." *Technological Forecasting and Social Change* 152: 17. https://doi.org/10.1016/j.techfore.2019.119894. <Go to ISI>://WOS:000514757800005.

Goyal, Nihit. 2019. "Promoting policy innovation for sustainability: Leaders, laggards and learners in the indian electricity transition." Ph.D. in Public Policy Doctoral Dissertation, National University of Singapore. https://scholarbank.nus.edu.sg/handle/10635/166349.

———. 2021a. "explaining policy success using the multiple streams framework: Political success despite programmatic failure of the solar energy policy in Gujarat, India."

Politics and Policy. https://doi.org/10.1111/polp.12426. www.scopus.com/inward/record.uri?eid=2-s2.0-85111120572&doi=10.1111%2fpolp.12426&partnerID=40&md5=81cf9769965fed8aba982159f53e925f.
———. 2021b. "Limited demand or unreliable supply? A bibliometric review and computational text analysis of research on energy policy in India." *Sustainability* 13(23): 13421. www.mdpi.com/2071-1050/13/23/13421.
———. 2021c. "Policy diffusion through multiple streams: The (non-)adoption of energy conservation building code in India." *Policy Studies Journal* n/a (n/a). https://doi.org/https://doi.org/10.1111/psj.12415. https://doi.org/10.1111/psj.12415.
Goyal, Nihit, and Michael Howlett. 2018. "Technology and instrument constituencies as agents of innovation: Sustainability transitions and the governance of urban transport." *Energies* 11(5): 1198. www.mdpi.com/1996-1073/11/5/1198.
———. 2020. "Who learns what in sustainability transitions?" *Environmental Innovation and Societal Transitions* 34: 311–321. https://doi.org/https://doi.org/10.1016/j.eist.2019.09.002. https://doi.org/10.1016/j.eist.2019.09.002.
———. 2021. "Conceptualizing energy democracy using the multiple streams framework: Actors, public participation, and scale in energy transitions." In *Routledge Handbook of Energy Democracy*, 66–81. Routledge.
Goyal, Nihit, Michael Howlett, and Namrata Chindarkar. 2020. "Who coupled which stream(s)? Policy entrepreneurship and innovation in the energy—water nexus in Gujarat, India." *Public Administration and Development* 40(1): 49–64. https://doi.org/https://doi.org/10.1002/pad.1855. https://doi.org/10.1002/pad.1855.
Goyal, Nihit, Michael Howlett, and Araz Taeihagh. 2021. "Why and how does the regulation of emerging technologies occur? Explaining the adoption of the EU General Data Protection Regulation using the multiple streams framework." *Regulation & Governance* n/a (n/a). https://doi.org/https://doi.org/10.1111/rego.12387. https://doi.org/10.1111/rego.12387.
Greene, David L., Janet L. Hopson, and Jia Li. 2006. "Have we run out of oil yet? Oil peaking analysis from an optimist's perspective." *Energy Policy* 34(5): 515–531. https://doi.org/https://doi.org/10.1016/j.enpol.2005.11.025. www.sciencedirect.com/science/article/pii/S0301421505003083.
Gutierrez, R., R. Gutierrez-Sanchez, and A. Nafidi. 2009. "Modelling and forecasting vehicle stocks using the trends of stochastic Gompertz diffusion models: The case of Spain." *Applied Stochastic Models in Business and Industry* 25(3): 385–405. https://doi.org/10.1002/asmb.754. <Go to ISI>://WOS:000267275300012.
Hahn, H., D. Hau, C. Dick, and M. Puchta. 2017. "Techno-economic assessment of a subsea energy storage technology for power balancing services." *Energy* 133: 121–127. https://doi.org/10.1016/j.energy.2017.05.116. <Go to ISI>://WOS:000407655700013.
Han, Shiyu, Runming Yao, and Nan Li. 2021. "The development of energy conservation policy of buildings in China: A comprehensive review and analysis." *Journal of Building Engineering* 38: 102229. https://doi.org/https://doi.org/10.1016/j.jobe.2021.102229. www.sciencedirect.com/science/article/pii/S2352710221000851.
Hansen, T., and L. Coenen. 2015. "The geography of sustainability transitions: Review, synthesis and reflections on an emergent research field." *Environmental Innovation and Societal Transitions* 17: 92–109. https://doi.org/10.1016/j.eist.2014.11.001. <Go to ISI>://WOS:000218922000008.
Hayes, D. 1977. "Coming energy transition." *Futurist* 11(5): 303–309. <Go to ISI>://WOS:A1977EE24200006.
Hellstrom, M., A. Tsvetkova, M. Gustafsson, and K. Wikstrom. 2015. "Collaboration mechanisms for business models in distributed energy ecosystems." *Journal of Cleaner Production* 102: 226–236. https://doi.org/10.1016/j.jclepro.2015.04.128. <Go to ISI>://WOS:000356741200020.
Hess, D. J. 2019. "Coalitions, framing, and the politics of energy transitions: Local democracy and community choice in California." *Energy Research and Social Science* 50: 38–50. https://doi.org/10.1016/j.erss.2018.11.013. www.scopus.com/inward/record.uri?eid=2-s2.0-85057496237&doi=10.1016%2fj.erss.2018.11.013&partnerID=40&md5=c0e53d55074af807d0d566159c70ec34.
Heyen, Dirk Arne, Klaus Jacob, Julia Teebken, and Franziska Wolff. 2021. "Spillovers between policy-transfer and transitions research." *Environmental Innovation and Societal Transitions* 38: 79–81. https://doi.org/https://doi.org/10.1016/j.eist.2020.11.005. www.sciencedirect.com/science/article/pii/S2210422420301325.
Hojnik, J., M. Ruzzier, S. Fabri, and A. L. Klopcic. 2021. "What you give is what you get: Willingness to pay for green energy." *Renewable Energy* 174: 733–746. https://doi.org/10.1016/j.renene.2021.04.037. <Go to ISI>://WOS:000654248500011.

Hölsgens, Rick. 2019. "Resource dependence and energy risks in the Netherlands since the mid-nineteenth century." *Energy Policy* 125: 45–54. https://doi.org/https://doi.org/10.1016/j.enpol.2018.10.020. www.sciencedirect.com/science/article/pii/S0301421518306761.

Hoppe, Thomas, Frans Coenen, and Maya van den Berg. 2016. "Illustrating the use of concepts from the discipline of policy studies in energy research: An explorative literature review." *Energy Research & Social Science* 21: 12–32. https://doi.org/https://doi.org/10.1016/j.erss.2016.06.006. www.sciencedirect.com/science/article/pii/S2214629616301359.

Howlett, Michael. 2014. "Why are policy innovations rare and so often negative? Blame avoidance and problem denial in climate change policy-making." *Global Environmental Change* 29: 395–403. https://doi.org/https://doi.org/10.1016/j.gloenvcha.2013.12.009. www.sciencedirect.com/science/article/pii/S0959378013002392.

Ibarra-Yunez, A. 2015. "Energy reform in Mexico: Imperfect unbundling in the electricity sector." *Utilities Policy* 35: 19–27. https://doi.org/10.1016/j.jup.2015.06.009. <Go to ISI>://WOS:000362382900002.

Jakob, M., R. Soria, C. Trinidad, O. Edenhofer, C. Bak, D. Bouille, D. Buira, H. Carlino, V. Gutman, C. Hubner, B. Knopf, A. Lucena, L. Santos, A. Scott, J. C. Steckel, K. Tanaka, A. Vogt-Schilb, and K. Yamada. 2019. "Green fiscal reform for a just energy transition in Latin America." *Economics-the Open Access Open-Assessment E-Journal* 13: 11. https://doi.org/10.5018/economics-ejournal.ja.2019-17. <Go to ISI>://WOS:000461275800001.

Joensuu, T., H. Edelman, and A. Saari. 2020. "Circular economy practices in the built environment." *Journal of Cleaner Production* 276. https://doi.org/10.1016/j.jclepro.2020.124215.

Jordan, A., and D. Huitema. 2014. "Policy innovation in a changing climate: Sources, patterns and effects." *Global Environmental Change* 29: 387–394. https://doi.org/10.1016/j.gloenvcha.2014.09.005. www.scopus.com/inward/record.uri?eid=2-s2.0-84898491276&doi=10.1016%2fj.gloenvcha.2014.09.005&partnerID=40&md5=f8b18a661ea940176b4427650a2c9b85.

Karapin, Roger. 2019. "Federalism as a double-edged sword: The slow energy transition in the United States." *The Journal of Environment & Development* 29(1): 26–50. https://doi.org/10.1177/1070496519886001. https://doi.org/10.1177/1070496519886001.

Kebede, K. Y., and T. Mitsufuji. 2017. "Technological innovation system building for diffusion of renewable energy technology: A case of solar PV systems in Ethiopia." *Technological Forecasting and Social Change* 114: 242–253. https://doi.org/10.1016/j.techfore.2016.08.018. <Go to ISI>://WOS:000390743900021.

Kern, F., and M. Howlett. 2009. "Implementing transition management as policy reforms: A case study of the Dutch energy sector." *Policy Sciences* 42(4): 391–408. https://doi.org/10.1007/s11077-009-9099-x. www.scopus.com/inward/record.uri?eid=2-s2.0-72449152852&doi=10.1007%2fs11077-009-9099-x&partnerID=40&md5=eb64f1517be82e0e24b423d52d638a03.

Kern, F., and K. S. Rogge. 2018. "Harnessing theories of the policy process for analysing the politics of sustainability transitions: A critical survey." *Environmental Innovation and Societal Transitions* 27: 102–117. https://doi.org/10.1016/j.eist.2017.11.001. www.scopus.com/inward/record.uri?eid=2-s2.0-85034441414&doi=10.1016%2fj.eist.2017.11.001&partnerID=40&md5=3488d10efe56a04603672332d7c64cfd.

Kern, Florian, and Adrian Smith. 2008. "Restructuring energy systems for sustainability? Energy transition policy in the Netherlands." *Energy Policy* 36(11): 4093–4103. https://doi.org/https://doi.org/10.1016/j.enpol.2008.06.018. www.sciencedirect.com/science/article/pii/S030142150800308X.

Kim, Sei-wan, Kihoon Lee, and Kiseok Nam. 2010. "The relationship between CO2 emissions and economic growth: The case of Korea with nonlinear evidence." *Energy Policy* 38(10): 5938–5946. https://doi.org/https://doi.org/10.1016/j.enpol.2010.05.047. www.sciencedirect.com/science/article/pii/S0301421510004180.

Kittner, Noah, Felix Lill, and Daniel M. Kammen. 2017. "Energy storage deployment and innovation for the clean energy transition." *Nature Energy* 2(9): 17125. https://doi.org/10.1038/nenergy.2017.125. https://doi.org/10.1038/nenergy.2017.125.

Kivimaa, P., W. Boon, S. Hyysalo, and L. Klerkx. 2019. "Towards a typology of intermediaries in sustainability transitions: A systematic review and a research agenda." *Research Policy* 48(4): 1062–1075. https://doi.org/10.1016/j.respol.2018.10.006. <Go to ISI>://WOS:000460714300017.

Kivimaa, P., and F. Kern. 2016. "Creative destruction or mere niche support? Innovation policy mixes for sustainability transitions." *Research Policy* 45(1): 205–217. https://doi.org/10.1016/j.respol.2015.09.008. <Go to ISI>://WOS:000367484100016.

Kopiske, J., S. Spieker, and G. Tsatsaronis. 2017. "Value of power plant flexibility in power systems with high shares of variable renewables: A scenario outlook for Germany 2035." *Energy* 137: 823–833. https://doi.org/10.1016/j.energy.2017.04.138. <Go to ISI>://WOS:000414879400073.

Kung, C. C., and B. A. McCarl. 2020. "The potential role of renewable electricity generation in Tai-wan." *Energy Policy* 138: 13. https://doi.org/10.1016/j.enpol.2019.111227. <Go to ISI>://WOS: 000526116500060.

Kuramochi, Takeshi, Niklas Höhne, Michiel Schaeffer, Jasmin Cantzler, Bill Hare, Yvonne Deng, Sebas-tian Sterl, Markus Hagemann, Marcia Rocha, Paola Andrea Yanguas-Parra, Goher-Ur-Rehman Mir, Lindee Wong, Tarik El-Laboudy, Karlien Wouters, Delphine Deryng, and Kornelis Blok. 2018. "Ten key short-term sectoral benchmarks to limit warming to 1.5°C." *Climate Policy* 18(3): 287–305. https:// doi.org/10.1080/14693062.2017.1397495. https://doi.org/10.1080/14693062.2017.1397495.

Langer, K., T. Decker, J. Roosen, and K. Menrad. 2016. "A qualitative analysis to understand the accept-ance of wind energy in Bavaria." *Renewable & Sustainable Energy Reviews* 64: 248–259. https://doi. org/10.1016/j.rser.2016.05.084. <Go to ISI>://WOS:000381833200017.

Lazowski, B., P. Parker, and I. H. Rowlands. 2018. "Towards a smart and sustainable residential energy cul-ture: Assessing participant feedback from a long-term smart grid pilot project." *Energy Sustainability and Society* 8: 21. https://doi.org/10.1186/s13705-018-0169-9. <Go to ISI>://WOS:000444653300001.

Le, Minh Thong, Huu Tung Do, Thanh Thuy Nguyen, and Thi Kim Ngan Nguyen. 2021. "What prospects for shale gas in Asia? Case of shale gas in China." *The Journal of World Energy Law & Busi-ness* 13(5–6): 426–440. https://doi.org/10.1093/jwelb/jwaa037. https://doi.org/10.1093/jwelb/ jwaa037.

Li, Lili, and Araz Taeihagh. 2020. "An in-depth analysis of the evolution of the policy mix for the sustainable energy transition in China from 1981 to 2020." *Applied Energy* 263(February): 114611– 114611. https://doi.org/10.1016/j.apenergy.2020.114611. https://linkinghub.elsevier.com/retrieve/ pii/S0306261920301239.

Li, R. L., G. Dane, C. Finck, and W. Zeiler. 2017. "Are building users prepared for energy flex-ible buildings?-A large-scale survey in the Netherlands." *Applied Energy* 203: 623–634. https://doi. org/10.1016/j.apenergy.2017.06.067. <Go to ISI>://WOS:000412379300046.

Liu, J., H. Liu, X. L. Yao, and Y. Liu. 2016. "Evaluating the sustainability impact of consolidation policy in China's coal mining industry: A data envelopment analysis." *Journal of Cleaner Production* 112: 2969– 2976. https://doi.org/10.1016/j.jclepro.2015.08.011.

Liu, Lanjian, Tian Zhang, Anne-Perrine Avrin, and Xianwen Wang. 2020. "Is China's industrial policy effective? An empirical study of the new energy vehicles industry." *Technology in Society* 63: 101356. https://doi.org/https://doi.org/10.1016/j.techsoc.2020.101356. www.sciencedirect.com/science/ article/pii/S0160791X20303705.

Llamosas, C., P. Upham, and G. Blanco. 2018. "Multiple streams, resistance and energy policy change in Paraguay (2004–2014)." *Energy Research and Social Science* 42: 226–236. https://doi.org/10.1016/j. erss.2018.03.011.

Lockwood, Matthew, Caroline Kuzemko, Catherine Mitchell, and Richard Hoggett. 2017. "Historical institutionalism and the politics of sustainable energy transitions: A research agenda." *Environment and Planning C: Politics and Space* 35(2): 312–333. https://doi.org/10.1177/0263774x16660561. https:// journals.sagepub.com/doi/abs/10.1177/0263774X16660561.

Loorbach, Derk. 2010. "Transition Management for Sustainable Development: A Prescriptive, Com-plexity-Based Governance Framework." *Governance* 23(1): 161–183. https://doi.org/10.1111/ j.1468-0491.2009.01471.x.

Lugovoy, O., S. Gao, J. Gao, and K. J. Jiang. 2021. "Feasibility study of China's electric power sector transi-tion to zero emissions by 2050." *Energy Economics* 96: 40. https://doi.org/10.1016/j.eneco.2021.105176. <Go to ISI>://WOS:000634096700002.

Lutjen, H., C. Schultz, F. Tietze, and F. Urmetzer. 2019. "Managing ecosystems for service innova-tion: A dynamic capability view." *Journal of Business Research* 104: 506–519. https://doi.org/10.1016/ j.jbusres.2019.06.001. <Go to ISI>://WOS:000484647500041.

Lutz, L. M., L. B. Fischer, J. Newig, and D. J. Lang. 2017. "Driving factors for the regional implementa-tion of renewable energy – A multiple case study on the German energy transition." *Energy Policy* 105: 136–147. https://doi.org/10.1016/j.enpol.2017.02.019. <Go to ISI>://WOS:000400532900014.

Ma, Chao-Qun, Jiang-Long Liu, Yi-Shuai Ren, and Yong Jiang. 2019. "The impact of economic Growth, FDI and energy intensity on China's manufacturing industry's CO2 emissions: An empirical study based on the fixed-effect panel quantile regression model." *Energies* 12(24): 4800. www.mdpi. com/1996-1073/12/24/4800.

Maji, P., Z. Mehrabi, and M. Kandlikar. 2021. "Incomplete transitions to clean household energy reinforce gender inequality by lowering women's respiratory health and household labour productivity." *World Development* 139: 10. https://doi.org/10.1016/j.worlddev.2020.105309. <Go to ISI>:// WOS:000643731700004.

Markard, J., M. Suter, and K. Ingold. 2016. "Socio-technical transitions and policy change–Advocacy coalitions in Swiss energy policy." *Environmental Innovation and Societal Transitions* 18: 215–237. https://doi.org/10.1016/j. eist.2015.05.003. www.scopus.com/inward/record.uri?eid=2-s2.0-84959560409&doi=10.1016%2fj. eist.2015.05.003&partnerID=40&md5 = 67e8cb01e88905397c1631ab84910c34.

Marquardt, Jens. 2016. *How Power Shapes Energy Transitions in Southeast Asia: A Complex Governance Challenge*. Routledge.

Marra, A., P. Antonelli, L. Dell'Anna, and C. Pozzi. 2015. "A network analysis using metadata to investigate innovation in clean-tech – Implications for energy policy." *Energy Policy* 86: 17–26. https://doi. org/10.1016/j.enpol.2015.06.025. <Go to ISI>://WOS:000364246100002.

Martinez-Jaramillo, J. E., A. van Ackere, and E. R. Larsen. 2020. "Towards a solar-hydro based generation: The case of Switzerland." *Energy Policy* 138: 11. https://doi.org/10.1016/j.enpol.2019.111197. <Go to ISI>://WOS:000526116500063.

Martiskainen, M. 2017. "The role of community leadership in the development of grassroots innovations." *Environmental Innovation and Societal Transitions* 22: 78–89. https://doi.org/10.1016/j.eist.2016.05.002. <Go to ISI>://WOS:000400269900007.

Matinga, M. N., J. S. Clancy, and H. J. Annegarn. 2014. "Explaining the non-implementation of health-improving policies related to solid fuels use in South Africa." *Energy Policy* 68: 53–59. https://doi.org/10.1016/j. enpol.2013.10.040. www.scopus.com/inward/record.uri?eid=2-s2.0-84895541621&doi=10.1016%2fj. enpol.2013.10.040&partnerID=40&md5 = 30c6158550093978958a0da51ec0e439.

Mauro, G. 2019. "The new "windscapes" in the time of energy transition: A comparison of ten European countries." *Applied Geography* 109: 15. https://doi.org/10.1016/j.apgeog.2019.102041. <Go to ISI>://WOS:000500199100010.

McConnell, Allan. 2010. "Policy success, policy failure and grey areas in-between." *Journal of Public Policy* 30(3): 345–362. https://doi.org/10.1017/S0143814X10000152. https://doi.org/10.1017/S0143814X10000152.

Meister, T., B. Schmid, I. Seidl, and B. Klagge. 2020. "How municipalities support energy cooperatives: Survey results from Germany and Switzerland." *Energy Sustainability and Society* 10(1): 20. https://doi. org/10.1186/s13705-020-00248-3. <Go to ISI>://WOS:000522103200002.

Mlecnik, E. 2013. "Development of the passive house market: Challenges and opportunities in the transition from innovators to early adopters." In *Innovation Development for Highly Energy-Efficient Housing: Opportunities and Challenges Related to the Adoption of Passive Houses, in Sustainable Urban Areas*, 119–140. Amsterdam: Ios Press.

Moore, Sharlissa. 2018. *Sustainable Energy Transformations, Power, and Politics: Morocco and the Mediterranean*. Routledge.

Negro, S. O., F. Alkemade, and M. P. Hekkert. 2012. "Why does renewable energy diffuse so slowly? A review of innovation system problems." *Renewable & Sustainable Energy Reviews* 16(6): 3836–3846. https://doi.org/10.1016/j.rser.2012.03.043. <Go to ISI>://WOS:000306860700026.

Negro, S. O., M. P. Hekkert, and R. E. Smits. 2007. "Explaining the failure of the Dutch innovation system for biomass digestion—A functional analysis." *Energy Policy* 35(2): 925–938. https://doi. org/10.1016/j.enpol.2006.01.027. <Go to ISI>://WOS:000244170000017.

Nguyen, T. T., T. T. Nguyen, V. N. Hoang, C. Wilson, and S. Managi. 2019. "Energy transition, poverty and inequality in Vietnam." *Energy Policy* 132: 536–548. https://doi.org/10.1016/j.enpol.2019.06.001. <Go to ISI>://WOS:000483425800051.

Oshiro, Ken, Shinichiro Fujimori, Yuki Ochi, and Tomoki Ehara. 2021. "Enabling energy system transition toward decarbonization in Japan through energy service demand reduction." *Energy* 227: 120464. https:// doi.org/https://doi.org/10.1016/j.energy.2021.120464. www.sciencedirect.com/science/article/ pii/S0360544221007131.

Patala, Samuli, Jouni K. Juntunen, Sarianna Lundan, and Tiina Ritvala. 2021. "Multinational energy utilities in the energy transition: A configurational study of the drivers of FDI in renewables." *Journal of International Business Studies* 52(5): 930–950. https://doi.org/10.1057/s41267-020-00387-x. https:// doi.org/10.1057/s41267-020-00387-x.

Perelman, L. J. 1980. "Speculations on the transition to sustainable energy." *Ethics* 90(3): 392–416. https:// doi.org/10.1086/292170. <Go to ISI>://WOS:A1980JY27400005.

Peters, A. M., E. van der Werff, and L. Steg. 2018. "Beyond purchasing: Electric vehicle adoption motivation and consistent sustainable energy behaviour in The Netherlands." *Energy Research & Social Science* 39: 234–247. https://doi.org/10.1016/j.erss.2017.10.008. <Go to ISI>://WOS:000429911000028.

Pitt, Hannah, and Mat Jones. 2016. "Scaling up and out as a pathway for food system transitions." *Sustainability* 8 (10): 1025. www.mdpi.com/2071-1050/8/10/1025.

Pregger, T., T. Naegler, W. Weimer-Jehle, S. Prehofer, and W. Hauser. 2020. "Moving towards socio-technical scenarios of the German energy transition-lessons learned from integrated energy scenario building." *Climatic Change* 162(4): 1743–1762. https://doi.org/10.1007/s10584-019-02598-0. <Go to ISI>://WOS:000568322600001.

Qu, H., W. Chen, M. Y. Niu, and X. D. Li. 2016. "Forecasting realized volatility in electricity markets using logistic smooth transition heterogeneous autoregressive models." *Energy Economics* 54: 68–76. https://doi.org/10.1016/j.eneco.2015.12.001. <Go to ISI>://WOS:000371942000007.

Quirapas, M. A. J. R., and A. Taeihagh. 2021. "Ocean renewable energy development in Southeast Asia: Opportunities, risks and unintended consequences." *Renewable and Sustainable Energy Reviews* 137(September): 110403–110403. https://doi.org/10.1016/j.rser.2020.110403. https://doi.org/10.1016/j.rser.2020.110403 https://linkinghub.elsevier.com/retrieve/pii/S1364032120306912.

Rahut, D. B., B. Behera, A. Ali, and P. Marenya. 2017. "A ladder within a ladder: Understanding the factors influencing a household's domestic use of electricity in four African countries." *Energy Economics* 66: 167–181. https://doi.org/10.1016/j.eneco.2017.05.020. <Go to ISI>://WOS:000412033900017.

Roberts, Margaret E, Brandon M Stewart, and Dustin Tingley. 2014. "Stm: R package for structural topic models." *Journal of Statistical Software* 10(2): 1–40.

Sabatier, Paul A. 1988. "An advocacy coalition framework of policy change and the role of policy-oriented learning therein." *Policy Sciences* 21(2/3): 129–168. https://doi.org/10.1007/BF00136406.

Salm, Sarah, Stefanie Lena Hille, and Rolf Wüstenhagen. 2016. "What are retail investors' risk-return preferences towards renewable energy projects? A choice experiment in Germany." *Energy Policy* 97: 310–320. https://doi.org/https://doi.org/10.1016/j.enpol.2016.07.042. www.sciencedirect.com/science/article/pii/S0301421516304037.

Schar, S., and J. Geldermann. 2021. "Adopting multiactor multicriteria analysis for the evaluation of energy scenarios." *Sustainability* 13(5): 19. *https://doi.org/10.3390/su13052594.* <Go to ISI>://WOS:000628635600001.

Schmid, Nicolas, Sebastian Sewerin, and Tobias S. Schmidt. 2019. "Explaining advocacy coalition change with policy feedback." *Policy Studies Journal* n/a (n/a). https://doi.org/10.1111/psj.12365. https://doi.org/10.1111/psj.12365.

Schroeder, H., S. Burch, and S. Rayner. 2013. "Novel multisector networks and entrepreneurship in urban climate governance." *Environment and Planning C-Government and Policy* 31 (5): 761–768. https://doi.org/10.1068/c3105ed. <Go to ISI>://WOS:000326502500001.

Sloot, D., L. Jans, and L. Steg. 2019. "In it for the money, the environment, or the community? Motives for being involved in community energy initiatives." *Global Environmental Change-Human and Policy Dimensions* 57: 10. https://doi.org/10.1016/j.gloenvcha.2019.101936. <Go to ISI>://WOS:000480375400018.

Sovacool, Benjamin K., Christopher Cooper, and Patrick Parenteau. 2011. "From a hard place to a rock: Questioning the energy security of a coal-based economy." *Energy Policy* 39 (8): 4664–4670. https://doi.org/https://doi.org/10.1016/j.enpol.2011.04.065. www.sciencedirect.com/science/article/pii/S030142151100348X.

Steffen, Bjarne. 2018. "The importance of project finance for renewable energy projects." *Energy Economics* 69: 280–294. https://doi.org/https://doi.org/10.1016/j.eneco.2017.11.006. www.sciencedirect.com/science/article/pii/S0140988317303870.

Stirling, A. 2019. "How deep is incumbency? A 'configuring fields' approach to redistributing and reorienting power in socio-material change." *Energy Research & Social Science* 58. https://doi.org/10.1016/j.erss.2019.101239. <Go to ISI>://WOS:000498882200011.

Strauch, Yonatan. 2020. "Beyond the low-carbon niche: Global tipping points in the rise of wind, solar, and electric vehicles to regime scale systems." *Energy Research & Social Science* 62: 101364. https://doi.org/https://doi.org/10.1016/j.erss.2019.101364. www.sciencedirect.com/science/article/pii/S2214629619300684.

Tavana, M., A. Shaabani, F. J. Santos-Arteaga, and I. R. Vanani. 2020. "A review of uncertain decision-making methods in energy management using text mining and data analytics." *Energies* 13(15): 23. https://doi.org/10.3390/en13153947. <Go to ISI>://WOS:000559157700001.

Tian, Yongxiao, and Xianming Yang. 2020. "Asymmetric effects of industrial energy prices on carbon productivity." *Environmental Science and Pollution Research* 27(33): 42133–42149. https://doi.org/10.1007/s11356-020-10204-5. https://doi.org/10.1007/s11356-020-10204-5.

Trondle, T. 2020. "Supply-side options to reduce land requirements of fully renewable electricity in Europe." *Plos One* 15 (8): 19. https://doi.org/10.1371/journal.Pone.0236958. <Go to ISI>://WOS:000561027200057.

van Soest, Heleen L., Harmen Sytze de Boer, Mark Roelfsema, Michel G. J. den Elzen, Annemiek Admiraal, Detlef P. van Vuuren, Andries F. Hof, Maarten van den Berg, Mathijs J. H. M. Harmsen, David E. H. J. Gernaat, and Nicklas Forsell. 2017. "Early action on Paris Agreement allows for more time to change energy systems." *Climatic Change* 144(2): 165–179. https://doi.org/10.1007/s10584-017-2027-8. https://doi.org/10.1007/s10584-017-2027-8.

van Vuuren, Detlef P., Elke Stehfest, David E. H. J. Gernaat, Maarten van den Berg, David L. Bijl, Harmen Sytze de Boer, Vassilis Daioglou, Jonathan C. Doelman, Oreane Y. Edelenbosch, Mathijs Harmsen, Andries F. Hof, and Mariësse A. E. van Sluisveld. 2018. "Alternative pathways to the 1.5 °C target reduce the need for negative emission technologies." *Nature Climate Change* 8(5): 391–397. https://doi.org/10.1038/s41558-018-0119-8. https://doi.org/10.1038/s41558-018-0119-8.

Vedung, Evert. 2006. "Evaluation research." In B. G. Peters and J. Pierre, *Handbook of Public Policy*, 397–416. London: SAGE.

Villo, S., M. Halme, and T. Ritvala. 2020. "Theorizing MNE-NGO conflicts in state-capitalist contexts: Insights from the Greenpeace, Gazprom and the Russian state dispute in the Arctic." *Journal of World Business* 55(3): 12. https://doi.org/10.1016/j.jwb.2019.101068. <Go to ISI>://WOS:000528198300007.

Walker, Jack L. 1969. "The diffusion of innovations among the American states." *The American Political Science Review* 63(3): 880–899. https://doi.org/10.2307/1954434. www.jstor.org.libproxy1.nus.edu.sg/stable/1954434.

Wang, Hui, Xin-gang Zhao, Ling-zhi Ren, Ji-cheng Fan, and Fan Lu. 2021. "The impact of technological progress on energy intensity in China (2005–2016): Evidence from a geographically and temporally weighted regression model." *Energy* 226: 120362. https://doi.org/https://doi.org/10.1016/j.energy.2021.120362. www.sciencedirect.com/science/article/pii/S0360544221006113.

Wang, Q., M. M. M'Ikiugu, I. Kinoshita, and Y. Y. Luo. 2016. "GIS-based approach for municipal renewable energy planning to support post-earthquake revitalization: A Japanese case study." *Sustainability* 8(7): 20. https://doi.org/10.3390/su8070703. <Go to ISI>://WOS:000380760400109.

Win, Z. C., N. Mizoue, T. Ota, T. Kajisa, and S. Yoshida. 2018. "Consumption rates and use patterns of firewood and charcoal in urban and rural communities in yedashe township, Myanmar." *Forests* 9(7): 11. https://doi.org/10.3390/f9070429. <Go to ISI>://WOS:000440018600060.

Yanosek, Kassia. 2012. "Policies for financing the energy transition." *Daedalus* 141(2): 94–104. https://doi.org/10.1162/DAED_a_00149. https://doi.org/10.1162/DAED_a_00149.

Yergin, D. 1979. "United-states energy-policy —transition to what." *World Today* 35(3): 81–91. <Go to ISI>://WOS:A1979GM99900001.

Zhang, P. P., L. X. Zhang, Y. Chang, M. Xu, Y. Hao, S. Liang, G. Y. Liu, Z. F. Yang, and C. Wang. 2019. "Food-energy-water (FEW) nexus for urban sustainability: A comprehensive review." *Resources Conservation and Recycling* 142: 215–224. https://doi.org/10.1016/j.resconrec.2018.11.018. <Go to ISI>://WOS:000457659400022.

Zhang, Z. X. 2015. "Carbon emissions trading in China: The evolution from pilots to a nationwide scheme." *Climate Policy* 15: S104-S126. https://doi.org/10.1080/14693062.2015.1096231. <Go to ISI>://WOS:000364543500006.

Zhou, Y., X. G. Zhao, X. F. Jia, and Z. Wang. 2021. "Can the renewable portfolio standards improve social welfare in China's electricity market?" *Energy Policy* 152: 14. https://doi.org/10.1016/j.enpol.2021.112242. <Go to ISI>://WOS:000636054400029.

Zimm, Caroline. 2021. "Improving the understanding of electric vehicle technology and policy diffusion across countries." *Transport Policy* 105: 54–66. https://doi.org/https://doi.org/10.1016/j.tranpol.2020.12.012. www.sciencedirect.com/science/article/pii/S0967070X20309495.

PART IV

Strategic and Deliberate Transitions

21

INTENTIONAL AND RESPONSIBLE ENERGY TRANSITIONS

Integrating Design Choices in the Pursuit of Carbon-Neutral Futures

Clark A. Miller, Yiamar Rivera-Matos, Angel Echevarria, and Gary Dirks

1. Introduction

The year 2020 was a watershed year for the transition to a carbon-neutral energy future. In January, BlackRock, the world's largest investment manager, with a $7 trillion portfolio, announced that the company would put "the transition to a low-carbon economy" at "the center of [their] investment approach" (Fink 2020). In the months that followed, a parade of the world's largest firms followed suit, as did many governments, including the European Union, China, and the US under the president-elect Joe Biden (Carbon Intelligence 2022; Shetty 2021). By late 2020, the UN announced, over 70% of the global economy was subject to targets of carbon neutrality by mid-century (Guterres 2020).

In this chapter, we ask, *what is next for energy transition leadership?* Many have proposed that now is the time to transform carbon-neutrality targets into concrete plans of action (Oakes 2021), but exactly what that means remains uncertain. Sophisticated models can now provide optimized, least cost scenarios, under different assumptions, of how to configure low-carbon energy systems at regional and national scales.[1] But that is still a long way from mapping out actual transitions, with real infrastructures, workforces, organizations, economies, industries, and communities. We propose, therefore, that *now is the time for leadership in the public and private sectors and civil society to build collaborative initiatives to intentionally and responsibly design regional transitions to carbon-neutral economies.*

- *By intentional and responsible design*, we mean systematically mapping what transitions to carbon neutrality will require, pathways to achieving them, and their implications for different groups, encompassing both the goals of the transition and its process. This will not be easy nor follow simple scripts, given the complex systems, intertwined social, economic, political, and technological dynamics, and polycentric decision-making involved in transitions. Transitions will entail navigating uneven and unknown terrain, with uncertain knowledge (Miller et al. 2013). Best practices in planning leverage flexible, iterative, and adaptive approaches to handle complex, emergent, and dynamic problems (Woodruff 2016), but transitions to carbon neutrality will likely challenge even the most sophisticated methodologies. And no coordination would be worse. Disconnected, decentralized decision-making will likely be too slow, lead to costly mistakes, and fall short of ensuring a just transition.

- *By regional transitions*, we suggest that regions, defined as major urban areas and their rural neighbors, offer a salient, manageable, and scalable focal point for decarbonizing energy systems.[2] Electricity and mobility systems largely operate and are governed on this scale. Hence, such regions will be where the bulk of transition work gets done (Späth and Rohracher 2010; Coenen et al. 2021). Regional systems are also more easily managed and coordinated than countries or the globe. Larger regions, such as California, New York, and Germany, which possess both social coherence and economic and policy power, have already demonstrated the ability to lead renewable energy transitions (Breetz et al. 2018) – as have major cities (Rutherford and Coutard 2014). Others, such as Texas, the Bakken, Alaska, Alberta, or the Middle East, have historically led resource extraction and production (Grubert 2018). We do not deny the potential significance of national governments in guiding and supporting transitions, although many are currently struggling to exercise leadership. But no matter what progress occurs in Paris, London, or Washington, DC, or at the Conferences of Parties, regional entities like electric utilities and metropolitan transportation agencies will still need to do a lot of work to make transitions happen. In our view, therefore, advancing regional transitions is critical and, if it can be done in parallel, across the world's regions, offers a pathway to plausible decarbonization. Dividing the Earth into 1,000 regions with roughly similar populations yields regions of ~3–10 million people each.[3] We propose, within each such region, that public, private sector, civil society, and research/educational leaders come together to plan and implement decarbonization.

- *By economy-wide transitions*, we highlight the deep cross-system and cross-sectoral demands of regional energy transitions. By itself, the energy sector is a major industry and locus of economic development. Energy systems are also critical infrastructures that support every facet of modern economies and are interdependent with other critical infrastructures, including food, water, transport, manufacturing, and communications. Transitioning these complex systems-of-systems will require an economy-wide approach, especially as energy systems converge via the electrification of transportation and industry and the use of electricity to manufacture renewable fuels like green hydrogen. Piecemeal approaches risk a variety of unanticipated, unhappy outcomes (Mikellidou et al. 2018), as has been illuminated in the recent wildfires in California (Ajaz and Bernell 2021; Wong-Parodi 2020), the near collapse of the electricity grid in Texas (Busby et al. 2021), the yellow vest protests against a proposed carbon tax in Paris, and political opposition to phasing out fossil fuels in carbon-dependent regions (Jetten et al. 2020).

In the rest of this chapter, we offer some initial reflections on how regions can pursue intentional and responsible energy transitions. There are unfortunately few simple answers. We organize the chapter around four questions: (1) What does the complexity of energy transitions imply for their intentional and responsible design? (2) To what ends should energy transitions be organized? (3) Through what processes should they be governed? And (4) how inclusively should those processes be designed?

To summarize our conclusions, we suggest (1) that regions take a nested, polycentric approach to breaking down the complexity of systems-of-systems reconfiguration into manageable pieces; (2) that regional governance be as inclusive as possible, reflecting the need for widespread public engagement in the transition for it to be successful; (3) that regions adopt analytic approaches and outcome targets that fully reflect the intertwined social, ecological, and technological dimensions of transitions; and (4) that they practice adaptive processes that allow regions to anticipate, chart, navigate, and evaluate the complexities of change on an iterative, ongoing basis.

This will necessarily entail time, care, thoughtfulness, capacity building, and engagement. Considerable social science research indicates that durable and transformative change is much more likely to occur when governance follows these practices (Bauwens 2017), and our analysis suggests that this is likely to be true for energy transitions. Objections to such approaches typically assert that it will be too slow for energy transitions, which need to happen rapidly. Yet simply pushing for faster action without sufficient attention to inclusive processes is often counterproductive (Johnston et al. 2011). We have three decades, which is more than sufficient for inclusive, adaptive approaches, especially if they begin now, are planned strategically, and occur in parallel across many regional geographies.

2. The Complexities of Energy Transitions and the Challenges of Intentional Governance

From our perspective, the transition of global energy systems to carbon neutrality is the most complex and dangerous project ever undertaken by humanity. This is true for many reasons. It is imperative that energy transitions happen rapidly to have a hope of minimizing the destructive impacts of anthropogenic climate change. Energy systems are also critical infrastructures, providing essential support for the global economy and other critical infrastructures. Hence, they must remain highly reliable and affordable throughout. Finally, throughout history, energy transitions have deeply reshaped human and environmental systems, altering how people live and earn livelihoods; urban and rural geographies; patterns of social and environmental risks, insecurities, and injustices; and the distribution of wealth and power at every scale from the village to the globe. Achieving rapid, stable transformations of this depth and complexity will be difficult, pose deep risks for society and the economy, and raise profound questions of justice. Being intentional will therefore be important.

The complexity of energy transitions is rooted in the scope of what is involved. Energy transitions are too often defined in terms of changes in the sources of energy or the technologies that provide energy services (Laird 2013). For example, across history, boats have been powered by human labor, via the technology of oars, wind via sails, and wood, coal, oil, and/or nuclear via engines and propellers. This definition is incomplete, however. When historians have looked in detail at energy transitions, they have found much larger transformations in human behavior and social practices (Nye 1992; Cowan 1976; Kline and Pinch 1996), distributions of power and wealth (Clark 1990), the forms of economic (Chandler 2009; Hughes 1983) and social organization (Jones 2014), fundamental values and expectations, and forms of culture and imagination (Hecht 2009). Underlying these interconnections are the complex yet foundational ways that energy systems become layered into and interwoven with diverse forms of human life, labor, and play, on scales from individual households to the political economy and governance of the globe. Many facets of today's cities, for example, like suburban sprawl, high-rise skylines, lit-up nights, and 24/7/365 cultures and economies, get taken for granted as "natural" features of cities but are, in reality, products of the electricity and fuels revolutions of the early 20th century (Miller et al. 2021). *To alter energy systems, therefore, has the potential to be as disruptive to human systems and societies as climate change.* Carbon neutrality may well usher in fundamental social and economic transformations as powerful as those that have been driven by other new technologies, such as the automobile, airplane, radio, telephone, television, and social media (Miller et al. 2015).

Given this depth of transformation, one key challenge will be to maintain energy security. Today's societies and economies are built on the premise of reliable and inexpensive energy and on particular forms of energy, such as electricity or gasoline, that can't always be easily substituted for one another. The metropolitan sprawl of Phoenix, Arizona (USA), simply wouldn't exist without reliable and widely available supplies of gasoline for its ubiquitous SUVs and pickup trucks; electricity to pump water from the Colorado River and to power cooling systems during the city's extremely hot summers; and diesel to fuel trucks to transport food into the city, day in and day out (Scott and Pasqualetti 2010). Nor is Phoenix alone. Phoenix wears its dependencies on highly reliable energy systems on its sleeves. Yet every city suffers from similar addictions to energy, even if they are not always visible.

Unfortunately, the world is already experiencing heightened instabilities in energy systems, driven by diverse factors, including the recent collapse of the electricity system in Texas (due to deregulation, coupled with extreme weather; see Busby et al. 2021), deep challenges in maintaining grid stability in Germany (due to decarbonization and denuclearization; see Stürm 2018 and now the exigencies of war in Ukraine and Russia), fuel supply disruptions recently experienced by the UK (due to interdependent, COVID-19-related workforce, supply chain, and economic shifts; see Axon and Darton 2021; Gross 2021), and the year-long collapse across multiple energy systems experienced by Puerto Ricans after Hurricane Maria (Santos-Lozada and Howard 2018). The extreme social, economic, and health risks posed by such failures are stark and real – and the resulting social backlash has the potential to slow or even undermine progress toward carbon neutrality (Abramsky 2010).

With energy dependencies also come concentrations of wealth and power (Kuzemko et al. 2019). Energy companies – including oil and gas, electricity, and automobiles – have historically been among the largest private enterprises. They are often major contributors to regional and national economies and labor forces and wield significant power and influence. For energy transitions, this poses very real trade-offs.[4] Existing energy institutions typically have extensive knowledge, skills, resources, and control over systems that can accelerate transitions and ensure continued operations of energy systems (De Laurentis 2021). Yet they can also suffer lock-in to modes of thinking and forms of organization that slow transitions. They may be slow to take actions that threaten business models, workforces, or positions of influence and control (Graffy and Kihm 2014). Working with existing institutions may also raise difficult questions about perpetuating forms of historical exploitation of people or resources that have done deep damage in the past and call out for justice (Routledge et al. 2018). Yet explicitly dismantling not only carbon-intensive energy technologies and systems but also the organizations that have thus far run them involves exercising considerable political, social, or economic power against those entities and the diverse risks such exercises of power always entail. And it threatens the loss of access to the knowledge, skills, and resources held by such organizations in efforts to further the transition.

At least four additional complexities confound energy transition planning. First, patterns of decision-making about different facets of energy and other interdependent systems are heterogeneous, combining both more and less centralized and decentralized elements. A wide range of actors make decisions across systems of resource extraction, transport, production, distribution, sale, and use of energy. For example, the management, operation, and governance of any given system (e.g., regional electricity grids) typically involve multiple organizations (utilities, systems operators, independent power plants, etc.). Going up a level, each interdependent system (electricity, water, fuels, transportation) is typically organized and regulated within its own institutions and professions (Gim 2019). Decision-making about many parts of all of these systems, such as the use of energy, and the purchase of technologies for using energy (such as cars or air conditioners) is decentralized to the level of individual households and businesses. Systems also often cross jurisdictional boundaries that further complicate coordination.

Second, interdependencies among energy and other infrastructure systems, as well as their relationships to culture, politics, economics, and geography, also vary from place to place. Although most societies rely on a very limited number of types of energy (wood, hydrocarbon fuels, electricity), local manifestations of these systems vary enormously. Similar heterogeneities occur in water, transit, food, resource, and other systems.

Third, energy transitions will need to balance speed and complexity. Given their depth and complexity, many energy transitions have historically been slow endeavors (Sovacool 2016; Sovacool and Geels 2016), as social practices and social, economic, and infrastructural systems reconfigure around the introduction of new technologies. Old energy systems often persist for long periods of times in sectors of economies or societies that do not see the need for change, in an example of technology lock-in and momentum (MacFadyen and Watson 2018; Hughes 1983). Both pose challenges to the rapid pace of change required to tackle climate change – and efforts to accelerate deep transformation in interdependent social, economic, and technological systems create very real risks.

Finally, the opportunities offered by energy transitions to improve human outcomes are also significant, as we will describe in the next section of the chapter. As discussed earlier, energy systems are deeply tied to social and economic arrangements and, both positively and negatively, to myriad human outcomes. Redesigning energy systems to create better social and economic outcomes (however one chooses to define and measure such goals) is an unprecedented opportunity in human history.

Navigating all of these complexities to get energy transitions "right", therefore, is a worthwhile ambition – and the heart of what we mean by intentionality and responsibility in transition governance. Of course, there is no clear and unambiguous definition of or approach to being "right" in the pursuit of energy transitions. That's an inevitable feature of what Allenby and Sarewitz term "Level 3" or "Earth systems" problems in their book *The Techno-Human Condition* (2011). Level 1 problems are hammer-and-nail problems: clearly defined, narrowly scoped problems with effective technology solutions. Level 2 problems are systems-level problems, such as the redesign of an electricity grid or an air transportation system. Level 2 problems integrate multiple technologies and networks of people in complex systems, but they still have clear objectives and the ability to orchestrate and optimize techno-human relationships and dynamics.

Level 3 problems, on the other hand, are systems-of-systems problems, in which changes ripple across diverse systems to create unanticipated feedbacks in unexpected locations, often with deleterious effects. Uncertainty and contingency are endemic in Level 3 problems, rendering linear causal logics deeply problematic. Consequently, attempts to create predictive models of causality in Level 1 and Level 2 processes are not only inevitably inaccurate but even potentially dangerously wrong in Level 3 contexts.

Transforming global energy systems to carbon neutrality is a Level 3 problem. Even just shifting electricity systems to low-carbon power is extraordinarily difficult, especially when it has to be coordinated with other challenges, such as cyberattacks, denuclearization, deregulation, digitalization, shifts in extreme weather patterns, and so on. Yet planners do not have the luxury of just tackling electricity. They must consider the integrated transformation of diverse technological and human systems, acknowledging both the risks that poses to regional economic vitality and security and the opportunities it contains for societal progress (Araújo and Shropshire 2021).

The transition to carbon neutrality involves not just electricity and fuels but transportation, housing, and banking, as well as other critical infrastructure systems that depend, utterly, on energy, like food, water, and manufacturing. In such transitions, the potential for unanticipated disruptions becomes something like Charles Perow described in *Normal Accidents* (1984) for the operation of complex nuclear power plants. As noted earlier, such disruptions have been fully on display in recent months, including major electricity grid failures (Wolf 2021) in California (wildfires), Texas (cold snap), and Puerto Rico and Florida (hurricanes) and rapidly rising energy prices (and therefore heating and electricity crises; Smith 2022) and fuel supply chain disruptions across carbon markets (e.g., coal in Germany and China, natural gas in Europe, gasoline in the United Kingdom, heating oil in the northeastern US; Valdmanis 2021). As a consequence, energy transitions will inevitably be uncertain (i.e., we will never have full information), contingent (i.e., involve navigating unanticipated surprises), and emergent (i.e., involve evolving goals, expectations, and dynamics that respond to transition dynamics).

To imagine deliberately transitioning global energy systems, therefore, all together, according to some sort of master plan, is laughable. The complexities far exceed the limits of human reasoning and computational modeling, even if one could imagine some sort of shared global will. The level of challenge is not dissimilar to the widespread problems currently bedeviling global supply chains, in the wake of COVID-19, as well as the deep incapacity of global institutions to manage them (Al-Mansour and Al-Ajmi 2020). Even national or supranational transitions in the United States, China, India, or the European Union seem likely to prove too complex, involving too many systems, too diverse economic and political actors, and too interwoven institutional layers to effectively coordinate.

By contrast, regions have been at the cutting edge of energy transitions. Germany, California, and Texas (Golding 2021), for example, have made notable progress, even if they have also faced challenges. It is relatively straightforward to imagine how these regions, and others like them, can continue to navigate pathways toward carbon neutrality. Indeed, it is increasingly plausible, drawing on lessons from their experiences, to envision how regional transitions might be pursued even more competently – and how to scale regional transition leadership globally.

To be sure, regional transitions remain a Level 3 challenge. Nonetheless, the degrees of complexity are simplified in key ways (Termeer et al. 2011). Jurisdictionally, there are fewer layers of overlapping institutions, and their boundaries line up better with important system boundaries. Regional electricity grids, for example, are often operated by a small number of utilities and governed by regional entities. Those boundaries, in turn, often map reasonably to state, county, and/or city jurisdictions. The institutional problem, in other words, is finite. At the same time, because food, energy, water, transport, industry, and communications systems interpenetrate deeply at regional scales, local experts often share at least some cross-cutting knowledge and networks. Regions must still manage their boundaries and interdependencies with larger, national and global systems dynamics, of course, but even these, too, are more manageable, taken region by region.

3. To What Ends an Energy Transition?

The transition to carbon neutrality is often framed as a discrete process with clear end points: today's carbon-rich energy system will be replaced with a new carbon-neutral energy system by 2050. This is a mistake. In reality, the evolution of the energy sector over the rest of the 21st century will be a continuous process, occurring within the context of an already shifting energy system. Since the beginning of the 20th century, the design and regulation of electricity systems, grids, and markets have evolved considerably, as have their sources of energy (Hughes

1983). So, too, has the extraction and refining of oil (Grätz 2012). And those changes continue today. Since the 1990s, for example, energy markets have seen dramatic changes in primary fuel supplies, such as with the rapid rise of unconventional natural gas, deep debates over the future of nuclear energy (Markard et al. 2020), and widespread deregulation in the electricity sector (Hirsh 2002).

In other words, energy transitions tend to involve complex, continuous shifts in the organization, governance, and meaning of energy systems, often with no clear end point. This has two important implications for decarbonization. First, it means that, whatever designs are planned or achieved for carbon-neutral energy systems by 2050, those designs are likely to be temporary, at best. Electric vehicles, for example, may well dominate the next 30 years of automobile sales. If a robust hydrogen economy emerges in the heavy transportation, chemicals, and electricity industries, however, and lithium or other critical materials supply chains for electric vehicle manufacturing run into environmental, security, justice, or other challenges, it could be that hydrogen fuel cell vehicles evolve to be dominant over longer time periods. Or perhaps modern societies will ultimately give up their fascination with cities designed for individual automobiles and turn to alternative modalities of mobility.

Second and even more importantly, as significant as carbon neutrality will remain as a core outcome criterion for the designs of future energy systems, it will also remain only one among a mix of significant design variables, alongside rising concerns with resilience, climate adaptation, and environmental justice, as well as historically significant values such as reliability, affordability, and security. This draws back into focus the central theme of this chapter: intentionality and responsibility in energy transition governance. In designing and building tomorrow's energy systems, we cannot simply focus on carbon neutrality. Energy systems are fundamental and critical infrastructures with deep importance in and implications for the future of economies and societies, at all scales, from the neighborhood to the community.

Toward what ends or outcomes should energy transitions be designed or guided? As we observed, energy systems profoundly shape many aspects of the organization and dynamics of economies and societies, from the geographies of cities to the creation of 24/7/365 cultures to the geopolitical economies and forms of security that circulate around oil, gas, and coal. Arguably, future energy systems will be just as influential in shaping the future of human affairs, but they will layer into the fabrics of human economies and societies in different ways, driving new patterns of social, economic, cultural, and political change across the planet.

In other words, as rich a set of criteria as has already evolved for guiding energy design – reliability, affordability, security, equity, carbon neutrality, and resilience – even wider considerations of the future of human societies and economies may be at stake and, therefore, appropriate to attempt to anticipate and draw into conversations about how to navigate current energy transitions. Given the scale of global investment in new energy systems required to achieve carbon neutrality – over $100 trillion, by some estimates (IEA 2021a) – and the depth of implications that investments of that scale will likely have for social and economic transformation, no single, comprehensive solution of "the ends to which energy transitions should be put" is plausible. We offer, here, therefore, an array of considerations that could, and arguably should, be brought into discussions of energy transitions to attempt to account for their full social and economic design:

- *Patterns of sociotechnical integration*: The phrases "low-carbon" or "carbon-neutral" hide a multiplicity of possible future energy technologies, such as solar, wind, nuclear, hydrogen, biofuels, and carbon capture, each with different future implications for economies and societies. As Langdon Winner observed a half century ago, future worlds built around different technologies have the potential to be radically divergent (Winner 1986). Differences will occur in numerous dimensions, from material life cycles and ownership patterns to security requirements and risk profiles. Questions of environmental justice, waste disposal, threats of cyberattacks or other forms of terrorism, and other issues play out very differently.[5] Nor is it just technology versus technology. Within each technology, significant design variations (small modular vs. large, highly centralized nuclear plants; distributed vs. utility-scale solar) could create profoundly different futures. *Our choices about which energy systems to design and build and how to layer them into the fabrics of our lives are also fundamentally choices about what kinds of societies and economies we propose to inhabit. What will the post-carbon cities, markets, and societies of tomorrow look like? How will they be shaped by the design choices we make in configuring tomorrow's clean energy systems?* (Eschrich and Miller 2019, 2021).

- *Political economies*: Modern political economies have circulated around energy for at least two centuries, and energy transitions offer the opportunity to redesign them, such as by altering patterns of ownership and revenue across resources, systems, and end-uses. At stake are long-standing questions of public versus private ownership, as well as structures of private ownership for key energy technologies (e.g., individually owned automobiles vs. fleets of collectively owned autonomous vehicles; centralized ownership of electricity generation replaced by distributed solar or hydrogen). Both will have deep implications far beyond the energy sector, and the governance rules that shape them are up for grabs in the years to come (Miller 2022a, b).

- *Human security, freedom, and livelihood*: As numerous recent disasters have illuminated, access to reliable and resilient energy systems has immediate implications for human security (Chen et al. 2022). Throughout the last century, access to individually owned automobiles and infrastructures built to facilitate their open travel has been considered a central contributor to expanding human freedom. In many parts of the world, access to modern energy services is increasingly viewed as a human right. Yet across all three dimensions, the past two centuries of energy development have been anything but inclusive and equitable (Sovacool and Dworkin 2015). An open question for future energy system designs, therefore, is how they distribute security, freedom, and economic opportunity across disparate groups (a point to which we will return later in the chapter).

- *Defining a "just transition"*: The idea that the transition to carbon neutrality must be "just" is now broadly accepted. Two definitions are common. The first focuses on the end of carbon-based industries and supply chains and the workers, communities, and regions dependent on them (Cha 2020). The second concerns communities who have suffered environmental and health impacts from past energy systems, especially where those are unequally distributed along racial or socioeconomic lines (Baker 2021). Each of these definitions is critical but also too narrow. More people will be harmed by energy transitions than just those in carbon industries. More injustices have occurred than just pollution. The question, then, is how to scope a just energy transition. Any approach has to start by taking these two principles, mapping the full array of implicated communities, and working to reduce both sets of harms. To this, we would propose adding other aspects of future energy systems design, including the distribution of benefits and risks from future energy systems; the establishment of inclusive practices of energy democracy, ownership, and/or sovereignty; and the justice of societies and economies built on or transformed by future energy systems. Scoping these issues in their entirety will be an enormous undertaking yet not impossible and, for the world's wealthiest societies, with their vast analytic capabilities, worth the modest investment required. Even prior to such scoping, however, we highlight the importance of simply asking, for each decision in a transition, "How is this choice an opportunity to enhance social justice?"

A lot is ultimately at stake in future energy systems designs. To treat transitions narrowly, as mere questions of technological change, is to risk creating, in the search for carbon neutrality, new social and economic dislocations and injustices that add to the historical legacies of the energy industry. It also risks missing out on an historically rare opportunity to improve the human condition and create more just and thriving human futures. To be sure, energy systems design cannot presume to merely set key design criteria and expect they will automatically be achieved. Conflicts over the potential aims and outcomes of energy transitions are too deep and endemic in the energy sector, too enshrined in alternative paradigms of how best to proceed, even among those who proclaim to share the ends of carbon neutrality.[6] And energy transitions remain too complex, contingent, and emergent as processes of change, with far too many decision-makers and stakeholders, operating in too decentralized of decision-making arrangements. Yet the importance of what is at stake in transitions is too high to simply proceed without strategic design and navigation. All of which emphasizes the importance of developing innovative strategies for governing transitions, to which we turn next.

4. Via What Means?

Given all that is at stake and their deep complexity, how should energy transitions be governed? We have already argued for *a region-first approach*. The idea of breaking up the global transition to carbon neutrality into heterogeneous regional parts is consistent with several of Allenby and Sarewitz's suggested strategies for governing Level 3 processes of change. They recommend, for example, strengthening "pluralism" and "lower[ing] the amplitude and increas[ing] the frequency of decision-making". Both can be facilitated by focusing on smaller subsystems with manageable dynamics and risks. Incremental decision-making can navigate the challenges, contingencies, and trade-offs of transformation in ways consistent with local values and expectations, while confronting less complexity and creating simpler (although still difficult enough) problems of knowledge and uncertainty than attempting to shift supranational or global systems all at once.

Such an approach is also consistent with theories of sociotechnical systems change, which highlight *the centrality of sociotechnical regimes* (of which, for example, electric utilities are a prime example, as are metropolitan transportation systems) as critical nodes in processes of multi-scalar transformation (Geels 2004; Hughes 1983). And it aligns with the recommendations by Ostrom and Keohane to pursue *polycentric approaches* that build up to large-scale change through the coordination of decentralized action by local and regional jurisdictions (Ostrom 2010; Keohane and Victor 2016). In the end, our bet is that navigating roughly 1,000 transitions, in parallel, encompassing 3–10 million people each, will be more feasible, faster, less complex, and less prone to dangerous or catastrophic errors than trying to transition 10 billion people all at once.

Our second recommendation is to tackle transitions via approaches that are *anticipatory, adaptive, and emergent*. This perspective is shared by innovative planning methodologies, such as in adaptive management of socio-ecological systems (Chaffin et al. 2014), as well as Allenby and Sarewitz. Alas, it is still rare in energy planning (Peluso 2021). Anticipatory, adaptive, and emergent approaches stress the inadvisability of mapping out and expecting to follow prescribed trajectories in transitions of the sort required to achieve carbon neutrality (Malekpour and Newig 2020). As we have noted, the systems-of-systems dynamics are simply too complex to be predicted because of both interdependencies across systems and the pervasiveness of human choices. Instead, they emphasize recursive and iterative processes that are frequently subject to critical review, reassessment, and revision.

This does not mean eschewing attempts to anticipate plausible or potential futures. Indeed, adaptive approaches leverage a variety of futures methodologies to explore potential societal, economic, and environmental changes (Buehring and Liedtka 2018). In this paper, we discuss the idea of design as serving a similar purpose: to present options and pathways and open them up for pluralistic review and deliberation. Such approaches are consistent with *anticipatory strategies for governing the introduction of new and emerging technologies* (and, certainly, low-carbon technologies like solar, wind, hydrogen, and electric vehicles count in this category; Guston 2014). It is also worth drawing lessons from the theory and practice of adaptive, responsive, and experimental approaches to environmental management and governance that acknowledge the importance of incremental and experimental decision-making, as well as monitoring, evaluation, and revision in the face of unexpected and surprising events within multi-system dynamics (Allen et al. 2011).

One of the central rationales for taking an adaptive and emergent approach to transitions is the *sociotechnical flexibility* that remains inherent in future energy systems design (Pinch and Bijker 1984). This flexibility stems in part from high rates of technological innovation still occurring within key potential pathways of energy futures. While the stabilization of certain key technologies – such as solar, wind, and nuclear electricity generation and battery electric vehicles as replacements for gasoline-powered cars – has made the very idea of a transition possible, others, like hydrogen and carbon capture, are still undergoing rapid innovation with high and still-growing levels of public and private investment. In addition, there is growing flexibility in how energy technologies are being integrated into new sociotechnical configurations, such as distributed solar and batteries (Clack et al. 2020), electric bikes and scooters (Flores and Jansson 2021), and home hydrogen solutions (Lavo 2022), that diverge significantly from prior systems models. Predicting which future technologies and socio-energy arrangements will be viable and/or desirable across diverse publics, even five or ten years from today, is a fool's game, and rigid governance approaches that do not allow for significant shifts in pathways between now and 2050 have the potential to encounter considerable difficulties.

Finally, we recommend *a justice-centered approach to transition governance* that foregrounds the perspectives of marginalized communities. As we suggested earlier and will discuss in more detail in the next section, the question of a just energy transition engages many different kinds of communities (frontline, low-income, indigenous, labor, disability, etc.). While such communities exist in all regions, their particular contexts, needs, and circumstances vary considerably from place to place, meaning that governance processes will need to foreground efforts to engage relevant communities, both to hear from them to understand the challenges they face and to collaborate with them to find meaningful solutions. Unfortunately, regional governance institutions also have the very real potential to be captured by locally powerful actors with little or no interest or capacity to tackle what are often enduring problems that they have historically either ignored or contributed to. That argues for the need to push for transition governance processes that are as inclusive as possible, to which we turn our attention next.

4. How Inclusive?

Historically, energy politics resonated broadly across nations and polities. In the US, for example, when Samuel Insull, president of Chicago Edison, first proposed the model of a regulated electric utility, he organized speaking events with many of Chicago's prominent civil society organizations, seeking support for the legislative changes that would be needed to implement it (Insull 1915). The subsequent debate over public power versus private power was one of the most resonant and recurrent political themes of Progressive Era and New Deal eras. When David Lilienthal built the Tennessee Valley Authority, he wrote a popular book, *TVA: Democracy on the March* (1944), to similarly explain its model to the public at large.

Today, by contrast, large swaths of energy decision-making occur without much public notice. When electricity regulators meet to debate electricity policy, for example, most citizens pay little attention, taking note only when the electricity goes out or their bill goes up. This shift results from two broad developments in energy governance. First, energy systems have become increasingly complex, which has tended to support narrow, expert-led decision-making in forums designed to minimize public input. Second, energy systems have become increasingly central and critical to the functioning of modern industrial societies and economies, such that decisions about their governance have carried increasingly high stakes.

Does the need to rapidly transition energy systems away from fossil fuels argue for continued expert-led decision-making or, by contrast, more open, inclusive, and participatory approaches? Arguments for the latter tend to fall out along three lines:

- *The values of energy experts and insiders increasingly seem out of line with wider publics who support faster transitions to renewable energy* (many argue, for example, for governments to actively set and enforce mid-century carbon neutrality targets; Armstrong 2021).
- *The stakes of energy transitions go well beyond the normal stakes of energy system management, thus justifying the inclusion of a much wider array of stakeholders in decision-making* (this includes, for example, the need to coordinate decision-making across many sectors, as well as the concern that stakeholders left out of the process may object and thus slow change; Skjølsvold and Coenen 2021).
- *Many communities have been historically excluded from decision-making and disadvantaged by the design of energy systems and therefore must be included both to make governance processes equitable and to ensure that the design of future energy systems is just* (this is often referred to as procedural justice and decolonization of energy systems; Stephens 2019).

The question of inclusion is further complicated by the question of how to institutionalize transition governance. As noted earlier, existing governance of energy systems is heterogeneous and fragmented, as is the governance, more broadly, of interdependent systems-of-systems. Developing strategies for ensuring that diverse stakeholders and communities have an effective voice in contributing to and shaping transition decision-making in such a landscape is extremely difficult, especially if not all forums support inclusive processes.

One approach would be to develop institutional mechanisms for inclusion that have broad reach across energy governance, such as a citizens' assembly or advisory body with broad powers to intervene across diverse sites. By creating a single point of convergence, this approach would simplify participation and engagement by diverse groups. The analogy is to legislatures. Legislatures have broad remit to govern the affairs of their jurisdiction and so create a single point of engagement for diverse civil society groups. And indeed, legislatures have played a key role, such as in California and the state of Washington, in establishing equity principles and requirements for inclusive participation in energy governance in recent years (CPUC 2021; Washington State Department of Public Health 2021).

Because legislatures have comprehensive policy authorities over multiple aspects of energy systems, they have served in these cases as both sites of inclusion, where diverse participants are able to contribute, as well as promulgators of rules favoring inclusion in other, subsidiary regulatory institutions. Cases such as these are rare, however, at least thus far, and not all legislatures are committed to open and inclusive processes. In places where transition governance remains fragmented, greater inclusiveness may be difficult to achieve and even seen as detrimental to agile decision-making, unless persistently advocated for by powerful actors.

Nonetheless, our view is that inclusive governance is essential to effective regional transitions to carbon neutrality. Several reasons support this idea:

- *Trust and transparency*: Energy transitions are going to involve and impact broad populations. As the world witnessed in France's Yellow Vest protests, those left out of planning may lose trust and even challenge proposed pathways, especially where they feel the bite of higher costs, choices imposed on them, or the loss of employment (Donadio and Meyer 2018). Indeed, deep decarbonization cannot be achieved without the active participation of all people, each of whom must alter their own energy technologies and consumption routines and practices to create a carbon-neutral future. Inclusivity means finding diverse and innovative approaches to engaging people in processes of change, especially by ensuring that they are made better off by it in ways they consider valuable.

- *Autonomy, sovereignty, and security*: The principle of democracy holds that people have a right to participate in and shape decisions that impact their lives, grounded in a recognition of their autonomy as individuals. Given the depth of significance of energy systems for security and economic thriving and the degree to which both may be impacted by energy transitions, it is hardly surprising that people would feel strongly about their rights to participate when critical energy decisions are at stake (Abramsky 2010). Around the world, we see growing agitation by grassroots communities for greater autonomy and sovereignty over energy futures, stimulated by the perceived availability of decentralized energy options, such as solar and microgrids (and also diesel generators), and by perceived failures of regional and national institutions to deliver reliable and affordable energy services. Yet we also see deep concentrations of power and wealth in the energy sector, across public and private institutions, that actively suppress and undermine public sovereignty and autonomy. One of the most divisive versions of this conflict today is in Puerto Rico, where multiple levels of colonial state institutions retain high degrees of control over an increasingly dysfunctional electricity system, fueling a diverse array of community solar initiatives, as well as angry social protests against the newly privatized electric utility (Rivera-Matos et al. 2021).

- *Collective world-building*: As we observed earlier, when Samuel Insull invented the regulated monopoly model for electric utilities at Chicago Edison in the early 20th century, he traveled throughout the city to present the idea and generate discussion among the city's leaders (Insull 1915). He understood that he was not just designing a new technology; he was inventing a whole new idea of what a city would look like, in which every aspect of people's lives would be reshaped into new forms and possibilities, and that people would care about that. The same scale of transformation is happening today: a once-in-many-generations transformation of systems-of-systems that will reshape everyone's lives and livelihoods. Understanding and shaping the imagination of what is possible – and the world that will be built – will inevitably become the subject of society-wide dialogue, debate, and perhaps conflict. Inclusive institutions can potentially help ensure that diverse voices are engaged and heard in that conversation earlier and with greater purchase, mediate and negotiate conflicting viewpoints, and reduce delays in bringing about needed change.

- *Decolonization of energy systems*: Energy is a prominent domain within which technological systems and ways of life have colonized geographies, resources, and communities (Barandiarán 2019). At its most benign, this has meant the worldwide reconfiguration of social values, identities, practices, behaviors, relationships, and institutions around modern energy systems and services (e.g., especially oil, automobiles, and electricity; Miller 2021). Much of humanity now inhabits electrified residences and businesses and urban geographies that cater to automobiles, even as the precise details of what that means may vary across places and cultures. Far more insidiously, at the other end of the spectrum, energy system colonization (e.g., especially, of resource geographies) has transformed Hobbes' line about lives being "nasty, brutish, and short" (Hobbes 1651) into social, economic, and environmental sacrifice zones all across the planet (Hernández 2015), not as a state of nature but as a product of global technological systems design. In between, for many of world's least well-off, energy system colonization has created a monthly financial drag on incomes to pay for expensive energy services that routinely perpetuates and exacerbates extreme poverty (Biswas 2020; Biswas et al. 2022), even as it has left significant populations outside the reach of modern energy services, especially in rural areas that are expensive to reach (Sovacool 2012). In the absence of more open and inclusive forms of energy governance that enable the integration of, for example, the social, economic, and ecological consequences of energy systems life cycles, supply chains, consumption patterns, and waste streams, these critical dimensions of just energy transitions will remain underprivileged in energy decision-making and future energy system design. Just as importantly, more inclusive governance regimes offer an opportunity to give voice and precedence to the people and communities who have been ill-served by existing modern energy systems to help shape the design of future energy systems that contribute to the goals of restorative justice and human thriving.

- *Vulnerable workers and communities*: Last but certainly not least, a key aspect of energy system colonization has been the creation of resource-dependent workforces and communities that are now highly vulnerable to the transition itself (Cha 2020). Inclusive governance processes can help ensure that the long-term concerns and needs of these communities received appropriate attention, especially in helping them to create more economically diversified futures that avoid further resource dependencies, as well to address short-term challenges with layoffs, loss of government financing for critical social services, declines in property values, and other key facets of transition adjustment.

5. Synthesis: Building Intentional and Responsible Regional Energy Transitions

The International Energy Agency has developed a broad initiative aimed at ensuring that transitions to carbon neutrality are "people-centered" (IEA 2021b). A core aim of this chapter has been to flesh out what that means. For us, it means attending to *the design of energy transitions*, understood as transitions in technology but also in the fabrics of people's lives, economies, and societies. More specifically, it means grappling with the question of what it means to intentionally and responsibly design and govern a techno–human transformation of the scope and scale required to end carbon emissions from global energy systems.

At the same time, we fully acknowledge the difficulty of pursuing intentional energy transitions. Energy systems are heterogenous. They are managed, operated, and regulated by a wide range of organizations and subject to the legal rules and institutions of manifold local, national, and international jurisdictions. For transition decision-making to be broadly coordinated, guided by fully shared goals and values, along predetermined pathways, is unimaginable. Yet there are real drawbacks to not trying and simply allowing the transition to evolve along whatever pathways processes of complex, socio-technological change happens to proceed. Failure to pursue intentional transitions could easily lead to deeply suboptimal and perhaps disastrous outcomes. Perhaps most importantly, it would likely forgo the opportunity to use the world's

$100 trillion in forthcoming investments in clean energy to achieve not only carbon neutrality but also significant additional improvements to human well-being, thriving, equality, and justice.

To that end, we propose that leaders seek out ways to work together to design transitions, via intentional processes of distributed and collective decision-making, that give due weight to concerns about *the human ends of energy transitions, the human processes of navigating from where we currently are to a world of carbon neutrality, and the inclusion of diverse human voices in shaping how and where transitions take us there.*

A key lesson that we draw in the chapter is that regional transitions will be especially critical. Regardless of national or global policies, much of the day-to-day work of transitioning energy systems must be done by regional actors who manage or govern regional infrastructures and who need to orchestrate the shift from one set of socio-energy systems to another, given the particularities of regional contexts. This will include significant numbers of ordinary people who are owner/operators of a rich diversity of energy producing and consuming technologies, as well as potential participants in future distributed energy infrastructures. Regions will also be, in many cases, where questions of just transitions hit home for communities inhabiting past or potential future sacrifice zones, workers and communities in the energy sector, and so on.

Regions offer key opportunities for developing the kinds of collaborative and polycentric governance required to pursue carbon neutrality, as well as the knowledge and research infrastructures that can inform them. Key aspects of energy and other relevant systems are already organized regionally, and many regions already have some existing capacity for collective knowledge-making and governance around energy.

Building on these, regions will need to add important new capabilities, including the capacity to create knowledge that anticipates, tracks, and evaluates transitions as integrated social, technical, and environmental processes. They will need to leverage that knowledge in new ways to inform coordinated decision-making across scales, systems, and institutions that usually operate in independent silos. And they will need to acknowledge that coordinated efforts to make knowledge and decisions will struggle to effectively understand and manage the complexities, contingencies, and emergent dynamics of the next several decades.

In that effort, regions may also be scales on which it is possible not simply to muddle through but to do so with grace, forgiveness, adaptability, and a growing degree of inclusivity over time. Or in cases where major regional actors simply refuse to play, regions are potentially scales on which active politics may be able to work to either force them to the table, to reorganize how they are controlled or governed, or to go around them through alternative pathways to carbon neutrality. One way of potentially achieving many of these goals, albeit beyond the scope of this chapter, could be the establishment of regional energy transition hubs that combine collaborative governance facilitation, transition justice planning, mechanisms for enhancing participation and inclusion of diverse voices, and supportive research infrastructures linking together regional universities and centers with industry, government, and civil society.

Notes

1 Two US examples of such modeling efforts are the Princeton Net Zero America study (Larson et al. 2021) and the Electric Power Research Institute Low-Carbon Resources Initiative (LCRI) modeling effort (www.epri.com/lcri).

2 In this, we are guided by histories of the powerful role of cities in organizing modern techno-economic systems and supply chains, including Cronon (1991), Hughes (1983), and Mumford (1934).

3 This is an estimate, obviously. The world population is currently 7.9 billion. Divided by 1,000 yields a size of ~8 million, although there will obviously be some variation. Major island archipelagos, like

Puerto Rico, with a population of 3 million, tend to be a bit smaller. Arizona, which is slightly larger than the national average for a US state, has 7.3 million people. Only a few major urban areas worldwide (e.g., New York, Los Angeles, Shanghai, Tokyo) have significantly larger populations than 10 million.

4 These issues are addressed broadly in the chapter on energy democracy by Keahey, Nadesan, and Pasqualetti.

5 Some might question whether terrorism is a risk for solar energy, but large solar power plants are vulnerable to bombings, while collections of solar plants are now operated from centralized facilities via smart inverters and telecommunications networks, which are vulnerable to cyberattacks like many other grid elements.

6 Witness, for example, the illustrative encounter between youth climate activist Lauren MacDonald and Shell CEO Ben van Beurden at a pre-COP 26 event in late 2021 (Dunn 2021).

References

Abramsky, Kolya. 2010. *Sparking a Worldwide Energy Revolution: Social Struggles in the Transition to a Post-Petrol World*. Chico: AK Press, 2010.

Ajaz, Warda, and David Bernell. 2021. "California's adoption of microgrids: A tale of symbiotic regimes and energy transitions." *Renewable and Sustainable Energy Reviews* 138: 110568.

Allen, Craig et al. 2011. "Adaptive management for a turbulent future." *Journal of Environmental Management* 92, no. 5: 1339–1345.

Allenby, Braden, and Daniel Sarewitz. 2011. *The Techno-Human Condition*. Cambridge: MIT.

Al-Mansour, Jarrah, and Sanad Al-Ajmi. 2020. "Coronavirus' COVID-19'-supply chain disruption and implications for strategy, economy, and management." *The Journal of Asian Finance, Economics, and Business* 7, no. 9: 659–672.

Araújo, Kathleen and Shropshire, D. 2021. "A meta-level framework for evaluating resilience in net-zero carbon power systems with extreme weather events in the United States." *Energies* 14, no. 14: 4243.

Armstrong, John H. 2021. "People and power: Expanding the role and scale of public engagement in energy transitions." *Energy Research & Social Science* 78: 102136.

Axon, C., and R. Darton. 2021. "Measuring risk in fuel supply chains." *Sustainable Production and Consumption* 28: 1663–1676.

Baker, Shalanda. 2021. *Revolutionary Power: An Activist's Guide to the Energy Transition*. Island Press.

Barandiarán, Javiera. 2019. "Lithium and development imaginaries in Chile, Argentina and Bolivia." *World Development* 113: 381–391.

Bauwens, Thomas. 2017. "Polycentric governance approaches for a low-carbon transition: The roles of community-based energy initiatives in enhancing the resilience of future energy systems." In *Complex Systems and Social Practices in Energy Transitions*, edited by Nicola Labanca. Cham: Springer. 119–145.

Biswas, Saurabh. 2020. "Creating social value of energy at the grassroots: Investigating the energy-poverty nexus and co-producing solutions for energy thriving." PhD Diss., Arizona State University.

Breetz, Hanna, Matto Mildenberger, and Leah Stokes. 2018. "The political logics of clean energy transitions." *Business and Politics* 20, no. 4: 492–522.

Buehring, Joern Henning, and Jeanne Liedtka. 2018. "Embracing systematic futures thinking at the intersection of strategic planning, foresight and design." *Journal of Innovation Management* 6, no. 3: 134–152.

Busby, Joshua et al. 2021. "Cascading risks: Understanding the 2021 winter blackout in Texas." *Energy Research & Social Science* 77: 102106.

California Public Utilities Commission. 2021. *Environmental and Social Justice Action Plan 2.0*. Accessed Feb. 7, 2022. www.cpuc.ca.gov/news-and-updates/newsroom/environmental-and-social-justice-action-plan

Carbon Intelligence. 2022. "Companies that have set net zero targets." Accessed Feb. 7, 2022. https://carbon.ci/insights/companies-with-net-zero-targets.

Cha, J. Mijin. 2020. "A just transition for whom? Politics, contestation, and social identity in the disruption of coal in the Powder River Basin." *Energy Research & Social Science* 69: 101657.

Chaffin, Brian et al. 2014. "A decade of adaptive governance scholarship: Synthesis and future directions." *Ecology and Society* 19, no. 3.

Chandler, Alfred. 2009. *Scale and Scope: The Dynamics of Industrial Capitalism*. Cambridge: Harvard.

Chen, Chien-fei et al. 2022. "Extreme events, energy security and equality through micro-and macro-levels: Concepts, challenges and methods." *Energy Research & Social Science* 85: 102401.

Clack, Christopher et al. 2020. *Why Solar for All Costs Less*. Boulder: Vibrant Clean Energy. Accessed Feb. 7, 2022. www.vibrantcleanenergy.com/wp-content/uploads/2020/12/WhyDERs_TR_Final.pdf

Clark, John Garretson. 1990. *The Political Economy of World Energy: A Twentieth-Century Perspective*. Chapel Hill: UNC Press.

Coenen, Lars et al. 2021. "Regional foundations of energy transitions." *Cambridge Journal of Regions, Economy and Society* 14, no. 2: 219–233.

Cowan, Ruth Schwartz. 1976. "The 'industrial revolution' in the home: Household technology and social change in the 20th century." *Technology and Culture* 17, no. 1: 1–23.

Cronon, William. 1991. *Nature's Metropolis: Chicago and the Great West*. New York: W. W. Norton.

De Laurentis, Carla. 2021. "What are the regionally specific institutions that matter for renewable energy deployment and how can they be identified? Some insights from Italian regions." *Local Environment* 26, no. 5: 632–649.

Donadio, R. and R. Meyer. 2018. "France's 'Yellow Vest' protesters aren't against climate action," *The Atlantic*, Dec. 7, 2018. www.theatlantic.com/international/archive/2018/12/france-yellow-vest-climate-action/577642.

Dunn, Kate. 2021. "Young climate activist confronts Shell CEO at pre-COP26 conference," *Fortune*, Oct. 14, 2021, https://fortune.com/2021/10/14/climate-activist-confronts-shell-ceo-cop26.

Eschrich, J., and C. Miller. 2019. *The Weight of Light: A Collection of Solar Futures*. Tempe: Center for Science and the Imagination.

Eschrich, J., and C. Miller. 2021. *Cities of Light: A Collection of Solar Futures*. Tempe: Center for Science and the Imagination.

Fink, Larry. 2020. "A fundamental reshaping of finance." Accessed Feb. 7, 2022. www.blackrock.com/corporate/investor-relations/2020-blackrock-client-letter

Flores, Phil Justice, and Johan Jansson. 2021. "The role of consumer innovativeness and green perceptions on green innovation use: The case of shared e-bikes and e-scooters." *Journal of Consumer Behaviour* 20, no. 6: 1466–1479.

Geels, Frank. 2004. "From sectoral systems of innovation to socio-technical systems: Insights about dynamics and change from sociology and institutional theory." *Research Policy* 33, no. 6–7: 897–920.

Gim, Changdeok. 2019. *Institutional Management for Infrastructure Resilience*. PhD Diss., Arizona State University.

Golding, Garrett. 2021. "Surging renewable energy capacity in texas prompts electricity generation adequacy questions." *Dallas Federal Reserve Bank*, Aug. 17, 2021, www.dallasfed.org/research/economics/2021/0817.aspx

Graffy, Elisabeth, and Steven Kihm. 2014. "Does disruptive competition mean a death spiral for electric utilities." *Energy Law Journal* 35.

Grätz, Jonas. 2012. "Unconventional resources: The shifting geographies and geopolitics of energy." In *Strategic Trends 2012*. Zurich: Center for Security Studies (CSS), ETH Zürich. 79–102.

Gross, Jenny. 2021. "Britain's gas crisis, explained." *New York Times*, Oct. 14, 2021. www.nytimes.com/2021/09/28/world/europe/why-uk-fuel-shortage.html

Grubert, Emily. 2018. "The Eagle ford and Bakken shale regions of the United States: A comparative case study." *The Extractive Industries and Society* 5, no. 4: 570–580.

Guston, David H. 2014. "Understanding 'anticipatory governance'." *Social Studies of Science* 44, no. 2: 218–242.

Guterres, Antonio. 2020. "Carbon neutrality by 2050: The world's most urgent mission." Accessed Feb. 7, 2022. www.un.org/sg/en/content/sg/articles/2020-12-11/carbon-neutrality-2050-the-world's-most-urgent-mission

Hecht, Gabrielle. 2009. *The Radiance of France: Nuclear Power and National Identity after World War II*. Cambridge: MIT.

Hernández, Diana. 2015. "Sacrifice along the energy continuum: A call for energy justice." *Environmental Justice* 8, no. 4: 151–156.

Hirsh, Richard 2002. *Power Loss: The Origins of Deregulation and Restructuring in the American Electric Utility System*. Cambridge: MIT.

Hobbes, Thomas. *Leviathan* (1651).

Hughes, Thomas. 1983. *Networks of Power: Electrification in Western Society, 1880–1930*. Baltimore: Hopkins.

Insull, Samuel. 1915. *Central-station Electric Service: Its Commercial Development and Economic Significance as Set Forth in the Public Addresses (1897–1914) of Samuel Insull*. Priv. Print.

International Energy Agency. 2021a. *World Energy Outlook*. Paris: IEA.

International Energy Agency. 2021b. *Recommendations of the Global Commission on People-Centred Clean Energy Transitions*. Paris: IEA. www.iea.org/reports/recommendations-of-the-global-commission-on-people-centred-clean-energy-transitions

Jetten, Jolanda et al. 2020. "How economic inequality fuels the rise and persistence of the Yellow Vest movement." *International Review of Social Psychology* 33, no. 1.

Johnston, Erik et al. 2011. "Managing the inclusion process in collaborative governance." *Journal of Public Administration Research and Theory* 21, no. 4: 699–721.

Jones, Christopher. 2014. *Routes of Power: Energy and Modern America*. Cambridge: Harvard University Press.

Keohane, Robert, and David Victor. 2016. "Cooperation and discord in global climate policy." *Nature Climate Change* 6, no. 6: 570–575.

Kline, Ronald, and Trevor Pinch. 1996. "Users as agents of technological change: The social construction of the automobile in the rural United States." *Technology and Culture* 37, no. 4: 763–795.

Kuzemko, Caroline et al. 2019. "New directions in the international political economy of energy." *Review of International Political Economy* 26, no. 1: 1–24.

Laird, Frank. 2013. "Against transitions? Uncovering conflicts in changing energy systems." *Science as Culture* 22, no. 2: 149–156.

Larson, E. et al. 2021. *Net-Zero America: Potential Pathways, Infrastructure, and Impacts*. Princeton University.

Lavo. 2022. "Lavo Hydrogen Battery." Accessed Feb. 7, 2022. https://lavo.com.au/lavo-hydrogen-battery.

Lilienthal, David. 1944. *TVA: Democracy on the March*. New York: Harper and Brothers.

MacFadyen, Joshua, and Andrew Watson. 2018. "Energy in a woodland-livestock agroecosystem: Prince Edward Island, Canada, 1870–2010." *Regional Environmental Change* 18, no. 4: 1033–1045.

Malekpour, Shirin, and Jens Newig. 2020. "Putting adaptive planning into practice: A meta-analysis of current applications." *Cities* 106: 102866.

Markard, Jochen et al. 2020. "Destined for decline? Examining nuclear energy from a technological innovation systems perspective." *Energy Research & Social Science* 67: 101512.

Mikellidou, Cleo Varianou et al. 2018. "Energy critical infrastructures at risk from climate change: A state of the art review." *Safety Science* 110: 110–120.

Miller, Clark. 2021. "Decolonizing technology." In *Cities of Light: A Collection of Solar Futures*, edited by Joey Eschrich and Clark Miller. Tempe: Center for Science and the Imagination.

Miller, Clark. 2022a. "Redesigning political economy: The promise and peril of a green new deal for energy." In *The Green New Deal and the Future of Work in America*, edited by B. Fong and C. Calhoun. New York: Columbia.

Miller, Clark. 2022b. "Energy, ownership, and the future of democracy." In *Democratizing Energy: Imaginaries, Transitions, Risks*, edited by J. Keahey, M. Nadesan, and M. Pasqualetti. Amsterdam: Elsevier.

Miller, Clark, Alastair Iles, and Christopher F. Jones. 2013. "The social dimensions of energy transitions." *Science as Culture* 22, no. 2: 135–148.

Miller, Clark, Jennifer Richter, and Jason O'Leary. 2015. "Socio-energy systems design: A policy framework for energy transitions." *Energy Research & Social Science* 6: 29–40.

Miller, Clark, Patricia Romero-Lankao, Andrew Dana Hudson, Joey Eschrich, and Ruth Wylie. 2021. "Imagined cities." In *Cities of Light: A Collection of Solar Futures*, edited by Joey Eschrich and Clark Miller. Tempe: Center for Science and the Imagination.

Mumford, Lewis. 1934. *Technics and Civilization*. Chicago: Chicago.

Nye, David. 1992. *Electrifying America: Social Meanings of a New Technology*. Cambridge: MIT.

Oakes, Kelly. 2021. "What makes a good net zero carbon emissions pledge?" *BBC*, Nov. 1, 2021. www.bbc.com/future/article/20211028-why-not-all-net-zero-emissions-targets-are-equal

Ostrom, Elinor. 2010. "Beyond markets and states: Polycentric governance of complex economic systems." *American Economic Review* 100, no. 3: 641–672.

Peluso, Nina. 2021. "Evolving paradigms in state-level integrated resource planning." PhD Diss., Massachusetts Institute of Technology. https://dspace.mit.edu/handle/1721.1/139486.

Perow, Charles. 1984. *Normal Accidents*. Princeton: Princeton.

Pinch, Trevor, and Wiebe Bijker. 1984. "The social construction of facts and artefacts: Or how the sociology of science and the sociology of technology might benefit each other." *Social Studies of Science* 14, no. 3: 399–441.

Rivera-Matos, Yiamar et al. 2021. *The Evolving Solar Energy Innovation System in Puerto Rico*. Tempe: Center for Energy & Society. http://cohemis.uprm.edu/solar2020/pdf/EvolvingSolarEnergy_March2021.pdf

Routledge, Paul et al. 2018. "States of just transition: Realising climate justice through and against the state." *Geoforum* 88: 78–86.

Rutherford, Jonathan, and Olivier Coutard. 2014. "Urban energy transitions: Places, processes and politics of socio-technical change." *Urban Studies* 51, no. 7: 1353–1377.

Santos-Lozada, Alexis, and Jeffrey Howard. 2018. "Use of death counts from vital statistics to calculate excess deaths in Puerto Rico following Hurricane Maria." *JAMA* 320, no. 14: 1491–1493.

Scott, Christopher, and Martin Pasqualetti. 2010. "Energy and water resources scarcity: Critical infrastructure for growth and economic development in Arizona and Sonora." *Natural Resources Journal* 50: 645.

Shetty, Disha. 2021. "A fifth of world's largest companies committed to net zero target." *Forbes*, Mar. 24, 2021. www.forbes.com/sites/dishashetty/2021/03/24/a-fifth-of-worlds-largest-companies-committed-to-net-zero-target/?sh=798f21e0662f

Skjølsvold, Tomas Moe, and Lars Coenen. 2021. "Are rapid and inclusive energy and climate transitions oxymorons? Towards principles of responsible acceleration." *Energy Research & Social Science* 79: 102164.

Smith, Ian. 2022. "Cost of living crisis: What are European countries doing to avoid soaring energy bills?" *Euronews*, Feb. 3, 2022. www.euronews.com/next/2022/02/03/cost-of-living-crisis-how-are-european-countries-responding-to-soaring-energy-bills

Sovacool, Benjamin. 2012. "The political economy of energy poverty: A review of key challenges." *Energy for Sustainable Development* 16, no. 3: 272–282.

Sovacool, Benjamin. 2016. "How long will it take? Conceptualizing the temporal dynamics of energy transitions." *Energy Research and Social Science* 13 (March): 202–215.

Sovacool, Benjamin, and Michael Dworkin. 2015. "Energy justice: Conceptual insights and practical applications." *Applied Energy* 142: 435–444.

Sovacool, Benjamin, and Frank Geels. 2016. "Further reflections on the temporality of energy transitions: A response to critics." *Energy Research & Social Science* 22: 232–237.

Späth, Philipp, and Harald Rohracher. 2010. "'Energy regions': The transformative power of regional discourses on socio-technical futures." *Research policy* 39, no. 4: 449–458.

Stephens, Jennie. 2019. "Energy democracy: Redistributing power to the people through renewable transformation." *Environment: Science and Policy for Sustainable Development* 61, no. 2: 4–13.

Stürm, Christine. 2018. "Germany's energy transition experiment: A case study about guiding decisions and steering large socio-technical systems in desired directions." PhD Diss., Arizona State University.

Termeer, Catrien et al. 2011. "The regional governance of climate adaptation: A framework for developing legitimate, effective, and resilient governance arrangements." *Climate Law* 2, no. 2: 159–179.

Valdmanis, Richard. 2021. "Global energy shortage or a coincidence of regional crises?" *Reuters*, Oct. 1, 2021. www.reuters.com/business/energy/global-energy-shortage-or-a-coincidence-regional-crises-2021-09-29.

Washington State Department of Public Health. 2021. *Clean Energy Transformation Act – Cumulative Impact Analysis.* Accessed Feb. 7, 2022. www.doh.wa.gov/DataandStatisticalReports/WashingtonTrackingNetworkWTN/ClimateProjections/CleanEnergyTransformationAct

Winner, Langdon. 1986. *The Whale and the Reactor.* Chicago: U. Chicago Press.

Wolf, Zachary. 2021. "It's time to pay serious attention to the power grid." *CNN*, Sept. 1, 2021. www.cnn.com/2021/08/31/politics/climate-change-power-grid-explainer/index.html

Wong-Parodi, Gabrielle. 2020. "When climate change adaptation becomes a "looming threat" to society: Exploring views and responses to California wildfires and public safety power shutoffs." *Energy Research & Social Science* 70: 101757.

Woodruff, Sierra 2016. "Planning for an unknowable future: Uncertainty in climate change adaptation planning." *Climatic Change* 139, no. 3: 445–459.

22

REDUCING MOBILITY-RELATED ENERGY USE IN FUTURE CITIES

The Planning Process for Urban Mobility in the City-District of Dietenbach in Freiburg, Germany

Arian Mahzouni

1. Introduction

Urban transport is a major driver of greenhouse gas emissions and currently stands for about 40% of total transport energy consumption. It is projected by 2050 two-thirds of the world population will live in cities, which will double mobility-related energy use in our cities despite rapid technological advancement in vehicle industry (ICLEI, 2022). Therefore, urban planners, policymakers and practitioners will have a major role in reducing mobility-related energy use.

This original research analyzes mobility transition pathways, policies, and practices in the new city-district of Dietenbach in Freiburg characterized by territorially bounded factors, such as planning policy framework, infrastructural artifact, architectural design, institutional arrangement, and the actor constellation. The aim of the urban design and infrastructure planning in Dietenbach has been to create a walkable mixed-use city-district that provides future residents access to daily services by non-motorized human-powered transport modes such as biking, walking, and small-wheeled vehicles (e.g., bike trailer, skates, and scooters). The Dietenbach project will be one of the largest construction projects in Germany, which will keep the City of Freiburg busy over the next decades, creating uncertainty about its feasibility of implementation. This chapter, as an early study of urban mobility development with respect to low-carbon change, will address the following two research questions:

- How could urban planning enable low-carbon mobility practices in the future cities through an integrated approach to urban mobility (e.g., urban design and the infrastructure planning for non-motorized transport modes and green spaces)?
- In what way(s) could new approaches to planning policies and practices for low-carbon urban mobility overcome path dependencies?

To address these questions, this study will undertake an interdisciplinary approach to the phenomenon of low-carbon urban mobility by combining insights from the literature on urban

DOI:10.4324/9781003183020-26

form (Dantzing and Saaty, 1973; Anderson et al., 1996; Banister et al., 1997; Dieleman et al., 1999; Jabareen, 2006; Rickwood et al., 2008; Clifton et al., 2008; Ewing and Rong, 2008), sSocio-technical transitions in urban mobility (Sovacool and Griffiths, 2020; Holden et al., 2020), and the emerging field of social practice theory (Spaargaren, 2003; Shove, 2010; Shove et al., 2012; Gram-Hanssen, 2014; Madsen and Gram-Hanssen, 2017; Julsrud and Farstad, 2020).

The remainder of this chapter is organized as follows. In the Section 2, the key elements of low-carbon urban mobility are conceptualized and at the end of the section an analytical framework is presented to better link the theoretical framework with case study results, highlighting the role of planning in enabling low-carbon urban mobility. In the Section 3, the methodological approach and empirical basis of this contribution is briefly explained. The Section 4 presents the case study including key information about the ongoing planning process and outcomes of urban development in Dietenbach. In the Section 5, the two research questions are discussed. Finally, Section 6 makes some remarks including an outlook for future research.

2. Theoretical Framework

2.1 The Links Between Urban Form and Mobility-Related Energy Use

There is no agreement on a standard definition of urban form. Williams (2014, 6) offers a more comprehensive definition of urban form as "the physical characteristics that make up built-up areas, including the shape, size, density and configuration of settlements. It can be considered at different scales: from regional, to urban, neighborhood, 'block' and street".

Three types of urban forms (urban sprawl, compact city, and polycentric city) have been discussed extensively in the literature. European urban scholars (Dieleman et al., 1999; Næss, 2005; Geurs and van Wee, 2006) indicate that urban sprawl, due to car dependence of households, results in higher energy use than a compact city or polycentric city. Furthermore, while a compact city consists of only one center with high-density development, a polycentric city has more than one center of activity and is characterized by sub-centers and "multinucleated form of development, with both population and employment concentrated around existing urban nodes" (Anderson et al., 1996, 26–27). Jabareen (2006, 48) believes that sustainable urban form is characterized by "a high density and adequate diversity, compact with mixed land-uses, and its design is based on sustainable transportation, greening, and passive solar energy". However, sustainability discourse in the last decades has raised disputed debates about the impacts of both compactness (Newman and Kenworthy, 1989; Ewing, 1997) and sprawl (Gordon and Richardson, 1989; Gordon and Richardson, 1997).

For many urban scholars the compact city concept (density and mix of uses) has been a physical response to many urban sustainability challenges, such as environmental protection by reducing car use and air pollution and social justice by providing citizens access to public transport (Neuman, 2005). However, most scholars support compact city design as an efficient urban form for sustainable mobility in many ways. For instance, it is economically viable because in densely populated area, infrastructure for different modes of transport (e.g., roads and cycle lane) can be provided cost-effectively per capita (Williams et al., 2000; Jabareen, 2006). The focus of this study is more on a "compact city", which was first coined by Dantzing and Saaty (1973). Their vision was to enhance the quality of life by efficient use of resources, which includes strategies to create compactness and density. The main elements of the compact city concept are density, proximity, availability, mixing of land uses, provision of local services that connect work, and leisure facilities together in order to reduce mobility-related energy use. Furthermore, it is proposed that in more compact cities, travel distance is shortened, and rural area is protected from development.

The work of Australian researchers Newman and Kenworthy (1989), which is still one of the most comprehensive international studies on the link between urban form and mobility-related energy use, supports the compact city concept. There is broad consensus about the three key elements/measures of urban form: density, mixed land use, and neighborhood accessibility. But much of the current debate consider density as the most important element of sustainable urban form (Rickwood et al., 2008), which is defined as "the ratio of people or dwelling units to land area" (Jabareen, 2006, 41). The relationship between densely populated cities and the mobility-related energy use was first established by Newman and Kenworthy (1991). They argued that sprawled and low-density populated areas due to limited or no connection to public transport are consequently characterized by car-dominated infrastructure systems, car ownership, and therefore, high rate of per capita mobility-related energy use. Also Clift et al. (2015, 26) state that "people who are spread out in a city, living further apart, must travel further distances to interact in their daily urban activities and hence use more energy".

According to Newman and Kenworthy (1989, 33), substantial amounts of energy can be saved by designing and implementing a mix of density policies such as "strengthening the city center; extending the proportion of a city that has inner-area land use; providing a good transit option; and restraining the provision of automobile infrastructure". About the link between urban form and energy use, Newman and Kenworthy (1989) argue that increased density decreases automobile dependency and increases the use of public transport and therefore gasoline consumption is lower in high-density cities. In subsequent writing, Kenworthy and Newman (1990) came to a similar conclusion that density is the key element to measure mobility-related energy use.

Another key element of compact city is mixed land use, which "indicates the diversity of functional land uses such as residential, commercial, industrial, institutional, and those related to transportation" (Jabareen, 2006, 41). Mixed land use can reduce car dependency when services related to housing, working, shopping, health, and leisure activities are reached within a reasonable distance by non-motorized human-powered transport modes, such as biking and walking. Furthermore, mobility is essentially tied to the building envelope, which might influence the use of alternative modes of transport, for instance, in the form of parking facilities for various non-motorized transport modes or charging facilities for electromobility. Also, O'Brien et al. (2010, 1003) indicate that "from an energy standpoint, compact buildings reduce heat loss and promote walking, bicycling, and cost-effective mass transportation". Thus, creating infrastructure for non-motorized transport modes and public transport systems are preconditions to reduce mobility-related energy use in cities.

2.2 Mobility-Related Energy Use as a Set of Social Practices

There is a trend away from car ownership to car usership among young people, for instance, in the large US cities (Davis et al., 2012; Davis and Baxandall, 2013). Therefore, as argued by Clift et al. (2015, 55), "understanding the consumption patterns, attitudes, values and potential for behavioural change among the urban young is an important challenge for policy-makers and researchers". The work of Giddens (1984) has provided a solid basis for considering consumption as a set of social practices that are shaped by both social norms and lifestyle choices and by the institutions and structures of society (Jackson, 2005; Shove et al., 2012; Clift et al., 2015). A key strategy to understand and change unsustainable consumption behaviors, as the Giddens' structuration theory proposes, is making a clear distinction between practical consciousness and discursive consciousness.

Practical consciousness is the everyday knowledge of people to do things in certain ways. Giddens (1984) believes that human agency mostly uses practical consciousness in routinized situations and behavioral contexts. In contrast, in discursive consciousness, social actors try to build a form of their awareness and action, based on communication or debate. So agency is largely created through "repetitive, routine practices of everyday life" (Jackson, 2005, 91). In addition, as stated by Jackson (2005, 90), actors must have access to "the 'transformative capacity' of historical social structures, such as language, rules, norms, meanings and power" to create adequate agency for change. So according to Giddens model, social structure is both the medium and the outcome of people's daily social practices. Furthermore, in Giddens' structuration theory, social interaction is characterized by recursiveness and constant reproduction of social practices in which social actors are involved (Giddens, 1984).

Drawing on the discussion by Giddens (1991), Spaargaren (2003, 689) argues that

> agency is analyzed not in terms of the isolated individual . . . but rather in terms of the twin concepts of lifestyles and social practices. The lifestyle of each individual is constructed from a series of building blocks – corresponding to the set of social practices an individual invokes when pursuing his or her daily life.

So our behavior in any particular situation is influenced partly by our attitudes and intentions and partly by the situational constraints and conditions under which we act. Thus, promoting sustainable living requires a transition from individual lifestyle (practical consciousness) to community-driven social practices (discursive consciousness). Individual behaviors and actions are constrained by social practices and remain mostly as practical consciousness. Therefore, pro-environmental behavioral change will be achieved largely by converting behaviors from the level of practical consciousness to the level of discursive consciousness (Spaargaren and van Vliet, 2000; Jackson, 2005). A successful example of scaling up individual practices to community-driven social practices is the creation of the community-driven district heating infrastructure in rural town of St. Peter in Germany (Mahzouni, 2019), in which agency for the promotion of low-carbon heating infrastructure has come from different scales (individual, household, neighborhood, city, and national) shaping a particular local energy transition.

2.3 Analytical Framework: Transition to Low-Carbon Urban Mobility

This study sheds new lights on two key elements of low-carbon urban mobility (low-mobility societies and collective transport), discussed by Holden et al. (2020). In a low-mobility society "due to higher population density and generally shorter distances to private and public services, people in cities have reasonable alternatives to cars, such as walking, cycling, and well-functioning public transport systems" (Holden et al., 2020, 6). Compact city concepts and practices have enabled car-free zones in European cities (e.g., Madrid, Freiburg, Strasbourg, Brussels, Utrecht, Dublin, Glasgow, Bergen, Copenhagen, Stockholm, and Helsinki) by banning cars from residential and commercial areas in certain city-districts. The exception has been vehicles that provide emergency, medical care, waste collection, and freight and delivery services (Hagen et al., 2017; Mahzouni, 2018; Holden et al., 2020; Sareen et al., 2021).

Collective transport includes both public transport and shared mobility. Holden et al. (2020, 3–4) argue that the major transport patterns are currently dominated by cars and airplanes and need to be replaced by "an affordable and well-functioning public transport system" (buses, trams, and trains), which are more energy-efficient, especially regarding occupancy rates. An efficient and affordable public transport can also facilitate access to shared mobility options, limiting the use of single-occupancy vehicle. Thus, local governments should promote the use of public transport (e.g., by subsidizing fares) to achieve the twin objectives of providing access to basic mobility needs for people without cars and encouraging car owners to shift to public transport. Shaheen et al. (2015, 3) describe shared mobility in this way

> the shared use of a vehicle, bicycle, or other mode – is an innovative transportation strategy that enables users to gain short-term access to transportation modes on an "as-needed" basis [which] includes various forms of carsharing, bikesharing, ridesharing (carpooling and vanpooling), and on-demand ride services.

The figure presents the role of planning in creating the complementary elements for socio-technical transition to a low-carbon urban mobility in the Dietenbach city-district. Each element influences and is influenced by three neighboring elements directly, which are illustrated by double-headed arrows to show mutual relationship between them. Low-carbon mobility practices influence and are influenced by all other elements directly. The planning policy framework of local government provides a solid platform for shaping both hard and soft elements of socio-technical transitions in urban mobility. The two hard elements of urban form and mobility infrastructure and the two soft elements of regulations and actor constellation will help citizens to change their mobility practices, enabling transition from car ownership to car usership. Planning policy framework through physical elements of transition (urban form and mobility infrastructure) can establish a solid base to make necessary regulations and to enable an appropriate actor constellation aimed at supporting both elements of low-mobility societies and collective transport and thereby discouraging car ownership and use.

Regulations, as a key social element of transition, can support the diffusion of collective transport. Here, the key role of regulations is illustrated by an example of a recent mobility transition effort from the City of Freiburg. The annual fees for residential parking until now have been only €30 in Freiburg. But the Freiburg City Council made a decision in December 2021 to increase the fees very drastically depending on the size of the vehicle, €240 for small, €360 for medium, and €480 for large cars. So, on average one euro per day, which is a reasonable price for the use of public space. The political parties, such as the the Green Party (Die Grünen), among others, which backed this decision, claim that sustainable mobility requires large investments, and the revenue from the fees will be used to expand the pedestrian and cycle paths to enable non-motorized human-powered transport modes. The new law, which will enter into force by April 2022, is a major step toward low-carbon urban mobility in Freiburg. The city has recently revised its climate action plan to be carbon-neutral by 2038, which is 12 years earlier than previously planned (Amtsblatt, 2021).

Furthermore, depending on a range of territorially based factors (e.g., governance modes, policy arrangements, and institutional logics) an appropriate actor constellation to shape particular roles and power relations can play a vital role in supporting socio-technical transitions (Späth & Rohracher, 2010; Becker et al., 2017; Bjerkan et al., 2021). In particular, intermediary organizations can play a vital role in shaping new constellations, power relations, and roles. They can be from all sectors (public, private, and civil society) and can shape "the traditional relationships between utilities, regulators, and consumers" (Moss, 2009, 1489).

To sum up, the Figure 22.1 shows that achieving low-carbon urban mobility requires far-reaching sociotechnical transition, engaging all sectors of the economy and society. The role of planning is to establish a systematic and effective intersection between the sectoral policy of urban form, mobility infrastructure, regulations, and actor constellation toward low-carbon

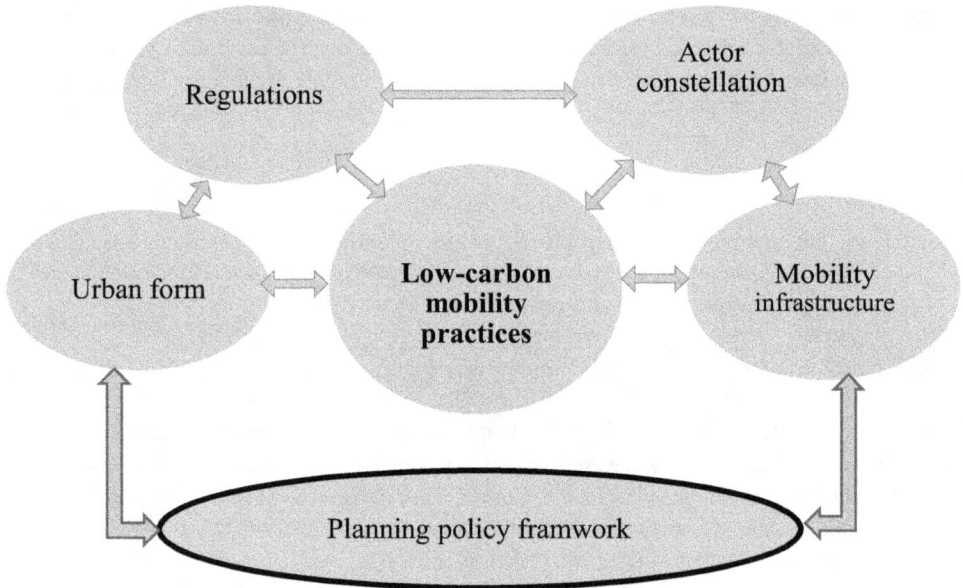

Figure 22.1 An integrated socio-technical approach to low-carbon urban mobility.

mobility practices. This framework might help the planners and policymakers to consider low-carbon mobility as an interdisciplinary platform that requires achieving synergy between different sectors, and therefore designing and implementing a mix of policy instruments accordingly. Based on territorially based factors, they can design and implement a right combination of elements to support low-carbon mobility practices. The framework illustrated in the figure has been applied to address the two research questions in the discussion section.

3. Methods

A qualitative case study methodology (Eisenhardt, 1989; Stake, 1995; Yin, 2003, 2009; Merriam, 2009) has been applied to analyze the phenomenon of low-carbon urban mobility in the particular context of the case study area. Miles and Huberman (1994, 25) define a case as "a phenomenon of some sort occurring in a bounded context", and Yin (2009,18) defines a case study as "an empirical inquiry that investigates a contemporary phenomenon in depth and within its real-life context, especially when the boundaries between phenomenon and context are not clearly evident". Here, low-carbon urban mobility in Dietenbach is defined as the main unit of analysis and the planning policy framework to enable low-carbon urban mobility as the embedded unit of analysis. Both units are studied within the historical and urban planning context of Freiburg and in relation to other infrastructure for land use planning, housing, and digitalization in the area.

As part of desk study, the information sources have been research articles, policy documents (e.g., Projektgruppe Dietenbach, 2020; City of Freiburg, 2021, 2022; EGS-plan, 2021) and local newspapers (e.g., *Amtsblatt*, *Badische Zeitung*, and *Immobilien Zeitung Freiburg*). Furthermore, the author has lived in Freiburg for many years and observed the sustainability events in Freiburg very closely. In recent years, he has participated in numerous public events on the Dietenbach discourse. Thus, the reliability of obtained information has been cross-checked through the active observations of the author. In the next section, the case study results are presented.

4. Planning for a Low-Carbon Urban Development in Dietenbach

Since 2013, the City of Freiburg in southwest Germany has been working on the development of a new city-district on the greenfield site of Dietenbach in West Freiburg to meet the twin objectives of providing a broad range of the population with affordable housing and achieving a low-carbon urban development. Against this background, this section will present the results of planning policies and practices for urban development in Dietenbach until now. It will first explain the key elements of urban design competition and land use planning for urban development, including the master and development plans. Then, the infrastructure development plan for green spaces and collective transport will be presented that might shape the mobility practices of future residents in Dietenbach in a particular way.

4.1 Urban Design Competition for a New City-District

In summer 2017, a Europe-wide invitation to tender for urban design of a city-district on the greenfield site of Dietenbach was announced. The main aim was to create a city-district of short distances and living room for people above all for lower- and middle-income groups. About 28 architectural firms submitted their design concepts of which four design drafts were selected in the first stage. In March 2018, the four selected architectural firms presented their concepts in a citizens' forum, in which around 300 residents used the opportunity to exchange information with the architectural firms. The suggestions from the residents were included in the four revised drafts. Then, in October 2018, the Jury of Freiburg City Council announced the winner draft of the architectural and urban design competition for the new city-district, which was produced by the team of K9 Architects. The Jury presented many arguments for the selection:

- The winning team and design concept make the living conditions in Dietenbach more visible characterized by urbanity and being green and socially diverse, ensuring a colorful and livable city-district. The concept was developed in line with the urban design, transport planning, and economic and ecological requirements of the Freiburg City Council and in many respects is a future-oriented model for urban development (City of Freiburg, 2022).[1]
- The design accounted many key issues, such as climate neutrality, species and nature conservation, and plans to build social infrastructure, such as schools, kindergartens, and the youth center. In addition, the team presented a clear concept for designing green spaces in the form of two parks along the streams of Dietenbach and Käserbach with a central located center (ibid.).

    The winning design concept is in line with the decision of the Freiburg City Council to provide 50% of residential area with subsidized rental housing (City of Freiburg, 2021).
- As required by the land policy instruments, the design relies on high density and different tenancy-related living forms (senior housing, multi-generational housing, student housing) and physically related building types (single-family house, apartment building, semi-detached house, high-rise buildings) to create a socially mixed city-district. Drawing on the experiences from the neighboring city-district of Rieselfeld, each courtyard will consist of four smaller building blocks grouped around a collective neighborhood space, which forms the basis for a variety of different construction projects and thereby a desired mix of living forms and building types. So the design concept supports the creation of the

building blocks in a particular way, which allow the involvement of many different housing associations like the municipal housing company Freiburger Stadtbau, housing cooperatives, co-building groups (Baugemeinschaften), and syndicate of tenements (Mietshäuser Syndikat). According to Mahzouni (2017, 394),

> A co-building group consists of different people (families, single, young, elderly people) who collectively buy land plots to design and build their own houses together. The group diversity provides access to a wide range of resources, experiences and networks for creative thinking and innovation.

The syndicate of tenements is a model for collectively owned houses as an alternative to the mainstream market and state provision of housing. It creates a common pool for sharing resources, skills, and responsibilities for collectively managed housing; it encourages a sense of collective belonging (Hurlin, 2018). These housing associations represent a wide range of societal groups, which will play an important role in creating affordable housing and in responding to the current housing shortage in Freiburg due to recent dramatic increases in real estate and rent prices. Socially mixed housing is anticipated to also create different lifestyles and values and thereby generating a range of mobility practices.

4.2 The Planning Process to Convert the Master Plan into the Development Plan

The planning process carries the design to implementation. For the purposes here, a master plan is an informal planning tool to outline the development potential and prospects for the future use of an area, which will be gradually converted into a legally binding development plan subject to standardized procedures. The development plan concretizes the previous plans including the master plan, indicating how the land in Dietenbach should be used in the future. It will indicate for whom and by whom the new city-district will be built, which will specify the mobility infrastructure in Dietenbach too. In each building block, the possibility for different uses including different living forms (e.g., multi-generational housing, senior housing, student housing) and tenancy and physically related building types (e.g., single-family house, apartment building, semi-detached house, high-rise buildings) will be defined.

Specific to the design, the Dietenbach greenfield area with a size of 110 hectares has been earmarked for different land use purposes. The design concept leaves around 61 hectares for residential buildings (including private gardens) and mix-use construction (City of Freiburg, 2022). This is a good basis for economic viability, which together with the high building density fulfills the prerequisites for creating a compact city-district up to 6,500 apartments for around 15,000 people. Furthermore, 21 hectares are needed for green spaces, and 21 hectares for mobility infrastructure and public places. A further 7 hectares are available for public buildings, such as day-care centers, primary and secondary schools, high schools, and sports halls. Other social facilities, such as shops, small commercial units, the community center, and church center, are to be integrated into the development plan.

Turing to the planning process, from the beginning of 2019, efforts have been made to develop the design concepts into a master plan for Dietenbach, in which public authorities have played a vital role. The master plan was finalized in 2021, which will be gradually transferred into a development plan as a basis for land use planning. The development plan should be adopted by 2022 as a standing rule to provide details about key issues, such as Freiburg's climate protection objectives, the energy standards of Freiburg (Freiburger Energiestandards), the share of social rented housing, and the concepts of cooperative housing and co-building groups (Baugemeinschaften). However, several architectural elements and building design features cannot be regulated in a development plan. This is then the task of the marketing concept (Vermarktungskonzepts) and the purchase contracts. Furthermore, around the year 2023, the first plots for the first construction site can come into the market. But until then, the city needs to conduct preparatory work related to basic infrastructure, such as high-voltage cables, gas lines, and transmitters.

4.3 Green Infrastructure

In the winning design concept, the new city-district is divided into three similarly structured quarters which are gradually consolidated from the margins to the center (ibid.). The quarters are oriented toward the large open spaces in the area adapting to the landscape setting of Freiburg. In addition to a small network of paths, a ring-formed boulevard connects the three quarters. The new center acts as a buttress between the two landscape areas of the Dietenbach-park in the east and the Mooswald forest in the west of the city-district.

Despite high building density, the city-district should be provided with at least 21 hectares of green space, as mentioned previously. The framework for urban green-space planning is characterized by two parks as green corridors: the Auenpark Dietenbach, which runs along the Dietenbach stream and will be under the greatest possible preservation of the trees for creating leisure areas, and the Mooswaldpark, which includes structures of the Mooswald forest and horticultural use. As a result, three roughly equal-sized quarters emerge that are linked by an urban center (City of Freiburg, 2022). These parks make access to the neighboring open spaces of the Dietenbachpark in the east and the Mooswald forest in the west of the city-district possible. At the same time, the two parks due to their location contribute to the improvement of the urban climate and offer future residents immediate access to green spaces, offering great recreational qualities. Also, short distances to the green areas through two large parks will ensure the immediate access to cooler vegetation zones (City of Freiburg, 2021).

4.4 Infrastructure for Collective Transport

This section will present infrastructure planning for collective transport (public transport, shared mobility). The Dietenbach city-district is characterized by a variety of mobility options that discourages car use and promotes walkability in the city-district. Efforts are made to create non-motorized infrastructure for a walkable mixed-use city-district to provide future residents access to daily services by bicycle or on foot.

An essential feature of the master plan is a ring-formed boulevard that connects the sub-centers in the city-district, which will create a traffic calming city-district center with services, restaurants, and retails. A bicycle lane along the ring-formed boulevard connects the public spaces in all quarters with each other. The center can be reached on foot and by bike from the areas within the ring. Car drivers reach the city-district from three sides of the city-district: over the new roundabout on heavily trafficked Besançonallee street in the southeast, through the federal highway B31 in the north and northeast, and over the existing connection Mundenhof/ Lehen in the west to the feeder road (City of Freiburg, 2022). A feeder road is a secondary road which "feeds" traffic to main highways. In addition, the construction of noise barrier walls along high-trafficked ways of the federal highway B31 and Besançonallee street has been included in the master plan (City of Freiburg, 2021). In the following, more details about collective transport, in the form of public transport and shared mobility, will be presented.

Public Transport

The last stop of the Tram Line 5 in the neighboring city-district of Rieselfeld will be extended into Dietenbach with three stops throughout the city-district to ensure fast connections to the city center. The extension is done via a ring-formed boulevard running across the central square. The inner quarters are accessed via residential streets designed as loops. So transport links with the ring-formed boulevard and the Tram Line 5 from Rieselfeld is ensured (City of Freiburg, 2021, 2022)

Shared Mobility: From Car Ownership to Car Usership

In order to achieve low-carbon urban mobility in the city-district, the parking space is by design set low, only 0.8 cars per apartment and no parking space will be provided on residents' own property. However, 12 collective garages are planned to be built along the ring-formed boulevard in the outskirt of the city-district, in which other important infrastructure for shared mobility services (e.g., stations for bike-sharing and car-sharing) will be provided. The street network is hierarchically constructed and relies on the ring-formed boulevard, where all quarter garages are located. In addition, all major traffic routes will lead to the central square of the city-district, which is an important social meeting place to provide local residents with many activities, such as weekly market or city-district festivals (City of Freiburg, 2022).

Cyclists and pedestrians will be provided with new connections via the bicycle highway along the Dreisam River and with continuation to other parts of Freiburg. The cycle traffic surrounds the quarter and crosses the quarter on a cycle lane via the central square with connection to the neighboring city-district of Rieselfeld in the south, an underpass across the Besançonallee highway in the northeast, and the new footbridge via the highway B31 in the north, which will lead to the bicycle highway along the Dreisam River (ibid.).

5. Discussion

Freiburg is internationally recognized for a range of urban sustainability policies and pathways. This study draws on a synthesis of urban design, socio-technical transitions in urban mobility, and social practice theory to provide early practical and scholarly insights into the ongoing planning process for the new city-district of Dietenbach in West Freiburg. In this section, efforts will be made to address the two research questions listed in the introduction section by reflecting on the analytical framework in the theory section and based on the case study results.

How Could Urban Planning Enable Low-Carbon Mobility Practices in the Future Cities Like Dietenbach?

The planning policy framework provides a common base for low-carbon mobility practices by encouraging future residents in Dietenbach to act more environmentally. It has supported both hard elements (urban form and mobility infrastructure) and soft elements (regulations and actor constellation) that might enable adequate agency and broad public support for enhancing low-carbon mobility practices.

Compact Urban Form

Living in compact cities is usually less energy-intensive than living in sprawling rural areas mainly because in the latter travel distances are likely to be longer and public transport less available, leading to high rate of car ownership and energy use. Thus, a mixed-use development and a city of short distances, a key element of compact urban form, might support the use of non-motorized human-powered transport modes such as biking, walking, and small-wheeled vehicles (e.g., bike trailer, skates, and scooters) and thereby institutionalizing low-carbon mobility practices in the Dietenbach city-district. However, the technological advancement for telecommuting can reduce the correlation between compact urban form and mobility-related energy use. A study by O'Brien et al. (2010, 1018–1019) shows that a "telecommuter during a five-day workweek use 11.5% less transport than non-telecommuters". As a result, the total potential benefit of compact urban form might be diminished by the fact that telecommuting enables people to work from home in less compact areas who do not use their car for work-related travel. A good example is increased home offices during COVID-19. However, working from home can lead to increased nonwork and holiday trips, which are usually longer than travel to work, strengthening the so-called rebound effects (Bieser et al., 2021).

Mobility Infrastructure

The planned infrastructure for urban mobility in the Dietenbach city-district is supposed to reduce car ownership, thereby shaping residents' mobility-related energy use in a particular way. Infrastructure for cyclists and pedestrians has been prioritized in order to allow access to services of daily living within a reasonable distance by non-motorized transport modes. In addition, green infrastructure planning will provide immediate access to green spaces, encouraging the residents to use non-motorized human-powered transport modes for their recreational activities. This will, in turn, reduce car use to access green spaces in the surrounding countryside. So the Dietenbach city-district is characterized by a variety of mobility options encouraging a transition from car ownership to car usership.

Regulations

Both urban form and mobility infrastructure, as physical elements of socio-technical transitions in urban mobility, can provide a solid platform for local government to enact and enforce necessary regulations that fosters collective transport (public transport, shared mobility) and discourages car ownership. The drastic increased annual fees for residential parking in the Freiburg city center in 2021, as described previously in the Section 2.3, might decrease car use given the fact that there is a relatively functioning infrastructure for non-motorized human-powered transport modes in Freiburg.

Actor Constellation

The people are the main agent in transition process to a low-mobility future (Holden et al., 2020). As discussed in the Section 4.1, the involvement of a range of housing associations in the construction process in Dietenbach (e.g., the municipal housing company, housing cooperatives, co-building groups, and syndicate of tenements) will create new business models for providing affordable housing. In addition, they are representing various societal groups, lifestyles, and preferences. Therefore, an early involvement of these associations in the planning process is key to gain a broad public acceptance for enacting and enforcing necessary regulations, which might gradually lead to institutionalizing low-carbon mobility practices.

How Could New Approaches to Planning Policies and Practices for Low-Carbon Urban Mobility Overcome Path Dependencies?

The concept of path dependence explains the process of change, a "dynamical process whose evolution is governed by its own history [and] characterized by positive feedbacks and self-reinforcing dynamics" (David, 2007, 92). According to Pierson (2000, 259), path dependence occurs when "previous choices often are relevant to current action", which might cause conflicting interests among actors from different spheres of society: public, private, and civil society. Therefore, planners, policymakers, and practitioners aimed at stimulating low-carbon urban mobility should recognize that mobility system, like all other complex urban systems (e.g., energy and housing systems), is characterized by inertia that occurs when past choices crowd out future, and it is therefore difficult to transform it from its existing state (Araújo, 2017; Barazza and Strachan, 2021).

However, relying only on physical elements of transition (urban form and mobility infrastructure) will unlikely enable the transformation of residents' behaviors from the level of practical consciousness (lifestyles) to the level of discursive consciousness (community practices). Residents may not be required to apply sustainable practices in their daily life. As Holden et al. (2020, 8) emphasizes, "it is ultimately the people that are key to creating the credibility and the acceptability of the narratives. People decide where and when to travel. People decide to travel by bike or bus. People decide which car to buy or not to buy". In addition, Araújo et al. (2019) have identified a range of demographic and territorially based factors (such as income, education, age, political orientation, population density, and urban form) as critical niches influencing the early-staged adoption or diffusion of clean technologies, such as electric vehicles and solar photovoltaics. For instance, the car-dependent nature of suburban and rural areas can requires more investment in electric vehicle charging stations. Similarly, in contrast to high-density urban areas, "the availability of space in suburban and rural contexts offers greater potential for Solar PV adoption by single home owners" (ibid., 10). Thus, a change in existing mobility practices, above all, requires personal adaption to new infrastructures, technologies, and urban forms.

It is hard to alter a sociotechnical mobility system which has become a self-sustaining and self-regulating system over a long period of time (Hughes, 1983). Past experiences from the Vauban city-district in Freiburg show that it might be difficult to achieve a car-free neighborhood in Dietenbach due to free rider problem caused by car owners who refused to register their car ownership and pay for a parking space (Mahzouni, 2018). They refused to adapt their mobility practices despite investing large resources in providing infrastructure for public transport and non-motorized human-powered transport modes. Nevertheless, constructing a new city-district in the greenfield area of Dietenbach might enable path-creation for both planning policies and with residential mobility practices by providing appropriate architectural design and infrastructure for low-carbon mobility. This will help the future residents in Dietenbach not to be locked into their existing choices and practices. Therefore, policymakers and planners should apply "adaptive expectations" as key self-reinforcing mechanisms to generate path-creation for urban mobility transition (Arthur, 1994). Path-creation strategies for low-carbon urban mobility should not only target infrastructural, regulative, and economic instruments at the policy level but more soft elements, such as lifestyles and consumption behavior at the household and personal level. Low-carbon mobility practices will need to garner broad acceptance and to meet public expectations. Achieving low-carbon mobility in Dietenbach, the future residents should adapt their actions and behaviors by using more walking, cycling, and public transport and thus minimizing car use.

To sum up, only relying on physical elements of transition will unlikely enable the transformation of residents' behaviors from the level of practical consciousness (lifestyles) to the level of discursive consciousness (community practices). So as illustrated in the Figure 22.1, planning policies and practices should also support soft elements, such as by enacting and enforcing regulations and involving a wide range of actors in planning process at a very early stage in order to gain public acceptance for transition toward low-carbon urban mobility.

6. Conclusions

This chapter addresses key approaches for low-carbon mobility in future cities by conducting a case study of current planning policies and practices of the City of Freiburg for constructing a new city-district in the greenfield area of Dietenbach. Based on the research results and analysis, several conclusions can be drawn, which urban planners, policymakers, and practitioners might consider in their effort to design and implement context-based strategies for low-carbon urban mobility, as a major step toward energy-efficient and climate-neutral cities.

First, path dependency can occur not only in planning policies and practices by local authorities but also in mobility practices by residences in future cities who might refuse to change their lifestyle and find it hard to shift from car ownership to car usership. So besides strategies to enable technological artifacts and architectural design, strategies to modify "ideological beliefs", which shape the mobility practices and choices of future residents, must be included (North, 1990). Planners and policymakers should encourage lifestyles and habits that might support the socio-technical transition toward low-carbon urban mobility. *Second*, the strategic plans for mobility infrastructure supports low-carbon urban mobility in Dietenbach. However, whether individual lifestyles may be expanded into broad social practices remains an open question, as a major step to reducing mobility-related energy use. *Third*, path dependency both in planning policies and mobility practices may continue to impede a transition to low-carbon urban mobility. Therefore, future research might explore how to address the role of path dependency in mobility transition policies and practices. Also, in light of increased energy use at home, partly due to recent trends in teleworking, future research might study the actions and policies of the combined effect of in-dwelling and mobility-related energy use in future cities.

Finally, in the literature there is a large debate on the role electromobility in urban mobility transitions. However, due to the scope of this study, the discourse of electromobility has been excluded from the analysis mainly because of two reasons. First, there is no clear data about infrastructure planning for electromobility in Dietenbach. Second, electromobility might also impose negative impacts on the outcome of the two key elements of low-carbon urban mobility (low-mobility societies and collective transport) discussed in this chapter. As stated by Holden et al. (2020, 8),

> More electromobility in terms of electric private cars could well mean more private car ownership and use, and subsequently lead to less collective transport and counteract the promotion of car-free societies. Thus, promoting generous incentive packages for EVs (e.g. as they do in Norway) encourages people to buy and drive privately owned cars instead of taking the bus or participating in shared mobility schemes.

Also, Clift et al. (2015) argue that the use of driverless vehicles in future cities might lead to increased mobility-related energy use once citizens switch from non-motorized human-powered transport modes (e.g., walking, biking and small-wheeled vehicles) to driverless vehicles for short trips.

Note

1 For details about master/development and mobility plans for Dietenbach, see City of Freiburg (2022).

References

Amtsblatt. (2021). Anwohnerparken drastisch teurer [Costs of residential parking will increase drastically]. 17. Dezember 2021, Nr. 806, Jahrgang 34, S. 1, Stadt Freiburg im Breisgau.

Anderson, W. P., Kanaroglou, P. S., & Miller, E. J. (1996). Urban form, energy and the environment: A review of issues, evidence and policy. *Urban Studies, 33*(1), 7–36. https://doi.org/10.1080/00420989650012095.

Araújo, K. (2017). *Low Carbon Energy Transitions: Turning Points in National Policy and Innovation.* Oxford University Press, New York.

Araújo, K., Boucher, J. L., & Aphale, O. (2019). A clean energy assessment of early adopters in electric vehicle and solar photovoltaic technology: Geospatial, political and socio-demographic trends in New York. *Journal of Cleaner Production, 216*(2019), 99–116. https://doi.org/10.1016/j.jclepro.2018.12.208.

Arthur, W. B. (1994). *Increasing Returns and Path Dependence in the Economy.* Ann Arbor: University of Michigan Press.

Banister, D., Watson, S., & Wood, C. (1997). Sustainable cities: Transport, energy, and urban form. *Environment and Planning B: Planning and Design, 24,* 125–143.

Barazza, E., & Strachan, N. (2021). The key role of historic path-dependency and competitor imitation on the electricity sector low-carbon transition. *Energy Strategy Reviews, 33*(2021), 100588. https://doi.org/10.1016/j.esr.2020.100588.

Becker, S., Naumann, M., & Moss, T. (2017). Between coproduction and commons: Understanding initiatives to reclaim urban energy provision in Berlin and Hamburg. *Urban Research and Practice, 10*(1), 63–85. https://doi.org/10.1080/17535069.2016.1156735.

Bieser, J. C. T., Vaddadi, B., Kramers, A., Höjer, M., & Hilty, L. M. (2021). Impacts of telecommuting on time use and travel: A case study of a neighborhood telecommuting center in Stockholm. *Travel Behaviour and Society, 23*(2020), 157–165. https://doi.org/10.1016/j.tbs.2020.12.001.

Bjerkan, K. Y., Ryghaug, M., & Skjølsvold, T. M. (2021). Actors in energy transitions: Transformative potentials at the intersection between Norwegian port and transport systems. *Energy Research and Social Science, 72*(2021), 101868. https://doi.org/10.1016/j.erss.2020.101868.

City of Freiburg. (2021). Neuer Stadtteil Dietenbach Wird Klimaneutral [New City-District of Dietenbach Will Be Climate-Neutral]. Press Release 13 July. www.freiburg.de/pb/1734171.html, Accessed 12 October 2021.

City of Freiburg. (2022). Der Rahmenplan für Dietenbach [The master and development plans for Dietenbach]. www.freiburg.de/pb/1631506.html, Accessed 24 January 2022.

Clift, R., Druckman, A., Christie, I., Kennedy, C., & Keirstead, J. (2015). Urban Metabolism: A Review in the UK Context. Future of Cities: Working Paper, Government Office for Science, London).

Clifton, K., Ewing, R., Knapp, G., & Song, Y. (2008). Quantitative analysis of urban form: A multidisciplinary review. *Journal of Urbanism, 1*(1), 17–45. https://doi.org/10.1080/17549170801903496.

Dantzing, G. B., & Saaty, T. L. (1973). *Compact City: A plan for a Livable Urban Environment.* San Francisco: W.H. Freeman.

David, P. A. (2007). Path dependence: A foundational concept for historical social science. *Cliometrica, 1,* 91–114. https://doi.org/10.1007/s11698-006-0005-x

Davis, B., & Baxandall, P. (2013). *Transportation in Transition: A Look at Changing Travel Patterns in America's Biggest Cities.* Boston: US PIRG Education Fund/Frontier Group.

Davis, B., Dutzik, T., & Baxandall, P. (2012). *Transportation and the New Generation: Why Young People Are Driving Less and What It Means for Transportation Policy.* Boston: US PIRG Education Fund/Frontier Group.

Dieleman, F. M., Dijst, M. J., & Spit, T. (1999). Planning the compact city: The randstad Holland experience. *European Planning Studies, 7*(5), 605–621. https://doi.org/10.1080/09654319908720541.

EGS-plan. (2021). Fortschreibung Energiekonzept Stadtteil Dietenbach, Freiburg *[Updated energy concept for Dietenbach city-district in Freiburg]*. 28 Juni 2021, EGS-plan Ingenieurgesellschaft für Energie-, Gebäude- und Solartechnik mbH, Stuttgart. Beauftragt von Stad Freiburg.

Eisenhardt, K. M. (1989). Building theories from case study research. *Academy of Management Review, 14*(4), 532–550.

Ewing, R. (1997). Is Los Angeles-style sprawl desirable? *Journal of the American Planning Association, 63*(1), 107–126.

Ewing, R., & Rong, F. (2008). The impact of urban form on US residential energy use. *Housing Policy Debate, 19*(1), 1–30. https://doi.org/10.1080/10511482.2008.9521624.

Geurs, K., & van Wee, B. (2006). Ex-post evaluation of thirty years of compact urban development in the Netherlands. *Urban Studies, 43*(1), 139–160. https://doi.org/10.1080/00420980500409318.

Giddens, A. (1984). *The Constitution of Society: Outline of the Theory of Structuration.* Cambridge: Polity Press.

Giddens, A. (1991). *Modernity and Self-Identity.* Cambridge: Polity Press.

Gordon, P., & Richardson, H. W. (1989). Gasoline consumption and cities: A reply. *Journal of the American Planning Association, 55*(3), 342–346. https://doi.org/10.1080/01944368908975421.

Gordon, P., & Richardson, H. W. (1997). Are compact cities a desirable planning goal? *Journal of the American Planning Association, 63*(1), 95–106. https://doi.org/10.1080/01944369708975727.

Gram-Hanssen, K. (2014). New needs for better understanding of household's energy consumption – behaviour, lifestyle or practices? *Architectural Engineering and Design Management, 10*(1–2), 91–107. https://doi.org/10.1080/17452007.2013.837251.

Hagen, O. H., Tønnesen, A., & Fossheim, K. (2017). *Car-Free City Solutions in Three Nordic Cities.*

Holden, E., Banister, D., Gössling, S., Gilpin, G., & Linnerud, K. (2020). Grand Narratives for sustainable mobility: A conceptual review. *Energy Research and Social Science, 65*, 101454. https://doi.org/10.1016/j.erss.2020.101454.

Hughes, T. (1983). *Networks of Power: Electrification in Western Society 1880–1930.* Baltimore, MD: The Johns Hopkins University Press.

Hurlin, L. (2018). Mietshäuser syndikat: Collective ownership, the 'housing question' and degrowth. In A. Nelson & F. Schneider (Eds.), *Housing for Degrowth: Principles, Models, Challenges and Opportunities.* London: Routledge.

ICLEI. (2022). Urban transport and climate change. https://sustainablemobility.iclei.org/urban-transport-and-climate-change/, Accessed 25 March.

Jabareen, Y. R. (2006). Sustainable urban forms: Their typologies, models, and concepts. *Journal of Planning Education and Research, 26*(1), 38–52. https://doi.org/10.1177/0739456X05285119.

Jackson, T. (2005). Motivating sustainable consumption: A review of evidence on consumer behaviour and behavioural change. A Report to the Sustainable Development Research Network. London, January 2005.

Julsrud, T. E., & Farstad, E. (2020). Car sharing and transformations in households travel patterns: Insights from emerging proto-practices in Norway. *Energy Research and Social Science, 66*(2020), 101497. https://doi.org/10.1016/j.erss.2020.101497.

Kenworthy, J. R., & Newman, P. W. G. (1990). Cities and transport energy: Lessons from a global survey. *Ekistics, 34*(4/5), 258–268.

Madsen, L. V., & Gram-Hanssen, K. (2017). Understanding comfort and senses in social practice theory: Insights from a Danish field study. *Energy Research & Social Science, 29*, 86–94. https://doi.org/10.1016/j.erss.2017.05.013.

Mahzouni, A. (2017). The role of co-building groups in creating sustainable buildings and neighbourhoods: Lessons from Freiburg in Germany. In M. Young (Ed.), *AMPS Proceedings Series 9. Living and Sustainability: An Environmental Critique of Design and Building Practices, Locally and Globally* (pp. 394–402). London: London South Bank University, February 08–09, 2017. http://architecturemps.com/proceedings/

Mahzouni, A. (2018). Urban brownfield redevelopment and energy transition pathways: A review of planning policies and practices in Freiburg. *Journal of Cleaner Production, 195*(2018), 1476–1486. https://doi.org/10.1016/j.jclepro.2017.11.116.

Mahzouni, A. (2019). The role of institutional entrepreneurship in emerging energy communities: The town of St. Peter in Germany. *Renewable and Sustainable Energy Reviews, 107*(2019), 297–308. https://doi.org/10.1016/j.rser.2019.03.011.

Merriam, S. B. (2009). *Qualitative Research: A Guide to Design And Implementation.* San Francisco: John Wiley and Sons.

Miles, M. B., & Huberman, A. M. (1994). *Qualitative Data Analysis: An Expanded Sourcebook*. Thousand Oaks, CA: Sage Publications.

Moss, T. (2009). Intermediaries and the governance of sociotechnical networks in transition. *Environment and Planning A, 41*, 1480–1495.

Næss, P. (2005). Residential location affects travel behavior—but how and why ? The case of Copenhagen metropolitan area. *Progress in Planning, 63*(2005), 167–257. https://doi.org/10.1016/j.progress.2004.07.004.

Neuman, M. (2005). The compact city fallacy. *Journal of Planning Education and Research, 25*, 11–26. https://doi.org/10.1177/0739456X04270466.

Newman, P., & Kenworthy, J. (1989). Gasoline consumption and cities: A comparison of U.S. cities with a global survey. *Journal of the American Planning Association, 55*(1), 24–37. https://doi.org/10.1080/01944368908975398.

Newman, P., & Kenworthy, J. (1991). *Cities and Automobile Dependence: An International Sourcebook*. Avebury: Aldershot.

North, D. C. (1990). *Institutions, Institutional Change and Economic Performance*. Cambridge University Press, Cambridge.

O'Brien, W. T., Kennedy, C. a., Athienitis, A. K., & Kesik, T. J. (2010). The relationship between net energy use and the urban density of solar buildings. *Environment and Planning B: Planning and Design, 37*, 1002–1021. https://doi.org/10.1068/b36030.

Pierson, P. (2000). Increasing returns, path dependence, and the study of politics. *The American Political Science Review, 94*(2), 251–267.

Projektgruppe Dietenbach. (2020). *Dietenbach: Städtebaulicher Rahmenplan- Erläuterungsbericht* [*Urban Development Master Plan of Dietenbach: Explanatory Report*]. 210 Seiten, November 2020, Stadt Freiburg.

Rickwood, P., Glazebrook, G., & Searle, G. (2008). Urban structure and energy – A review. *Urban Policy and Research, 26*(1), 57–81. https://doi.org/10.1080/08111140701629886.

Sareen, S., Remme, D., Wågsæther, K., & Haarstad, H. (2021). A matter of time: Explicating temporality in science and technology studies and Bergen's car-free zone development. *Energy Research and Social Science, 78*(2021), 102128. https://doi.org/10.1016/j.erss.2021.102128.

Shaheen, S., Chan, N., Bansal, A., & Cohen, A. (2015). *Shared Mobility: Definitions, Industry Developments, and Early Understanding*. 3 November 2015, Transportation Sustainability Research Center, University of California, Berkeley.

Shove, E. (2010). Beyond the ABC: Climate change policy and theories of social change. *Environment and Planning A, 42*(6), 1273–1285.

Shove, E., Pantzar, M., & Watson, M. (2012). *The Dynamics of Social Practice: Everyday life and How it Changes*. London: SAGE Publication.

Sovacool, B. K., & Griffiths, S. (2020). The cultural barriers to a low-carbon future: A review of six mobility and energy transitions across 28 countries. *Renewable and Sustainable Energy Reviews, 119*(2020), 1–12. https://doi.org/10.1016/j.rser.2019.109569.

Spaargaren, G. (2003). Sustainable consumption: A theoretical and environmental policy perspective. *Society and Natural Resources, 16*, 687–701. https://doi.org/10.1080/08941920309192.

Spaargaren, G., & van Vliet, B. (2000). Lifestyle, Consumption and the environment: The ecological modernisation of domestic consumption. *Society and Natural Resources, 9*(1), 50–76.

Späth, P., & Rohracher, H. (2010). 'Energy regions': The transformative power of regional discourses on socio-technical futures. *Research Policy, 39*(4), 449–458. https://doi.org/10.1016/j.respol.2010.01.017.

Stake, R. E. (1995). *The Art of Case Study Research*. Thousand Oaks, CA: SAGE Publications.

Williams, K. (2014). Urban form and Infrastructure: A Morphological Review. June 2014, Technical Report. Foresight, Government Office for Science. University of the West of England, Bristol. http://eprints.uwe.ac.uk/24989.

Williams, K., Burton, E., & Jenks, M. (2000). Conclusions. In K. Williams, E. Burton, & M. Jenks (Eds.), *Achieving Sustainable Urban Form* (pp. 347–355). London: E & FN Spon.

Yin, R. K. (2003). *Case Study Research: Design and Methods*. 3th ed. Thousand Oaks: Sage Publications.

Yin, R. K. (2009). *Case Study Research: Design and Methods*. 4th ed. Thousand Oaks: Sage Publications.

23

ENERGY-SERVICES-LED TRANSFORMATION

C. Wilson, A. Grubler, and C. Zimm

Definition of Key Terms and Acronyms

decent living standards (DLS): a bundle of goods and services that constitute minimum thresholds for satisfying human needs

energy services: useful functions, such as mobility and thermal comfort, that meet human needs and provide for well-being

GHGs: greenhouse gases

ICTs: information and communication technologies

SDGs: (United Nations) Sustainable Development Goals

service provisioning: combinations of technologies, infrastructures, markets, and business models that make energy services available to final users

sufficiency: consuming fewer services without adversely impacting well-being

sufficiency corridor: the range of service levels that ensure decent living for all while remaining within planetary boundaries

---

In line with the focus of this *Handbook*, we define an *energy transition* as a notable qualitative, quantitative, or geospatial shift in how energy is sourced, delivered, or utilized. It can also include the formation, evolution, or phase-out of energy industries or clusters. By focusing on energy services, we emphasize the importance to energy transitions of the final conversion step of energy into useful functions like mobility or thermal comfort that meet human needs.

Introduction: The Building Blocks of Energy-Services-Led Transformation

Providing Useful Services to Improve Well-Being While Reducing Resource Use

This chapter uses "services" as an analytical entry point into the challenge of rapid and sustainable energy transitions to meet global climate targets and the wider United Nations Sustainable Development Goals

DOI:10.4324/9781003183020-27

(SDGs).[1] Services include nutrition, mobility, shelter, thermal comfort, entertainment, and social inter-action, all of which provide for human needs and well-being. The provisioning systems which deliver these services require energy, material, and land resource inputs. The use of these resources – par-ticularly in the form of fossil fuel combustion, agricultural practices, and forest conversion – generate greenhouse gas (GHG) emissions and other forms of pollution, which negatively impact well-being. By crossing planetary boundaries that define "the safe operating space for humanity with respect to the earth system" (Rockstrom et al. 2009), excessive resource use also undermines the future capacity of provisioning systems to continue to meet human needs. Services are therefore a bridge between human needs and well-being on the one hand, and resource use and GHG emissions on the other.

From this perspective, the challenge for sustainable energy transitions is this: How can human needs be satisfied and well-being improved while using fewer resources to provide useful energy services? Answering this question is a multidisciplinary endeavor, drawing on methods and data from energy systems analysis and engineering, concepts and tools from sustainability science and international development, and insights from welfare economics and psychology.

Focusing on useful services is important for various reasons. First, services are a direct antecedent of needs satisfaction and well-being fulfillment, which is the ultimate purpose of the energy system. More traditional economic concepts of demand apply to bundles of goods (and services) acquired through market or nonmarket transactions. But demand is an intermediate not a final step. Human needs are met by the useful services provided by demand (Grubler et al. 2012). Second, services as an entry point follows sustainable design principles by first exploring how to size the system appro-priately before then exploring how to make that system function sustainably and effectively. Third, services recognize that how energy is used (to provide useful functions) creates pressures and needs for how energy is sourced and delivered. Fourth, services help broaden the search for solutions beyond dominant supply-side decarbonization and technological mitigation options.

Figure 23.1 shows the basic organizing framework we use in this chapter. It describes human needs being satisfied by useful services made available by service-provisioning systems, which require resource inputs. These are the four main building blocks of our argument.

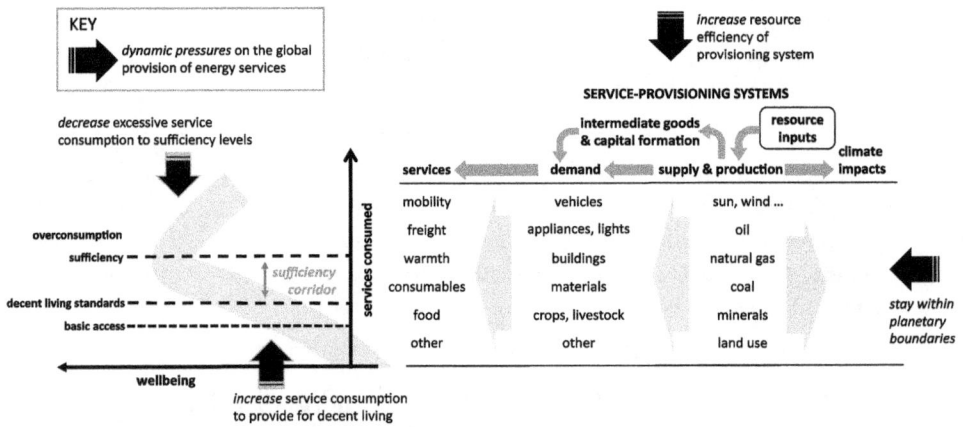

Figure 23.1 Transforming how services are provided to meet human needs with fewer resource inputs. The left panel shows a stylized relationship between levels of service provision and human well-being. The right panel shows the resource processing flows (Sankey diagram) through the global service provisioning system. Thick arrows show dynamic pressures to ensure sus-tainable energy transition.

Sources: Adapted from TWI2050 (2020).

The thick black arrows in Figure 23.1 shows the dynamic pressures that define the challenge of sustainable energy transition: *downward* pressure on service-provisioning systems to use fewer resources, *upward* pressure on the provision of services to satisfy basic needs and achieve decent living for all, and *downward* pressure on excessive consumption of services that do not improve well-being.

These pressures are dynamic for two reasons. First, they are shaped by cultural, political, and technological contexts that are constantly changing. In one context, poverty eradication may create pressure to increase access to services and consumption levels, whereas in a different context, tackling obesity may create pressure to decrease harmful overconsumption of services. Second, they define directions of travel, not fixed destinations. Although planetary boundaries for pollution, climate change, and global nutrient cycles define absolute constraints or upper thresholds, negative impacts are reduced at lower levels of resource use.

Human Needs and Well-Being

Providing for human needs is at the heart of sustainable development. There are many different conceptual and empirical approaches to needs. Maslow (1943)'s widely used hierarchy of needs distinguishes *basic* needs for food, shelter, and security, *intermediate* needs for belonging, connectedness, and self-esteem, and *higher* needs for self-actualization including through learning, creativity, and morality. Max-Neef (1991) developed a similar typology of fundamental human needs consisting of subsistence, protection, creation, freedom, leisure, identity, understanding, and participation. How these needs are satisfied is determined differently by individuals and groups, and with widely varying implications for resource consumption and sustainability. Using Max-Neef's typology of needs, Vita et al. (2019) found that meeting human needs for subsistence and protection accounted for nearly half of global GHG emissions, with most of the remaining emissions accounted for by freedom, identity, creation, and leisure needs. Meeting needs for understanding and participation accounted for less than 4% of global emissions.

Satisfying needs ensures human well-being and quality of life. There are two distinct conceptualizations of well-being: hedonic and eudaemonic. Hedonic well-being is a subjective state of motivation and pleasure, linked to happiness and life satisfaction, and is commonly measured (subjectively) through self-reports. Eudaemonic well-being is associated with fulfillment and self-realization, and places greater emphasis on the social and material context enabling people to flourish. Eudaemonic well-being is embedded in development strategies to enhance human capabilities (Sen 2000) or otherwise meet human needs (Doyal and Gough 1991).

The left panel of Figure 23.1 presents a stylized representation of the relationship between services and human well-being. Its exact shape will vary according to the type of service and the indicator of well-being being considered. Increases in well-being per additional unit of services consumed are high at low levels of service provision, declining to a peak at the upper bound of decent living standards, after which excessive service consumption causes well-being to decline.

Basic Access to Services and Decent Living Standards

Ensuring basic access to services is the most important step for ensuring human well-being. Lack of access to safe drinking water and sanitation, undernourishment, or lack of access to basic health care services are examples of inadequate or absent service provisioning that negatively impact well-being.

Globally, basic needs are not universally met, capabilities are insufficiently developed, and inequalities persist along multiple dimensions including health, relationships, safety, ability to have influence, knowledge, and financial security (UNDP 2019). Inequalities undermine resilience and place greater risk and exposure to external shocks on those with limited resources (Leach et al. 2018). The COVID-19 pandemic has amplified inequalities, for example, by disproportionately endangering people living in conditions that do not allow for social distancing or access to health care facilities.

Although the provision of access to basic services is essential for survival, it does not necessarily allow full participation in economic activities, society, and self-fulfillment. Rao and Min (2017) developed a quantitative framework of the bundle of services and material resource prerequisites for human well-being that they called decent living standards (DLS). We discuss these further in the next sections. As shown in the left panel of Figure 23.1, moving from basic access to DLS yields large benefits for human well-being.

Sufficiency and Overconsumption

Beyond DLS is a level of service provision labelled "sufficiency" in Figure 23.1. Here, there are diminishing returns to well-being from ever-higher levels of service provision and consumption. Sufficiency levels are difficult to define *ex ante* as they are subject to values and expectations that vary widely across individuals, social groups, and cultures. Yet there are two important social and environmental constraints that influence sufficiency levels. First, Pareto efficiency principles mean that improvements in service consumption for one individual should not diminish anyone else's consumption. Second, planetary boundaries mean there are absolute limits on resource use that should not be breached (Rockstrom et al. 2009).

With significant technological, social, and behavioral innovations to minimize resource use and environmental impacts, it is possible to provide sufficient levels of service while remaining within planetary boundaries and also ensuring equitable access to natural resources (Grubler et al. 2018). This is represented as the sufficiency corridor in Figure 23.1. Beyond sufficiency, high levels of service consumption, labelled "overconsumption" in Figure 23.1, diminish well-being either at the level of the individual or at the level of society.

Consider motorized transport with privately owned vehicles. At high levels of vehicle ownership and use, particularly in densely populated urban environments, the benefits of a convenient and flexible individual mode of transport are quickly counterbalanced by congestion, urban air pollution, and noise that diminishes not only the utility of vehicle ownership and use, but also the well-being of the individual car user and all urban residents. A high level of consumption that is therefore neither efficient in terms of individual well-being nor in terms of Pareto efficiency is classified as "overconsumption".

Nutrition is another example.[2] Food consumption exceeding physiological needs (especially high-energy food like fats and sugars) is associated with obesity, especially when combined with a lack of exercise. In 2016, approximately two billion adults (~40%) worldwide were considered overweight, and 650 million (13%) were classified as obese (WHO 2020). At the same time, over 800 million people go hungry and over 150 million children are stunted through malnutrition (WHO 2020). Obesity rates in countries like the US, Saudi Arabia, and South Africa approach or even surpass one-third of the child and adult population. This is a major health risk for individuals, reducing life expectancy and well-being. It also creates pressure for the provisioning systems of health care services.

Service-Provisioning Systems

Useful services are delivered by provisioning systems. Service-provisioning systems are (alternative) combinations of technologies, infrastructures, markets, and business models that make useful services available at the point of consumption. They include a wide variety of actors, from manufacturers to local governments and from entrepreneurs to standards organizations, operating in particular institutional and physical contexts, from regulatory frameworks to transport networks.

The service concept is well established in some sectors like education, with a provisioning system of schools, national curricula, and teachers, or like health, with a provisioning system of hospitals, medical treatments, nurses, and safety regulators. Service concepts were first applied to environmental and sustainability challenges in the energy literature in the 1990s (Nakićenović, Gilli, and Kurz 1996; Nakicenovic et al. 1993). But they are still not widely taken up in sectors like transport or consumer goods, in which the misleading idea still persists that well-being is derived from buying and owning a product like a car or a computer rather than from the useful services like mobility or entertainment that those products can provide.

The resource flow (Sankey) diagram in the right panel of Figure 23.1 shows how the global provisioning system consumes energy, materials, biomass, and other resources to deliver services, including mobility, warmth, and nutrition in domestic and other settings (Cullen and Allwood 2010). Also shown on the far right of Figure 23.1 are the climate impacts caused by GHG emissions caused by this use of resources, dominated by fossil fuel combustion and land-use change.

Not all resources are converted into final services. Mobility, not cars, meets people's needs to move around in order to work, socialize, and recreate. Passenger-kilometers (p-km) of mobility is a direct measure of useful service rather than numbers of cars or quantities of fuel. However, the delivery of mobility clearly requires technologies (vehicles), infrastructure (roads, bridges), and resources (fuels) in service-provisioning systems. Demand for cars, roads, and fuel is therefore an intermediate step toward providing the useful service of mobility.

Intermediate goods, including the formation of capital, such as roads, buildings, and broadband networks, facilitate the provision of final services and so meet human needs indirectly rather than directly. Other examples of intermediate goods required upstream (rather than downstream at the point of consumption) include construction materials, freight transport, manufacturing operations, and so on.

Improving the Resource Efficiency of Service-Provisioning Systems

Service-provisioning systems are highly inefficient and wasteful. Globally, only 14% of the energy resources harvested from nature deliver useful services to final users (72 of 511 EJ) – the remaining 86% is wasted (De Stercke 2014). Figure 23.2 (panels A–D) represents the cascading of losses at each conversion stage from energy resources (primary energy) all the way down to final services. Although there are thermodynamic limits on how much this wastage can be reduced, there are enormous potentials for improving the efficiency of the global energy system as a service-provisioning system to meet human needs (Cullen, Allwood, and Borgstein 2011).

More recently, the narrow emphasis on energy has broadened to include materials and other resources. A world on its way to net-zero GHG emissions means declining fossil fuel combustion. This remains the case even if carbon capture, utilization, and storage expand dramatically. Consequently, the consumption of materials, biomass, land, and other resource inputs will become more and more significant as a cause of environmental degradation. The term "resource-intensive services" better captures this broadened focus, although "energy services" is often used interchangeably.

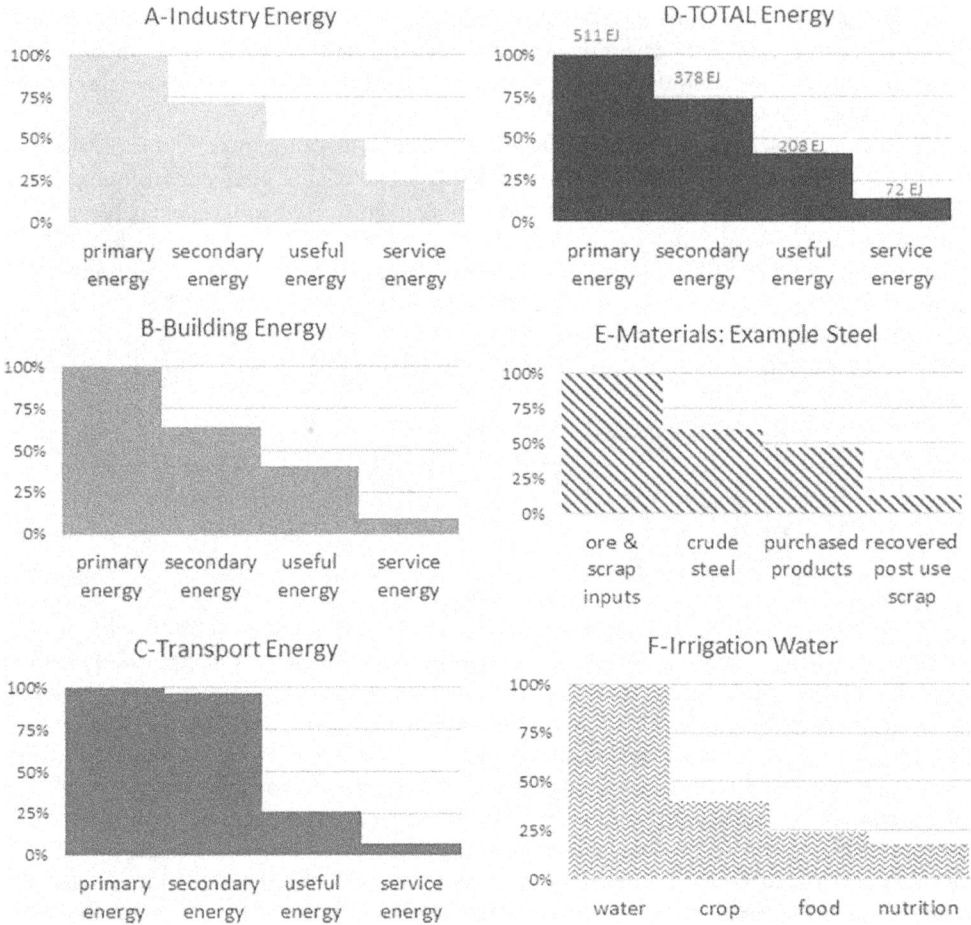

Figure 23.2 Global resource efficiency cascades throughout the service provisioning system for energy both by sector (panels A–C) and overall (panel D), water embodied in food (panel E), and materials in steel manufacturing, use and recycling (panel F) for 2020. Each step of the cascade shows the percentage of the extracted primary resource remaining after conversion losses. Estimates for energy from De Stercke (2014). Estimates for embodied water in food production, processing, and consumption based on Lundqvist, De Fraiture, and Molden (2008) and Sadras, Grassini, and Steduto (2011). Estimates for are steel from Allwood et al. (2012) and Cullen (pers. comm.), as well as Ayres and Simonis (1994) and Fischer-Kowalski et al. (2011).

Source: Reproduced from TWI2050 (2018).

Figure 23.2 (panels E and F) shows similar conversion cascades in the service-provisioning systems for water resources (in food) and materials (in steel). Invariably, accumulated conversion losses are substantial, ranging from 83% (water) to 86% (energy) and 87% (steel) of primary resource inputs. In all cases less than one-fifth of the primary resources extracted from nature remain in the final service that meets human needs. The remainder is dissipated to the environment. As Figure 23.2 shows, a substantial portion of these losses happens at the level of end-use in the way final services are delivered and consumed (60% of aggregate system losses for energy, 47% for steel, and 23% for water embodied in food).[3]

The central challenge for services-led transformation is to reduce resource inputs to service-provisioning systems while expanding the capacity of those systems to deliver useful services to meet human needs. There are two types of strategy:

- *Demand-side strategies* change how services are delivered (e.g., more energy-efficient end-use technologies and infrastructure, digitalization, business models to increase efficient utilization of resources).
- *Supply-side strategies* change how resources are converted (e.g., precision agriculture, decarbonization of power production) and how intermediate goods and capital are formed (e.g., net-zero energy buildings, circular economies, dematerialization, urban design).

This chapter focuses on demand-side strategies, while also recognizing the interdependencies between how useful services are consumed and how they are provided. Evidence to date shows that demand-side strategies have stronger synergies and fewer trade-offs with SDGs than supply-side strategies (Rogelj et al. 2018).

Ensuring Access to Basic Services and Providing for Decent Living Standards

Access to Basic Services

The UN 2030 Agenda and SDGs provide a broad framework that targets inclusive, equitable, and universal access to basic infrastructure and services for human needs, acknowledging a multidimensional perspective of poverty and human needs. Energy access enables many of these basic services. Indicators of access show progress has been insufficient, and there is wide variation in access to basic services within countries and regions, and between urban, rural, and other population subgroups.

The UN Sustainable Development Goals Report in 2019 highlights the remaining gap between SDG ambitions and current achievements.

With respect to basic energy services, over three billion people lack access to clean cooking or essential health services, and around 800 million lack access to basic electricity – according to data from the UN, WHO, and World Bank that reveal these large inequalities (Zimm 2019).

Achieving the SDGs will require additional capacities, improved governance, and increased investment. In the developing countries of Asia-Pacific alone, additional annual investments of $1.5 trillion are needed to end extreme poverty through provision of power stations, universal health coverage, quality education, and enabling infrastructure, such as roads, rail, ports, and sanitation facilities (UNESCAP 2019).

Bridging the gap between SDG ambition and achievement requires new approaches that go beyond extending supply infrastructure. Integrated planning for service provision requires linking supply to broader developmental and socioeconomic objectives. Providing access to basic services is a fundamental first step to enhancing well-being. But extending access to energy, education, health care, water, sanitation, food, housing, and other basic goods and services is not on its own sufficient unless it is also affordable, reliable, of good quality, and sustainable.

Although efforts to expand access to electricity in sub-Saharan Africa have not kept pace with population growth in the region, more than 115 million people gained access over the period of 2014–2019. At a subnational scale, this progress has been uneven (Falchetta et al. 2019) and the access to electricity that has been achieved varies widely in quality (Falchetta et al. 2020).

In countries like Kenya, estimated final services among newly electrified households remain very limited (Fobi et al. 2018). Low service levels indicate that people have not moved beyond subsistence uses for lighting and phone charging to levels of demand that provide a means of livelihood through enhanced employment, education, and income-earning opportunities.

Accelerating efforts to provide reliable and affordable access to basic services requires rapid innovation in technologies, financing, and distribution models. Disruptive and digital technologies can increase access to basic goods and services and directly contribute to achieving many of the SDGs (TWI2050 2019). Recent ICT innovations, particularly mobile phones and virtual financial services, help deliver essential services like e-governance, energy, education, health, water, and financial inclusion (Alstone, Gershenson, and Kammen 2015). Satellite datasets and mobile phones have been used to monitor access and estimate road quality (Cadamuro, Muhebwa, and Taneja 2019), electricity supply outages (Correa, Klugman, and Taneja 2018), and latent demand for electricity (Falchetta et al. 2020; Fobi et al. 2018). Providing services through decentralized infrastructure in combination with highly efficient end-use appliances and ICTs is also quicker to scale up and diffuse (Wilson et al. 2020). For example, direct current (DC) appliances and productive equipment in conjunction with off-grid solar energy technologies can provide reliable and affordable energy services even to remote rural communities (Phadke et al. 2017).

Decent Living Standards

Going beyond providing access to basic services, everyone should have the means to pursue a decent life and avoid harm from extreme weather, disease, and pollution (Doyal and Gough 1991). Decent living means access to amenities, including safe and un-cramped shelter, adequate nutrition and water supply, clothing, health care, and basic comforts in the home, such as lighting and thermal comfort (including water heating), refrigerators, and clean cooking. Decent living also means the ability to engage with society, communicate with others, and seek knowledge about the world, which in turn means access to education, mobility, devices in the home to communicate (e.g., mobile phones), and broadcast media (e.g., television).

The components of a decent life are reasonably well established in literature (Alkire and Robles 2017; Alkire and Santos 2014). The exact quantities of such living standards are subjective and vary both within and between populations, cultures, time periods, and places. Yet there is enough commonality across humanity to identify a minimum set of core requirements (Doyal and Gough 1991). To provide a first generalization, Rao and Min (2017) developed a framework and quantitative estimates for decent living standards (DLS) covering physical and social well-being and the necessary goods and services to be accessible at household and community level.

Household-level DLS cover the following:

- *Nutrition, including food and cold storage.* People need the opportunity to follow an adequate, nutritious, and balanced diet with a limited share of meat derived calories, with access to a refrigerator for food storage. DLS food requirements are specified in kcal per capita per day and aggregated based on the population structure of each country (men, women, and children have different calorie requirements). More detailed studies have also specified micronutrient levels (e.g., DeFries et al. [2018] for India).
- *Shelter and living conditions.* Durable walls and roof, and a minimum size of 30 m² per household (or 10 m² per person) to accommodate a bathroom and kitchen with adequate space. Basic comforts include electricity and clean cooking devices, inhouse sanitation and

freshwater access, and devices to provide lighting, hot water, and thermal comfort if applicable (e.g., fans, heating, or air-conditioning).

- *Clothing.* Minimal material requirements to dress oneself suitable for the prevailing climate and culture.
- *Information and communication.* One mobile phone per adult and one television set per household to enable communication, information, and engagement with society.
- *Mobility.* Access to either private or public motorized transport depending on local conditions to ensure livelihood opportunities.

These household-level goods and services are supplemented by community-level DLS necessary for people to develop their capabilities and thrive. Community-level DLS cover accessible and adequate health care services, primary and secondary education, and public infrastructure (electricity, water, sewage, public transport and road infrastructure, ICT networks).

The Resource Requirements of Decent Living Standards

The DLS are universal minimum requirements but can be adapted to local conditions when applied to estimate country-specific thresholds. For example, a universal standard for adequate floor space, durable housing, and thermal comfort translates to regionally differentiated construction materials (depending on building practices and standards) and space heating and cooling requirements (depending on climatic conditions). Similar differences apply to modal choice (mobility) or diets (nutrition). Quantitative service thresholds for DLS in India, Brazil, and South Africa were estimated by Rao, Min, and Mastrucci (2019) as follows:

Nutrition: An average of 2,500 kcal per person (accounting for age and gender) with the share of animal protein differing across countries. This is in contrast with current household data that shows food deprivation and micronutrient deficiency (e.g., zinc).

Shelter and living conditions: An annual energy demand of 1 MWh per household, which would cover an LPG cookstove (45% efficiency), an LCD television set, phone charging, electric lighting with a minimum illuminance level of 150 lux, and a 150-liter refrigerator. Additional thermal energy requirements for water heating, air-conditioning, or space heating would depend on climatic conditions. Improved efficiency and longer lifetimes for lighting and appliances are needed to expand service provision while reducing energy and material demand. For example, currently only 25–32% of households in India, Brazil, and South Africa use LEDs for lighting.

Sanitation and water: Sanitation facilities at home and a minimum quantity of 65 liters of running water per capita per day (based on Gleick 1996).

Clothing and footwear: A requirement of 1.3 kg of clothing and 0.9 kg of footwear in India, compared with 2.3 kg of clothing and 1.4 kg of footwear in Brazil (estimated from observed quantities in household surveys for these two countries). Clothing needs vary across climates.

Shelter: Single-story masonry structures in rural areas and four-story reinforced concrete structure in urban areas as low-energy building archetypes (Mastrucci and Rao 2019) using durable materials and local construction practices (Bansal, Singh, and Sawhney 2014; Paulsen and Sposto 2013).

Mobility: A minimum of 10,000 p-km (including for leisure travel), which is roughly the average service in an efficient, densely populated, affluent economy like Japan (Rao and Baer 2012). Service-provisioning systems for mobility vary widely based on existing preferences, public transport options, and historical accumulation of infrastructure. In India, two-wheelers make up around three-quarters of the total stock of light-duty vehicles, but as incomes rise, passenger cars are growing in share. For more mobility to be sustainable, public and shared modes need to play a larger role.

These minimum thresholds for DLS reveal the prevailing extent of multidimensional poverty (Rao, Min, and Mastrucci 2019). Many more people lack DLS than are defined as income poor by the World Bank's International Poverty Line (2011 PPP $1.90 per day). In India, 15–93% of the population lack various elements of DLS, in contrast to 20% of income poor. As an example, more than 93% of Indians (over a billion people) lack access to space cooling to avoid heat-induced adverse health effects, 70% of households do not own a refrigerator, and 4% have no access to any source of electricity (Mastrucci et al. 2019).

It is also important to note that the DLS are a floor or minimum threshold, not a target or a ceiling. They are far below the service levels prevailing in the Global North and (increasingly) wealthier populations elsewhere. But they are also well above anything resembling poverty. Average future living standards may exceed the DLS, and in a changing technological and social landscape, what is considered decent or what can be provided in a sustainable manner will itself change.

How much energy would be required to provision universal DLS? In a stylized scenario experiment, Millward-Hopkins et al. (2020) show that, at least in theory, global energy demand would be significantly lower than today. Historically, however, the lifestyles people adopt as their income increases and their opportunities expand have led to material and energy demand growth (Creutzig et al. 2018; Semieniuk et al. 2021). The services-led transformation set out in this chapter as the basis for a sustainable energy transition implies dramatically higher resource efficiency into the future. As examples, future energy and material demands could be minimized if slums and poor-quality rural homes were upgraded with energy-saving construction practices, and mobility demand was met through public and collective transportation (Mastrucci and Rao 2019). Means of social affiliation, including basic education and access to broadband and social media, require just a few gigajoules of energy per capita.

Reducing Overconsumption Toward Sufficiency of Service Consumption

Overconsumption of Useful Services, Sufficiency, and Well-Being

Sufficiency concepts are based on evidence for decreasing marginal returns to consumption of services and goods that suggests an ultimate upper floor to consumption (represented stylistically in Figure 23.1). The statistician Ernst Engel first demonstrated that the share of food expenditure decreases with rising incomes (Engel 1857). More recently, Easterlin (1973) and Easterlin et al. (2010) observed that happiness or self-reported well-being saturates or even diminishes beyond certain income thresholds. It is important to note that these empirical findings typically apply to income expenditure on material goods rather than, for example, time spent consuming services associated with the higher levels of Maslow's hierarchy, which can also be resource-intensive (e.g., long-distance travel for cultural or recreational purposes). The sufficiency concept is also consistent with limits to growth arguments originally proposed by Thomas Malthus (1798) and popularized more recently in the 1970s, with growing concerns over overconsumption of resources and potential environmental collapse (Meadows et al. 1972). Similar ideas are embedded in the planetary boundaries framework (Rockstrom et al. 2009) and doughnut economics (Raworth 2017).

Diminishing marginal utilities of consumption imply an upper bound on the ability of more services to meet more needs or provide more well-being. There are two further reasons for an upper bound on service consumption. First, if the resource costs and environmental impacts of service-provisioning systems are factored into the well-being calculus, then ever-higher levels of services imply ever more degradation of natural systems, which undermines well-being for all (Knight and Rosa 2011). Second, well-being is experienced in relative and absolute terms. As the others' well-being cannot be observed, it is perceived through proxies such as income, material wealth, and service consumption. Inequalities in any of these measures can therefore undermine subjective well-being (Wilkinson and Pickett 2010). Global information flows mean that referent social groups against which people compare their own service consumption are increasingly distant elites. This further polarizes perceived inequality.

Sufficiency means consuming less in absolute terms or placing upper limits on consumption and is an intentional contrast to efficiency, which implies doing more with less in relative terms. First introduced by Sachs (1993) and popularized by Princen (2005), sufficiency is a common articulation of the need for downward pressure on high levels of service consumption, considering human needs and well-being in the broader context of planetary boundaries (Rijnhout et al. 2018). Sufficiency builds on long-standing arguments for resource conservation and curtailment to avoid unnecessary waste but has a stronger normative meaning in which more becomes too much and is not only undesirable but also morally wrong and unjust.

Sufficiency Levels in Energy Transition Scenarios

Various national and global energy transition scenarios have put forward sufficient levels for energy service consumption, particularly thermal comfort and mobility. At the national level, one of the first energy transition scenarios that explicitly defined sufficiency levels linked to human well-being was the French négaWatt scenario developed in 2003 and regularly updated since (négaWatt 2017). This scenario halves final energy demand while maintaining high-energy service levels to provide for human needs. Other national studies have taken a sectoral focus. One example is a sustainable mobility scenario for Scotland in which a combination of lifestyle changes and sufficiency limits constrain annual mobility service consumption to under 10,000 p-km per capita (Brand, Anable, and Morton 2019).

Global scenario studies either explicitly or implicitly assuming sufficiency levels are more problematic because of the diversity of economic, geographic, and cultural context shaping how much service is sufficient. To explore this space, Grubler et al. (2018) developed a global low-energy-demand (LED) scenario with dramatically higher resource efficiencies allowing service consumption to expand in the Global South without overshooting the 1.5°C warming limit. Millward-Hopkins et al. (2020) developed a much more stringent (and hypothetical) global scenario for decent living with minimum energy inputs.

The sufficiency thresholds in these studies vary as a function of scenario assumptions and geographic scope. The French négaWatt scenario assumes sufficiency in service consumption of 42 m^2/capita (housing space), 15 kWh/m^2/year (residential heating), 10 kWh/m^2/year (residential cooling), and 15,000 p-km/year (mobility) (négaWatt 2017). The global LED scenario assumes upward trends in the Global South converging on downward pressure on overconsumption in the Global North toward sufficiency bounds of 29 m^2/capita (housing space), 21 kWh/m^2/year (heating and cooling), and 9,500–17,000 p-km/year (mobility) (Grubler et al. 2018). The much more stringent (and hypothetical) global decent living scenario places sufficiency thresholds at 15 m^2/capita, 10–13 kWh/m^2/year (heating), 10–14 kWh/m^2/year (cooling), and 4,900–15,000 p-km/year (mobility) (Millward-Hopkins et al. 2020).

How quantitative lower thresholds for "decent living" and upper thresholds for "sufficiency" are perceived will depend on current status, social expectations, and development stage. Decent living for some may mean little more than minimal subsistence while sufficiency for others may mean opportunities and choices over a wide set of goods and services. Setting expectations in sustainable energy transitions for the "sufficiency corridor" shown in the left panel of Figure 23.1 necessarily requires consideration of social and cultural identities in patterns of product choice and service consumption. Perceptions of needs and wants are shaped by available opportunities communicated through social networks, media, advertising, travel, and cultural transmission. Distributional equity and fairness also shape perceptions, which are shaped by observations of referent groups or aspirational others.

What constitutes decent living and sufficiency is an ongoing process of assessment and adjustment from local to global scales. Open platforms for societal debates can help facilitate this process, revealing areas of synergies and trade-offs and emphasizing the future as open possibility not deterministic outcome of historical development trajectories (Pereira et al. 2018).

Conclusions

We end with the question we posed at the beginning: How can human needs be satisfied and well-being improved while using fewer resources to provide energy services? This is the essential challenge for sustainable energy transitions.

The Pareto efficiency principle means that increasing one person's service consumption and resulting well-being should not diminish others' service consumption given planetary boundaries. This leaves the resource efficiency of provisioning systems as the only malleable variable for ensuring universal access to the "sufficiency corridor" of service consumption (Figure 23.1). So as a pragmatic answer to this question, we propose the following:

> Across different social and spatial scales, a minimum threshold or floor for service consumption, defined by decent living standards (DLS), is first satisfied. Once this level is reached, any remaining resource use that would not violate planetary boundaries is made available to increase service consumption and resulting wellbeing in line with cultural, social, regional and historical circumstances under some equity principle. O'Neill et al. (2018), for example, define a desirable social threshold for distributional equity as a Gini coefficient lower than 0.3, which is seen in countries with more equitable income distributions. (Gini coefficients range from 0–1 with smaller numbers representing more equitable distributions, usually of income, but can also be applied to access to energy services). While planetary boundaries define an upper bound to resource use, the context-specific distribution of services provided by those resources is guided by equity principles. These two sets of constraints leave efficiency of service provision as the principle 'lever' for driving sustainable energy transitions. The global LED scenario sets out one such vision for how behavioral, organizational, and technological innovations can be marshalled to maximize service provision while remaining within planetary boundaries.
>
> (Grubler et al. 2018)

Notes

1 The ideas and arguments presented in the chapter draw on our earlier exposition of energy-services-led transformation in Chapter 3 of *The World In 2050* (2020). Innovations for Sustainability: Pathways to an efficient and sufficient post-pandemic future. Laxenburg, Austria, International Institute for Applied Systems Analysis (IIASA). doi.org/10.22022/TNT/07–2020.16533.

2 Food systems transitions are covered in depth in the chapter by Pereira et al. on "The Role of Food Systems in Energy Transitions".

3 Material transitions are covered in depth in the chapter by Zotin et al. on "Material transitions".

References

Alkire, Sabina, and Gisela Robles. 2017. "Multidimensional poverty index summer 2017: Brief methodological note and results." *OPHI Methodological Notes* 45.

Alkire, Sabina, and Maria Emma Santos. 2014. "Measuring acute poverty in the developing world: Robustness and scope of the multidimensional poverty index." *World Development* 59:251–274.

Allwood, Julian M., Jonathan M. Cullen, Mark A. Carruth, Daniel R. Cooper, Martin McBrien, Rachel L. Milford, Muiris C. Moynihan, and Alexandra C. H. Patel. 2012. *Sustainable Materials: With Both Eyes Open*. UIT Cambridge Cambridge.

Alstone, Peter, Dimitry Gershenson, and Daniel M. Kammen. 2015. "Decentralized energy systems for clean electricity access." *Nature Climate Change* 5 (4):305–314. doi: 10.1038/nclimate2512 www.nature.com/nclimate/journal/v5/n4/abs/nclimate2512.html#supplementary-information.

Ayres, Robert U., and Udo Ernst Simonis. 1994. *Industrial Metabolism: Restructuring for Sustainable Development*. Vol. 376: United Nations University Press Tokyo.

Bajželj, Bojana, Julian M. Allwood, and Jonathan M. Cullen. 2013. "Designing climate change mitigation plans that Add Up." *Environmental Science & Technology* 47 (14):8062–8069. doi: 10.1021/es400399h.

Bansal, Deepak, Ramkishore Singh, and R. L. Sawhney. 2014. "Effect of construction materials on embodied energy and cost of buildings – A case study of residential houses in India up to 60 m2 of plinth area." *Energy and Buildings* 69:260–266.

Brand, Christian, Jillian Anable, and Craig Morton. 2019. "Lifestyle, efficiency and limits: Modelling transport energy and emissions using a socio-technical approach." *Energy Efficiency* 12 (1):187–207. doi: 10.1007/s12053-018-9678-9.

Cadamuro, Gabriel, Aggrey Muhebwa, and Jay Taneja. 2019. "Street smarts: Measuring intercity road quality using deep learning on satellite imagery." In Proceedings of the 2nd ACM SIGCAS Conference on Computing and Sustainable Societies.

Correa, Santiago, Noah Klugman, and Jay Taneja. 2018. "How Many Smartphones Does It Take To Detect A Power Outage?" Proceedings of the Ninth International Conference on Future Energy Systems.

Creutzig, Felix, Joyashree Roy, William F. Lamb, Inês M. L. Azevedo, Wändi Bruine de Bruin, Holger Dalkmann, Oreane Y. Edelenbosch, Frank W. Geels, Arnulf Grubler, Cameron Hepburn, Edgar G. Hertwich, Radhika Khosla, Linus Mattauch, Jan C. Minx, Anjali Ramakrishnan, Narasimha D. Rao, Julia K. Steinberger, Massimo Tavoni, Diana Ürge-Vorsatz, and Elke U. Weber. 2018. "Towards demand-side solutions for mitigating climate change." *Nature Climate Change* 8 (4):268–271. doi: 10.1038/s41558-018-0121-1.

Cullen, Jonathan M., and Julian M. Allwood. 2010. "The efficient use of energy: Tracing the global flow of energy from fuel to service." *Energy Policy* 38 (1):75–81. doi: http://dx.doi.org/10.1016/j.enpol.2009.08.054.

Cullen, Jonathan M., Julian M. Allwood, and Edward H. Borgstein. 2011. "Reducing energy demand: What are the practical limits?" *Environmental Science & Technology* 45 (4):1711–1718. doi: 10.1021/es102641n.

De Stercke, S. 2014. *Dynamics of Energy Systems: A Useful Perspective*. Laxenburg, Austria: International Institute for Applied Systems Analysis (IIASA).

DeFries, Ruth, Ashwini Chhatre, Kyle Frankel Davis, Arnab Dutta, Jessica Fanzo, Suparna Ghosh-Jerath, Samuel Myers, Narasimha D Rao, and Matthew R Smith. 2018. "Impact of historical changes in coarse cereals consumption in India on micronutrient intake and anemia prevalence." *Food and Nutrition Bulletin* 39 (3):377–392.

Doyal, Len, and Ian Gough. 1991. *A Theory of Human Need*. Macmillan International Higher Education.

Easterlin, Richard A. 1973. "Does money buy happiness?" *The Public Interest* 30:3.

Easterlin, Richard A., Laura Angelescu McVey, Malgorzata Switek, Onnicha Sawangfa, and Jacqueline Smith Zweig. 2010. "The happiness – income paradox revisited." *Proceedings of the National Academy of Sciences* 107 (52):22463. doi: 10.1073/pnas.1015962107.

Engel, E. 1857. "Die Productions- und Consumtionsverhältnisse des Königreichs Sachsen." *Zeitschrift des Statistischen Bureaus des Königlich Sächsischen Ministerium des Inneren* 8–9:28–29.

Falchetta, Giacomo, Shonali Pachauri, Edward Byers, Olha Danylo, and Simon C Parkinson. 2020. "Satellite observations reveal inequalities in the progress and effectiveness of recent electrification in sub-Saharan Africa." *One Earth* 2 (4):364–379.

Falchetta, Giacomo, Shonali Pachauri, Simon Parkinson, and Edward Byers. 2019. "A high-resolution gridded dataset to assess electrification in sub-Saharan Africa." *Scientific Data* 6 (1):1–9.

Fischer-Kowalski, Marina, Fridolin Krausmann, Stefan Giljum, Stephan Lutter, Andreas Mayer, Stefan Bringezu, Yuichi Moriguchi, Helmut Schütz, Heinz Schandl, and Helga Weisz. 2011. "Methodology and indicators of economy-wide material flow accounting: State of the art and reliability across sources." *Journal of Industrial Ecology* 15 (6):855–876.

Fobi, Simone, Varun Deshpande, Samson Ondiek, Vijay Modi, and Jay Taneja. 2018. "A longitudinal study of electricity consumption growth in Kenya." *Energy Policy* 123:569–578.

Gleick, Peter H. 1996. "Basic water requirements for human activities: Meeting basic needs." *Water International* 21 (2):83–92.

Grubler, A., T. B. Johansson, L. Mundaca, N. Nakicenovic, S. Pachauri, K. Riahi, H.-H. Rogner, and L. Strupeit. 2012. "Energy primer." In *Global Energy Assessment*. Cambridge: Cambridge University Press.

Grubler, Arnulf, Charlie Wilson, Nuno Bento, Benigna Boza-Kiss, Volker Krey, David McCollum, Narasimha D. Rao, Keywan Riahi, Joeri Rogelj, Simon De Stercke, Jonathan Cullen, Stefan Frank, Oliver Fricko, Fei Guo, Petr Havlík, Matt Gidden, Daniel Huppmann, Gregor Kiesewetter, Peter Rafaj, Wolfgang Schoepp, and Hugo Valin. 2018. "A Low energy demand scenario for meeting the 1.5oC target and sustainable development goals without negative emission technologies." *Nature Energy* 3:515–527. doi: DOI.org/10.1038/s41560-018-0172-6.

Knight, Kyle W., and Eugene A. Rosa. 2011. "The environmental efficiency of well-being: A cross-national analysis." *Social Science Research* 40 (3):931–949. doi: https://doi.org/10.1016/j.ssresearch.2010.11.002.

Leach, Melissa, Belinda Reyers, Xuemei Bai, Eduardo S Brondizio, Christina Cook, Sandra Díaz, Giovana Espindola, Michelle Scobie, Mark Stafford-Smith, and Suneetha M. Subramanian. 2018. "Equity and sustainability in the Anthropocene: A social – ecological systems perspective on their intertwined futures." *Global Sustainability* 1.

Lundqvist, Jan, Charlotte De Fraiture, and David Molden. 2008. Saving water from field to fork: Curbing losses and wastage in the food chain. In *SIWI Policy Brief*. Stockholm, Sweden: Stockholm International Water Institute (SIWI).

Malthus, Thomas. 1798. *An Essay on the Principle of Population. An Essay on the Principle of Population, as it Affects the Future Improvement of Society with Remarks on the Speculations of Mr. Godwin, M. Condorcet, and Other Writers.* St. Paul's Churchyard: J. Johnson.

Maslow, A. H. 1943. "A theory of human Motivation." *Psychological Review* 50 (4):370–396. doi: https://doi.org/10.1037/h0054346.

Mastrucci, Alessio, Edward Byers, Shonali Pachauri, and Narasimha D. Rao. 2019. "Improving the SDG energy poverty targets: Residential cooling needs in the Global South." *Energy and Buildings* 186:405–415.

Mastrucci, Alessio, and Narasimha D. Rao. 2019. "Bridging India's housing gap: Lowering costs and CO2 emissions." *Building Research & Information* 47 (1):8–23.

Max-Neef, Manfred A. 1991. *Human-Scale Development: Conception, Application and Further Reflections.* New York: Apex.

Meadows, Donella H., Dennis L. Meadows, Jørgen Randers, and William W. Behrens III. 1972. *The Limits to Growth: A Report for the Club of Rome's Project on the Predicament of Mankind.* New York: Universe Books.

Millward-Hopkins, Joel, Julia K. Steinberger, Narasimha D. Rao, and Yannick Oswald. 2020. "Providing decent living with minimum energy: A global Scenario." *Global Environmental Change* 65:102168. doi: https://doi.org/10.1016/j.gloenvcha.2020.102168.

Nakicenovic, N., D. Victor, A. Grubler, and L. Schrattenholtzer. 1993. "Long-term strategies for mitigating global warming." *Energy* 18:403–409.

Nakićenović, Nebojsa, Paul Viktor Gilli, and Rainer Kurz. 1996. "Regional and global exergy and energy efficiencies." *Energy* 21 (3):223–237.

négaWatt. 2017. *Scénario négaWatt 2017–2050: Hypothèses et Résultats.* Paris, France: négaWatt Association.

O'Neill, Daniel W, Andrew L. Fanning, William F. Lamb, and Julia K. Steinberger. 2018. "A good life for all within planetary boundaries." *Nature Sustainability* 1 (2):88–95.

Paulsen, Jacob Silva, and Rosa Maria Sposto. 2013. "A life cycle energy analysis of social housing in Brazil: Case study for the program "MY HOUSE MY LIFE"." *Energy and Buildings* 57:95–102.

Pereira, Laura M., Tanja Hichert, Maike Hamann, Rika Preiser, and Reinette Biggs. 2018. "Using futures methods to create transformative spaces." *Ecology and Society* 23 (1).

Phadke, A., A. Jacobson, W. Park, G. Lee, P. Alstone, and A. Khare. (2017). "Powering a home with just 25 watts of solar PV: Super-efficient appliances can enable expanded off-grid energy service using small solar power systems." Lawrence Berkeley National Laboratory. LBNL Report #: LBNL-175726. https://escholarship.org/uc/item/3vv7m0x7

Princen, Thomas. 2005. *The Logic of Sufficiency.* Vol. 30. Cambridge, MA: MIT Press.

Rao, N. D., and P. Baer. 2012. ""Decent Living" emissions: A conceptual framework." *Sustainability Science* 4:656–681.

Rao, N. D., and J. Min. 2017. "Decent living standards: Material prerequisites For human wellbeing." *Social Indicators Research*. doi: doi: 10.1007/s11205-017-1650-0.

Rao, Narasimha D., Jihoon Min, and Alessio Mastrucci. 2019. "Energy requirements for decent living in India, Brazil and South Africa." *Nature Energy*. doi: https://doi.org/10.1038/s41560-019-0497-9.

Raworth, Kate. 2017. *Doughnut Economics: Seven Ways to Think Like a 21st-Century Economist*. White River Junction, VT: Chelsea Green Publishing.

Rijnhout, L, R. Mastini, J. Potocnik, J. Spangenberg, B. Alcott, V. Kiss, A. Coote, A. Reichel, S. Lorek, and M. Mathai. 2018. *Sufficiency—Moving Beyond the Gospel of Eco-Efficiency*. Brussels, Belgium: Friends of the Earth Europe.

Rockstrom, Johan, Will Steffen, Kevin Noone, Asa Persson, F. Stuart Chapin, Eric F. Lambin, Timothy M. Lenton, Marten Scheffer, Carl Folke, Hans Joachim Schellnhuber, Bjorn Nykvist, Cynthia A. de Wit, Terry Hughes, Sander van der Leeuw, Henning Rodhe, Sverker Sorlin, Peter K. Snyder, Robert Costanza, Uno Svedin, Malin Falkenmark, Louise Karlberg, Robert W. Corell, Victoria J. Fabry, James Hansen, Brian Walker, Diana Liverman, Katherine Richardson, Paul Crutzen, and Jonathan A. Foley. 2009. "A safe operating space for humanity." *Nature* 461 (7263):472–475.

Rogelj, J., D. Shindell, K. Jiang, S. Fifita, P. Forster, V. Ginzburg, C. Handa, H. Kheshgi, S. Kobayashi, E. Kriegler, L. Mundaca, R. Séférian, and M. V. Vilariño. 2018. "Mitigation pathways compatible with 1.5°C in the context of sustainable development." In *Global Warming of 1.5°C: An IPCC Special Report.*, edited by V. Masson-Delmotte, P. Zhai, H. O. Pörtner, D. Roberts, J. Skea, P. R. Shukla, A. Pirani, W. Moufouma-Okia, C. Péan, R. Pidcock, S. Connors, J. B. R. Matthews, Y. Chen, X. Zhou, M. I. Gomis, E. Lonnoy, T. Maycock, M. Tignor and T. Waterfield, 93–174. Geneva, Switzerland: World Meteorological Organization.

Sachs, Wolfgang. 1993. "Die vier E's: Merkposten für einen maß-vollen Wirtschaftsstil."

Sadras, Victor O., Patricio Grassini, and Pasquale Steduto. 2011. *Status of Water use Efficiency of Main Crops.* Rome, Italy: Food and Agricultural Organization (FAO).

Semieniuk, Gregor, Lance Taylor, Armon Rezai, and Duncan K. Foley. 2021. "Plausible energy demand patterns in a growing global economy with climate policy." *Nature Climate Change* 11 (4):313–318. doi: 10.1038/s41558-020-00975-7.

Sen, Amartya. 2000. *Development as Freedom.* New York: Anchor Books.

TWI2050. 2018. *Transformations to Achieve the Sustainable Development Goals. Report prepared by The World in 2050 initiative.* Laxenburg, Austria: International Institute for Applied Systems Analysis (IIASA).

TWI2050. 2019. *The Digital Revolution and Sustainable Development: Opportunities and Challenges.* Laxenburg, Austria: International Institute for Applied Systems Analysis (IIASA).

TWI2050.2020. *Innovations for Sustainability: Pathways to an Efficient and Sufficient Post-Pandemic Future.* Laxenburg, Austria: International Institute for Applied Systems Analysis (IIASA).

UNDP. 2019. *Human Development Report 2019: Beyond Income, Beyond Averages, Beyond Today – Inequalities in Human Development in the 21st Century.* New York, US: United Nations Development Program.

UNESCAP. 2019. *Economic and Social Survey 2019: Ambitions Beyond Growth.* Bangkok, Thailand: United Nations Economic and Social Commission for Asia and the Pacific (ESCAP).

Vita, Gibran, Edgar G. Hertwich, Konstantin Stadler, and Richard Wood. 2019. "Connecting global emissions to fundamental human needs and their satisfaction." *Environmental Research Letters* 14 (1):014002. doi: 10.1088/1748-9326/aae6e0.

WHO. 2020. *World Health Statistics 2020.* Geneva, Switzerland: World Health Organisation.

Wilkinson, Richard, and Kate Pickett. 2010. *The Spirit Level: Why Equality is Better for Everyone.* Penguin UK.

Wilson, C., A. Grubler, N. Bento, S. Healey, S. De Stercke, and C. Zimm. 2020. "Granular technologies to accelerate decarbonization." *Science* 368 (6486):36–39. doi: 10.1126/science.aaz8060.

Zimm, Caroline. 2019. "Methodological issues in measuring international inequality in technology ownership and infrastructure service use." *Development Studies Research* 6 (1):92–105.

24

THE EMERGING DEMAND-SIDE PARADIGM IN THE POWER SECTOR

Fereidoon Sioshansi

1. Introduction

Traditionally, the power sector *predicted aggregate customer demand* and adjusted *generation* resources to meet it in real time. Utilities, most of which were vertically integrated[1] and regulated monopolies, maintained a portfolio of plants – baseload, intermediate, and peaking[2] – to meet demand as it varied from hour to hour, day to day, and across the seasons. Moreover, demand was treated as a "given" – namely, the sum of the load from all electricity-using devices connected to the network. There was little or no attempt to manage either the quantity or *when* and *where* it was consumed. Finally, all consumers bought all the kilowatt-hours (kWh) from the network and paid a regulated bundled tariff.

As we move forward, this business paradigm is changing. As states, countries, and entire continents – notably the EU – aim to reach carbon neutrality by mid-century, if not sooner, the generation mix is rapidly shifting toward renewables, many of which are variable and/or non-dispatchable.[3] Consequently, grid operators will increasingly be *predicting the output of variable renewable generation* and – when possible – adjusting *demand* to match it. This is likely to happen in more places as the percentage of renewables reaches higher levels as many countries are aiming for 100% low-carbon electricity generation by 2045–2050.

In this context, the future will be one where flexible demand and/or adjustable loads complement the inherent variability of renewable resources – with prices seamlessly flowing to smart, connected behind-the-meter[4] (BTM) assets turning virtually all electricity using devices, distributed generation, and storage into flexible or controllable loads. In time, this paradigm shift will lead to the emergence of a *demand-side merit order*.

Just as the supply-side merit order historically determined which power plants were dispatched to meet the load, in the future, sophisticated intermediaries or orchestrators are likely to determine which BTM assets will be dispatched to minimize the cost of keeping supply and demand in balance. This, it is argued, can be accomplished by managing large portfolios of BTM assets, which will increasingly be intelligent, connected, and remotely managed through digitalization and sophisticated software using artificial intelligence (AI) and machine learning (ML).

This chapter describes the obvious and the not-so-obvious reasons for this fundamental paradigm shift and its implications for the operability, resilience, and security of future power

DOI:10.4324/9781003183020-28

networks. The chapter contributes to the aims of the handbook, which takes a cross-sectoral narrative in the context of energy transition by examining concepts, theory, methods, practices, and fundamental evidence.

The chapter is organized as follows:

- Section 2 describes some of the fundamental changes taking place on the demand side, which taken together, may lead to the aforementioned *paradigm shift*.
- Section 3 describes the emergence of the *demand-side merit order* and its significance.
- Section 4 describes advances including digitalization and load aggregation leading to schemes such as *virtual power plants* (VPPs), *peer-to-peer* (P2P) *trading*, and *transactive energy* (TE).
- Section 5 highlights some of the game-changing developments including the emergence of new players with business models that suggest how the evolution of the demand side may play out followed by the chapter's conclusions.

2. Changes Taking Place on the Demand Side

The digitalization of electricity sector is driven by the falling cost of communication and the fact that progressively more electricity using devices are getting smart and increasingly connected (IEA, 2017), the so-called internet of things (IoT). Increasingly, thermostats controlling HVAC systems are smart and able to communicate with other household devices wirelessly and effortlessly. Likewise, ubiquitous personal assistants such as Amazon's Alexa and its competitors from Google and Apple allow effortless, hands-free management of many devices within homes. The concept of *smart home energy management system*, or SEMS (Shaw-Williams, 2020), is gradually moving from science fiction to practical reality as energy management, security, and other useful functions are integrated and managed through easy-to-use apps on mobile phones.

There are several powerful forces pushing more households and commercial customers toward digitalization, automation, and integration:

- First is the rapid proliferation of distributed generation[5] (DG), primarily through rooftop solar photovoltaics (PVs) on many homes and commercial establishments.
- Second is the emergence of distributed storage options – not just batteries – as their cost continues to fall and their performance continues to improve.
- Finally, the rapid expected penetration of electric vehicles (EVs) are mandated and/or subsidized in a number of key markets.

As described in the literature (e.g., Sioshansi, 2020), these important developments have resulted in the gradual transformation of a number of consumers to *prosumers* – that is consumers who produce or self-generate a portion of their needs (the first item in the previous list).

Going a step further, a *prosumer* can become a *prosumager*[6] by investing in storage, either a stationary battery, an electric water heater with a storage tank, or a number of other devices that can absorb, store, and release energy for use at a later time. EVs, of course, can be viewed as batteries on wheels.

The gradual migration of consumers to being less dependent on the electricity services delivered by the retailers through the regulated distribution network is reflected in the growth of distributed energy resources, or DERs. What is more interesting is that as the scale and significance of DERs increases they become more mainstream and can no longer be considered alternative.

Broadly speaking, DERs consist of energy efficiency (EE), demand response (DR), distributed generation (DG), distributed storage (DS), and efforts to monitor and manage all sorts of BTM assets, be it electric vehicles or smart thermostats that control heating ventilation and air-conditioning (HVAC) systems. As the industry moves away from its historical centralized business paradigm, the next phase in the evolution of DERs is their *integration* into the upstream utility functions to provide a host of services to customers, in what is called *integrated* DERs or iDERs.

According to Peter Asmus (2020), the key operating word moving forward is the "i" in iDERs – both upstream with the utility and grid operations and downstream with customers' BTM assets.

> Navigating the energy transformation requires new approaches to integrating DER into the grid and onsite energy networks to bolster resiliency and foster sustainability. It also requires innovative business models that recognize previously hidden value for both end users and stakeholders across the broader energy system.
>
> (Ibid.)

Once the integration is achieved, the resulting system will be "better attuned to evolving demand and customer needs and more compatible with a rapidly changing policy and regulatory environment" (ibid.).

The significance of the "i" in the iDERs is that DERs are moving mainstream and dominating new additions/investments in networks globally as the electricity business is increasingly moving toward a decentralized model.

By one estimate, the State of New York by 2040 could have

- 100,000 buildings offering flexible DR loads,
- 5,000 MW of distributed energy storage devices deployed behind and in front of the meter,
- 9,000 MW of offshore wind,
- 250,000 solar plus storage systems installed at residences, and
- 100,000 EVs plugged into distribution networks (ibid.).

Likewise, California and a number of states are also striving to develop and integrate DERs into their traditional networks. The white paper argues that the key to successful iDERs is to harness their full value:

> The value of DER assets is typically measured by the capital expense attached to their purchase. However, the value of each asset to the commonly shared power grid grows exponentially if integrated with other DER assets in a smart and nimble way, even as the capital cost declines.
>
> (Ibid.)

An integrated vision of the DERs envisions a host of players, potentially collaborating and/or competing for a slice of the growing DER pie. This is beginning to happen as new players are joining the incumbents in what is likely to remain a highly fragmented space. Many new players are beginning to appear in what promises to lead to significant disruptions in the way energy services will be delivered and paid for in a future enabled by the rapid proliferation of smart connected devices and the connectivity of BTM assets enabled by the cloud.

There are signs that the gradual transition of consumers to *prosumers* is already quite pronounced in places such as Australia, with an abundance of sunshine, relatively high retail tariffs, and supporting polices that have resulted in roughly one out of four household to have solar roofs (Swanston, 2020). It is not a stretch to claim that many of the same households will eventually become *prosumagers* as they invest in storage, stationary, or mobile (Sioshansi, 2021a).

In fact, in places like Australia, where customers predominantly live in detached homes in suburbs with big roofs, many are likely to become *nonsumers* – defined as consumers who buy little or no *net* kilowatt-hour from the network (Ben-David, 2019). For anyone who is inclined to dismiss such trends as unrealistic, the state of California, the most populous in the US (World Population Review, 2022), adopted a zero net energy building code for all *new* residential buildings starting in 2020 (CEC, 2020). As time goes on and more new homes are added, this building code – and similar ones elsewhere – will begin to turn many new households into virtual *nonsumers* – that is they will generate as much power as they consume on an annual basis.

The notion of *nonsumer* is already a reality, at least anecdotally, as individual homeowners make the necessary investments to turn their households into net-zero energy homes even in the absence of any building codes or mandates (Schlesinger, 2020). An example of how and why some customers may choose to go this route is described in Box 24.1.

Box 24.1: From Consumer to *Nonsumer*[7]

What ordinary electricity consumers can do with some spare cash and know-how is amazing. The main motivation is not necessarily to save money or the planet but to achieve a level of energy self-sufficiency and resilience they have never had before. Some have other motives such as enjoying service during power outages; others like to live with a low-carbon footprint. As it happens, all of these are technically feasible and are financially viable.

A case in point is Andrew Wilson, who lives in Brisbane, Australia, and who explains his main motivation as described here (Wilson, 2020):

> The goal was to be as self-reliant with our own renewable energy as possible – directly consuming solar generation during the day and charging the battery for use overnight. On this metric we're pretty happy with the result – achieving 92.2% self-sufficiency over the year.

To get there, the following are required:

- Tesla Powerwall 2 battery – 5 kW/13.5 kWh
- 20x SunPower P19 315 W modules – 6.3 kWp
- 20x SolarEdge P320 power optimizers
- SolarEdge HD-Wave single-phase inverter – 5 kWac
- SolarEdge Immersion Hot Water Heater Controller – 3.6 kWac

(Ibid.)

Like most innovators, there was a lot of experimentation and learning – some gadgets worked better than advertised; others, less so. Wilson's first surprise was the discovery of the usefulness of the

---

humble electric water heater. "I've long been saying that most solar households are sitting on top of the best battery there is without realizing it – their hot water system".

Once equipped with a controller, the electric water heater could function as a smart device dynamically diverting the excess PV generation into hot water instead of exporting it to the grid. By doing so, the electric water heater could store a lot of the midday solar generation as hot water in the tank for later use, working as an inexpensive battery.

The battery takes up any residual excess solar generation, discharging the stored energy in the evening after the sun has set. The hot water in the tank could be used at any time. Together, the two devices absorb most of the excess solar output, reducing the need to export the excess generation into the grid[8].

The next step for many *prosumagers* may be to invest in an electric vehicle – which offers even more storage capacity and more demand flexibility. Once the functionality of vehicle-to-grid (V2G) or vehicle-to-home (V2H) or vehicle-to-vehicle (V2V) technology becomes available, the fully charged EV batteries can provide services to the grid, run appliances at home, or – potentially some day – charge up a neighbor's EV.

Averaged over a full year, such a household could be nearly self-sufficient or a net producer. The details, of course, vary by the key parameters – the size of the solar panels, the battery, the EV, the water tank, and overall consumption patterns within the household. Moreover, there are external variables, such as how many sunny days there are in a year, the location of the system, and of course, the prevailing weather. The same system in a more northerly or cloudy part of the world would not perform as well especially in winter.

Another critical variable is the prevailing tariffs. If grid-supplied electricity is expensive, solar generation will be more attractive, encouraging self-consumption. Australia, which had only three solar roofs in 2001, now has three million and is projected to reach six million by 2030 (ibid.), one in two houses. It meets all the necessary conditions for *prosumagers*.

Another important variable is how much is paid for the excess power exported to the grid. In some places *prosumers* get paid as much for the power exported as for the kilowatt-hour bought from the grid. If retail prices are high, as they are in California or Hawaii or Germany, then customers are incentivized to export as much as they can. Finally, in remote areas with unreliable service, *prosumers* will have strong incentives to become *prosumagers*. For the Wilson family, the solar panels generated 7,834 kWh over the year compared to 5,092 consumed – that is, 35% of the solar generation was exported to the grid.

A perennial issue facing all self-generators is the seasonal variations in solar output, typically too much in sunny months and not enough in the winter and cloudy ones. For the Wilsons who live in the Southern Hemisphere, November and January tied for the best result in sunny Queensland, with 99.3% self-sufficiency, while June had the worst outcome with only 74.9% self-sufficiency. There is not much one can do about solar insolation, the amount of solar radiation on a specific surface for a period of time. But even in the worst case, roughly 75% of the required power could be obtained from the solar panels – even more if the house is well-insulated and has efficient appliances.

Needless to say, the sizing of the panels, batteries, and water heater matters. The key factor is the battery's *utilization factor* – what portion of the battery's storage capacity is discharged each night on average over the year. In the case of Wilson's, the answer was 6.21 kWh – or 46% of the battery's nameplate capacity of 13.5 kWh. "If I were to make any change it would be to add more PV capacity to help with generation shortfalls during winter".

Another important factor is the ability to ride through power outages. If properly configured, the Tesla Powerwall can function as an uninterrupted power supply (UPS). The family was spared from the inconvenience of at least 13 outages between October and February 2019. Wilson has not had to reset the oven clock since the system was commissioned, adding, "I'm a firm believer that this is an area where home batteries can very quickly deliver immeasurable value to their owners". With more extreme weather, such as wildfires and hurricanes, many consumers will find the UPS functionality among the most compelling reason to invest in such systems.

Of course, ultimately most customers want to know the amount of investment and its payback. In this particular case, the system cost was roughly $15,000, the net cost after factoring for a grant and an interest free loan. Figure 24.1 compares the annual energy bill under three scenarios:

- Consumer: fully reliant on grid-supplied kilowatt-hour
- *Prosumer* with flexible load: solar + hot water controller only
- *Prosumager* with solar + hot water controller + battery

The results are specific to each installation, local tariffs, available incentives, and how well the components are integrated and operated. In this case, a solar only scenario delivers a 67% saving compared to grid power alone, but the solar + battery option is even better, delivering 86% saving versus grid power and a net cost less than half of the solar only scenario. The solar only option would have a payback period of 7.2 years, and 12.5 years for the solar + battery system.

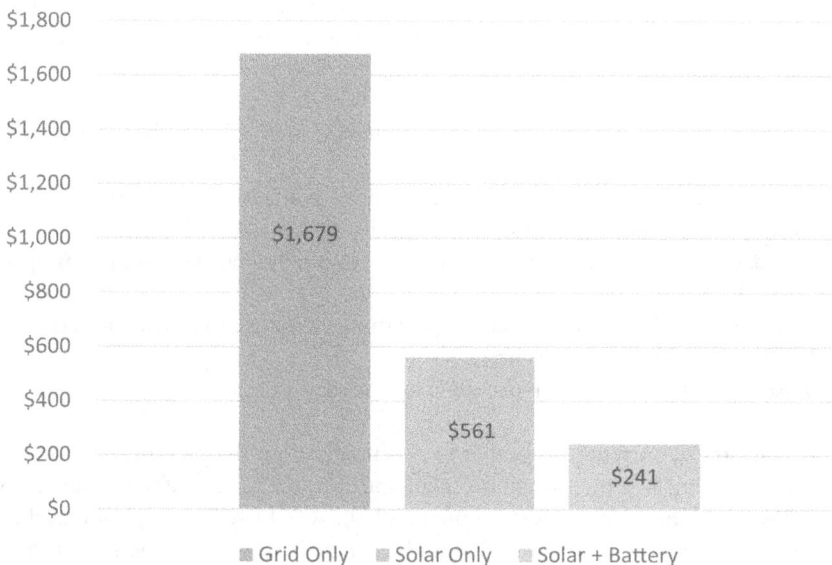

Figure 24.1 Annual energy cost outcomes for three scenarios for residential solar, battery, and water heater system in Queensland, Australia.

Source: Wilson, 2020.

Would more consumers follow a similar path? It depends on three critical factors:

- Costs – both hardware and installation
- Retail tariffs
- How much can be earned for exporting the surplus energy into the grid

The early adopters are appearing in increasing numbers all over the world. Saving on energy bills helps but is not necessarily the only or the main motivator. Payback period is important but does not appear to be the driver. Achieving some degree of self-sufficiency by not relying on the dirty power provided by the grid and the added security of riding through localized power outages are all important factors, but clearly, it is not for everybody.

The same phenomenon, called by other names such as *passive homes* in Europe, is catching on across the globe as concerns about climate change and sustainability gain momentum. It should also be mentioned that the concept of *nonsumer* is also emerging in some of the least developed parts of the world with nonexistent and/or unreliable electricity services such as rural parts of Africa, Arctic Canada, Alaska, remote parts of Australia, and virtually all remote islands, basically anywhere where the cost of serving customers is excessive and/or the service quality is poor (Couture et al., 2019).

As the preceding discussion explains, the traditional definition of consumers buying all their kilowatt-hour from the local distribution network and paying a *flat bundled-regulated tariff*, usually based on *volumetric usage*, is likely to be challenged with the proliferation of the new options identified (Sioshansi, 2021b). How likely will this transition be and how soon will vary from place to place depending on the following:

- The prevailing retail rates, taxes, and add-ons therein, such as levies for social and/or environmental objectives
- The presence or absence of supportive/prohibitive policies and regulations allowing the export of the excess solar generation to the grid
- The abundance of sun and the prevalence of single-family detached homes (as opposed to multi-story high-rises)
- The overall economics of self-generation (and storage) relative to buying from the network, which depends on the prior two items
- Of course, if the house is owner-occupied or rented

Returning to Australia, currently an outlier in decentralization, the prevailing conditions favor self-generation, making it economic to go solar, and not just for the affluent but also for the middle and lower income customers (Mountain, 2019). At 644 watts per person at the end of 2019, Australia is ahead of Germany at 589 and Japan at 500 watts per person in distributed solar generation (ibid.).

The same is happening in other places with high retail tariffs such as California and Hawaii (EIA, 2021), pushing more consumers to self-generate and/or store the excess generation from solar PVs for use at later times. If the distribution grid is *not* reliable and prone to frequent outages due to storms, hurricanes, wildfires, and the like – as in parts of California or Puerto Rico – then the motivation to self-generate and store becomes even more compelling. The recent well-publicized outages in Texas (https://energy.utexas.edu/ercot-blackout-2021), for example, may compel some customers to invest not only in self-generation but self-reliance and resilience. Finally, there is growing interest in P2P trading and sharing, including the formation of energy communities or micro-grids, concepts that allow better utilization of localized generation, consumption, and storage with demand flexibility.

3. The Emergence of the Demand-Side Merit Order

The supply-side merit order is well-known to anyone familiar with the operational dispatch of the power plants (Sioshansi, 2021a). As the demand rises on a given network, more expensive generation is dispatched to meet the demand; the reverse happens as the electricity load falls off. This results in minimizing the overall cost of power generation while serving customers' needs. The concept was well-suited to an industry long focused on the supply side and mostly oblivious to the demand side.

This chapter argues that the time has arrived to seriously question the efficacy of relying solely on the supply side resources to balance supply and demand for at least two reasons:

- First, as time goes on, many networks will be dominated by non-dispatchable renewables, notably solar and wind. Unlike most thermal plants, solar plants or wind farms cannot be turned on or off or adjusted in the same way without storage, regional smoothing, or curtailment.
- Second, nearly all customer loads have certain degree of *flexibility* (Sioshansi, 2021b), which can be monitored and controlled, which, in turn, means that they can be aggregated and adjusted at will – in other words dispatched just as supply-side resources were in the past.

Among customer loads with particular flexibility are HVAC, electric water heaters, pool pumps, washers, and dryers, to name a few. The addition of EVs plus other forms of distributed storage means that a significant portion of many customers' loads can be adjusted in response to price signals, which in turn depend on the supply and demand conditions on the grid. A typical EV, for example, may require to be charged for around two hours during a 13-hour window, from 6:00 p.m. to 7:00 a.m., where many EVs are typically parked at home (Sioshansi, 2021c). If the demand on the network peaks between 4:00–9:00 p.m., as it currently does in California (US DOE, 2017), for example, most EVs can be charged *outside* this window, when price tends to be low and supplies plentiful.

In the same vein, most buildings can be pre-cooled prior to the peak demand period. Similarly, electric water heaters can be programmed to heat and store water during off-peak hours. The flexible opportunities are not limited to the residential sector but are equally applicable to commercial and the industrial sectors (Lobbe et al., 2021).

While the challenge of aggregating millions of small loads into large blocks of flexible demand on the scale of large thermal power plants is not trivial, the technology to digitalize and manage large portfolios of BTM assets is rapidly advancing offering promising potential (Barrager & Cazalet, 2020), as further described in the following section.

The point, however, is that instead of relying solely on the supply side, the demand side will be increasingly deployed to play a constructive role in balancing supply and demand. The emerging demand-side merit order works exactly as the supply-side merit order has done in the past, which means that the least-cost demand-side options are utilized in such a way as to minimize the cost of balancing supply and demand while optimizing the integration of renewable generation resources when they are available.

4. Digitalization, Load Aggregation, and Virtual Power Plants

There are a number of promising developments making it possible to monitor and manage how much electricity is used and when in any end-use device. This is broadly referred to as the digitalization of the BTM assets. As more devices become smart and are connected to the internet, the task of remotely monitoring and optimizing their consumption, storage, or generation – as the case may be – becomes easier. The motivations, the means, the benefits, and the challenges of digitalization in the electricity sector are further described in *AI Insights: The Power Sector in a Post Digital Age* by the Eurelectric's Beyond Digital Platform (2020). This report examines the opportunities for utilities to accelerate the energy transition by facilitating the role of AI and digitalization in the power sector. It says the power sector can enable an efficient integration of variable sources of energy, address reliability and intermittency, support grid stability, and improve forecasting, all leading to an optimal match between supply and demand.

> The ongoing volatility of the grid remains a key challenge as the share of renewable generation grows. AI-driven data analysis will look to counteract this by ascertaining the optimal location for low carbon technologies by predicting the ongoing impact on network capacity.
>
> (Ibid.)

The report suggests that one of the most important ways digitalization and AI can drive the energy transition will be in optimizing distributed energy resources, battery storage and vehicle-to-grid technology. Virtual power plants (VPPs) will aggregate DERs, electric vehicles, and flexible loads to modulate aggregated power exchanged with the grid and ensure grid stability, delivering increased flexibility and support.

AI offers enormous potential for the energy industry and is a fundamental building block of the clean energy transition. It will lead to greater automation, save money and drive operating efficiencies, develop new products and services, and enable a reduction in carbon emissions.

The International Data Corporation estimated that worldwide revenues for AI, including software, hardware, and services, were as much as $156 billion in 2020 and are projected to rise to $300 billion by 2024 (ibid.). If Europe continues at its current pace of adoption, AI revenue could add around $3.26 trillion, or 20% to its combined economic output by 2030 (ibid.). But if it accelerates to catch up with the US, as much as $4.35 trillion could be added to GDP over this period (ibid.). These are clearly non-trivial numbers.

Time will tell whether digitalization of the power sector and AI lives up to the expectations. Google's CEO, Sundar Pichai, has identified AI as one of the most important areas humanity is working on, potentially more profound than electricity or fire (Petroff, 2018).

In the short term, the immediate applications of AI are likely to be around optimizing existing assets, but the highest potential may be at the intersection of different sectors. Examples include the electrification of transport and heat, smart homes to maximize energy efficiency through integration and autonomous optimization, and flexibility services for grids and edge computing[9] to enable the rapid response of decentralized energy assets with limited intervention.

Since individual devices tend to be small, millions of them in thousands of households have to be aggregated before the impact of managing them becomes noticeable and worth the bother. Monitoring and managing large numbers of small devices has historically not been practical or cost-effective. This, however, is changing as new means for communicating and control, including transmitting variable prices to smart devices, are deployed (Barrager & Cazalet, 2020).

Once large numbers of customers and their BTM assets are digitalized, connected, and aggregated, truly exciting possibilities emerge, which take advantage of the full range of flexibilities of the aggregated portfolio of customers, assets, and smart devices in three stages:

- First, digitalization and connectivity of individual customers and BTM devices
- Second, aggregation of large numbers of customers and their devices into portfolios of flexible loads, generation, and storage
- Third, managing the aggregated portfolio as a virtual power plant (VPP)[10]

A number of existing and new players have emerged in the last few years to develop and deliver on the components of the items highlighted (Poplavskaya & de Vries, 2020). Some are focused on digitalization and aggregation, while others are developing different versions of the VPP model including offering VPP-as-a-service. Not surprisingly, a number of new business models are emerging to capitalize and monetize on the opportunities as explained in the next section.

5. The Emergence of New Business Models

The rapid pace of change especially in the BTM space offers risks for the incumbent stakeholders and opportunities for the newcomers. Since the utility value chain is for the most part a zero-sum game, one stakeholder's gain usually comes at the expense of a loss for others. In this sense, the incumbent utilities *potentially* have a lot to lose especially in places where they dominate the entire space, such as states in the US where vertically integrated monopoly utilities still exist. On the other hand, some of these same incumbents may be able to transform themselves into new viable entities while they still can.

There is growing evidence for the burgeoning non-conventional services and the threat they imply for the incumbents. Given these developments, it is not surprising that both the incumbents and newcomers are trying to identify where the value lies in the long utility value chain and capture it, a topic examined by Peter Fox-Penner in his book *Power after Carbon* (2020).

Fox-Penner focuses on the power sector and how it may face up to the challenges of the transition to a low-carbon future, which is being shaped by three drivers:

- The rise of energy-independent cities, energy communities, and micro-grids powered by clean – and local – energy resources
- The gradually deregulated, transaction-centric marketplace of *prosumers* who make some of their own power, buy the rest, and continue to rely on the regulated grid as a trading platform
- The transformation of current utilities into advanced service providers, still regulated and relying on large and small generation resources

(2020)

Examining the pros and cons of these drivers, one could consider the existing infrastructure that generates, transmits and distributes power to millions of customers and surmise that the infrastructure has many useful attributes that will be hard to beat (ibid.). Despite shortcomings, current infrastructure delivers reasonably reliable service at reasonable cost most of the time.

Gradually, however, the focus is shifting to the *downstream* where the network delivers the electricity and related services to the customers and the BTM assets. Here is where new service options are emerging including smart-connected devices remotely monitored and managed by aggregators using sophisticated software, AI, and ML to reduce costs and improve service while increasing the efficiency of the network.

Two competing business/regulatory models are evident (ibid.): (1) smart integrator (SI) and (2) energy service utility (ESU). The middle portion of the utility sector's infrastructure, which includes high-voltage transmission and low-voltage distribution network, is and will continue to remain a *natural monopoly* and therefore will remain regulated. Moreover, some services – such as the delivery of energy efficiency – may fall by the wayside under some business/regulatory scenarios. Making matters even more complicated is the vexing issue of how to set tariffs to preserve fairness and equity among different customer segments as the service needs of consumers, *prosumers*, and *prosumagers* diverge.

Additional complexities arise as groups of consumers form energy communities and begin to trade and/or share energy – P2P transactions – and so on. Who will pay for the grid in such a future? Will electricity evolve into a subscription service – similar to Amazon's Prime or Netflix's business model where subscribers get a bundle of services for a fixed monthly fee and pay for extra services? Will future *prosumers* prefer to buy bundled services from a single provider – which is so much simpler – or to buy bits and pieces from competing providers – which takes time and effort to figure out the optimal rebundled package at reasonable cost? There are, of course, many possible variations to the two basic models. Fox-Penner recommends envisioning a *prosumer* who says to the energy service company (ESCO):

> "I am giving you control of my thermostat, vehicle chargers, ventilation systems, and other things, with the proviso that you keep all my services available within the limits I set. In this case, if the ESCO can generate savings, the customer gets a portion – and it is win–win. But as we already know, ESCOs have not managed to succeed as many had previously predicted. Would they succeed in the future?"

Even if the ESCO model succeeds, there are likely to be winners and losers – the electric equivalent of grocery chains not serving low-income neighborhoods, a phenomenon prevalent in low-income American inner cities. Regulators are not likely to tolerate such outcomes.

Because of these complicated factors, it is not clear which regulatory/business model is likely to prevail. Perhaps different utilities in different parts of the world facing different regulatory regimes and constraints will try and fail until they succumb to a reasonable solution. There are too many moving parts, and the technology is advancing faster than regulators can follow.

While the incumbent players jockey for position in the emerging energy services marketplace, newcomers including many start-ups are moving ahead with plans of their own. Many have decided that the BTM space is the most lucrative with the best potential. Others are targeting alternative means of aggregation and management or are offering electronic platforms for trading or acquiring flexibility and so on.

But with the proliferation of so many BTM devices, how can anyone manage the complicated transactions among and between the customers, service providers, and retailers? Historically, the power sector read the customers' meters once a month or a quarter and sent them a bill. The utilities may, arguably, not be among the most customer-focused companies. But they are adapting to the changing times and the new demands of their customers and new regulatory requirements to offer innovative rates, especially in places where customers have retail choice.

Another promising area is the growing interest to better manage customers' demand, not only how much they use but when. Since nearly all customer demand has some flexibility, concepts such as demand response (DR) are becoming more widespread.

Finally, there is speculation about how long it will take for many of the ideas presented in the preceding to become widely practiced within the power sector. The speed of change and adaption, many believe, will be accelerated by the intense pressures of the energy transition.

6. Conclusions

As stated at the outset, this *Handbook* aims to provide cross-sectoral insight on energy transitions by examining concepts, theory, methods, practices, and fundamental evidence. This chapter endeavored to contribute to the book's goal by describing how developments on the demand side are likely to result in a paradigm shift where digitalized and connected smart loads, distributed generation and storage enabled by intelligent software and sophisticated intermediaries can play a greater role in balancing supply and demand.

The chapter's main insight is that the traditional definition of consumers buying all their kilowatt-hour needs from the local distribution network and paying a *flat bundled-regulated tariff*, usually based on *volumetric usage* is likely to be challenged with the proliferation of the new demand-side technologies offering customers new service options and leading to the emergence of new business models.

Notes

1 The term *generally* refers to companies where generation, transmission, distribution, and retailing services are handled by the same.
2 These terms refer, respectively, to plants that operate around the clock at full capacity, those that adjust output based on variations in load, and units that only operate during peak demand hours, respectively.
3 Not all renewables are variable or non-dispatchable, of course, but the two that are growing the fastest, wind and solar, generally are.
4 The term generally refers to all energy-consuming, energy-generating, or energy-storing devices on the customer side of the meter (Sioshansi, 2020) in contrast to the rest of the utility infrastructure, which sits *in front of the meter.*
5 Distributed generation is a subset of distributed energy resources, or DERs, which broadly includes to generation, storage, and enhanced control and management of consumption behind the meter.
6 The term *prosumager*, further described in Sioshansi (2019), is hereafter used to broadly refer to *prosumers* with storage and hence increased flexibility to manage their self-generation, storage, and consumption.
7 This draws upon https://onestepoffthegrid.com.au/analysis-of-my-home-battery-solar-systems-first-year-performance and www.usaee.org/usaee2019/submissions/Presentations/Schlesinger_USAEE19.pdf;
http://ahmadfaruqui.blogspot.com.
8 The issue of how much the "excess" solar generation is worth is debated. By self-absorbing it, *nonsumers* can minimize exports to the grid.
9 This is a distributed computing construct that draws computation and data storage closer to the sources of data.
10 Among the most successful examples of VPPs is Next Kraftwerke, based in Cologne, Germany, which was acquired by Shell in February 2021.

References

Asmus, P. (2020). Integrated DERs: Orchestrating the Grid's Last Mile, *Guidehouse*, https://guidehouse.com/insights/energy/2020/ider

Barrager, S. & E. Cazalet, (2020). *Transactive Energy in California: A Platform for 100 Percent Clean Energy and Electrification*. Baker Street Publishing, San Francisco, CA.

Ben-David, R. (2019). Preface, in: Sioshansi, F. (Ed.), *Consumer, Prosumer, Prosumager: How Service Innovations Will Disrupt the Utility Business Model*. Academic Press, Oxford.

California Energy Commission (2020). Zero net Energy Building Code at www.cpuc.ca.gov/zne.

Couture, T. et al. (2019). Off-grid Prosumers: Electrifying the Next Billion with PAYGO Solar, in: Sioshansi, F. (Ed.), *Consumer, Prosumer, Prosumager: How Service Innovations Will Disrupt the Utility Business Model*. Academic Press, Oxford.

Energy Information Administration (EIA) (2021). State Electricity Profiles, Release date November 4, www.eia.gov/electricity/state/, Accessed February 27, 2022.

Eurelectric, (2020). AI Insights: The Power Sector in a Post-Digital Age, Report, November 26, www.eurelectric.org/media/5016/ai-insights-final-report-26112020.pdf

Fox-Penner, P. (2020). *Power after Carbon*, Building a Clean, Resilient Grid. Harvard Univ. Press, Cambridge, MA.

IEA (2017). *Digitalisation and Energy*, Report, IEA, Paris, France, www.iea.org/reports/digitalisation-and-energy

Lobbe, S. et al. (2021). Industrial Demand Flexibility: A German Case Study, in: Sioshansi, F. (Ed.), *Variable Generation, Flexible Demand*. Academic Press, Oxford.

Mountain, B. (2019). Do I have a Deal for you? buying well in Australia's Contestable Retail Electricity Markets, in: Sioshansi, F. (Ed.), *Consumer, Prosumer, Prosumager: How Service Innovations will Disrupt the Utility Business Model*. Academic Press, Oxford.

Petroff, A. (2018). Google CEO: AI is More Profound than Electricity or Fire, *CNN Business*, January 24, https://money.cnn.com/2018/01/24/technology/sundar-pichai-google-ai-artificial-intelligence/index.html.

Poplavskaya, K. & L. de Vries (2020). Aggregators Today and Tomorrow: From Intermediaries to Local Orchestrators? in: Sioshansi, F. (Ed.), *Consumer, Prosumer, Prosumager: How Service Innovations will Disrupt the Utility Business Model*. Academic Press, Oxford.

Schelsinger, B. (2020). It is not science fiction: Going zero net energy and loving it, in: Sioshansi, F. (Ed.), *Behind & Behind the Meter: Digitalization, Aggregation, Optimization, Monetization*. Academic Press, Oxford.

Shaw-Williams, D. (2020). The Expanding Role of Home Energy Management Ecosystem: An Australian case study, in: Sioshansi, F. (Ed.), *Behind & Behind the Meter: Digitalization, Aggregation, Optimization, Monetization*. Academic Press, Oxford.

Sioshansi, F. (2021a). What Is Flexible Demand: What Demand Is Flexible?, in: Sioshansi, F. (Ed.), *Variable Generation, Flexible Demand*. Academic Press, Oxford.

Sioshansi, F. (2021b). How Can Flexible Demand Be Aggregated and Delivered?, in: Sioshansi, F. (Ed.), *Variable Generation, Flexible Demand*. Academic Press, Oxford.

Sioshansi, F. (2021c). Electric Vehicles: The Ultimate Flexible Demand, in: Sioshansi, F. (Ed.), *Variable Generation, Flexible Demand*. Academic Press, Oxford.

Sioshansi, F. (Ed.) (2020). *Behind & Behind the Meter: Digitalization, Aggregation, Optimization, Monetization*. Academic Press, Oxford.

Sioshansi, F. (Ed.) (2019). *Consumer, Prosumer, Prosumager: How Service Innovations will Disrupt the Utility Business Model*. Academic Press, Oxford.

Swanston, M. (2020). Two Million Plus Solar Roofs: What's in it for the Consumers? in: Sioshansi, F. (Ed.), *Behind & Behind the Meter: Digitalization, Aggregation, Optimization, Monetization*. Academic Press, Oxford.

US Department of Energy (US DOE) (2017). Confronting the Duck Curve, October 12, 2017, www.energy.gov/eere/articles/confronting-duck-curve-how-address-over-generation-solar-energy, Accessed February 27, 2022.

Wilson, A. (2020). "My Home Battery and Rooftop Solar System: How it Performed in its First Year," *Renew Economy*, Augugust 14, https://reneweconomy.com.au/my-home-battery-and-rooftop-solar-system-how-it-performed-in-its-first-year-86315.

World Population Review, US States Ranked by Population – 2022, https://worldpopulationreview.com/states, Accessed February 27, 2022.

25

RETHINKING ENERGY DEMOCRACY

Jennifer Keahey, Majia Nadesan, and Martin J. Pasqualetti

Introduction

In *Energy: A Human History*, Richard Rhodes (2018, xiii) argues that contemporary energy and climate challenges are the "legacies of historical transitions" from wood to coal and oil to gas, nuclear, and renewables. Rhodes argues that conversations regarding these challenges are technical and esoteric, with the discourse making little reference to past realities or current opportunities for public engagement. In a similar vein, Sabeel Rahman (2011) finds that the energy sector is guided by a technocratic impulse that prioritizes the knowledge, autonomy, and authority of energy experts within industry and state governance. This technocratic impulse devalues community participation and suppresses concerns regarding the human-nature relationship, exacerbating the risk of regulatory capture and ineffectiveness. Hyper-rationalized extraction regimes, like that of energy, generate inequitable and irrational outcomes, in part by assuming that the detached logic of the scientific expert is the only valid means of knowing. Angela Roothaan (2019) clarifies the role that techno-rational science has played in producing the global environmental crisis by showing how influential European Enlightenment scholars, such as René Descartes came to limit moral existence to the individual and rational mind. This human-nature rupture has enabled conventional Western science to perceive most life as morally dead matter that one may extract and destroy without consequence.

If broader Enlightenment scholarship sparked the formation of democratic states in Europe and its Western affiliates, overreliance upon techno-rational logic has sown the seeds of democracy's demise. According to Richard Sclove (1995, 7), many of our most foundational industrial technologies perpetuate "antidemocratic power relations" and erode social avenues for "expressing citizenship". The post–World War II industrialized global economy was founded in oil, a form of energy that Timothy Mitchell (2011, 235) argues produced a "dematerialized and denatured" politics. The unique sociotechnical features of cheap oil did not empower workers, unlike with coal, whose 19th-century flows, Mitchell contends, bolstered worker demands for labor rights. Petrochemical extraction and distribution have produced technical, natural, and human entanglements that reinforce distinctions between experts and laypersons, not just in terms of energy production but also in attendant areas, such as science, warfare, industrial management, public health, law, and more. The technocratic impulse has also empowered global finance, causing nature to become denatured, or in other words, stripped of calculations of

DOI:10.4324/9781003183020-29

social and biological risk. Financial faith in the "limitless horizon of growth" has created a "peculiar orientation towards the future" that is empirically false given our collective placement on a planet of finite resources (ibid.).

The idea of a limitless horizon of growth predicated on cheap energy governed by socio-technical experts is a grand narrative that has lost fidelity. Yet it remains entrenched in public policy. If a grand narrative is a master story that organizes and orders representations of what was, what is, and what will be, the master narrative of energy "too cheap to meter" illustrates how the limitless-growth narrative has dominated debates for far too long (Cohn 1997). Despite its failure to accurately describe the costs of oil and nuclear production, this master narrative is now being applied to renewable energy. For example, a June 2020 press release by the International Renewable Energy Agency (IRENA, 2020) claims, "Renewables Increasingly Beat Even Cheapest Coal Competitors on Cost", merely substituting "renewables" for coal, oil, or nuclear retains the master narrative, deflecting attention from material resource challenges and social/health/environmental externalities, or costs that privatized firms shift to the taxpayer after securing profit for their shareholders.

The master narrative of endless cheap energy is co-produced by industry and state regulatory experts. While the energy sector as a whole is failing to account for resource limits, including carbon-era limits on oil and renewable constraints on lithium and rare earths, these limits are making news media headlines. Recent climate disasters are eroding public confidence in a horizon of limitless growth (Reuters 2019). Astronomical electricity bills delivered to Texas citizens in the wake of climate-induced power outages in February of 2021 have dramatized the potential impacts of projected weather irregularities, showing voters that Texas is incapable of anticipating and mitigating disasters (Blunt and Gold 2021). Although the myth of limitless growth has begun to yield to criticism regarding unsustainable false premises, decision-making at all institutionalized levels continues to cling to this master narrative.

The assumption of a perfect, well-ordered, and grid-like society founded on denatured technologies is as unreal as it is unachievable. Therefore, this chapter does not offer yet another material narrative of cheap energy and limitless horizons of optimistic growth. Instead, it adapts the concept of heterotopia – developed by the French theorist Michel Foucault – to revisit the question of energy democracy in an inequitable world. Defined as an alternative sociotechnical order, heterotopia provides a theoretical foundation for examining the interplay between conventional and alternative communities (Foucault and Miskowiec 1986). In the context of energy, heterotopias are distinct from big energy but embedded within existing social and technical geographies. One case in point is the first London neighborhood to have developed a community-based heating system (Johnson 2016). While the British energy policy landscape has continued to prioritize consumption and profit, this neighborhood heterotopia has resisted the status quo by redefining energy as a social good and by developing an organizational infrastructure that enshrines community ownership.

Johnson's example is but a single representation of alternative energy possibilities signified by the heterotopia. The concept of heterotopia gives visibility to social spaces that operate according to alternative social logics and material technologies, clarifying the empirical lack of uniformity in energy solutions and displacing techno-rational management as the sole mode of governance. In this chapter, we engage the concept of the heterotopia to unpack techno-rational grand narratives in relation with the counternarratives of energy democracy, making visible the hidden power dynamics hindering energy transitions to more sustainable futures.

Techno-Rational Gridlock

If energy, in material form, shapes social conditions and possibilities, then it is important to differentiate between renewable and nonrenewable energy sources. When comparing the case of solar energy technologies with oil, it is clear that solar production is more compatible with democratic governance. While oil requires expensive infrastructures to extract, process, and distribute, solar energy is capable of skipping these steps, as it harnesses the power of the sun, a plentiful and locally available source of energy in many parts of the world. For oil producers, meeting the present demand of 90+ million barrels of oil per day, year after year, is a gargantuan effort, enacted by millions of workers – including thousands of highly trained and expensive specialists – in increasingly remote areas of the world. Solar does not require this level of management. Oil-based development is densely centralized, controlled by small groups of investor-owned companies and self-serving government agencies. Solar-based technologies do not require organizational centralization. This is why sustainable development organizations tout solar energy as the ideal energy source for Global South countries lacking in oil reserves, capital, or foreign direct investment (Bugaje 2006). Yet solar is not without its particular shortcomings. Intermittency remains an obvious challenge in solar production and other challenges include the environmental and health impacts of toxic materials that are produced during manufacturing, the need to develop recycling strategies to limit solid-waste disposal when modules are decommissioned and the ecological impact of placing massive solar farms in natural habitats rather than on rooftops.

A heterotopian reading of solar energy moves beyond technological discussions by reminding us that solar production is not isolated from the dominant social order. As discussed in the following section, renewable technologies are unlikely to generate sustainable outcomes in societies experiencing techno-rational gridlock. Dominant social orders maintain power and control by assimilating potentially threatening technologies and narratives into institutionalized practice, as has been seen in the context of sustainability standards in agricultural production and trade (Jaffee and Howard 2010). In the case of energy, the democratic potential of renewable energy sources may be erased by the agents of big energy who are only interested in incorporating renewable energy as an add-on to conventional practice. Indeed, solar energy still remains cost-prohibitive for many, as individual households rather than public or corporate entities may assume the cost of converting to solar infrastructure.

Techno-Rational Assimilation of Renewable Energy

As mentioned earlier, contemporary energy policies are guided by a centralizing and techno-cratic impulse that is reflected in energy ownership and decisions about energy futures. This trend is by no means limited to the oil industry: it is a pervasive element of the entire energy industry. Aggregating and technocratic impulses inform the production of energy commodities ranging from coal, natural gas, and hydropower to nuclear power, where rationalized production and management technologies are especially noticeable. Industry and state actors justify this approach to governance by claiming need for grid stability. They have bought into the theory that the electricity supply would be less efficient, more expensive, and more chaotic without centralized governance. This theoretical assumption has resulted in a world where energy – a life-sustaining good historically produced by communities and households – has become a global commodity extracted and governed by a small number of commercial and governmental entities who are not beholden to meaningful public control.

Once touted as a liberating off-grid technology, solar energy development increasingly is managed by techno-rational interests positioned at the center of the grid despite the potential for decentralized production. If regulatory mandates have encouraged big energy to begin investing in solar developments that allow for economies of scale, these mandates have failed to consider the social dimension of sustainable development, enabling firms to maintain centralized and autocratic control of the generated power. In the United States, many utility companies have worked to repress localized energy democracy by lobbying regulatory interests and donating to political campaigns. In Arizona, the Phoenix-based Public Service (APS) and Salt River Project (SRP) utilities have hindered the democratic development of solar energy in distributed systems such as individually or community-owned solar rooftops by brokering deals with state interests; indeed, such power plays are rooted in the making of modern America (Jones, 2014).

The public remains unaware of the degree of centralized control within the energy sector, as several smokescreens obscure the nature of governance. One smokescreen is the assumption of an effective regulatory apparatus. Within the United States, public utilities operate according to a regulated-business model that was instated in the early 20th century to prevent public utilities from seeking profit (Kibbey 2021). Within this model, utilities pass on the cost of operating expenses to customers, but they cannot raise rates to secure profit. According to Kibby, however, public utilities are allowed to pass on the cost plus an additional percentage for profit to customers when investing in physical infrastructure. This capital expense allowance has incentivized public utilities to make expensive infrastructural investments, even when cleaner and lower-cost solutions are available, perversely engendering a high-cost energy system.

A second smokescreen that limits the democratic potential of renewable energy is the failure of governance to factor in external costs and subsidies of fossil and nuclear fuels. Going into detail on this point is beyond the scope of our discussion here, but the discrepancy is simple to illustrate. When considering the external costs of nuclear energy, including issues related to future decommissioning responsibility, disaster potential, and weapons proliferation, several nations have enshrined liability limits in laws, as in the US case of the Price-Anderson Nuclear Industries Indemnity Act (1957). In the coal industry, external costs include health hazards of air pollution, coal ash disposal, black lung disease, and greenhouse gases, and within the context of oil, a report by the International Monetary Fund (IMF 2013) estimated the global fossil fuel subsidies amounted to $1.9 trillion annually, of which $1.4 trillion was due to externalities and another $800 billion due to climate change.

Although solar power does have some external costs, such as the consequences involved in mining lithium to power batteries and broader costs related to decommissioning and recycling, these are by no means comparable to those found in nuclear, coal, and oil. If the external costs of nuclear, coal, and oil were included in the sale price of the electricity produced from their use, consumption costs would be much higher than presently positioned, putting solar at an even more significant cost advantage than it already enjoys. Rather than accounting for costs and removing subsidies to support market-based transitions to renewable energy, global and national policymakers have pursued a path of deregulation, essentially releasing energy monopolies from financial responsibility for the disasters they cause while empowering them to retain monopolistic control into post-oil futures.

From Renewable to Sustainable Energy

According to Kathleen M. Araújo (2017), the semantic distinction between "renewable" and "sustainable" reflects divergent understandings of how energy is harnessed and used, drawing attention to the social and cultural challenges of sustainable energy transitions. As commonly defined, renewable energy refers to resources that "replenish themselves naturally without being depleted in the earth" and may include sources such as hydropower, geothermal, solar, wind, and ocean energy (Owusu and Asumadu-Sarkodie 2016). Renewable energy provides energy security in the most basic sense of providing a continuous supply of energy for economic infrastructures (Kruyt, van Vuuren, de Vries, and Groenenberg 2009). However, renewable energy is not inherently socially, biologically, or ecologically sustainable if its supply and utilization chains fail to promote access, health, and welfare. In the introduction to this volume, Araújo broadly defines energy transitions "as a notable qualitative, quantitative, or geo-spatial shift in how energy is sourced, delivered, or utilized as well as the emergence or decline of an energy industry", suggesting that social shifts need to occur in tandem with energy sourcing shifts. To address social issues pertaining to access and equity, we argue that renewable energy must be implemented as but one component of a broader sustainable energy project.

Beyond the materially grounded definition of renewability, the concept of sustainability incorporates what philosopher Michel Foucault called biopolitics (Burchell, Davidson, and Foucault 2008). In essence, biopolitics denotes governance that elicits and administers vital (biological) forces. There are many historical examples of biopolitical campaigns and institutions that have sought to define and govern life. These range from reactionary and destructive campaigns, such as the early-20th-century eugenics movement – which attempted to create a master Aryan race by eradicating ethnic and racial difference – to more socially progressive movements, such as mid-20th-century public nutrition campaigns for poor children.

Contemporary sustainability movements are employing a *biopolitics of regeneration* to contest the socially and environmentally destructive impacts of centralized techno-rational systems. These movements offer alternative models of decentralized energy production and control, often envisioned within loosely coupled systems that offer resilience against localized failures. The empirical diversity of sustainability initiatives occurring across product sectors ranging from food and manufactured goods to energy and rural livelihoods illustrates the profound biopolitical differences. Indigenous heterotopias provide critical insight into human/nature relationships. Calling for the demilitarization of conservation practices, these spaces of resistance argue that we cannot effectively protect the environment by banishing people from nature. Not only are humans part of nature, but such protocols can only inhibit the public from seeing the impacts of degradation and holding experts accountable for poor management. Thus, in indigenous conservation heterotopias, the goal is to heal the human/nature rupture through the development of securitization practices that cultivate protective and caring human/nature relationships (Kashwan et al. 2021).

As illustrated in this definition of sustainability drawn from the US EPA (n.d.) website, there has been a historical trend toward the development of more inclusive, interdependent, and ecologically formulated biopolitical visions:

> Sustainability is based on a simple principle: Everything that we need for our survival and well-being depends, either directly or indirectly, on our natural environment. To pursue sustainability is to create and maintain the conditions under which humans and nature can exist in productive harmony to support present and future generations.

According to the global definition provided by the World Commission on Environment and Development (WCED 1987), sustainability emphasizes social, ecological, and economic harmony for present and future generations. It also recognizes that human well-being cannot be ensured without ensuring the flourishing of other species, with whom we are fundamentally interconnected. More specific definitions of sustainability are culturally and temporally situated but increasingly recognize that interspecies ethics is integral to achieving sustainability's social justice mandate (Probyn-Rapsey et al. 2016). Although enacted in different contexts and often diluted by the dominant social order, the discourse on sustainability writ large represents a utopian biopolitical vision of security, whose realizations will be contingent upon technological, political, and sociocultural innovations.

A sustainability framework sheds light on the complex challenges hindering energy transitions. While there is an urgent need to innovate biophysical technologies that expand energy storage capacity and reduce reliance upon toxic rare earths, alternative technologies are in development. As a case in point, the magnetization of nickel and other common metals are being innovated to reduce reliance upon the rare-earth permanent magnets used to produce energy-efficient cars, air-conditioning systems, and other goods (Matizamhuka 2018). However, a social sustainability reading also emphasizes the importance of democratizing energy production and control. To achieve such an agenda, organizational innovations, such as the development of loosely coupled energy producer networks, may facilitate decentralized control while engaging in a practice of energy sharing to reduce the risk of energy outages when local sources become overloaded. Democratization also cannot occur without investing in political innovations that open room for community-based production and control. Finally, there is a need to consider what Horlings (2015, 163) term "the inner dimension of sustainability" or the personal and cultural values that either hinder or inspire change. Although movement toward sustainable energy requires broad-scale systemic investments by nation-states, system-wide change should not be confused with universal design principles: the concept of the heterotopia requires prioritization of local conditions and needs over homogenized infrastructural development (Foucault and Miskowiec 1986).

As the following section discusses, energy transitions are unlikely to be sustainable unless development approaches prioritize community ownership. Inspired storytelling is one example of a critically needed driver of change because this cultural practice enables communities to develop a collective vision of a better future. In "Environmental Justice Storytelling: Angels and Isotopes at Yucca Mountain, Nevada", Donna Houston (2012) demonstrates how storytelling can prompt the hard work of transformation by building conceptual bridges between the unsustainable present and more ecological and democratic futures. This process of creating visions for change must be in significant part defined by grassroots constituencies who make sustainability meaningful and achievable within specific socio-ecological spaces. Bidtah Becker and Dana E. Powell (2022) add to the burgeoning discourse in this subject by using the case of the Navajo Nation to illustrate how storytelling infrastructure is being created to mitigate the dual impacts of climate change and COVID-19 on Navajo communities in the US Southwest.

Energy Utopias and Heterotopias

Utopian visions of energy futures are grounded in the material possibilities of particular places and times, where they either languish in the philosophical realm of ideals or become enacted as key drivers of change. Energy utopias often emerge in connection with newly identified sources of energy, as Mark Schrope (2001) has argued in "Which Way to Energy Utopia?" Although idealistic visions struggle to overcome systemic barriers, such as public resistance to change and political opposition, energy utopias can and do prompt meaningful transitions. Providing a case in point, Abraham Tidwell and Jacqueline Hettel Tidwell (2022) show how the current dominant energy system in the United States was once a public service utopia. However imperfect this model has proven, the ideal of public service company led a nation of people to transition from reliance upon in-home coal usage to a publicly regulated power grid in the early 20th century.

Heterotopias represent a trajectory away from the status quo toward an envisioned alternative that guides efforts to transform existing institutional arrangements and forms of life. Yet heterotopias are rarely utopian in practice. The concept of the heterotopia provides a framework for problematizing the interplay between a perfectly imagined energy utopia and an imperfectly actualized alternative energy assemblage. If assemblages are material sites that join various components into a unified whole, "utopias are sites with no real place" and are therefore "fundamentally unreal spaces" (Foucault and Miskowiec 1986, 24). Joining these two concepts, heterotopia denotes a materially enacted utopia, or a counter-site assemblage constructed in accordance with a particular utopian vision. The concept of the heterotopia therefore raises the question of the relationship between normative ideals and achieved sociotechnical relations.

The scale and social inclusivity of transformative efforts often explain the gap between ideals and achievements. Heterotopias differ in terms of breadth of geographic scale and degree and characteristics of formal organization. On one hand, they include self-sufficient collectives for which residents follow a profoundly different mode of existence, as in Amish communities, where the daily rhythm is defined by agrarian and religious precepts. On the other hand, they also include inchoate spaces carved out by alternative societies at the fringes of the existing social system, as illustrated by off-grid survivalist communities.

A defining feature of heterotopia is its paradoxical emplacement outside/within the existing social order. Heterotopias are situated outside in so far as they are driven by counterculture logic that expressly differs from the logic of the status quo. Yet these spaces are physically surrounded by and in communication with surrounding institutional spaces. In other words, the transformative knowledges and practices of heterotopias are, to some degree, bounded by the knowledges and practices of the dominant system in which they mirror. Given their tangential position to an existing social order, the actualized alternatives that heterotopias offer are a product of the collision between paradoxical forces. Within the context of energy, heterotopian systems both upset and reflect the institutionalized geographies and techno-rational logics of energy production, distribution, and consumption.

Heterotopian Production Networks

Energy democracies have been developing in tandem with sustainability and social justice movements in other product sectors, whose experiences provide productive instruction. Among these, the case of fair trade provides critical insight into the challenges and prospects of carving out space for alternative production networks in a neoliberal world system. Now operating within the food, handicrafts, and manufacturing sectors, the global fair trade movement originated the utopian vision of Edna Ruth Byler, a Mennonite social entrepreneur who began selling crafts produced by a Puerto Rican women's group after a 1946 visit to the island (Keahey, Littrell, and Murray 2011). By the end of the 20th century, Byler's enterprise had become Ten Thousand Villages, a nonprofit business selling artisanal goods through a network of shops in Canada and the US. Edna Byler was not alone in her vision, for other fair trade networks developed under different governance structures in the latter half of the 20th century.

Some of the fair trade heterotopias that have been established are radical in orientation, with devotees such as Equal Exchange establishing a worker-owned cooperative retail center that buys goods directly from democratically organized small-scale producer cooperatives in accordance with fair pricing agreements and long-term trade arrangements. Other fair trade heterotopias build bridges to conventional markets, with Fairtrade International (FTI), instituting a third-party product certification system that brings fairly traded goods into conventional retail spaces that have more purchasing power. While divergent in practice, these fair trade systems have remained loosely coupled in mission, in part through adherence to the International Fair Trade Charter (2018), which defines fair trade as

> a trading partnership, based on dialogue, transparency and respect, that seeks greater equity in international trade. It contributes to sustainable development by offering better trading conditions to, and securing the rights of, marginalized producers and workers – especially in the [Global] South.

In "Fair Trade: Social Justice and Production Alternatives", Raynolds and Keahey (2014, 167) argue that alternative production networks such as fair trade operate both within and against markets by "working through market channels to create alternative commodity networks . . . and simultaneously working against the conventional market forces that create and uphold inequalities". If this paradoxical approach has made the fair trade concept vulnerable to neoliberal market assimilation, the more radical heterotopias within the movement serve as a moral voice for deepening commitments to justice and sustainability, and more broadly embracing a biopolitics of regeneration.

A fair trade assemblage has not yet come into existence within the energy sector, but sustainable energy heterotopias share in common the strategy of working both within and against conventional energy markets. Figure 25.1 provides a visual map of renewable energy approaches, illustrated as heterotopian spaces situated in greater or lesser opposition to conventional energy, with degree of permeability depicted by dashes. In the following section, we discuss how the energy democracy movement is seeking to embed equity, justice, and democracy standards into a variety of renewable energy heterotopias. These range from more conventional initiatives that are working within centralized energy infrastructures to scale up renewable energy to radical delinking approaches that are involved in establishing grassroots energy cooperatives. While centralized initiatives are positioned to transition the energy sector writ large to renewable technologies, partnership with big energy threatens the democratic and justice aims of energy sustainability. In contrast, radical energy heterotopias may have less impact in terms of ameliorating pressing environmental problems because these are situated outside centralized systems that have the capacity to achieve widespread transitions. Yet the establishment of community-based sustainable energy cooperatives are of central import to 21st-century energy transitions, for these extend the normative boundaries of what is and what could be due to their revolutionary emphasis on institutional transformation.

Centralized Transitions

As electrical grids are digitalized and renewable technologies, such as solar and wind power, become more affordable, utilities now have the capacity to develop multidirectional energy systems. This technological shift has enabled energy consumers to become co-producers, sending excess energy back to the grid. Within the terrain of centralized transitions, renewable energy advocates work with conventional utilities to reduce reliance upon fossil fuels; thus, smart grid transitions represent a critical step in moving renewable energy from the margins to the center of the energy sector. The energy democracy movement also has had an impact upon smart grid campaigns, which often are marketed as socially inclusive energy transitions that empower users to become actively involved in the transition to sustainability. Yet Tarasova and Rohracher (2022) uncover a more troubling story when interrogating smart grid rollout in Sweden.

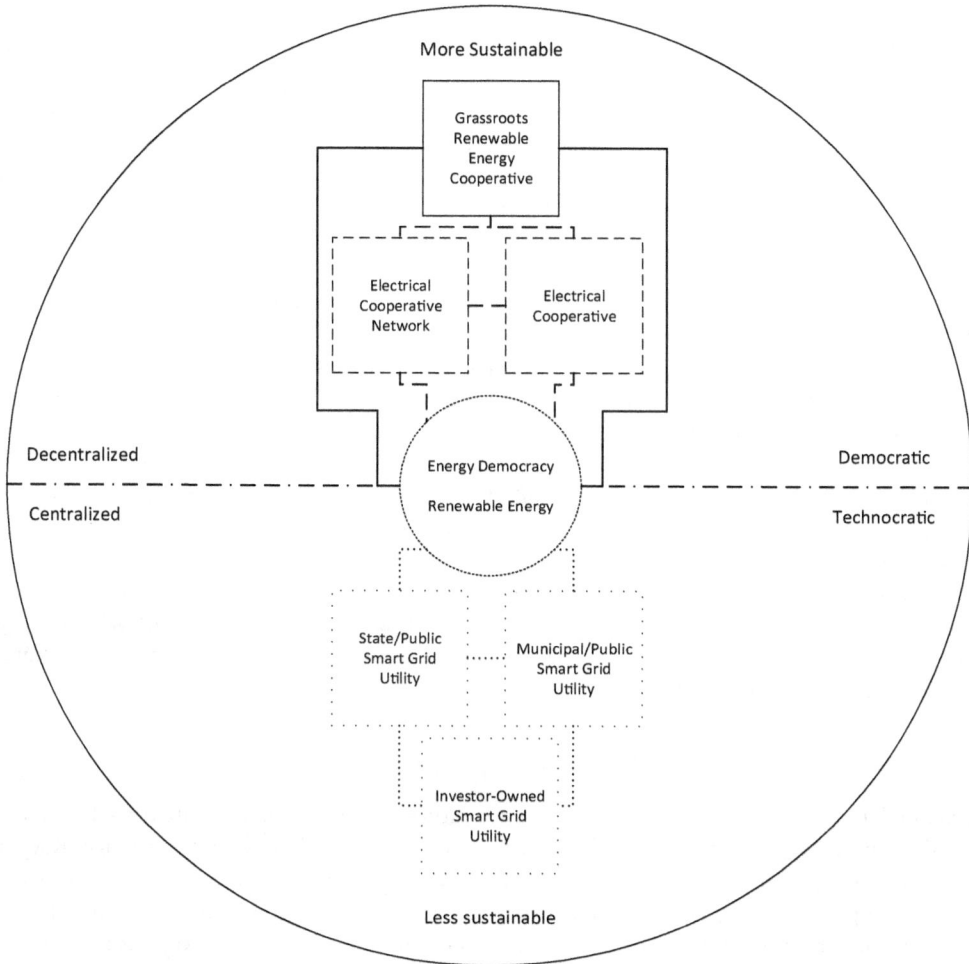

Figure 25.1 Map of renewable energy approaches.

Although Sweden's program employed public funding and sought active user engagement, the techno-rational configuration and economic dictates of the existing energy infrastructure have generated hard and soft forms of social exclusion, ultimately privileging middle-class users while marginalizing the poor. This is because the elite actors driving Sweden's smart grid transition have made little effort to acknowledge the diversity of energy users, let alone take steps to incorporate diverse perspectives into governance and practice.

On the other side of the world, Japan has instituted its own large-scale transition to smart grid technology. Rather than taking an active-user approach, as in the case of Sweden, the Japanese model was solely a government-led and business-driven affair (Mah et al. 2013). While the rollout of four large projects achieved functionality and ecological benefit, the techno-rational focus of Japan's smart grid transition prevented these projects from achieving any higher-level or third-order transformations, such as changes to regulations or the emergence of new business models. Indeed, these projects failed to deliver any social benefits to their customers or broader user communities.

It is clear that smart grid transitions are capable of reducing reliance upon fossil fuels by harnessing state and commercial resources to drive large-scale transitions. When implemented in connection with an active-user agenda, as in the Swedish case, prospects for social sustainability become plausible. Yet it is also clear that centralized endeavors are falling short of their promise, even when socially inclusive agendas are centered in planning. The vast differences between material outcomes and marketing suggest the assimilation of energy democracy rhetoric in centralized smart grid transitions. This issue crosscuts different geopolitical landscapes. According to Lourdes Alonso-Serna and Edgar Talledos-Sánchez (2022), wind farm developments in Mexico are undergoing a similar pattern of assimilation, causing renewable energy to become entrenched within a conventional energy sector that has no interest in democratizing governance or ensuring energy access to all.

Despite these trends, the energy democracy movement is becoming more vocal in its demands for change, and this is helping to reshape what is politically possible. As new energy actors move from the margins to the center by entering the political realm of policymaking, they are igniting a political struggle that speaks to the potential of organizing energy democracy at multiple scales (Brisbois 2022). As the collective voice of alternative energy heterotopias, the energy democracy movement is pushing centralized smart grid systems to consider participatory governance mechanisms. If centralized smart grid transitions provide one point of action for the energy democracy movement, the realization of more foundational sustainability transitions will require investment in "distributed, decentralized, and heterogeneous renewable-energy systems" that are capable of addressing deeper issues pertaining to political, social, and cultural resilience (Stephens 2019, 341).

Decentralized Transitions

Energy democracy may be a buzzword in conventional renewable energy initiatives, but in the outsider spaces of decentralized transition, it is a revolutionary call for systemic change. Kacper Szulecki (2018) defines energy democracy as a three-dimensional governance system enacted through popular sovereignty, participatory governance, and civic ownership. In this conception, energy democracy strategically challenges the detached rationality underlying both statist and capitalist logics. Within the utopian ideal of the decentralized energy cooperative, the social relations of production are inverted from the conventional norm. Here, the user is a prosumer-citizen empowered by their co-ownership of the means of production to be critically informed, actively involved, and democratically engaged (2018, 12). By liberating themselves from energy oligopolies, prosumers can become involved in efforts to establish a more sustainable energy system, comprising dispersed community-based cooperatives that bring even the most disadvantaged users in the community into ownership and governance. If this utopian vision shares ties with socialist visions of social change, as Szulecki notes, it is by no means a statist vision, for energy democracy conceptualizes decentralized and community-based approaches to energy democracy.

The notion of cooperative-based governance is by no means new to the energy sector. Not only was the idea of public control a key driver in the formation of publicly owned utility companies within US states and municipalities, but the progressive New Deal politics of the early 20th century enabled rural communities to invest in electric cooperatives, bringing electricity into rural spaces that for-profit energy utilities ignored (Pacyniak 2020). Electric cooperatives also have been established in the Global South, where rural communities likewise lacked access to energy grids. According to Yadoo and Cruickshank (2010), electric cooperatives in Nepal emerged in the wake of structural adjustment in the 1980s and 1990s, when international finance forced Global South nations to privatize services and cut public spending in accordance with neoliberal economic philosophy. Driven by capitalist logic, private companies neglected rural areas that were deemed too unprofitable for investment, and in response to this problem, Nepal turned to an age-old tradition of forming community-based organizations (CBOs). With funding from the United Nation's Development Program, Nepalese activists instituted CBO-led grid extensions by forming into energy cooperatives, later using profits generated from electricity sales to provide micro-financing for cooperative members to invest in small-scale income generation activities.

As heterotopian counter-sites to centralized energy, these examples of electric cooperatives demonstrate the potential of decentralized transitions. Yet they also illustrate the challenges involved in ensuring the kind of support and oversight that enables cooperatives to achieve environmental and social agendas. Although rural electric cooperatives located in wealthy areas are proving to be more willing to innovate, with two Colorado associations encouraging the adoption of electric vehicles (Best 2018), most rural American cooperatives actually have been slower to convert to renewable energy than have municipal and for-profit utilities (Pacyniak 2020). In the case of rural Nepal, the formation of CBO cooperatives succeeded in achieving the energy democracy imperatives of civic ownership and participatory governance but likewise failed to center renewable energy (Yadoo and Cruickshank 2010). Just as there is no guarantee that renewable energy will generate more socially just energy systems, decentralized transitions do not necessarily enshrine environmentally responsible practices. Thus, a key question facing energy democracy heterotopias is how to develop socio-legal infrastructures that support both social and environmental aims without reinscribing top-down control.

Sustainable Energy Challenges and Prospects

The utopian ideal of renewable energy democracies offers considerable promise to a world that is grappling with interlocking environmental, social, and political crises. There has been a tendency to consider local renewable energy cooperatives, especially solar cooperatives, as potentially disruptive for driving sustainable and democratic change, yet key technical and social challenges remain (Hufen and Koppenjan 2015). While the renewable energy cooperative may be broadly perceived by energy democracy activists as the gold standard of energy transitions, a heterotopian reading of energy alternatives illustrates the difficulties involved in inverting existing energy narratives and systems. The challenges facing renewable energy democracy go beyond any one dimension of materiality, for if transition to renewable energy is technologically challenging, energy democracies additionally must confront the economic, political, and socio-cultural dictates of institutionalized energy. In part, this task involves working against techno-rational assumptions regarding limitless horizons and the human/nature relationship.

At the most foundational level, sustainable energy demands a moral recognition of the inter-dependence of all life, and a commitment to inverting destructive systems for the sake of genera-tions to come. Yet sustainable energy must also confront a wide range of material issues as well, including ongoing technological challenges with renewables, entrenched geopolitical interests, and economic barriers.

A key renewable energy challenge is that the construction of solar panels and wind turbines continue to rely on nonrenewable rare earths. Yet rare earth mining produces environmental damage and refinement is largely controlled by one nation, China, which has declared will-ingness to restrict exports in response to trade conflicts (Reuters 2019). A 2007 report on the geopolitics of renewables observed that conflict may arise over capital for investment and tech-nology, technology transfer, asymmetries in dependencies between producers and consumers leading to escalating geopolitical conflicts, new resource curses, and increased inequalities (O'Sullivan, Overland, and Sandalow 2017). Efforts to develop rare earth alternatives are underway, but much work remains. More broadly, many skeptics argue that no renewable energy source could permit the continued expansion of Global North lifestyles and consumption, with Boggs (2012, 114–115) going so far to claim that the failure to acknowledge this material truth is but a "liberal delusion".

A key political challenge is decentralization. It is unlikely that significant changes in energy landscapes will occur in the absence of state involvement; however, centralized energy systems have only emerged through democratic engagement on rare occasions and, apart from such instances, have failed to provide opportunities for democratic decision making and community wealth-building (Fairchild and Weinrub 2017). Human security is closely tied to energy security; thus, energy future narratives produced by institutional actors typically are encumbered with a dominant security paradigm rooted in a Hobbesian worldview of scarcity and conflict. Energy's connection to national-security assemblages has been well documented (Bridge et. al. 2018; Pascual and Elkind 2010). Kacper Szulecki (2022) suggests that the democratization of energy hinges on new and unencumbered conceptions of security. Hence, energy security will figure largely in alternative visions of renewables, with specific heterotopias varying significantly in terms of the values informing specific articulations of security (Cherp 2012). These divergences will shape preferences for everything from selected energy forms to production and distribution practices and governance modalities. If biodiverse ecologies are more resilient than monocultures in the face of shock, then it follows that decentralized energy networks designed to function in richly diverse ways will strengthen human resilience to the sweeping challenges of climate change.

A key economic challenge is how to redirect market logic from the path of capitalist accumulation to one of social provision. David Harvey (2018) reminds us that the logic of capitalism is inherently flawed. By requiring never-ending growth (defined in biology as cancer) to ensure its survival, not only capitalism does end up harming people and planet, but it also sows the seeds of its own demise. While capital is ingenious at circumventing barriers to economic growth, it is incapable of responding to systemic problems with anything but short-term solutions. For example, when people become too indebted to buy the goods that capitalism offers, the only way for capital to survive is by finding new consumers or by lowering production costs. As a deregulated model, neoliberal capitalism accelerates this process by forcing societies around the world to participate in an economic race to the bottom. Compelled to offer ever cheaper sources of labor and raw goods to a ravenous market, societies diminish their own consumer purchasing power.

Within the global fair trade movement, actors have achieved some success in halting this process by reregulating production and trade through the use of third-party certifications that operate according to stringent social and environmental standards (Raynolds and Greenfield 2015). Inverting industrial conventions that prioritize cost and quantity, fair traders use market logic to redefine product quality in terms of social and environmental welfare. As new energy actors enter into previously closed commercial and political spaces where they are disrupting the status quo, there is potential for introducing third-party auditing systems that add value to renewable energy produced by worker-owned cooperatives and unionized utilities through the use of fair trade certifications that appeal to green consumers.

Despite the technological challenges of renewable energy described earlier, the advent of modern solar and wind power does bring with it the option of delinking from non-sustainable political and economic spaces. Within the modern industrial US, off-grid energy movements have existed at least since the mid-20th century, when a generation of back-to-the-landers broke from society to build off-grid homes. The spirit of this environmental movement remains alive in

21st-century energy heterotopias, where grassroots activists are harnessing renewable technologies to construct off-grid energy grids maintained by community-based cooperatives (Brennan 2022). While highly localized, the effects of these efforts may yet be compounded through the development of loosely coupled regional networks that build mutual resilience. Rather than pushing big energy to invest in renewable and democratic energy systems, these insurrectional energy actors are seeking to create regional zones where reliance upon big energy simply becomes obsolete.

Conclusion

Renewables are not necessarily sustainable, but this chapter has demonstrated how a transition to renewable energy can be guided by a biopolitics of sustainability that decentralizes and democratizes governance. We have examined the concept of the heterotopia to show how sustainable spaces can be carved out of existing institutional systems of energy production as a strategy for inverting systemic inequalities and externalities. However, the capacities to carve out such sustainable communities is limited by resource and cultural constraints. The concept of the heterotopia is distinguished from utopia by this essential acknowledgment of the gap between the achievement and the ideal; hence, we see in these resource constraints how social transformation is complicated by material and cultural challenges, such as barriers to localized infrastructural investments. Changes in energy policy are needed to support the development of ethical supply chains that deliver sustainable outcomes. Additionally, the narrative of endless cheap energy must be shifted to enable new community value orientations and future visions of energy to emerge. Recent research suggests that storytelling will be a critical tool for developing the collective visions needed to assemble renewable technologies in connection with socially sustainable practices. Community-based energy cooperatives illustrate avenues for building envisioned energy heterotopias, but there remain significant cost and resource barriers that require substantial infrastructural investments by governments, particularly if community-based energy cooperatives are not to become an option solely available to affluent communities with the financial means to establish a distributed network.

This chapter offers a framework for envisioning change at distinct scalar levels, in part modeled on sustainability work in fair trade. Accordingly, it recognizes that neither centralized nor decentralized energy transitions are likely to achieve systemic change on their own. Where centralized efforts fall short of the social mandate of energy sustainability and, more broadly, are incapable of achieving energy democracy writ large, decentralized efforts fall short of the renewable energy mandate and lack the resources needed to achieve change at scale. Yet when viewed as two prongs of an interrelated renewable energy and energy democracy movement, these centralized and decentralized heterotopias hold promise for by working against and within the conventional energy sector, and they may yet generate a world where energy production is increasingly diversified and dynamic.

References

Alonso-Serna, Lourdes and Edgar Talledos-Sánchez. 2022. Fossilizing Renewable Energy: The Case of Wind Power in the Isthmust of Tehuantepec, Mexico. In: Majia H. Nadesan, Martin Pasqualetti and Jennifer Keahey (eds). *Energy Democracies for Sustainable Futures*, New York: Elsevier.

Araújo, Kathleen M. 2017. *Low Carbon Energy Transitions: Turning Points in National Policy and Innovation.* Oxford: Oxford University Press.

Becker, Bidtah N. and Dana E. Powell. 2022. Just Energies: Storytelling Infrastructure, Climate, and COVID in the Navajo Nation. In: Majia H. Nadesan, Martin Pasqualetti and Jennifer Keahey (eds). *Energy Democracies for Sustainable Futures*, New York: Elsevier.

Best, Allen. 2018. In Colorado, Two Rural Co-ops Are Leading the Charge on Electric Vehicles. *Energy News Network*, September 19, 2018. https://energynews.us/2018/09/19/in-colorado-two-rural-co-ops-are-leading-the-charge-on-electric-vehicles.

Blunt, Katherine and Russell Gold. 2021. The Texas Freeze: Why the Power Grid Failed. *The Wall Street Journal*, February 19, 2021. www.wsj.com/articles/texas-freeze-power-grid-failure-electricity-market-incentives-11613777856?mod=hp_lead_pos7.

Boggs, Carl. 2012. *Ecology and Revolution: Global Crisis and Political Challenge*. New York: Palgrave.

Brennan, Ry. 2022. Technoregions of Insurrection: Decentralizing Energy Infrastructures and Manifesting Change at Scale. In: Majia H. Nadesan, Martin Pasqualetti and Jennifer Keahey (eds). *Energy Democracies for Sustainable Futures*, New York: Elsevier.

Bridge, Gavin, Stewart Barr, Stefan Bouzarovski, Michael Bradshaw, Ed Brown, Harriet Bulkeley and Gordon Walker. 2018. *Energy and Society: A Critical Perspective*. London: Routledge.

Brisbois, Marie Claire. 2022. Power Relations and the Possibilities of Participatory Energy Governance. In: Majia H. Nadesan, Martin Pasqualetti and Jennifer Keahey (eds). *Energy Democracies for Sustainable Futures,* New York: Elsevier.

Bugaje, I. M. 2006. Renewable Energy for Sustainable Development in Africa: A Review. *Renewable and Sustainable Energy Reviews* 10: 603–612.

Burchell, Graham, Arnold Davidson and Michel Foucault. 2008. *The Birth of Biopolitics: Lectures at the Collège de France, 1978–1979*. New York: Springer.

Cherp, A. 2012. Defining Energy Security Takes More Than Asking Around. *Energy Policy* 48: 841–842.

Cohn, Steven Mark. 1997. *Too Cheap to Meter: An Economic and Philosophical Analysis of the Nuclear Dream*. Albany: SUNY.

Fairchild, Denise and Al Weinrub, eds. 2017. *Energy Democracy: Advancing Equity in Clean Energy Solutions*. Washington: Island Press.

Foucault, Michel and Jay Miskowiec. 1986. Of Other Spaces. *Diacritics* 16(1): 22–27.

Harvey, David. 2018. *The Limits to Capital*. London: Verso.

Horlings, L. G. 2015. The Inner Dimension of Sustainability: Personal and Cultural Values. *Current Opinion in Environmental Sustainability* 14: 163–169.

Houston, Donna. 2012. Environmental Justice Storytelling: Angels and Isotopes at Yucca Mountain, Nevada. *Antipode* 45(2): 417–435.

Hufen, J. A. M. and J. F. M. Koppenjan. 2015. Local Renewable Energy Cooperatives: Revolution in Disguise? *Energy, Sustainability and Society* 5: 18. Doi 10.1186/s13705-015-0046-8.

IMF. 2013. *Energy Subsidy Reform: Lessons and Implications*. London: International Monetary Fund.

International Fair Trade Charter. 2018. "The International Fair Trade Charter." www.fair-trade.website/the-charter-1. Accessed May 30, 2021.

International Renewable Energy Agency. 2020. Renewables Increasingly Beat Even Cheapest Coal Competitors on Cost. June 02, 2020. www.irena.org/newsroom/pressreleases/2020/Jun/Renewables-Increasingly-Beat-Even-Cheapest-Coal-Competitors-on-Cost

Jaffee, Daniel and Philip H. Howard. 2010. Corporate Cooptation of Organic and Fair Trade Standards. *Agriculture and Human Values* 27: 387–399.

Johnson, Charlotte. 2016. District Heating as Heterotopia: Tracing the Social Contract Through Domestic Energy Infrastructure in Pimlico, London. *Economic Anthropology* 3(1): 94–105.

Jones, Christopher F. 2014. *Routes of Power*. Cambridge: Harvard University Press.

Kashwan, Prakash, Rosaleen V. Duffy, Francis Massé, Adeniyi P. Asiyanbi and Esther Marijnen. 2021. From Racialized Neocolonial Global Conservation to an Inclusive and Regenerative Conservation. *Environment: Science and Policy for Sustainable Development* 63: 4–19.

Keahey, Jennifer, Mary A. Littrell and Douglas L. Murray. 2011. Business with a Mission: The Ongoing Role of Ten Thousand Villages within the Fair Trade Movement. In: Alain Epp Weaver (ed). *A Table of Sharing: Mennonite Central Committee and the Expanding Networks of Mennonite Identity,* 265–283. Telford, PA: Cascadia Publishing House.

Kibbey, J. C. 2021. Utility Accountability 101: How Do Utilities Make Money? National Resource Defense Council January 20, 2021. www.nrdc.org/experts/jc-kibbey/utility-accountability-101-how-do-utilities-make-money

Kruyt, Bert, D. P. van Vuuren, H. J. M de Vries and H. Groenenberg. 2009. Indicators for Energy Security. *Energy Policy* 37(6): 2166–2181,

Mah, Daphne Ngar-yin, Yun-Ying Wu, Jasper Chi-man Ip, and Peter Ronald Hills. 2013. The Role of the State in Sustainable Energy Transitions: A Case Study of Large Smart Grid Demonstration Projects in Japan. *Energy Policy* 63: 726–737. doi: https://doi.org/10.1016/j.enpol.2013.07.106.

Matizamhuka, Wallace. 2018. The Impact of Magnetic Materials in Renewable Energy-Related Technologies in the 21st Century Industrial Revolution: The Case of South Africa. *Advances in Materials Science and Engineering* 3149412.

Mitchell, Timothy. 2011. *Carbon Democracy: Political Power in the Age of Oil*. London: Verso.

O'Sullivan, Meghan, Indra Overland and David Sandalow. 2017. The Geopolitics of Renewable Energy. HKS Working Paper No. RWP17-027, Available at SSRN: https://ssrn.com/abstract=2998305 or http://dx.doi.org/10.2139/ssrn.2998305.

Owusu, Phebe Asantewaa and Samuel Asumadu-Sarkodie. 2016. A Review of Renewable Energy Sources, Sustainability Issues and Climate Change Mitigation. *Cogent Engineering* 3(1): 1167990. doi: 10.1080/23311916.2016.1167990.

Pacyniak, Gabriel. 2020. Greening the Old New Deal: Strengthening Rural Electric Cooperative Supports and Oversight to Combat Climate Change. *Missouri Law Review* 85(2): 409–494.

Pascual, Carlos and Jonathan Elkind, eds. 2010. *Energy Security: Economics, Politics, Strategies and Implications*. Washington, DC: Brookings.

Probyn-Rapsey, Fiona, Sue Donaldson, George Ioannides, Tess Lea, Kate Marsh, Astrida Neimanis, Annie Potts, Nik Taylor, Richard Twine and Dinesh Wadiwel. 2016. A Sustainable Campus: The Sydney Declaration on Interspecies Sustainability. *Animal Studies Journal* 5(1): 110–151.

Rahman, K. Sabeel. 2011. Envisioning the Regulatory State: Technocracy, Democracy, and Institutional Experimentation in the 2010 Financial Reform and Oil Spill Statutes. *Harvard Journal on Legislation* 48(2): 555–590.

Raynolds, Laura T. and Nicholas Greenfield. 2015. Fair Trade: Movement and Markets. In: Laura T. Raynolds and Elizabeth A. Bennett (eds). *Handbook of Research on Fair Trade*, 24–41. Northampton, MA: Edward Elgar.

Raynolds, Laura T. and Jennifer Keahey. 2014. Fair Trade: Social Justice and Production Alternatives. In: Martin Parker, George Cheney, Valérie Fournier and Chris Land (eds). *The Routledge Companion to Alternative Organization,* 165–181. New York: Routledge.

Reuters. 2019. China's Rare Earth Supplies Could Be Vital Bargaining Chip in U.S. Trade War. May, 29, 2019. www.reuters.com/article/us-usa-china-rareearth-explainer/explainer-chinas-rare-earth-supplies-could-be-vital-bargaining-chip-in-u-s-trade-war-idUSKCN1T00EK

Rhodes, Richard. 2018. *Energy: A Human History*. New York: Simon & Schuster.

Roothaan, Angela. 2019. *Indigenous, Modern and Postcolonial Relations to Nature: Negotiating the Environment*. New York: Routledge.

Schrope, Mark. 2001. Which Way To Energy Utopia? *Nature* 414(6865): 682–684. doi: 10.1038/414682a.

Sclove, Richard. 1995. *Democracy and Technology*. New York: Guildford.

Stephens, Jennie C. 2019. Assessing Resilience in Energy System Change Through an Energy Democracy Lens. In: Matthias Ruth and Stefan Goessling-Reisemann (eds). *Handbook on Resilience of Socio-Technical Systems*, 341–359. Northampton: Edward Elgar.

Szulecki, Kacper. 2022. Does Security Push Democracy Out of Energy Governance? In: Majia H. Nadesan, Martin Pasqualetti and Jennifer Keahey (eds). *Energy Democracies for Sustainable Futures*, New York: Elsevier.

Szulecki, Kacper. 2018. Conceptualizing Energy Democracy. *Environmental Politics* 27(1): 21–41. doi: 10.1080/09644016.2017.1387294.

Tarasova, Ekaterina and Harald Rohracher. 2022. Democratizing Energy through Smart Grids? Discourses of Empowerment vs Practices of Marginalization. In: Majia H. Nadesan, Martin Pasqualetti and Jennifer Keahey (eds). *Energy Democracies for Sustainable Futures*, New York: Elsevier.

Tidwell, Abraham and Jacqueline Hettel Tidwell. 2022. Samuel Insull and the Public Service Utility Imaginary. In: Majia H. Nadesan, Martin Pasqualetti and Jennifer Keahey (eds). *Energy Democracies for Sustainable Futures*, New York: Elsevier.

US EPA. n.d. Learn About Sustainability. US Environmental Protection Agency. www.epa.gov/sustainability/learn-about-sustainability#what. Accessed October 5, 2021.

WCED. 1987. *Our Common Future*. Oxford: Oxford University Press.

Yadoo, Annabel and Heather Cruickshank. 2010. The Value of Cooperatives in Rural Electrification. *Energy Policy* 38(6): 2941–2947. doi: https://doi.org/10.1016/j.enpol.2010.01.031.

26

WHOSE TRANSITION?
A REVIEW OF CITIZEN
PARTICIPATION IN THE
ENERGY SYSTEM

Niall P. Dunphy and Breffní Lennon

1. Introduction

Energy, and indeed the energy system, is tightly intertwined with people's everyday lives – albeit as Ambrose (2020) observes, it is perhaps often somewhat invisible. At the same time, as Axon et al. (2018) suggest, energy-related culture, practices, and behaviors can be said to "constitute a powerful human factor in the energy system". Therefore, it can be understood that the current moves to decarbonize energy systems will both necessitate and result in significant changes to the way in which people go about their daily lives. Of course, there is no one single route to decarbonizing societies, and indeed, there is no single vision as to the intended destination[1]. There are many decisions yet to be made, which will shape and influence what decarbonized energy systems will look like and how they may be achieved.

There is a growing realization that the success of the ongoing energy transition is dependent on the perceived legitimacy and social acceptability of the societal transformation required for its actualization. Consequently, citizen participation in the energy discourse is increasingly seen as an important aspect of the energy transition. Although as Mullally et al., (2018, 71) note, while there is some agreement that energy policy should "no longer be the exclusive concern of public institutions and utilities", the nature and extent of this participation remains contested.

Citizen participation in the energy domain has traditionally been thought of in terms of their role as a consumer. Increasingly, citizens are requested to be active consumers and to use their purchasing power and consumption practices to both reduce their environmental impact and to influence the market (Fox, Foulds, and Robison 2017). More recently, citizens, under certain circumstances, have also been invited to become producer-consumers (so-called prosumers) producing renewable energy (typically electricity) for their own consumption and, where allowed, for sale back to the centralized grid. While this represents somewhat of a shift in expectations of the role of the citizenry in the energy domain – these forms of participation (albeit an evolution on the traditional role of passive consumer) are still quite limited and rather limiting.

There is, however, a move toward more expansive and inclusive visions of the type of roles citizens are permitted to play within the energy system. The growing discourse on energy democracy, for example, illustrates the emergent social movement calling for changes in the sociopolitical dynamics around the energy domain (Burke and Stephens 2017) and for

DOI:10.4324/9781003183020-30

"restructuring energy systems technologies and governance for greater democracy and inclusivity" (Burke and Stephens 2018, 90). These emergent citizen roles around energy can manifest as either individualist or collectivist, depending on circumstances and objectives. The modes of participation in emergent citizens roles differ but focus on one or more elements of the energy domain *e.g.*, consumption, production, policy and planning, regulatory enforcement, and advocacy (Lennon et al. 2020; Mullally, Dunphy, and O'Connor 2018).

Of course, the types of roles citizens wish, or are permitted, to play in the energy sphere will vary according to sociocultural norms, socioeconomic contexts, influence of the energy incumbents, and the policy and regulatory environment. The latter of which, in particular, is dependent on the economic ideological orientations of those in decision-making positions. This chapter examines how under-theorized and contested concepts like the "energy citizen" are already framing our collective experience(s) of the energy transition and asks for whom is the emerging energy system designed?

2. Understanding the Energy System as a Social System

This section posits that the energy system, long considered a technological construct is in fact a sociotechnical system and, moreover, first and foremost inherently social. The energy system has traditionally been seen through a techno-centric lens, and there has arguably been a reluctance, among some at least, to acknowledging its intrinsic social dimension. For the most part, it has been understood as corresponding with Bruckner et al.'s (2014) definition of the energy supply sector – that is, comprising the various processes that extract, convert store, and transmit and distribute energy for use. In this perspective, energy is considered a technical domain and an important one at that, which provides society with an essential product, the energy that is so vital to our lives. In this traditional view of the energy system, citizens were considered, if at all, as the fortunate recipients and beneficiaries of a key element of modern life from what was portrayed as an almost benevolent industry (Dunphy and Lennon 2020).

Of course, technology on its own does not provide the entire picture of the acquisition and supply of energy. Society's energy needs are met through the interaction of technologies and what might be termed the human factor. This human dimension is to be found both within and without the industrial structures traditionally considered to constitute the energy sector. This expanded view of the energy domain comprises what is termed a sociotechnical system[2] of interacting social and technical elements. Sociotechnical systems are described as "a cluster of aligned elements, e.g. artefacts, knowledge, user practices and markets, regulation, cultural meaning, infrastructure, maintenance networks and supply networks" (Geels 2005, 445). Geels (2004) comments that these systems do not function independently from, but rather are the results of and are dependent on, people's actions – emphasizing an important social aspect of such systems.

Walker et al. (2008) suggest that the interrelatedness of the social and technical dimensions is key to understanding the performance of (and by extension realizing change within) sociotechnical systems. They argue that the differences in their behavior means that focusing optimization attempts on just one dimension (usually technical) is likely to be detrimental to performance – this can be seen in, for example, community opposition to new energy infrastructure arising from top-down exclusionary planning and development processes (Argenti and Knight 2015) or the so-called rebound effect in energy efficiency projects, where energy savings arising from technical measures can be (at least partially) negated by increased consumption (Sorrell and Dimitropoulos 2008). All of these highlights the importance of acknowledging and appreciating the human and societal dimension in sustainable energy transitions.

While energy is a sociotechnical system, it can also be considered a human system. As described earlier, the energy system is designed, built, and used by people, but moreover, it is an integral part of the way people live their lives. But as Hunt and Ryan (2015, 274) observe, energy in and of itself is "not of intrinsic value", but rather it is the energy-related services that people desire. Nobody consciously *chooses* to use energy, but rather its use enables and supports the practices that form the basis of the way we live, including heating, lighting, cooking, cleaning, transportation, entertainment, and so on (Dunphy et al. 2017). In this way, energy consumption as a separate activity can be seen as a somewhat artificial construct. Thus, although the energy system and energy reach into all aspects of people's daily lives, it is often invisible (especially since the reduced use of solid fuels) and is usually only made present by its absence through power outage or unaffordability. Ambrose (2020) suggests that the "contemporary relationship with energy is arguably one characterised by complete dependency and almost complete ignorance".

Energy systems, as they exist today, developed in tandem with society, both responding to and actuating societal needs and demands. The configuration of such systems is a product of and closely intertwined with the societies that depend so much on them. Because of this, decarbonizing energy systems can be seen to be not just about changing energy technologies but rather entailing substantial societal transformation. Not only must the technologies change, but so too must people's relationship with the energy system and with energy – both to facilitate the new technologies and as a result of their realization. Miller et al. (2013, 135) put it quite succinctly commenting that transforming energy systems is not only about changes to technology "but also to the broader social and economic assemblages that are built around energy production and consumption". Indeed, given that much of the required technology is already available, it is plausible that the real challenge of the current energy transition (and the implicit societal transformation, which will accompany it) is actually sociopolitical.

3. Justice Within the Energy Domain

Fairness and perceptions of fairness are key to realizing the societal acceptability required for the ongoing energy transitions. There is a concern in particular that the burden of social, financial, and environmental costs of the energy system are disproportionately placed on those with less social and economically privilege, particularly on those at the margins of society, while it is those with more privilege who benefit most. In response, Sovacool and Dworkin (2015) describe energy justice as a system where both the benefits and costs of the provision of energy are fairly distributed and in which decision-making processes are representative and unbiased. McCauley et al.(2013) suggest there are three principal themes or tenets within energy justice: distributional justice (equity in the allocation of both the benefits and detriments of the energy system), recognition justice (which can perhaps be best understood as the absence of cultural domination, non-recognition, and disrespect in processes), and procedural justice (inclusive, transparent, and informed decision-making processes) (Jenkins et al. 2016).

Wenz (1988, 4) posits that the chief topics relating to environmental justice are concerned with distributional equities. *Distributional justice* is concerned with the fairness of how benefits and ills of the energy system are shared across the population. Distributional understandings of justice are not just about unfairness in impacts but also inequities in "distribution of responsibilities and the spatialities that are implicated within these" Walker (2009, 615). However, as Velasco-Herrejón and Bauwens (2020) observe, although a core element of energy justice, ideas of distributional justice need to be supplemented by other justice concepts to appreciate the causes underlying energy injustice and related maldistribution. A second key concept is that of *recognition justice*, which argues for an appreciation of the diversity of stakeholders and their experiences, both individually and collectively. Lack of recognition devalues some people, cultural groups, and place identities (Walker 2009); it is demonstrated by "various forms of insults, degradation and devaluation" of both individuals and groups (Schlosberg 2004, 519). Recognition injustice of itself directly causes harm, but moreover, such disrespect is the basis for distribution injustices. The third energy justice tenet is concerned with process. *Procedural justice* calls for fair, equitable, and inclusive processes around energy-related decision-making. Walker (2009, 627) notes that procedural justice fundamentally is about the interaction of "information, access and people". Perceptions on procedural fairness in decision-making consider opportunities for all groups to participate, adequacy of engagement, quality of information disclosure, impartiality, and trust (Tyler 2000; McCauley et al. 2013).

Hazrati and Heffron (2021) note a number of later additions to the energy justice framework. Along with distribution, recognition, and procedural justice, there is *cosmopolitan justice* (which acknowledges a common humanity and a responsibility to consider global implications), and *restorative justice*, which aims to repair (and proactively prevent) harm caused by decisions and actions around the energy system. Taking a cosmopolitan justice perspective, Sovacool et al. (2016) suggest a reframing of decisions on energy in terms of ethical and justice concerns. Suggesting energy systems can be thought of as a (pseudo) polity in which competing preferences play out. Their eight principles for a more sustainable energy system are availability of energy for people's daily needs, affordability of energy, due process in energy system planning and realization, transparency and accountability in energy-related decision-making, sustainability and conservation of energy resources, intragenerational equity in benefits and burdens between peoples, intergenerational equity and rights of future generations, and responsibility (of all countries to protect the natural environment). Emerging in the context of criminal justice studies, *restorative justice* is concerned with repairing the harm done to victims. Restorative justice has been proposed as relevant to the energy domain, with the victim in this context conceived as people, society, or nature (Heffron and McCauley 2017). It is intended as a means of correcting existing and preventing new, distributional, recognition, and procedural injustices through governmental (or other) interventions (Hazrati and Heffron 2021).

The ongoing moves to decarbonize our energy systems will inherently result in winners and losers, albeit this is something that is not always fully acknowledged in the transitions discourse and related policy dialogues. Energy justice is vital to ensure the new energy system is developed and realized in a fair and just manner. However, there is an additional requirement for ensuring that the energy transition and wider decarbonization of society is done in a manner that is fair and equitable. Healy and Barry (2017) argue that building support for decarbonization requires greater recognition of (and discussion about) the potential associated socioeconomic costs. The idea of a just transition originated in the trade union movement in response to the implications transitioning away from fossil fuels has on jobs in coal and other carbon-intensive regions. This explains the emphasis on workers' rights, particularly those associated with fossil fuels in much of the discourse. However, while such labor-driven perspectives are indeed correct, there are few who would argue that this was the only valid perspective on a just transition. Heffron and McCauley (2018) argue that integrating the concepts of and ideas from related justice scholarships provide a more holistic view on what justice is and what it can be in the context of the energy transition. They suggest this offers an opportunity to develop an integrated framework for developing more understanding and promoting fairness in the transition through a fusion of the concept of energy justice (as discussed earlier) with concepts of climate justice (seen, for example, by Caney 2014 as a combination of fairness in how climate impact burdens are shared, and fairness through people seeking to avoid or limit climate-related harm, which threatens the rights of others) and environmental justice (described, for example, by Wenz 1988 as fairness in the allocation of burdens of environmental protection between those with socioeconomic privilege and those without, fairness in the sharing of natural resources between developed and lesser developed countries, fairness between the present and future generations, and fairness between humanity and other species).

Mascarenhas-Swan (2017, 38) argues that a deep democracy "in which workers and communities have control over the decisions that affect their daily lives" is vital for realizing a just transition. She posits that any energy democracy must adhere to five central principles: diversification, democratization, decentralization, reduction in consumption, and provision of reparations to ensure that "energy systems . . . do not exacerbate or create new forms of social inequity and the consequential ecosystem erosion" (ibid., 48).

4. Democratizing Energy

Citizen participation is increasingly seen as an essential element of the transition to a decarbonized energy future, although the nature and scale of the desired participation remains contested. Some envision a decarbonized energy system to be structured much the same as today's, with centralized grids, seeing renewable energy sources replacing carbon-intensive fuels. In such a scenario, citizen participation would seem to be quite restricted. However, the ongoing energy transition enables (and some would suggest requires) a far more extensive restructuring and reimagining. Wittmayer et al. (2022) argue that greater citizen participation, and particularly collective participation in energy projects not only means new ways of realizing energy projects and organizing but also leads to a new way of thinking about the energy system. The ongoing transitions debate including support for decentralization and localization of energy systems is closely linked to concepts such as energy democracy and energy citizenship (Wahlund and Palm 2022).

Democratizing energy is often forwarded as the solution to the sociopolitical challenges associated with the energy transition (see Keahey et al., this volume, for a treatment of energy democracy). This campaign, which evolved into Germany's Energiewende (energy transition), originally had a strong democracy focus, but as Morris and Jungjohann (2016, 4) note, by the 1990s, "climate change had clearly taken center stage as the main goal". The term "energy democracy" originated in the German climate justice movement,[3] with a call for socialization, decarbonization, and democratization of the energy system, it quickly gained currency among non-governmental organizations and researchers, particularly across Europe and the USA (Angel 2016). Much of the early writings on the concept were found in so-called gray literature, and it is really only since around 2015 that the term appears in peer-reviewed literature (van Veelen and van der Horst 2018). While the idea of restructuring the energy system to be more democratic is increasingly accepted, energy democracy as a concept remains somewhat underdefined. Energy democracy may be seen as a process (challenging incumbents), an outcome (or by-product) of decarbonization, and/or a normative goal (Szulecki and Overland 2020). Burke and Stephens (2018, 90) posit that the concept "represents a contemporary expression of ongoing struggles for social and environmental justice through engagement with technological systems". Wahlund and Palm (2022) posit that there are three overarching goals of energy democracy: opposition to fossil fuels and support of decarbonization of the energy mix, socialized control of the energy system, and a restructuring of energy system technologies and governance to support more inclusive participatory democratic processes. In this way, they see energy democracy as a means of envisioning transitions to renewable energies as "pathways for democratic development" (ibid., 90). In this regard, Angel (2019) suggests that people see energy democracy in two ways: there are those for whom it is about increasing agency within the energy sphere for those with little or none, while for others, energy democracy is a means of bringing energy system governance under the democratic control of energy users (and to an extent workers).

Szulecki and Overland (2020) observe that while originating as a term identifying a larger role for citizens in decision-making on energy (and the energy transition), energy democracy has evolved into a term more concerned with energy governance. Szulecki (2018, 21) suggests there are three key dimensions that make up energy democracy: popular sovereignty, where citizens (re)claim authority to shape their energy systems "in ways that are culturally relevant and ecologically sustainable" (Laldjebaev, Sovacool, and Kassam 2015, 98); participatory governance, where there is inclusive, informed, and transparent decision-making at all levels relating to energy and the energy system; and civic ownership, where there is increased non-state and non-corporate involvement – that is, greater "citizen, community, co-operative and municipal ownership" (Hall et al. 2016, 5) of energy generation and transmission systems. However, energy democracy, like all social movements, has a wide diversity of subgroups, each with its own goals and priorities, which do not necessarily align. Examples of differing opinions on energy democracy include the role of the state (see, for example, Angel 2017 for an exploration of energy politics in, against, and beyond the state); oppositional versus alternative positionality of arguments (see, for example, Wahlund and Palm 2022); and ownership and control (level of citizen involvement) (see, for example, Van Veelen 2018). Involvement of citizens across these three dimensions may be conceived as a form of citizenship within the energy domain.

5. Energy Citizenship

The terms "energy citizenship" and "energy citizen" have increasing currency in discourse around the energy transition (albeit perhaps concentrated within academic and policy circles). Understanding energy citizenship first requires an appreciation of citizenship of which there are two main traditions: the liberal tradition, which focuses on the entitlement of citizens to fundamental rights (Schuck 2002), and the civic republican tradition, which is based on duties and responsibilities (Richard 2002). While traditional views of citizenship involved membership of a particular polity, more inclusive perspectives include ideas of cosmopolitan citizenship, which acknowledges a shared humanity and Dobson's (2003) idea of a post-cosmopolitan citizenship in his theoretical considerations on the ecological citizenship concept, which adds a perceived obligation of justice (of righting a wrong) to the idea of common humanity. Citizenship in the energy domain could seem to relate to one or more of these perspectives depending on context.

The concept of energy citizenship can be understood as a social construct; it is a sociotechnical vision[4] conceptualized by activists, academics, and increasingly, policymakers of the potential roles that citizens could, or perhaps should, play in the energy system (Pel et al. 2021). While not claiming it as a novel idea, Devine-Wright (2007) forwards the concept of energy citizenship as fundamentally about citizens becoming active stakeholders in an energy system where equity in addressing the effects of energy consumption is foregrounded. His idea of the energy citizen was positioned in stark contrast to the traditional passive role filled by consumers. Energy citizenship, however, remains a nebulous idea and one that means different things to different people – as Lennon et al. (2020, 184) observe, what it "might involve in practice remains open to interpretation".

Aronson and Stern (1984, 16) forward four representations on the nature of energy: energy as a commodity, energy as an ecological resource, energy as a social necessity, and energy as a strategic material. Each of these perspectives inherently reflects a set of values and beliefs resulting in differing framing of the citizenry, their potential roles within the energy domain, and the level of empowerment to be afforded them (Lennon et al. 2020; Pel et al. 2021). Mullally et al. (2018) suggest citizenship within in the energy domain involves a combination of rights and responsibilities underpinned by important principles of sustainability and social justice. They posit that while there is growing acceptance that increasing citizen participation is warranted, there is a lack of consensus as to the nature and depth of the roles that citizen should be invited to (or indeed permitted to) play in the energy domain.

Lennon et al. (2020) argue that discussions on citizenship in energy transitions discourse have been skewed toward normative descriptions of how a "good citizen" can contribute to energy conservation and decarbonization by being an active consumer in the energy sphere. Energy citizenship is often framed with an emphasis of the responsibilities of citizens (particularly with regard to their consumer activities), with little if any acknowledgment of the rights accorded to them. This privatization of responsibility acts to shift the onus away from those with real power and influence (e.g., energy companies and state bodies) to individuals whose agency is far more bounded. It places the responsibility on reducing energy consumption on individuals, thereby disregarding the need for wider societal and structural changes arising from the strong influence sociotechnical contexts have on social practices (Dunphy and Lennon 2020). This focus on citizens' private consumption activities disregards the duality of the citizen who exists "simultaneously as a 'communal being' and as a private individual" (Rosenow 1992, 45). The consumerist conceptualization of energy citizenship greatly restricts the types of roles that citizens might play in the energy system and, by doing so, limits their potential contribution to the energy transition.

Substantial societal transformation will both be necessary to achieve the transition away from carbon-intensive energy sources and will, in turn, result in even more societal transformation. A successful energy transition and the associated social changes in the way we live our lives requires citizens to not just acquiesce to certain decisions but rather to wholeheartedly embrace the transition not as something foisted upon them but as something of which they are part. This requires inclusive and participatory governance forms that enable and support citizens' full participation in the energy system and provide for them with an equitable share in its benefits (Dunphy et al. 2018).

6. Citizen Participation

As we have seen, there is a growing acceptance of the need for a more inclusive energy system; this has meant that greater roles for the citizenry is increasingly to the fore in energy transition discourses. The traditional role of citizens in the energy system was that of a passive consumer, a supposedly grateful recipient of a much-needed product – this was especially the case as electricity and, to a lesser extent, as gas networks were rolled out. Energy was viewed very much in technocratic terms, with an elite of technical experts guiding, if not directing, energy policy. The citizen's relationship with the energy domain was seen very much in transactional terms and their expected – and really their only permitted – role was that of a customer and preferably one that was deferential to those who knew what was best. However, this is changing in response to a range of geopolitical (e.g., oil crises, gas blockades), economic (i.e., cost of energy), and environmental (e.g., climate impact) challenges associated with energy consumption and production. As a result, in recent decades there has been an increasing acceptance (albeit hesitant in places) of new roles for citizens in the energy domain, whatever form they may take.

Among attempts to develop categories of citizen participation, the most prominent perhaps is Arnstein's seminal work on societal power structures, where she differentiates between "the empty ritual of participation and having the real power needed to affect the outcome of the process" (1969, 216). She forwards a typology comprising three main categories of participation. The lowest levels are classified as non-participation and involve manipulating/curing participants by "educating" and engineering them into supporting a particular position. They are not meant to enable true participation but rather designed so that those with power can orchestrate support. The second category involves informing, consulting, and placating participants. While information will be provided and views exchanged, they may not necessarily be any serious consideration of participants' contributions. In its most advanced forms, engagement can be quite inclusive, involving those not ordinarily included; however, in all cases, power is retained by the existing power holders. At the highest levels comprise different degrees of citizen power, ranging from partnership with other actors, enabling negotiation and real input in decisions, to delegated power and citizen control where the citizenry have real ability to make decisions themselves.

The energy system (in most countries) is somewhat of a public-private hybrid, with various levels of public sector oversight and control. It involves (multi-level) state authorities licensing, highly regulating, in some instances granting a monopoly franchise, and occasionally owning individual components (see, for example, Osofsky and Wiseman 2014 for a treatment of hybrid energy governance). Within the energy sector, while citizen participation was originally (principally) construed as a process for citizens to feed into decision-making processes on infrastructure projects (and more recently on energy policy), the term is now understood more expansively, and citizens have increasing social, political, and economic agency in the energy domain. The quasi-public (at least) nature of energy governance, combined with the technological changes and business opportunities arising from the energy transition, makes it an appropriate sector for citizens to potentially participate across a range of activities. When once limited to a passive consumption role, citizens individually or collectively are increasingly contributing to and participating in the energy system through a growing number of activities, involving the exercise of social, political, and economic power (Dunphy and Lennon 2020).

In a review of a consultation process on energy policy in Ireland, Mullally, Dunphy, and O'Connor (2018) identified six distinct narratives of citizens participation in the energy system. These narratives (paternalist, majoritarian, consumerist, constitutionalist, communitarian, and deliberative) provide some insights into different perspectives of how energy citizenship might (or perhaps should) be realized.

- The *paternalist* narrative sees the citizens' role as facilitating and enabling energy system developments in a way determined by a politico-technical elite. This narrative imagines the citizenry as ill-informed and requiring education and information to be persuaded to behave in the right way (however that might be portrayed).
- In the *majoritarian* narrative, the views of the politico-technical elite are supported by the so-called silent majority. The citizenry is viewed in almost a binary fashion – a vocal minority who obstruct the required changes to the energy system and a supporting majority whose views are not heard.
- The *consumerist* narrative is quite a limited perspective: it sees citizens only through their traditional role as consumers. Its focus is on their consumption and production activities, and there is little if any participation beyond that envisaged by law.
- The *constitutionalist* narrative sees everything through right-based legalistic perspective. It maintains that citizens have legally mandated rights (including consultation and contribution to consent processes) which must be upheld (e.g., in energy infrastructure projects) but does not really envisage participation outside of such official mechanisms.

- A more bottom-up perspective is given in the *communitarian* narrative, which acknowledges that citizens exist within communities and that they act collectively and individually. It holds that citizens and the communities that they form have a right to participate in and to benefit from energy projects.
- The final viewpoint identified in this paper is the *deliberative* narrative where citizens participate across all levels of governance and practice on decisions on energy and the energy system. Citizens are considered to be capable of and have the right to deliberate with other stakeholders to identify and resolve issues around energy and the energy system.

Dunphy and Lennon (2020) used this framework of narratives to characterize a number of modes of energy participation, based around consumption, market influence (and power), access to information, enforcement of rights, individual and collective production, and challenging power holders through formal and informal politics. Building on and drawing from this work and informed by other analyses of participation around energy (e.g., public participation in the UK energy system by Pallett, Chilvers, and Hargreaves 2019; emergent public engagement in societal transitions by Revez et al. 2022). The following paragraphs outline some examples of energy citizenship expression across a range of potential modes of participation. Some of these expressions of citizenship are quite passive (in keeping with the idea that not all participation has to be active), while others are inherently active. They encompass both individualistic and collective approaches.

The first category are those manifestations of energy citizenship, which are fundamentally about access to energy and energy resources. These include the *dispossessed*, indigenous peoples and other marginalized groups from whom energy resources have been unjustly taken, such as tar sands production on Canadian First Nations land (Parson and Ray 2018) and/or the extraction of which has resulted in their displacement, such as the Three Gorges Dam (Jackson and Sleigh 2000); the *excluded*, those who are prevented from connecting to energy grids due to sociopolitical and/or economic reasons, such as the electrification of rural communities in sub-Saharan Africa (Falchetta et al. 2020); the *energy-vulnerable*, those for whom affordability of energy is a key issue and who are at risk of energy poverty – that is, being unable to afford energy required for adequate heat, lighting, cooking, and appliance use (Bouzarovski 2014). Another group impacted by energy access are those who live in what might be termed energy deserts – areas with limited choice of and access to energy suppliers (analogous to the concept of food desert; see, for example, Hamidi 2020).

The second categorization of energy citizenship is consumption-orientated. It includes the traditional *passive consumer* role in which, citizens have little or no expected participation beyond being simply recipients of a product; the *active consumer* persona, so beloved of government energy efficiency programs, which uses their purchasing power to influence the market; the narrative of the *good citizen*, which involves changing energy consumption habits for environmental and other public rationales in response to the persuasion of government information campaigns; the *digital native*, those early adopters of technology and for whom the energy system is mediated through a combination of smart meters, smartphone apps, connected appliances, and a multitude of gadgets (see, for example, Strengers et al. 2022 on visions or imaginaries of future home life); the interesting narrative of the *energy champion*, who is somewhat of an evangelist for sustainable energy practices and acts as a source of knowledge and advice for her neighbors; and an emerging citizen consumption narrative, that of the *collectivist-consumer*, the energy consumer that combines with others to group purchase, manage, and/or consume energy (sometimes in the form of heat).

The third category for energy citizenship is production-orientation and encompasses the various means by which citizens become involved in and contribute to the production of energy. Narratives of citizen energy production include the *prosumer*, a citizen that is a hybrid producer and consumer of energy. Such prosumption is a bottom-up phenomenon where the prosumer that produces, sells, trades, or stores while typically linking with an energy company to sell excess production. The *self-consumer* may be considered a variant of the prosumer – she seeks a level of individual sovereignty, to be autonomous of centralized energy systems by producing all the energy required for her own use. The *collectivist-producer* narrative is where citizens combine in community energy projects, typically in the form co-operatives or similar structures, for the purpose of generating energy for sale (or indeed for collective self-consumption see, for example, Reis et al. 2022). Another production narrative involves the citizen as a *citizen-investor* in energy projects, companies, or crowdsourcing initiatives. These citizen-investors have a varying mixture of economic and environmental (and occasionally social) motives.

The final category of energy citizenship is within the political sphere; this can take a number of different forms, all of which are concerned with decision-making: The *citizen-litigator* is concerned with procedural correctness; she argues her rights through established processes and work to ensure that laws about information provision, consultation, and permitting are strictly followed. In doing so, she contributes to better energy policy development and regulation. The *citizen-challenger* is active in political processes; she envisages the energy transition as an implementation challenge and, therefore, a sociopolitical one. In light of this, she combines with others to challenge the status quo and enact change through the political system by means of public awareness, political campaigning, lobbying, electoral politics, and the like. The citizen-challenger is typically well-informed, motivated, and organized and necessarily has sociopolitical agency. The final example type is the *citizen-activist*, who overlaps somewhat with the challenger, but she works more on the political margins. The citizen-activist also wishes to change the status quo, the difference being she does not fully trust the political system to deliver such change. She is involved in social mobilization, protest movements, and other radical action – her aim is not so much to achieve change through the system but rather to change the system itself.

Energy citizens may display one or more of these expressions of energy citizenship in the energy domain concurrently or at different times. The different expressions undoubtedly speak to different levels of socioeconomic privilege and to different life experiences. Certain sections of society are either locked in (e.g., those residing in areas with a monopoly electricity franchise) or locked out (e.g., citizens without the resources to take part in collective initiatives) of aspects of the energy system, which limits their ability to participate. Pallett, Chilvers, and Hargreaves note, "All forms of participation – whether invited or uninvited, insider or outsider – are always orchestrated and framed in powerful and highly partial ways, and are thus subject to exclusions" (2017, 607). Thus, the big questions for the realization of energy citizenship and indeed the future shape of energy systems are as follows: Where in the system will participation be permitted? Who will be allowed to participate? On what basis, and to what end? The answers to these queries will be system-specific and will be reveled over time as various ad hoc interest groups representing different narratives (what Mullally, Dunphy, and O'Connor 2018 referred to as discourse coalitions) compete/collaborate/negotiate to reach a resolution across a range of sociocultural, sociopolitical, and sociotechnical battlegrounds.

7. Conclusion

Devine-Wright's (2007) early conceptualization of energy citizenship offered a hint of an energy future in which citizens meaningfully participate in the governance of the energy system. This concept, albeit ambiguously defined, has facilitated (and continues to facilitate) visions of citizen participation around systems of energy consumption, production, and supply. Energy citizenship can be thought of as a form of citizenship beyond the state, but a citizenship of energy ought not be thought of as a legal-political conception; it is not to be understood in terms of a political membership but rather (principally) as a democratic function focused on decision-making around energy. However, as described in the preceding section, the energy citizen is not just interested in policymaking and decision-making; one can also express citizenship through action in both public and private aspects of life. In this way, energy citizenship shares many post-cosmopolitan citizenship attributes found in ideas, such as environmental citizenship (non-territorial in nature, involved in public and private spheres, and with focus on virtues; see Dobson 2003).

Chilvers and Longhurst (2016, 601) conceive of participation in the energy transition as an emergent, relational, and co-produced phenomena, expanded beyond just deliberation to take multiple forms "including activism, grassroots innovation, and interactions with more mundane technologies in everyday life". The labels of "citizen" and "citizenship" are useful metaphors for considering such energy participation, invoking the language of citizenship helps appreciation of the responsibilities and rights involved (particularly in the context of the energy transition). In this chapter, we have discussed the multiple existing and emerging modes of participation, which embody expressions of energy citizenship. These include the excluded and dispossessed, non-engaged passive consumers, and various modes of more active individual and group consumption; production by individuals and through collectives; and a range of political and campaigning roles trying to inform and/or shape the energy system to align with their priorities.

Not all perspectives on what energy citizenship could or should be are equally supported. Governments, energy companies, and other energy stakeholders all have preferences for the type of roles they wish citizens to play in the energy system. There is (growing) support among traditional energy system power holders for certain expressions of energy citizenship. The more "acceptable" expressions are really those that do not threaten the status quo, such as active consumerism (citizens are encouraged to use their purchasing power to send signals to the market) or prosumerism (citizens both produce and consume energy). Other expressions of energy citizenship that challenge incumbents or government policy are not so welcome, and indeed, such energy citizens are often vilified and/or marginalized by the incumbent power holders (see, for example, Mullally, Dunphy, and O'Connor 2018, for a discussion of the discourse coalitions that coalesce around different conceptualizations of energy citizenship).

There has been a tendency in the discourse on citizenship around energy to focus on cases of active participation, with citizens typically construed as economic actors. However, economics is a rather exclusory basis for citizenship. In the domestic sphere, for example, energy citizens are imagined as making informed, rational decisions on energy consumption, assuming levels of resources, financial and otherwise, that many householders simply do not have. Similar assumptions of resources and agency are to be found in productivist manifestations of energy citizenship, whether individual or collective. As Lennon et al. (2020, 189) observe, such perspectives ignore "issues of unequal access to energy, limited financial resources, educational privilege and expertise, or differential levels of control over one's environment and practices". There is notably a classism inherent in many expressions of energy citizenship (which in some principled expressions overlaps with an environmental classism; see, for example, Bell 2020), which can be quite exclusory. If energy citizenship remains predominately focused on economic modes of participation, those with less economic privilege will at best be quasi-citizens in the energy future. Energy citizenship should not be just about active involvement, it ought not be something that one becomes, but rather, it should exist by virtue of our existing close relations with energy and the energy system. This conceptualization of citizenship in the energy domain is not only more inclusive but arguably could form the basis for a comprehensive framework of people's relationship with energy, establishing rights and responsibilities for a continuum of expressions of energy citizenship.

Notes

1 This work has been carried out in the context of two research projects. ENCLUDE, funded under the EU's Horizon 2020 Research and Innovation Program, under grant agreement no. 101022791, is seeking to understand the diversity of expressions of energy citizenship; EnergyPolities, funded by the SEAI R&RD program, 18/RRD/356, is exploring social mobilization around energy.
2 It is notable that the concept of sociotechnical systems originated within the energy domain in attempts to improve the efficiency of the British coal industry in the 1950s (Pasmore et al. 2019).
3 An early example of a grassroots movement to make decisions on energy more democratic was the 1970s campaign against nuclear power plants in Breisach and Wyhl, West Germany (Kalb 2012).
4 Such visions are often referred to as sociotechnical imaginaries, described by Jasanoff and Kim (2009, 120) as "collectively imagined forms of social life and social order reflected in the design and fulfillment of nation-specific scientific and/or technological projects".

References

Ambrose, Aimee R. 2020. "Walking with Energy: Challenging Energy Invisibility and Connecting Citizens with Energy Futures through Participatory Research." *Futures* 117: 102528. https://doi.org/10.1016/j.futures.2020.102528.

Angel, James. 2016. *"Strategies of Energy Democracy."* Brussels: Rosa Luxemburg Foundation.

Angel, James. 2017. "Towards an Energy Politics In-Against-and-Beyond the State: Berlin's Struggle for Energy Democracy." *Antipode* 49 (3): 557–576. https://doi.org/10.1111/anti.12289.

Angel, James. 2019. "Urban Energy Democracy Contesting the Energy System in London and Barcelona (Doctoral Thesis)." King's College London.

Argenti, Nicolas, and Daniel M. Knight. 2015. "Sun, Wind, and the Rebirth of Extractive Economies: Renewable Energy Investment and Metanarratives of Crisis in Greece." *Journal of the Royal Anthropological Institute* 21 (4): 781–802. https://doi.org/10.1111/1467-9655.12287.

Arnstein, Sherry R. 1969. "A Ladder of Citizen Participation." *Journal of the American Institute of Planners* 35 (4): 216–224.

Aronson, E, and P.C. Stern. 1984. *Energy Use: The Human Dimension.* Washington DC: The National Academies Press. https://doi.org/10.17226/9259.

Axon, Stephen, John E. Morrissey, Rosita Aiesha, Joanne Hillman, Alexandra Revez, Breffní Lennon, Mathieu Salel, Niall P. Dunphy, and Eva Boo. 2018. "The Human Factor: Classification of European Community-Based Behaviour Change Initiatives." *Journal of Cleaner Production* 182: 567–586. https://doi.org/10.1016/j.jclepro.2018.01.232.

Bell, Karen. 2020. *Working-Class Environmentalism: An Agenda for a Just and Fair Transition to Sustainability.* Cham: Palgrave Macmillan/Springer Nature.

Bouzarovski, Stefan. 2014. "Energy Poverty in the European Union: Landscapes of Vulnerability." *Wiley Interdisciplinary Reviews: Energy and Environment* 3 (3): 276–289. https://doi.org/10.1002/wene.89.

Bruckner, Thomas, Igor A. Bashmakov, Yacob Mulugetta, Helena Chum, Angel de la Vega Navarro, James Edmonds, Andre Faaij et al. 2014. "Energy Systems." In: O. Edenhofer, R. Pichs-Madruga, Y. Sokona, E. Farahani, S. Kadner, K. Seyboth, A. Adler et al. (eds). *Climate Change 2014: Mitigation of Climate Change. Contribution of Working Group III to the Fifth Assessment Report of the Intergovernmental Panel on Climate Change*, 511–597. Cambridge and New York: Cambridge University. https://doi.org/10.4324/9780203223017-18.

Burke, Matthew J., and Jennie C. Stephens. 2017. "Energy Democracy: Goals and Policy Instruments for Sociotechnical Transitions." *Energy Research and Social Science* 33: 35–48. https://doi.org/10.1016/j.erss.2017.09.024.

Burke, Matthew J., and Jennie C. Stephens. 2018. "Political Power and Renewable Energy Futures: A Critical Review." *Energy Research and Social Science* 35: 78–93. https://doi.org/10.1016/j.erss.2017.10.018.

Caney, Simon. 2014. "Two Kinds of Climate Justice: Avoiding Harm and Sharing Burdens." *Journal of Political Philosophy* 22 (2): 125–149. https://doi.org/10.1111/jopp.12030.

Chilvers, Jason, and Noel Longhurst. 2016. "Participation in Transition(s): Reconceiving Public Engagements in Energy Transitions as Co-Produced, Emergent and Diverse." *Journal of Environmental Policy & Planning* 18 (5): 585–607. https://doi.org/10.1080/1523908X.2015.1110483.

Devine-Wright, Patrick. 2007. "Energy Citizenship: Psychological Aspects of Evolution in Sustainable Energy Technologies." In: J. Murphy (ed.). *Governing Technology for Sustainability*, 63–86. Earthscan.

Dobson, Andrew. 2003. *Citizenship and the Environment.* Oxford University Press.

Dunphy, Niall P., and Breffní Lennon. 2020. Citizen Participation in the Energy System at the Macro Level. Cork: EnergyPolities project.

Dunphy, Niall P., Breffní Lennon, Estibaliz Sanvicente, Suzi Tart, Daniele Kielmanowicz, and John E Morrissey. 2018. "Innovative Business Models to Foster Transition." *Entrust* H2020 Project. https://doi.org/10.5281/zenodo.3479235.

Dunphy, Niall P., Alexandra Revez, Christine Gaffney, Breffní Lennon, Ariadna Ramis Aguilo, John E. Morrissey, and Stephen Axon. 2017. "Intersectional Analysis of Energy Practices." Cork: ENTRUST H2020 Project. https://doi.org/10.5281/zenodo.3479295.

Falchetta, Giacomo, Shonali Pachauri, Edward Byers, Olha Danylo, and Simon C. Parkinson. 2020. "Satellite Observations Reveal Inequalities in the Progress and Effectiveness of Recent Electrification in Sub-Saharan Africa." *One Earth* 2 (4): 364–379. https://doi.org/10.1016/j.oneear.2020.03.007.

Fox, Emmet, Chris Foulds, and Rosie Robison. 2017. "Energy & the Active Consumer – a Social Sciences and Humanities Cross-Cutting Theme Report." Cambridge: Shape Energy. https://shapeenergy.eu/wp-content/uploads/2017/07/SHAPE-ENERGY_ThemeReports_ENERGY-THE-ACTIVE-CONSUMER.pdf.

Geels, Frank W. 2004. "From Sectoral Systems of Innovation to Socio-Technical Systems." *Research Policy* 33 (6–7): 897–920. https://doi.org/10.1016/j.respol.2004.01.015.

Geels, Frank W. 2005. "The Dynamics of Transitions in Socio-Technical Systems: A Multi-Level Analysis of the Transition Pathway from Horse-Drawn Carriages to Automobiles (1860–1930)." *Technology Analysis and Strategic Management* 17 (4): 445–476. https://doi.org/10.1080/09537320500357319.

Hall, Stephen, Timothy J. Foxon, and Ronan Bolton. 2016. "Financing the Civic Energy Sector: How Financial Institutions Affect Ownership Models in Germany and the United Kingdom." *Energy Research and Social Science* 12: 5–15. https://doi.org/10.1016/j.erss.2015.11.004.

Hamidi, Shima. 2020. "Urban Sprawl and the Emergence of Food Deserts in the USA." *Urban Studies* 57 (8): 1660–1675. https://doi.org/10.1177/0042098019841540.

Hazrati, Mohammad, and Raphael J. Heffron. 2021. "Conceptualising Restorative Justice in the Energy Transition: Changing the Perspectives of Fossil Fuels." *Energy Research and Social Science* 78: 102115. https://doi.org/10.1016/j.erss.2021.102115.

Healy, Noel, and John Barry. 2017. "Politicizing Energy Justice and Energy System Transitions: Fossil Fuel Divestment and a 'Just Transition.'" *Energy Policy* 108 (November 2016): 451–459. https://doi.org/10.1016/j.enpol.2017.06.014.

Heffron, Raphael J., and Darren McCauley. 2017. "The Concept of Energy Justice across the Disciplines." *Energy Policy* 105: 658–667. https://doi.org/10.1016/j.enpol.2017.03.018.

Heffron, Raphael J., and Darren McCauley. 2018. "What Is the 'Just Transition'?" *Geoforum* 88: 74–77. https://doi.org/10.1016/j.geoforum.2017.11.016.

Hunt, Lester C., and David L. Ryan. 2015. "Economic Modelling of Energy Services: Rectifying Misspecified Energy Demand Functions." *Energy Economics* 50: 273–285. https://doi.org/10.1016/j.eneco.2015.05.006.

Jackson, Sukhan, and Adrian Sleigh. 2000. "Resettlement for China's Three Gorges Dam: Socio-Economic Impact and Institutional Tensions." *Communist and Post-Communist Studies* 33 (2): 223–241. https://doi.org/10.1016/S0967-067X(00)00005-2.

Jasanoff, Sheila, and Sang Hyun Kim. 2009. "Containing the Atom: Sociotechnical Imaginaries and Nuclear Power in the United States and South Korea." *Minerva* 47 (2): 119–146. https://doi.org/10.1007/s11024-009-9124-4.

Jenkins, Kirsten, Darren McCauley, Raphael J. Heffron, Hannes Stephan, and Robert Rehner. 2016. "Energy Justice: A Conceptual Review." *Energy Research & Social Science* 11 (9): 174–182. https://doi.org/10.1016/j.erss.2015.10.004.

Kalb, Martin. 2012. "'Rather Active Today Than Radioactive Tomorrow!' Environmental Justice and the Anti-Nuclear Movement in 1970s Wyhl, West Germany." *Global Environment* 5 (10): 156–183. https://doi.org/10.3197/ge.2012.051009.

Laldjebaev, Murodbek, Benjamin K. Sovacool, and Karim-Aly S. Kassam. 2015. "Energy Security, Poverty, and Sovereignty Complex Interlinkages and Compelling Implications." In: Lakshman Guruswamy and Elizabeth Neville (eds). *International Energy and Poverty: The Emerging Contours*, 97–112. Abingdon & New York: Routledge. https://doi.org/10.4324/9781315762203.

Lennon, Breffní, Niall P. Dunphy, Christine Gaffney, Alexandra Revez, Gerard Mullally, and Paul O'Connor. 2020. "Citizen or Consumer? Reconsidering Energy Citizenship." *Journal of Environmental Policy & Planning* 22 (2): 184–197. https://doi.org/10.1080/1523908X.2019.1680277.

Mascarenhas-Swan, Michelle. 2017. "The Case for a Just Transition." In: *Energy Democracy*, 37–56. Washington, DC: Island Press/Center for Resource Economics. https://doi.org/10.5822/978-1-61091-852-7_3.

McCauley, Darren, Raphael J. Heffron, Hannes Stephan, and Kirsten Jenkins. 2013. "Advancing Energy Justice: The Triumvirate of Tenets." *International Energy Law Review* 32 (3): 107–110.

Miller, Clark A., Alastair Iles, and Christopher F. Jones. 2013. "The Social Dimensions of Energy Transitions." *Science as Culture* 22 (2): 135–148. https://doi.org/10.1080/09505431.2013.786989.

Morris, Craig, and Arne Jungjohann. 2016. "Energiewende: The Solution to More Problems Than Climate Change." In: *Energy Democracy: Germany's Energiewende to Renewables*, 1–14. Springer International Publishing. https://doi.org/10.1007/978-3-319-31891-2.

Mullally, Gerard, Niall P. Dunphy, and Paul O'Connor. 2018. "Participative Environmental Policy Integration in the Irish Energy Sector." *Environmental Science and Policy* 83: 71–78. https://doi.org/10.1016/j.envsci.2018.02.007.

Osofsky, Hari M., and Hannah Jacobs Wiseman. 2014. "Hybrid Energy Governance." *University of Illinois Law Review* 2014 (1): 1–66.

Pallett, Helen, Jason Chilvers, and Tom Hargreaves. 2017. "Mapping Energy Participation: A Systematic Review of Diverse Practices of Participation in UK Energy Transitions, 2010–2015," no. April: 115. www.ukerc.ac.uk/publications/mapping-energy-participation-a-systematic-review-of-diverse-practices-of-participation-in-energy-transitions-2010–1015.html.

Pallett, Helen, Jason Chilvers, and Tom Hargreaves. 2019. "Mapping Participation: A Systematic Analysis of Diverse Public Participation in the UK Energy System." *Environment and Planning E: Nature and Space* 2 (3): 590–616. https://doi.org/10.1177/2514848619845595.

Parson, Sean, and Emily Ray. 2018. "Sustainable Colonization: Tar Sands as Resource Colonialism." *Capitalism, Nature, Socialism* 29 (3): 68–86. https://doi.org/10.1080/10455752.2016.1268187.

Pasmore, William, Stu Winby, Susan Albers Mohrman, and Rick Vanasse. 2019. "Reflections: Sociotechnical Systems Design and Organization Change." *Journal of Change Management* 19 (2): 67–85. https://doi.org/10.1080/14697017.2018.1553761.

Pel, Bonno, Ariane Debourdeau, René Kemp, Adina Dumitru, Martina Schäfer, Edina Vadovics, Frances Fahy, Aurore Fransolet, and Thomas Pellerin-Carlin. 2021. "Conceptual Framework Energy Citizenship." *EnergyProspects* H2020 Project.

Reis, Inês F. G., Ivo Gonçalves, Marta A. R. Lopes, and Carlos Henggeler Antunes. 2022. "Collective Self-Consumption in Multi-Tenancy Buildings – To What Extent Do Consumers' Goals Influence the Energy System's Performance?" *Sustainable Cities and Society* 80: 103688. https://doi.org/10.1016/j.scs.2022.103688.

Revez, Alexandra, Niall P. Dunphy, Clodagh Harris, Fionn Rogan, Edmond Byrne, Connor McGookin, Paul Bolger, Brian Ó Gallachóir, John Barry, Geraint Ellis, Barry O'Dwyer, Evan Boyle, Stephen Flood, James Glynn and Gerard Mullally. 2022. "Mapping Emergent Public Engagement in Societal Transitions: A Scoping Review." *Energy, Sustainability and Society* 12 (2). https://doi.org/10.1186/s13705-021-00330-4.

Richard, Dagger. 2002. "Republican Citizenship." In: Engin F. Isin and Byran S. Turner (eds). *Handbook of Citizenship Studies*, 145–158. Sage Publications.

Rosenow, Eliyahu. 1992. "Bourgeois or Citoyen? The Democratic Concept of Man." *Educational Philosophy and Theory* 24 (1): 44–50. https://doi.org/10.1111/j.1469-5812.1992.tb00218.x.

Schlosberg, David. 2004. "Reconceiving Environmental Justice: Global Movements and Political Theories." *Environmental Politics* 13 (3): 517–540. https://doi.org/10.1080/0964401042000229025.

Schuck, Peter H. 2002. "Liberal Citizenship." In Engin F. Isin and Byran S. Turner (eds). *Handbook of Citizenship Studies*, 131–144. Sage Publications.

Sorrell, Steve, and John Dimitropoulos. 2008. "The Rebound Effect: Microeconomic Definitions, Limitations and Extensions." *Ecological Economics* 65 (3): 636–649. https://doi.org/10.1016/j.ecolecon.2007.08.013.

Sovacool, Benjamin K., and Michael H. Dworkin. 2015. "Energy Justice: Conceptual Insights and Practical Applications." *Applied Energy* 142: 435–444. https://doi.org/10.1016/j.apenergy.2015.01.002.

Sovacool, Benjamin K., Raphael J. Heffron, Darren McCauley, and Andreas Goldthau. 2016. "Energy Decisions Reframed as Justice and Ethical Concerns." *Nature Energy* 1 (5): 16024. https://doi.org/10.1038/nenergy.2016.24.

Strengers, Yolande, Kari Dahlgren, Sarah Pink, Jathan Sadowski, and Larissa Nicholls. 2022. "Digital Technology and Energy Imaginaries of Future Home Life: Comic-Strip Scenarios as a Method to Disrupt Energy Industry Futures." *Energy Research & Social Science* 84 (May 2021): 102366. https://doi.org/10.1016/j.erss.2021.102366.

Szulecki, Kacper. 2018. "Conceptualizing Energy Democracy." *Environmental Politics* 27 (1): 21–41. https://doi.org/10.1080/09644016.2017.1387294.

Szulecki, Kacper, and Indra Overland. 2020. "Energy Democracy as a Process, an Outcome and a Goal: A Conceptual Review." *Energy Research and Social Science* 69: 101768. https://doi.org/10.1016/j.erss.2020.101768.

Tyler, Tom R. 2000. "Social Justice: Outcome and Procedure." *International Journal of Psychology* 35 (2): 117–125. https://doi.org/10.1080/002075900399411.

Veelen, Bregje Van. 2018. "Negotiating Energy Democracy in Practice: Governance Processes in Community Energy Projects." *Environmental Politics* 27 (4): 644–665. https://doi.org/10.1080/09644016.2018.1427824.

Veelen, Bregje van, and Dan van der Horst. 2018. "What Is Energy Democracy? Connecting Social Science Energy Research and Political Theory." *Energy Research and Social Science* 46 (February): 19–28. https://doi.org/10.1016/j.erss.2018.06.010.

Velasco-Herrejón, Paola, and Thomas Bauwens. 2020. "Energy Justice from the Bottom up: A Capability Approach to Community Acceptance of Wind Energy in Mexico." *Energy Research and Social Science* 70 (July): 101711. https://doi.org/10.1016/j.erss.2020.101711.

Wahlund, Madeleine, and Jenny Palm. 2022. "The Role of Energy Democracy and Energy Citizenship for Participatory Energy Transitions: A Comprehensive Review." *Energy Research & Social Science* 87 (1): 102482. https://doi.org/10.1016/j.erss.2021.102482.

Walker, Gordon. 2009. "Beyond Distribution and Proximity: Exploring the Multiple Spatialities of Environmental Justice." *Antipode* 41 (4): 614–636. https://doi.org/10.1111/j.1467-8330.2009.00691.x.

Walker, Guy H., Neville A. Stanton, Paul M. Salmon, and Daniel P. Jenkins. 2008. "A Review of Sociotechnical Systems Theory: A Classic Concept for New Command and Control Paradigms." *Theoretical Issues in Ergonomics Science* 9 (6): 479–499. https://doi.org/10.1080/14639220701635470.

Wenz, Peter S. 1988. *Environmental Justice*. Albany: State University of New York Press.

Wittmayer, Julia M., Inês Campos, Flor Avelino, Donal Brown, Borna Doračić, Maria Fraaije, Swantje Gährs et al. 2022. "Thinking, Doing, Organising: Prefiguring Just and Sustainable Energy Systems via Collective Prosumer Ecosystems in Europe." *Energy Research & Social Science* 86: 102425. https://doi.org/10.1016/j.erss.2021.102425.

27

THE HUMAN DEVELOPMENT PARADIGM AND SOCIAL VALUE OF ENERGY

Saurabh Biswas, Faheem Hussain, and Mary Jane Parmentier

1. Introduction

Energy systems and their transitions innately possess the potential to undermine human development, disproportionately affecting the well-being of marginalized populations and places, irrespective of whether such communities are in the Global South or North. When off-grid electrification projects under-deliver on promised outcomes causing a slowdown in community development (see Ikejemba and Schuur, 2020), or grid expansion for universal access deepens exclusionary social structures and social divides in India (see Patnaik and Jha, 2020; Mayer, Ghosh Banerjee and Trimble, 2015; Nelson and Kuriakose 2017), or national green goals override local priorities in European community energy initiatives (see van Bommel and Höffken, 2021), or votes-for-electricity politics in Indonesia undermine democratic processes (Mohsin, 2014) – sacrifice zones of energy transitions arise. What may be considered as sustainable outcomes from one standpoint might be deeply unsustainable and unjust from other perspectives. Case in point is the "sustainable" logging for industrial-scale wood pellets production in the states of North Carolina and Virginia. The predominantly black and less affluent families in the communities around the production sites are reported to suffer from a range of negative outcomes to the health and overall quality of life (de Puy Camp, 2021). Any possible economic benefit to the community from the export of renewable fuel is steadily being replaced by air and noise pollution, increase in property taxes, marginalization of political voice, and as some ecologists point out, a steady loss of ecological balance.

Thus, climate goals of future energy systems are not separable from their purpose of enabling socioeconomic, cultural, and spiritual well-being of people, particularly of those in existing or potential sacrifice zones of energy transitions. This implies that the supply chains and politics of energy infrastructure must not replicate the systematic injustices and extractive practices of the past that deepen vulnerabilities and undermine coping capabilities in already marginalized communities (Smith, 2009; Sachs, 2008). Furthermore, energy transitions will need to be instrumental drivers of restoring agency and freedoms among marginalized communities by creating new opportunities and conditions for re-envisioning sustainable futures (McNeish et al., 2015).

This chapter explores the questions of human development, equity, justice, and safeguards for marginalized communities undergoing localized energy transitions. The concept of social value of energy mapping is first introduced as a heuristic approach to chart the current and future benefits, burdens, risks, and trade-offs to individuals and communities experiencing shifts

DOI:10.4324/9781003183020-31

in their energy system. The concept is then applied to a review of local energy and human development initiatives in a selection of communities of Bolivia, Nepal, and Bangladesh. The findings demonstrate that for decarbonizing energy transitions to be equitable and sustainable, they must be deliberately designed to pursue social value creation alongside other aims like emissions reductions and energy access in the communities. Key drivers and dynamics are discussed, forming the basis for framing the human development ambition and design guidelines for energy transitions in marginalized communities.

2. Methods: Social Value of Energy Mapping

Social value of energy mapping is the systemic exploration of the energy and human well-being relationship, characterizing positionally and temporally differentiated benefits, costs, burdens, trade-offs, and risks to people and places created by the services, actors, and their organization and managerial practices of the energy system (Biswas, 2020). At its center is the conception that there is a consistent production of social value for energy users, when multiple forms of benefits are generated through the enhancement of user capabilities and the simultaneous lowering of burdens, barriers, and risks to social value (Biswas et al., 2020; Miller, Altamirano-Allende and Agyemang, 2015).

As a methodological exploration of sustained human development co-located with energy transitions, social value of energy mapping is underpinned by three key ideas. First is an ethical perspective of social marginalization, described as the people and places that are largely excluded from benefits and comforts normally available, disproportionately increasing their hardships, deprivation, and vulnerabilities (Hennink et al., 2012; Peace, 2001). Second is a sociologically broad view of human development, where individual opportunities, capabilities and freedoms to pursue a better life co-exists and depends upon the structural ecosystem (infrastructure, governance, financing, and policy), which can directly influence and create conditions by widening set of choices, capabilities, and freedoms (Ponzio and Ghosh, 2016; Sen, 2006; Alkire, 2005). Finally, in this layered structure with human, ecological, and infrastructural attributes, the acknowledgment that selectively aimed interventions not addressing a multiplicity of attributes will eventually create counterproductive outcomes. The sociocultural subversion of marginalized spaces through simplistic biodiversity and wildlife conservation interventions in several African countries illustrate this risk (Banerjee and Linstead, 2004; Akama, 2004; Wall and McClanahan, 2015).

Figure 27.1 represents the logical process of posing interrogative questions to generate a profile of capabilities and freedoms affected by the energy system as it resides and functions within the social, cultural, economic, and ecological structural elements of the community. Mapping social value of energy identifies and characterizes capabilities and freedoms achievable and affected by the interactions of the various sociotechnical elements constituting the energy system, including its infrastructure, services, actors and their organization, transactions, and managerial practices. Thus, advancing to a value paradigm for energy transitions, the social value of energy mapping enables the framing of a vision of capabilities and freedoms for individuals and communities, which can then inform goals for structural change, techno-human capacities, and governing principles for the socio-technical system to transition in ways that value is generated for the people and communities at the center of the transitions. Used as an evaluative process, social value of energy maps can be analyzed from the perspectives of multiple stakeholders and interests in society, illustrating how sociotechnical arrangements of the energy systems delivers value and whether energy transitions generate cycles of value that overcome marginalization or if they lead to extractive cycles that continue to sustain and deepen marginalization.

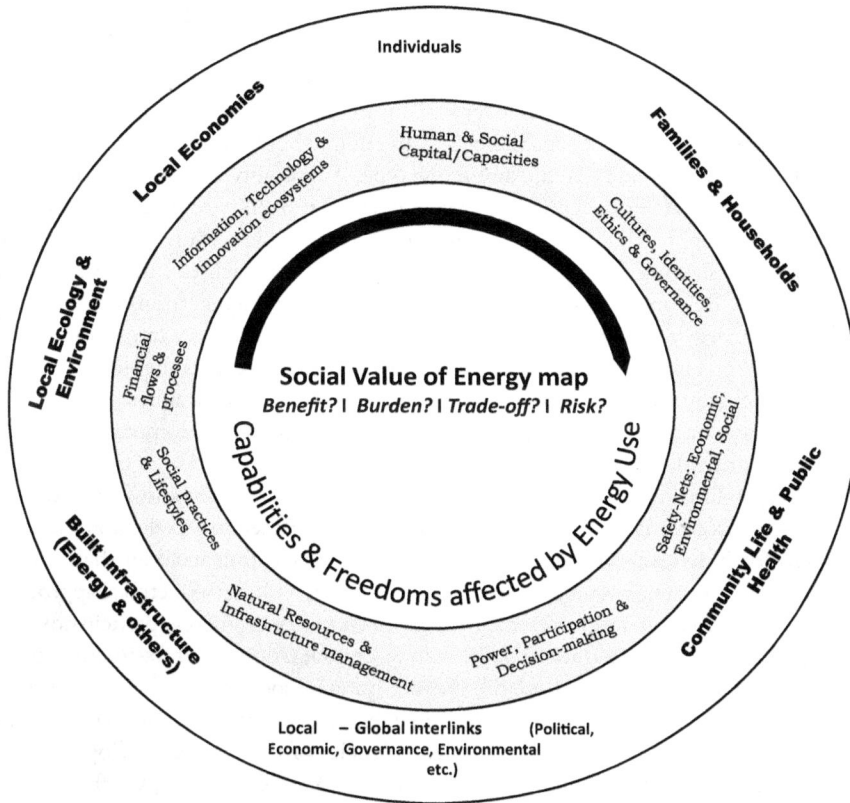

Figure 27.1 Social value of energy mapping (process schematic).

The outermost level depicts the actors and structural elements of a community formed at the intersection of social, cultural, economic, and ecological attributes. The next layer depicts the techno–human variables and functional attributes of the energy intervention or project that link the technical and managerial aspects of infrastructure to the community. The interactions and feedbacks among the various elements of the two layers form the socio-energy system. The innermost layer represents the outcomes in terms of a variety of capabilities and freedoms resulting from or affected by the dynamic interactions in the socio-energy system. A social value of energy map parses the system dynamics and outcomes thus produced, as beneficial or burdensome or creating risks and trade-offs for each of the actors and structural elements of the community, adding to or limiting human development and well-being.

Identifying Capabilities and Freedoms as Social Value of Energy

Prior to delving further into the dynamics of social value of energy, it is important to understand the interpretation of the capabilities approach as a concept and its methodological integration here. The ability of any individual to lead a satisfactory life and to progressively pursue enhanced well-being is a cumulative outcome of ways of *doing and being*, collectively called *functioning*, and the various combinations of functioning(s) producing *capabilities* to pursue well-being goals (Nussbaum and Sen, 1993). Alongside capabilities, there needs to exist an ecosystem of freedoms that enable the application and expression of one's capabilities. The combination of resources,

447

like income, public services, goods, and commodities, coupled with the appropriate capabilities and under conditions of necessary freedoms, results in consistent realization of human development and well-being outcomes (Alkire, 2005; Robeyns, 2005; Cohen, 1990), overcoming current deficits and unlocking the discovery of future potential (Hennink et al., 2012). It is important to note that a suite of capabilities and freedoms should co-exist for human development, and they are neither additive (meaning one type of capability and freedom does not compensate for others) and are not irreducible (meaning partial capability or freedom does not exist).

The capabilities approach offers a rigorous and adaptable framework for investigating the human condition, allowing energy researchers to apply the approach to study the complexities of energy systems, human development, and poverty. However, applications of the capabilities approach to the energy and human development relationship, particularly in quantifying impacts of energy services on energy poverty, appear to selectively focus on capabilities linked to the direct consumption of energy services (see Day, Walker and Simcock, 2016; Sadath and Acharya, 2017). From a standpoint of just and equitable transitions of energy systems, consumption-focused investigations tend to prioritize and isolate the problem-access and consumption of energy for improving selected capabilities of energy users (e.g., quantity, quality, affordability) from the broader sociotechnical environment of the energy system and their effects on a collection of capabilities and freedoms for individuals and communities that go beyond (but include) their identities as energy users (e.g., transactions and human resource practices, cultural and ethical appropriateness of technology, information and feedback mechanisms, politics of energy, participatory governance and goal setting, local economy and ecology considerations in infrastructure decisions). Although studies continue to investigate the broader systemic sociotechnical challenges in energy transitions, their grounding in an energy service–energy user lens limits the exploration of the co-located systemic issues and complex interdependencies of the human development problem (e.g., the complexities of forced migration and humanitarian energy access; Grafham, 2019).

In order to overcome such gaps and fully utilize the conceptual depth of the capabilities approach, it is important to shift the investigative lens to primarily focus on the systemic assemblage of well-being, capabilities, and freedoms in society through which energy access, consumption and infrastructures are viewed to characterize the energy-human development relationship. An example of this shift can be found in a recent study from Zimbabwe, identifying patriarchal institutions, social ordering, and differential ability to influence the national sociopolitical narrative of electrification as strong influence on freedoms and capabilities of individuals in the community, causing temporal and spatial variations in how energy poverty is experienced differently among social groups (Chipango, 2021).

Social value of energy mapping presents a methodological approach that incorporates capabilities and freedoms as a conceptual framework for linking human well-being to energy services and systems. It follows an exploratory heuristic to identify place-based drivers of well-being, the necessary set of capabilities, and the enabling conditions of freedoms, including the role of energy services and systems. By recognizing the structural elements of community well-being and mapping a broad set of capabilities and freedoms, energy systems can be positioned within a polycentric conceptualization of human development. This is critical to ensure that an artificial isolation of energy system–energy user interactions are avoided and energy transitions are not decontextualized from the broader questions of sustainable human development.

3. Reviews: Mapping Social Value of Energy in Marginalized Communities

Capabilities and freedoms associated with energy services, the rules and norms around energy use, and the structure of the energy system can be explored from multiple human and social development perspectives, as discussed earlier. In this section, the social value of energy mapping methodology explores such possibilities in places and among people engaged in localized efforts to transition their existing energy system (Table 27.1). These transitions are part of greater

Table 27.1 Mapping Social Value of Energy in Communities Undergoing Energy Transitions

	Benefit Pathways	*Burdens, Risks, and Trade-offs*	*Social Value of Energy Potential*
Case 1: Socioeconomic value creation in Amazon and Altiplano communities of Bolivia	• Enhanced incomes and profitable livelihood • Entrepreneurship opportunities • Co-existing formal and informal economies • Cooperative management of resources and activities • Preserving indigenous identity, culture, and ecological traditions • Lower carbon footprint	• Uncertain institutional support • Vulnerabilities at the informal-formal economy interface • Inclusion in participatory governance and management • Redistribution of benefits	• Enhancements in benefit redistribution pathways • Stronger autonomy of informal economy and lowered vulnerability • Enhanced coordination between local and national entities for broadening scope of Sustainable Development Goals
Case 2: Livelihoods, drinking water, and local governance in climate-vulnerable communities of Nepal	• Reliable supply of piped water to homes and farmland • Replacing diesel generators • Increased resilience to natural disasters and climate change • Equitable access and affordable service • Strengthening technical, self-governance, co-production, and impact assessment capacities in the community	• Tensions in participation, end-user energy needs, and ability to pay • Elevated risk of new burdens for women and children • Confluence of inequal distribution of benefits and pre-existing social divides risks project sustainability • Uncertainty from lack of transparency in national grid expansion plans	• Adaptive electricity service to match evolving energy needs due to economic, demographic, and climate changes • Coupling energy transition with interventions and institutions for gender rights, social justice, education, and skill development
Case 3: Reviving hope and dignity in refugee camps of Bangladesh	• Access to cooking fuel and reduced dependency on firewood • Domestic and public-area lighting for physical safety • Cell phone charging for communication and information access • Business opportunities in repairs for portable solar devices, cell phones, and charging stations • Emerging solidarity network and sharing economy within the camps and host communities	• Accidents and fire damages due to unfamiliarity with safe operation of cooking gas • Increased vulnerability to thefts in households with solar lights • Tensions with host communities over economic opportunities and resource distribution	• Safeguards to reduce and avoid tensions with host community and exploitative opportunities for vested interests • Investing in skill and innovation capacities, businesses opportunities and linkage to policy interventions for harmonious local economies and reintegration with society

visions of sustainable futures, encompassing changes in the problematic state of their socioeconomic well-being, livelihoods, environment, and political agency. Three primary case studies are examined, drawing from the longitudinal engagements of the co-authors with the communities in each case. In the Bolivian Amazon and Altiplano regions, the co-authors (M.J.P. and S.B.) have studied the social value of energy dynamics in 11 community-based off-grid energy projects in 2018–2019, with assistance from the global nonprofit Practical Action. A parallel study was conducted in various locations of Nepal, facilitated by SunBridge Solar, a Kathmandu-based off-grid solar project developer specializing in mini-grids and water-pumping projects in rural communities. The third case study is situated in the social value of energy dynamics intertwined with interventions for quality of life and dignity in the refugee camps of Cox's Bazar in Bangladesh, where over 800,000 displaced Rohingya families and individuals from the Rakhine State of Myanmar are housed. This is a continuing study since 2018 by one of the co-authors (F.H.) facilitated by two local nonprofits, Bangladesh Rural Advancement Committee (BRAC) and Young Power in Social Action (YPSA). Each of the studies are conducted using a multimodal data process that includes field visits and project evaluations, direct community engagement using interviews and focus group discussions, and reflection workshops with the facilitating organizations to deconstruct their experiences and project data. This section demonstrates a variety of social value of energy possibilities and how these are dynamically shaped by local and global factors.

Case 1

Poverty alleviation, informal economy, and cultural identity as social value of energy in remote communities in Bolivia

Background

Since 2000, Bolivia has recognized poverty as the systematic exclusion of groups of people from the benefits of natural and built resources, economic growth, social agency, and accountable governance (BPRSP, 2001). Rural poverty, in particular, is considered to be exclusions based on gender and ethnicity, caused by limited participation in decision-making and in the formal mechanisms of state policy. As a result, policies and programs for poverty alleviation were formulated and implemented, to develop human and social capital alongside infrastructures. Interventions like creating access to electricity especially for the rural poor (e.g., Programa Electicidad para Vivir con Dignidad) (PEVD, 2014), off-grid energy projects to increase productivity and incomes in the informal economy with particular attention to restoring, and reinforcing dignity and agency of the historically excluded groups (SmartVillages, 2016) are some examples. During the same timeframe, a general improvement in the country's economy and enhanced social spending helped increase wages and reduce income inequalities (Vargas and Garriga, 2015). However, it has been pointed out that legal frameworks are in need of greater flexibility and customizability to facilitate sustained integration of renewable energy technologies with rural development needs (Zegada, 2016). Opportunities exist in further empowering rural entrepreneurs, incentivizing user-centric and use-inspired business models, better-directed state support and finances, capacity development support, and platforms of engagement and deliberation on situating distributed energy systems to further stimulate structural improvements in local informal economies, environmental preservation, household well-being, and community identity (Eras-Almeida, 2019).

With this backdrop, we employ the social value of energy mapping to explore the systemic dynamics in selected off-grid projects implemented by several local communities, together with the international nonprofit Practical Action UK. Key informant interviews in the communities and with the project developers, focus group discussions, participant observations, and project planning and operating document reviews were conducted over a period of four weeks.

Planned and Emerging Benefit Pathways

In the Amazon region, near the town of Rurrenabaque on the Beni River, several small-scale solar-powered refrigeration units are being used to store fish before it is transported to the local market, about an hour away by canoe. Women entrepreneurs from the riverside communities have formed a cooperative entity to manage and operate the refrigeration units. Economic benefits have been in the form of improved and regular incomes since acquiring the refrigerators, as the businesswomen can now sell more fish than a day's catch and set up shop regularly, irrespective of fluctuations in the catch volume. Other households have supplemented their incomes through small commercial activities, like using solar ovens to dry beads for making jewelry sold to tourists and drying cacao beans for producing chocolate. Equally important is the freedom from being dominated by the formal economy and retaining control over their capability to pursue artisanal production of commodities. This is a culturally significant freedom, directly related to the sustenance of rural and ethnic identities, resisting forced choices of urban migration or industrial-scale production.

Likewise, in the Altiplano region near Lake Titicaca, a small solar water-pumping project improves incomes of livestock herders. The alpaca and llama herds can better survive the long dry winters, improving the availability of food, clothing, and additional income from selling the wool and meat. One farmer reported that before the solar pump, they would lose dozens of animals due to lack of water. Similarly, on the other side of the lake, a lakeside community installed a donated solar pump to irrigate the ancient Incan terraces cut into the steep hillsides around the community. The farming terraces were long dormant, with a drying climate and salination of Lake Titicaca. The pump enables dozens of terrace rows to remain green with crops of fava beans, potato, and other native tubular plants. The community now often has surplus produce, which they sell in the big market in La Paz, and plans to develop the community as a sustainable tourism destination where visitors could learn indigenous traditions and eat locally produced food. The pump will also potentially bring fresh groundwater (not from the lake) to buildings for drinking, cooking, and washing. In both communities, the energy transition and its economic benefits are moderated by ethnic values grounded in identity and relationship with nature. From a sustainability standpoint, these communities can therefore be considered to possess greater capabilities and freedoms in articulating and operationalizing the ethic of sustainable development through their energy transitions.

A notable dynamic of value creation from the off-grid renewable energy systems is the local community's governance and collective action to ensure that economic and infrastructural gains do not create unwanted trade-offs for social, environmental, and cultural values. Interviews and project plan reviews with the Practical Action team and community leadership clearly identified the central role of the deliberative process involving community representatives and domain experts like engineers and social scientists. Social capacity in technical and governance domains are collaboratively developed, with consideration to social practices, preferences, and existing norms of collaborations. Other key value propositions are the lowering of financial barriers to ownership of the technology and the intentional exploration of balance between formal and informal economies. For the communities, informal economic activities and transactions are key values embedded in their social fabric of indigeneity, with its roots in the ethic of harmonious co-existence with nature. Admittedly, these values are pressured by the stronger economic and material incentives of the formal economy. Hence, the intentional boundary-setting process for integration of renewable energy projects with formal structures (e.g., markets and banking) and defining domains of autonomy for the informal economic structures (e.g., rules for the cooperatives, self-consumption, and intensity of production) assume critical importance. For instance, the Altiplano wool collection to garment production aided by solar energy remains within cooperative norms set by the local artisans, whereas the finished product follows market forces, finding its way even to department stores in France.

Finally, the interactions of community renewable energy projects with local, regional, or national government capacities were reported to influence social value creation. Anecdotally, involvement of government agencies improved capabilities to finance and scale projects. On the contrary, project delays and procedural barriers are reportedly common when the linkage is weak. Further research is required to better understand this dynamic.

Burdens, Risks, and Trade-Off Dynamics

The variations and uncertainties due to the range of actors and contextual factors result in a profile of burdens, risks, and limitations to value creation in each case. In the Beni River community, while the refrigeration units create beneficial economic returns for the community members, there remains a recognized lack of electricity to power the streetlights along the paths from homes in the forest to the center of the community. This issue was addressed in some of the other nearby communities in similar situations through a solar light ownership program. The process of selecting beneficiaries did not meet the community's expectations of transparency, causing some discontent and diminished trust in the intent of institutional efforts to consistently address social value creation. Other burdens, although temporary, were sometimes caused due to delays in implementation timelines as initial consultation and training activities took longer than expected. The most prescient and dominant forms of risk are associated to the intersection of the formal and informal economies, where a large dependency on revenue from markets exposes the vulnerabilities of informal structures to global market oscillations. It was also possible to discern inequities within the communities in the governance of the technologies. For implementing the refrigeration project, the local women formed a cooperative; however, not all were initially able to join for financial and other reasons, and the cost of joining after its implementation was significantly higher. Thus, even at a small community level, equitable inclusion is not a given.

Social Value of Energy Potential

The extent of off-grid renewable energy system's role in sustainable futures for Bolivian communities relies heavily on sustained efforts at discovering new social value creation pathways and strengthening the existing ones through appropriate policy and financial safety nets. As pointed out before, many Bolivians live in rural areas, and their human development needs are multiple, sometimes unique and closely tied to their culture, place, and relationship to nature (BPRSP, 2001). A variety of technologies and application of clean energy is uniquely positioned to deliver social value in ways that overcome historic deficits and rapidly enhance development. These would have to contend with the established business as usual paradigms (e.g., the formal economy and grid extension for electricity access) that offer an easier path to economic benefits (Zegada, 2016). A just and sustainable energy transition for the Bolivian communities will therefore be one that deliberately includes the diversity of factors that constitutes their visions of human development and utilizes the social value offered by energy technologies and services to achieve the visions.

Case 2

Water, livelihoods, and managing socio-ecological vulnerabilities as social value of energy in communities in Nepal

Background

Small towns and rural areas of Nepal are endeavoring to radically improve their water supply situation, with the goal of lowering community vulnerabilities to health, irrigation, food supply, and drinking water shocks (Nepal et al., 2021). In the past two decades, several cities, towns, and rural areas have adopted the community operated water infrastructure model that follows a cost-sharing arrangement between the communities and funding agencies like the Asian Development Bank (ADB) to develop water pumping, storage, filtration, and distribution infrastructure(ADB, 2019; Vijayraghavan and Kilroy, 2017). Despite operational challenges and varying degrees of success, the community-operated model remains critical to water supply resilience in Nepal, particular for recovery post the devastating earthquake of 2015 (Gautam and Dahal, 2020). In agrarian communities, water supply and livelihood vulnerabilities are linked through irrigation capacities that are severely affected due to changes in rainfall patterns, flow rate variation of rivers, and insufficient electricity supply to operate pumps (He et al., 2018; Uprety et al., 2017). Solar energy is playing an increasingly important role here, by reliably and resiliently powering the pumps for local drinking water and irrigation systems.

The Ministry of Energy, Water Supply, and Irrigation estimated a growth of 176% in solar irrigation pumps and 30% in solar powered drinking water projects, in the 2018–2019 period (APEC, 2019). Off-grid solar PV installations for water pumping, along with rooftop installations and mini-grids, constitutes the majority of Nepal's 60 MW of installed solar capacity (Lohani and Blakers, 2021). The social value of energy mapping studies three such projects – of a fully operational drinking water supply project in the Gorkha municipality, a recently commissioned solar pumping project for irrigation in a village of the Nawalpur district, and a solar-diesel hybrid mini-grid at planning and design stage in the challenging terrain of Ghurmi Bazar village of Udayapur district – to illustrate the dynamics of this transition. Key informant interviews and project document reviews were used for the Gorkha and Nawalpur cases, whereas the Ghurmi Bazar case additionally involves focus group discussions, participant observations, and energy utilization surveys.

Planned and Emerging Benefit Pathways

The drinking water project in Gorkha adopts the Small Towns Water Supply and Sanitation Sector Project (STWSSSP) model (ADB, 2019), enabling layers of social value in the community through the organization and design of the coupled energy and water supply systems. The primary benefits of improved water supply include (1) creating equitable access to safe drinking water, consequently lowering health and financial burdens of contaminated drinking water especially for children; (2) reliable piped supply to minimize disruptions to household activities; and (3) improved sanitation outcomes. Coupling the solar photovoltaic power creates a second layer of benefits: (1) provide the technological choice of switching seamlessly between solar power and grid power based on assessments of cost and reliability benefits and (2) reduce the carbon footprint of the infrastructure. Similar social value of energy benefits emerged in the evaluation of the solar powered irrigation project in the Nawalpur district and in the design scoping for the solar-diesel hybrid mini-grid in Ghurmi Bazar. The benefits and lowering of burdens in Nawalpur were predominantly linked to improvements in agricultural livelihood and reducing vulnerability to seasonal rainfall for irrigation. At Ghurmi Bazar, they are linked to supplementing livelihood activities and meeting the critical electricity demand of the community through an energy system that is sufficiently resilient to natural hazards in an ecologically fragile community. An important design feature of the Ghurmi Bazar mini-grid is the blueprint for phased social value upgrades. The mapping exercise illuminated that the expansion of electricity service will sustain value creation by subsequently encompassing water pumping, increased electric cooking, school electrification, and gradual lowering of energy cost to benefit selected population groups in the village.

For the benefits to be realized and sustained, it is critical that the projects are embedded with enabling structures that engage the social capital, are culturally appropriate and align with social practices, and balance technological sophistication with capacities to operate and govern them. Each project relies fundamentally on an empowered local governance platform with capabilities to deliberate upon and set standards for delivery of service, as well as design and implement a context-sensitive billing and dues collection mechanisms. Agency in making financial decisions creates the necessary freedom for the project managers to respond to uncertainties, such as waiving penalties and deferring payments for a household facing temporary financial difficulties. The capabilities and freedoms to envision and realize the direct benefits and lower burdens are only possible through the exploration of socio-technical capacities and limits of actors involved. Thus, the limits of social capital and governance capacities of the community have to be necessarily supported by technical, financial, legal, and knowledge interventions. The role of multilateral institutional funding, therefore, becomes critical in achieving financial sustainability through not only the capital investment but also the regulatory, accounting, and management practices they introduce. Similarly, technical personnel, end-user awareness improvements, and consultative process during the design and construction phases are reported as key mechanisms of cultural integration of the projects. SunBridge Solar Nepal, a technology service provider involved in the projects and a key facilitator of the studies, played an important role in the formative community engagement processes.

Burdens, Risks, and Trade-Off Dynamics

The participation barrier due to the initial cost of subscription is a common burden. Irrespective of the energy tariff and membership dues for the cooperative, a proportion of community members find it difficult to afford the fee. This can be either temporary or persistent, especially for households where energy services may not add to their incomes (directly or indirectly) to balance the increased expense. If left unrecognized and unaddressed, inclusion barriers morph into a variety of risks, such as increased cost burden, disconnections, and shrinkage in end-user numbers, disproportionate distribution of benefits, and fractures in the relationship among social groups. Such risks have negative consequences for energy users and the sustenance of the project, destroying social value over time.

Mapping social value of energy in Ghurmi Bazar through interviews and focus groups offered insights on the foundations and evolution of such risks. Residential homes expected incremental growth in energy use starting with indoor comfort and reduction in manual effort for daily chores, without a significant increase in energy costs. On the other hand, businesses preferred aggressive utilization to power revenue-generating applications, driven by their assessment of being able to recover additional costs from their customers. The competing preferences were contested at the community forum and needed moderation by the project developer, resulting in differentiated pricing strategy. Other risks and trade-offs emerge from sociocultural issues, like underrepresentation of women in community governance and stunted educational outcomes for children due to increased propensity of employing them in growing business activities as a result of improved energy supply. Another risk to off-grid projects in general is the uncertainty of the national grid extension to off-grid communities. In the absence of clear plans or timelines for grid expansion (or not), decentralized project developers undertake a significant financial risk and design aggressive tariffs to recover investments in shorter timeframes. This also discourages supplementary investments in capacity building and scaling community projects due to the possibility of becoming obsolete.

Social Value of Energy Potential

The three cases here demonstrate that the situational and future oriented understanding of social value of energy is critical in providing direction to solar energy projects and design them to create sustained social value. Solar projects integrated with the goals of drinking water, irrigation, and climate-resilient community services and livelihoods carry a significantly higher value proposition and are able to bypass certain barriers in policy and regulations of the electrification sector, by being data-driven and distributing the cost of electricity planning through localized discovery of socially valuable uses of electricity (IRENA, 2016). Targeted programs like these are also better positioned to cater to specific needs of disadvantaged population groups and indigenous people in Nepal by directing solar energy to meet social equity goals (APEC, 2019).

However, for the energy transition to sustain, strategic interventions will have to be equally attentive to the internalized sociopolitical risks and trade-offs. Plans would have to identify and build safety nets for emergent burdens and risks, like those caused by the inclusion barriers. This would have to include the ability to implement adaptive fixes by local stakeholders with limited resources, as well as institutional interventions to transform root causes over the long term. Sociocultural drivers of negative consequences require involvement of specialized actors, like governmental and non-governmental entities specializing in gender and child rights, indigenous rights, economic justice, transparency and accountability in local governance, and so on. External actors, like developers and researchers, might be able to identify challenges but do not possess the social and political legitimacy to address those. Interestingly, this present a social value proposition where energy transitions can be a proactive instrument to pursue social justice goals.

Case 3

Hope, agency, and revival as social value of energy in refugee camps in Bangladesh

Background

Nearly 800,000 refugees find shelter in Cox's Bazar district of Bangladesh, creating the world's largest refugee camp administered under the supervision of the government of Bangladesh with support from multilateral agencies and nonprofits. The refugees of the Rohingya ethnicity have been escaping decades of political persecution and more recently violent ethnic cleansing in Myanmar (UNHCR, 2020), and have settled in the Cox's Bazar refugee settlement since 2017. A massive humanitarian crisis is underway, presenting a challenge not only for providing relief to the refugee population but also for sustaining and improving the quality of life for the host population. The situation is particularly critical as Cox's Bazar is a region with relatively higher incidence of poverty and weak basic infrastructure, including energy. Limited electricity and cooking fuel services are strained by the additional demand from the camps, spurring multiple interventions to mitigate the infrastructural, environmental, and humanitarian challenge. Social value of energy mapping is utilized to gather information through interviews and ethnographic observations over several years. This data is analyzed to understand the interlinkages of energy and human distress, intended and unintended effects of solutions implemented, and potential future pathways for energy transitions in this highly fluid and resource deficient situation.

Planned and Emerging Benefit Pathways

Unsurprisingly, the foremost forms of social value of energy are in mitigating the extremely deficient conditions and preventing cascading negative impacts in a situation where thousands struggle to meet basic survival needs. The first forms of energy deficiency appeared in cooking fuel access, which rapidly compounded into a cascade of economic and environmental consequences. Firewood, as the least costly and easily accessible source of cooking fuel, led to an alarming depletion of the forest cover and the local ecology in the vicinity of the refugee camps. The resulting reduction in availability and increase in prices of firewood, combined with the loss of traditional livelihoods dependent on the forest, has diminished economic capabilities of the local host population. To reduce the dependency of the refugee population on firewood, bottled cooking gas was introduced to the camps. As a result, a significant portion of the refugee population are no more entirely dependent of firewood and some of the cascading consequences reportedly reduced in intensity.

The second key dimension of the energy deficiency is the lack of electricity for domestic use and public amenities. With no grid supply, solar panels and diesel generators are used for basic lighting, mobile phone charging, and other essential infrastructure at the camp site, including lighting of public places for safety. A limited number of solar lanterns, solar charging devices, and solar panels exist at the campsite, distributed by humanitarian aid agencies. Physical security at night is a critical value proposition of lighting. However, the most important function is reported to be mobile phone charging. Mobile phones are the only medium for communication, access to information and educational content, and entertainment in the absence the internet. Given the value of mobile phone devices and keeping them powered, an entrepreneurial ecosystem has emerged in the camps, sometimes extending to the host community. This consists of businesses that repair devices and sell multimedia content on memory cards, as well as a network of charging points and stations. Similarly, a sharing ecosystem of solar lights and charging devices has developed among the residents of the camp. Certain human and social capacities emerge as residents mobilize around the energy interventions, creating some semblance of a safety net for the communities. First is the sense of hope and some level of agency experienced by the refugees, as they feel better informed and not entirely helpless when they engage with the relief agencies to discuss ways to improve living conditions in the camps. Second is the emergence of the solidarity networks among fellow campers and some host community members, supporting the entrepreneurial ecosystem and sharing economy built around mobile phones and solar devices. This has helped develop a degree of trust with the host community, which is an important avenue of material and technical expertise inflows into the micro-economy of the refugee camps.

Burdens, Risks, and Trade-Off Dynamics

Given the intensely deficient and overwhelmed infrastructure, ecology, economy, and governance capacities, negative consequences and risks from energy interventions were not unexpected. So while bottled cooking gas mitigates the fuel availability situation and deforestation problem to some extent, it increased the risk of fires in the camps. A recent blaze in the camps was found have been caused by a cooking gas accident. This may be attributed to the unfamiliarity with the fuel and unsafe cooking devices and practices. Also, it was reported that break-ins became common in predominantly female-led households at a certain point in time, as these could be identified at night by the solar lights distributed to this selected group. Another critical risk is the increasing dissatisfaction among the host community, first owing to the strain on the local infrastructure and environment, then subsequently due to the feeling of losing out on the cooking gas and solar interventions. Although an informal economy of exchange and trade has emerged among the two groups, there remains very real possibilities of exploitative transactions and shocks to the rehabilitation efforts arising from conflicting socioeconomic interests. These patterns indicate social value gaps and opportunities in the governance, safety net formulation, and explorations of resource and benefits distribution among the two communities.

Social Value of Energy Potential

The realized social value and emerging challenges from energy access interventions at Cox's Bazar, viewed from the overarching perspective of ensuring dignified relief and rehabilitation to distressed people, highlight potential pathways for transitions. First priority would be to mitigate some of the most pertinent resource deficits and depletion challenges. Energy interventions integrated with food security (supply, cooking, storage) with minimal impact on the local environment is one such pathway. Similarly, enhancing education and vocational skills of youth from both the camps and host communities enables a pathway to livelihood opportunities and social integration. For the medium term, pathways to developing social capacity and innovation in energy systems can stabilize and strengthen the informal economic exchanges between host and settler populations. Improved energy access, harmonious co-existence founded on mutual economic value creation, conflict management, and ecological revitalization are key areas for strengthening this relationship. More critically, a large-scale transition needs to happen where the refugee population can move out of camps and rebuild dignified and fulfilling lives. With the role of hope, agency, and entrepreneurial spirit proving to be critical to rebuilding lives, the energy transition can become a vehicle for advancing these elements through targeted application areas like digital platforms for education, vocational training, and employment.

4. Discussion

Energy systems and their social contexts co-exist, co-evolve, and are mutually shaped by the dynamics emerging from the broad interactions of the infrastructural, sociocultural, and organizational elements of the socio-energy system (Miller and Richter, 2014). The case studies illustrate that advancing human development in communities require energy systems to engage with multiple societal levels, technical and environmental attributes, and human facets of well-being, a complex relationship known as the energy-poverty nexus (Biswas, 2020). Mapping the social value of energy illustrates the possible pathways of change for the energy-poverty nexus, presenting choices for socio-technical design of energy transitions. Such transitions can be instrumental in creating, advancing, and sustaining human development in communities, a combination of meeting material needs to sustain healthy and productive individuals while creating conditions of dignified agency, choice, and opportunities for satisfactory co-existence in society (Alkire, Roche, and Vaz, 2017; Stewart, 2013). At the same time, risks, trade-offs, and negative consequences reinforced or created due to fundamental weaknesses in infrastructure and social organization of the community, can be addressed proactively by the energy system. Mapping the social value of energy in each of the three cases of energy-poverty nexus points to three essential characteristics of energy transitions for human development.

i.   *Social purpose of energy transitions – filling gaps, discovering new value, and preventing trade-offs in human well-being*

Designs for energy transitions must be predicated on a social purpose that includes addressing well-being needs and mechanisms for identifying failure modes, mapping trade-offs, and sensitivity to dynamics that could offer new value propositions. This can include layers of goals, such as energizing homes, productive uses, and livelihood opportunities that strengthen the local economy, social safety nets, strategic adaptations to socioecological risks, and multidimensional resilience. Mapping social value of energy at the outset helps establish early and intermediate tangible outcomes, as witnessed in the case of Bolivian communities and refugee camps of Cox's Bazar. Integrating social value of energy mapping over the phases of the transition helps rediscover and adapt goals to the changing social, economic, and ecological conditions. For instance, in the Nawalpur solar irrigation project, the community is exploring utilizing excess solar electricity for other farm and residential uses, and optimizing pumping hours for balancing irrigation needs with underground water conservation concerns.

Allocating social purpose to energy transition must be a deliberate design process founded in contextual evidence and co-produced visions of thriving futures. Merely the fact that they create instantaneous access to clean and affordable energy and are community-based does not automatically make energy transitions just or sustainable. Gaps in the right mix of socio-technical capacities, trust deficits between stakeholders at different scales, or weak mechanisms to track realized societal benefits are some key factors that lead to underwhelming human development outcomes (e.g., see Ikejemba and Schuur, 2020), deepen social strife when large-scale energy transitions lack safeguards against existing exclusionary structures (Patnaik and Jha, 2020; Mayer, Ghosh Banerjee, and Trimble, 2015) and foster new barriers for selected groups, such as women (Nelson and Kuriakose 2017).

ii.   *Grounding energy transitions in ethics and sociocultural values*

Energy decisions are usually technological and economic choices made by empowered actors and entities, the implications of which transcend boundaries, and the negative externalities disproportionately absorbed by weaker sections of society (Miller, 2012; Sachs, 2008). Just and equitable human development as a goal of energy transitions will not be realized unless ethics and sociocultural values are incorporated in decision-making (Miller et al., 2013; Sovacool and Dworkin, 2014) through contextual exploration of what is ethical in terms of cultural and spiritual values of the people (Biswas et al., 2020; Bethem et al., 2020) and formulate a contextual knowledge basis for energy decisions going beyond universal assumptions (Biswas and Miller, 2022). This includes the exploration of the ethical norms for self-governance, authority and boundaries of decision-making among stakeholders, and processes of identifying and negotiating contestations in the community projects. In the communities in Bolivia and Nepal, safeguarding ethnic identities, cultural norms, and practices of collective action were non-negotiable values from the community perspective. Solidarity and values around the human–nature relationship strongly influence rules of transactions and production. In the case of the refugee camps of Cox's Bazar, energy interventions are expected to evolve and match the ethical expectation of restoring and strengthening human dignity.

The process of grounding energy decisions in place-based values and knowledge is not linear and requires deliberations over time and other changes. The resulting operational tensions for project implementation can have implications on stakeholder commitments to the process of transition. A critical paradigm shift is therefore required here, recognizing that the suite of social value in energy transitions is an evolving opportunity space. For instance, the UN refugee agency's strategy paper (UNHCR, 2019) points out in the context of energy transitions for refugee populations – to begin with energy access is a key bridge to short-term humanitarian response, but it is also a work in progress for long-term development goals through sustained energy interventions guided by the ethic of "protection" that includes safety and dignity, peaceful co-existence, avoiding harms, and mitigating potential risks. Thus, viewing energy transitions as exploration of social value of energy for human development instead of a one-off project helps recognize potential negative consequences for the marginalized, like burdens of disproportionate distribution of roles and agency (van Bommel and Höffken, 2021), dominance of sociocultural values by market forces (Ockwell et al., 2017, 2018), and the subjugation of political freedoms in exchange for infrastructure investments (Mohsin, 2014).

iii. *Diversity of engagement processes, actors, and roles*

Curated social engagement platforms and self-governance capabilities accompanying the energy projects stand out as the fundamental driver for the formulation of social purpose and ethical grounding of the energy project, as well as for generating a positive externality in terms of strengthening local democratic cultures and values. The processes and forms of consultation throughout the life cycle of the project, capturing perspectives, intentions, and capacities of the multiple stakeholders, enables reflexive and adaptive planning, designs, and operations. Engagement platforms and processes should therefore result in enabling conditions that sustain value creation through knowledge and negotiation capabilities to deliberate upon human, social, economic, technological, and environmental attributes of projects. They should reveal issues that could be fundamental barriers to freedoms of choice and reasoned agency (Sen, 2004), recognition of marginalized groups, power differentials, and consent in participatory processes (Farhar, Osnes, and Lowry, 2014; Rojas and Kindornay, 2014; Sovacool and Dworkin, 2014).

Given that any credible project roadmaps for a sustained creation of social value would pursue a suite of development goals and span multiple levels of complexities (Yillia, 2016), social engagement platforms require explicit attention. They should include actors and capacities of governance, skills for planning and evaluation, knowledge processes for converging social-environmental-indigenous-technological knowledge, and risk anticipation and mitigation capacities (Carley, Engle, and Konisky, 2021; Romero-Lankao et al., 2021). Moreover, engagement needs to be financially and operationally sustained over time, beginning with goal setting, developing and implementing designs, and evaluating outcomes (Bazilian et al., 2011; Cummings & Worley, 2014; McMichael and Weber, 2020).

The case studies and other reports show that even the most marginalized communities highly value participation and agency to shape energy interventions, as much as they appreciate partners who bridge the gaps in capacity (see UNHCR, 2019; Betts at al., 2014). Assuming otherwise will undercut the sanctity and effectiveness of consultative processes and platforms through processes of disproportionate control over the goal setting process (Levidow and Raman, 2020), political legitimacy (van Bommel and Höffken, 2021) or techno-centric worldview (Mengolini and Masera, 2021), and exploiting the absence of public mobilization and intermediaries that facilitate mobilization (Fedele, Locatelli, and Djoudi, 2017; Eaton and Kinchy, 2016).

5. Conclusion: Social Value of Energy and Transitions for Human Development

The human development and well-being paradigm requires energy transitions to produce a set of services, products, and forms of capital – the social value of energy that empowers communities and individuals to leverage this value to improve and sustain their social, economic, and ecological situation. This transition should be driven by informed agency, the values and ethics of identity, cooperation and altruism for the people and planet instead of insecurities and power struggles. In marginalized communities facing simultaneous social, economic, and ecological challenges, planned energy transitions are an opportunity for purpose-driven and creative co-evolution of energy systems with human and social aspirations for a good life.

Exploring the drivers, conditions, and interactions of energy projects and human development, maps of social value of energy in the cases discussed earlier puts forth three key considerations for equitable and just human development outcomes from energy transitions. First is the question of the *social purpose* – exploring what human and social development goals energy transitions can facilitate, in addition to creating sufficient access to clean and environmentally benign energy services. This includes current and future aspirations about the quality of life, conditions of material and sociocultural thriving without negative trade-offs, and the aesthetics of inhabited spaces. Second are considerations of *ethics and values* – grounding energy decisions in principles that ensure existing forms of marginalization are mitigated and not perpetuated. These would include exploring tensions in power differentials, identity and cultures, knowledge making capabilities, the human-nature relationship, restoration and enhancements in agency, and the like. Achieving ethical grounding requires a paradigmatic shift in framing energy transitions as layered and phased pursuits of multiple goals and not as one-off energy projects. From a planning standpoint, this shift revisits how the questions of timelines, resources, facilitating structures, and rules are traditionally framed from a project perspective. Finally, to match the plural and multidimensional structure of energy transitions, considerations of *actors and processes* require creative and sustained efforts. Legitimate actors, mechanisms of engagement in the discovery of social purpose and ethical values of the transitions, and complementarity in capacities are essential considerations for resilient and adaptive transitions.

Energy transitions for human development requires going beyond minimal energy solutions and a singular goal. Instead, energy transitions must be reimagined as exploration and expansion in social value of energy through context-relevant processes, facilitating participation of marginalized populations as decision-making stakeholders and not merely as beneficiaries of energy services.

Acknowledgments

The field studies have been possible due to the continued support and facilitation of the collaborators – Practical Action UK in Bolivia, SunBridge Solar Nepal in Nepal, and Bangladesh Rural Advancement Committee (BRAC) and Young Power in Social Action (YPSA) in Bangladesh. M.J.P. and S.B. received financial support from the Global Consortium for Sustainability Outcomes (GCSO) to conduct research in Bolivia and Nepal. Dr. Clark A. Miller, Dr.

Witold-Roger Poganietz, and Mr. Davi Ezequiel François have provided critical inputs to this research.

References

Asian Development Bank (ADB). 2019. Completion Report. "Nepal: Second Small Towns Water Supply and Sanitation Sector Project." October. www.adb.org/sites/default/files/project-documents/41022/41022-022-pcr-en.pdf

Akama, John S. 2004. "Neo-Colonialism, Dependency and External Control of Africa's Tourism industry." In: *Tourism and Postcolonialism: Contested Discourses, Identities and Representations*. London: Routledge, 140–152.

Alkire, Sabina. 2005. *Valuing Freedoms: Sen's Capability Approach and Poverty Reduction*. Oxford University Press on Demand.

Alkire, Sabina, Jose Manuel Roche, and Ana Vaz. 2017. "Changes over Time in Multidimensional Poverty: Methodology and Results for 34 Countries." *World Development 94*: 232–249. https://doi.org/10.1016/j.worlddev.2017.01.011.

Alternative Energy Promotion Centre (APEC). 2019. "Progress at a Glance: A Year in Review FY 2075/76 (2018/19)." November 4. www.aepc.gov.np/documents/annual-progress-report-aepc

Banerjee, Subhabrata Bobby, and Stephen Linstead. 2004. "Masking Subversion: Neocolonial Embeddedness in Anthropological Accounts of Indigenous Management." *Human Relations 57*, no. 2: 221–247. https://doi.org/10.1177/0018726704042928.

Bazilian, Morgan, Holger Rogner, Mark Howells, Sebastian Hermann, Douglas Arent, Dolf Gielen, Pasquale Steduto et al. 2011. "Considering the Energy, Water and Food Nexus: Towards an Integrated Modelling Approach." *Energy Policy 39*, no. 12: 7896–7906. https://doi.org/10.1016/j.enpol.2011.09.039.

Bethem, Jacob, Giovanni Frigo, Saurabh Biswas, C. Tyler DesRoches, and Martin Pasqualetti. 2020. "Energy decisions within an Applied Ethics Framework: An Analysis of Five Recent Controversies." *Energy, Sustainability and Society 10*, no. 1: 1–6. https://doi.org/10.1186/s13705-020-00261-6.

Betts, Alexander, Louise Bloom, Josiah David Kaplan, and Naohiko Omata. 2014. *Refugee Economies: Rethinking Popular Assumptions*. University of Oxford, Refugee Studies Centre.

Biswas, Saurabh. 2020. "Creating Social Value of Energy at the Grassroots: Investigating the Energy-Poverty Nexus and Co-Producing Solutions for Energy Thriving." PhD Diss., Arizona State University.

Biswas, Saurabh, and Clark A. Miller. 2022. "Deconstructing Knowledge and Reconstructing Understanding: Designing a Knowledge Architecture for Transdisciplinary Co-creation of Energy Futures." *Sustainable Development 30*, no. 2: 293–308.

Biswas, Saurabh, Jennifer Richter, Clark A. Miller, Carlo Altamirano Allende, Mary Jane Parmentier, Nalini Chhetri, Netra Chhetri, Stacia Dreyer°, and Davi E. François. 2020. "Eradicating Poverty through Energy Innovation." *Proceedings of the 26th International Sustainable Development Research Society Conference*, 802–813.

Bolivia Poverty Reduction Strategy Paper (BPSRP) 2001. "Government of Bolivia," March. www.imf.org/external/NP/prsp/2001/bol/01/Index.htm

Carley, Sanya, Caroline Engle, and David M. Konisky. 2021. "An Analysis of Energy Justice Programs across the United States." *Energy Policy 152*: 112219.

Chipango, Ellen Fungisai. 2021. "Beyond Utilitarian Economics: A Capability Approach to Energy Poverty and Social Suffering." *Journal of Human Development and Capabilities 22*, no. 3: 446–467.

Cohen, Gerald A. 1990. "Equality of what? On Welfare, Goods and Capabilities." *Recherches Économiques de Louvain/Louvain Economic Review 56*, no. 3–4: 357–382.

Cummings, Thomas G., and Christopher G. Worley. 2014. *Organization Development and Change*. Cengage Learning.

Day, Rosie, Gordon Walker, and Neil Simcock. 2016. "Conceptualising Energy use and Energy Poverty using a Capabilities Framework." *Energy Policy 93*: 255–264. https://doi.org/10.1016/j.enpol.2016.03.019.

de Puy Camp, Majlie. 2021. "How Marginalized Communities in the South are Paying the Price for 'Green Energy' in Europe." *CNN*, July 9. www.cnn.com/interactive/2021/07/us/american-south-biomass-energy-invs.

Eaton, Emily, and Abby Kinchy. 2016. "Quiet Voices in the Fracking Debate: Ambivalence, Nonmobilization, and Individual Action in Two Extractive Communities (Saskatchewan and Pennsylvania)." *Energy Research & Social Science 20*: 22–30. https://doi.org/10.1016/j.erss.2016.05.005.

Eras-Almeida, Andrea A., Miguel Fernandez, Julio Eisman, Jose G. Martin, Estefania Caamano, and Miguel A. Egido-Aguilera. 2019. "Lessons Learned from Rural Electrification Experiences with third Generation Solar Home Systems in Latin America: Case Studies in Peru, Mexico, and Bolivia." *Sustainability 11*, no. 24: 7139. https://doi.org/10.3390/su11247139.

Farhar, Barbara C., Beth Osnes, and Elizabeth A. Lowry. 2014. "Energy and Gender." In: Antoine Halff, Benjamin K. Sovacool, and Jon Rozhon (eds.), *Energy Poverty: Global Challenges and Local Solutions*. Oxford: Oxford Academic. https://doi.org/10.1093/acprof:oso/9780199682362.003.0008

Fedele, Giacomo, Bruno Locatelli, and Houria Djoudi. 2017. "Mechanisms Mediating the Contribution of Ecosystem Services to Human Well-Being and Resilience." *Ecosystem Services 28*: 43–54. https://doi.org/10.1016/j.ecoser.2017.09.011.

Gautam, G., and K. R. Dahal, 2020. "Issues and Problems of Community Water Supply Schemes with Special Reference to Nepal." *Journal of Civil, Construction and Environmental Engineering, 5* no. 5: 114.

Grafham, Owen, ed. 2019. *Energy Access and Forced Migration*. Routledge.

He, Lulu, Jonathan C. Aitchison, Karen Hussey, Yongping Wei, and Alex Lo. 2018. "Accumulation of Vulnerabilities in the Aftermath of the 2015 Nepal Earthquake: Household Displacement, Livelihood Changes and Recovery Challenges." *International Journal of Disaster Risk Reduction 31*: 68–75. https://doi.org/10.1016/j.ijdrr.2018.04.017.

Hennink, Monique, Ndunge Kiiti, Mara Pillinger, and Ravi Jayakaran. 2012. "Defining Empowerment: Perspectives from International Development Organisations." *Development in Practice 22*, no. 2: 202–215. https://doi.org/10.1080/09614524.2012.640987.

Ikejemba, Eugene C.X., and Peter C. Schuur. 2020. "The Empirical Failures of Attaining the Societal Benefits of Renewable Energy Development Projects in Sub-Saharan Africa." *Renewable Energy 162*: 1490–1498. https://doi.org/10.1016/j.renene.2020.08.052.

IRENA. 2016. *Solar Pumping for Irrigation: Improving Livelihoods and Sustainability*. Abu Dhabi: The International Renewable Energy Agency.

Levidow, Les, and Sujatha Raman. 2020. "Sociotechnical Imaginaries of Low-Carbon Waste-Energy Futures: UK Techno-Market Fixes Displacing Public Accountability." *Social Studies of Science 50*, no. 4: 609–641. https://doi.org/10.1177/0306312720905084.

Lohani, Sunil Prasad, and Andrew Blakers. 2021. "100% Renewable Energy with Pumped-Hydro-Energy Storage in Nepal." *Clean Energy 5*, no. 2: 243–253. https://doi.org/10.1093/ce/zkab011.

Mayer, Kristy, Sudeshna Ghosh Banerjee, and Chris Trimble. 2015. Elite *Capture: Residential Tariff Subsidies in India*. Washington, DC: World Bank. doi:10.1596/978-1-4648-0412-0.

McMichael, Philip, and Heloise Weber. *2020. Development and Social Change*. Sage Publications.

McNeish, John-Andrew, Axel Borchgrevink, and Owen Logan (eds.). 2015. *Contested Powers: The Politics of Energy and Development in Latin America*. Chicago, IL: The University of Chicago Press. doi: 10.5040/9781350219366.

Mengolini, Anna, and Marcelo Masera. 2021. "EU Energy Policy: A Socio-Energy Perspective for an Inclusive Energy Transition." In *Shaping an Inclusive Energy Transition*, 141–161. Springer, Cham. https://doi.org/10.1007/978-3-030-74586-8_7.

Miller, Clark. 2012. "Energy Justice: Ensuring Human Dignity in the Post-Carbon Future." *Cairo Review of Global Affairs 5*: 46–59.

Miller, Clark A., and Jennifer Richter. 2014. "Social Planning for Energy Transitions." *Current Sustainable/Renewable Energy Reports 1*, no. 3: 77–84. https://doi.org/10.1007/s40518-014-0010-9.

Miller, Clark A., Carlo Altamirano-Allende, Nathan Johnson, and Malena Agyemang. 2015. "The Social Value of Mid-Scale Energy in Africa: Redefining Value and Redesigning Energy to Reduce Poverty." *Energy Research & Social Science 5*: 67–69. https://doi.org/10.1016/j.erss.2014.12.013.

Miller, Clark A., Alastair Iles, and Christopher F. Jones. 2013. "The Social Dimensions of Energy Transitions." *Science as Culture 22*, no. 2: 135–148. https://doi.org/10.1080/09505431.2013.786989.

Mohsin, Anto. 2014. "Wiring the New Order: Indonesian Village Electrification and Patrimonial Technopolitics (1966–1998)." *Sojourn: Journal of Social Issues in Southeast Asia 29*, no. 1: 63–95.

Nelson, Sibyl, and Anne T. Kuriakose. 2017. "Gender and Renewable Energy: Entry Points for Women's Livelihoods and Employment." *Climate Investment Funds*. https://www.climateinvestmentfunds.org/sites/cif_enc/files/knowledge-documents/gender_and_re_digital.pdf

Nepal, Santosh, Nilhari Neupane, Devesh Belbase, Vishnu Prasad Pandey, and Aditi Mukherji. 2021. "Achieving Water Security in Nepal Through Unravelling the Water-Energy-Agriculture Nexus." *International Journal of Water Resources Development 37*, no. 1: 67–93. https://doi.org/10.1080/07900627.2019.1694867.

Nussbaum, Martha, and Amartya Sen, eds. 1993. *The Quality of Life*. Clarendon Press.

Ockwell, David, Rob Byrne, Ulrich Elmer Hansen, James Haselip, and Ivan Nygaard. 2018. "The Uptake and Diffusion of Solar Power in Africa: Socio-Cultural and Political Insights on a Rapidly Emerging Socio-Technical Transition." *Energy Research & Social Science 44*: 122–129. https://doi.org/10.1016/j.erss.2018.04.033.

Ockwell, David, Rob Byrne, Kevin Urama, Nicholas Ozor, Edith Kirumba, Adrian Ely, Sarah Becker, and Lorenz Gollwitzer. 2017. "Debunking Free Market Myths: Transforming Pro-Poor, Sustainable Energy Access for Climate Compatible Development." In *Making Climate Compatible Development Happen*, 130–150. Routledge.

Patnaik, Sasmita, and Shaily Jha. 2020. "Caste, Class and Gender in Determining Access to Energy: A Critical Review of LPG Adoption in India." *Energy Research & Social Science 67*: 101530. https://doi.org/10.1016/j.erss.2020.101530.

Peace, Robin. 2001. "Social Exclusion: A Concept in Need of Definition?" *Social Policy Journal of New Zealand 16*: 17–36.

Programa Electicidad para Vivir con Dignidad (PEVD). 2014. "Supreme Decree No 2061, Financing Agreement, Law 581 Approval Credit Agreement." www.pevd.gob.bo/informacion-publica/infraestructura-descentralizada-para-la-transformacion-rural-idtr-ii

Ponzio, Richard, and Arunabha Ghosh. 2016. *Human Development And Global Institutions: Evolution, Impact, Reform*. Routledge.

Robeyns, Ingrid. 2005. "Selecting Capabilities for Quality of life Measurement." *Social Indicators Research 74*, no. 1: 191–215.

Rojas, Cristina and Kindornay Shannon. 2014. "The Politics of Governing Development." In: Heloise Weber (ed.), *The Politics of Development. A Survey*, 10–30. London; Routledge. https://doi.org/10.4324/9780203804919.

Romero-Lankao, Patricia, Alana Wilson, Joshua Sperling, Clark Miller, Daniel Zimny-Schmitt, Benjamin Sovacool, Chris Gearhart, et al. 2021. "Of Actors, Cities and Energy Systems: Advancing the Transformative Potential of Urban Electrification." *Progress in Energy 3*, no. 3: 032002.

Sachs, Jeffrey. 2008. *Common Wealth: Economics for a Crowded Planet*. New York: Penguin.

Sadath, Anver C., and Rajesh H. Acharya. 2017. "Assessing the Extent and Intensity of Energy Poverty using Multidimensional Energy Poverty Index: Empirical Evidence from Households in India." *Energy Policy 102*: 540–550. https://doi.org/10.1016/j.enpol.2016.12.056.

Sen, Amartya. 2004. *Rationality and Freedom*. Harvard University Press.

Sen, Amartya. 2006. "Development as Freedom: An India perspective." *Indian Journal of Industrial Relations 42*, no. 2: 157–169.

Smart Villages. 2016. "Sustainable Energy Sources for Off-Grid Rural Communities in Bolivia: Opportunities, Challenges, and Perspectives." *Workshop Report 17 LA PAZ, BOLIVIA*, April. https://e4sv.org/wp-content/uploads/2017/02/WR17-Sustainable-energy-sources-for-off-grid-rural-communities-in-Bolivia-opportunities-challenges-and-perspectives-1.compressed.pdf

Smith, James. 2009. *Science and Technology for Development*. London: Bloomsbury Publishing.

Sovacool, Benjamin K., and Michael H. Dworkin. 2014. *Global Energy Justice*. Cambridge University Press. https://doi.org/10.1017/CBO9781107323605.

Stewart, Frances. 2013. "Capabilities and Human Development: Beyond the Individual-the Critical Role of Social Institutions and Social Competencies." *UNDP-HDRO Occasional Papers* 2013/03.

UNHCR. 2020. "Rohingya Refugee Crisis Timeline." www.unrefugees.org/news/rohingya-refugee-crisis-timeline.

United Nations High Commissioner for Refugees (UNHCR). 2019. "Global Strategy for Sustainable Energy: A UNHCR Strategy 2019–2024.". www.unitar.org/ptp/sustainable-energy

Uprety, Sital, Juliet Iwelunmor, Nora Sadik, Bipin Dangol, and Thanh H. Nguyen. 2017. "A Qualitative case Study of Water, Sanitation, and Hygiene Resources after the 2015 Gorkha, Nepal, Earthquake." *Earthquake Spectra 33*, no. 1_suppl: 133–146. https://doi.org/10.1193/112916eqs212m

van Bommel, Natascha, and Johanna I. Höffken. 2021. "Energy Justice within, between and Beyond European Community Energy Initiatives: A Review." *Energy Research & Social Science 79*: 102157. https://doi.org/10.1016/j.erss.2021.102157.

Vargas, Mr Mauricio, and Santiago Garriga. 2015. *Explaining Inequality and Poverty Reduction in BOLIVIA.* International Monetary Fund.

Vijayraghavan, Maya, and Garret Kilroy. 2017. "Impact Evaluation of Small Towns Water Supply and Sanitation Sector Project in Nepal." *Asian Development Bank Evaluation Approach Paper*, March 2017. www.adb.org/sites/default/files/evaluation-document/237456/files/eap-ie-stwsssp-nepal.pdf

Wall, Tyler, and Bill McClanahan. 2015. "Weaponising Conservation in the "Heart of Darkness": The War on Poachers and the Neocolonial Hunt." *Environmental Crime and Social Conflict: Contemporary and Emerging Issues*: 221–240.

Yillia, Paul T. 2016. "Water-Energy-Food Nexus: Framing the Opportunities, Challenges and Synergies for Implementing the SDGs." *Österreichische Wasser-und Abfallwirtschaft 68*, no. 3–4: 86–98. https://doi.org/10.1007/s00506-016-0297-4.

Zegada, Maria Elena. 2016. "The Integration of Renewables into the Bolivian Energy Mix-Legal Framework and Policies." *Renewable Energy Law and Policy Review 7*, no. 1: 71–80. www.jstor.org/stable/26256482.

28

BUILD BACK SOLAR

Designing Solar Energy for a Just Transition

Robert Ferry and Elizabeth Monoian

Introduction

In response to the consensus projections of the climate science community, there is widespread planning underway targeting a transition to a low-carbon economy by mid-century in an effort to restrict global temperature rise to 2°C or less (IPCC, 2021). In a 100% clean energy economy, which includes electrified transportation and heating, solar power could be a significant contributor, providing 25–40% of total energy (Jacobson, 2017; Jenkins, 2021; Griffith, 2020; IRENA, 2021). A study by the US Department of Energy (USDOE SETO, 2021) estimates that solar photovoltaic (PV) generation could account for as much as 45% of US electricity supply by 2050. To achieve 40% of total worldwide energy from solar PV will require the installation of around 40,000 GW of solar capacity or 80 billion commercial solar modules globally – ten 72-cell PV modules for every person on the planet. As of 2020, the world has installed less than 1.5 billion modules, and solar power provides less than 2% of total energy (EIA, 2021).

Yet we are already seeing pressures on land use and obstacles to permitting of new solar landscapes and the requisite power transmission and distribution infrastructures. Solar energy landscapes are increasingly competing with other land use interests such as agriculture, recreation, visual resources stewardship, land conservation, forest preservation, and biodiversity. It is already difficult to get approval for renewable energy developments, with more than four out of five proposed solar projects never reaching commercial operation (Rand, 2021). The reasons for withdrawal of grid interconnection requests are varied, but local objections and land use conflicts often play a role. In 2021, what was to be the United States' largest solar farm, the one-billion-dollar Battle Born Solar Project, was scrapped, in part because it would have destroyed the visual context for Michael Heizer's epic work of land art, *Double Negative* (Angeleti, 2021).

With so much pushback and local objection, it is essential that new solar installations be designed with intention, mindful of their relationship to nature, place, and people. To avoid unnecessary disruptions to deployment, we may consider "rethinking energy policy as socio-energy systems design" (Miller, 2015), recognizing that technology does not live in a vacuum. It must appeal to the desires of people if it is to be deployed at scale.

The more solar generation we can bring into our cities the more we can ease the burden on rural landscapes. Early and consistent community engagement throughout the design process is critically important, especially for solar installations within or near to population centers.

DOI:10.4324/9781003183020-32

Answers to the questions of where we should install solar panels, who will be impacted by them, and who will benefit from them are fundamental to the success of the energy transition.

The Urban Versus Rural Debate

Over the past decade there has been a robust debate regarding the most beneficial mix of distributed energy (smaller solar installations within or near population centers) versus centralized energy infrastructure (large and remote power plants) to provide the most economical and reliable electricity grid for a post-carbon economy (Hansen, 2013; Martin-Martinez, 2017; Barbose, 2017; USDOE FERC, 2020; Khatib, 2021). This debate has benefitted recently from grid modeling tools developed by Vibrant Clean Energy (VCE) (Clack et. Al., 2020) that, according to VCE, consider a more complete set of variables and feedback loops than older modeling tools that do not model local distribution at a high enough level of detail. The results of VCE's analysis are challenging long-held assumptions regarding the costs and benefits of distributed energy resources.[1]

Before the emergence of solar PV technology, centralized infrastructures were more cost-effective than distributed infrastructures within population centers. Thermal power systems experience major benefits from economies of scale that make large centralized systems preferable to distributed systems. Solar PV systems do not require operational economies of scale (a coal-fired boiler operates at greater efficiency when it is larger, but a single solar panel does not operate more efficiently when placed next to other solar panels).

Distributed energy resources (DER) also tend to have higher soft costs and require the duplication of many small balance-of-system components, such as inverters, which can be efficiently consolidated in centralized solar power arrays. But the added marginal costs of DER must be considered against the cost of the construction, operation, and maintenance of the distribution and transmission infrastructure required to deliver remotely generated electricity to population centers. Another important consideration is the raft of political and regulatory barriers to permitting new overhead transmission lines and establishing new rights of way (Reed et. al. 2020). Such barriers have led to the study of non-transmission alternatives (NTA), of which distributed solar is one of the most important (Kasam-Griffith et. al., 2020).

Pathways to the kind of deep decarbonization warranted by the findings of the IPCC will require rapid electrification of transportation, domestic heating, and heavy industry, which in turn will require at least a doubling of existing electrical generation infrastructure (Griffith et. al., 2020). The more this added generation capacity is located in remote landscapes, the more long-distance transmission infrastructure will be required. When considered together, all these factors indicate a prioritization of DER over centralized energy systems may provide a more rapid pathway to decarbonization.

Large, centralized solar landscapes in remote areas will always be a significant part of the solar energy portfolio because centralized solar farms allow developers to deploy large amounts of generation capacity under one permit. According to VCE, however, the mix of distributed energy resources may be as much as ten times more than previous assumptions to arrive at the most cost-effective and reliable grid (Clack et.al., 2020). Assuming the validity of VCE's grid modeling and assuming that policy will provide the proper incentives to shift solar development toward more DERs, what are the social and cultural ramifications that follow from a prioritization of distributed solar plus storage over centralized solar energy generation infrastructure? What do all those solar panels look like when installed in our cities? Will there be enough viable rooftop area? Deploying ten times more solar energy infrastructure within the built environment than we had previously assumed may require new models for design and development that can expand permissible site opportunities. This is where community engagement can play an important role. The rest of this chapter proposes a model for public-private partnership that could establish a best practice standard for the design and implementation of distributed solar deployment in population centers.

Challenges to Urban Deployment

The standard model of solar energy development on vacant parcels is often characterized by ground-mounted modules surrounded by chain-link fencing and security cameras. Community engagement is often limited to the most basic level required to obtain permits after design has been completed, and in many cases, there may not be any community engagement at all. If the land use is allowed by right, there is a limit to what design concessions the public can expect to receive from solar development projects. Expanding this conventional model to meet the level of deployment necessary to establish a 100% renewable energy grid, we can expect to find solar developers meeting localized resistance to new deployments, especially for sites that are considered significant (from an urban planning perspective) or for historical significance. In general, the design outcomes that result from the conventional model of solar development do not serve to advance from a political or cultural perspective the long-term goals of the energy transition. Who wants to see rows of flat blue panels on gravel behind a barbed-wire chain-link fence in their neighborhood?

The utilitarian design standards of some solar developers may pose a danger to the long-term success of decarbonization strategies. If the perception of solar energy in the minds of the public carries negative connotations related to aesthetics and real estate values, then it will be difficult to build the political will necessary to implement pro-solar public policy, and at the local level, we may see more projects rejected due to public opposition.

Potential Opportunities and Intersectional Benefits

Providing carbon-free energy in the fight against climate change is the most obvious benefit of solar power. What may be overlooked are the benefits of distributed solar for local economies and DER's potential to increase social equity and quality of life. Thoughtfully planned and implemented, city-integrated solar infrastructure can provide a wide array of such co-benefits.

The application of solar infrastructure can increase the energy resilience and independence of municipalities. Cities are already taking advantage of long-term power purchase agreements for clean energy to establish pricing stability. By bringing solar installations into urban neighborhoods, cities can accomplish the same goal while also contributing to greater resilience and economic efficiency by limiting reliance on monopoly utilities and reducing maintenance costs for remote distribution infrastructures. Localized supply chains can emerge to provide good-paying jobs while locally produced solar components may decrease the embodied environmental footprint of PV modules.

When designed in collaboration with communities, solar installations can improve the quality of life and the beauty of public spaces. PV modules can provide aesthetic shade, creating new microclimates that expand the comfort zone in hotter regions during the summer months for human activities in outdoor public spaces and reduce heat island effects (see Figure 28.1).

Cities can define themselves and instill a sense of pride and ownership of shared energy infrastructures by creating new visual icons and cultural markers of what solar looks like for their city. Solar technology is a versatile design medium that can take shape in a variety of styles and

Figure 28.1   Solar canopy provides shade over the Orange Mall of the Arizona State University campus in
Tempe, Arizona. Design and installation by Power Parasol.

Source: Author.

expressions that capture the vibe of local cultures (see Figure 28.2). More vibrant and livable
cities attract more business and investment. They retain people over generations and attract new
long-term residents. Improvements to interstitial or liminal spaces in the margins of the city
(think of the neglected slivers of property near highway on ramps) can make productive use of
neglected land.

By considering innovative shared land uses for solar development projects, the benefits to
cities of distributed solar will not only include urban sustainability and resilience but could
also increase livability; contribute to cultural expression, storytelling, and food security; help
address racial and economic injustice; and support economic development. These co-benefits
can potentially improve overall social equity and quality of life.

Solar PV energy production on reservoirs reduces losses from evaporation. The body of
water also acts as a heat sink, keeping the solar modules cool and therefore operating more efficiently. Similarly, solar energy generation can make urban gardens more productive, help harvest
stormwater, and reduce the need for artificial irrigation.

Community solar projects can provide opportunities for small individual investors to see
low-risk rates of return and help tackle issues of energy poverty when designed with an eye on
social return on investment. Community solar with storage can offer additional revenue streams
through energy services like demand management that keep the grid stable through periods of
peak demand or remote outages.

Figure 28.2 *La Monarca*, a Land Art Generator Solar Mural installation. Visual graphic designed by San Antonio artist Cruz Ortiz with creative direction by Penelope Boyer. Shown on display at the San Antonio Zoo (reprinted with permission by Penelope Boyer). Recent technology allows functioning solar panels to be hidden behind fields of color or even graphic images. Conversion efficiency of the solar cells is impacted, but only slightly.

Source: http://solarmural.com.

Creative uses of solar power technology can even enhance public spaces as works of civic art using innovations such as organic PV (translucent, colorful, flexible) and custom laminations that can be applied to the face glass of standard PV modules, transforming solar panels into any color or even displaying printed images. The authors of this chapter are the co-founders of the Land Art Generator Initiative, a nonprofit that has been hosting international design competitions for the design of renewable energy landscapes since 2010. The outcomes of these open-call design challenges point the way to what could be possible if policymakers take a holistic approach to city planning that fuses the thoughtful design of civic spaces with distributed energy deployment strategies (see Figure 28.3).

To incentivize solar development that adds value to public space, cities could adopt more stringent standards for design and community engagement. But strengthening design and engagement standards will present additional barriers to energy developers.

Therefore, it is critical that cities proactively identify land areas for solar aggregation within their strategic plans and zoning ordinances so that solar deployments within urban contexts can be accomplished wholesale in megawatt capacity installations rather than piecemeal with a few kilowatts here and there. The goal should be to present a development environment for urban contexts that can come close to parity with rural installations with regard to the levelized cost of energy.

Figure 28.3   Left: *Nest*, by Robert Flottemesch, generates 6,633 MWh per year using monocrystalline bifacial PERC solar modules (LAGI 2019 Abu Dhabi). Reprinted with permission from Robert Flottemesch. Right: *Light Up* – by Martin Heide, Dean Boothroyd, Emily Van Monger, David Allouf, Takasumi Inoue, Liam Oxlade, Michael Strack, and Richard Le (NH Architecture); Mike Rainbow and Jan Talacko (Ark Resources); John Bahoric (John Bahoric Design); and Bryan Chung, Chea Yuen Yeow Chong, Anna Lee, Amelie Noren (RMIT students) – generates 2,220 MWh per year using laminated monocrystalline Si solar (first place winner, LAGI 2018, Melbourne).

Sources: Reprinted with permission from Robert Flottemesch and Martin Heide.

For the reasons stated in the previous section, turning this vision into a reality will likely require an evolution from the current development paradigm for solar infrastructure and the coordination of public and private capital. Presently, the business models of private solar development companies are designed to profit from strategies that limit soft costs related to community engagement and project design complexity. A more holistic life-cycle model for DER development might internalize considerations related to the range of social benefits possible through well-planned community solar projects and recognize the value streams created by site activation and creative placemaking that are familiar to mixed-use real estate developers and city planners.

Looking at a broader spectrum of co-benefits besides electrons can also lead to more inclusive models for financing solar deployment in cities and can expand the base of the public who are personally invested in a clean energy future. Community energy projects and energy cooperatives can provide a low-risk means of wealth building for marginalized populations who have been intergenerationally excluded from investment opportunities. By working in partnership with cities, solar developer business models may begin to bring more externalities into the calculation for return on investment. Innovative revenue models may be established that provide income streams from sources in addition to kilowatt-hours.

By breaking down the silos of developer models (energy developer versus community developer) we can begin to expand the realm of what is possible for urban solar integration. Cities and energy developers can work together to improve public perceptions of renewable energy infrastructure. By doing so, we might accelerate renewable energy deployment, meet the most beneficial mix of distributed versus centralized infrastructure, increase support for pro-solar policy agendas, help to lessen the intergenerational wealth gap, and make cities more vibrant, livable, resilient, and sustainable.

The energy of the sun can be considered a universal natural resource from a political perspective. Throughout history, there are examples of a "universal property" approach to the distribution of natural resource benefits. The most often cited example is the Alaska Permanent Fund. Writing in *Scientific American*, James Boyce (2020, November) summarizes the program:

> In 1976, as oil production commenced on Alaska's North Slope, the state amended its constitution to create a new entity called the Alaska Permanent Fund. The idea was the brainchild of Republican governor Jay Hammond, who believed that Alaska's oil wealth belonged to all its residents, and that all should receive equal annual dividends from its extraction.

The idea extends to the principle of the "solar energy commons" – a recognition that the sun's energy belongs to all of us if it belongs to any of us (Milun, 2020). Especially when installed on public land or community land trusts, solar power offers an opportunity for an equitable distribution of a solar dividend, a policy mechanism that can be a powerful tool to combat the cycles of poverty. At the national scale, the sale of solar electricity could even pay out in universal basic income and provide new opportunities for private wealth generation through distributed energy ownership, a democratization of what is today dominated by monopoly ownership.

Technologies such as virtual net metering, which assigns and exchanges credits for electricity associated with small-scale solar and wind installations, make it possible for individuals to become energy prosumers selling kilowatt-hours directly. As privatization of solar infrastructure becomes more widespread, it will be important to make sure that the financial benefits are not limited to the upper class but intentionally provide diversity of ownership as a means to close the wealth gap.

Community Co-design and Creative Placemaking

Incorporating community and stakeholder engagement into the energy developer business model can provide lasting benefits throughout the life cycle of projects and instill positive associations with renewable energy infrastructure at the neighborhood level. This can be accomplished by learning from best practices in urban planning and creative placemaking, which include community and stakeholder engagement from the pre-planning stage of development.

There are many resources available to developers who see the benefit of a multifaceted approach to energy projects. These include the Project for Public Places, which provides guidance through *Eleven Principles For Creating Great Community Places (How to Turn a Place Around,* 1999); the Trust for Public Land, which published *The Field Guide for Creative Placemaking In Parks* (2017); the American Planning Association; EcoDistricts®, which provides protocols and certification for city planners and urban designers; the American Society of Landscape Architects; the Landscape Architecture Foundation, which hosts the Landscape Performance Series of online resources; the National Endowment for the Arts; and Art Place America, a collaboration of foundations with an excellent set of resources, including the field survey *Farther, Faster, Together: How Arts And Culture Can Accelerate Environmental Progress* (2018).

Considerations of Social Justice

Assuming that we will be installing a significant percentage of solar power infrastructure in cities, it will be important to learn lessons from the unintended consequences of past infrastructure development that contributed to racially motivated spatial and environmental injustices (Caro, 1974; Jacobs, 1992). The destruction of inner-city neighborhoods through mid-20th century interstate highway planning is one example of the kind of top-down planning that should be avoided during the deployment of new energy infrastructures in the 21st century. Communities can instead be engaged in the decision-making processes for significant solar installations proposed in their neighborhoods so that the resulting energy landscapes are designed to add value to the built environment, contribute to economic development, and provide co-benefits to people beyond kilowatt-hours.

Other social risks of the energy transition include the furtherance of structural wealth disparities. Centralized infrastructures are typically financed by institutional investors with the returns accruing over time to benefit the already wealthy. With trillions of dollars in near future investment and low-risk returns for solar deployments, it is critically important that the opportunity for wealth generation be offered to a broad array of stakeholders and not be limited to Wall Street investors.

The Idea of a PV City

Cities are well-positioned to take the lead in the energy transition. As a starting point for strategic planning, every city can aspire to become a "PV city" by 2030. A PV city in this context is one that has installed and interconnected to the grid at least one commercial solar module per resident within the city limits. For a smaller city of 250,000 inhabitants, this would equate to roughly 115 MW of installed capacity. For a city of a million inhabitants, this would equate to roughly 460 MW. By implementing the following model proposed, cities can kickstart their path to achieving this goal, advance the energy transition, and help to conserve rural landscapes.

Bringing It All Together: Build Back Solar

Build Back Solar (BBS) is a proposal (Land Art Generator Initiative, 2021) to advance the installation of utility-scale solar in cities by co-designing shared land uses and improvements to the built environment with the communities where the solar is installed.

Under the proposal, the costs related to the design of new solar power projects and the marginal costs related to project social co-benefits would be paid for by the public sector under a program designed to incentivize solar developers to pursue sites other than rural greenfields.

The program has the added benefit of expanding public awareness and increasing the quality of solar infrastructure design. While the pre-development costs of community engagement and design are proposed to be borne by the public sector, the hard costs of installation and operations will remain the purview of private developers of any ownership model.

In order to be eligible, solar development projects would need to meet specific criteria related to population density around the site, and developers would need to engage a licensed design professional, such as a landscape architect, to facilitate the co-design process with the local community. The construction costs of elements of the design that provide verifiable co-benefits to the community could also be covered under related programs, perhaps implemented at the local level.

The goal of BBS is to leverage public investment in the most effective way possible on solar infrastructure while ensuring that we are proactively addressing issues related to land use, public pushback, equity, and resilience by incentivizing solar developers to look at sites other than rural greenfields and to think about ways their projects can support social good.

Community engagement through the design process may take place in a variety of ways, such as a public design challenge, invited competition, participatory design, or consistent public meetings throughout pre-design, design development, and construction documents. With the soft costs of development carried by the public sector and with multiple small- and medium-size sites aggregated into one large bid package, developers will be more readily able to invest in urban solar projects while the BBS process ensures that an equitable and just transition is a centerpiece of each project plan.

Process Overview

The following section illustrates what the implementation of a design competition by a city seeking BBS funding might look like. In this example, the lead is the municipal government, but a lead applicant could also be a solar developer, cooperative, nonprofit, or land-owning institution.

The first phase of work includes a spatial analysis and solar survey, identification of legal constraints, outreach to solar developers and market experts, meetings with city officials, and recommendations for inclusive financing models. Partnerships will be established with local community representative organizations, who will assist in coordinating public events and design workshops.

Through community collaboration and geospatial analysis, the BBS project identifies several suitable sites over a total of 50–100 acres and in multiple neighborhoods that aggregates in a coordinated design of between 20 and 40 MW of solar power infrastructure, assuming 400 kW of solar per acre. By aggregating a utility-scale project (45,000–90,000 commercial solar panels) across multiple sites, BBS projects can meet the kind of economies of scale that are standard for exurban installations while bringing greater resilience and reliability to the city grid, directly conserving 50–100 acres of countryside, and bringing a suite of co-benefits to neighborhoods.

Community engagement begins with an initial meeting with community leaders, followed by a public presentation of the idea and open discussion. Subsequent meetings include interactive mapping, review of local histories, discussion of anecdotes of relevance, site selection and analysis, co-benefit ideation workshops, procurement strategy, solar financing and equity workshops, and the co-creation of the design brief documents with key members of the community. At any step of the process, a community can decide how to participate in the project or not to participate at all.

For the communities who decide to participate, the BBS program launches an open-call design challenge. Unique design briefs for each of the sites reflect the stated aspirations of each community for shared land uses and co-benefits of community solar. In parallel, some communities may opt instead to engage in a participatory design project with a local artist or designer.

Responding to the open-call design challenge, hundreds of participating design teams submit concept design proposals for one or more of the site-specific design briefs, illustrating their design ideas and outlining the co-benefits of the shared land use.

At the conclusion of the design challenge, the community juries select their preferred project for each of the selected sites. Short-listed design proposals are displayed through exhibitions, publications, and media coverage, presenting an exemplary blueprint to the world for equitable urban solar development and further engaging the public in the host city. Projects are then selected by community representatives to receive detailed design commissions.

By supporting the development of community-driven bid packages, the soft costs and risks of development are removed from the balance sheets of the solar energy developers who will come forward to install the projects that are selected by local stakeholders.

Phase 1

For a BBS project, the first phase of work after establishing the local consultant partnerships and qualifying for BBS funding will include the following:

- **City-wide geographic information system (GIS) spatial analysis and solar survey:**
  Using a series of geographic overlays related to available solar energy resources and tree inventory, as well as demographics, land values, land utilization/vacancy, land ownership, walkability, food scarcity, access to open space, and other considerations, a detailed analysis will be conducted to identify BBS sites within a range from possible to preferred. These lists of sites will be used as a starting point for communications with local communities.
- **Identification of legal framework and constraints:**
  Working with legal and policy consultants produces a "state of regulatory frameworks" report that points to the various barriers and pathways to implementation within the specific city and within the regulatory framework of the state. The report will look at community solar, solar cooperatives, power purchase agreements, municipal utilities, and other aspects related to solar development in the region. Findings will inform the design briefs and the public engagement frameworks, as well as the financial model for implementation.
- **Outreach to solar developers and market experts:**
  Develop a list of solar developers and consultants and engage them in outreach with the goal of learning what incentive structures will be required to make BBS projects successful. How can the bid be made attractive to developers and result in a competitive process? What is the comfort level with solar developers regarding projects that share land use with civic functions and amenities, like community gardens and recreational uses? How can the premium capital costs for public human-centered solar projects be offset by creative income streams that reflect the social value of the installations? How can the ownership be shared for equitable returns on investment? This same network of developers will comprise the short list for invitations to bid during the implementation stage of the project.
- **Meetings with city officials and commissions:**
  Early and frequent meetings with city administrators, commission members, and city council will be key to the success of the project and ensure that no one feels left out of the process. It will be important to establish what role the city will play in the BBS project,

ranging from passive support helping to facilitate where appropriate to active partnership providing resources and perhaps even purchasing the power from completed solar installations.

Municipal responsibilities might include the following:

o   Access to neighborhood planners
o   BBS city representative
o   Sharing of GIS data
o   Provision of city-owned land
o   Power purchasing of energy generated on BBS sites
o   Coordination with commissions and departments
o   Zoning and approvals
o   Coordination with city staff overseeing diversity, equity, and inclusion

- **Identify options for inclusive ownership:**
  Working with a project equity consultant, compile a comprehensive planning document that sets forth the aspirational goals and the minimum requirements for the inclusive and diverse ownership of the BBS outcomes so that the project provides the greatest benefit to those most in need. Identify best practices for community ownership of solar infrastructure that can help to address energy poverty in city neighborhoods. Coordination with the legal framework and constraints will be important.
- **Establish community partnerships:**
  Identify the district boundaries that will guide community engagement. Together, they may comprise the entirety of the city.

Phase 2

The second phase of the project is the primary public outreach and co-design process, including the following:

- **Initial meetings with community leaders:**
  Initial feedback and reception to a presentation of the program will inform the next steps and begin to customize the approach for each neighborhood.
- **Public presentation (one presentation per neighborhood group):**
  These presentations will be open to the public and will take place in venues within each neighborhood.
- **Interactive mapping and co-benefit ideation:**
  Over the course of one or two additional public meetings and utilizing the findings of the city-wide GIS analysis, each community will work collaboratively to identify the most viable sites that they consider to be within their neighborhood. Characteristics, use cases, and cultural context for each site will be identified. The outcomes of these meetings will be collected in a report issued by project consultants to all participating communities. During these meetings, focus groups and committees will be established to oversee various aspects of the planning process in the time between open public meetings.
- **Determination of the design delivery model:**
  A second series of community planning meetings will determine whether each site will be a part of the headline BBS open-call design competition or whether the site would be better served by engaging in a co-design process with a local design team.

- **Co-create design guidelines document with community:**
  For those sites that decide to participate in the open-call design competition, a series of working sessions will establish the parameters of the design guidelines that pertain to their specific site(s). These local guidelines will form a part of a larger BBS design challenge document. For those sites that decide to engage a local design team in a community co-design process, a series of working sessions will establish a request for proposal (RFP) process to identify the facilitating designer.
- **All-neighborhood networking meetings:**
  At key stages in the process, neighborhood leaders will come together to share ideas and experiences and engender a spirit of city-wide collaboration and community. Details regarding the aggregation of total capacity will be discussed at these convenings.

At the conclusion of phase 2, the selected sites for aggregation will be determined and their capacity for solar energy generation will be known.

Subsequent Phases

Subsequent phases of the project will include the development and management of the design challenge for participating sites, the selection process led by communities, public display of the short-listed design proposals, educational outreach activities, and the local artist-led co-design for sites that opted out of the design challenge process.

Following the selection of the concept designs for each site, detailed design proposals are solicited, and the architecture and engineering (A/E) firms are chosen to develop documents for fabrication and construction. At each step in the detailed design process, the community will be re-engaged through a series of public presentations and workshops coordinated by the A/E.

With all the soft costs of community engagement and detailed design covered by the public sector and with allowances provided to cover construction costs of co-beneficial features, solar developers will bid on a package that can be categorized at this stage as "notice to proceed", the lowest risk investment for solar development. Tenders for development will include the specific requirements for solar developers to meet the standards of equity, inclusion, and co-benefit of land use established by each community, along with detailed working drawings for the solar installation and co-beneficial features.

Conclusion: Impacts and Potential Benefits

By merging the fields of community development and energy development, BBS proposes that policymakers and city leaders can take advantage of opportunities presented by the energy transition to create co-benefits for cities that are greater than the sum of their constituent parts. Employing the best practices of creative placemaking and the human-centered design of public spaces to the design of distributed solar energy installations can increase the value that the public places on solar energy, expand the range of viable urban site typologies for solar energy landscapes, and increase the profile of cities to attract businesses and talent.

Neighborhoods will be the beneficiaries of well-designed solar infrastructure that adds value to public space, providing new places for recreation, outdoor enjoyment, education, urban farming, and open-air markets. STEAM[2] education, project-based learning, and continuing education can be incorporated around the design, implementation, and operation of these new shared energy landscapes.

Equitable financing models can bring wealth-building opportunities to residents who sign up for and/or buy into the program.

For each BBS project, 20 MW or more of power capacity will be added to the distributed energy generation network of the city, providing opportunities for expanding energy services through interconnected microgrids with battery storage.

BBS projects can help to advance the deployment of other sustainable installations throughout the city, such as rooftop solar.

The solar infrastructure, because it is designed to share land uses with civic amenities, may increase the useability of outdoor spaces (protecting them from urban heat island effects and from winter weather) and may create lasting cultural landmarks for future generations.

The hosting city's leadership role in advanced urban solar integration may help to raise the positive profile of the city for the climate-conscious public nationally and internationally, attracting new people and businesses and increasing retention of the existing population.

The BBS model – by focusing on the social aspects of solar deployment at the city and neighborhood scales – may establish an attractive blueprint for replication, eventually leading to the proliferation of thousands of megawatts of solar installations within cities across the country and the world for land areas that would have previously been considered impractical for solar development.

Initially, the BBS model can be implemented within the kinds of urban living labs detailed in the chapter by Sharp et al. (this volume).

If scaled nationally and internationally, the overall impact of the project could contribute significantly to rapid decarbonization goals in line with the consensus recommendations of climate scientists while preserving thousands of acres of rural land and providing substantial co-benefits to people in support of a thriving democratic society.

Notes

1 See the ICF report for the US Department of Energy (Meyer, 2018) for a meta-analysis of 15 energy modeling studies showing only slight marginal benefits or even additional costs associated with net energy metering of distributed energy resources. These analyses – often paid for by utility companies – have been cited by some state-level regulations that have disincentivized rooftop solar. The January 2022 controversy around the California Public Utilities Commission proposal is one example (Sforza, 2022).
2 STEAM is an acronym for science, technology, engineering, art, and math. STEM education is more widely known. The addition of A (art) in STEAM was pioneered by John Maeda at Rhode Island School of Design. See www.risd.edu/steam.

References

Angeleti, Gabriella. 2021. "Plans Scrapped for Solar Project that Would Disrupt Michael Heizer's Double Negative." *The Art Newspaper*. www.theartnewspaper.com/2021/07/26/plans-scrapped-for-solar-project-that-would-disrupt-michael-heizers-double-negative.

Barbose, Galen. 2017. *Putting the Potential Rate Impacts of Distributed Solar into Context*. Online: Energy Analysis and Environmental Impacts Division of the Lawrence Berkeley National Laboratory. www.michigan.gov/documents/mpsc/LaRoy_lbnl-1007060_605390_7.pdf.

Boyce, James. 2020. "The Case for Universal Property." *Scientific American*. www.scientificamerican.com/article/the-case-for-universal-property.

Caro, Robert A. 1974. *The Power Broker: Robert Moses and the Fall of New York*. New York: Knopf.

Clack, C.T.M., A. Choukulkar, B. Coté, and S. McKee. 2020. *Why Local Solar for All Costs Less: A New Roadmap for the Lowest Cost Grid*. Online: Vibrant Clean Energy. www.vibrantcleanenergy.com/wp-content/uploads/2020/12/WhyDERs_TR_Final.pdf

EIA (United States Energy Information Administration). 2021. *Table 1.2: Primary Energy Production by Source*. Online. www.eia.gov/totalenergy/data/monthly/pdf/sec1_5.pdf

Griffith, Saul, Sam Calisch, and Laura Fraser. 2020. *The Rewiring America Handbook: A guide to Winning the Climate Fight*. Online. www.rewiringamerica.org.

Hansen, Lena, Virginia Lacy, and Devi Glick. 2013. *A Review of Solar PV Benefit & Cost Studies*. Online: Rocky Mountain Institute. https://pscdocs.utah.gov/electric/14docs/14035114/263609ExASierraClubInitCommEx1RMIMetaAnaly2-6-2015.pdf

IPCC. 2021. *Climate Change 2021: The Physical Science Basis*. Contribution of Working Group I to the Sixth Assessment Report of the Intergovernmental Panel on Climate Change [Masson-Delmotte, V., P. Zhai, A. Pirani, S.L. Connors, C. Péan, S. Berger, N. Caud, Y. Chen, L. Goldfarb, M.I. Gomis, M. Huang, K. Leitzell, E. Lonnoy, J.B.R. Matthews, T.K. Maycock, T. Waterfield, O. Yelekçi, R. Yu, and B. Zhou (eds.)]. Cambridge University Press.

IRENA. 2021. *World Energy Transitions Outlook: 1.5°C Pathway*. Abu Dhabi: International Renewable Energy. www.irena.org/publications/2021/Jun/World-Energy-Transitions-Outlook

Jacobs, Jane. 1992. *The Death and Life of Great American Cities*. New York: Vintage Books.

Jacobson, Mark et al. 2017. "100% Clean and Renewable Wind, Water, and Sunlight All-Sector Energy Roadmaps for 139 Countries of the World." *Joule, Issue* 1 (September 6, 2017): 108–121. http://dx.doi.org/10.1016/j.joule.2017.07.005.

Jenkins, Jesse D., Erin N. Mayfield, Eric D. Larson, Stephen W. Pacala, and Chris Greig. 2021. "Mission net-zero America: The nation-building path to a prosperous, net-zero emissions economy." *Joule*, 5(11): 2755–2761. ISSN 2542–4351. https://doi.org/10.1016/j.joule.2021.10.016.

Kasam-Griffith, Alisha, Natasha S. Turkmani, Martin J. Wolf, Nina C. Peluso, and Tomas W. Green. 2020. "Transmission Transition: Modernizing U.S. Transmission Planning to Support Decarbonization." In: Ruaridh Macdonald and Anthony Tabet (eds). *MIT Science Policy Review*, Vol 1 (December): 87. https://sciencepolicyreview.org/downloads/2020/12/Vol1no11_Kasam-Griffith.pdf

Khatib, Tamer, and Lama Sabri. 2021. "Grid Impact assessment of centralized and decentralized photovoltaic-based distribution generation: A Case study of power distribution network with high renewable energy penetration." *Mathematical Problems in Engineering, 2021*. https://doi.org/10.1155/2021/5430089.

Land Art Generator Initiative. 2021. *Build Back Solar*. Online. http://buildbacksolar.org.

Martin-Martinez, F., A. Sánchez-Miralles, M. Rivier, and C.F. Calvillo. 2017. "Centralized vs distributed generation. A model to assess the relevance of some thermal and electric factors. Application to the Spanish case study." *Energy*, 134: 850–863. ISSN 0360–5442. https://doi.org/10.1016/j.energy.2017.06.055.

Meyer, David. 2018. "Review of Recent Cost-Benefit Studies Related to Net Metering and Distributed Solar." *ICF*. (Report Prepared for The U.S. Department of Energy). www.energy.gov/sites/prod/files/2020/06/f75/ICF%20NEM%20Meta%20Analysis_Formatted%20FINAL_Revised%208-27-18.pdf

Miller, Clark, Jennifer Richter, and Jason O'Leary. 2015. "Socio-energy systems design: A policy framework for energy transitions." *Energy Research & Social Science*, 6:29–40.10.1016/j.erss.2014.11.004.

Milun, Kathryn. 2020. "Solar commons: A "commons option" for the 21st Century." *The American Journal of Economics and Sociology*, 79. https://doi.org/10.1111/ajes.12348.

Rand, Joseph, Mark Bolinger, Ryan Wiser, and Seongeun Jeong. 2021. *Queued Up: Characteristics of Power Plants Seeking Transmission Interconnection as of the End of* 2020. Online: Lawrence Berkeley National Laboratory.

Reed, Liza, Michael Dworkin, Parth Vaishnav, and M. Granger Morgan. 2020. "Expanding transmission capacity: Examples of regulatory paths for five alternative strategies." *The Electricity Journal*, 33(6): 106770. https://doi.org/10.1016/j.tej.2020.106770.

Sforza, Teri. 2022. "California's Rooftop Solar War Intensifies as Regulators Pull Reform Proposal." *The Mercury News*. www.mercurynews.com/2022/01/24/rooftop-solar-war-intensifies-as-regulators-pull-reform-proposal.

USDOE (United States Department of Energy) Federal Energy Regulatory Commission (FERC). 2020. *Participation of Distributed Energy Resource Aggregations in Markets Operated by Regional Transmission Organizations and Independent System Operators*. Online. https://ferc.gov/sites/default/files/2020-09/E-1_0.pdf.

USDOE (United States Department of Energy) Solar Energy Technologies Office (SETO). 2021. *Solar Futures Study*: 197. Online. www.energy.gov/sites/default/files/2021-09/Solar%20Futures%20Study.pdf.

Author Bios

Robert Ferry is a LEED-accredited licensed architect whose practice centers on regenerative design and urban energy landscapes. As founding co-director of the nonprofit Land Art Generator Initiative, he supports the role of design as part of a comprehensive solution to climate change.

Elizabeth Monoian (MFA, Carnegie Mellon University) is the founding co-director of the Land Art Generator, a nonprofit working to provide models of sustainable community-centered infrastructures that inspire people about the beauty of a world without fossil fuel and educate the next generation of designers.

Conclusion

29

RECONCEPTUALIZING THE NEXT FRONTIER IN ENERGY TRANSITIONS

Kathleen M. Araújo, Timothy J. Foxon, Jochen Markard,
Rob Raven, and Roberto Schaeffer

1. Introduction

The *Routledge Handbook of Energy Transitions* takes stock of the evolving field of energy transitions. As has been shown in earlier chapters, the field has undergone considerable growth in recent years, as the subject came into mainstream use (see Araújo, this volume, Chapters 1–2).

Alongside exponential growth in writing about the practice and theory of this subject, the *Handbook* considered dimensions that have become increasingly explored in the field. In a period when strategies are more fully focusing on net-zero energy systems that are resilient and secure (ibid. and Markard and Rosenbloom, this volume), aspects like how to account for geopolitics and resilience in addition to traditional carbon-intensive industries are also being more fully explored. The aim in this perspective chapter is to reflect on the frontier for further analysis and action that is relevant for the choices, context, transition practice, and research.

2. Observations of and from the Field

Broadly speaking, energy transitions that are underway or yet to occur require *more extensive consideration of how everyday life, the societal context, and the co-benefits relate to energy use.* Among the many factors that matter are issues associated with resilience and jobs, health impacts and fairness, industrial reorientation, as well as impacts on nature. Related considerations include land use, waste and/or recycling, and affordability. Here, decision-makers have a key role to play in approaching the choices and process more holistically, prioritizing benefits (not just costs), and balancing the trade-offs.

In connection with energy decision-making and planning, there is a need for a more *integrated systems approach.* To date, most assessments of energy transitions tend to focus on technological solutions and supply-side issues within key energy sectors – electricity, transport, and heating and cooling. To achieve energy security with *net-zero carbon (and more fully, net-zero greenhouse gas) energy systems* – arguably, a paradigm shift for today's society – major changes are also necessary on the demand side. This includes mobility practices, consumption patterns, and household dynamics, all of which are deeply embedded in societal norms and values as well as in what matters to people. These changes can be difficult to achieve and politically sensitive

DOI:10.4324/9781003183020-34

and/or contested, but also have the potential to significantly improve human well-being, while at the same time as reducing emissions (Creutzig et al., 2022).

Questions related to societal change, including community-informed decision-making, the role of cities and local governments as transition managers, as well as social innovation in energy transitions, are increasingly addressed in different ways across geographies. Further research is needed to understand better practices as well as the *interactions between individual behaviors, consumption practices, lifestyles, and drivers of energy systems change.*

Technological paths, including the build-out of hydrogen across sectors and the decarbonizing of heating that seemed primarily conceptual until recently, are now priorities or high-level objectives for countless public and private sector agendas (Bischof-Niemz and Creamer, this volume; Kiruja and Barrera, this volume). Looking ahead, much more attention will need to be paid to the *direct costs of technology paths* alongside other measures to reduce high levels of energy demand. In tandem, the *direct benefits to the users and the wider co-benefits* should receive similar attention in areas, such as quality of life and of the fuller environment, as well as lower-carbon job creation and economic regeneration.

To date, shifts in decarbonizing industrial economies have often been achieved largely through substitution of coal-fired electricity generation by renewables and other lower-carbon forms of electricity generation. This has been done, at times, with little direct involvement of energy users. Moreover, many individual decisions that influence energy use are made without consideration for the energy implications. Instead, the focus may be on priorities, such as well-being, social status, careers, and so on.

There is a need to more fully leverage the *strengths and insights from social sciences and humanities* to inform on societal influences and the flows of knowledge that often fall outside of energy studies. In doing so, novel approaches may be developed to overcome difficult challenges and to inclusively account for participation in energy transitions. New approaches integrating *design and design anthropology* (Ferry and Monoian, this volume; Sharp et al., this volume), as well as consumption *sociology and social practice* approaches (Raven et al., 2021) with insights from energy and sustainability transitions research can be promising venues for future fuller research into the demand side of energy transitions.

Energy transitions may increasingly be framed in terms of public policy imperatives and fairness, which will be at the center of *politics and contention*. Politics, like that evident in Japan, following the Fukushima Daiichi accident (Matsuura, this volume; Trencher et al., 2020) can stymie expected transformations. By contrast, politics can also enable shifts, as seen in periods of support with Denmark's wind energy cluster (Karnøe and Raghu, this volume; Araújo, 2017) or the coal phase-outs in Ontario and the UK (Rosenbloom, 2018; Isoaho and Markard, 2020). This area holds strong potential for expanded study.

As energy transitions priorities increasingly move into the mainstream, so will their influence be increasingly felt beyond the usual *actors*, such as early adopters and coalitions of the willing. This raises new questions and policy challenges in relation to engaging more deliberately with people and communities whose lives may be deeply affected by energy transitions (Johnstone and Hielscher, 2017) but are harder to reach or may deliberately resist social, economic, material, or cultural changes related to energy transitions. What are new approaches for engaging? How can slowly changing and emotionally important impacts of energy transitions be accounted for more fully in deep or rapid decarbonization? What new types of lifestyles, business models, or social contracts between the state, market, and community still need to be developed?

Debates associated with the recovery from the COVID-19 pandemic offer added insight for transitions studies (Cohen, 2020; Markard and Rosenbloom, 2020). The pandemic highlighted the importance of accounting for wider societal and community values. Proponents of strong action to support a net-zero energy transition point to *lessons from the pandemic* on the scale of government action that is possible in the face of a perceived urgency. Meanwhile, opponents of action may point to evidence of inflation as industrialized economies start to recover. Increasing consumption and constraints on supply chains for key energy and material resources add further challenges to energy transitions (Araújo, this volume, Chapter 1; Sovacool et al., 2020).

The net-zero debate also brings scrutiny to additional sectors that are energy-intensive and difficult to decarbonize (Davis et al., 2018). Examples include aviation, shipping, steel, cement, and chemicals. Analysts and policymakers need to pay *attention to critical, underlying systems*, such as energy finance, education, administration, or policymaking (European Commission, 2019, n.d.). These are important for providing not only necessary services but also well-functioning institutions/regulations.

The *gap between planned and unplanned transitions* reflects an area that is ripe for expanded thinking and analysis. A planned transition pathway may end up in an unplanned domain, given the pace of sociotechnical changes or unforeseen developments and disruptions, such as those from the COVID-19 pandemic or shifting geopolitics. Eventually an unplanned transition may need to become course-corrected into a "newer", planned transition. That is why some experts have suggested that pathways have to encompass multiple goals concurrently, weighting different objectives and managing their interactions (Pahle et al., 2021; see also Araújo, Chapter 2 for *readiness*).

Rethinking and repurposing infrastructure should also be factored more holistically in planning for energy transitions. As projects reach their end of expected life, can they be retooled/retrofitted or should they be decommissioned? Better practices in permitting and licensing are needed, here, to account more fully for change in technology and societal growth, geospatial aspects, preferences and learning, as well as infrastructure abandonment. For planners and policymakers, in addition to educators and industry, tapping existing skills for project decommissioning, retrofitting, and managing reclamation of abandoned oil and gas wells, for example, could be prioritized in infrastructure and decarbonization investment.

Current energy transitions are being significantly driven by social and environmental goals, notably the aim to have a decarbonized planet by the middle of the 21st century with resilient and secure energy. This will require high levels of public and private *investment* in innovation and diffusion of a range of energy technologies and measures to promote energy efficiency and reduce waste.

Greater private sector investment will be needed in both industrialized and industrializing countries. This is starting to happen as initial public investment and incentives have helped to bring down the costs of the low-carbon energy supply and end-use technologies. Yet a more rapid reorientation of investment by large private sector organizations, such as banks, pension funds, and insurance companies, is needed (Geddes and Schmidt, 2020). Arguably, one of the outcomes of the Conference of Parties 26 Summit in 2021 was the formation of the Glasgow Finance Alliance for Net-zero (GFANZ), which aimed to align $130 trillion of private finance to science-based net-zero targets and near-term milestones (GFANZ, n.d.). Its progress remains to be seen, as there is much work to be done to translate these commitments into action, and further research is needed to understand the means to achieve this within the financial system.

In industrialized economies, measures to promote a net-zero energy transition are often framed in terms of their benefits for green growth and job creation in new energy sectors. However, opponents of such measures typically point to the dangers of public spending contributing to increasing price inflation, as well as to the loss of jobs in traditional energy industries. Here, decision-makers and analysts can facilitate broader understanding with objective and transparent reporting.

Thinking beyond purely economics, a net-zero transition can be achieved in the context of a *wider sustainability transition*, with countries committing to meet the UN Sustainable Development Goals (SDGs) by 2030, including eradicating poverty and hunger and promoting good health and well-being, alongside economic and environmental goals. This means that energy transitions may look different in developing and emerging economies, with distinct priorities and different resources and institutional contexts. SDG 7 sets the goal of ensuring not only access to affordable and reliable energy but also sustainable and modern energy for all. There may, however, be competing pathways to achieve this. While some developing countries are pursuing approaches to electrification based on large-scale centralized generation technologies, others are aiming to "leapfrog" to more decentralized energy technologies and nature-based and circular economy solutions. It is important that citizens have the opportunity to participate in making energy choices and that environmental goals are recognized with social development priorities.

As industrialized and emerging economies advance from the pandemic, the prices of oil and gas in global markets have increased significantly, due to shifting *geopolitics* and mismatches between supply and demand. This is pushing up the cost of energy services for households and businesses in many countries, which can be politically sensitive. At the same time, the US, many European countries, and other countries continue to reduce their dependence on imported Russian oil and gas (Araújo, this volume, Chapter 2), creating intense political debates around the direction and speed of energy transitions that aim to move toward net-zero while maintaining energy security and resilience. Here, the connection between energy security, resilience, and decarbonization presents a critical point of focus for decision-makers, communities, and industry.

As the next frontier of energy transitions emerges, there is an opportunity to improve the process of change across the sectors. Sustainable solutions will need to properly factor for environmental impacts, costs, co-benefits, and political interests. For energy transitions to account for varied sets of values and positive aims, they will need to be strategically coordinated to balance different objectives, to manage the interactions, and to navigate new terrain.

Bibliography

Afonso, T., Marques, A., Fuinhas, J., 2021. Does energy efficiency and trade openness matter for (the) energy transition? *Environment, Development and Sustainability*, 23 (9), 13569–13589.

Araújo, K., this volume. The Evolving Field of Energy Transitions, chapter in: Araújo, K. (Ed.) *Routledge Handbook of Energy Transitions*. Routledge: Abingdon.

Araújo, K., this volume. A Roadmap for Concepts and Theory of Energy Transitions, chapter in: Araújo, K. (Ed.) *Routledge Handbook of Energy Transitions*. Routledge: Abingdon.

Araújo, K., 2017. *Low Carbon Energy Transitions: Turning Points in National Policy and Innovation*. Oxford University Press: New York.

Bischof-Niemz, T., Creamer, T., this volume. South Africa: Building a Green Hydrogen Products Superpower, chapter in: Araújo, K. (Ed.) *Routledge Handbook of Energy Transitions*. Routledge: Abingdon.

Calvert, K., Greer, K., Maddison-MacFayden, M., Theorizing energy landscapes for energy transition management. *Geoforum*, 102, 191–201.

Cohen, M., 2020. Does the COVID-19 outbreak mark the onset of a sustainable consumption transition? *Sustainability: Science, Practice and Policy* 16 (1), 1–3.

Cozzi, L., Motherway, B., 2021. The Importance of Focusing on Jobs and Fairness in Clean Energy Transitions. *IEA*, July 6, www.iea.org/commentaries/the-importance-of-focusing-on-jobs-and-fairness-in-clean-energy-transitions

Creutzig, F., Niamir, L., Bai, X., Calaghan, M., Cullen, J. et al., 2022. Demand-side solutions to climate change mitigation consistent with high levels of well-being. *Nature Climate Change*, 12, 36–46.

Davis, S.J., Lewis, N.S., Shaner, M., Aggarwal, S., Arent, D., Azevedo, I.L., Benson, S.M., Bradley, T., Brouwer, J., Chiang, Y.-M., Clack, C.T.M., Cohen, A., Doig, S., Edmonds, J., Fennell, P., Field, C.B., Hannegan, B., Hodge, B.-M., Hoffert, M.I., Ingersoll, E., Jaramillo, P., Lackner, K.S., Mach, K.J.,

Mastrandrea, M., Ogden, J., Peterson, P.F., Sanchez, D.L., Sperling, D., Stagner, J., Trancik, J.E., Yang, C.-J., Caldeira, K., 2018. Net-zero emissions energy systems. *Science* 360, 1419.

European Commission, 2019. The European Green Deal. COM (2019) 2640 final, Brussels.

European Commission, n.d. Circular Economy Strategy, https://research-and-innovation.ec.europa.eu/research-area/environment/circular-economy/circular-economy-strategy_en, Accessed June 10, 2022.

Ferry, R., Monoian, E., this volume. Planning and Designing Inclusive Solar Energy Infrastructures, chapter in: Araújo, K. (Ed.) *Routledge Handbook of Energy Transitions*. Routledge: Abingdon.

Geddes, A., Schmidt, T.S., 2020. Integrating finance into the multi-level perspective: Technology niche-finance regime interactions and financial policy interventions. *Research Policy* 49, 103985.

Geels, F., 2011. The multi-level perspective on sustainability transitions: Responses to seven criticisms. *Environmental Innovation Societal Transitions* 1 (1), 24–40.

Ghosh, B., Kivimaa, P., Ramirez, M., Schot, J., Torrens, J., 2021. Transformative outcomes: Assessing and reorienting experimentation with transformative innovation policy. *Science Public Policy*.

Glasgow Finance Alliance for Net-zero, n.d. www.gfanzero.com/, Accessed March 2, 2022.

Grin, J., Rotmans, J., Schot, J., 2011. On patterns and agency in transition dynamics: Some key insights from the KSI programme. *Environmental Innovation Societal Transitions* 1 (1), 76–81.

Grin, J., Rotmans, J., Schot, J., Geels, F.W., Loorbach, D., 2010. *Transitions to Sustainable Development: New Directions in the Study of Long-Term Transformative Change*. Routledge: London.

Grubler, A., 2012. Energy transitions research: Insights and cautionary tales. *Energy Policy* 50, 8–16.

Grubler, A., Wilson, C., Nemet, G., 2016. Apples, oranges, and consistent comparisons of the temporal dynamics of energy transitions. *Energy Research & Social Science* 22, 18–25.

Isoaho, K., Markard, J., 2020. The politics of technology decline: Discursive struggles over coal phase-out in the UK. *Review of Policy Research* 37, 342–368.

Johnstone, P., Hielscher, S., 2017. Phasing out coal, sustaining coal communities? Living with technological decline in sustainability pathways. *The Extractive Industries and Society* 4, 457–461.

Kanger, L., Schot, J., 2019. Deep Transitions: Theorizing the long-term patterns of socio-technical change. *Environmental Innovation Societal Transitions* 32, 7–21.

Karnøe, P., Raghu, G., this volume. Transformation of the Danish Wind Turbine Industry through Path Creation, chapter in: Araújo, K. (Ed.) *Routledge Handbook of Energy Transitions*. Routledge: Abingdon.

Kern, F., Rogge, K., Howlett, M., 2019. Policy mixes for sustainability transitions. *Research Policy* 48 (10), 103832.

Kiruja, J., Barrera, F., this volume. Decarbonization of the Heating Sector through Direct Use of Geothermal Energy, chapter in: Araújo, K. (Ed.) *Routledge Handbook of Energy Transitions*. Routledge: Abingdon.

Markard, J., Rosenbloom, D., 2020. A tale of two crises. *Sustainability: Science, Practice and Policy*, 16(1).

Markard, J., Rosenbloom, D. this volume. Phases of the Net-Zero Energy Transition and Strategies to Achieve, chapter in: Araújo, K. (Ed.) *Routledge Handbook of Energy Transitions*. Routledge: Abingdon.

Matsuura, M., this volume. The Policies and Politics of Missed Opportunity for a Post-Fukushima Energy Transition, chapter in: Araújo, K. (Ed.) *Routledge Handbook of Energy Transitions*. Routledge: Abingdon.

Pahle, M., Schaeffer, R., Pachauri, S., Eom, J., Awasthy, A., Chen, W., Di Maria, C., Jiang, K., He, C., Portugal-Pereira, J., Safonov, G., and Verdolini, E. 2021. The crucial role of complementarity, transparency and adaptability for designing energy policies for sustainable development, *Energy Policy* 159, 112662, https://doi.org/10.1016/j.enpol.2021.112662.

Papachristos, G., Sofianos, A., Adamides, E., 2013. System interactions in socio-technical transitions. *Environmental Innovation Societal Transitions* 7, 53–69.

Raven, R.P.J.M., Reynolds, D., Lane, R., Lindsay, J., Kronsell, A., Arunachalam, D., 2021. Households in sustainability transitions: A systematic review and new research avenues. *Environmental Innovation and Societal Transitions* 40, 87–107.

Rosenbloom, D., 2018. Framing low-carbon pathways: A discursive analysis of contending storylines surrounding the phase-out of coal-fired power in Ontario. *Environmental Innovation and Societal Transitions* 27, 129–145.

Sharp, D., Pink, S., Raven, R., Farrelly, M., this volume. People, Politics and Place: An Interdisciplinary Agenda for the Governance of Urban Energy Transitions, chapter in: Araújo, K. (Ed.) *Routledge Handbook of Energy Transitions*. Routledge: Abingdon.

Smil, V., 2016. Examining energy transitions: A dozen insights based on performance. *Energy Research and Social Science* 22, 194–197.

Sovacool, B., Geels, F., 2016. Further reflections on the temporality of energy transitions. *Energy Research and Social Science* 22, 232–237.

Sovacool, B.K., Ali, S.H., Bazilian, M., Radley, B., Nemery, B., Okatz, J., Mulvaney, D., 2020. Sustainable minerals and metals for a low-carbon future. *Science* 367, 30–33.

Tainter, J.A., 2011. Energy, complexity, and sustainability: A historical perspective. *Environmental Innovation Societal Transitions* 1 (1), 89–95.

Trencher, G., Rinscheid, A., Duygan, M., Truong, N., Asuka, J., 2020. Revisiting carbon lock-in in energy systems: Explaining the perpetuation of coal power in Japan. *Energy Research & Social Science* 69, 101770.

Van der Vleuten, E., 2019. Radical change and deep transitions: Lessons from Europe's infrastructure transition 1815–2015. *Environmental Innovation Societal Transitions* 32, 22–32.

Verbong, G., Loorbach, D., 2012. *Governing the Energy Transition?* Routledge: Oxfordshire.

INDEX

For Product Safety Concerns and Information please contact our EU
representative GPSR@taylorandfrancis.com
Taylor & Francis Verlag GmbH, Kaufingerstraße 24, 80331 München, Germany